Student Solutions Manual to Accompany

Physics

The Nature of Things

Susan M. Lea
San Francisco State University

John Robert Burke
San Francisco State University

Peter Salzman
University of California-Davis

Chris Kelly
San Francisco State University

Brooks/Cole Publishing Company
West Publishing Company

I(T)P® An International Thomson Publishing Company

Pacific Grove • Albany • Belmont • Bonn • Boston • Cincinnati • Detroit
Johannesburg • London • Madrid • Melbourne • Mexico City
New York • Paris • Singapore • Tokyo • Toronto • Washington

WEST'S COMMITMENT TO THE ENVIRONMENT

In 1906, West Publishing Company began recycling materials left over from the production of books. This began a tradition of efficient and responsible use of resources. Today, 100% of our legal bound volumes are printed on acid-free, recycled paper consisting of 50% new fibers. West recycles nearly 27,700,000 pounds of scrap paper annually—the equivalent of 229,300 trees. Since the 1960s, West has devised ways to capture and recycle waste inks, solvents, oils, and vapors created in the printing process. We also recycle plastics of all kinds, wood, glass, corrugated cardboard, and batteries, and have eliminated the use of polystyrene book packaging. We at West are proud of the longevity and the scope of our commitment to the environment.

West pocket parts and advance sheets are printed on recyclable paper and can be collected and recycled with newspapers. Staples do not have to be removed. Bound volumes can be recycled after removing the cover.

Production, Prepress, Printing and Binding by West Publishing Company.

an International Thomson Publishing company•**ITP**
610 Opperman Drive
P.O. Box 64526
St. Paul, MN 55164–0526

Printed in the United States of America
04 03 02 01 00 99 8 7 6 5 4 3 2

ISBN 0–314–20731–7

Contents

Chapter 0

The Roots of Science

0.1 We should calculate the cost per cm^3 for each package.

$$\text{Brand X: } \frac{\$4.50}{750\ \text{cm}^3} = 0.60\ \text{¢}/\text{cm}^3$$

$$\text{Brand Y: } \frac{\$0.80}{120\ \text{cm}^3} = 0.67\ \text{¢}/\text{cm}^3$$

The price per cm^3 is less for Brand X, which therefore offers the best value.

0.3 The loop of string passes around both thumbtacks (at the foci) and the pencil (on the ellipse). Thus its total length must be $r_1 + r_2 + 2ae$. But $r_1 + r_2 = 2a$, so the total length of string is:

$$\ell = 2a(1+e) = (24\ \text{cm})(1+0.30) = 31.2\ \text{cm}$$

The distance between the thumbtacks should be:

$$2ae = 2(12\ \text{cm})(0.30) = 7.2\ \text{cm}.$$

0.5 According to Kepler's second law, a line drawn from the Sun to the comet sweeps out equal areas in equal times. Thus the comet sweeps out the largest arc when its distance from the Sun is smallest. The comet's speed is greatest at point A and smallest at point C.

0.7 The apple shares the horizontal motion of the train, but so do you! Neglecting air resistance, the apple will appear to fall straight down, which is where you should look.

0.9 We use Galileo's law of fall:

$$\frac{\ell_1}{\ell_2} = \left[\frac{t_1}{t_2}\right]^2 \Rightarrow \frac{\ell_1}{\frac{1}{4}\,\text{shaft}} = \left[\frac{2\ \text{s}}{1\ \text{s}}\right]^2 = 4$$

Thus $\ell_1 = 4(\frac{1}{4}\ \text{shaft}) = 1$ shaft. The rock falls all the way to the bottom.

0.11 According to Kepler's second law, a line drawn from the Sun to the comet sweeps out equal areas in equal times. Thus we must estimate the area of each segment and compare them. Each segment is approximately a triangle with area $= \frac{1}{2}\,(\text{base})\,(\text{height})$.

Segment	Radial line (cm)	arc length (cm)	product (cm^2)
AB	1.3	0.5	0.65
CD	2.2	0.9	2.0
EF	3.7	0.6	2.2
GH	3.4	0.25	0.85

Thus the region that most closely represents the same time interval as AB is GH.

0.15 The greatest and least distances from the center of the ellipse to the curve are the semimajor axis and the semiminor axis. With $a = 5.0$ cm, an ellipse of eccentricity 0.093 has semiminor axis b where:

$$a^2 = b^2 + (ae)^2$$

Thus: $b = a\sqrt{1-e^2} = (5.0\ \text{cm})\sqrt{1-0.093^2} = 4.98$ cm. The difference is only 0.2 mm, or less than the width of the pencil line. You cannot tell that the curve is not a circle.

Chapter 1

Introducing the Language of Physics

1.1 The smallest division on a meter stick would be a millimeter. Any measurements made would be precise to $\pm\frac{1}{2}$ of the smallest division. In this case, the precision would be ± 0.0005 m or ± 0.5 mm. When declaring the result, it would probably be given in cm and we would say something like "this paper is 20.5 cm long."

1.3 a) The precision of a number is the uncertainty, assumed to be one half of the last figure quoted. Here that is

$$0.5(0.001) \times 10^{-4}\ \text{s} = 5 \times 10^{-8}\ \text{s}$$

The accuracy is 4 significant figures.

b) The precision is 50 kg and the accuracy is 1 significant figure.

c) The precision is 5×10^{-6} and the accuracy is 3 significant figures.

d) The precision is 0.00005 m = 50 μm and the accuracy is 3 significant figures.

1.5 $1\ \text{Ts} = 10^{12}\,\text{s}\left(\dfrac{10^6\ \mu\text{s}}{1\ \text{s}}\right) = 10^{18}\ \mu\text{s}$

1.7 a) $[\text{density}] : \dfrac{[\text{mass}]}{[\text{volume}]} = \dfrac{[\text{mass}]}{[\text{length}]^3} = \dfrac{M}{L^3}$

b) angle: dimensionless

c) $[\text{Area}] : [\text{length}]^2 = L^2$

d) $[\text{volume}] : [\text{length}]^3 = L^3$

1.9 The displacement vector depends only on the child's starting point and ending point—in this case the school and park respectively. The vector would have a magnitude of 15 m and point north.

1.11

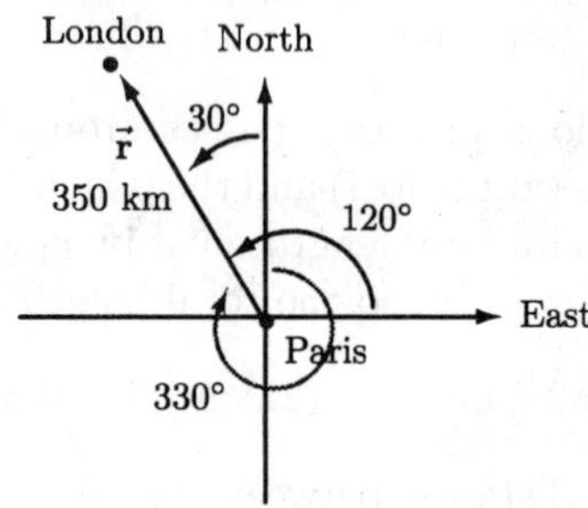

Vector $\vec{\mathbf{r}}$ is the position vector from Paris to London. It has a magnitude of 350 km and points 120° CCW from the east axis.

1.13 We can simply read the components off the graph. Remember that if a component vector points along the negative coordinate axis, that component is negative.

$$\begin{aligned} a_x &= -1, & a_y &= 2 \\ b_x &= 3, & b_y &= 2 \\ c_x &= 1, & c_y &= -3 \\ d_x &= 2, & d_y &= 2 \\ e_x &= -2, & e_y &= -1 \end{aligned}$$

Each unit vector has length 1. Thus $\hat{\mathbf{a}}$ has unit length and is parallel to $\vec{\mathbf{a}}$, etc.

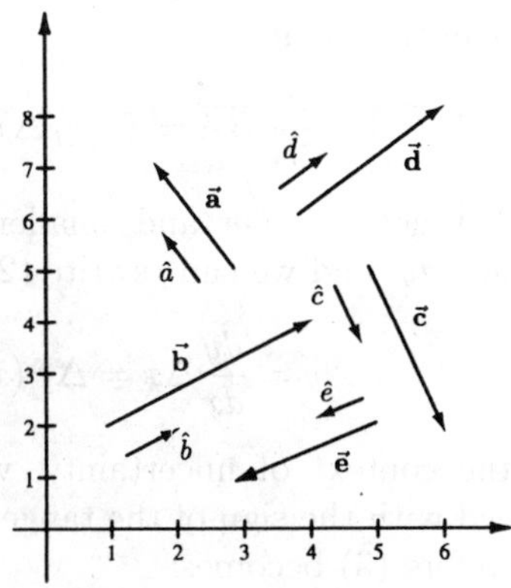

1.15 Now we'll calculate the components of $\vec{\mathbf{A}} + \vec{\mathbf{B}}$:

$$\vec{\mathbf{A}} + \vec{\mathbf{B}} = (3,2) + (-1,1) = (2,3)$$

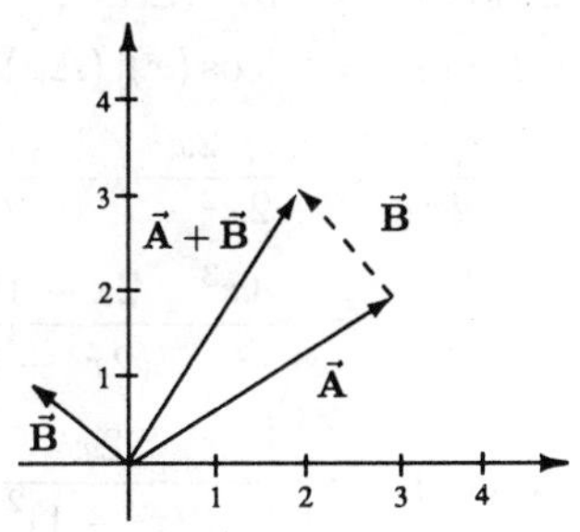

Looking at the graph, this is precisely what we constructed.

1.19 **a)** The precision is dependent on your measuring instrument, so it would be ±0.5 mm. If you measure to the nearest millimeter, the accuracy would be to *four* decimal places. The uncertainty arising from marking off four points on the wall (using one ruler four times) will be much greater, probably several millimeters.

b) Considering that you will have to measure around the cabin and the mast, the uncertainty is even higher than it was in (a). The precision would be the same: ±0.5 mm. You could generate a number with perhaps 5 digits but this wouldn't mean much considering the uncertainty. The figure below shows a possible scenario (the dashed lines are your estimate of where to begin measuring).

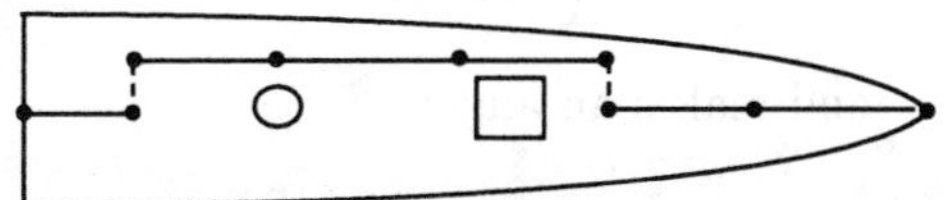

c) You will be able to determine halfway between a second, but no better. Your precision would be ±0.5 sec. A typical measurement would be something like 1:41, which has an accuracy of 3 digits.

d) This should be exactly the same as part (c) with the additional observation that the precision of the measurement is totally unsuited for its purpose.

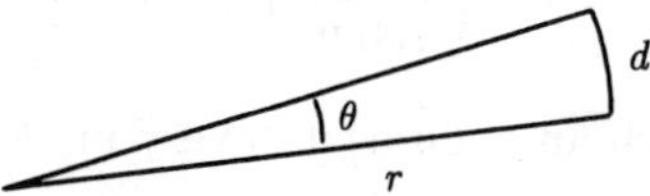

1.23 Let d be the distance between Alexandria and Syene (the arc length between A and S.

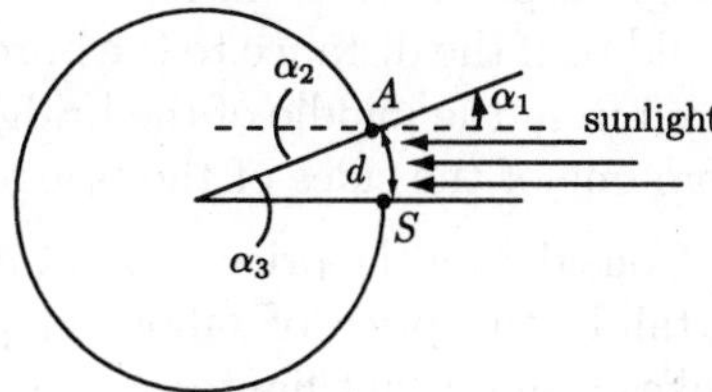

Let C denote the Earth's circumference. The angle α_1, which is given as 7°, is the angle between the sun's rays and a radial line from the center of the Earth extending through Alexandria. By the "opposite angle theorem" $\alpha_2 = \alpha_1 = 7°$. Then by the "alternate interior angle theorem", $\alpha_3 = \alpha_2 = 7°$. The ratio of the Earth's circumference to the distance d between Alexandria and Syene is:

$$\frac{C}{d} = \frac{360°}{7°} = 51$$
$$C = \left(\frac{360°}{7°}\right) 800 \text{ km}$$
$$= 4 \times 10^4 \text{ km}$$

1.27 **a)** The precision of the value painted on the bridge is ±0.0005 mi = ±0.8 m.

b) It is extremely improbable that there is another bridge with a distance to the border equal to this one plus or minus 0.8 m. This precision is probably unnecessary.

c) For the measurement to be correct for both ends of the bridge, the bridge needs to be situated so that either end lies—within 0.8 m—at the same distance from the border.

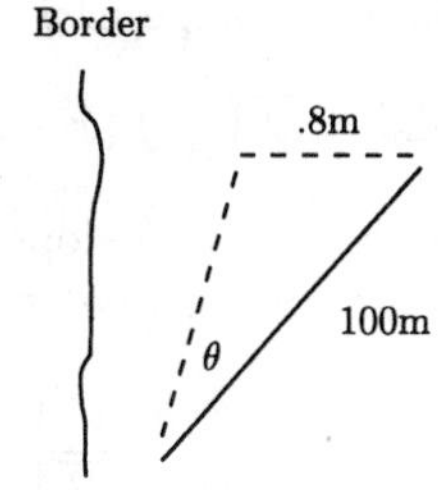

That means that (in this diagram—which shows a worst case scenario) θ needs to be a maximum of

$$\sin^{-1}\left(\frac{0.8\text{ m}}{100\text{ m}}\right) = (0.458°) = 8\text{ mrad}$$

Such an alignment is very unlikely!

d) Ridiculous—the bridge is certainly wider than 0.8 m which makes the purported precision meaningless. It may be true for one side of the bridge, but if the bridge is wider than 0.8 m, the given precision is misleading. An exception would be if the distance to the border were measured from the middle of the bridge rather than from one of the sides of the bridge.

e) Considering the price of paint it may be, but certainly the price of labor (in person-hours) makes 4 significant figures not cost effective.

1.31 Example: my height is 5′7″ which is 67 in.

$$67\text{ in.}\left(\frac{0.0254\text{ m}}{1\text{ in.}}\right) = 1.7\text{ m}$$

1.35 The speed of light is

$$\left(\frac{3.0\times10^{8}\text{ m}}{\text{s}}\right)\times\left(\frac{3600\text{ s}}{1\text{ h}}\right)\times\left(\frac{24\text{ h}}{1\text{ day}}\right)\times$$
$$\left(\frac{14\text{ days}}{1\text{ fortnight}}\right)\times\left(\frac{3.28\text{ ft}}{1\text{ m}}\right)\times\left(\frac{1\text{ furlong}}{660\text{ ft}}\right)$$
$$= 1.8\times10^{12}\text{ furlong/fortnight.}$$

1.43 y is some arbitrary function of x. $\overline{BC}$ defines a line tangent to y at the point $B(x, f(x))$.

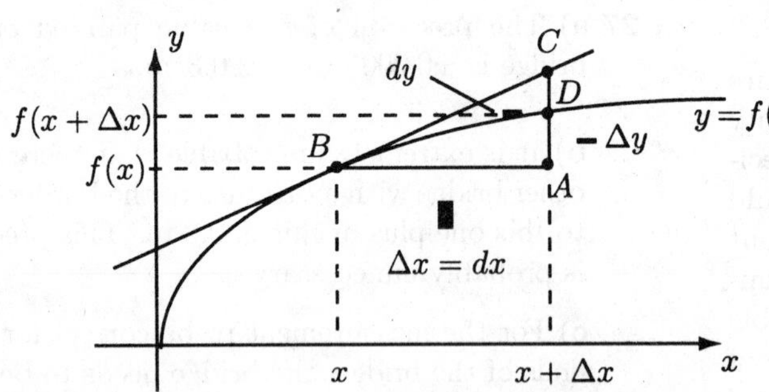

If we change x by an amount $\Delta x = dx$, the corresponding change in y is

$$\Delta y = \overline{DA} = f(x+\Delta x) - f(x)$$

From the definition of slope we can express the slope of the tangent line as:

$$m = \frac{dy}{dx} = \frac{\overline{CA}}{\overline{BA}} = f'(x) \qquad (1)$$

We can then write

$$\overline{CA} = \frac{dy}{dx}\overline{BA} = f'(x)\,\Delta x = dy \qquad (2)$$

As Δx gets smaller and smaller, Δy will approach dy, and we can rewrite (2) as:

$$\Delta y = \frac{dy}{dx}\Delta x = \Delta f(x) \qquad (3)$$

In the context of uncertainty, we aren't concerned with the sign of the tangent line's slope. Therefore (3) becomes:

$$\Delta f(x) = \left|\frac{dy}{dx}\right|\Delta x \qquad (4)$$

$$\begin{aligned}
\Delta f_1(x) &= 3x^2(\Delta x)\\
\Delta f_2(x) &= |\cos(x)|(\Delta x)\\
\frac{df_3}{dx} &= \frac{2x}{2x^2-1} - \frac{(x^2+5)(4x)}{(2x^2-1)^2}\\
&= \frac{4x^3-2x-4x^3-20x}{(2x^2-1)^2}\\
\Delta f_3(x) &= \left|-\frac{22x}{(2x^2-1)^2}\right|(\Delta x)
\end{aligned}$$

and with $x = \frac{\pi}{3} \pm 0.020$

$$\begin{aligned}
\Delta f_1(x) &= 6.6\times10^{-2}\\
\Delta f_2(x) &= 1.0\times10^{-2}\\
\Delta f_3(x) &= 0.32
\end{aligned}$$

1.47 The answer becomes apparent if we redraw the diagram.

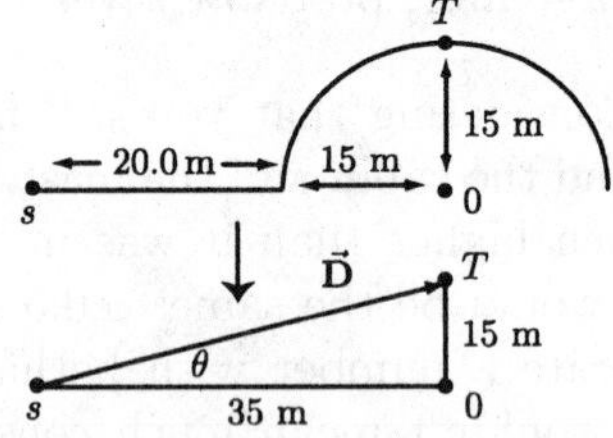

Your displacement vector has a magnitude of

$$\left|\vec{D}\right| = \sqrt{(35\text{ m})^2 + (15\text{ m})^2} = 38\text{ m}$$

and makes an angle

$$\theta = \tan^{-1}\left(\frac{15}{35}\right) = 23°$$

from the ground.

1.51 See the diagram. (e) most closely represents the result.

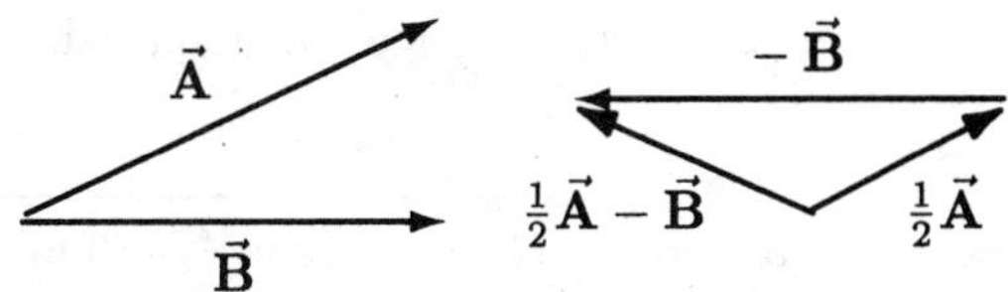

1.55 a) Since $|\vec{a}| = |\vec{b}|$, the diagonal $\vec{s}$ exactly bisects the angle between $\vec{a}$ and $\vec{b}$. This angle between the two vectors is equal to $\theta_b - \theta_a$. Since $\vec{s}$ bisects the angle between $\vec{a}$ and $\vec{b}$ we can say that

$$\begin{aligned}\theta_s &= \theta_a + \frac{1}{2}\left(\text{angle between } \vec{a} \text{ and } \vec{b}\right) \\ &= \theta_a + \frac{1}{2}(\theta_b - \theta_a) = \frac{\theta_a + \theta_b}{2}\end{aligned}$$

The components of $\vec{a}$ and $\vec{b}$ which are $\perp$ to $\vec{s}$ point in opposite directions and are equal in magnitude—thus they cancel. The components of $\vec{a}$ and $\vec{b}$ which are parallel to $\vec{s}$ are in the same direction and therefore they add.

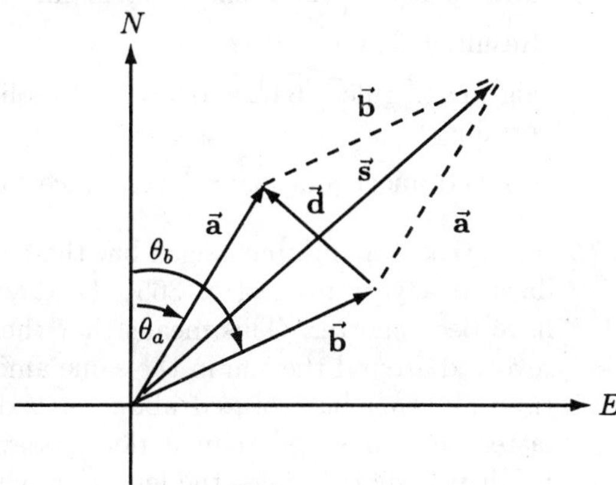

Computing the component of each vector parallel to $\vec{s}$, we find:

$$\begin{aligned}|\vec{s}| &= |\vec{a}|\cos\left(\frac{\theta_b - \theta_a}{2}\right) + |\vec{b}|\cos\left(\frac{\theta_b - \theta_a}{2}\right) \\ &= \left(|\vec{a}| + |\vec{b}|\right)\cos\left[\frac{1}{2}(\theta_b - \theta_a)\right]\end{aligned}$$

Since we were told that $|\vec{a}| = |\vec{b}|$,

$$|\vec{s}| = 2|\vec{a}|\cos\left[\frac{1}{2}(\theta_b - \theta_a)\right]$$

or

$$|\vec{s}| = 2|\vec{b}|\cos\left[\frac{1}{2}(\theta_b - \theta_a)\right]$$

b) Computing the components perpendicular to $\vec{s}$, we find the difference $\vec{d}$ to have magnitude

$$|\vec{d}| = 2|\vec{a}|\sin\left(\frac{\theta_b - \theta_a}{2}\right)$$

$\vec{d}$ is perpendicular to $\vec{s}$, so its direction from north is

$$\frac{\theta_a + \theta_b}{2} - \frac{\pi}{2} = \theta_d$$

1.59 We can't apply the cosine law to find $|\vec{D}|$ until we find the angle α.

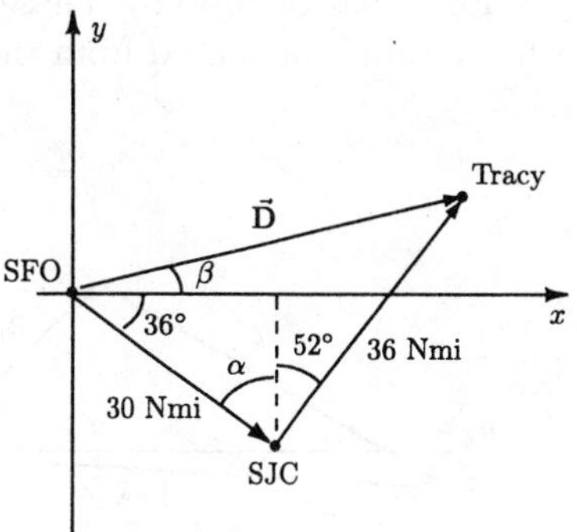

Note the perpendicular from SJC to the x-axis forms 2 right triangles. We can find α:

$$\alpha = 180° - 90° - 36° = 54°$$

Now we can apply the cosine law:

$$\begin{aligned}|\vec{D}| &= \sqrt{\begin{array}{l}(30\ \text{Nmi})^2 + (36\ \text{Nmi})^2 \\ -2(30\ \text{Nmi})(36\ \text{Nmi})\cos(54° + 52°)\end{array}} \\ &= 53\ \text{Nmi (about 61 mi)}\end{aligned}$$

Now we will find β, the angle $\vec{D}$ makes with the x-axis using the sine law:

$$\frac{\sin(106°)}{53\ \text{Nmi}} = \frac{\sin(36° + \beta)}{36\ \text{Nmi}} \Rightarrow \beta = 5°\ \text{N of E.}$$

1.63 Measuring from the figure

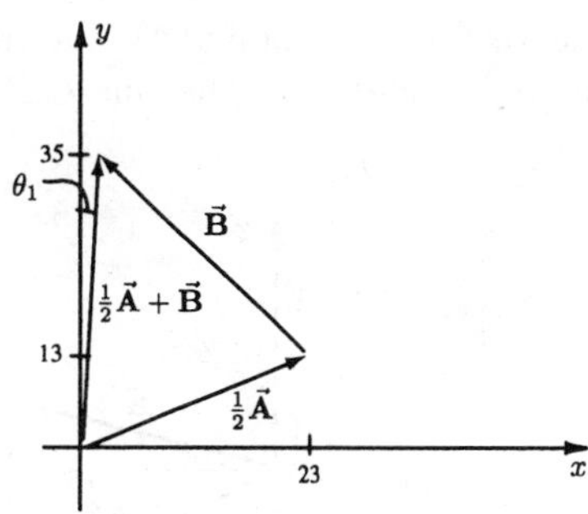

$A_x = 46$ m, $A_y = 26$ m; $B_x = -22$ m, $B_y = 22$ m.

Using components, we have:

$$\begin{aligned}\left(\frac{1}{2}\vec{A} + \vec{B}\right)_x &= \frac{1}{2}A_x + B_x \\ &= 23\ \text{m} - 22\ \text{m} = 1\ \text{m}. \\ \left(\frac{1}{2}\vec{A} + \vec{B}\right)_y &= \frac{1}{2}A_y + B_y \\ &= 13\ \text{m} + 22\ \text{m} = 35\ \text{m}\end{aligned}$$

Thus

$$\left|\frac{1}{2}\vec{\mathbf{A}}+\vec{\mathbf{B}}\right| = \sqrt{(1\text{ m})^2+(35\text{ m})^2} = 35\text{ m}$$

and

$$\tan\theta_1 = \frac{1\text{ m}}{35\text{ m}} = 0.029$$

$\theta_1 = 1.6°$ from the y-axis. These results agree with the values measured from the figure: 35 m and 2°.

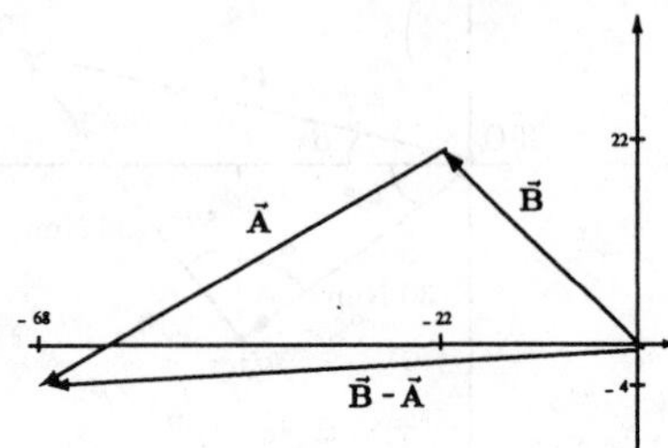

Similarly, for $\vec{\mathbf{B}}-\vec{\mathbf{A}}$:

$$\begin{aligned} B_x - A_x &= -22\text{ m} - 46\text{ m} \\ &= -68\text{ m} \\ B_y - A_y &= 22\text{ m} - 26\text{ m} = -4\text{ m} \end{aligned}$$

Thus

$$\left|\vec{\mathbf{B}}-\vec{\mathbf{A}}\right| = \sqrt{(68\text{ m})^2+(4\text{ m})^2} = 68\text{ m}$$

and

$$\tan\theta_2 = \frac{4\text{ m}}{68\text{ m}} = 0.059;\ \theta_2 = 3°$$

Again these values agree with the measurements.

1.67 The vector $\vec{\mathbf{a}}$ (to point A), the line AB ($\perp$ to the x-y plane), and the line OB form a right triangle.

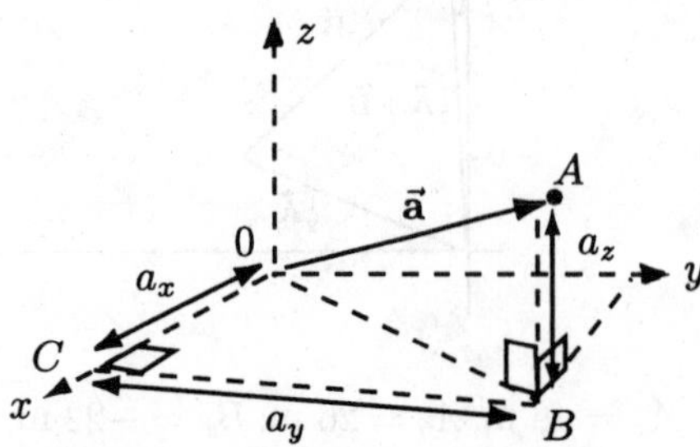

So

$$|\vec{\mathbf{a}}|^2 = \left(\overline{OB}\right)^2 + \left(\overline{AB}\right)^2$$

by Pythagoras' Theorem. But $\overline{AB} = a_z$, and using Pythagoras' Theorem again for ΔOBC, we have

$$\left(\overline{OB}\right)^2 = \left(\overline{OC}\right)^2 + \left(\overline{CB}\right)^2 = a_x^2 + a_y^2.$$

Thus $|\vec{\mathbf{a}}|^2 = a_x^2 + a_y^2 + a_z^2$, as required.

$$\begin{aligned} |\vec{\mathbf{a}}| &= \sqrt{(1\text{ m})^2+(2\text{ m})^2+(3\text{ m})^2} \\ &= \sqrt{14}\text{ m} \\ \left|\vec{\mathbf{b}}\right| &= \sqrt{(2\text{ m})^2+(2\text{ m})^2+(2\text{ m})^2} \\ &= \sqrt{12}\text{ m} \\ \vec{\mathbf{a}}-\vec{\mathbf{b}} &= (1\text{ m}, 2\text{ m}, 3\text{ m}) - (2\text{ m}, 2\text{ m}, 2\text{ m}) \\ &= (-1\text{ m},\ 0\text{ m}, 1\text{ m}) \\ \left|\vec{\mathbf{a}}-\vec{\mathbf{b}}\right| &= \sqrt{(1\text{ m})^2+0+(1\text{ m})^2} = \sqrt{2}\text{ m} \end{aligned}$$

1.71 $d = 2.33 \pm 0.02$ cm

Volume $= \frac{4}{3}\pi r^3 = \frac{4}{3}\pi\left[\frac{d}{2}\right]^3 = \frac{\pi d^3}{6}$

"Best" value $= \frac{\pi}{6}(2.33\text{ cm})^3 = 6.62\text{ cm}^3$

Max value $= \frac{\pi}{6}(2.35\text{ cm})^3 = 6.80\text{ cm}^3$

Min value $= \frac{\pi}{6}(2.31\text{ cm})^3 = 6.45\text{ cm}^3$

Result: $6.62\text{ cm}^3 \pm 0.17\text{ cm}^3$

Sig. Dig. rules: $6.62 \pm 0.05\,\text{cm}^3$ (implied uncertainty)

Implied uncertainty is $\approx \frac{1}{3}$ true uncertainty.

1.75 If the year remains unchanged but there are 400 days in a year instead of 365, the days must have been shorter. This means that the Earth revolved around the sun in the same amount of time (1 year) but rotated about its axis at a faster angular speed than it does today. The result is a shorter day—the length of which we will find. We'll set up a ratio between days and years to find out how long a day was (in seconds).

$$\frac{400\text{ days}}{1\text{ yr}}\left(\frac{1\text{ yr}}{8760\text{ hr}}\right)\left(\frac{1\text{ hr}}{3600\text{ s}}\right) = \frac{1.27\times10^{-5}\text{ days}}{1\text{ s}}$$

So there were 7.88×10^4 sec in 1 day. This is about 21.9 hr. (We can do this calculation because 1 hr was and is 60 min, and 1 min was and is 60 sec.) The rate of change of the length of Earth's day would be expressed by

$$\frac{\Delta\text{ seconds}}{500\text{ million years}}$$

where Δ seconds is the difference between the number of seconds per day *now* and the number of seconds per day *then.*

$$\begin{aligned} \Delta\text{ seconds} &= 24\,\text{h}\left(\frac{3600\,\text{s}}{1\,\text{h}}\right) - 7.88\times10^4\,\text{s} \\ &= 7600\,\text{s} \end{aligned}$$

So the rate of change is:

$$\frac{7600 \text{ seconds}}{500 \text{ million years}}\left(\frac{1 \text{ yr}}{3.16\times 10^7 \text{ sec}}\right)=5\times 10^{-13}$$

If we define the day to be exactly 24 hr rather than the time it takes for the Earth to rotate completely once, we can express the rate of change in seconds per day:

$$\frac{7600 \text{ seconds}}{500\times 10^6 \text{ years}}\left(\frac{1 \text{ yr}}{365 \text{ day}}\right)=4\times 10^{-8} \text{ sec/day}$$

1.79 Traveling at 0.001 c implies that you'll cover 10^{-3} light years in one year (if you travelled at c you would cover 1 light year in a year). If there are 10 light years between stars a trip to another star (the closest one) would take time

$$t=\frac{x}{v}=\frac{10 \text{ ly}}{10^{-3} \text{ ly}/\text{y}}=10^4$$

If a simple colony takes 20000 y to develop both technologically and socially into an industrialized planet with a strong economy to support space travel, then civilized space should expand its radius (assuming it is spherical) by 10 ly every 30,000 years. To achieve a radius of 3×10^4 ly, it would take time

$$t=3\times 10^4 \text{ ly}\left(\frac{30,000 \text{ y}}{10 \text{ ly}}\right)=9\times 10^7 \text{ y}$$

The universe is about 10 billion years old (10^{10} y). The question "where are they?" is unanswered ... they should be here by now.

Chapter 2

Kinematics

2.1 $S = \dfrac{170\ \text{km}}{2.0\ \text{h}} = 85\ \text{km/h}$

2.3

$$\begin{aligned} v_x &= (3.0\ \text{cm/s})(\cos 150°) = -2.6\ \text{cm/s} \\ v_y &= (3.0\ \text{cm/s})(\sin 150°) = +1.5\ \text{cm/s} \end{aligned}$$

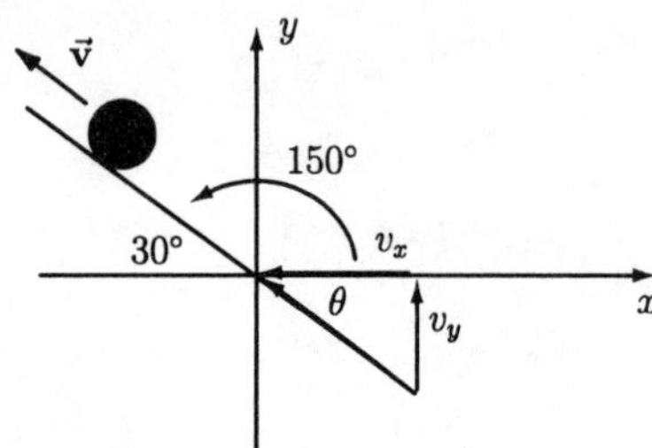

2.5 Since we are asked to *estimate* we shouldn't get hung up about whether to measure using both time intervals around a point, the time interval before a point or after a point. I will measure both intervals and estimate curves as well as possible.

Point A:

$$(1.25\ \text{cm})\left(\frac{1.0\ \text{m}}{0.55\ \text{cm}}\right)\frac{1}{2.0\ \text{s}} = 1.1\ \text{m/s}$$

parallel to the x-axis

Point B:

$$(1.95\ \text{cm})\left(\frac{1.0\ \text{m}}{0.55\ \text{cm}}\right)\frac{1}{2.0\ \text{s}} = 1.8\ \text{m/s}$$

parallel to the y-axis

Point C:

$$(0.85\ \text{cm})\left(\frac{1.0\ \text{m}}{0.55\ \text{cm}}\right)\frac{1}{2.0\ \text{s}} = 0.77\ \text{m/s}$$

toward 30° clockwise from the y-axis.

2.7 The particle follows a curved path so we expect acceleration toward the center of curvature. This leaves us to consider only choices **a** and **b**. If you can imagine the path being perfectly circular, vector **b** points more nearly to its center than vector **a** which is a little too angled. Thus, choice **b** is the correct answer.

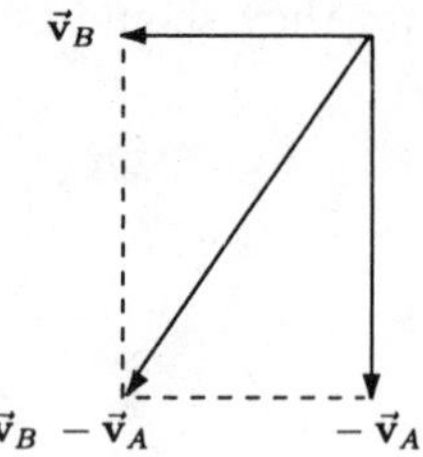

In the figure, note that $\vec{v_B} - \vec{v_A}$ = choice **b**. This shows that the qualitative argument is correct.

2.9 This would be true only when the initial velocity were zero. Indeed $v_{\text{avg}} = \frac{1}{2}(v_f + v_i) = \frac{1}{2}v_f$ only when $v_i = 0$!

2.11 The acceleration of the object is constant between $t = 0$ and 5 s.

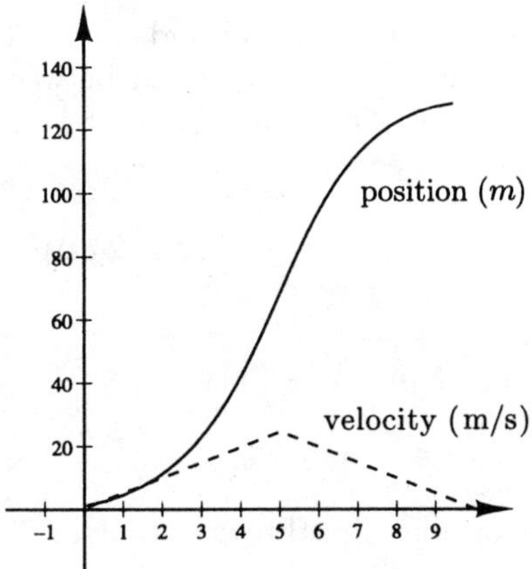

Between $t = 0$ and $t = 5$ s,

$$a_x = +\frac{5\ \mathrm{m}}{\mathrm{s}^2}$$

Thus

$$v_x = v_0 + a_x t = \left(5\ \mathrm{m/s^2}\right) t$$

and

$$\begin{aligned} x &= 0 + v_0 t + \frac{1}{2} a_x t^2 \\ &= \left(2.5\ \mathrm{m/s^2}\right) t^2 \end{aligned}$$

Between $t = 5$ s and 10 s the acceleration has the constant value $a_x = -5\ \mathrm{m/s^2}$. The starting values of position and velocity are:

$$\begin{aligned} x &= 62.5\ \mathrm{m} \\ v_x &= 25\ \mathrm{m/s} \end{aligned}$$

Thus

$$v_x = 25\ \mathrm{m/s} - \left(5\,\mathrm{m/s^2}\right)(t - 5\ \mathrm{s})$$

and

$$\begin{aligned} x = &\ 62.5\ \mathrm{m} + (25\ \mathrm{m/s})(t - 5\,\mathrm{s}) \\ &- \left(2.5\ \mathrm{m/s^2}\right)(t - 5\ \mathrm{s})^2 \end{aligned}$$

2.13 We will use the formula $v_f^2 - v_i^2 = 2as$ to solve for the car's acceleration.

$$(80\ \mathrm{km/h})^2 - (100\ \mathrm{km/h})^2 = 2a(1.5\ \mathrm{km})$$

$$\begin{aligned} a &= -1200\ \mathrm{km/h^2}\left(\frac{1\ \mathrm{h}}{3600\ \mathrm{s}}\right)^2\left(\frac{10^3\ \mathrm{m}}{1\ \mathrm{km}}\right) \\ &= -9.3\times 10^{-2}\ \mathrm{m/s^2} \end{aligned}$$

2.17 The figure shows the displacement vector for the car.

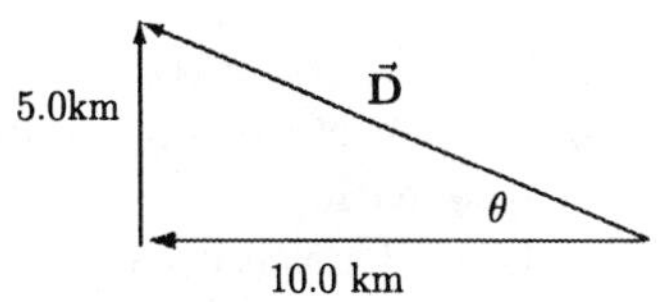

Its magnitude is given by:

$$\left|\vec{D}\right| = \sqrt{(10\ \mathrm{km})^2 + (5\ \mathrm{km})^2} = 11\ \mathrm{km}$$

Thus the magnitude of the car's average velocity is:

$$\begin{aligned} |\vec{v}_{\mathrm{avg}}| &= \frac{11\ \mathrm{km}}{13\ \mathrm{min}} \\ &= 0.86\ \mathrm{km/min}\left(\frac{10^3\ \mathrm{m}}{\mathrm{km}}\frac{1\ \mathrm{min}}{60\ \mathrm{s}}\right) \\ &= 14\ \mathrm{m/s} \end{aligned}$$

The angle θ is:

$$\theta = \tan^{-1}\left(\frac{5\ \mathrm{km}}{10\ \mathrm{km}}\right) = 27°$$

Since the given data have only one significant figure, we quote the result as $\vec{v}_{\mathrm{avg}} = (10\ \mathrm{m/s}, 30°$ north of west).

2.21 The magnitude of $\vec{D}$ is given by:

$$\left|\vec{D}\right| = \sqrt{(5\ \mathrm{km})^2 + (3\ \mathrm{km})^2} = 5.8\ \mathrm{km}$$

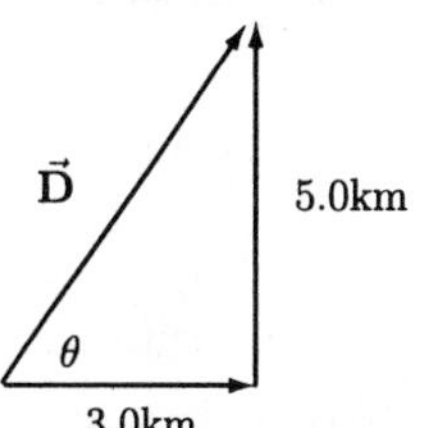

Thus, $|\vec{v}_{\mathrm{avg}}|$ is given by:

$$\begin{aligned} |\vec{v}_{\mathrm{avg}}| &= \frac{5.8\ \mathrm{km}}{11.0\ \mathrm{min}} \\ &= 0.53\ \frac{\mathrm{km}}{\mathrm{min}}\cdot\left(\frac{10^3\ \mathrm{m}}{1\ \mathrm{km}}\right)\left(\frac{1\ \mathrm{min}}{60\ \mathrm{s}}\right) \\ &= 8.8\ \mathrm{m/s} \end{aligned}$$

The angle θ is:

$$\theta = \tan^{-1}\left(\frac{5\ \mathrm{km}}{3\ \mathrm{km}}\right) = 59°$$

So $\vec{v}_{\mathrm{avg}} = (8.8\ \mathrm{m/s}, 59°$ north of east).

2.25 $|\vec{a}_{\mathrm{avg}}| = \dfrac{25\ \mathrm{m/s}}{5.0\ \mathrm{s}} = 5.0\ \mathrm{m/s^2}$

2.29 Choice **b** cannot be correct. Acceleration is the rate of change of velocity, represented not by the area beneath a velocity *vs* time curve, but by its slope. Choice **c** cannot be correct since if $\frac{dv}{dt}$ = constant the slope of the tangent line to this graph should be constant everywhere within the time interval t, which it obviously is not. Only choice **a** remains, and it is correct, since *displacement*, and not distance traveled, is represented by the area beneath a velocity *vs* time curve.

2.33 The slope of a graph is the ratio of change in dependent variable to change in independent variable.

$$\text{Slope} = \frac{\Delta \text{ position}}{\Delta \text{ time}} = \text{velocity}$$

Since you have the choice of scales to use in plotting meters and seconds on your graph, the lengths that correspond to the same Δ position and Δ time on your graph are arbitrary. A protractor would measure

$$\theta = \tan^{-1}\left(\frac{\text{length that represents } \Delta\text{position}}{\text{length that represents } \Delta\text{time}}\right)$$

Depending on your scale choices, θ could have any value between $+90°$ and $0°$ for the same (positive) value of slope.

2.35 Since no times were given, the relation

$$v_f^2 - v_i^2 = 2a\Delta y$$

should be used (with down being the positive direction):

$$\begin{aligned} (0\ \mathrm{m/s})^2 - (20\ \mathrm{m/s})^2 &= 2a(2 \times 10^{-2}\ \mathrm{m}) \\ \Rightarrow a &= -10^4\ \mathrm{m/s^2} \end{aligned}$$

2.37 First convert everything to m and s:

$$\begin{aligned} 80.0\ \mathrm{km/h} &= 22.2\ \mathrm{m/s} \\ 1.5\ \mathrm{km} &= 1.5 \times 10^3\ \mathrm{m} \end{aligned}$$

Next plug all the given values into the general distance formula. We'll end up with a quadratic in t (the unknown). Take point F to be the origin.

$$1.5 \times 10^3\,\mathrm{m} = (22.2\,\mathrm{m/s})t + \frac{1}{2}(2\,\mathrm{m/s^2})t^2$$

$$t = \frac{-22.2\,\mathrm{m/s} \pm \sqrt{492.8\,\mathrm{m^2/s^2} - 4(\mathrm{m/s^2})(-1.5\times10^3\,\mathrm{m})}}{2\ \mathrm{m/s^2}}$$

The only positive answer is $t = 29$ s. The velocity at point E is

$$\begin{aligned} v &= v_0 + at \\ &= 22.2\ \mathrm{m/s} + (2\ \mathrm{m/s^2})(29\ \mathrm{s}) \\ &= 80\ \mathrm{m/s} = 290\ \mathrm{km/h} \end{aligned}$$

parallel to the track.

2.41 We know an initial and final velocity and the distance the car traveled between the two velocities. This is all the information we need to use the equation:

$$v_f^2 - v_i^2 = 2a\,\Delta y$$

$$\begin{aligned} (50.0\ \mathrm{m/s})^2 - (30.0\ \mathrm{m/s})^2 &= 2a(200\ \mathrm{m}) \\ \Rightarrow a &= 4.00\ \mathrm{m/s^2} \end{aligned}$$

2.45 Consider Jim throwing to John first.

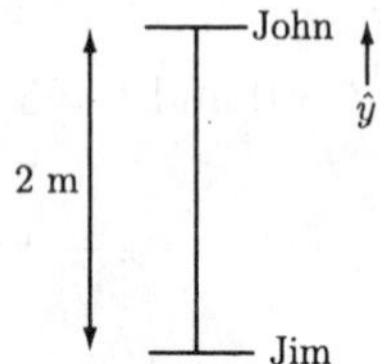

We'll take Jim to be at the origin. To find the time it takes the ball to reach John we'll use the general distance formula and solve it for t.

$$2.0\,\mathrm{m} = (7.0\,\mathrm{m/s})t - \frac{1}{2}(9.8\,\mathrm{m/s^2})t^2$$

$$\begin{aligned} t &= \frac{-7.0\ \mathrm{m/s} \pm \sqrt{49\,\mathrm{m^2/s^2} - 4(-49\,\mathrm{m/s^2})(-2\,\mathrm{m})}}{-9.80\,\mathrm{m/s^2}}\ \mathrm{s} \\ &= 0.39\ \mathrm{s}, 1.0\ \mathrm{s} \end{aligned}$$

The result 0.39 s is the one we're looking for. 1.0 s represents the time it takes the ball to reach John on the way back down from reaching the maximum height of its trajectory. The speed of the ball at $t = 0.39$ s is:

$$v = (7\ \mathrm{m/s}) - (9.80\ \mathrm{m/s^2})(.39\ \mathrm{s}) = 3.1\ \mathrm{m/s}$$

Now we'll consider John throwing the ball.

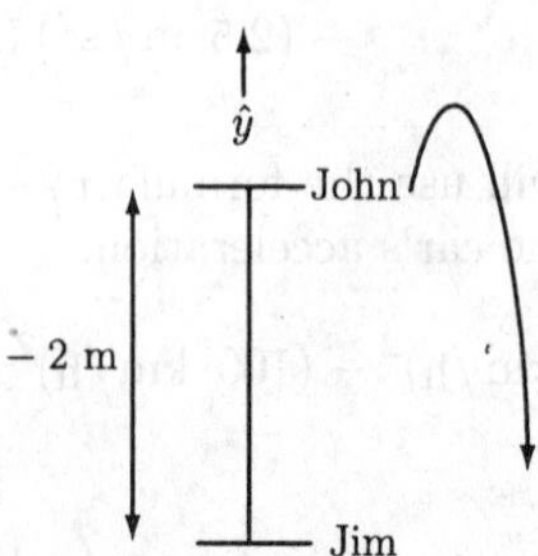

The symmetric nature of the ball's trajectory tells us that John will throw the ball upwards at 7 m/s, the ball will reach some high point and then start to come back down—and when it passes John it will be going at a velocity

$-7\ \text{m/s}\ \hat{\mathbf{j}}$. Take John to be the origin. The method will be the same as before:

$$\begin{aligned} -2\ \text{m} &= (+7\ \text{m/s})t - \frac{1}{2}\left(9.80\ \text{m/s}^2\right)t^2 \\ \Rightarrow t &= -0.244\ \text{s}, 1.67\ \text{s} \end{aligned}$$

This time we want the 2nd root. Jim catches the ball 1.67 s after John throws it. At this point, the ball has a velocity of:

$$\begin{aligned} v &= (7\ \text{m/s})\hat{\mathbf{j}} - (9.8\ \text{m/s}^2)(1.67\ \text{s})\hat{\mathbf{j}} \\ &= -9.4\ \text{m/s}\ \hat{\mathbf{j}} \end{aligned}$$

and a speed of 9.4 m/s.

2.49 First convert the given acceleration to km/h^2:

$$\begin{aligned} a &= -8\ \text{m/s}^2\left(\frac{1\ \text{km}}{1000\ \text{m}}\right)\left(\frac{3600\ \text{s}}{1\ \text{h}}\right)^2 \\ &= -1.04\times10^5\ \text{km/h}^2 \end{aligned}$$

Thus the minimum distance from G is: $v_f^2 - v_i^2 = 2as$

$$\begin{aligned} (80\ \text{km/h})^2 - (240\ \text{km/h})^2 & \\ &= 2(-1.04\times10^5\ \text{km/h}^2)s \\ s &= .25\ \text{km} = 250\ \text{m} \end{aligned}$$

The total time to travel from E to G is:

$$\begin{aligned} T &= \frac{1.25\ \text{km}}{240\ \text{km/h}} + \frac{0.25\ \text{km}}{\frac{1}{2}(240+80)\ \text{km/h}} \\ &= 6.77\times10^{-3}\ \text{h} \\ &= 24\ \text{s} \end{aligned}$$

2.53 The word "heavy" indicates that air resistance is not to be considered. The ball dropped first must have been dropped from a higher point than the 2nd ball, since it takes more time to reach the ground. Using the distance formula $\Delta x = \frac{1}{2}at^2$, we find the two heights:

$$\begin{aligned} \Delta x &= \frac{1}{2}(9.8\ \text{m/s}^2)(4.0\,\text{s})^2 = 78\ \text{m} \\ \Delta x &= \frac{1}{2}(9.8\ \text{m/s}^2)(4.0\ \text{s} - 2.2\ \text{s})^2 = 16\ \text{m} \end{aligned}$$

2.57 a) To find the objects' speed we find first how long their fall took:

$$19.5\ \text{m} = \frac{1}{2}(9.80\ \text{m/s}^2)t^2 \Rightarrow t = 1.99\ \text{s}$$

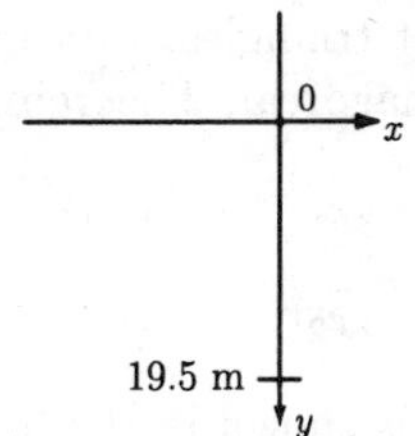

So at the end of their fall they're traveling at a speed of:

$$(+9.80\ \text{m/s}^2)(1.99\ \text{s}) = 19.6\ \text{m/s}$$

In $\frac{1}{16}$ s at this speed, the second object travels an approximate distance:

$$\Delta d = (19.6\,\text{m/s})\left(\frac{1}{16}\,\text{s}\right) = 1.2\,\text{m}$$

As the first object hits the ground the second object is still at a height of 1.2 m.

b) The first object's average speed is:

$$\frac{19.5\ \text{m}}{1.99\ \text{s}} = 9.80\ \text{m/s}$$

So:

$$\vec{\mathbf{v}}_{\text{avg1}} = (9.80\ \text{m/s, downward})$$

The second object falls for a time

$$1.99\ \text{s} + \frac{1}{16}\ \text{s} = 2.05\ \text{s}$$

so its average speed is

$$\frac{19.5\,\text{m}}{2.05\,\text{s}} = 9.51\ \text{m/s}$$

and

$$\vec{\mathbf{v}}_{\text{avg2}} = (9.51\ \text{m/s, downward})$$

So $\Delta\vec{\mathbf{v}}_{\text{avg}} = \vec{\mathbf{v}}_{\text{avg1}} - \vec{\mathbf{v}}_{\text{avg2}} = (.29\ \text{m/s, downward})$

c) From Example 2.10, we have the change in displacement caused by air resistance:

$$\delta y = \frac{\alpha}{12t^4}$$

to a good enough approximation, we may take δy as -1.2 m, the amount ball 2 lags behind ball 1 after $t = 1.99$ s. So

$$\alpha = \frac{-12\,\delta y}{t^4} = \frac{-12(-1.2\ \text{m})}{(2.0\ \text{s})^4} = 0.90\ \text{m/s}^4$$

2.61 *Question*: How long does it take for the lunch bag to reach the train window height (i.e., the time it takes the bag to fall 15 m)

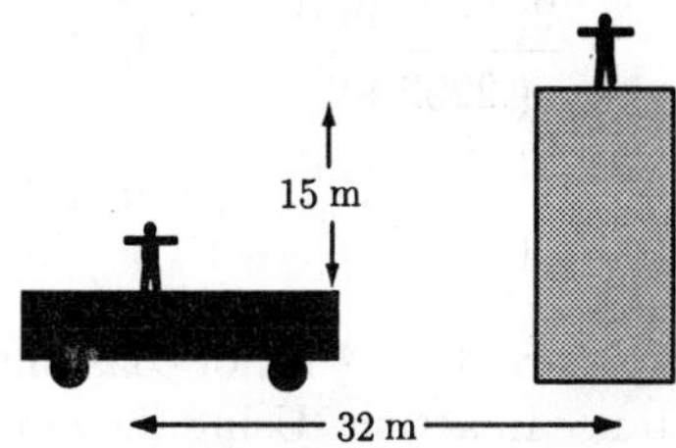

Answer:

$$15\text{ m} = \frac{1}{2}(9.8\text{ m/s}^2)t^2 \Rightarrow t = 1.75\text{ s}$$

Question: What speed does the train need to cover 32 m in 1.75 s?

Answer:

$$v = \frac{32\text{ m}}{1.75\text{ s}} = 18\text{ m/s}$$

2.65 We probably wouldn't get any improved accuracy. To see why, let's compare the size of the Apollo space craft with the distance between the Earth and Moon:

$$\frac{4\text{ m}}{3.84\times 10^5\text{ km}} = \frac{4\text{ m}}{3.84\times 10^8\text{ m}} \approx 10^{-8}$$

For comparison, if the Earth–Moon system were shrunk to the size of a person, the space craft would be 1/100 the size of a bacterium. Modeling the spacecraft as a particle is an excellent approximation.

2.69 Our knowledge of uniformly accelerated linear motion tells us that the ball's position should be described by $s = \frac{1}{2}at^2$. We would expect that

$$\frac{2s}{t^2} = \text{constant}$$

and that constant should be a, the acceleration due to Earth's gravity.

0 cm —	●	ball
		}5cm
10 cm —	•	
20 cm —	•	detectors
30 cm —	•	

When the first beam is interrupted by the edge of the sphere, its center has fallen 5.00 cm:

$$\frac{2(5.00\text{ cm})}{(.1010\text{ s})^2} = 980.3\text{ cm/s}^2$$

Similarly:

$$\begin{aligned}\frac{2(15.00\text{ cm})}{(.1751\text{ s})^2} &= 978.5\text{ cm/s}^2\\ \frac{2(25.00\text{ cm})}{(.2262\text{ s})^2} &= 977.2\text{ cm/s}^2\\ \frac{2(35.00\text{ cm})}{(.2677\text{ s})^2} &= 976.8\text{ cm/s}^2\end{aligned}$$

It appears that $\frac{2s}{t^2}$ is not constant; this is air resistance in action. Using the model of Example 2.10 for air resistance,

$$\begin{aligned}\frac{2s}{t^2} &= \frac{2}{t^2}\left(\frac{1}{2}gt^2 - \frac{\alpha}{12}t^4\right)\\ \frac{2s}{t^2} &= g - \frac{\alpha t^2}{6} \qquad (1)\end{aligned}$$

where α is our coefficient of air resistance. Given that our model of air resistance is an accurate one, the graph of $\frac{2s}{t^2}$ vs t^2 should be a straight line.

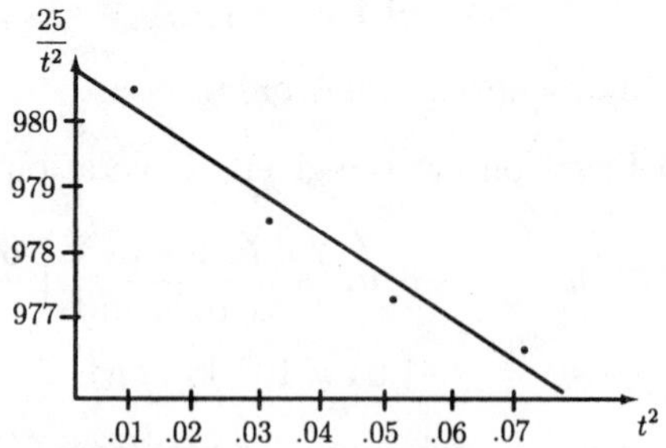

This line should have a slope of $\frac{-\alpha}{6}$ and y-intercept of g; since the equation

$$\frac{2s}{t^2} = \frac{-\alpha t^2}{6} + g$$

is of the form

$$y = mx + b$$

t^2 (s^2)	$\frac{2s}{t^2}$
.01020	980.3
.03066	978.5
.05117	977.2
.07166	976.8

There are a few ways you could find the slope of a line that fits through these points. We chose the least squares method and found

$$m = \frac{-\alpha}{6} = -57.59\text{ cm/s}^4$$

The y-intercept gives a value for g:

$$\begin{aligned}b &= g = 980.6\text{ cm/s}^2\\ \alpha &= 3.5\times 10^2\text{ cm/s}^4\end{aligned}$$

2.73 a) Let x_1 be the length of Daphne's path on land and x_2 be the length of her path on the water. Since x_1 and x_2 form hypotenuses of two right triangles, we can express them using the Pythagorean Theorem:

$$\begin{aligned}x_1 &= \sqrt{d^2+s^2}\\ x_2 &= \sqrt{(2d-s)^2+d^2},\end{aligned}$$

where s is shown in the figure.

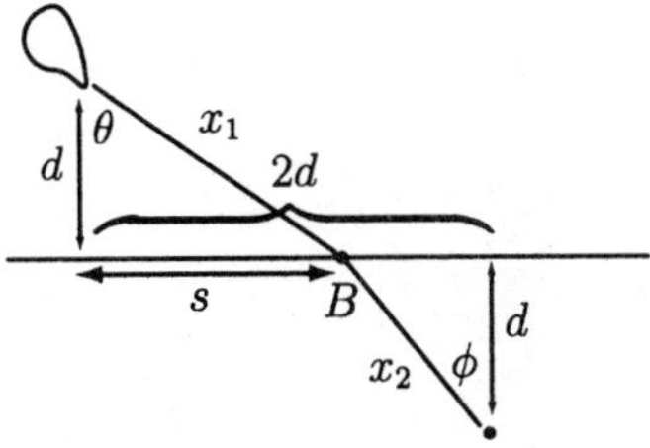

The total time that it takes Daphne to reach Paradise Island is t_1 (time on land) $+t_2$ (time in water). Thus

$$
\begin{aligned}
t_{\text{total}} &= t_1 + t_2 = \frac{x_1}{v_1} + \frac{x_2}{v_2} \\
&= \frac{\sqrt{d^2+s^2}}{v_1} + \frac{\sqrt{(2d-s)^2+d^2}}{v_2}
\end{aligned}
$$

Since we want to minimize the time it takes to reach Paradise Island by changing the point at which we reach the water (point B), we'll take the derivative of t total with respect to s.

$$
\begin{aligned}
\frac{dt}{ds} &= \frac{\frac{1}{2}\left[(d^2+s^2)\right]^{-\frac{1}{2}}(2s)}{v_1} \\
&\quad + \frac{\frac{1}{2}\left[(2d-s)^2+d^2\right]^{-\frac{1}{2}}(4d-2s)(-1)}{v_2}
\end{aligned}
$$

To find the minimum time with respect to point B we'll set $\frac{dt}{ds} \equiv 0$. This gives:

$$
\begin{aligned}
\frac{(d^2+s^2)^{-\frac{1}{2}}(2s)}{v_1} &= \frac{\left[(2d-s)^2+d^2\right]^{-\frac{1}{2}}(4d-2s)}{v_2} \\
\frac{2s}{\sqrt{d^2+s^2}}\frac{1}{v_1} &= \frac{2(2d-s)}{\sqrt{(2d-s)^2+d^2}}\frac{1}{v_2}
\end{aligned}
$$

However, notice that from the diagram,

$$\sin\theta = \frac{s}{\sqrt{d^2+s^2}}$$

and

$$\sin\phi = \frac{2d-s}{\sqrt{(2d-s)^2+d^2}}.$$

Using this information we can rewrite this equality in terms of θ and ϕ:

$$\frac{2\sin\theta}{v_1} = \frac{2\sin\phi}{v_2} \Rightarrow \frac{v_2}{v_1} = \frac{\sin\theta}{\sin\phi}$$

The conclusion, as was to be shown, is independent of specific numerical values. When $v_1 = 2v_2$ she reaches Paradise Island in minimum time when $\sin\theta = 2\sin\phi$.

b) A second relation between ϕ and ϕ is

$$2d = d\tan\theta + d\tan\phi$$

Thus

$$2 = \frac{\sin\theta}{\sqrt{1-\sin^2\theta}} + \frac{\frac{\sin\theta}{2}}{\sqrt{1-\left(\frac{\sin\theta}{2}\right)^2}} \equiv f(\theta)$$

We find by trial and error:

θ	$\sin\theta$	$f(\theta)$
40°	0.6428	1.179
50°	0.7660	1.606
45°	0.7071	1.378
60°	0.8660	2.212
55°	0.8192	1.877
58°	0.8480	2.07
57°	0.8387	2.002
56°	0.8290	1.938
56.5°	0.8339	1.970
56.75°	0.8363	1.986
56.875°	0.8375	1.994
56.9375°	0.8381	1.998
56.96881°	0.838373	1.99974

She should leave home at $\theta = 57°$.

Chapter 3

Advanced Kinematic Models

3.1 The flyball is a good example of projectile motion —constant horizontal motion and accelerated vertical motion. This acceleration which is associated with the ball's vertical motion is the acceleration of gravity and has an (approximately) constant value of 9.80 m/s^2. Since the acceleration is constant and nonzero, choices **a**, **b**, **c** and **e** are incorrect; choice **d** is the correct answer.

3.3 a) The go-kart travels an angle of 2π radians (makes 1 complete revolution) in 10.0 s. Thus its angular speed is:

$$\omega = \frac{2\ \text{rad}}{10.0\ \text{s}} = \frac{\pi}{5.00\ \text{rad}/\text{s}} = 0.628\ \text{rad}/\text{s}$$

when the go-kart makes one complete revolution, it travels a distance equal to the circumference of its track.

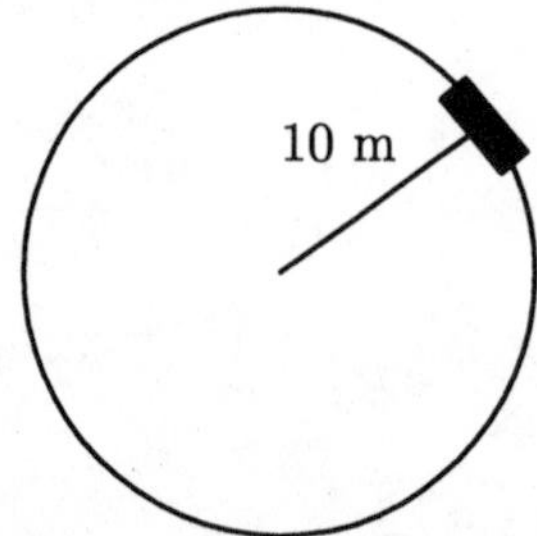

So, its speed is

$$v = \frac{2\pi r}{10.0\ \text{s}} = \frac{2\pi(10.0\ \text{m})}{10.0\ \text{s}} = 2\pi\ \text{m}/\text{s} = 6.28\ \text{m}/\text{s}$$

b) The frequency of the go-kart is the number of complete trips it makes about the track in 1 second. We can compute this by using the fact that $\omega = 2\pi f$.

$$f = \frac{\omega}{2\pi} = \frac{(\pi/5.00)\ \text{rad}/\text{s}}{2\pi\ \text{rad}} = 0.100\ \text{Hz}$$

The go-kart's acceleration (centripetal) has magnitude

$$a_{cent} = \frac{v^2}{r} = \frac{(2\pi\ \text{m}/\text{s})^2}{10.0\ \text{m}} = 3.95\ \text{m}/\text{s}^2$$

3.5 Although the car's speed is constant at 10 m / s, its velocity changes because the car changes direction. The car's velocity is always tangent to the track as shown in the figure.

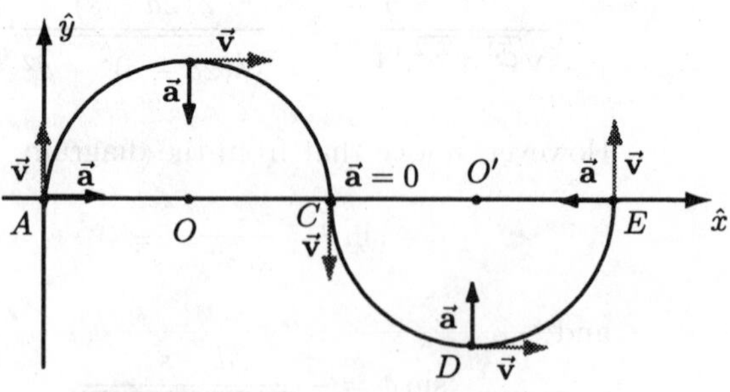

If we use the Cartesian coordinate system in the diagram the velocity vectors to be found are all $\perp$ or $\|$ to the coordinate axes:

$$\begin{aligned}
\vec{v}(A) &= 10\ \text{m}/\text{s}(\hat{\jmath}) \\
\vec{v}(B) &= 10\ \text{m}/\text{s}(\hat{\imath}) \\
\vec{v}(C) &= -10\ \text{m}/\text{s}(\hat{\jmath}) \\
\vec{v}(D) &= 10\ \text{m}/\text{s}(\hat{\imath}) \\
\vec{v}(E) &= 10\ \text{m}/\text{s}(\hat{\jmath})
\end{aligned}$$

Between A and C the car accelerates in a circle around point O and then, between C and E, around point O'. There is one point, however, where the car is circling around neither point O nor point O', that is at point C. There the track has zero curvature and the car instantaneously is not changing direction: $\vec{a}(c) = 0\ \mathrm{m/s^2}$. At all other points the centripetal acceleration has a magnitude of

$$\frac{v^2}{r} = \frac{(10\ \mathrm{m/s})^2}{5\ \mathrm{m}} = 20\ \mathrm{m/s^2}$$

and points inward along the radius towards the center of the track's curve, as shown.

$$\begin{aligned}\vec{a}(A) &= 20\ \mathrm{m/s^2}\,\hat{\imath} \\ \vec{a}(B) &= -20\ \mathrm{m/s^2}\,\hat{\jmath} \\ \vec{a}(D) &= 20\ \mathrm{m/s^2}\,\hat{\jmath} \\ \vec{a}(E) &= -20\ \mathrm{m/s^2}\,\hat{\imath}\end{aligned}$$

3.7 Your friend's and your own net displacements are each the sum of a displacement before lunch and a displacement after lunch.

a) you

$$\begin{aligned}\vec{D}_y &= (3.0\ \text{mi, north}) + (2.0\ \text{mi, northeast}) \\ &= (3.0\ \text{mi})\hat{\jmath} + (2.0\ \text{mi})\left(\tfrac{\sqrt{2}}{2}\hat{\imath} + \tfrac{\sqrt{2}}{2}\hat{\jmath}\right) \\ &= (4.4\ \text{mi})\hat{\jmath} + (1.4\ \text{mi})\hat{\imath}\end{aligned}$$

b) friend

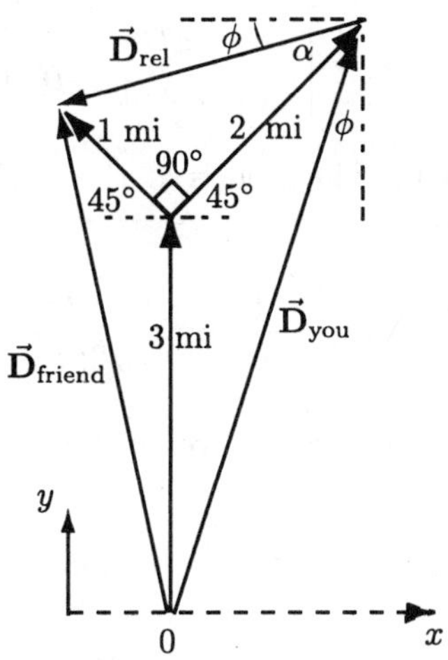

$$\begin{aligned}\vec{D}_f &= (3.0\ \text{mi, north}) + (1.0\ \text{mi, northwest}) \\ &= (3.0\ \text{mi})\hat{\jmath} + (1.0\ \text{mi})\left(-\tfrac{\sqrt{2}}{2}\hat{\imath} + \tfrac{\sqrt{2}}{2}\hat{\jmath}\right) \\ &= (3.7\ \text{mi})\hat{\jmath} - (0.71\ \text{mi})\hat{\imath}\end{aligned}$$

c) Your friend's displacement relative to you is the difference of your two individual displacements

$$\begin{aligned}\vec{D}_{\text{rel}} &= \vec{D}_f - \vec{D}_y \\ &= (1.0\ \text{mi})\left(-\frac{\sqrt{2}}{2}\hat{\imath} + \frac{\sqrt{2}}{2}\hat{\jmath}\right) \\ &\quad -(2.0\ \text{mi})\left(\frac{\sqrt{2}}{2}\hat{\imath} + \frac{\sqrt{2}}{2}\hat{\jmath}\right)\end{aligned}$$

(*Note*: The common term of $(3.0\ \text{mi})\hat{\jmath}$ subtracts out.) So

$$\vec{D}_{\text{rel}} = \left(1.0\frac{\sqrt{2}}{2}\ \text{mi}\right)\left[-3\hat{\imath} - \hat{\jmath}\right]$$

We may also express the result as a magnitude and direction:

$$\begin{aligned}\left|\vec{D}_{\text{rel}}\right| &= \left(1.0\frac{\sqrt{2}}{2}\ \text{mi}\right)\sqrt{3^2+1^2} \\ &= 1.0\sqrt{5}\,\text{mi} = 2.2\ \text{mi} \\ \tan\phi &= 1/3 \Rightarrow \phi = 18^\circ\end{aligned}$$

So $\vec{D}_{\text{rel}}$ is at 198° from the x-axis (or toward 252° in nautical terminology) (or 18° south of west).

3.9 We need the velocity of raindrops with respect to you, which is found from

$$\vec{v}_{\text{rain,you}} = \vec{v}_{\text{rain,earth}} + \vec{v}_{\text{earth,you}}$$

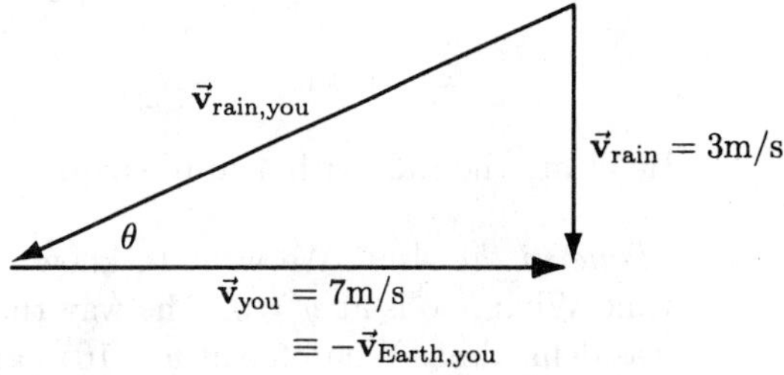

Let θ be the angle with the horizontal at which the rain drops hit you in the face. We can see that

$$\theta = \tan^{-1}\left(\frac{3\,\mathrm{m/s}}{7\,\mathrm{m/s}}\right) = 23^\circ$$

3.13 Using Eqn. (3.2) for the range R—the distance traveled horizontally to a landing point at the same height as the launch point, we have:

$$\begin{aligned}\sin 2\theta &= \frac{gR}{v_0^2} = \frac{(9.8\ \mathrm{m/s^2})(31.25\ \mathrm{m})}{(25\ \mathrm{m/s})^2} = .49 \\ \Rightarrow 2\theta &= \sin^{-1}(.49) = 29.4^\circ \\ \Rightarrow \theta &= 15^\circ, \text{or } 75^\circ\end{aligned}$$

3.17 Using the given information we can write equations for Whammo's position:

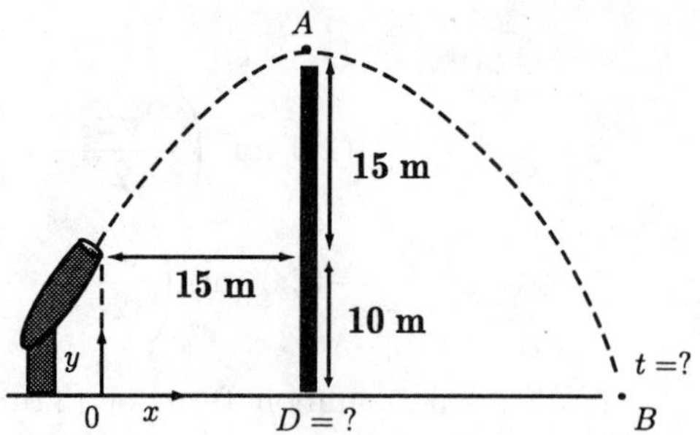

(taking point 0 to be the origin)

$$\begin{aligned} y &= 10.0 \text{ m} + (20.0 \text{ m/s})t - \frac{1}{2}gt^2 \\ x &= (10.0 \text{ m/s})t \end{aligned}$$

Does he miss the wall? Find t_A, the time in which he travels a horizontal distance of 15 m. At this time he has the same x-coordinate as the wall.

$$\begin{aligned} x &= v_x t \\ 15 \text{ m} &\overset{\text{set}}{\equiv} (10.0 \text{ m/s})t \\ \Rightarrow t &= 1.5 \text{ s} \end{aligned}$$

Now we'll find $y(1.5\,\text{s})$; We hope it's greater than 25 m!

$$\begin{aligned} y(1.5\,\text{s}) &= 10.0 \text{ m} + (20.0 \text{ m/s})(1.5 \text{ s}) \\ &\quad -\frac{1}{2}\left(9.81 \text{ m/s}^2\right)(1.5 \text{ s})^2 \\ &= 29 \text{ m} \end{aligned}$$

He clears the wall with 4 m to spare.

Time in the Air? We want to know at what time Whammo is at $y = 0$. The way the origin was defined has Whammo at $y = 10\,\text{m}$ at $t = 0$, so we'd expect that one of the times at which Whammo is at $y = 0$ will be negative; we want the other root.

$$y = 10.0 \text{ m} + (20.0 \text{ m/s})t - \frac{1}{2}(9.81 \text{ m/s}^2)t^2 \overset{\text{set}}{\equiv} 0$$

So

$$\begin{aligned} t &= -(20.0 \text{ m/s}) \\ &\quad \pm \frac{\sqrt{(400 \text{ m}^2/\text{s}^2) + 2(9.81 \text{ m/s}^2)(10.0 \text{ m})}}{-9.81 \text{ m/s}^2} \\ &= -0.45 \text{ s}, 4.5 \text{ s} \end{aligned}$$

Thus the time at which Whammo falls into the net is $t = 4.5$ s.

Where to place the net? We will calculate $x(t_B)$:

$$x(4.5 \text{ s}) = (10.0 \text{ m/s})(4.5 \text{ s}) = 45 \text{ m} = D$$

3.21 If we place our origin at the mouth of the cannon, Whammo's position at any time t is described by the equations

$$y = v_0 \sin\theta_0 t - \frac{1}{2}gt^2 \qquad (1)$$

$$x = v_0 \cos\theta_0 t \qquad (2)$$

Eliminating $t = \dfrac{x}{(v_0 \cos\theta)}$ in Eqn. (1):

$$y = x\tan\theta_0 - \frac{gx^2}{2v_0^2\cos^2\theta_0}$$

Solving the above equation for v_0 yields

$$v_0 = \left[\frac{gx^2}{[2\cos^2(\theta_0)][x\tan(\theta_0) - y]}\right]^{\frac{1}{2}}$$

We are told that Whammo is launched at $\theta = 45°$, and that the center of the ring is located at $x = 20.0$ m, $y = 10.0$ m. The initial speed necessary for Whammo to pass this point is

$$\begin{aligned} v_0 &= \left[\frac{(9.81 \text{ m/s}^2)(20.0 \text{ m})^2}{2\cos^2(45°)\,(20.0 \text{ m}\tan(45°) - 10.0 \text{ m})}\right] \\ &= 19.8 \text{ m/s} \end{aligned}$$

Putting the given dimensions into the algebraic result for v_0,

$$v_0 = \left[\frac{gD^2}{2\left(\frac{1}{2}\right)\left(D \cdot 1 - \frac{D}{2}\right)}\right]^{\frac{1}{2}} = \sqrt{2gD}$$

Then, Whammo is at the center of the ring at time

$$t = \frac{D}{v_0\cos\theta} = \frac{D}{\sqrt{2gD}\frac{1}{\sqrt{2}}} = \sqrt{\frac{D}{g}}.$$

At this time, the vertical velocity component is

$$v_y = v_0\sin\theta - gt = \sqrt{2gD}\frac{1}{\sqrt{2}} - g\sqrt{\frac{D}{g}} = 0,$$

as we were to show. If Whammo is moving horizontally as he passes through the ring, the vertical component of his velocity will be zero at that point.

3.25 a) We take our origin to be where the ball is at $t = 0$ s which means $x_i = y_i = 0$ m.

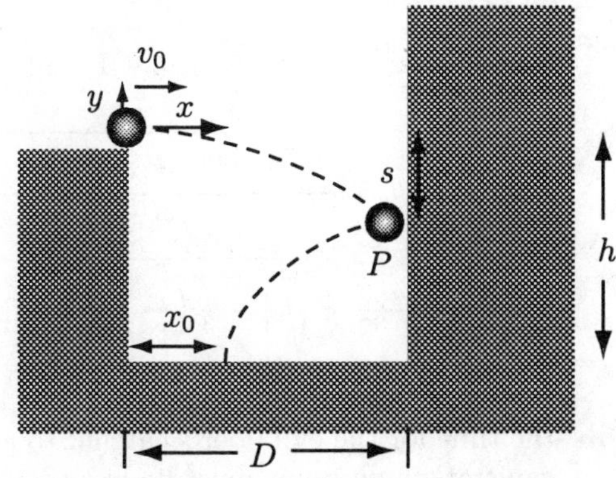

Initially the ball has no vertical velocity; its horizontal velocity is

$$v_{x_i} = v_0$$

So

$$\begin{aligned} y &= -\frac{1}{2}gt^2 & (3) \\ x &= v_0 t & (4) \end{aligned}$$

b) The time that it takes the ball to hit the opposite wall is the time that it takes for the ball to travel a horizontal distance D at a velocity v_0. This time is

$$t = \frac{D}{v_0}$$

In this time $t = \frac{D}{v_0}$ the ball falls downward a distance s. Using this time with Eqn. (3) gives us an expression for s.

$$s = \left| -\frac{1}{2}g\left(\frac{D}{v_0}\right)^2 \right| = +\frac{gD^2}{2v_0^2},$$

and the velocity v_y is given by:

$$v_y = -at = -g\frac{D}{v_0}.$$

c) The ball is now at point P. To find an expression for x_0 we need to know how much time t_f it will take the ball to drop a distance of $h-s$, with initial downward velocity $v_y = -\frac{gD}{v_0}$:

$$-(h-s) \overset{\text{set}}{\equiv} \left(-\frac{gD}{v_0}\right)t_f - \frac{1}{2}gt_f^2$$

$$\frac{1}{2}t_f^2 + \frac{D}{v_0}t_f - \left(\frac{h}{g} - \frac{D^2}{2v_0^2}\right) = 0$$

The quadratic has solutions

$$\begin{aligned} t_f &= -\frac{D}{v_0} \pm \frac{1}{v_0}\sqrt{D^2 + \frac{(2v_0^2 h - gD^2)}{g}} \\ &= -\frac{D}{v_0} \pm \sqrt{\frac{2h}{g}} \end{aligned}$$

Taking the positive root gives us the time in which the ball falls from $y = -s$ to $y = -h$, the distance $h - s$. If the ball has a horizontal velocity of $-v_0$, x_0 is given by:

$$\begin{aligned} x_0 &= D - v_0 t_f \\ &= 2D - v_0\sqrt{\frac{2h}{g}} \end{aligned}$$

Setting $x_0 = 0$ and solving for v_0 gives

$$v_0 = D\sqrt{\frac{2g}{h}}$$

The problem is much easier if you notice that the path after hitting the wall is the reverse of the path the ball would follow if the wall were not there. Then the question becomes: *What v_0 is needed to travel a distance $2D$ while falling a height h?*

3.29 The equations for the banana's position are given by:

$$\begin{aligned} y &= (v_0 \sin\theta)t - \frac{1}{2}gt^2 \\ x &= (v_0 \cos\theta)t \end{aligned}$$

Since we do not need to know the banana's time of flight, eliminate t from the eqns, as in Eqn. 3.3:

$$y = x\tan - \frac{gx^2}{2v_0^2}(1 + \tan^2\theta)$$

Rearranging terms gives a quadratic equation for $\tan\theta$

$$-\tan^2\theta + \frac{2v_0^2}{gx}\tan\theta - \frac{gx^2 + 2v_0^2 y}{gx^2} = 0,$$

with solutions:

$$\begin{aligned} \tan\theta &= \frac{v_0^2}{gx} \pm \sqrt{\frac{v_0^4}{g^2x^2} - \left(1 + \frac{2v_0^2}{g}\right)} \\ &= \frac{v_0^2}{gx}\left[1 \pm \sqrt{1 - \frac{g^2x^2}{v_0^4}\left(1 + \frac{2v_0^2 y}{gx^2}\right)}\right] \end{aligned}$$

We are told that the banana is thrown at $v_0 = 8.0$ m/s and it is caught at $x = 2.0$ m, $y = 3.0$ m. Then

$$\frac{v_0^2}{gx} = \frac{(8.0\ \text{m/s})^2}{[(9.8\ \text{m/s}^2)(3.0\ \text{m})]} = 2.18$$

and

$$\begin{aligned} \tan\theta &= 2.18\left[1 \pm \sqrt{1 - \frac{\left[1 + \frac{2(2.18)(2.0)}{3.0}\right]}{(2.18)^2}}\right] \\ &= 1.26, 3.09 \end{aligned}$$

The corresponding values of θ are $\theta = 52°, 72°$. The keeper would probably throw at 52° to the non-jumping monkey.

3.33 Since Rimtown and Copernicus Spaceport are at the same level, the Rimtown postmaster uses a launch speed v_i such that the canisters range at launch angle 30° is the given distance D to the spaceport:

$$D = \frac{v_i^2}{g}\sin 60° \ \Rightarrow\ v_i^2 = \frac{2gD}{\sqrt{3}}$$

The time for the canister to reach the top of its path is

$$t_{\text{top}} = \frac{v_i \sin 30°}{g} = \frac{\sqrt{\frac{2gD}{\sqrt{3}}}\left(\frac{1}{2}\right)}{g} = \sqrt{\frac{D}{2\sqrt{3}g}}$$

The height of the top of the path is found from

$$\Delta y = \frac{v_f^2 - v_i^2}{2a} = \frac{0 - (v_i \sin 30°)^2}{-2(g)} = \frac{\left(\frac{2gD}{\sqrt{3}}\right)\frac{1}{4}}{2g} = \frac{D}{4\sqrt{3}}$$

Now, the evildoers' problem is to fire the catcher at speed v_e and angle θ_e and have it pass through the point $x_e = D, y_e = \frac{D}{4\sqrt{3}}$. (The origin for x_e and y_e is at the evildoers' launch site). Then $x_e = v_e \cos\theta_e t$ and

$$y_e = v_e \sin\theta_e t - \frac{1}{2}gt^2$$

$$\Rightarrow y = x_e \tan\theta_e - \frac{g x_e^2}{2v_e^2\cos^2\theta_e}$$

Thus the evildoers' constraint is

$$\frac{D}{4\sqrt{3}} = D\tan\theta_e - \frac{gD^2}{2v_e^2\cos^2\theta_e}$$

or

$$1 = 4\sqrt{3}\tan\theta_e - \frac{2\sqrt{3}gD}{v_e^2\cos^2\theta_e} = 4\sqrt{3}\tan\theta_e - \frac{3}{\cos^2\theta_e}\frac{v_i^2}{v_e^2}$$

Now

$$\frac{1}{\cos^2\theta_e} = \sec^2\theta_e = 1 + \tan^2\theta_e$$

so

$$1 = 4\sqrt{3}\tan\theta_e - 3\left(\frac{v_i}{v_e}\right)^2(1+\tan^2\theta_e)$$

$$3\left(\frac{v_i}{v_e}\right)^2\tan^2\theta_e - 4\sqrt{3}\tan\theta_e + 1 + 3\left(\frac{v_i}{v_e}\right)^2 = 0$$

$$\Rightarrow \tan\theta_e = \frac{4\sqrt{3} \pm \sqrt{48 - 12\left(\frac{v_i}{v_e}\right)^2\left(1+3\left(\frac{v_i}{v_e}\right)^2\right)}}{6\left(\frac{v_i}{v_e}\right)^2}$$

$$= \left(\frac{v_e}{v_i}\right)^2\left[\frac{2\sqrt{3}}{3} \pm \sqrt{\frac{4}{3} - \frac{1}{3}\left(\frac{v_i}{v_e}\right)^2\left(1+3\left(\frac{v_i}{v_e}\right)^2\right)}\right]$$

If they are to wait until the postmaster launches, the time for the evil doer's catcher to get to the canister's high point must be less than or equal to the time for the canister to arrive there

$$t_e = \frac{D}{v_e\cos\theta_e} \le \sqrt{\frac{D}{2\sqrt{3}g}}$$

or

$$v_e\cos\theta_e \ \ge\ \sqrt{2\sqrt{3}gD}$$

$$\Rightarrow \frac{v_e}{v_i}\cos\theta_e \ge \frac{\sqrt{2\sqrt{3}gD}}{\sqrt{\frac{2gD}{\sqrt{3}}}} = \sqrt{3}$$

Since $\frac{v_e}{v_i} \le 2$, this requires $\cos\theta_e > \frac{\sqrt{3}}{2}$ or $\theta_e < 30°$. The evildoers need not rely on the postmaster to launch on schedule.

3.37 The average acceleration is defined as $\frac{\Delta\vec{\mathbf{v}}}{\Delta t}$. When a particle makes one complete revolution in a circular path it ends up at the same location from which it started with a velocity which is equal in both magnitude and direction to the velocity it had when it started. Thus $\Delta\vec{\mathbf{v}} = \vec{\mathbf{v}}_f - \vec{\mathbf{v}}_i = 0$; therefore $\vec{\mathbf{a}}_{\text{avg}}$ must be 0. Another argument would be to picture how $\vec{\mathbf{a}}_{\text{cent}}$ changes as the particle moves along its circular path.

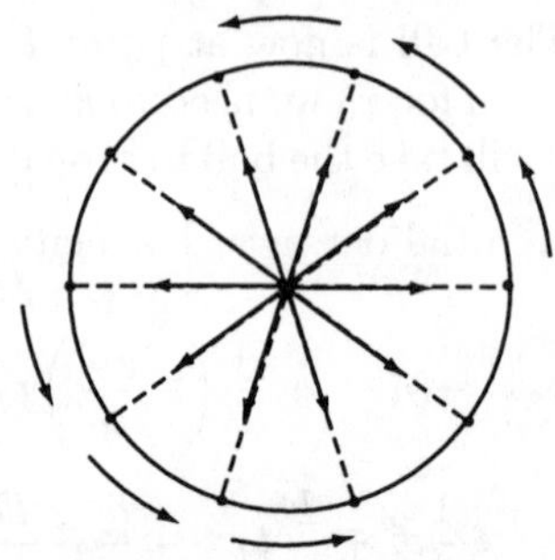

The picture shows just this. In one revolution $\vec{\mathbf{a}}_{\text{cent}}$ (which is constant in magnitude) moves through 2π. For each $\vec{\mathbf{a}}_{\text{cent}}$ pointing in some arbitrary direction $\hat{\mathbf{r}}$ there's an equal in magnitude $-\vec{\mathbf{a}}_{\text{cent}}$ which points in the opposite direction $-\hat{\mathbf{r}}$. The average of all these vectors must necessarily be 0.

3.41 Since $\vec{\mathbf{v}}_{\text{avg}} = \frac{\vec{\mathbf{D}}}{\Delta t}$, if the car travels from point A to point B, its displacement is represented best by choice **d**, as is its $\vec{\mathbf{v}}_{\text{avg}}$. Average acceleration is defined as $\vec{\mathbf{a}}_{\text{avg}} = \frac{\Delta\vec{\mathbf{v}}}{\Delta t}$.

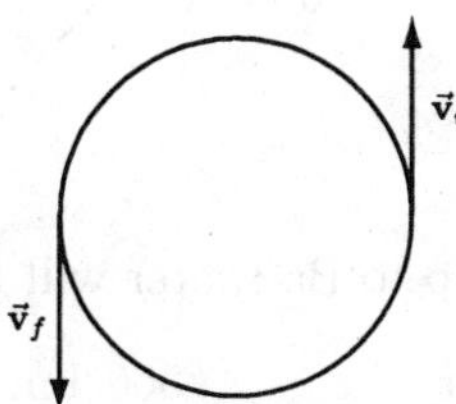

In the diagram below we can see what $\Delta\vec{\mathbf{v}}$ looks like by drawing $\vec{\mathbf{v}}_f - \vec{\mathbf{v}}_i$:

$\vec{\mathbf{v}}_f$ + $-\vec{\mathbf{v}}_i$ = $\Delta\vec{\mathbf{v}}$

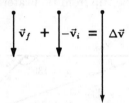

Since $\vec{\mathbf{a}}_{\text{avg}} = \frac{\Delta\vec{\mathbf{v}}}{\Delta t}$, where Δt is a scalar, the best vector to represent $\vec{\mathbf{a}}_{\text{avg}}$ is **b**, since dividing a vector by a scalar will change its magnitude but not its direction.

3.45 Since the train cruises at constant speed the accelerations to be considered arise from changing its direction around curves. On the smallest allowable curve the train's centripetal acceleration is

$$a_{\text{cent}} = .050\,g$$

So,

$$a_{\text{cent}} = \frac{v^2}{r} = .050g \Rightarrow r = \frac{v^2}{0.050g}$$

$$\begin{aligned} r &= \frac{\left[(216\ \text{km}/\text{h})\left(\frac{1000\ \text{m}}{1\ \text{km}}\right)\left(\frac{1\ \text{h}}{3600\ \text{s}}\right)\right]^2}{.050(9.8\ \text{m}/\text{s}^2)} \\ &= 7.3\ \text{km} \end{aligned}$$

3.49 The speed is given by

$$\begin{aligned} v_{\text{tip}} &= \omega r_{\text{tip}} \\ &= (2625\ \text{rpm})\left(\frac{2\pi\ \text{rad}}{1\ \text{rev}}\right)\left(\frac{1\ \text{min}}{60\ \text{s}}\right) \\ &\quad \cdot\left(\frac{2.08\ \text{m}}{2}\right) \\ &= 286\ \text{m}/\text{s} \end{aligned}$$

close to, but not quite equal to, the speed of sound. Propellers with tips moving near the speed of sound can be quite noisy. Airports now attempt to route light aircraft away from residential areas.

3.53 The problem states that the shuttle makes 1 complete orbit in 89.5 min, or 5370 s. Using this we can find the shuttle's centripetal acceleration, which is:

$$\begin{aligned} |\vec{\mathbf{a}}_{\text{cent}}| &= \omega^2 R = \left(\frac{2\pi}{T}\right)^2 R \\ &= \frac{4\pi^2}{(5370\ \text{s})^2}(6.63\times10^6\ \text{m}) \\ &= 9.08\ \text{m}/\text{s}^2 \end{aligned}$$

If we wanted to express this in terms of g, the acceleration at the Earth's surface due to gravity

$$x(9.81\ \text{m}/\text{s}^2) = (9.08\ \text{m}/\text{s}^2) \Rightarrow x = .926$$

The shuttle's acceleration is $.926g$ where $g = 9.81\ \text{m}/\text{s}^2$, and is directed to the center of the Earth. Using $r_{\text{earth}} = 6.38\times10^6$ m we can write the proportion as

$$\left(\frac{6.38\times10^6\ \text{m}}{6.63\times10^6\ \text{m}}\right)^n = \frac{.926g}{g} \Rightarrow \left(\frac{6.38}{6.63}\right)^n = .926$$

Taking the log of both sides gives

$$\begin{aligned} \ln\left(\frac{6.38}{6.63}\right)^n &= \ln(.926) \\ \Rightarrow\ n\cdot\ln\left(\frac{6.38}{6.63}\right) &= \ln(.926) \\ \Rightarrow\ n = \frac{\ln(.926)}{\ln\left(\frac{6.38}{6.63}\right)} &= 2.00 \end{aligned}$$

3.57 When you travel with the wind—when the wind blows in the same direction in which you're traveling—you feel it moving much less slowly relative to you.

observer from water — v_{wind} →, v_{boat} →

observer from boat — v_{wind} →, $v_{\text{boat}} = 0$

In fact, you can really only detect the difference between your velocity and the wind velocity. When you travel in the direction opposite to the wind's direction, you feel it moving much faster relative to you.

observer from water — v_{wind} ←, v_{boat} →

observer from boat — v_{wind} ←, $v_{\text{boat}} = 0$

In this case you'll feel the wind traveling at the sum of the wind's speed and your speed. Thus, even though the boat moves more slowly close-hauled, its speed *relative to the air* is greater.

3.61 Choose the y-direction across the river and the x-direction downstream, as shown.

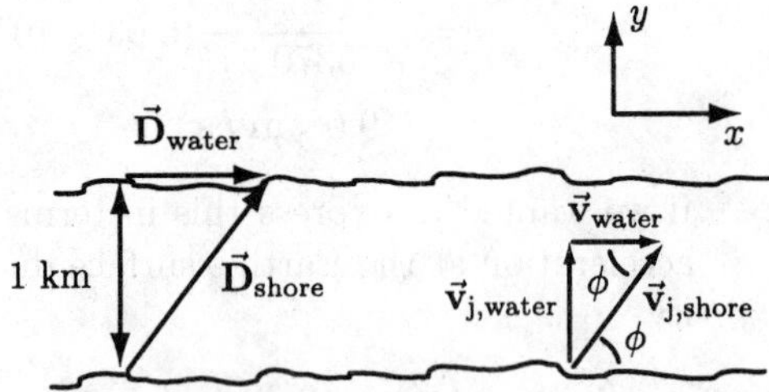

Taking the particle "p" as (John's boat), "1" as the shore reference frame and "2" as the water reference frame, John's velocity with respect to the shore is

$$\vec{v}_{shore} = (2.0 \text{ km/h})\hat{\imath} + (3.0 \text{ km/h})\hat{\jmath}$$

The triangles formed by John's velocity and displacement vectors are similar. John moves in the y-direction at 3 km/h whether the water moves or not. So, it takes him a time

$$\Delta t = \frac{\text{(displacement wrt shore)}}{\text{(velocity} \perp \text{shore)}} = \frac{(1 \text{ km})}{(3 \text{ km/h})} = \frac{1}{3} \text{ h}$$

according to either shore or water reference frames. John's displacement with respect to the shore has magnitude

$$\left|\vec{D}_{shore}\right| = \left|\vec{v}_{\hat{\jmath},shore}\right| \Delta t = \sqrt{(2 \text{ km/h})^2 + (3 \text{ km/h})^2}(1/3 \text{ h}) = \frac{\sqrt{13}}{3} \text{ km} = 1.2 \text{ km}$$

Its direction (ccw from the x-axis) is

$$\phi = \tan^{-1}\frac{|\vec{v}_{j,water}|}{|\vec{v}_{water}|} = \tan^{-1}\frac{3}{2} = 56°$$

John's displacement downstream is

$$\vec{D}_{water} = \vec{v}_{water}\Delta t = (2 \text{ km/h}\,\hat{\imath})\left(\frac{1}{3}\text{h}\right) = (0.67 \text{ km})\hat{\imath}.$$

3.65 A little geometry tells us that the 3 cities form 3 vertices of a square. On the trip from Aces to Badwater the wind blows parallel to the plane's path; their velocities are parallel and the sum of their speeds gives the plane's ground speed.

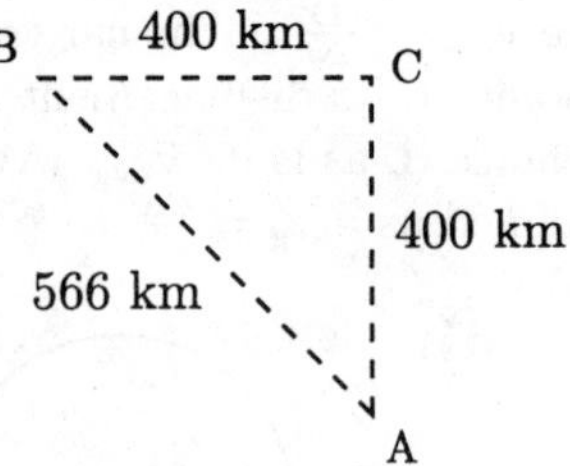

The trip to Badwater will take a time

$$t = \frac{x}{v} = \frac{566 \text{ km}}{300 \text{ km/h} + 60 \text{ km/h}} = 1.57 \text{ h}.$$

We now need to calculate $|\vec{v}_{p,ground}|$ to find how long the trip from Badwater to Clinch City will take.

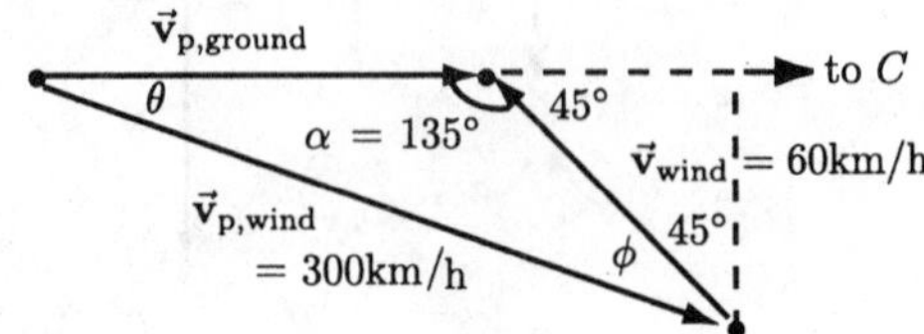

In the diagram we show $\vec{v}_{p,ground}$ pointing toward Clinch City, which is what the pilot wishes to achieve. We are given $\vec{v}_{wind}$; $\vec{v}_{p,wind}$ is what the plane must do. Since the wind is toward the NW, $\alpha = 135°$. Before calculating $\vec{v}_{p,ground}$, we need to compute θ which we can do with the sine law. *Note*: we have no way of calculating ϕ unless we calculate θ first.

$$\frac{\sin\theta}{60 \text{ km/h}} = \frac{\sin 135°}{300 \text{ km/h}} \Rightarrow \theta = 8.13°$$

Thus,

$$\phi = 180° - 135° - 8.13° = 36.9°.$$

We can now compute

$$\vec{v}_{p,ground} = [(300 \text{ km/h})^2 + (60 \text{ km/h})^2 - 2(300 \text{ km/h})(60 \text{ km/h})\cos 36.9°] = 255 \text{ km/h}$$

This trip will take a time

$$t = \frac{x}{v} = \frac{400 \text{ km}}{255 \text{ km/h}} = 1.57 \text{ h}$$

The total time it will take for the plane to go from Aces to Badwater to Clinch City is

$$1.57 \text{ h} + 1.57 \text{ h} = 3.14 \text{ h}$$

3.69 In getting to Treasure Cove Mei-Lin swims in the same direction as the current.

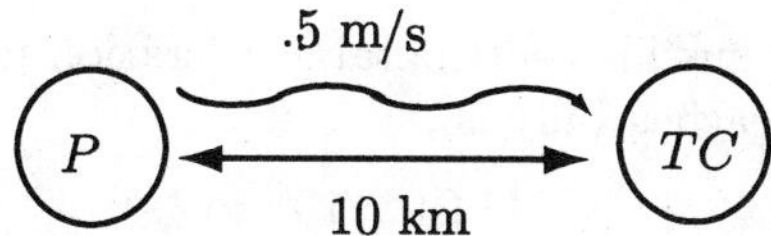

The current aids her in getting to Treasure Cove.

$$v_{\text{Mei-Lin}} = 3.0\ \text{m/s} + 0.5\ \text{m/s} = 3.5\ \text{m/s}$$

$$t_{\text{to TC}} = \frac{x}{v} = \frac{10.0\ \text{km}}{3.5\ \text{m/s}} = 0.79\ \text{h}$$

Swimming back to Paradise Island, she will be swimming against the current, which will impede her in getting back

$$\begin{aligned} v_{\text{Mei-Lin}} &= 3.0\ \text{m/s} - 0.5\ \text{m/s} = 2.5\ \text{m/s} \\ t_{\text{back}} &= \frac{x}{v} = \frac{10.0\ \text{km}}{2.5\ \text{m/s}} = 1.1\ \text{h} \end{aligned}$$

She's cutting it close; her round trip will take 1.9 hours, just 6 minutes short of her limit, but she should be able to do it. If the current were in the other direction, it would still take her 1.9 h for the round trip. All that would happen is that it would take her 1.1 h to get there and 0.8 h to get back.

3.73 The equations describing the position of both the package and friend are:

package	friend
$y_p = -\frac{1}{2}gt^2$	$y_f = -h$
$x_p = v_p t$	$x_f = d - v_f t$

where $v_f = 1.0\ \text{m/s}$.

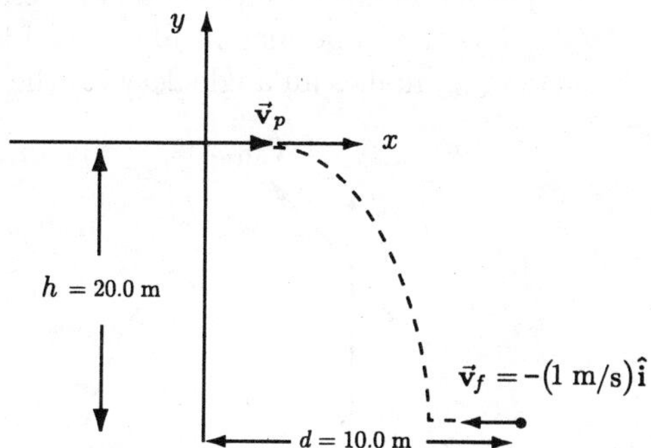

When the friend catches the package, $y_p = y_f$ and $x_p = x_f$:

$y_p = y_f$:

$$-\frac{1}{2}gt^2 = -h \Rightarrow t = \sqrt{\frac{2h}{g}}$$

$x_p = x_f$:

$$\begin{aligned} v_p t &= d - v_f t \\ v_p &= \frac{d}{t} - v_f = d\sqrt{\frac{g}{2h}} - v_f \\ &= (10.0\ \text{m})\sqrt{\frac{9.8\ \text{m/s}^2}{2(20.0\ \text{m})}} - 1.0\ \text{m/s} \\ &= 4.0\ \text{m/s} \end{aligned}$$

An initial horizontal velocity of 4.0 m/s is required if the friend is to catch the package.

3.77 The motion of a point on the rim is the combination of the linear motion of the bicycle down the street with the rotational motion of the rim about the center of the wheel.

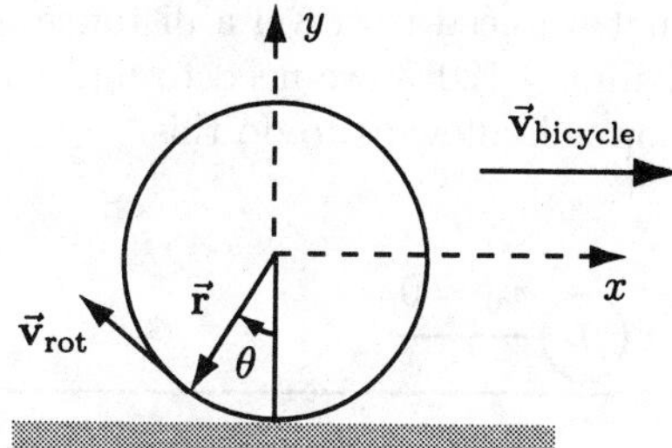

Since the wheel rolls without slipping, both contributions have the same magnitude

$$v_0 = v_{\text{bicycle}} = 5.0\ \text{m/s}$$

The angular speed of the wheel is

$$\omega = \frac{v_0}{r} = 1.0 \times 10^1\ \text{rad/s}$$

If we choose $t = 0$ when $\theta = 0$, then $\theta = \omega t$, and the vector sum in the next diagram gives the velocity of the rim point with respect to the ground.

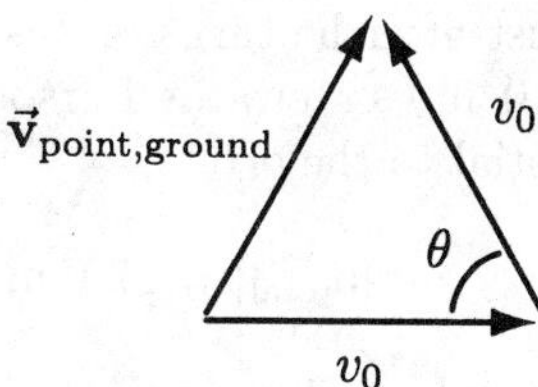

The required speed is the magnitude of this vector. From the law of cosines:

$$\begin{aligned} |\vec{v}_{\text{point,ground}}| &= \sqrt{v_0^2 + v_0^2 - 2v_0v_0\cos(\theta)} \\ &= v_0\sqrt{2(1-\cos\theta)} \\ &= (7.1\ \text{m/s})\sqrt{1 - \cos[(10\ \text{rad/s})t]} \end{aligned}$$

3.81 There are two things we can do to help us solve this problem.

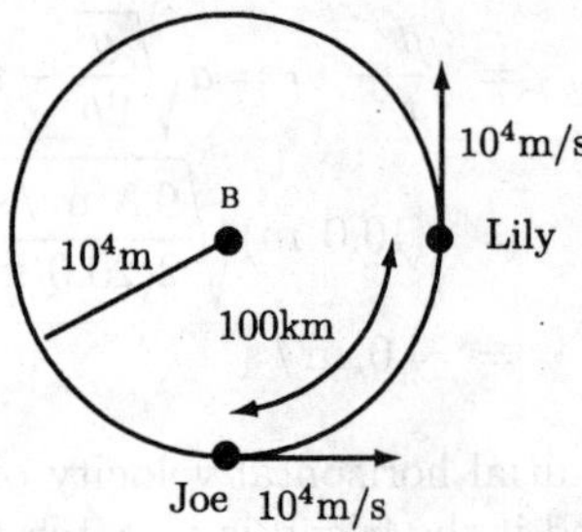

The first is a change of reference frame; if we picture ourselves on Joe's craft, we would be traveling at the same speed as Lilly. The second thing we can do is picture the path we are about to take as being "stretched out." We are given that we need to travel a distance of 100 km in 2 min = 120 s; we need to find what acceleration will allow us to do this.

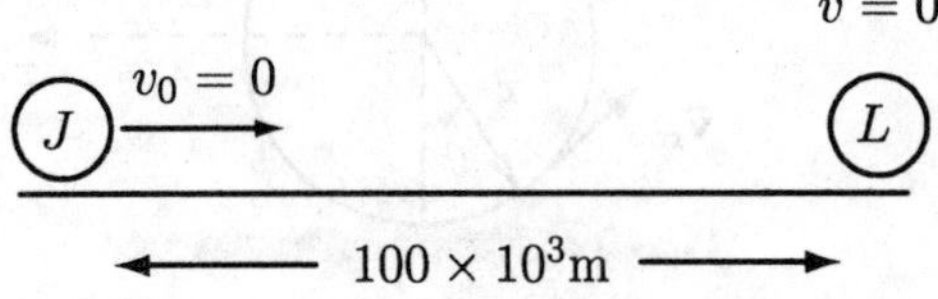

$$100 \times 10^3 \text{ m} = \frac{1}{2}a(120 \text{ s})^2 \Rightarrow a = 13.9 \text{ m/s}^2$$

This acceleration is parallel to Joe's orbit; it increases his speed in orbit, as opposed to his centripetal acceleration that actually keeps him in orbit. Just after he starts up his engines, he'll still be (approximately) traveling at $v_0 = 104$ m/s, so his centripetal acceleration will be

$$\vec{\mathbf{a}}_{\text{cent}} = \frac{(10^4 \text{ m/s})^2}{10 \text{ m}} = 10.0 \text{ m/s}^2$$

directed towards Barsoom. His total acceleration just after he turns on his engines is therefore (10 m/s^2 towards Barsoom) +(13.9 m/s tangential to the orbit),

$$|\vec{\mathbf{a}}_{\text{total}}| = 17.1 \text{ m/s}^2$$

Joe's total acceleration just before he reaches Lilly will still have the 13.9 m/s^2 component tangential to his orbit. His centripetal acceleration will have increased since he's now going at a faster speed. His speed just before he reaches Lilly is:

$$\begin{aligned} v_f^2 - (10^4 \text{ m/s})^2 &= 2(13.9 \text{ m/s}^2)(100 \times 10^3 \text{ m}) \\ &\Rightarrow v_f = 1.01 \times 10^4 \text{ m/s} \end{aligned}$$

Thus his centripetal acceleration just before he reaches Lilly is:

$$|\vec{\mathbf{a}}_{\text{cent}}| = \frac{(1.01 \times 10^4 \text{ m/s})^2}{10^7 \text{ m}} = 10.3 \text{ m/s}^2$$

and

$$|\vec{\mathbf{a}}_{\text{total}}| = 17.3 \text{ m/s}^2$$

3.85 The ball's velocity just before it hits the surface is:

$$v_f^2 - v_i^2 = 2gh \Rightarrow \vec{\mathbf{v}}_y = -\sqrt{2gh}\,\hat{\mathbf{j}}$$

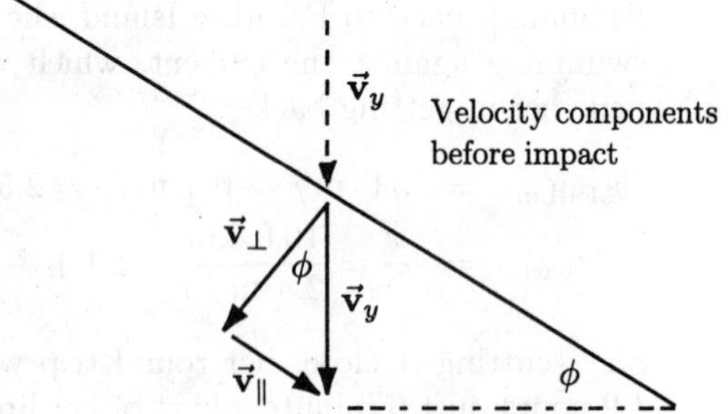

We need to analyze the ball's velocity components $\perp$ and $\parallel$ to the surface before the collision to see what happens after the collision.

After impact

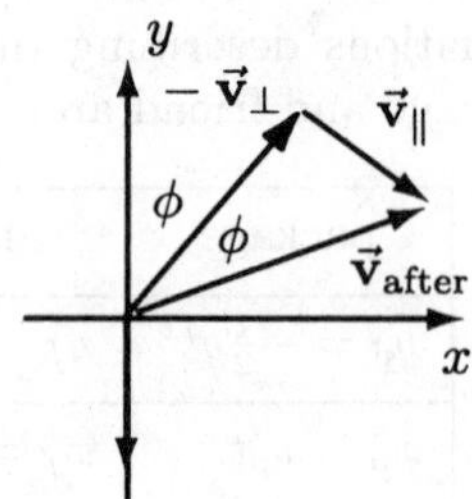

The parallel component of $\vec{\mathbf{v}}$ is unchanged, while the perpendicular component is reversed. So, $\vec{\mathbf{v}}_{\text{after}}$ has the same magnitude as $\vec{\mathbf{v}}_y$ before impact; $\vec{\mathbf{v}}_{\text{after}}$ makes an angle 2ϕ with the vertical.

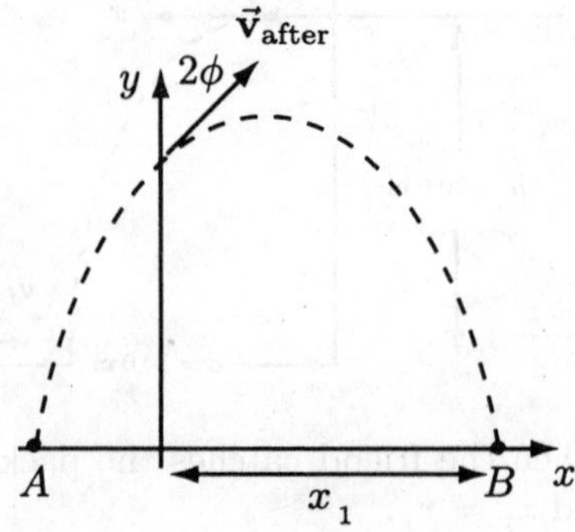

We can now restart the problem with this information: where

$$\begin{aligned} v_{oy} &= \sqrt{2gh}\cos 2\phi\,\hat{\mathbf{j}} \\ v_{ox} &= \sqrt{2gh}\sin 2\phi\,\hat{\mathbf{i}} \end{aligned}$$

We'll find the time it takes for the ball to reach $y = 0$:

$$\begin{aligned} 0 &= h + \sqrt{2gh}\cos 2\phi t - \frac{1}{2}gt^2 \\ t &= \frac{-\sqrt{2gh}\cos 2\phi \pm \sqrt{2gh\cos^2 2\phi + 2gh}}{-g} \\ &= \sqrt{\frac{2h}{g}}\left(+\cos 2\phi + \sqrt{\cos^2 2\phi + 1}\right) \end{aligned}$$

(Use the positive root—the negative root corresponds with point A.) In this time, the ball will travel a horizontal distance of

$$\begin{aligned} x &= v_{ox}t \\ &= \sqrt{2gh}\sin 2\phi\sqrt{\frac{2h}{g}}\left[\cos 2\phi + \sqrt{\cos^2 2\phi + 1}\right] \\ &= 2h\sin 2\phi\left[\cos 2\phi + \sqrt{\cos^2 2\phi + 1}\right] \end{aligned}$$

3.89 Case 1: $A \to B \to C$

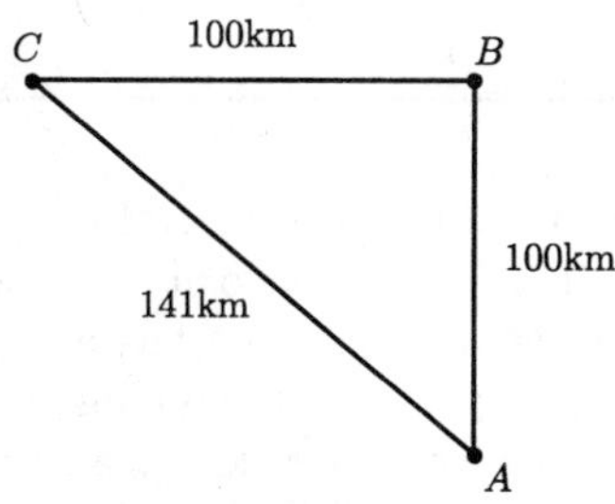

From $A \to B$,

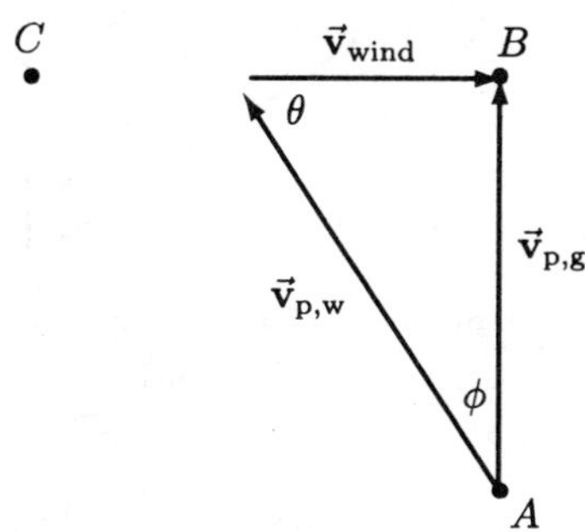

$\vec{v}_{\text{p,g}}$ is the plane's velocity over the ground needed to reach Bodie. $\vec{v}_{\text{p,w}}$ is the velocity of the plane through the air, and has a magnitude of 100 km / h. We cannot use the cosine law to find $|\vec{v}_{\text{p,g}}|$ directly since we don't know θ. We can get ϕ, though:

$$\sin\phi = \frac{70.7 \text{ km/h}}{100 \text{ km/h}} = \frac{1}{\sqrt{2}} \quad \Rightarrow \quad \phi = 45°$$

Since all triangles have an interior angle sum of $180°, \theta = 45°$. Thus the triangle is isosceles, and

$$|\vec{v}_{\text{p,g}}| = 70.7 \text{ km/h}$$

Thus, in going from A to B it will take the plane a travel time of:

$$t_{A\to B} = \frac{100 \text{ km}}{70.7 \text{ km/h}} = 1.41 \text{ h}$$

From $B \to C$

Here, the plane is flying directly into the wind. There is no angle correction in $\vec{v}_{\text{p,w}}$, but the plane will be traveling a lot slower.

$$\vec{v}_{\text{p,g}} = \vec{v}_{\text{p,w}} + \vec{v}_{\text{w}}$$

so

$$\begin{aligned} |\vec{v}_{\text{p,g}}| &= 100 \text{ km/h} - 70.7 \text{ km/h} \\ &= 29.3 \text{ km/h} \end{aligned}$$

In going from B to C it will take the plane a travel time of:

$$t_{B\to C} = \frac{100 \text{ km}}{29.3 \text{ km/h}} = 3.41 \text{ h}$$

The total time for the trip $A \to B \to C$ is

$$1.41 \text{ h} + 3.41 \text{ h} = 4.82 \text{ h}$$

Case 2: $A \to C \to B$

From $A \to C$

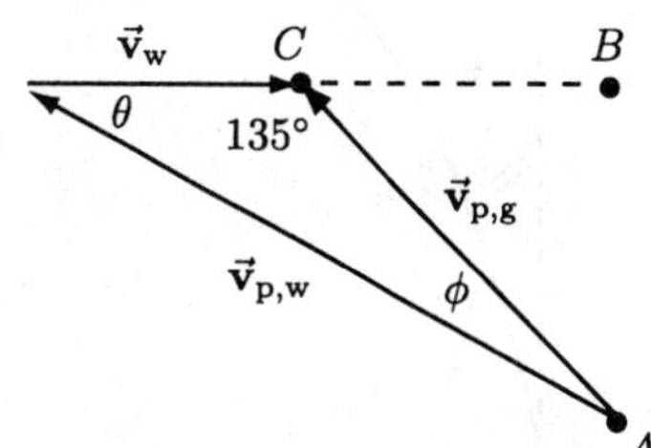

We would like to know θ to find $\vec{v}_{\text{p,g}}$ with the cos law; ϕ is obtained from the sin law.

$$\frac{\sin\phi}{70.7 \text{ km/h}} = \frac{\sin 135°}{100 \text{ km/h}} \Rightarrow \phi = 30°$$

Thus

$$\theta = 180° - 30° - 135° = 15°$$

$$\begin{aligned} |\vec{v}_{\text{p,g}}| &= \left[(100 \text{ km/h})^2 + (70.7 \text{ km/h})^2 \right. \\ &\quad \left. - 2(100 \text{ km/h})(70.7 \text{ km/h})\cos 15°\right]^{\frac{1}{2}} \\ &= 36.6 \text{ km/h} \end{aligned}$$

The time in going from A to C is:

$$t_{A\to C} = \frac{141 \text{ km}}{36.6 \text{ km/h}} = 3.85 \text{ h}$$

From $C \to B$

In this case the plane is flying parallel to the wind and in the same direction. Their velocities will add.

$$\begin{aligned}\vec{\mathbf{v}}_{\rm p,g} &= \vec{\mathbf{v}}_{\rm w} + \vec{\mathbf{v}}_{\rm p,w} \\ |\vec{\mathbf{v}}_{\rm p,g}| &= (70.7\ \mathrm{km/h}) + (100\ \mathrm{km/h}) \\ &= 170.7\ \mathrm{km/h}\end{aligned}$$

The time in going from C to B is:

$$t_{C\to B} = \frac{100\ \mathrm{km}}{170.7\ \mathrm{km/h}} = .586\ \mathrm{h}$$

The whole trip will take

$$3.85\ \mathrm{h} + .586\ \mathrm{h} = 4.44\ \mathrm{h}$$

It is faster for the plane to fly from $A \to C \to B$ than it is to fly from $A \to B \to C$. The wind has its biggest effect when you fly with it or against it. The $A \to C \to B$ route takes the head wind at an angle and gets full benefit from the tail wind.

3.97 **a)** To find the angle ϕ, we need vertical and horizontal velocity components for the gymnast as she leaves the horse.

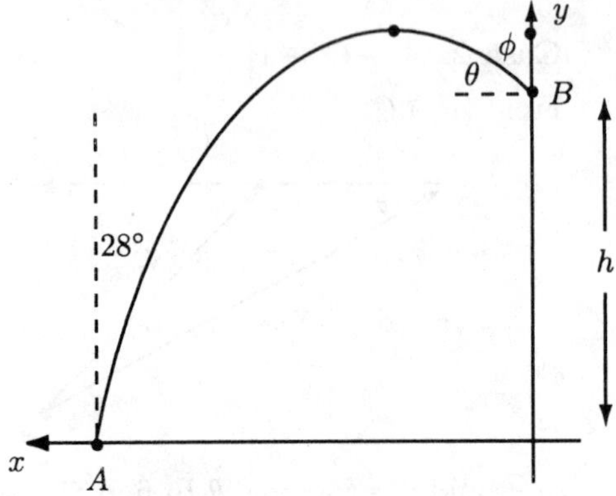

Her horizontal velocity component is constant, and so equal to its value on landing:

$$v_x = (8.0\ \mathrm{m/s}) \sin 28^\circ = 3.76\ \mathrm{m/s}$$

Upon landing, the gymnast is $(1.8\ \mathrm{m} - 0.7\ \mathrm{m}) = 1.1\ \mathrm{m}$ lower than when leaving the horse, so we have

$$(v_{y,\mathrm{landing}})^2 - (v_{y,\mathrm{leaves\ horse}})^2 = 2(-g)(-1.1\ \mathrm{m})$$

So

$$\begin{aligned}v_{y,\mathrm{leaves\ horse}} &= +\sqrt{\begin{matrix}[(8.0\ \mathrm{m/s})\sin 28^\circ]^2 \\ -2\,(9.81\ \mathrm{m/s^2})\,(1.1\ \mathrm{m})\end{matrix}} \\ &= 5.32\ \mathrm{m/s}\end{aligned}$$

So the angle ϕ is given by

$$\phi = \tan^{-1}\frac{v_h}{v_{y,\mathrm{leaves\ horse}}} = \tan^{-1}\frac{3.76\ \mathrm{m/s}}{5.32\ \mathrm{m/s}} = 35^\circ$$

b) The horizontal distance she travels is given by her horizontal speed and the time she travels. We obtain the time from the change in her vertical velocity:

$$\begin{aligned}\Delta x_2 &= v_h \Delta t = v_h \frac{\Delta v_y}{g} \\ &= v_h \left(\frac{v_{y,\mathrm{landing}} - v_{y,\mathrm{leaves\ horse}}}{-g}\right) \\ &= (3.76\ \mathrm{m/s})\left[\frac{(8.0\ \mathrm{m/s})\sin 28^\circ - 5.32\ \mathrm{m/s}}{-9.81\ \mathrm{m/s^2}}\right] \\ &= (3.76\ \mathrm{m/s})\,(1.26\ \mathrm{s}) = 4.8\ \mathrm{m}\end{aligned}$$

c) The gymnast's horizontal velocity is constant between her leap and her arrival at the horse.

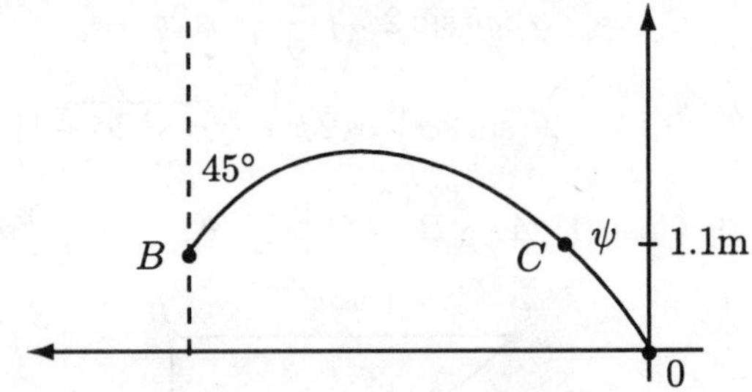

Since her velocity at arrival is at 45° to the vertical, her vertical velocity component has this same magnitude. (She has a negative velocity component, as we are labeling downward as minus.) Her vertical velocity at arrival must also be given by her initial vertical velocity and her vertical displacement.

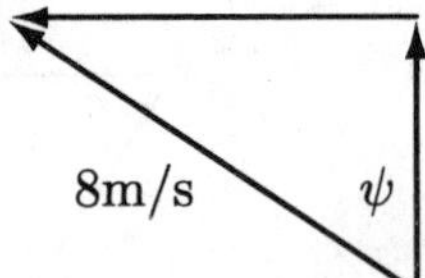

So, letting $v_0 \equiv 8.0\ \mathrm{m/s}$ be her initial speed, and $h \equiv 1.1\ \mathrm{m}$ be her vertical displacement on arrival at the horse, we have

$$v_{y,\mathrm{arrival}} = -v_0 \sin\psi$$

and

$$v_{y,\mathrm{arrival}}^2 - (v_0 \cos\psi)^2 = -2gh$$

so

$$v_0^2\left(\sin^2\psi - \cos^2\psi\right) = -2gh$$

or

$$\begin{aligned}\cos^2\psi - \sin^2\psi &= \cos 2\psi = \frac{2gh}{v_0^2} \\ &= \frac{2\,(9.81\ \mathrm{m/s^2})\,(1.1\ \mathrm{m})}{(8.0\ \mathrm{m/s^2})^2} \\ &= 0.337 \\ \psi &= \frac{1}{2}\cos^{-1}(0.337) = 35^\circ\end{aligned}$$

Again, we find horizontal distance from velocity and time

$$\begin{aligned}\Delta x_1 &= v_0 \sin\psi \frac{-v_0 \sin\psi - v_0\cos\psi}{-g} \\ &= \frac{v_0^2}{g}\sin\psi\left(\sin\psi + \cos\psi\right) = 5.2 \text{ m}\end{aligned}$$

d) In each case, average acceleration equals change of velocity divided by time interval:

Landing:

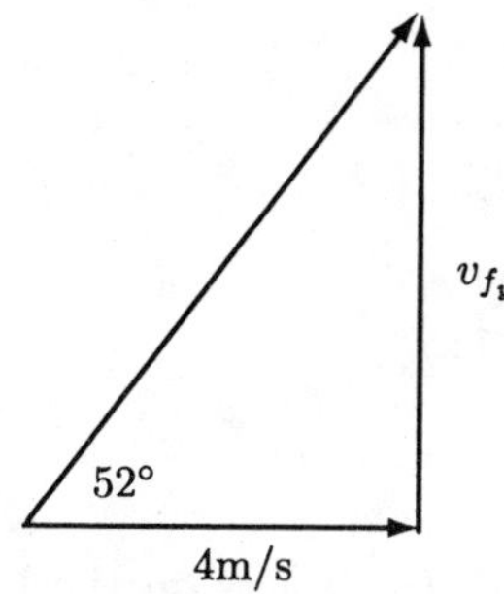

$$\begin{aligned}\vec{\mathbf{a}}_{AV,\text{land}} &= \frac{\vec{\mathbf{v}}_f - \vec{\mathbf{v}}_i}{\Delta t} = \frac{-\vec{\mathbf{v}}_i}{\Delta t} \\ &= \frac{-(8.0 \text{ m/s}) \text{ 28° left of downward}}{0.15 \text{ s}} \\ &= \left(53 \text{ m/s}^2, \text{ 28° right of upward}\right)\end{aligned}$$

On horse:

$$\begin{aligned}\vec{\mathbf{a}}_{AV,\text{horse}} &= \frac{(3.76 \text{ m/s}, 5.32 \text{ m/s}) - (v_0 \sin\psi, -v_0 \sin\psi)}{0.30 \text{ s}} \\ &= \left(-2.8 \text{ m/s}^2, 3.3 \text{ m/s}^2\right) \\ &= \left(33 \text{ m/s}^2, \text{ 5° right of upward}\right)\end{aligned}$$

e) The gymnast's average angular speed in either interval is the angle she turns through divided by the time interval.

Leap to horse:

$$\omega_1 = \frac{\left(180° - \psi - 45°\right)\left(\frac{\pi \text{ rad}}{180°}\right)}{\frac{\Delta x_1}{v_0 \sin\psi}} = 1.5 \text{ rad/s}$$

Horse to landing:

$$\omega_2 = \frac{\left(180° - \phi - 28°\right)\left(\frac{\pi \text{ rad}}{180°}\right)}{(4.8 \text{ m})\,(3.76 \text{ m/s})} = 1.6 \text{ rad/s}$$

Chapter 4

Force and Newton's Laws

4.1 There are normal or contact forces acting on the rock climber which are exerted by the wall.

Frictional forces ($\vec{f}$) act on the rock climber which keep him from sliding down. Lastly there's the gravitational force ($m\vec{g}$) which tries to pull him back to the ground.

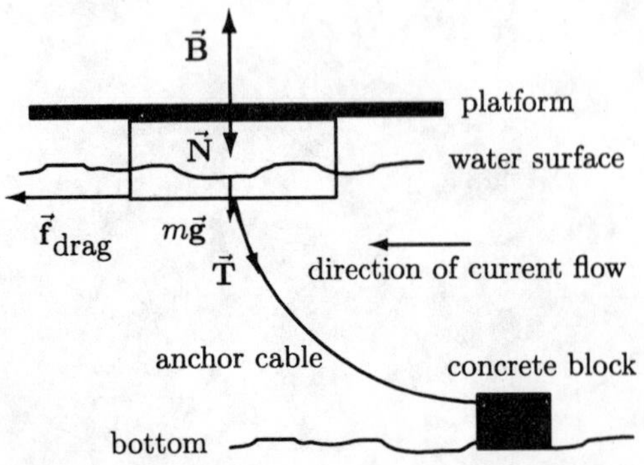

The cable exerts a tension force ($\vec{T}$). The current exerts a drag force ($\vec{f}$ drag) which pushes the buoy to the left. The gravitational force, or weight ($m\vec{g}$) acts on the buoy by pulling it downward. Finally, there's the buoyant force that the water exerts on the buoy upwards.

4.3 For $\vec{\mathbf{F}}_3$ to balance $\vec{\mathbf{F}}_1$ and $\vec{\mathbf{F}}_2$ the following condition must be satisfied:

$$\vec{\mathbf{F}}_1 + \vec{\mathbf{F}}_2 + \vec{\mathbf{F}}_3 = 0$$

or

$$\vec{\mathbf{F}}_1 + \vec{\mathbf{F}}_2 = -\vec{\mathbf{F}}_3$$

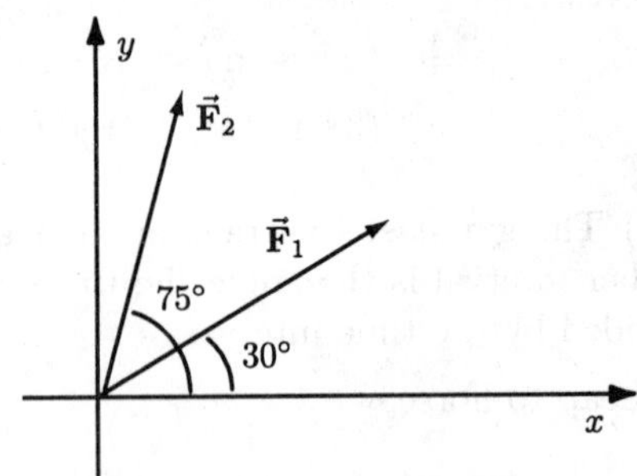

It might be easier to convert everything to Cartesian coordinates. We will add the x and y components of $\vec{\mathbf{F}}_1$ and $\vec{\mathbf{F}}_2$:

$$F_x \ : \ (2.0\text{ N})\cos 30° + (4.0\text{ N}\cos 75°) = 2.8\text{ N}$$
$$F_y \ : \ (2.0\text{ N})\sin 30° + (4.0\text{ N}\sin 75°) = 4.9\text{ N}$$

Thus

$$\vec{\mathbf{F}}_1 + \vec{\mathbf{F}}_2 = (2.8\text{ N})\,\hat{\mathbf{i}} + (4.9\text{ N})\,\hat{\mathbf{j}}$$

and to balance this:

$$\vec{\mathbf{F}}_3 = -(2.8\text{ N})\,\hat{\mathbf{i}} - (4.9\text{ N})\,\hat{\mathbf{j}}.$$

In polar coordinates,

$$\vec{\mathbf{F}}_3 = -(\vec{\mathbf{F}}_1 + \vec{\mathbf{F}}_2)$$

$$= \left[\sqrt{(-2.8\ \mathrm{N})^2 + (-4.9\ \mathrm{N})^2} \text{ at } \tan^{-1}\left(\frac{4.9\ \mathrm{F}}{2.8\ \mathrm{F}}\right)\right]$$

A calculator gives 60° as the result for the inverse tangent. However with both components negative the angle is in the third quadrant: $\theta = 240°$ Thus $\vec{\mathbf{F}}_3 = (5.6\ \mathrm{N}$, at $\theta = 240°$ from $+x$-axis), or (5.6 N, 60° below the $-x$-axis.)

4.5 Neglecting the SI prefixes we have a $g \cdot \mathrm{cm}/\mathrm{s}^2$ which is a mass · length/time² which are the dimensions of force. Considering the SI prefixes,

$$\frac{(1\times10^9\ \mathrm{g})\left(\frac{1\ \mathrm{kg}}{1000\ \mathrm{g}}\right)(1\ \mathrm{cm})\left(\frac{1\ \mathrm{m}}{10^2\ \mathrm{cm}}\right)}{(1\times10^{-12}\ \mathrm{s})^2} = 1\times10^{28}\ \mathrm{N}$$

4.7 Let M denote the mass of one box of nails. The maximum force that the table can withstand on the Earth is $F_{\mathrm{max}} = 6Mg$. If we let X be the maximum number of boxes the table can withstand on Mars we can write

$$F_{\mathrm{max}} = 6Mg_{\mathrm{earth}} = Xg_{\mathrm{mars}}$$

Since $0.38g_{\mathrm{earth}} = g_{\mathrm{mars}}$ we can then rewrite

$$\begin{aligned} F_{\mathrm{max}} &= 6Mg_{\mathrm{earth}} = X(0.38g_{\mathrm{earth}}) \\ X &= \frac{6Mg_{\mathrm{earth}}}{0.38g_{\mathrm{earth}}} = 15.7M \\ &\Rightarrow 15 \text{ boxes} \end{aligned}$$

The table can withstand the weight of 15 boxes of nails on Mars.

4.9 If I weigh 840 N, I am able to exert half of this on the lever: 420 N. Therefore I am able to compress the spring by an amount

$$s = \frac{F}{k} = \frac{420\ \mathrm{N}}{2.7\times10^3\ \mathrm{N/m}} = 15\ \mathrm{cm}$$

$$\vec{\mathbf{F}}_{\mathrm{on\ spring}} = +(420\ \mathrm{N})\,\hat{\mathbf{i}} = +k\vec{\mathbf{s}}$$

where $\vec{\mathbf{s}}$ is the displacement of the end of the spring.

$$\begin{aligned} \vec{\mathbf{F}}_{\mathrm{spring\ on\ you}} &= -(420\ \mathrm{N})\,\hat{\mathbf{i}} = -k\vec{\mathbf{s}} \\ \vec{\mathbf{F}}_{\mathrm{on\ mounting}} &= +(420\ \mathrm{N})\,\hat{\mathbf{i}} \end{aligned}$$

Note that there is *no* displacement of the bottom end of the spring.

4.11 Since the crate is at rest, there is no net force acting on it.

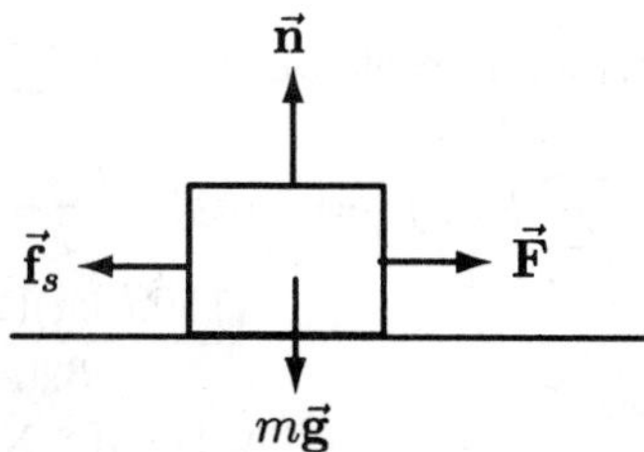

To balance the horizontal forces, the frictional force must be exactly equal but opposite in direction to the force you exert on the crate by pushing on it: $f_s = F$.

4.13 By considering the horizontal forces the engine must have applied a force $F_{\mathrm{bump}} = Ma$.

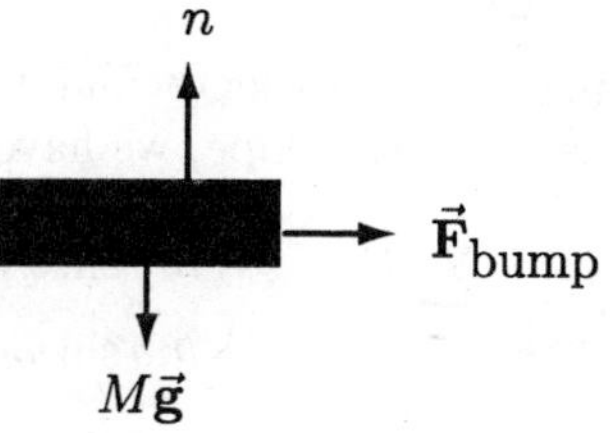

4.15 Since the car does not accelerate vertically, the sum $\sum F_y$ should be equal to 0; that is, $n = mg$ which means the gravitational force the Earth exerts on the car is balanced by the contact force that the road exerts on the car.

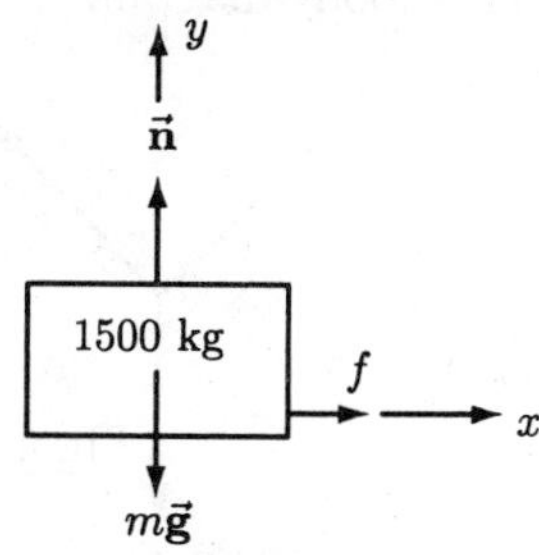

$$\begin{aligned} n &= mg = (1500\ \mathrm{kg})(9.81\ \mathrm{m/s^2}) \\ &= 1.5\times10^4\ \mathrm{N} \end{aligned}$$

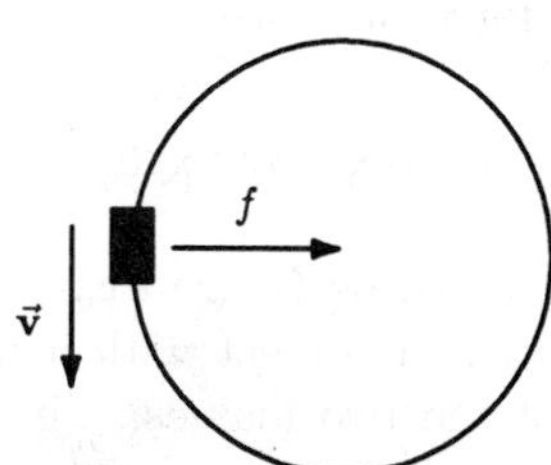

As we learned last chapter, the car is accelerating at $a = \frac{v^2}{r}$ inward along the radius of curvature. Frictional force is responsible for the cars centripetal acceleration; thus $\sum F_x \neq 0$

$$\begin{aligned} \sum F_x : f &= ma = \frac{mv^2}{r} \\ &= \frac{(1500\ \text{kg})(21\ \text{m/s})^2}{320\ \text{m}} \\ &= 2.1 \times 10^3\ \text{N} \end{aligned}$$

4.19 In each case the block is at rest, so the sum of all the forces acting is zero.

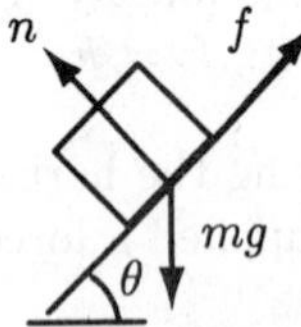

Taking y-axis perpendicular to the slope, and x-axis down the slope, we have, for each block:

$$\begin{aligned} \sum F_x &= mg\sin\theta - f = 0 \\ \Rightarrow f &= mg\sin\theta \end{aligned}$$

where θ is the angle the ramp makes with the horizontal. Since the values g and θ do not depend on the block we focus on, the values of f are directly proportional to the masses. Thus the friction force on the top block of mass $2M$ is the largest, and its magnitude is $2Mg\sin\theta$.

4.23 See the free body diagram.

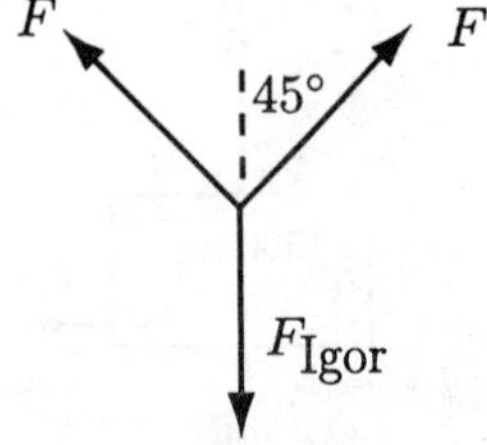

The opponents exert forces at equal angles to Igor's rope. For the force components perpendicular to the rope to balance, they must exert forces of equal magnitude. Then to balance Igor's force, we must have:

$$2F\cos 45° = 2.5 \times 10^3\ \text{N} \Rightarrow F = 1.8 \times 10^3\ \text{N}$$

4.27 When the objects are fastened together, they form a single object with a mass equal to the sum of the two masses. Since the same force acts in each case:

$$\begin{aligned} 20.0\ \text{N} &= F \\ &= m_1 a_1 = m_2 a_2 \\ &= (m_1 + m_2)a_3 \end{aligned}$$

We may use the given data for a_1 and a_2 to find the masses m_1 and m_2:

$$\begin{aligned} m_1 &= \frac{20.0\ \text{N}}{15.0\ \text{m/s}^2} = 1.33\ \text{kg} \\ m_2 &= \frac{20.0\ \text{N}}{10.0\ \text{m/s}^2} = 2.00\ \text{kg} \end{aligned}$$

Then

$$a_3 = \frac{20.0\ \text{N}}{1.33\ \text{kg} + 2.00\ \text{kg}} = 6.00\ \text{m/s}^2$$

4.31 We call the force unit the pound-force to distinguish is from the mass unit. In these units Newton's second law is:

$$\vec{\mathbf{F}} = km\vec{\mathbf{a}}$$

Now we use the definition of the lb-force .

$$1\ \text{lb-force} = k(1\ \text{lb-mass})(32\ \text{ft/s}^2)$$

Thus

$$k = (1\ \text{lb-force})/[(1\ \text{lb-mass})(32\ \text{ft/s}^2)]$$

The slug is defined by the relation:

$$1\ \text{lb-force} = k(1\ \text{slug})(1\ \text{ft/s}^2)$$

so

$$\begin{aligned} 1\ \text{slug} &= \frac{(1\ \text{lb-force})}{(1\ \text{ft/s}^2)} \cdot \frac{(1\ \text{lb-mass})(32\ \text{ft/s}^2)}{(1\ \text{lb-force})} \\ &= 32\ \text{lb-mass}. \end{aligned}$$

Using the conversion factor

$$1\ \text{lb-mass} = 0.4536\ \text{kg},$$

we have:

$$\begin{aligned} 1\ \text{slug} &= (32\ \text{lb-mass})(0.4536\ \text{kg/lb-mass}) \\ &= 14.5\ \text{kg} \end{aligned}$$

4.35 For the cup to remain at rest, the table must exert a normal force on the cup to balance the cup's weight; $n = Mg$.

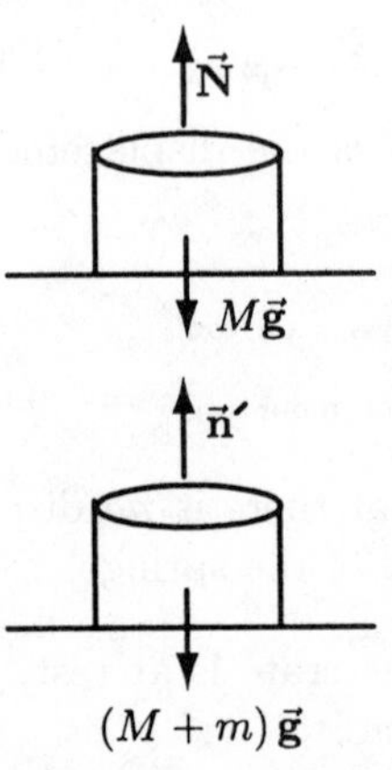

When the coffee is poured in the cup, the normal force that the table exerts on the cup must still balance the weight force of the combined coffee-cup system.

$$n' = (M + m)g$$

So the normal force increases by an amount mg.

4.39 Apply $F = ks$ to each of the four springs and solve for the spring constant k.

a)

$$Mg = ks,\ k = \frac{Mg}{s}$$

b)

$$\frac{1}{3}Mg = k\left(\frac{1}{3}s\right),\ k = \frac{Mg}{s}$$

c)

$$2Mg = k(2s),\ k = \frac{Mg}{s}$$

d)

$$\frac{4}{5}Mg = k\left(\frac{4}{5}s\right),\ k = \frac{Mg}{s}$$

The springs all have the same spring constant.

4.41 The free body diagram shows the forces acting on the crate.

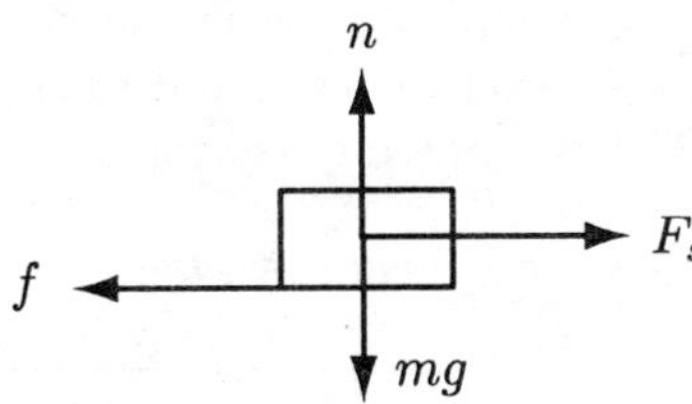

Since the crate remains at rest, the forces sum to zero. In particular, the sum of the horizontal forces must be zero, so $F_s = f$. Thus:

$$\begin{aligned} f &= ks \\ &= (6.43 \times 10^4 \text{ N/m})(2.00 \times 10^{-2} \text{ m}) \\ &= 1.29 \times 10^3 \text{ N} \end{aligned}$$

Since $f \lesssim \mu_s n$ and $n = mg$, we may check that this situation is possible.

$$\begin{aligned} \mu_s mg &= 0.65(2500 \text{ kg})(9.8 \text{ m/s}^2) \\ &= 1.5 \times 10^4 \text{ N} > f \end{aligned}$$

So we are OK.

4.43 The upper spring supports the weight mg and stretches by $s = \frac{mg}{k}$. Each of the two lower springs balances an equal amount of weight—$\frac{1}{2}mg$—so each stretches by $\frac{s}{2}$. The whole system stretches by $s + \frac{s}{2} = \frac{3s}{2}$ while supporting the weight mg. Thus the equivalent spring constant is

$$k' = \frac{mg}{\frac{3s}{2}} = \frac{2}{3}\frac{mg}{s} = \frac{2k}{3}$$

4.47 For the block to remain at rest, the two upward forces (tension and normal force) together balance the downward gravitational force. Thus by pulling harder on the string, the person reduces the normal force. If the person releases the tension in the string, n increases to mg.

4.51 In each case Newton's second law applies to the woman.

The scale is calibrated to read the normal force acting on it, in newtons. By Newton's third law this is also the magnitude of the normal force the scale exerts on the woman. Thus in each case:

$$n - mg = ma_{\text{up}}$$

a) Elevator stationary:

$$a_{\text{up}} = 0,$$

and

$$n = mg = (65 \text{ kg})(9.8 \text{ m/s}^2) = 640 \text{ N}$$

The scale reads 640 N.

b) Elevator accelerating upward at 2.0 m/s^2:

$$\begin{aligned} a_{\text{up}} &= +2.0 \text{ m/s}^2 \\ n &= m(g + a_{\text{up}}) \\ &= (65 \text{ kg})(9.8 \text{ m/s}^2 + 2.0 \text{ m/s}^2) \\ &= 770 \text{ N} \end{aligned}$$

The scale reads 770 N.

c) Elevator accelerating downward at 2.0 m/s^2:

$$\begin{aligned} a_{\text{up}} &= -2.0 \text{ m/s}^2 \\ n &= m(g + a_{\text{up}}) \\ &= (65 \text{ kg})(9.8 \text{ m/s}^2 - 2.0 \text{ m/s}^2) \\ &= 510 \text{ N} \end{aligned}$$

The scale reads 510 N.

d) Elevator moving at constant velocity: The woman is not accelerating, $a_{\text{up}} = 0$, and so the scale reads 640 N as in part (a).

e) Elevator in free fall:

$$\begin{aligned} a_{\text{up}} &= -9.8\,\text{m}/\text{s}^2 \\ n &= m(g + a_{\text{up}}) \\ &= (65\text{ kg})(9.8\text{ m}/\text{s}^2 - 9.8\text{ m}/\text{s}^2) \\ &= 0\text{ N} \end{aligned}$$

The scale reads zero!

4.55 The three forces on the balloon sum to zero:

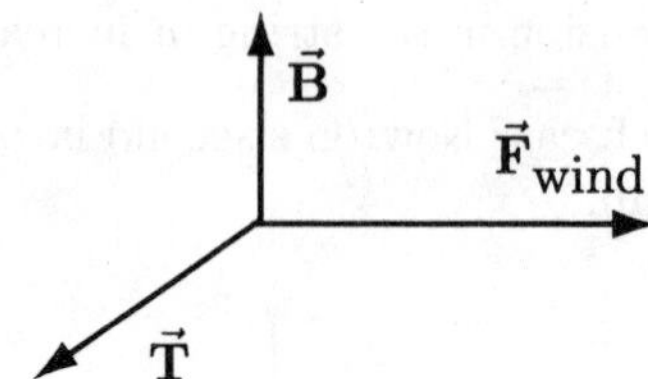

$$\vec{\mathbf{B}} + \vec{\mathbf{F}}_{\text{Wind}} + \vec{\mathbf{T}} = 0.$$

Since $\vec{\mathbf{B}}$ and $\vec{\mathbf{F}}_{\text{Wind}}$ are orthogonal,

$$\begin{aligned} \left|\vec{\mathbf{T}}\right| &= \sqrt{(B^2 + F_{\text{Wind}}^2)} \\ &= \sqrt{(12\text{ N})^2 + (15\text{ N})^2} \\ &= 19\text{ N} \end{aligned}$$

4.59 The forces acting on the coach are:

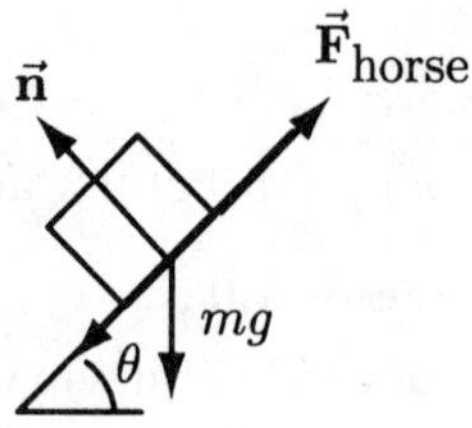

the pull of the horse, its weight, the normal force due to the slope. With the brakes off there is no force due to friction. We choose x and y-coordinates along and perpendicular to the slope. The net force acting is zero.

$$\begin{aligned} \sum F_x &= F_{\text{horse}} - mg\sin\theta = 0 \\ F_{\text{horse}} &= (1100\text{ kg})(9.8\text{ m}/\text{s}^2)\sin 10^\circ \\ &= 1900\text{ N} \end{aligned}$$

4.63 Before the ballast is dropped, the balloon is descending at 5.0 m/s. Since the speed is constant, the forces balance.

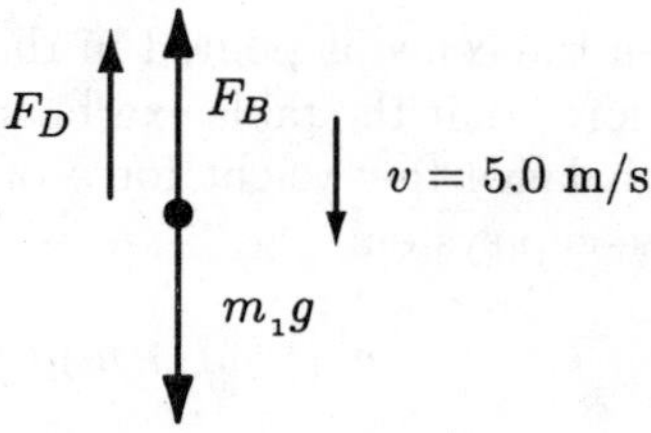

Remember that the drag force acts in the direction opposite the velocity since it's a fluid frictional force.

$$F_B + F_D = m_1 g$$

After the 25 kg of ballast are dropped, the ballon moves at the same constant speed, but upward, and F_D acts downward:

$$F_B = F_D + m_2 g$$

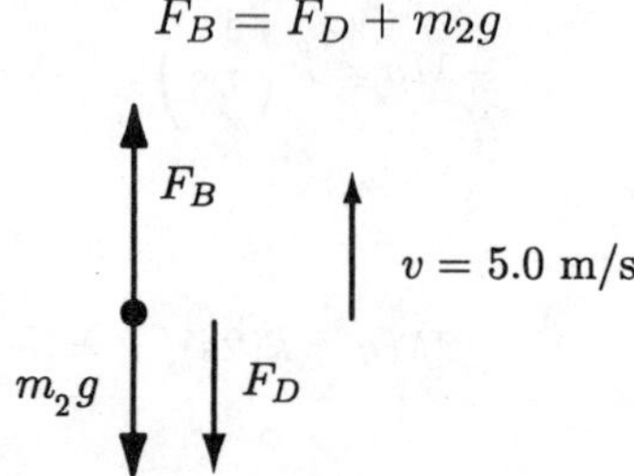

To find F_D, we eliminate F_B:

$$\begin{aligned} 2F_D &= (m_1 - m_2)g \\ &= (25\text{ kg})(9.8\text{ m}/\text{s}^2) \\ \Rightarrow F_D &= 120\text{ N} \end{aligned}$$

Immediately after the ballast is dropped, the balloon is still moving downward at 5.0 m/s, so $\vec{\mathbf{F}}_D$ still acts upward, but the mass is reduced by 25 kg. Applying Newton's second law:

$$\begin{aligned} F_B + F_D - m_2 g &= m_2 a \\ \Rightarrow a &= \frac{F_B + F_D}{m_2} - g \\ &= \frac{m_1 g}{m_2} - g \\ &= \frac{g(m_1 - m_2)}{m_2} \\ &= \frac{(9.8\text{ m}/\text{s}^2)(25\text{ kg})}{(730\text{ kg})} \\ &= 0.34\text{ m}/\text{s}^2 \end{aligned}$$

4.67 Frictional force between the car's tires and the road decelerates the car. The force acts opposite the car's velocity. We may find it with Eqn. (2.13):

$$\begin{aligned} v_f^2 - v_i^2 &= 2a\Delta x \\ \Rightarrow 0 - (113\text{ km}/\text{h})^2 &= 2a(56.4\text{ m}) \end{aligned}$$

Thus

$$a = \frac{-(113\ \text{km}/\text{h})^2 \left(\frac{1\ \text{h}}{3600\ \text{s}}\right)^2 \left(\frac{103\ \text{m}}{1\ \text{km}}\right)^2}{[2(56.4\ \text{m})]} = -8.73\ \text{m}/\text{s}^2$$

Thus the force has magnitude

$$m|a| = (1.40 \times 10^3\ \text{kg})(8.73\ \text{m}/\text{s}^2) = 1.22 \times 10^4\ \text{N}$$

and acts opposite the car's velocity.

4.71 After drawing the FBD, the choice of coordinate system should be straightforward.

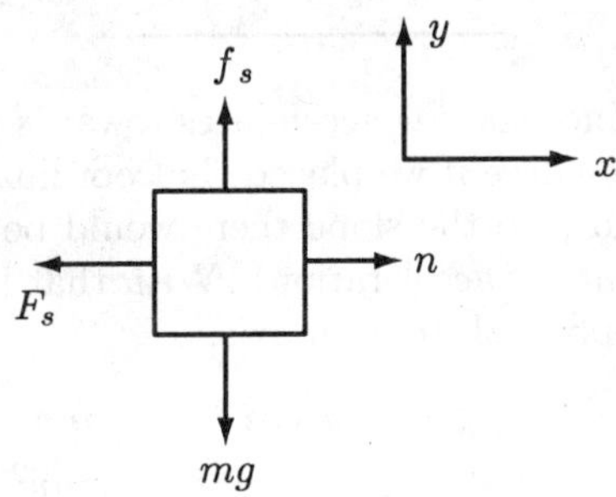

The spring as shown in the diagram is compressed by an amount d, and therefore exerts a force $F_s = kd$ in the $-\hat{\mathbf{i}}$ direction. Under the force of gravity, the block wants to accelerate downward, so the frictional force exerted by the wall on the block acts upward in the $\hat{\mathbf{j}}$ direction. When friction takes its maximum value, we have:

$$F_y : f_s = \mu_s n = mg$$
$$F_x : F_s = k_{\text{min}} d = n$$

Solving the above equations for k in terms of known quantities gives

$$k_{\text{min}} = \frac{n}{d} = \frac{mg}{\mu_s d}$$

If

$$k = \frac{2mg}{\mu_s d} = 2k_{\text{min}},$$

then $n = \dfrac{2mg}{\mu_s d}$ and $f = mg$. Note that f supports the block's weight and does not increase as n increases.

4.75 For the system to be at rest the forces acting on the ball must balance.

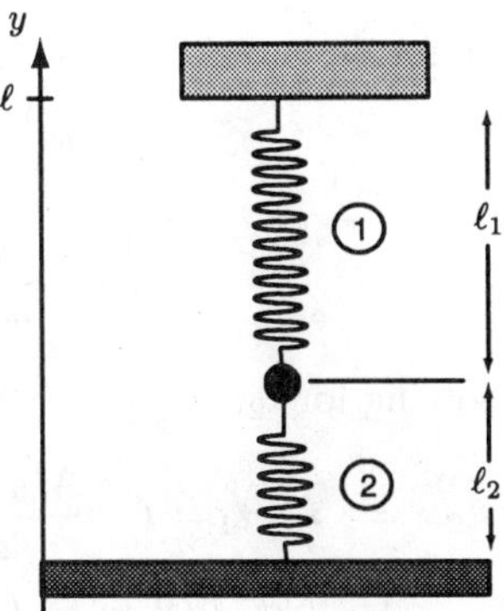

N2L tells us

$$\vec{\mathbf{F}}_{S1} + \vec{\mathbf{F}}_{S2} + M\vec{\mathbf{g}} = 0 \qquad (1)$$

The ball finds equilibrium somewhere below $\frac{1}{2}\ell$ since the top spring must be stretched and the bottom spring must be compressed as shown.

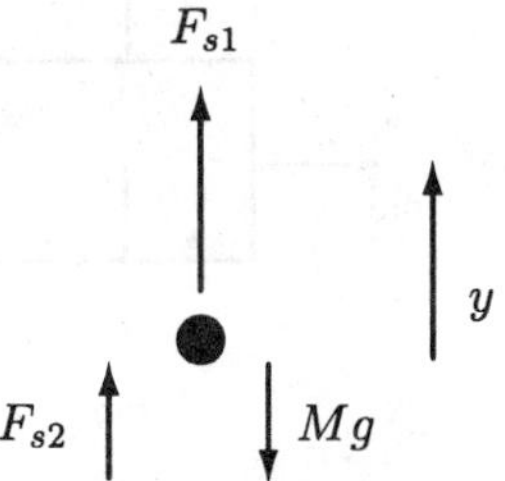

Let ℓ_1 be the distance between the ceiling and the ball, and ℓ_2 be the distance between the ball and ground; thus $\ell_1 + \ell_2 = \ell$. We will define $\hat{\mathbf{j}}$ to be upwards so we would like $F_{S1,y} > 0$ and $F_{S2,y} > 0$. Since

$$\vec{\mathbf{F}}_s = F_s\hat{\mathbf{j}} = -k\vec{\mathbf{s}}$$

where $\vec{\mathbf{s}}$ is the displacement of the spring's end, then:

$$F_{S1,y} = k\left(\ell_1 - \frac{\ell}{2}\right)$$

since $\ell_1 > \frac{1}{2}\ell$, and

$$F_{S2,y} = k\left(\frac{\ell}{2} - \ell_2\right)$$

since $\ell_2 < \frac{1}{2}\ell$. The quantities $\left(\ell_1 - \frac{\ell}{2}\right)$ and $\left(\frac{\ell}{2} - \ell_2\right)$ represent the magnitude of the displacement for the first and second springs respectively. Rewriting Eq. (1),

$$k\left(\ell_1 - \frac{\ell}{2}\right) + k\left(\frac{\ell}{2} - \ell_2\right) = Mg$$
$$k\ell_1 - k\ell_2 = Mg$$

using the fact that $\ell_1 + \ell_2 = \ell$

$$k\ell_1 - k(\ell - \ell_1) = Mg$$
$$2k\ell_1 - k\ell = Mg$$

Solving for ℓ_1,

$$\begin{aligned}\ell_1 &= \frac{Mg + k\ell}{2k} \\ &= \frac{1}{2}\ell\left(1 + \frac{Mg}{k\ell}\right)\end{aligned}$$

Now solving for ℓ_2,

$$\begin{aligned}\ell_2 &= \ell - \ell_1 = \ell - \frac{Mg + k\ell}{2k} \\ &= \frac{k\ell - Mg}{2k} = \frac{1}{2}\ell\left(1 - \frac{Mg}{k\ell}\right)\end{aligned}$$

4.79 While the plane is traveling at constant speed, there are no unbalanced forces, that is: $n = mg$ and $F_{\text{drag}} = F_{\text{thrust}}$.

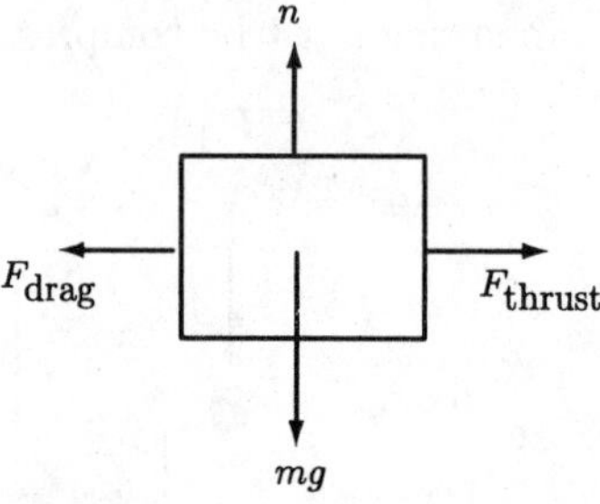

Now n, the lift force exerted by air on the wings, is increased. The result is a net force acting up. Horizontal velocity remains constant, but the unbalanced net force causes acceleration perpendicular to $\vec{\mathbf{v}}$. From our work in Chapter 3, we know that the plane's path will curve upward.

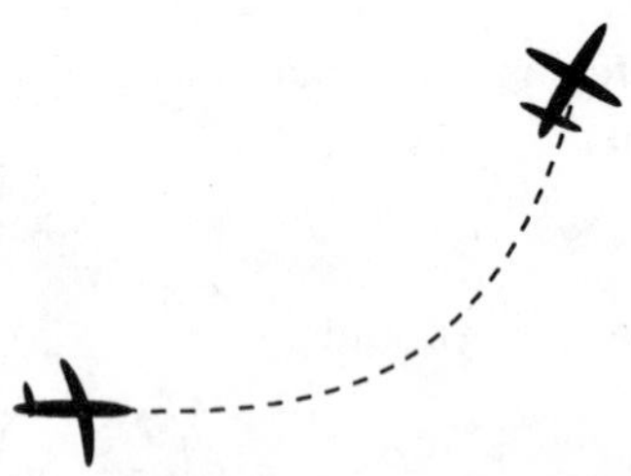

4.83 An unbalance in vertical forces produces an acceleration. This centripetal acceleration is what makes the car travel in a circular manner.

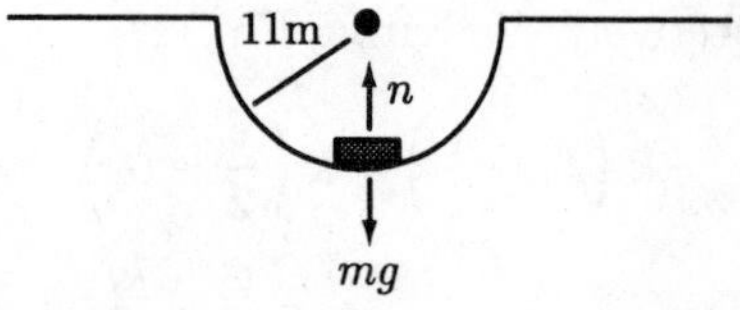

Using N2L,

$$n - mg = ma_{\text{cent}} \Rightarrow n = m\left(\frac{v^2}{r}\right) + mg$$

Using the values given we can find n.

$$\begin{aligned}n &= \frac{1000\ \text{kg}(32\ \text{m}/\text{s})^2}{11\ \text{m}} + (1000\ \text{kg})(9.81\ \text{m}/\text{s}^2) \\ &= 1.0 \times 10^5\ \text{N}\end{aligned}$$

4.87 a) We chose a coordinate system so that the car's acceleration is along one axis:

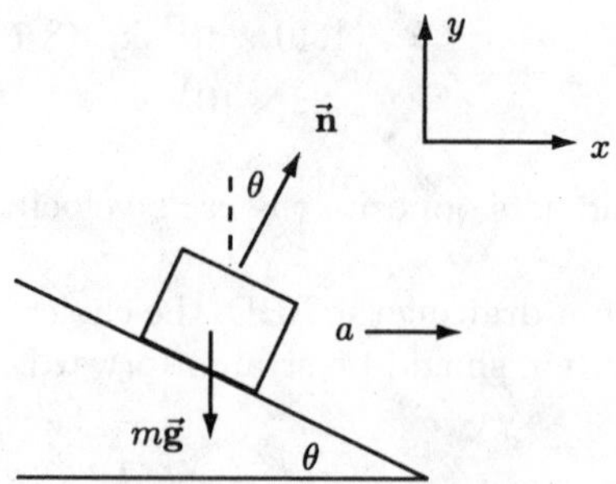

(Since the car accelerates towards the center of the curve if we placed the coordinate system $\perp$ and $\parallel$ to the slope there would be two components of acceleration). With that in mind we'll apply N2L to each axis:

$$F_y : n\cos\theta = mg \qquad (2)$$

$$F_x : n\sin\theta = \frac{mv_1^2}{r} \qquad (3)$$

from Eq. (2), $n = \dfrac{mg}{\cos\theta}$ and using this to eliminate n in Eq. (3) we can solve Eq. (3) for v_1:

$$\begin{aligned}v_1 &= \sqrt{gr\tan\theta} \\ &= \sqrt{(9.81\ \text{m}/\text{s}^2)(1.0 \times 10^3\ \text{m})\tan 5°} \\ &= 29\ \text{m}/\text{s}\end{aligned}$$

b) Now we are to consider the maximum speed the car can travel around the turn. In this scenario, friction between the car and road will act on the car in the direction of downslope.

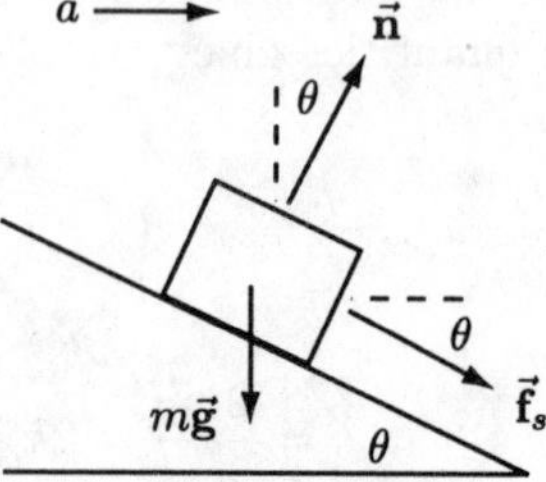

As with the previous case, the coordinate system was chosen so that the car has only one component of accelerated motion. Applying N2L to each direction

$$F_y : n\cos\theta = f_s\sin\theta + mg \qquad (4)$$

$$F_x : n\sin\theta + f_s\cos\theta = \frac{mv_m^2}{r} \qquad (5)$$

The frictional force is given by $f_s \leq \mu n$, but we are considering the maximum speed the car can have without flying off the track. This requires friction to be maximized, so: $f_s = \mu n$. Using this to rewrite Eq. (4) and Eq. (5)

$$n\cos\theta = \mu n \sin\theta + mg \qquad (6)$$

$$n\sin\theta + \mu n\cos\theta = \frac{mv_{\max}^2}{r} \qquad (7)$$

Solving Eq. (6) for n,

$$n = \frac{mg}{\cos\theta - \mu\sin\theta}$$

We can use this expression to replace n in Eq. (7) and solve for $v_{\max}$

$$v_{\max} = \left[\frac{gr(\sin\theta + \mu\cos\theta)}{\cos\theta - \mu\sin\theta}\right]^{1/2}$$

Using $\mu = .4$ and $\theta = 5°$ we can find a numerical value for $v_{\max}$:

$$\begin{aligned} v_{\max} &= \left[\frac{(9.81\ \mathrm{m/s})(1.0\times10^3\ \mathrm{m})(\sin 5° + (.4)\cos 5°)}{\cos 5° - (.4)\sin 5°}\right]^{1/2} \\ &= 7.0\times10^1\ \mathrm{m/s} \end{aligned}$$

For level ground, $\theta = 0$ so the sin terms are 0 and cos terms are 1.

$$\begin{aligned} v_m &= \left[\frac{(9.81\ \mathrm{m/s})(1.0\times10^3\ \mathrm{m})(0+.4)}{1-(.4)(0)}\right]^{1/2} \\ &= 63\ \mathrm{m/s} \end{aligned}$$

c) If $v < v_1$, friction acts up the slope to reduce the force component toward the center of the circle. There is a minimum speed if $\tan\theta > \mu$. The free body diagram is the same as in part b) except that f now acts up the slope. Thus Eqns. (6) and (7) become:

$$\begin{aligned} n\cos\theta &= f\sin\theta + mg \qquad (8) \\ n\sin\theta - f\cos\theta &= \frac{mv^2}{r} \qquad (9) \end{aligned}$$

and

$$f \leq \mu n$$

Multiply (8) by $\cos\theta$ and (9) by $\sin\theta$ and add. The terms in f cancel. Then since $\cos^2\theta + \sin^2\theta = 1$, we have:

$$n = mg\cos\theta + \left(\frac{mv^2}{r}\right)\sin\theta$$

Similarly, if we multiply (8) by $\sin\theta$ and (9) by $\cos\theta$ and subtract, we find:

$$f = mg\sin\theta - \left(\frac{mv^2}{r}\right)\cos\theta.$$

Then applying the limit on f:

$$\begin{aligned} & mg\sin\theta - \left(\frac{mv^2}{r}\right)\cos\theta \\ \leq\ & \mu\left(mg\cos\theta + \left(\frac{mv^2}{r}\right)\sin\theta\right) \end{aligned}$$

or

$$\left(\frac{v^2}{r}\right)(\mu\sin\theta + \cos\theta) \geq g(\sin\theta - \mu\cos\theta)$$

Thus

$$v^2 \geq gr\frac{\tan\theta - \mu}{\mu\tan\theta + 1}$$

These results can be compared with Exercise 4.11. As $\theta \to 90°$, we get back the results for a wall:

$$v_{\min} = \sqrt{\frac{gr}{\mu}}$$

4.93 When a climber falls, he plummets to Earth at an acceleration $g = 9.81\ \mathrm{m/s^2}$. He can accelerate to a pretty fast speed after falling a short distance. For the first few meters, his rope is slack and exerts no force on him. The problem lies when he has fallen a distance equal to the length of his rope.

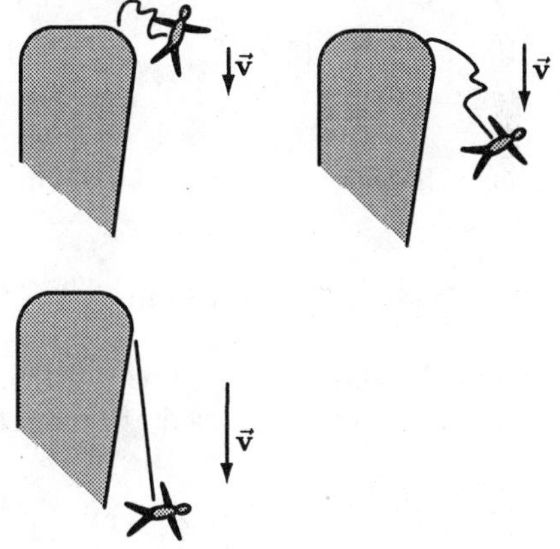

If the rope doesn't stretch much when it goes taught, the rope will have to provide a large enough force to decelerate him from a large speed to zero in a few centimeters! If the rope stretches, when he has fallen a distance equal to the length of the rope the rope isn't going to exert the tremendous force needed to bring his speed to 0 m/s in a short distance. The rope will act much the way a spring does. It will exert a force proportional to the distance stretched. The overall effect is similar to comparing the force needed to decelerate a car from 20 m/s to 0 m/s in a centimeter with the force needed to decelerate the same car in 40 meters. The first case is called a crash and is likely to turn the car into a pancake while the second case is called an emergency stop. So if the rope doesn't stretch, the climber suffers a major injury!

4.97 **a)** Since Igor reached a terminal speed, he is no longer accelerating, and must therefore have no net force acting on him. Before he pops his parachute, the only two forces acting on him are his weight (mg) and drag force (F_D).

Since $\sum F = 0$,

$$\begin{aligned} F_D &= mg = (85 \text{ kg})(9.81 \text{ m/s}^2) \\ &= 830 \text{ N} \end{aligned}$$

b) If we assume he accelerates uniformly, we can use his initial and final velocities to find his acceleration. During the first $\frac{1}{2}$ s he travels $(100 \text{ m/s})(\frac{1}{2} \text{ s}) = 50$ m, so he accelerates over a 150 m distance.

$$\begin{aligned} v_f^2 - v_i^2 &= 2as \\ (100 \text{ m/s})^2 - (5 \text{ m/s})^2 &= 2a(150 \text{ m}) \\ &\Rightarrow a = 33 \text{ m/s}^2 \end{aligned}$$

Thus, the force that the parachute exerts on Igor is

$$F = ma = (85 \text{ kg})(33 \text{ m/s}^2) = 2800 \text{ N}$$

4.101 The FBD shows the sled with a force that the explorer exerts on the sled at an angle of θ.

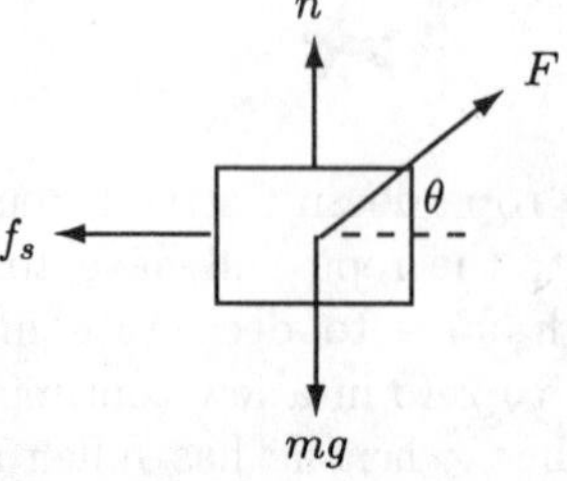

The minimum force required to start the sled will be just barely larger than that which results in maximum possible friction force. With this in mind we'll apply N2L along each coordinate axis.

$$F_x : \quad F\cos\theta = f_s \tag{10}$$

$$F_y : \quad n + F\sin\theta = mg \tag{11}$$

As stated, $F\cos\theta$ will just barely match the maximum frictional force to start the sled moving.

$$F\cos\theta = f_{s,\max} = \mu n \tag{12}$$

From Eq. (11) we see that $n = mg - F\sin\theta$, so Eq. (12) becomes

$$F\cos\theta = \mu(mg - F\sin\theta)$$

(Note that we are considering F for a given angle θ. F will change as θ changes). Solving for F,

$$F = \frac{\mu mg}{\cos\theta + \mu\sin\theta}$$

Finding a minimum or maximum involves taking a derivative.

$$\begin{aligned} \frac{dF}{d\theta} &= \frac{d}{d\theta}\left[\mu mg(\cos\theta + \mu\sin\theta)^{-1}\right] \\ &= \mu mg(-1)(\cos\theta + \mu\sin\theta)^{-2}(-\sin\theta + \mu\,\mathrm{cc} \\ &= \frac{\mu mg(\sin\theta - \mu\cos\theta)}{(\cos\theta + \mu\sin\theta)^2} \end{aligned}$$

Setting the derivative equal to 0, we find:

$$\begin{aligned} \mu &= \tan\theta \\ \Rightarrow \theta &= \tan^{-1}\mu \\ &= \tan^{-1}(0.40) = 22^\circ \end{aligned}$$

Thus the angle which results in the minimum value for f is 22° in which case

$$F_{\min} = \frac{(0.4)(110 \text{ kg})(9.8 \text{ m/s}^2)}{\cos 22^\circ + 0.4\sin 22^\circ} = 0.40 \text{ kN}$$

On a 30° slope:

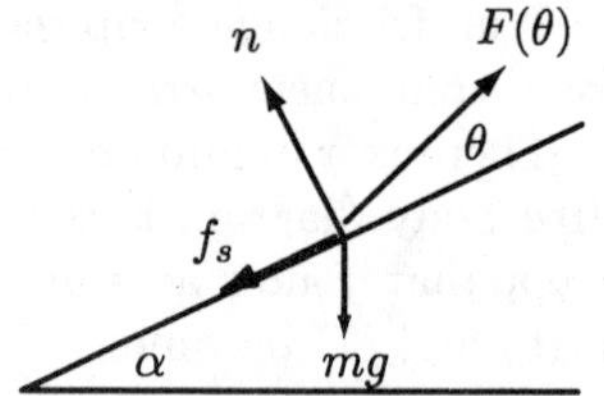

$$\begin{aligned} F\cos\theta &= f_s + mg\sin\alpha \\ F\sin\theta + n &= mg\cos\alpha \\ f &\le \mu n = \mu(mg\cos\alpha - F\sin\theta) \\ F\cos\theta - mg\sin\alpha &\le \mu(mg\cos\alpha - F\sin\theta) \end{aligned}$$

at the limit:

$$F[\cos\theta + \mu\sin\theta] = \mu(mg\cos\alpha) + mg\sin\alpha$$

Since F has the same dependence on θ, the slope doesn't change the value of θ at the limit. The minimum value is:

$$\begin{aligned} F &= mg\frac{[\mu\cos\alpha + \sin\alpha]}{[\mu\sin\theta + \cos\theta]} \\ &= (110 \text{ kg})(9.8 \text{ m/s}^2)\frac{(0.40\cos 30^\circ + \sin 30^\circ)}{(0.40\sin 22^\circ + \cos 22^\circ)} \\ &= 850 \text{ N} \end{aligned}$$

4.105 We will apply NL2 along each coordinate axis:

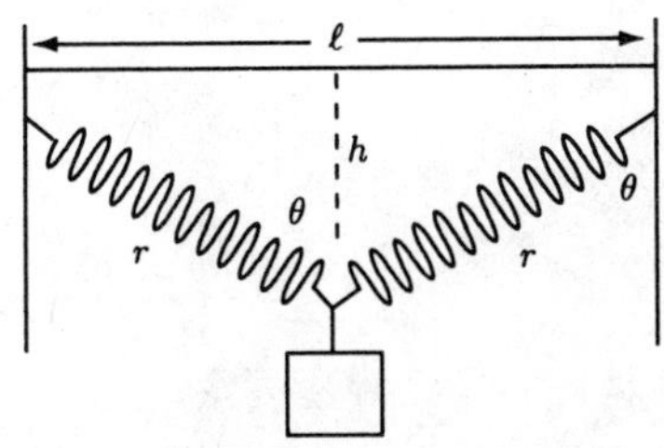

and make sure to balance forces since the mass is at rest.

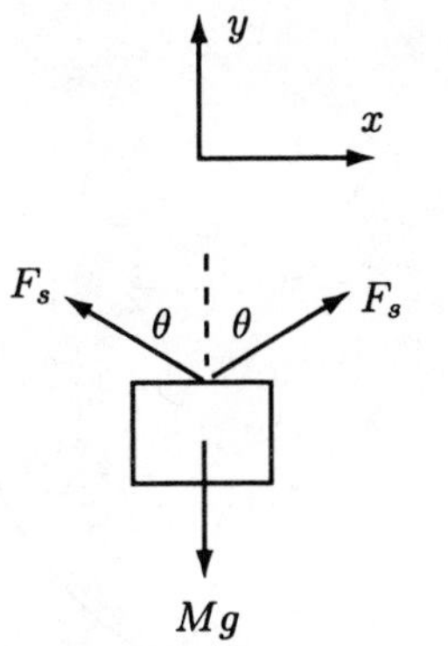

$$F_x \ : \ F_s \sin\theta = F_s \sin\theta \qquad (13)$$

$$F_y \ : \ 2F_s \cos\theta = Mg \qquad (14)$$

Eq. (13) doesn't tell us much; since the mass hangs from the middle of the spring the two spring forces are equal in magnitude and are directed at equal angles from the vertical. Eq. (14) is what we need to work with. The next thing to do is to obtain an expression for F_s in terms of known quantities and solve the resulting equation for θ. To do this we need to know the spring's stretched length.

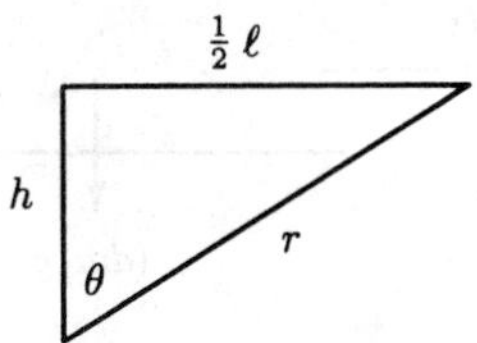

From the geometry of the problem, the springs length is:

$$\text{spring's length} = 2r = 2\left(\frac{\ell}{2\sin\theta}\right) = \frac{\ell}{\sin\theta}$$

The spring's stretch is then

$$\Delta\ell = \frac{\ell}{\sin\theta} - \ell = \ell\left[\csc\theta - 1\right]$$

and thus

$$F_s = k\ell\left[\csc\theta - 1\right]$$

[Equivalently, each half of the spring is stretched $\frac{\Delta\ell}{2}$ and exerts force $2k\left(\frac{\Delta\ell}{2}\right) = k\Delta\ell$.] From Eq. (14),

$$\begin{aligned} 2(k\ell)\left(\frac{1}{\sin\theta} - 1\right)\cos\theta &= Mg \\ \left(\frac{2k\ell}{Mg}\right)(1 - \sin\theta) &= \tan\theta \\ 1 - \sin\theta &= \frac{Mg}{2k\ell}\tan\theta \qquad (15) \\ \text{where } \frac{Mg}{2k\ell} &= \frac{(0.453 \text{ kg})(9.81 \text{ m/s}^2)}{2(78.5 \text{ N/m})(0.275 \text{ m})} \\ &= 0.103 \end{aligned}$$

The figure shows graphs of the two sides of Eq. (15). They intersect at $\theta = 57°$, which is the required solution.

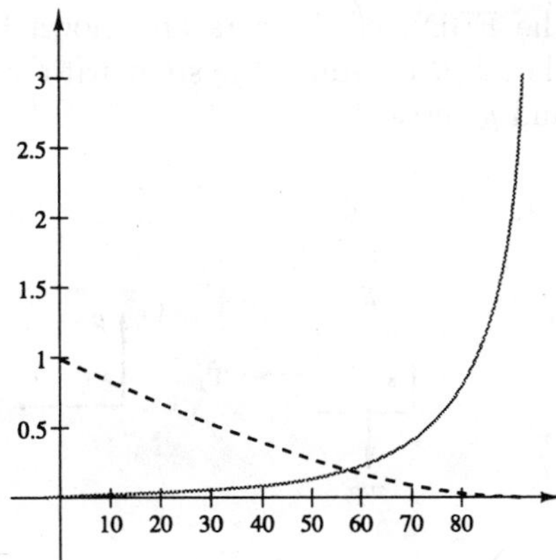

Chapter 5

Using Newton's Laws

5.1 The FBD's of the cars are shown below. We choose a coordinate system with x horizontal and y vertical.

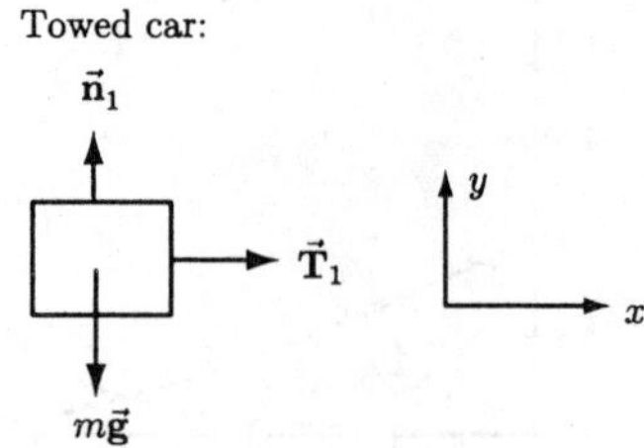

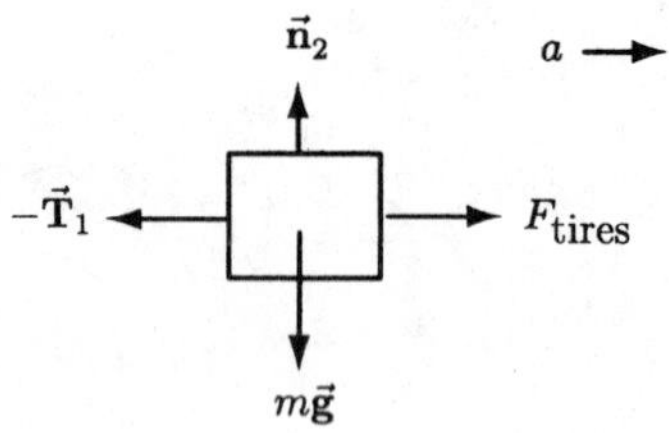

$$\sum F_y = 0: \quad n_1 = mg; \quad n_2 = mg$$
$$\sum F_x = 0: \quad T_1 = ma; \quad F_{\text{tires}} - T_1 = ma$$

Thus,

$$T_1 = ma = (850 \text{ kg})\left(2.0 \text{ m/s}^2\right) = 1700 \text{ N}$$

and

$$F_{\text{tires}} = 2T_1 = 3400 \text{ N}$$

5.3 We'll look at a typical element of a massive cable.

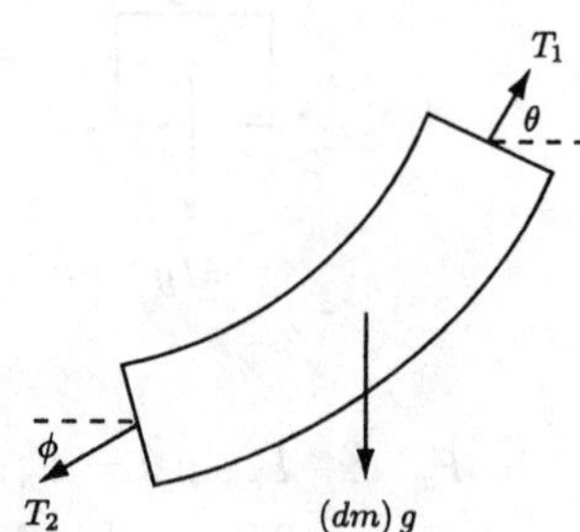

The necessary conditions for equilibrium are:

$$T_1 \sin\theta = T_2 \sin\phi + (dm)g$$
$$T_1 \cos\theta = T_2 \cos\phi$$

There are no contradictions in these equations. They are entirely plausible—however, let's look at what happens when the cable does not hang in a curve.

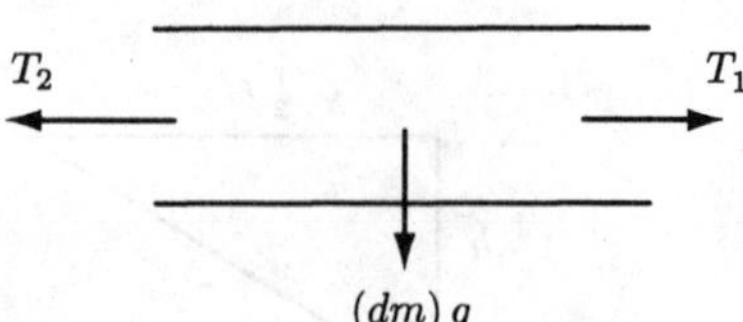

Then

$$\theta = \phi = 0$$

and there is no force to balance the cable's weight. If this were the case we'd never be able to hang cables—they'd all accelerate downwards.

5.5 Assume you and your friend have a mass of 65 kg and 75 kg respectively,

$$F_{\text{you on friend}} = \frac{Gm_{\text{you}}m_{\text{friend}}}{r^2} = \frac{(7\times10^{-11}\ \text{N}\cdot\text{m}^2/\text{kg}^2)(65\ \text{kg})(75\ \text{kg})}{(3\ \text{m})^2} = 4\times10^{-8}\ \text{N}$$

Your friend's weight (gravitational force exerted by the Earth on your friend) is

$$mg = (75\ \text{kg})(9.81\ \text{m}/\text{s}^2) = 740\ \text{N}$$

Its ratio to the force you exert on your friend is:

$$\frac{740\ \text{N}}{3.6\times10^{-6}\ \text{N}} = 2\times10^{10}$$

As we expected, the Earth's force on your friend is vastly larger than the force you exert on your friend.

5.7 The density of the Earth (and therefore the asteroid) is $5.52\times10^3\ \text{kg}/\text{m}^3$. Since mass = density × volume, if we model the asteroid as a sphere of radius 1.0 km we can find its mass.

$$M_{\text{ast}} = (5.52\times10^3\ \text{kg}/\text{m}^3)\,\frac{4}{3}\pi(1000\ \text{m})^3 = 2.3\times10^{13}\ \text{kg}$$

Acceleration on the surface of the asteroid is given by:

$$g_{\text{ast}} = \frac{GM_{\text{ast}}}{R_{\text{ast}}^2} = \frac{(6.67\times10^{-11}\ \text{N}\cdot\text{m}^2/\text{kg}^2)(2.3\times10^{13}\ \text{kg})}{(1000\ \text{m})^2} = 1.5\times10^{-3}\ \text{m}/\text{s}^2$$

To find the radius from the center of the asteroid at which its acceleration is half of that at its surface, set up the equality

$$\frac{1}{2}g_{\text{ast}} = \frac{GM_{\text{ast}}}{R^2}$$

and solve for R.

$$R = \left[\frac{2GM_{\text{ast}}}{1.5\times10^{-3}\ \text{m}/\text{s}^2}\right]^{\frac{1}{2}} = 1.4\times10^3\ \text{m}$$

(measured from the asteroid's center). Thus you need to be 400 m from the asteroid's surface for g_{ast} to be half that on the surface.

5.9 For m_3 to remain at rest, n_3 (exerted by the crate underneath) must balance the weight of m_3.

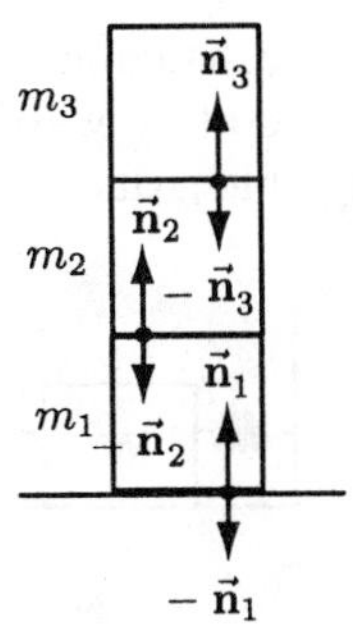

$$n_3 = Mg$$

However, by N3L, the third block must exert a force $n_3 = Mg$ back on the second block.

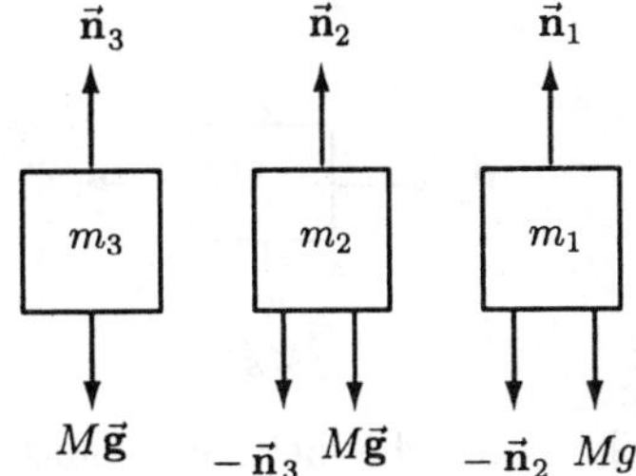

Thus, for the second block to remain at rest,

$$n_2 = n_3 + Mg = 2Mg$$

Again, by N3L, the second block must exert $n_2 = 2Mg$ down on the first block since the first block exerts $n_2 = 2Mg$ up on the second block. Equilibrium requires that

$$n_1 = n_2 + Mg = 3Mg$$

5.13 a) The free body diagrams for each crate are:

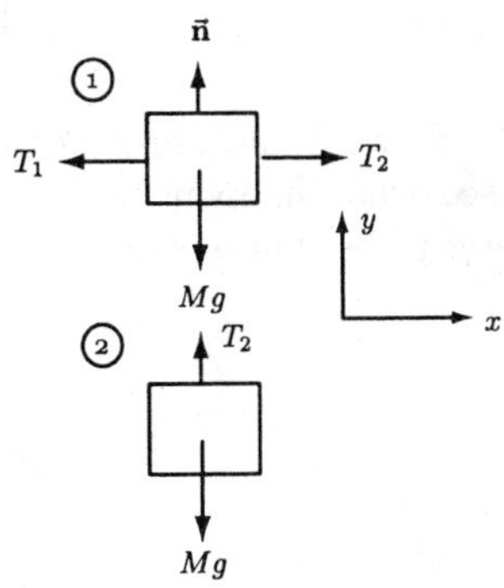

Using N2L we obtain:

$$\sum F_{x1} = T_2 - T_1 = 0$$
$$T_1 = T_2$$
$$\sum F_{y2} = T_2 - Mg = 0$$
$$T_1 = T_2 = Mg$$

Thus

b) Since the hanging orange crate mass pulls the orange crate towards the ledge, we expect the spring of length ℓ to be stretched to a length $\ell + \Delta\ell$.

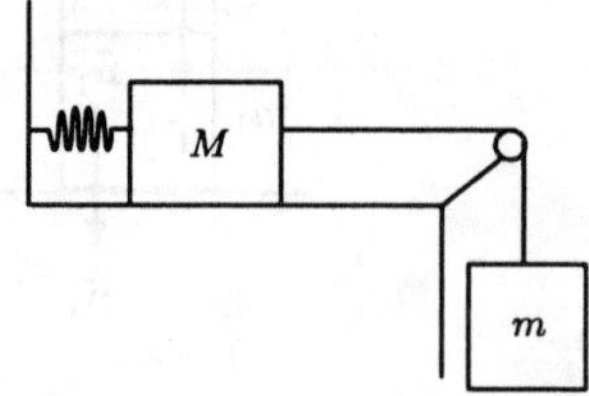

It must therefore, by Hooke's Law, exert a force $k(\Delta\ell)$ on the orange crate on the ledge that replaces T_1 in part a). Thus, $F_s = k(\Delta\ell) = Mg$.

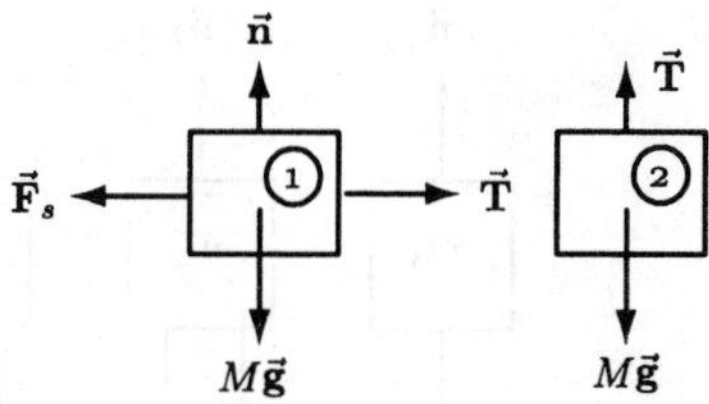

Solving for the distance from the ledge,

$$\ell + \Delta\ell = \frac{Mg}{k} + \ell$$

5.15 From watching physics professors we've learned a small trick that may reduce the work needed to solve this problem. The problem statement asks to analyze the problem first without friction and then with friction.

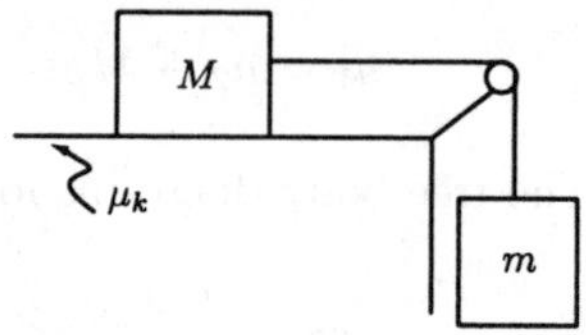

Instead, we'll analyze it with frictional forces. To see what happens without friction, simply take μ to be 0 in our answer.

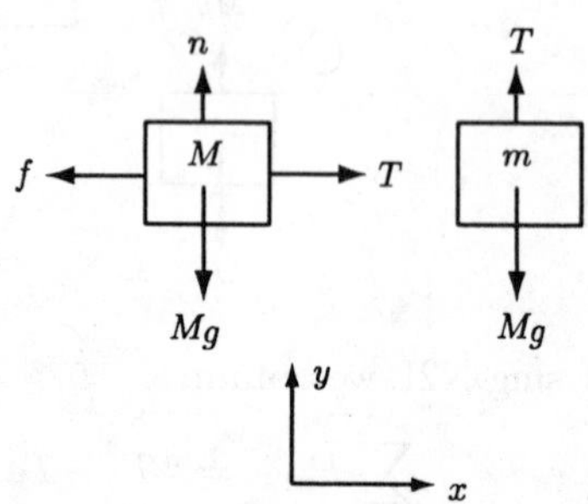

Now we'll apply N2L:

	Box on table	hanging box
F_y:	$n = Mg$ (i)	$T - mg = ma_{2y}$ (iii)
F_x:	$T - f = Ma_{1x}$ (ii)	no horizontal forces

Because the string cannot stretch, the hanging box moves down the same distance that M moves to the right. Thus

$$a_{2,y} = -a_{1,x} = -a$$

We wish to solve for acceleration, so we'll start by solving Eq. (iii) for T and use it to eliminate tension in Eq. (ii).

$$T = m(g + a_{2,y}) = m(g - a)$$

so:

$$mg - ma - f = Ma \Rightarrow mg - f = a\,(m + M)$$

Since $f = \mu_k n$ and $n = Mg$, we have

$$mg - \mu_k Mg = (m + M)\,a \Rightarrow a = \frac{(m - \mu_k M)}{(m + M)} g$$

When the table top is frictionless, the acceleration of both blocks is

a) $a = \dfrac{gm}{m + M}$

b) $a = \dfrac{g\,(m - \mu_k M)}{(m + M)}$.

If $\mu_k = .7$, their acceleration is $a = \dfrac{g\,(m - 0.7M)}{m + M}$.

5.17 Applying N2L:

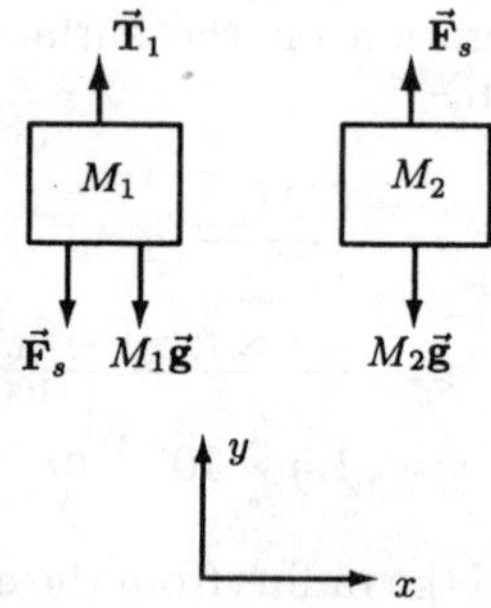

$$T_1 - F_s - M_1 g = 0 \quad (1)$$
$$F_s - M_2 g = 0 \quad (2)$$

We can find the stretch of the spring, s, directly from Eq. (2)

$$F_s = ks = M_2 g \Rightarrow s = \frac{M_2 g}{k}$$

The tension in the string can be obtained by substituting Eq. (2) into Eq. (1)

$$T_1 = M_2 g + M_1 g = g(M_1 + M_2)$$

5.21 The tension force T_1 keeps the object of mass m accelerating centripetally towards point 0.

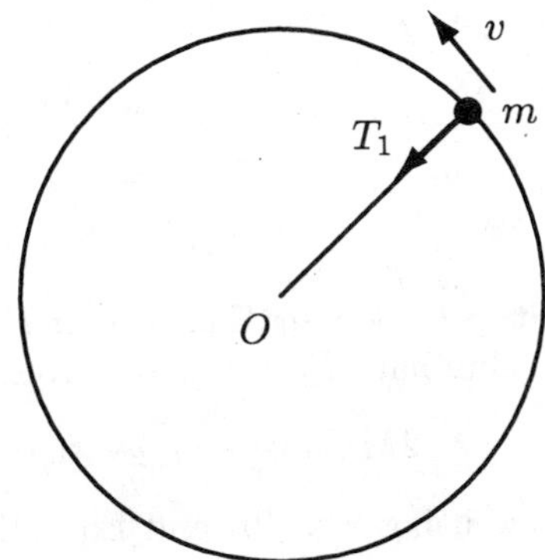

$$T_1 = \frac{mv^2}{r}$$

And this tension force must also balance the weight force acting on the hanging mass if this mass is to remain in equilibrium

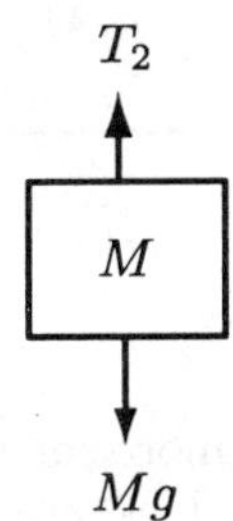

$$T_2 = Mg$$

If this is an ideal string, tension is constant throughout its length and thus, $T_1 = T_2$

$$\frac{mv^2}{r} = Mg$$

solving for v,

$$\begin{aligned} v &= \sqrt{\frac{Mgr}{m}} \\ &= \sqrt{\frac{(.100\ \text{kg})(9.81\ \text{m}/\text{s}^2)(.100\ \text{m})}{0.0100\ \text{kg}}} \\ &= 3.13\ \text{m}/\text{s} \end{aligned}$$

5.25 Our plan is to draw FBDs for both blocks, derive N2L equations and find the maximum acceleration possible for each block separately. If we find that the friction exerted on the smaller block can't become large enough to make it accelerate with an acceleration equal to that of the larger block then the smaller block must slide with respect to the larger block.

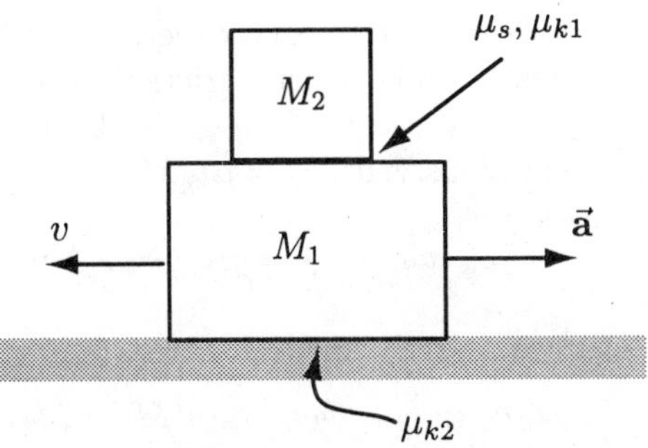

Some notes: the only thing that decelerates the small block is f_2; from physical considerations (the block must decelerate) its acceleration must point opposite the direction of motion and in the same direction as $\vec{a}$. f_2 and n_1 in the FBD for the larger block are the N3L pairs of those acting on the smaller block.

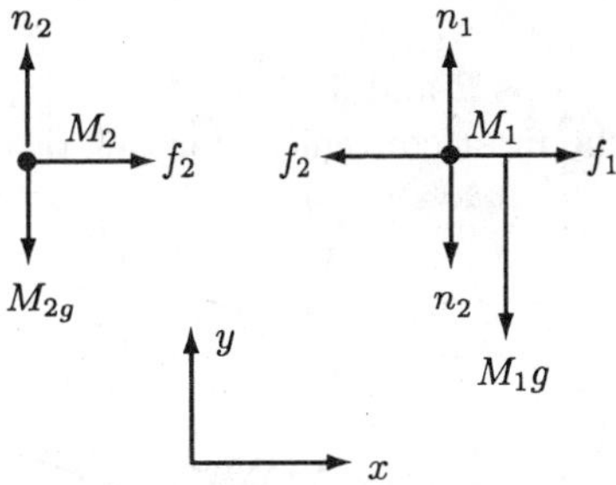

Applying N2L:

$$\begin{aligned} n_1 - n_2 - M_1g &= 0 && (3) \\ f_1 - f_2 &= M_1a_1 && (4) \\ n_2 - M_2g &= 0 && (5) \\ f_2 &= M_2a_2 && (6) \end{aligned}$$

First, as a trial solution, assume there is no sliding and the two blocks have the same acceleration $a_1 = a_2 = a$. Then $f_2 = M_2a$,

$$f_1 = \mu_{k1}n_1 = \mu_{k1}(n_2 + M_1g) = \mu_{k1}g(M_1 + M_2)$$

and

$$\mu_{k1}g(M_1 + M_2) - M_2a = M_1a \Rightarrow a = \mu_{k1}g$$

Now, the maximum static friction force on the small block is $f_{\text{max,static}} = \mu_s M_2 g$, and the maximum acceleration it can cause is

$$a_{\text{max}} = \frac{f_{\text{max,static}}}{M_2} = \mu_s g$$

But

$$\mu_s = 0.40 < \mu_{k1} = 0.50$$

Static friction cannot accelerate block 2 rapidly enough to avoid sliding. So the blocks do slide with respect to each other. Once block m begins to slide, kinetic friction takes over. Since $f_k < f_s$, we need to go back and rewrite the equations. With $f_2 = \mu_{k2} M_2 g$ we have

$$a_2 = \frac{f_2}{M_2} = \mu_{k2} g = 2.9 \text{ m/s}^2$$

And

$$M_1 a_1 = f_1 - f_2 = \mu_{k1} g (M_1 + M_2) - \mu_{k2} M_2 g$$

So

$$a_1 = g \left[\mu_{k1} + \frac{M_2}{M_1} (\mu_{k1} - \mu_{k2}) \right] = 5.3 \text{ m/s}^2$$

5.29 We will use the 6-step process for analyzing dynamical systems.

Step 1: The two significant particles are the boxes. The spring can be represented by forces it exerts.

Steps 2 and 3: Axes $\perp$ and $\parallel$ to the slope are the most convenient for this problem.

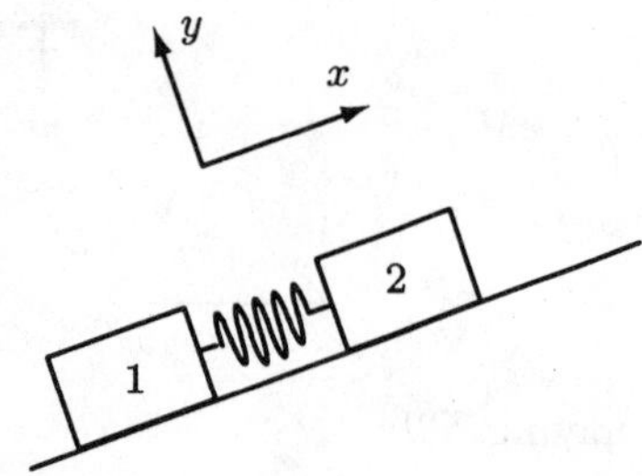

Steps 4 and 5: Note that $\mu_1 > \mu_2$ so we would expect that block 2 would try to accelerate faster than block 1.

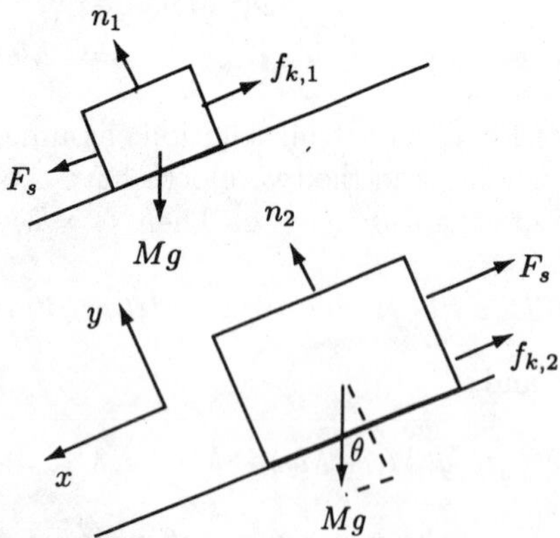

Because of this the spring will be compressed, exerting a spring force downslope on block 1 and upslope on block 2, forcing them to have the same acceleration.

Block 1:

$$\begin{aligned} n_1 &= Mg\cos\theta && (7) \\ F_s + Mg\sin\theta - f_{k,1} &= Ma && (8) \\ f_{k,1} &= \mu_1 n_1 = \mu_1 Mg\cos\theta && (9) \end{aligned}$$

Block 2:

$$\begin{aligned} n_2 &= Mg\cos\theta && (10) \\ Mg\sin\theta - F_s - f_{k,2} &= Ma && (11) \\ f_{k,2} &= \mu_2 n_2 = \mu_2 Mg\cos\theta && (12) \end{aligned}$$

Step 6: Solving Eq. (11) for F_s and using this to eliminate F_s in Eq. (8) gives:

$$2Mg\sin\theta - f_{k,2} - f_{k,1} = 2Ma$$

Now using Eq. (9) and Eq. (12), and solving for a gives:

$$a = g\sin\theta - \frac{g}{2}\cos\theta(\mu_1 + \mu_2)$$

Now we need to go back to Eq. (8) and solve for $s = \dfrac{F_s}{k}$

$$\begin{aligned} ks &= Mg\left[\sin\theta - \frac{\cos\theta}{2}(\mu_1 + \mu_2)\right] \\ &\quad + \mu_1 Mg\cos\theta - Mg\sin\theta \\ s &= \frac{Mg\cos\theta}{2k}(\mu_1 - \mu_2) \end{aligned}$$

Consistency check: Note that as $\mu_2 \to \mu_1$,

$$s \to 0$$

That is, the more equal the frictional forces acting on the block are the less compressed the spring is. This is what we would expect since if the blocks accelerated equally there would be no reason for the spring to stretch or compress. Also note that from our expression for acceleration it is possible to have a $\mu_1 + \mu_2$ s.t. the blocks don't accelerate at all. This happens when

$$0 = g\sin\theta - \frac{1}{2}g\cos\theta(\mu_1 + \mu_2)$$

or, when

$$\tan\theta = \frac{\mu_1 + \mu_2}{2}$$

the average of the two coefficients.

5.33 If the pulley system is ideal, then tension is constant throughout the string. Also, the lower pulley is massless. Then a free-body diagram for the object—lower pulley plus suspended mass—is as shown:

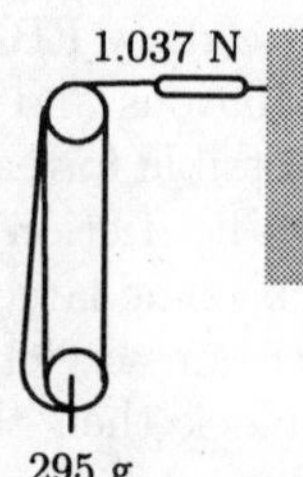

So

$$T_{\text{ideal}} = \frac{mg}{3} = \frac{(0.295 \text{ kg})(9.81 \text{ m/s}^2)}{3} = 0.965 \text{ N}$$

Since the measured tension differs from the ideal value by much more than the accuracy of the data, we conclude that the pulley system is not ideal.

5.37 **Step 1**: The significant object in the system is the bottom pulley. Since the block of mass m is fixed to the bottom pulley we can represent it as another weight force acting on pulley #2.

Steps 2 and **3**:

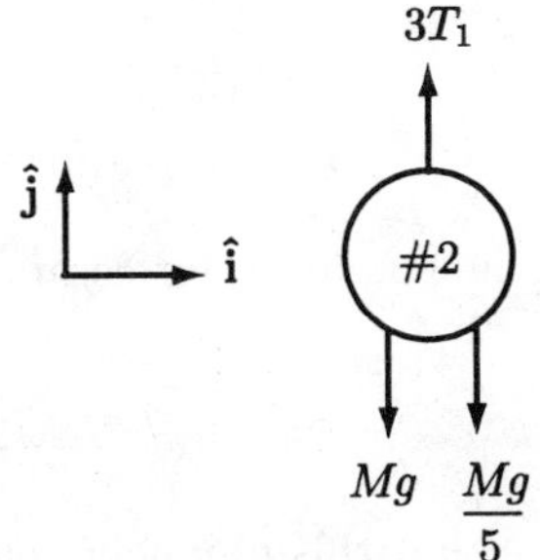

Step 4: We will use N2L on the FBD.

$$3T_1 = \frac{6Mg}{5} \tag{13}$$

Since tension along the length of string must be constant,

$$F = T_1 \tag{14}$$

Step 5: Using Eq. (13) and (14),

$$3F = \frac{6Mg}{5} \Rightarrow F = \frac{2Mg}{5}$$

5.41 We will draw a FBD for each significant portion of system and analyze it using N2L.

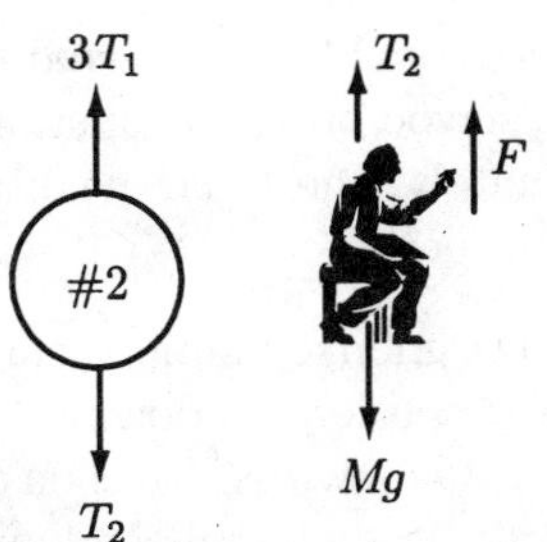

Let the upper pulley be pulley #1 and the lower pulley be pulley #2. Let the professor have a mass of M.

$$3T_1 = T_2 \tag{15}$$
$$T_2 + F = Mg \tag{16}$$

If we assume ideal string, the tension along the entire length of string is constant. At its end the string is being pulled on by a force F; thus $F = T_1$. From Eq. (16)and (15),

$$3T_1 + F = Mg \Rightarrow 3F + F = Mg \Rightarrow F = \frac{1}{4}Mg$$

5.45 **Step 1**: There are two elements of the system we can analyze: the spring with negligible unstretched length and the hanging mass.

Step 2 and **3**:

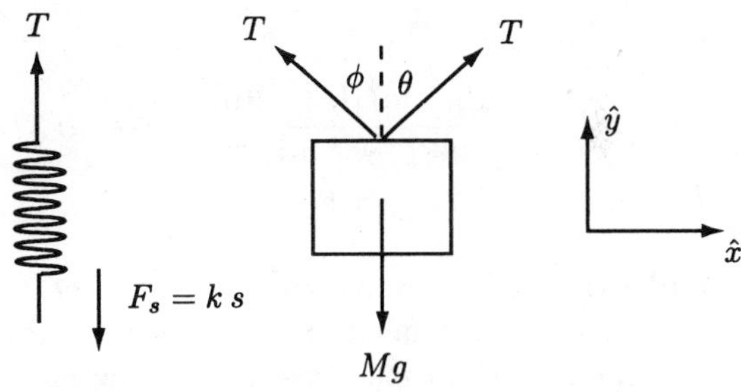

Step 4:

$$T = F_s = ks = k\,(\alpha\ell) \tag{17}$$
$$T\sin\phi = T\sin\theta \tag{18}$$
$$T\cos\phi + T\cos\theta = Mg \tag{19}$$

From Eq. (18), $\theta = \phi$ or $180 - \phi$. However, if $\theta = 180 - \phi$ then

$$T\cos\phi = -T\cos\theta$$

that would make Eq. (18) $0 = Mg$! Thus, we must conclude that $\theta = \phi$. Then,

$$2T\cos\theta = 2\alpha k\ell\cos\theta = Mg$$

Now we have two unknowns θ and M, but only one equation. The last equation comes from the geometry of the problem.

Step 5:

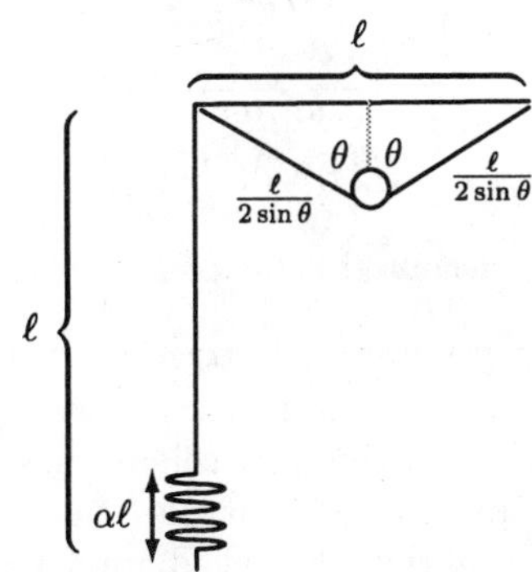

Consider half the triangle depicted in the figure. The hypotenuse is $\dfrac{\ell}{2\sin\theta}$. Since the path of the string is two such hypotenuses, we can find out what θ is:

$$\ell - \alpha\ell + \frac{\ell}{\sin\theta} = 2\ell$$

(2ℓ is the total length of the string.) So

$$\sin\theta = \frac{1}{1+\alpha}$$

So

$$M = \frac{2k\ell}{g}\alpha\cos\theta = \frac{2k\ell}{g}\alpha\sqrt{1-\left(\frac{1}{1+\alpha}\right)^2}$$

With the given numbers,

$$\begin{aligned}\theta &= \sin^{-1}\frac{1}{1.32} = 49^\circ\\ M &= \frac{(5.7\ \mathrm{m})\left(6.3\times10^4\ \mathrm{N/m}\right)}{9.8\ \mathrm{m/s^2}}(0.32)\cos 49^\circ\\ &= 7.7\times10^3\ \mathrm{kg}\end{aligned}$$

5.49 Her response would be very different. All objects in free fall accelerate at the same rate. This acceleration depends on how far you are from the Earth's center and is independent of the object's mass. Assuming she didn't throw the jar, it would remain right by her hand when released. On the other hand she might injure someone much more effectively by landing on the person herself!

5.53 The gravitational attraction between the Sun and Earth is the force that centripetally accelerates the Earth about the Sun. Using N2L,

$$\begin{aligned}F_{\text{Sun-Earth}} &= ma_{\text{cent}}\\ \frac{GM_{\text{Sun}}M_{\text{E}}}{R_{\text{ES}}^2} &= M_{\text{E}}\omega_{\text{E}}^2R_{\text{ES}} = \frac{M_{\text{E}}4\pi^2R_{\text{ES}}}{T_{\text{E}}^2}\end{aligned}$$

Solving this equation for M_{Sun},

$$\begin{aligned}M_{\text{Sun}} &= \frac{4\pi^2R_{\text{ES}}^3}{GT_{\text{E}}^2}\\ &= \frac{4\pi^2\left(1.5\times10^{11}\ \mathrm{m}\right)^3}{\left(6.67\times10^{-11}\ \mathrm{N\cdot m^2/kg^2}\right)\left[365\ \mathrm{d}\cdot\frac{24\ \mathrm{h}}{\mathrm{d}}\cdot\frac{3600\ \mathrm{s}}{\mathrm{h}}\right]}\\ &= 2.0\times10^{30}\ \mathrm{kg}\end{aligned}$$

The accepted value of M_{Sun} is 1.99×10^{30} kg.

5.57 The gravitational attraction between the Earth and satellite is the force that produces the necessary centripetal acceleration to keep the satellite in orbit at that particular height. By neglecting the difference between R_E and R_{sat}, we are to consider the satellite orbiting the Earth at the Earth's surface; $R_{\text{sat}} = R_E = 6400$ km

$$\begin{aligned}F_{E,\text{sat}} &= M_{\text{sat}}a_{\text{cent}}\\ \Rightarrow\ \frac{GM_EM_{\text{sat}}}{R_E^2} &= M_{\text{sat}}\omega_{\text{sat}}^2R_E^2\\ &= M_{\text{sat}}\left(\frac{2\pi}{T_{\text{sat}}}\right)^2R_E\end{aligned}$$

Solving for the satellite's period:

$$\begin{aligned}T_{\text{sat}} &= \sqrt{\frac{4\pi^2R_E^3}{GM_E}}\\ &= \sqrt{\frac{4\pi^2\left(6400\times10^3\ \mathrm{m}\right)^3}{\left(6.7\times10^{-11}\ \mathrm{N\cdot m^2/kg^2}\right)\left(5.98\times10^{24}\ \mathrm{kg}\right)}}\\ &= 5100\ \mathrm{s} = 1.4\ \mathrm{h}\end{aligned}$$

5.61 Now any point on the stalk rotates at the same angular speed that the moon does, so $v_{\text{stalk}} = \omega_{\text{moon}}r$ where

$$\begin{aligned}\omega_{\text{moon}} &= \frac{2\pi}{T_{\text{moon}}}\\ &= \frac{2\pi}{(27.5\ \mathrm{d})\left(\frac{24\ \mathrm{h}}{\mathrm{d}}\right)\left(\frac{3600\ \mathrm{s}}{\mathrm{h}}\right)}\\ &= 2.64\times10^{-6}\ \mathrm{s^{-1}}\end{aligned}$$

Now the condition for escape velocity says:

$$\frac{v_{\text{stalk}}}{v_{\text{esc}}} = \frac{\omega_{\text{moon}}r}{\sqrt{\frac{2GM_{\text{moon}}}{r}}} \overset{\text{set}}{\equiv} 1$$

Notice that this uniquely determines r:

$$r = \left[\frac{\sqrt{2GM_{\text{moon}}}}{\omega}\right]^{\frac{2}{3}}$$

Solving for r gives 1.1×10^8 m. Now the mysterious point is about $\frac{1}{9}$ of the Earth-Moon distance away from the Moon. However, the length of the Lunar beanstalk is about $\frac{1.1\times10^8}{3.8\times10^8}\approx\frac{1}{4}$ of the Earth-Moon distance.

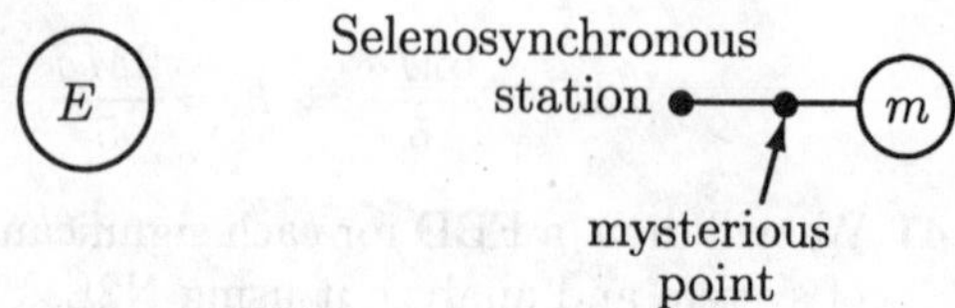

Gravitational forces exerted by the Earth would cause havoc on the beanstalk's operation. Unfortunately, the lunar beanstalk will not function properly.

5.65 a) If the monkey simply clutches the rope the monkey accelerates downward and the bananas accelerate upwards. Tension acts on the monkey upwards as shown in the figure.

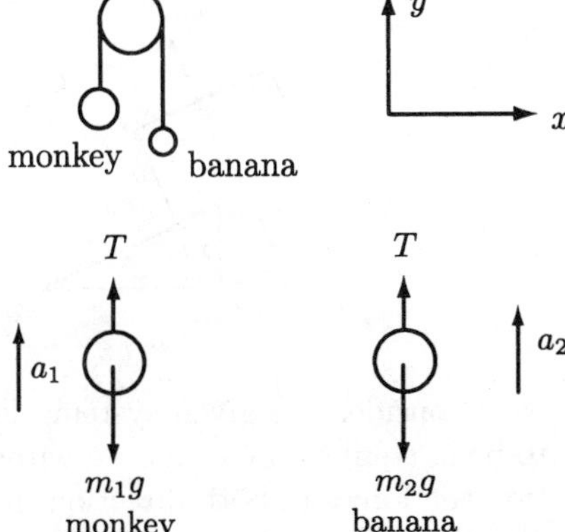

The magnitude of the tension is constant along the rope, and causes the bananas to be accelerated upwards (the magnitudes of the two accelerations must be equal because of conservation of string) but a_1 and a_2 are in opposite directions (someone needs to go up; the other goes down). Thus we have the relations

$$T - m_1 g = m_1 a_1 \quad (20)$$

$$T - m_2 g = m_2 a_2 \quad (21)$$

$$a_1 = -a_2 \quad (22)$$

but by multiplying Eq. (21) by -1 and adding it to Eq. (20) we get:

$$g(m_2 - m_1) = m_1 a_1 - m_2 a_2$$

and by Eq. (22)

$$\begin{aligned} g(m_2 - m_1) &= m_1 a_1 + m_2 a_1 \\ \Rightarrow a_1 &= \frac{g\,(m_2 - m_1)}{m_2 + m_1} \\ &= \frac{10g}{50} = \frac{g}{5} \end{aligned}$$

Thus the monkey accelerates at 1.96 $\mathrm{m/s^2}$ downward while the bananas accelerate at 1.96 $\mathrm{m/s^2}$ upward.

b) For the monkey not to accelerate he needs to climb up the rope. He must pull harder on the rope so that tension exactly balances his weight. From N2L,

$$T = m_1 g$$

Then the force acting on the bananas is

$$T - m_2 g = (m_1 - m_2)\,g = m_2 a_2$$

The bananas accelerate at

$$a_2 = \left(\frac{m_1 - m_2}{m_2}\right) g = \frac{1}{2} g,$$

faster than before. Notice that conservation of string does not apply in this case because the length of rope between monkey and bananas is changing.

c) Continuing the same line of reasoning,

$$\begin{aligned} T - m_1 g &= \frac{m_1 g}{3} \\ T &= m_1 \left[\frac{4}{3} g\right] \end{aligned}$$

and

$$\begin{aligned} T - m_2 g &= \left(\frac{4}{3} m_1 - m_2\right) g = m_2 a_2 \\ a_2 &= \left(\frac{4m_1 - 3m_2}{3m_2}\right) g = g \end{aligned}$$

The bananas accelerate up faster than they did in either case a) or b).

d) To determine how long it takes the monkey to reach the bananas we need to determine the acceleration of the monkey with respect to the bananas: $a_1 - a_2$.

Case a): The bananas are initially 10 m away and are accelerating at 3.92 $\mathrm{m/s^2}$ towards him.

$$10\,\mathrm{m} \overset{\text{set}}{\equiv} \frac{1}{2}(3.92\ \mathrm{m/s^2})t^2 \Rightarrow t = 2.26\ \mathrm{s}$$

The monkey will reach the banana in 2.26 s.

Case b)

$$\begin{aligned} a_2 - a_1 &= a_2 - 0 = a_2 \\ &= \frac{9.81\ \mathrm{m/s^2}\,(10\ \mathrm{kg})}{20\ \mathrm{kg}} = 4.9\ \mathrm{m/s^2} \end{aligned}$$

Thus the monkey will be able to reach the bananas in a time t, where

$$10\ \mathrm{m} = \frac{1}{2}(4.9\ \mathrm{m/s^2})t^2 \Rightarrow t = 2.02\ \mathrm{s}$$

In **Case c)** the monkey accelerates up at $\frac{1}{3}g$ and the bananas accelerate up at g. From the monkey's viewpoint, the bananas are accelerating towards him at $\frac{2}{3}g$.

$$\begin{aligned} 10\ \mathrm{m} &= \frac{1}{2}\left(\frac{2}{3}g\right)t^2 \\ \Rightarrow t &= \sqrt{\frac{30\ \mathrm{m}}{9.81\ \mathrm{m/s^2}}} = 1.75\ \mathrm{s} \end{aligned}$$

It appears that the monkey will minimize the time it takes for him to reach the bananas by Case c), accelerating up the rope.

e) If the bananas are initially above the monkey, then to reach them, the monkey must accelerate up and/or the bananas accelerate down.

But for the monkey to accelerate upward,

$$T > m_1 g > m_2 g$$

and the bananas would have to accelerate up too. This is not possible. If the monkey lets go of the rope, both monkey *and* bananas accelerate downward, but the monkey stays ahead of the bananas until they reach the ground! The monkey gets the bananas once they reach the ground.

5.69 This pulley was formally analyzed in Problem 5.4, but we'll briefly discuss it here. The argument hinges on the fact that tension is constant along the length of cord in the pulley system.

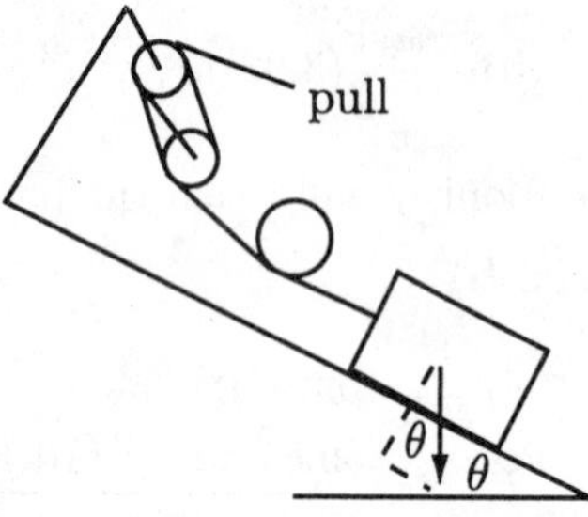

The box you wish to lift is fixed to the bottom pulley but note that there are 3 tension forces acting on this bottom pulley.

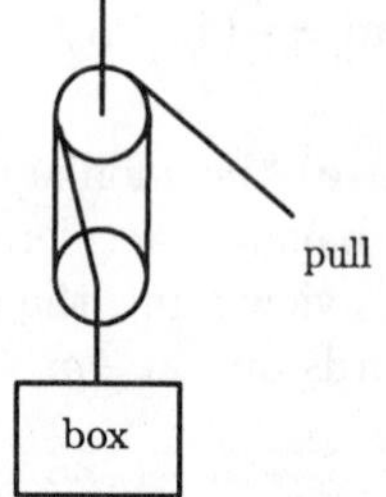

Each individual tension force is equal to the force with which you pull down on the cord. If the box has a weight mg this means you can support the box by applying a force of $\frac{1}{3}mg$. Thus the pulley allows you to support a given weight by applying only $\frac{1}{3}$ of the force you would have to apply without the pulley.

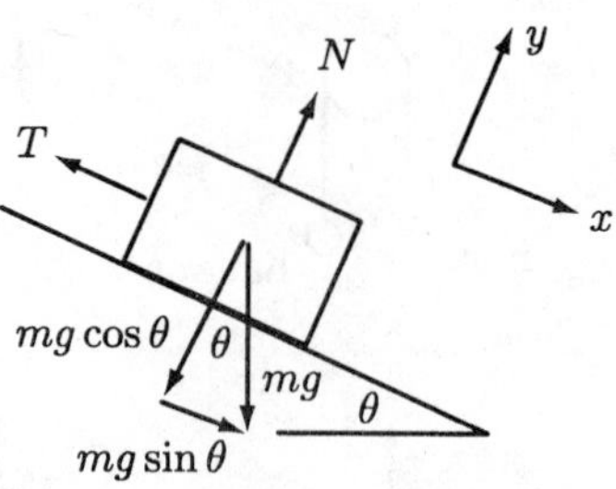

Now consider the given system. For the block to be in equilibrium or moved without acceleration the forces in both directions must balance. The floor of the ramp exerts a normal force on the box which balances the component of weight parallel to the ramp's surface;

$$n = mg\cos\theta$$

If it weren't for the tension force shown the component of the boxes' weight parallel to the ramp's surface, $mg\sin\theta$, would be unbalanced and cause an acceleration. To hoist the box slowly up the ramp (without acceleration) we must apply a tension force that equals $mg\sin\theta$. However, recall that using the pulley, we need only apply $\frac{1}{3}$ of this force. Thus to hoist the box of mass M slowly up the ramp we need only apply a force of $\frac{1}{3}Mg\sin\theta$ to the pulley. That gives us a mechanical advantage $\frac{W}{F}$ of

$$\frac{Mg}{\frac{1}{3}Mg\sin\theta} = \frac{3}{\sin\theta}$$

As discussed, using the pulley by itself, would require us to apply a force of $\frac{1}{3}Mg$ and using the ramp by itself would require us to apply a force of $Mg\sin\theta$ giving $\frac{W}{F}$ ratios of 3 and $\frac{1}{\sin\theta}$ respectively.

5.73 This is called an Atwood's machine.

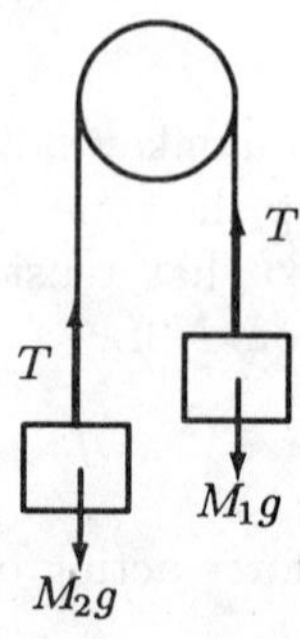

Step 1: The only two significant particles are the masses which hang. The frictionless rod provides a path over which the string travels.

Steps 2 and 3:

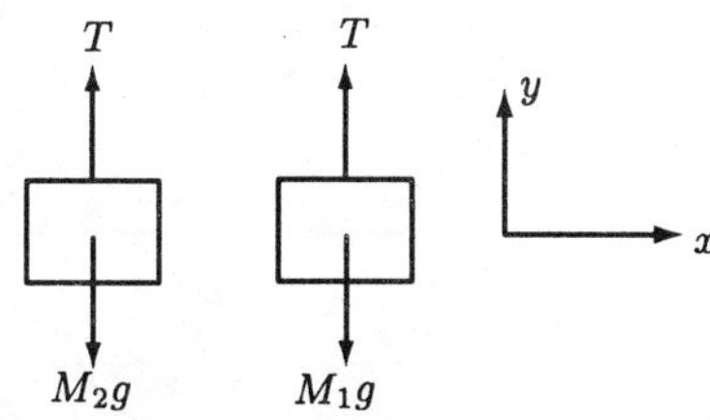

Steps 4 and **5**: Since $M_1 \neq M_2$, there will be accelerations $a_{1,y}$ and $a_{2,y}$ with $a_{1,y} = -a_{2,y}$ because of conservation of string

$$\begin{aligned} T - M_2 g &= M_2 a_{2,y} && (23) \\ T - M_1 g &= M_1 a_{1,y} && (24) \\ a_{1,y} &= -a_{2,y} && (25) \end{aligned}$$

Step 6: (abbreviated) Solving this system of equations gives:

$$a_{1,y} = \frac{g\,(M_2 - M_1)}{M_2 + M_1}$$

Initially

$$M_2 = M + m_2 = 1.003 \text{ kg}$$

and

$$M_1 = M + m_1 = 1.006 \text{ kg}$$

The acceleration of M_1 during the first 15 s is

$$a_i = -\frac{g\,(3 \times 10^{-3} \text{ kg})}{2.009 \text{ kg}} = \left(-\frac{3}{2009}\right) g$$

Now after the 6 g rider is removed, the acceleration of M_1 becomes

$$a_f = \left(\frac{3}{2003}\right) g$$

(Nearly the same magnitude but in the opposite direction). During the first 15 s the object on the left attains a velocity

$$v_1 = a_i t_1$$

and reaches a coordinate

$$y_1 = \frac{1}{2} a_i t_1^2$$

During the next 15 s, the object comes to rest at a final coordinate

$$y_f = -1.241 \text{ m}$$

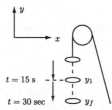

y
x
$t = 15$ s y_1
$t = 30$ sec y_f

We may use expressions for distance as a function of time to obtain the acceleration due to gravity from these data:

a)

$$y_f = y_i + v_i \Delta t + \frac{1}{2} a \Delta t^2$$

$$-1.241 \text{ m} = \frac{1}{2} a_i t_i^2 + a_i t_1 (t_f - t_1) + \frac{1}{2} a_f (t_f - t_1)^2$$

Now we are told that

$$\begin{aligned} t_f - t_1 &= t_1 = 15 \text{ s} \\ -1.241 \text{ m} &= a_i \frac{3t_1^2}{2} + a_f \frac{t_1^2}{2} \\ &= \frac{t_1^2}{2}\left[\frac{3}{2003} g + 3\left(\frac{-3}{2009} g\right)\right] \\ g &= \frac{(1.241 \text{ m})\,2}{(15 \text{ s})^2\, 3\left[\frac{-1}{2003} + \frac{3}{2009}\right]} \\ &= 3.70 \text{ m}/\text{s}^2 \end{aligned}$$

Comparing this with the astronomical data from the front endpapers (inside front cover) in the text, we can conclude that this experiment was conducted on Mars, where $g = 3.73$ m/s^2.

b) Now, to check whether removal of the rider caused a detectable disturbance, we can calculate g from observations of the system's velocity. Taking initial position at the removal of the rider and final position when the system comes to rest, we should have:

$$\begin{aligned} 2a_f (y_f - y_1) &= v_f^2 - v_1^2 = 0 - v_1^2 \\ -1.241 \text{ m} &= -\frac{(a_i t_1)^2}{2a_f} + \frac{1}{2} a_i t_1^2 \\ &= \frac{1}{2} a_i t_1^2 \left[1 - \frac{a_i}{a_f}\right] \end{aligned}$$

$$\begin{aligned} -1.241 \text{ m} &= \frac{1}{2}\left(\frac{-3}{2009} g\right)(15 \text{ s})^2 \left[1 - \frac{\left(\frac{-3}{2009}\right)}{\left(\frac{3}{2003}\right)}\right] \\ g &= \frac{2\,(1.241 \text{ m})}{(15 \text{ s})^2} \frac{(2009)}{3} \frac{1}{1 + \left(\frac{2003}{2009}\right)} \\ &= 3.70 \text{ m}/\text{s}^2 \end{aligned}$$

Since the two results are consistent, we may conclude that removing the riders did not significantly disturb the system.

5.77 a) Let some test mass, M_T be located beneath the Earth's surface at a distance r from the Earth's center ($r < R_E$).

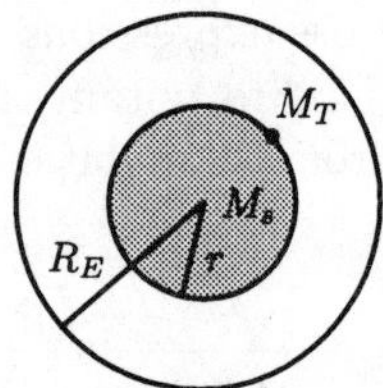

According to Newton, the unshaded portion of the Earth exerts no force on M_T—only the shaded part which we'll call M_s. This is the consequence of a spherical shell of mass not exerting any net force on an object inside the shell. Since a sphere of mass attracts objects outside of itself as a point mass located at the sphere's center we can write an equation for the force exerted on M_T as

$$F = \frac{GM_sM_T}{r^2} \tag{26}$$

However,

$$M_s = V_s\rho = \frac{4}{3}\pi r^3 \rho$$

where V_s is the volume of the shaded sphere and ρ is the density of the Earth. Now, Eq. (26) becomes

$$F = \frac{4\pi Gr\rho M_T}{3} = M_T g\,(r) \tag{27}$$

We are accustomed to seeing $F = mg$, but it looks strange to see $mg(r)$. In this problem we are asked to consider how g changes with radial distance from the Earth's center, thereby obtaining g as a function of "r" —$g(r)$. Now ρ, the Earth's density, is the mass of the whole Earth divided by the volume of the whole Earth:

$$\frac{M_E}{\frac{4}{3}\pi R_E^3}$$

Substituting this in to Eq. (27) and simplifying yields

$$g\,(r) = \frac{GM_E r}{R_E^3} = \left(\frac{GM_E}{R_E^2}\right)\frac{r}{R_E} \tag{28}$$

The expression in the parentheses is g_0, the acceleration due to gravity at the Earth's surface. Thus:

$$g\,(r) = \frac{g_0 r}{R_E}$$

Note that at the Earth's surface, $r = R_E$ and

$$g\,(r) = g_0 = 9.81 \text{ m}/\text{s}^2$$

as expected. Furthermore at the center of the Earth where $r = 0$, $g\,(r) = 0$, also as expected.

b) Now the tunnel is shown below.

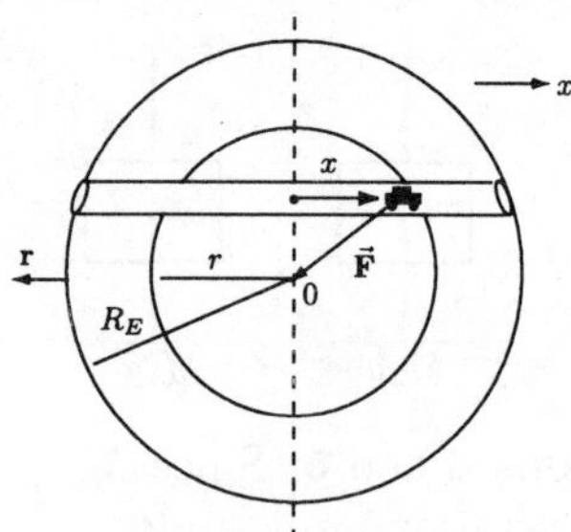

The train is a distance x from the axial line and a distance r from the Earth's center. The force acting on the car is

$$\begin{aligned}\vec{\mathbf{F}} &= -mg\,(r)\,\hat{\mathbf{r}} \\ &= -\frac{mg_0 r}{R_E}\hat{\mathbf{r}}\end{aligned}$$

where m is the mass of the train. However we need only the x component of the force acting on the train:

$$F_x = -F\cos\theta = -F\frac{x}{r}$$

where $\cos\theta = \frac{x}{r}$.

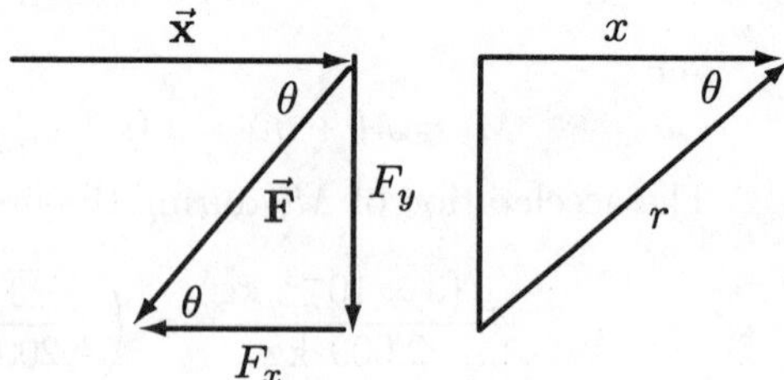

Thus,

$$\begin{aligned}F_x &= -F\cos\theta = -F\frac{x}{r} \\ &= -\frac{mg_0 r}{R_E}\frac{x}{r} \\ &= -\frac{mg_0 x}{R_E}\end{aligned}$$

c)

$$t = \pi\sqrt{\frac{R}{g_0}} = \pi\sqrt{\frac{R^3}{GM}} = \pi\sqrt{\frac{3}{4\pi G\rho}} = \sqrt{\frac{3\pi}{4G\rho}}$$

Since the time depends only on the density, it is the same on the asteroid as on Earth

$$t = \pi\sqrt{\frac{6.4\times10^6 \text{ m}}{10 \text{ m}/\text{s}^2}} = 0.7 \text{ h}$$

Travel on the asteroid would not be all that rapid, but it would be amazingly efficient.

5.81 If the cable's mass is concentrated at its center, then a total force of

$$(85 \text{ kg} + 31 \text{ kg})(9.8 \text{ m}/\text{s}^2) = 1137 \text{ N}$$

acts as depicted below (note that the diagram is not even close to being drawn to scale).

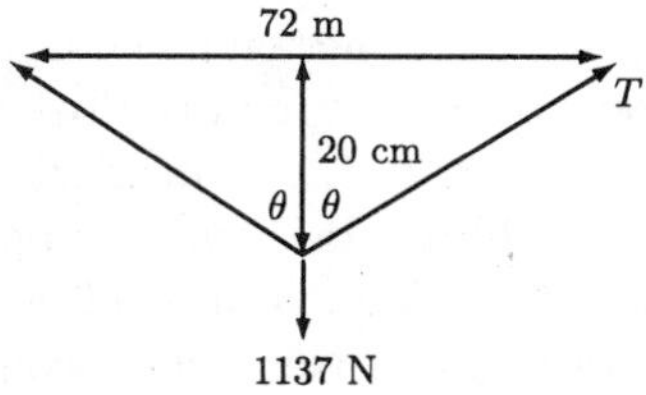

N2L gives

$$2T\cos\theta = 1137 \text{ N} \tag{29}$$

to ensure equilibrium. The highly symmetrical geometry of the problem is enough to assure that the two tension forces must be equal and the angles marked "θ" are equal. If we had assumed that the mass of the cable or the tightrope walker were at some position along the rope other than the middle, this assumption would not be true. We can find θ from the geometry of the problem:

$$\theta = \tan^{-1}\left(\frac{36 \text{ m}}{0.20 \text{ m}}\right) = 89.68^\circ$$

Thus from Eq. (29) the tension in the rope is 1.0×10^5 N.

b) Step 1: There are 3 masses—these must be the significant particles we need to analyze.

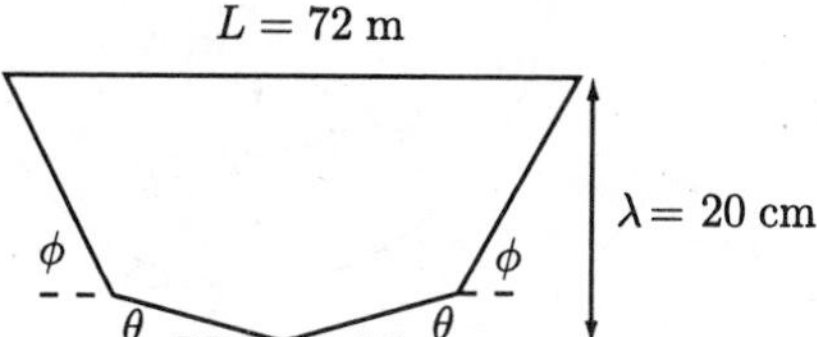

Steps 2 and **3**: The mass of the cable is distributed evenly between the 3 masses: 10.3 kg each. The middle mass includes the mass of the walker, 85 kg, so $m = 10.3$ kg and $M = 95.3$ kg. There are two things to notice here. First is the choice of coordinate system. The weight force for each particle is the only force that shares the same direction in each FBD, so we choose the axes to be $\perp$ and $||$ to it.

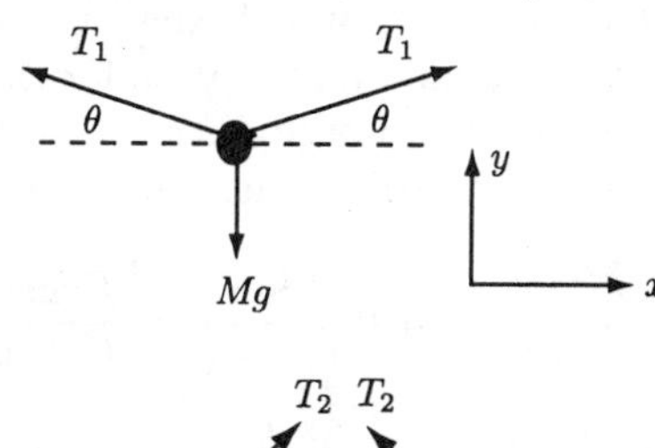

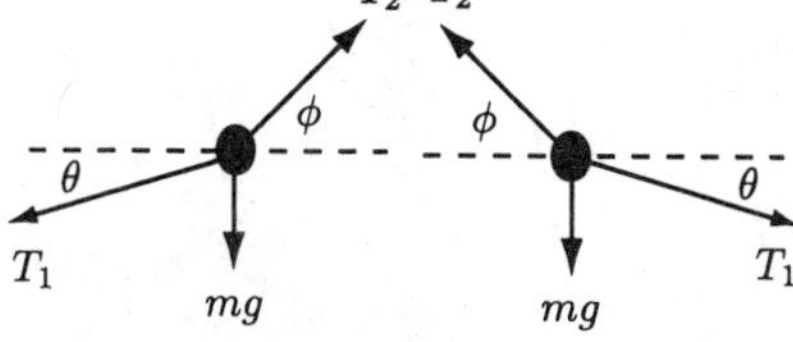

The two FBD for the two m are mirror images of each other so we'll get the same information from each.

Steps 4 and **5**:

$$\begin{aligned} 2T_1\sin\theta &= Mg & (30)\\ T_2\sin\phi &= T_1\sin\theta + mg & (31)\\ T_2\cos\phi &= T_1\cos\theta & (32)\end{aligned}$$

Oh oh! Trouble: We have 4 unknowns (T_1, T_2, θ, ϕ) but only 3 equations from N2L. The fundamental theorem of algebra says that we need one more equation to solve for each of the unknowns. We turn to the geometry of the problem to glean more equations. Since $L/\lambda = 360$, θ and ϕ must be incredibly small. Even the figure shown is a gross exaggeration.

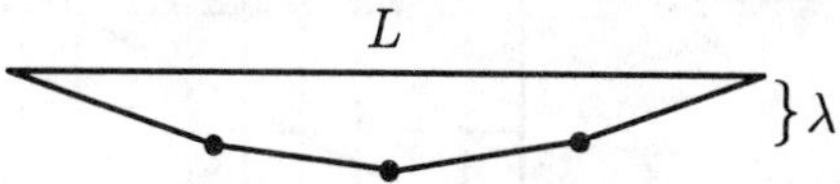

Thus the distance between any two points is $\approx \frac{1}{4}L$, (remember that the 3 points are equally spaced), and we can write

$$\frac{1}{4}L\sin\phi + \frac{1}{4}L\sin\theta = \lambda \tag{33}$$

We have the last equation we need.

Step 6: To make things easier we will utilize the fact that θ and ϕ are really small. Using small angle approximations and neglecting all terms greater than first order:

$$\begin{aligned} \cos\phi &= 1 - \frac{\phi^2}{2} + \frac{\phi^4}{6} - \cdots \approx 1 & (34)\\ \cos\theta &= 1 - \frac{\theta^2}{2} + \frac{\theta^4}{6} - \cdots \approx 1 & (35)\\ \sin\theta &= \theta - \frac{\theta^3}{3!} + \frac{\theta^5}{5!} - \cdots \approx \theta & (36)\\ \sin\phi &= \phi - \frac{\theta^3}{3!} + \frac{\theta^5}{5!} - \cdots \approx \phi & (37)\end{aligned}$$

Using Eq. (34) and (35) we can rewrite Eq. (32):

$$T_2 = T_1 = T \tag{38}$$

Thus according to our small angle approximation the tension is equal along the cable!

$$\begin{aligned} 2T\theta &= Mg & (39)\\ T\phi &= T\theta + mg & (40)\\ \phi + \theta &= \frac{4\lambda}{L} & (41)\end{aligned}$$

Now let's rewrite Eq. (40) and (41), and add them together.

Eq.(40): $\theta - \phi = -\frac{mg}{T}$
Eq. (41): $\theta + \phi = \frac{4\lambda}{L}$

Summing the two equations:

$$2\theta = \frac{4\lambda}{L} - \frac{mg}{T} \Rightarrow \theta = \frac{2\lambda}{L} - \frac{mg}{2T} \tag{42}$$

Now plugging Eq. (42) into Eq. (39),

$$2T\left[\frac{2\lambda}{L} - \frac{mg}{2T}\right] = Mg \Rightarrow T = \frac{g\,(M+m)\,L}{4\lambda}$$

Using the given numerical values,

$$T = 9.3 \times 10^4 \text{ N}$$

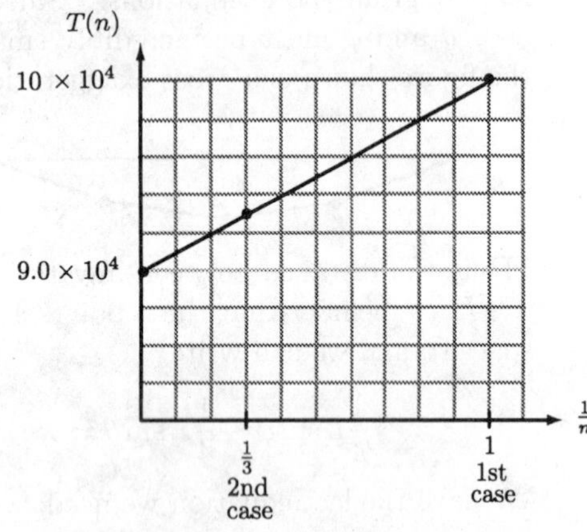

Now let's find the equation of the plotted line.

$$\text{slope} = \frac{\Delta y}{\Delta x} = \frac{7 \times 10^3 \text{ N}}{\frac{2}{3}} = 1.05 \times 10^4 \text{ N}$$

Now we'll choose one point to find the y-intercept

$$9.3 \times 10^4 \text{ N} \equiv (1.05 \times 10^4 \text{ N})\left(\frac{1}{3}\right) + b$$

$$\Rightarrow b = 8.95 \times 10^4 \text{ N}$$

Note that $\frac{1}{n} = 0$ corresponds to the mass of the cable being evenly distributed continuously. The tension when this happens is $T(0)$, the y-intercept which is $T = 9.0 \times 10^4$ N.

5.85 The typical element of cable is as shown in the figure.

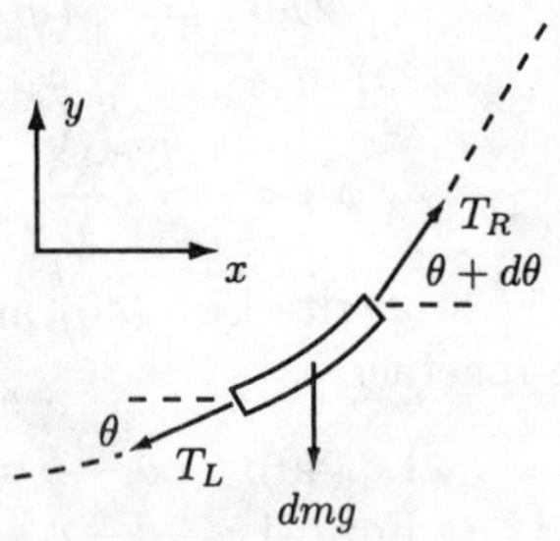

Tension forces act on its ends, and the weight of a portion of the roadway below the element is transmitted to the element via the vertical (relatively thin) cables which support the roadway. Since the bridge is in equilibrium, it is not accelerating, and Newton's second law gives: vertical components:

$$T_R \sin(\theta + d\theta) - T_L \sin\theta - dmg = 0$$

horizontal components:

$$T_R \cos(\theta + d\theta) - T_L \cos\theta = 0$$

The second equation states that the product tension × cosine of the angle is the same at both ends of the cable element. Extending this argument element by element to the ends of the cable implies

$$T\cos\theta = \text{constant} \equiv C$$

(a name!). The first equation gives the change in $T\sin\theta$ between the two ends of the cable element.

$$d\,(T\sin\theta) = dmg = g\lambda\, dx$$

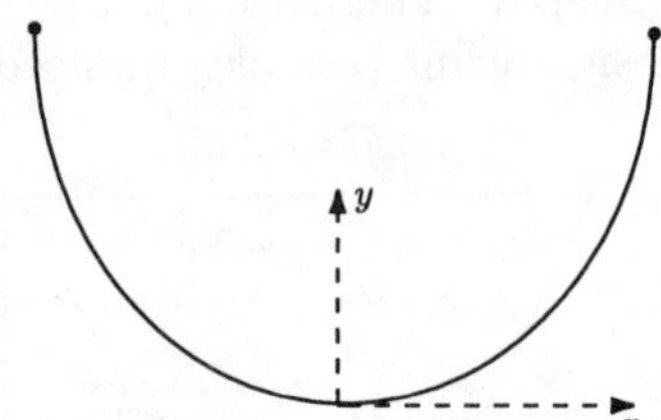

Now, if we take $x = 0$ at the low point of the cable, where $\theta = 0$, then we may integrate the equation as follows:

$$\int_{x=0}^{T\sin\theta} d\,(T'\sin\theta') = \int_{x=0}^{x} g\lambda\, dx$$

$$T\sin\theta = g\lambda x$$

{The integrated terms $T\sin\theta$ and $g\lambda x$ both vanish at the lower limit.} Now, we can convert our two results into an equation for the shape of the cable. If $y(x)$ describes the cable's y-coordinate as a function of x, then $\frac{dy}{dx} = \tan\theta$. So

$$\frac{dy}{dx} = \tan\theta = \frac{\sin\theta}{\cos\theta} = \frac{T\sin\theta}{T\cos\theta} = \frac{g\lambda x}{C}$$

Then

$$\int_{x=0}^{x} \frac{dy}{dx}\, dx = \int_{x=0}^{x} \frac{g\lambda x\, dx}{C}$$

or

$$y(x) - y|_{x=0} = \frac{g\lambda x^2}{2C}$$

Since $y(0) = 0$

$$y(x) = \frac{g\lambda x^2}{2C}$$

which is a parabola, as required. The sag of the cable, in our notation, is the value of y at the connection point to the towers, where $x = \frac{D}{2}$, half the distance between the towers:

$$\text{sag} = \frac{g\lambda D^2}{8C} \propto D^2$$

The constant C is still unknown. It can be found from the length of the cable ℓ, but finding it involves a rather nasty integration that we won't pursue.

Chapter 6

Linear Momentum

6.1 The ratio of the momenta of two objects with the same mass is the ratio of their speeds.

$$\begin{aligned} \frac{p_1}{p_2} &= \frac{mv_1}{mv_2} = \frac{v_1}{v_2} \\ &= \frac{30\ \mathrm{km/h}}{80\ \mathrm{km/h}} = \frac{3}{8} \end{aligned}$$

6.3 We can calculate the momentum of the system directly.

$$\vec{\mathbf{p}}_1 + \vec{\mathbf{p}}_2 = \frac{5}{4}Mv\hat{\mathbf{i}} - \frac{3}{4}Mv\hat{\mathbf{i}} = \frac{1}{2}Mv\hat{\mathbf{i}}$$

1 $\vec{\mathbf{v}}$ → ← $\vec{\mathbf{v}}$ 2

$\frac{5}{4}M$ $\frac{3}{4}M$

We have narrowed down the selection to *either* choices **c** or **e**. Momentum is a vector quantity; if the question wanted the magnitude of the system's momentum it would have asked for it. Choice **c** represents the correct magnitude and direction for the system's momentum.

6.5 By definition, the impulse I is equal to the change of the astronaut's momentum, Δp.

$$\begin{aligned} I &= p_f - p_i \\ &= (100\ \mathrm{kg})(3\ \mathrm{m/s}) - (100\ \mathrm{kg})(1\ \mathrm{m/s}) \\ &= 2 \times 10^2\ \mathrm{N \cdot s} \end{aligned}$$

6.7 The momentum of the system of three horses is the vector sum of the momenta of the individual horses.

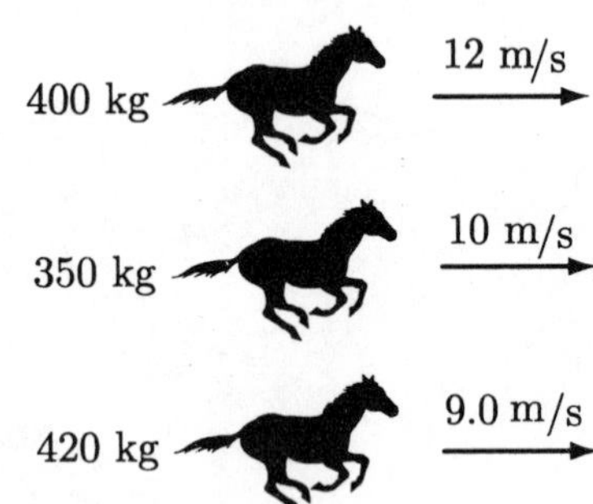

$$\begin{aligned} \vec{\mathbf{P}}_{\text{total}} &= \vec{\mathbf{p}}_1 + \vec{\mathbf{p}}_2 + \vec{\mathbf{p}}_3 \\ &= (390\ \mathrm{kg})(12\ \mathrm{m/s})\hat{\mathbf{i}} \\ &\quad +(350\ \mathrm{kg})(10.0\ \mathrm{m/s})\hat{\mathbf{i}} \\ &\quad +(420\ \mathrm{kg})(9.0\ \mathrm{m/s})\hat{\mathbf{i}} \\ &= 12 \times 10^3\ \mathrm{N \cdot s}\hat{\mathbf{i}} \end{aligned}$$

6.9 Momentum is a conserved quantity. Both skaters start with $v = 0$ m/s, so $\vec{\mathbf{p}}_i = 0$ and therefore $\vec{\mathbf{p}}_f = 0$.

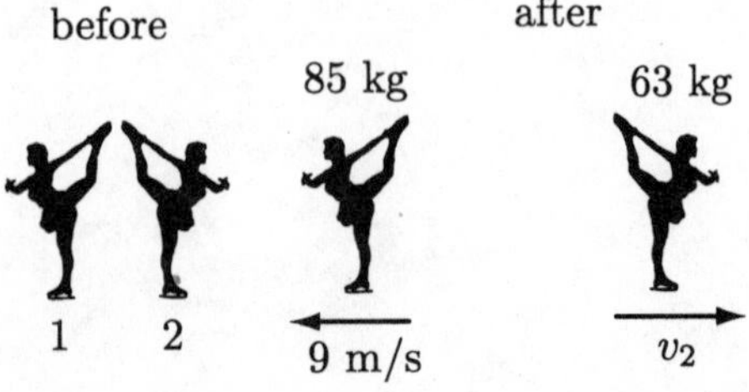

Their final velocities are in opposite directions along the "x-axis."

$$\vec{\mathbf{p}}_f = -(9\ \mathrm{m/s})(85\ \mathrm{kg})\hat{\mathbf{i}} + v_2(63\ \mathrm{kg})\hat{\mathbf{i}} = 0$$

Solving for $\vec{\mathbf{v}}_2$:

$$\vec{\mathbf{v}}_2 = \frac{(9\ \text{m/s})(85\ \text{kg})}{63\ \text{kg}}\hat{\mathbf{i}} = 12\ \text{m/s}\,\hat{\mathbf{i}}$$

The skater's speed is 12 m / s .

6.11 Assuming that no external forces act on the astronaut or satellite (so that we can consider them to comprise an isolated system) momentum must be conserved.

$$\begin{aligned} |\vec{\mathbf{p}}|_{\text{before}} &= (75\ \text{kg})(2.67\ \text{m/s}) \\ &\quad +(1248\ \text{kg})(0\ \text{m/s}) \\ &= 200\ \text{N}\cdot\text{s} \\ |\vec{\mathbf{p}}|_{\text{after}} &= (1323\ \text{kg})v_{\text{final}} \end{aligned}$$

where *before* and *after* refer to *before the collision* and *after the collision.* To conserve momentum, $\vec{\mathbf{p}}_{\text{before}} = \vec{\mathbf{p}}_{\text{after}}$

$$\begin{aligned} 200\ \text{N}\cdot\text{s} &= (1323\ \text{kg})v_{\text{final}} \\ v_{\text{final}} &= 0.15\ \text{m/s} \end{aligned}$$

The velocity is in the same direction in which the astronaut approached the satellite.

6.13 Intuitively, the ratio of the 2 momenta should be the ratio of the masses of the 2 vehicles since their speeds are the same

$$\begin{aligned} \frac{|\vec{\mathbf{p}}_t|}{|\vec{\mathbf{p}}_{vw}|} &= \frac{m_t v_t}{m_{vw} v_{vw}} \\ &= \frac{(170\ \text{kg})60\ \text{km/h}}{(850\ \text{kg})60\ \text{km/h}} \\ &= \frac{170\ \text{kg}}{850\ \text{kg}} = 0.20 \end{aligned}$$

6.17 If we call "down-alley" the "$\hat{d}$" direction,

$$\vec{\mathbf{p}}_{\text{ball}} = m\vec{\mathbf{v}} = (10\ \text{kg})(5\ \text{m/s})\hat{\mathbf{d}} = (50\ \text{N}\cdot\text{s})\,\hat{\mathbf{d}}$$

6.21 The impulse needed to stop the cricket ball equals the change of ball's momentum in going from 50 m / s to 0 m / s.

$$\begin{aligned} I &= \Delta p = p_f - p_i \\ &= 0 - (.250\ \text{kg})(50.0\ \text{m/s}) \\ &= -12.5\ \text{N}\cdot\text{s} \end{aligned}$$

The minus sign indicates that the impulse must be directed in the opposite direction from the ball's initial velocity.

6.25 Initially the car is at rest and has no momentum. The change in the car's momentum due to the force exerted by the tow truck is given by:

$$\begin{aligned} \Delta p_{\text{car}} &= F_{\text{on car}}\Delta t = (1000\ \text{N})(30\ \text{s}) \\ &= 30\,000\ \text{N}\cdot\text{s} \end{aligned}$$

Since its initial momentum is zero, this is its final momentum. From the definition of momentum its speed is:

$$v_{\text{car}} = \frac{p_{\text{car}}}{m_{\text{car}}} = \frac{30000\ \text{N}\cdot\text{s}}{1000\ \text{kg}} = 30\ \text{m/s}$$

6.29 The astronaut fires the pistol ⊥ to his initial velocity. From our work in Chapter 3 we know that, no matter what happens, his final velocity will have a component of 3 m / s$\hat{\mathbf{i}}$ and some component $v_y\hat{\mathbf{j}}$.

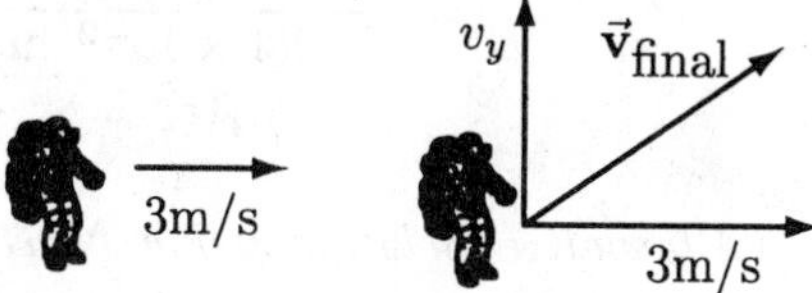

Impulse is the change in momentum, but since he started out with 0 momentum in the y-direction,

$$I_y = \Delta p_y = p_{f,y} - p_{i,y} = p_{f,y}$$

so his final momentum is

$$\begin{aligned} &(100\ \text{kg})(3\ \text{m/s})\hat{\mathbf{i}} + (800\ \text{N}\cdot\text{s})\,\hat{\mathbf{j}} \\ &= 100\ \text{kg}\left[(3\ \text{m/s})\,\hat{\mathbf{i}} + (8\ \text{m/s})\,\hat{\mathbf{j}}\right] \end{aligned}$$

Thus his new velocity is $(3\ \text{m/s})\,\hat{\mathbf{i}} + (8\ \text{m/s})\,\hat{\mathbf{j}}$.

6.33 The force exerted on the dummy by the seat belt must decelerate it from 30 m / s to 0 m / s in 0.1 s. In terms of momentum, the impulse delivered to the dummy must change its momentum from

$$(90\ \text{kg})(30\ \text{m}) = 2700\ \text{N}\cdot\text{s}$$

to 0 N· s. If we estimate the force acting on the dummy as being constant over the 0.1 s of crash, this force would be

$$\begin{aligned} F &= \frac{\Delta p}{\Delta t} = \frac{2700\ \text{N}\cdot\text{s}}{0.1\ \text{s}} \\ &= 27\times10^3\ \text{N} \simeq 3\times10^4\ \text{N} \end{aligned}$$

6.37 **a)** Since the ball has 0 final momentum (no velocity), its change in momentum ($\vec{\mathbf{I}} = \Delta\vec{\mathbf{p}}$) is minus its initial momentum:

$$\Delta\vec{\mathbf{p}} = -\vec{\mathbf{p}}_i$$

We can find the impulse if we know its velocity. We know its speed. From Eqn 2.13:

$$\begin{aligned} v_f^2 - v_i^2 &= 2gh \\ \Rightarrow v_f &= \sqrt{2gh} = \sqrt{2(5 \text{ m})(9.81 \text{ m/s}^2)} \\ &= 10 \text{ m/s} \end{aligned}$$

Thus:

$$\begin{aligned} \vec{\mathbf{I}} &= -\vec{\mathbf{p}}_i = (0.6 \text{ kg})(10 \text{ m/s}) \text{ up} \\ &= (6 \text{ N} \cdot \text{s upward}) \end{aligned}$$

b) Since the question asks for the force acting on the ball, $N2L$ might be a good thing to use.

$$\begin{aligned} F &= ma = \frac{mv^2}{2h} \\ &= \frac{(0.6 \text{ kg})(10 \text{ m/s})^2}{2(1 \times 10^{-2} \text{ m})} \\ &= 3 \times 10^3 \text{ N} \end{aligned}$$

(*Alternative Solution using Newton's original second law*):

$$F = \frac{\Delta p}{\Delta t}$$

Assuming the force is constant we can calculate the time it takes the ball to decelerate to zero in the sand

$$t = \frac{x}{v_{\text{AV}}} = \frac{10^{-2} \text{ m}}{\frac{0+10 \text{ m/s}}{2}} = 2 \times 10^{-3} \text{ s}$$

$$\therefore F = \frac{\Delta p}{\Delta t} = \frac{6 \text{ N} \cdot \text{s}}{2 \times 10^{-3} \text{ s}} = 3 \times 10^3 \text{ N}$$

6.41 Before the bull charges the momentum of the bull-railroad system is zero, since both are at rest. The train's brakes aren't set and if the train's wheels can roll without friction then the momentum of the system after the charge must equal the momentum of the system before the charge: 0 N·s. Thus the train has a velocity of 0 m/s after the crash.

6.45 In the reference frame in which the rocket is initially at rest, it has zero momentum; so the system of rocket and spent fuel also has zero total momentum:

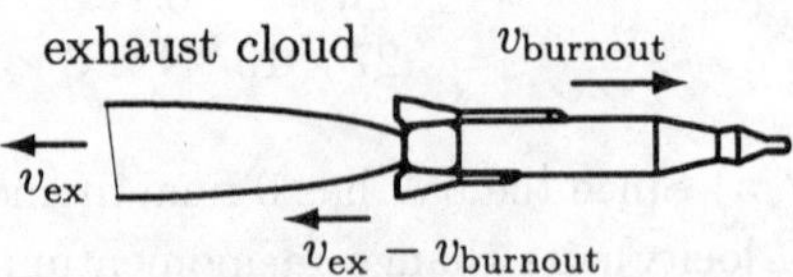

Since the payload momentum is to the right in the figure, the momentum of the exhaust cloud is to the left. Since the exhaust speed is constant *with respect to the rocket*, it is moving at v_{ex} with respect to the initial reference frame only at the left. Elsewhere in the cloud, the leftward velocity of the gas is less than v_{ex}. So, the total momentum of the exhaust has magnitude less than $m_{\text{exhaust}} v_{\text{ex}}$.

6.49 To a very good approximation there are no net external forces acting on the system comprised of 2 dogs; the momentum of this system must be conserved. When, the smart dog kicks the dumb dog it exerts an impulse on the dumb dog; the dumb dog slides across the ice with some momentum $\vec{\mathbf{p}}$ and velocity $\vec{\mathbf{v}}_i$.

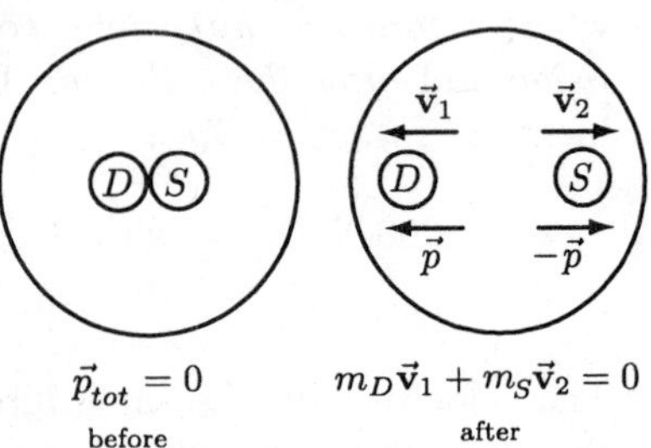

Because momentum is conserved, the smart dog must have an impulse applied to it from the dumb dog; this impulse gives the smart dog a momentum $-\vec{\mathbf{p}}$. As a result it slides off the ice in a direction opposite the dumb dog.

6.53 The problem states that the tool kit's speed relative to the astronaut is 5 m/s. This implies that the two speeds are related by:

$$v_1 + v_2 = 5 \text{ m/s} \tag{1}$$

v_1 v_2

10 kg 65 kg

toolkit astronaut

We also need to conserve momentum: since the astronaut/tool box system's momentum is initially 0, it must remain zero since there are no external net forces acting here. This says

$$(11 \text{ kg})v_1 - (65 \text{ kg})v_2 = 0 \tag{2}$$

Using Eq. (1) to obtain an expression for v_1 yields

$$v_1 = 5 \text{ m/s} - v_2$$

Using this expression to eliminate v_1 in Eq. (2) gives:

$$(11\ \text{kg})\,[(5\ \text{m/s}) - v_2] - 65\ \text{kg}(v_2) = 0$$

$$\Rightarrow v_2[65+11]\ \text{kg} = +11\times 5\ \text{kg m/s}$$
$$v_2 = 0.72\ \text{m/s}$$

and from Eq. (1), $v_1 = 4.3\ \text{m/s}$.

6.57 The skater and flowers end up with a final velocity of 0. Since ice is usually taken to be frictionless there are no net external forces acting on the skater–flower system:

60kg v_1 → −5 m/s ← 1kg

momentum is conserved and therefore the system must have had a net momentum of 0 to begin with.

$$(60\ \text{kg})(v_1) + (1\ \text{kg})(-5\ \text{m/s}) = 0$$
$$\Rightarrow v_1 = 8.3\times 10^{-2}\ \text{m/s}$$

A note on sign conventions: Although we didn't formally declare a coordinate system, that doesn't change the fact that the skater's and flower's velocities are clearly in opposite directions, which is what the minus sign means. The choice of which velocity is "negative" is arbitrary. We could have just as well written:

$$(60\ \text{kg})(-v_1) + (1\ \text{kg})(5\ \text{m/s}) = 0$$

which would have yielded the same result.

6.61 $\vec{\mathbf{v}}_1 = 3.5\ \text{m/s}\,\hat{\mathbf{i}}$ and $\vec{\mathbf{v}}_2 = 2.2\ \text{m/s}\,\hat{\mathbf{i}}$. The momenta are:

$$\vec{\mathbf{p}}_1 = (0.025\ \text{kg})(3.5\ \text{m/s})\hat{\mathbf{i}} = (0.0875\ \text{N}\cdot\text{s})\,\hat{\mathbf{i}}$$

and

$$\vec{\mathbf{p}}_2 = (0.025\ \text{kg})(2.2\ \text{m/s})\hat{\mathbf{i}} = (0.055\ \text{N}\cdot\text{s})\,\hat{\mathbf{i}}$$

The total momentum is

$$\begin{aligned}\vec{\mathbf{P}} &= \vec{\mathbf{p}}_1 + \vec{\mathbf{p}}_2\\ &= (0.0875\ \text{N}\cdot\text{s} + 0.055\ \text{N}\cdot\text{s})\,\hat{\mathbf{i}}\\ &= (0.1425\ \text{N}\cdot\text{s})\,\hat{\mathbf{i}}\end{aligned}$$

The final speed is

$$\frac{|\vec{\mathbf{P}}|}{2m} = \frac{0.1425\ \text{N}\cdot\text{s}}{0.050\ \text{kg}} = 2.85\ \text{m/s}$$

Then

$$\vec{\mathbf{P}} = (0.14\ \text{N}\cdot\text{s})\,\hat{\mathbf{i}}$$

and

$$\vec{\mathbf{v}}_f = (2.9\ \text{m/s})\,\hat{\mathbf{i}}$$

6.65 The first question we should ask is "how long does it take for 2×10^4 kg to be loaded from a hopper if the hopper releases ore at 1000×10^3 kg/h?"

$$\begin{aligned}2\times10^4\ \text{kg} &= (1000\times10^3\ \text{kg/h})(x\,\text{h})\\ x &= .02\ \text{h} = 1.2\ \text{min} = 72\ \text{s}\end{aligned}$$

Thus the freight train will fill completely in 72 s. The optimal loading scheme has the front of the freight car under the hopper just as loading begins and the back of the freight car under the hopper 72 s later as the last bit of ore is loaded. During this process the train will have moved a distance of 15 m (the length of a single freight car). So the train's speed is $\frac{15\ \text{m}}{72\ \text{s}} = 0.21\ \text{m/s}$. The necessary force is

$$F = \left(10^6\ \frac{\text{kg}}{\text{h}}\right)\left(\frac{1\ \text{h}}{3600\ \text{s}}\right)(0.21\ \text{m/s}) = 58\ \text{N}$$

6.67 Using Equation (6.6) for v_{burnout} on p. 212 of the text and the given information from the problem statement:

$$\begin{aligned}1000\ \text{m/s} &= -(2000\ \text{m/s})\ln\left(\frac{m_{\text{fuel}}}{m_{\text{launch}}}\right)\\ -\frac{1}{2} &= \ln\left(\frac{m_{\text{fuel}}}{m_{\text{launch}}}\right)\\ \Rightarrow \frac{m_{\text{fuel}}}{m_{\text{launch}}} &= e^{-\frac{1}{2}} \approx 0.6\end{aligned}$$

6.69 From the section on rocket equations we know that

$$v_{\text{burn}} = v_{\text{ex}}\ln\left[\frac{m_{\text{launch}}}{m_{\text{burn}}}\right]$$

but we are now asked to consider the effect of $2m_{\text{burn}}$ on the final speed of the rocket, v_{burn}.

$$\begin{aligned}v'_{\text{burn}} &= v_{\text{ex}}\ln\left[\frac{m_{\text{launch}}}{2m_{\text{burn}}}\right]\\ &= \left(\ln\left[\frac{m_{\text{launch}}}{m_{\text{burn}}}\right] + \ln\left(\frac{1}{2}\right)\right)v_{\text{ex}}\end{aligned}$$

The final speed is changed by $v_{\text{ex}}\ln\left(\frac{1}{2}\right) = -0.69\ v_{\text{ex}}$ (i.e. it is reduced.)

6.73 The rocket is shown in the figure. After the rocket is launched, all the fuel in the 1st stage is burned.

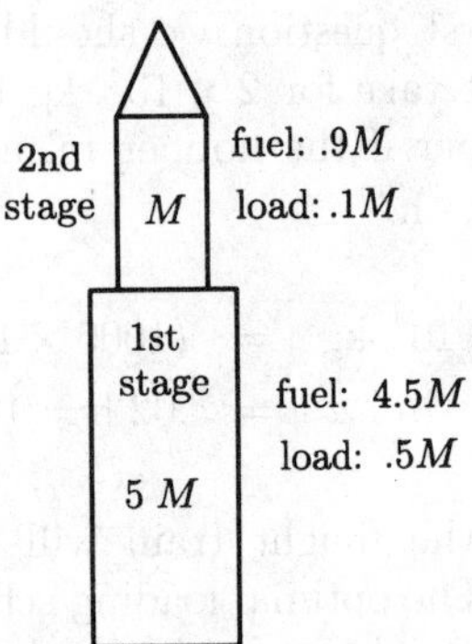

The rocket's speed, v_{B1} at this time is

$$\begin{aligned} v_{B1} &= -v_{\text{ex}} \ln \left[\frac{M_{\text{burn}}}{M_{\text{launch}}} \right] = -v_{\text{ex}} \ln \left[\frac{M + .5M}{M + 5M} \right] \\ &= -v_{\text{ex}} \ln \left[\frac{1.5}{6} \right] = -v_{\text{ex}} \ln[.25] \\ &= v_{\text{ex}} \ln(4) \end{aligned}$$

With the fuel of the 1st stage spent, we will now drop the $.5M$ payload structure and fire up the second stage engines. We are starting with a speed of $v_{\text{ex}} \ln(4)$, an initial mass M and once the fuel is spent our burnout mass will be $.1M$.

$$\begin{aligned} v_{B2} &= v_{B1} - v_{\text{ex}} \ln \left[\frac{.1M}{1M} \right] \\ &= v_{\text{ex}} \ln(4) - v_{\text{ex}} \ln(.1) \\ &= v_{\text{ex}} [\ln(4) + \ln(10)] \\ &= v_{\text{ex}} \ln(40) \end{aligned}$$

Now we will pretend that no stages are dropped, so that the rockets initial mass is $5M+M = 6M$ and final burnout mass is $.1M + .5M = .6M$

$$v_B = v_{\text{ex}} \ln \left[\frac{.6M}{6M} \right] = -v_{\text{ex}} \ln[.1] = v_{\text{ex}} \ln(10)$$

We achieve a faster burnout speed by dropping the 1st stage; the speed will be increased by a factor of $\frac{\ln(40)}{\ln(10)} = 1.6$.

6.77 When you stand on a (mechanical) bathroom scale, notice that its reading oscillates between overshooting and undershooting your weight, but eventually stabilizing. The box is being dropped so we'd expect the scale to over estimate the weight by quite a bit before the oscillations dampen.

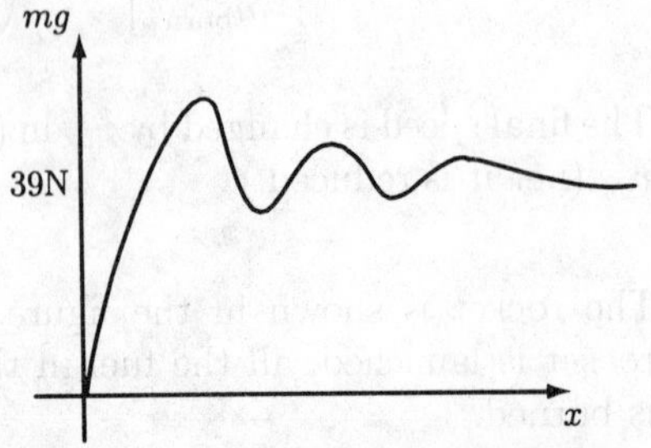

6.81 First we need to find the first twin's velocity when she reaches the stationary twin at a height of $s_1 = \frac{v_0^2}{4g}$. Let this velocity be v_1.

$$\begin{aligned} v_1^2 - v_0^2 &= 2(-g)\frac{v_0^2}{4g} \\ \Rightarrow v_1 &= \frac{v_0}{\sqrt{2}} \end{aligned}$$

The momentum of the system of the two twins just before the collision is $p_b = \frac{mv_0}{\sqrt{2}}$. Just after the collision, if the combined system has a speed v_a,

$$p_a = (2m)v_a$$

To conserve momentum we'll set $p_b = p_a$ and from that we can determine v_a.

$$\begin{aligned} p_b \overset{\text{set}}{\equiv} p_a \quad &\Rightarrow \quad \frac{mv_0}{\sqrt{2}} = (2m)v_f \\ &\Rightarrow \quad v_a = \frac{v_0}{2\sqrt{2}} \end{aligned}$$

Now we can determine how far the combined system of two twins travels from their current height of $s_1 = \frac{v_0^2}{4g}$

$$v_f^2 - v_i^2 = 2as_2$$

with $v_f = 0$ and $v_a = v_i$:

$$\begin{aligned} (0\,\mathrm{m/s})^2 - \left(\frac{v_0}{2\sqrt{2}} \right)^2 &= 2(-g)s_2 \\ \Rightarrow s_2 &= \frac{v_0^2}{16g} \end{aligned}$$

The total distance that the 1st twin travels is

$$s = s_1 + s_2 = \frac{v_0^2}{4g} + \frac{v_0^2}{16g} = \frac{5v_0^2}{16g}$$

6.85 The system of two skaters and a ball starts out with zero momentum. Since no net external forces act on this system, the momentum of the system must remain zero.

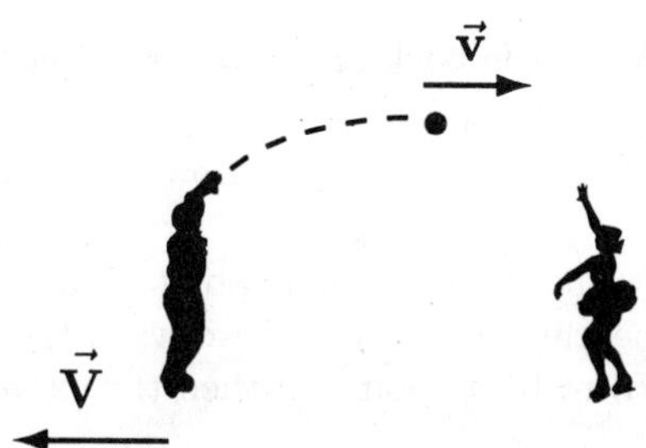

At this point it may be useful to change what we mean by "the system." Let's consider a new system consisting of a ball of mass m and velocity $\vec{\mathbf{v}}$ approaching a motionless skater of mass M. The initial momentum of this system is $\vec{\mathbf{p}}_i = m\vec{\mathbf{v}}$. Since no net external forces act on this system, momentum is conserved. Note that once the skater catches the ball, ball and skater will have equal velocity $\vec{\mathbf{v}}_f$.

$$\begin{aligned} \vec{\mathbf{p}}_i &= \vec{\mathbf{p}}_f \Rightarrow m\vec{\mathbf{v}} = (M+m)\vec{\mathbf{v}}_f \\ \Rightarrow \vec{\mathbf{v}}_f &= \left[\frac{m}{M+m}\right]\vec{\mathbf{v}} \end{aligned}$$

The skater/ball combination will be traveling in the same direction as the ball before it was caught and at a speed slower than the ball's initial speed. Since nothing is present to decelerate the skaters their velocities remain $-\frac{m}{M}\vec{\mathbf{v}}$ and $\left(\frac{m}{M+m}\right)\vec{\mathbf{v}}$ respectively. If

$$\left|\vec{\mathbf{I}}\right| = |\Delta\vec{\mathbf{p}}| = 20.0(\mathrm{N\cdot s})$$

then

$$20.0\ \mathrm{N\,s} = (0.65\ \mathrm{kg})v \Rightarrow v = 30.8\ \mathrm{m/s}$$

a) Then

$$v_1 = \frac{m}{M}v = \frac{0.65\ \mathrm{kg}}{75\ \mathrm{kg}}30.8\ \mathrm{m/s} = 0.27\ \mathrm{m/s}$$

b) and

$$\begin{aligned} v_2 &= \frac{m}{M+m}v = \frac{0.65\ \mathrm{kg}}{75.65\ \mathrm{kg}}(30.8\ \mathrm{m/s}) \\ &= 0.26\ \mathrm{m/s} \end{aligned}$$

Summarizing:

a) $v_1 = 0.27\ \mathrm{m/s}, v_2 = 0$

b) $v_1 = 0.27\ \mathrm{m/s}, v_2 = 0.26\ \mathrm{m/s}$

c) The skater's gliding speeds are not changed. But the velocity components perpendicular to the gliding direction are as given in (a) and (b) above.

6.89 The only force acting on the 1st released car is friction, which produces an acceleration

$$a = \frac{F}{m} = \frac{f_r}{m} = \frac{\mu_r n}{m} = \frac{\mu_r mg}{m} = \mu_r g$$

opposite the direction of v_0.

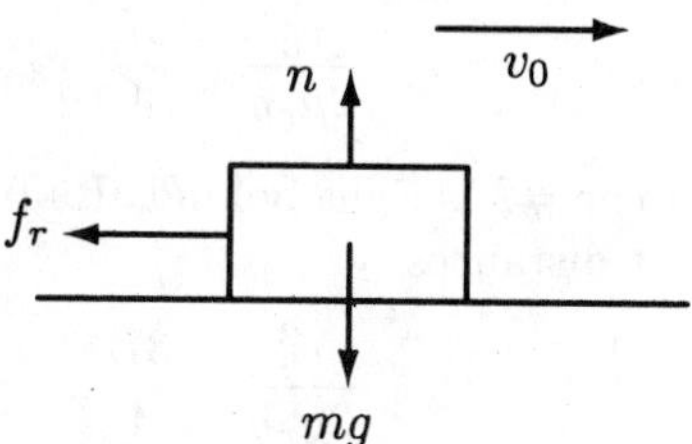

We will calculate the distance that the car rolls before coming to rest.

$$v_f^2 - v_0^2 = 2(-\mu_r g)s_1 \Rightarrow s_1 = \frac{v_0^2}{2\mu_r g}$$

s_1 is the distance travelled by the front end of the train from point A. Thus, the back of the train is a distance

$$d = \frac{v_0^2}{2\mu_r g} - L$$

from point A. Now the second car is released at a speed v_0. It would roll to rest at exactly the same place that car 1 did if car 1 was not in the way—the 2 cars will collide at a distance d from point A.

L

car #1

A $\quad \frac{v_0^2}{2\mu_r g} - L \quad \frac{v_0^2}{2\mu_r g}$

We need to know the second car's speed just before it collides with car 1.

$$v_f^2 - v_0^2 = 2(-\mu_r g)\left(\frac{v_0^2}{2\mu_r g} - L\right) \Rightarrow v_f = \sqrt{2\mu_r gL}$$

Thus, just before the collision the 2-car system has an initial momentum of $p_i = m\sqrt{2\mu_r gL}$. Just after the collision the system of two cars have a momentum of $p_f = 2mv_f$ where v_f is the speed of the two interlocked cars after the collision.

$$2mv_f = m\sqrt{2\mu_r gL} \Rightarrow v_f = \sqrt{\frac{\mu_r gL}{2}}$$

Now we can find the cars' displacement when they come to rest after the crash.

$$0 - \frac{\mu_r gL}{2} = 2(-\mu_r g)s_2 \Rightarrow s_2 = \frac{1}{4}L$$

The front and back of car #1 are displaced by a distance $\frac{1}{4}L$

front of car #1:

$$\frac{v_0^2}{2\mu_r g} + \frac{L}{4}$$

back of car #1:

$$\frac{v_0^2}{2\mu_r g} - \frac{3L}{4}$$

And car #2 is right behind. Its front and back are at distances

$$\frac{v_0^2}{2\mu_r g} - \frac{3L}{4}$$

and

$$\frac{v_0^2}{2\mu_r g} - \frac{7L}{4}$$

from point A, respectively. Now car 3 is released.

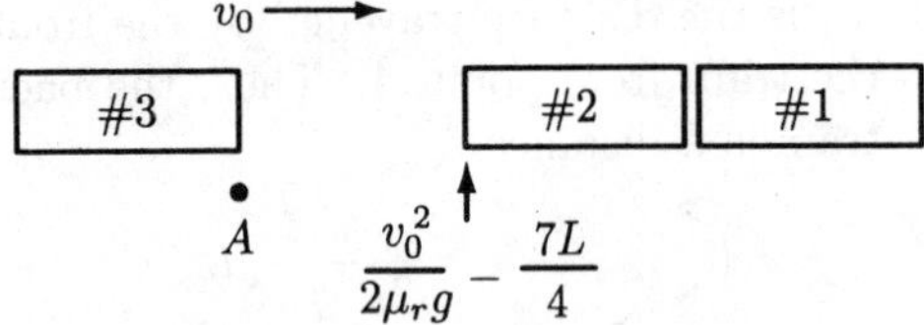

We will follow the same logic plan that we did before. The speed of car #3 just before the collision with car #2 is :

$$\begin{aligned} v_f^2 - v_0^2 &= 2(-\mu_r g)\left[\frac{v_0^2}{2\mu_r g} - \frac{7L}{4}\right] \\ v_f^2 &= \frac{7L}{4}(2\mu_r g) - v_0^2 + v_0^2 \\ \Rightarrow v_f &= \sqrt{\frac{7L}{2}\mu_r g} \end{aligned}$$

The momentum of the three trains before the crash is

$$p_i = m\sqrt{\frac{7L}{2}\mu_r g}$$

and after the crash, $(3m)v_f$. Conserving momentum during the crash and solving for v_f (the speed of the trains after the crash).

$$\begin{aligned} m\sqrt{\frac{7L}{2}\mu_r g} &= 3mv_f \\ \Rightarrow v_f &= \frac{1}{3}\sqrt{\frac{7L}{2}\mu_r g} \end{aligned}$$

At this speed the trains will move a distance

$$-\frac{1}{9}\left(\frac{7L}{2}\mu_r g\right) = 2(-\mu_r g)s_3 \Rightarrow s_3 = \frac{7L}{36}$$

Thus the front of car #3 is now at the position

$$\frac{v_0^2}{2\mu_r g} - \frac{7L}{4} + \frac{7L}{36} = \frac{v_0^2}{2\mu_r g} - \frac{56L}{36}$$

While the back of car #3 is at the position

$$\frac{v_0^2}{2\mu_r g} - \frac{56L}{36} - L = \frac{v_0^2}{2\mu_r g} - \frac{23L}{9}$$

We have been using point A as an origin to measure distances. If we want the back of car #3 to be at point A, then the expression

$$\frac{v_0^2}{2\mu g} - \frac{23L}{9}$$

which is the distance of the back of the train from point A must be equal to 0. Solving for v_0,

$$\begin{aligned} \frac{v_0^2}{2\mu g} - \frac{23L}{9} &= 0 \\ \Rightarrow v_0 &= \frac{1}{3}\sqrt{46\mu_r g L} = 2.3\sqrt{\mu_r g L} \end{aligned}$$

6.93 The impulse delivered to the object is given by $\vec{\mathbf{I}} = \int \vec{\mathbf{F}}(t)\, dt$. We can rewrite $\vec{\mathbf{F}}(t)$ to make it look like a typical tangent substitution in your calculus textbook.

$$F_x(t) = \frac{F_0}{1 + \frac{t^2}{t_0^2}} = \frac{F_0 t_0^2}{t_0^2 + t^2}$$

Although not specifically stated, F_0 and t_0 are both constants.

$$\vec{\mathbf{I}} = \int_{-\infty}^{\infty} \vec{\mathbf{F}}(t)\, dt = F_0 t_0^2 \hat{\mathbf{i}} \int_{-\infty}^{\infty} \frac{dt}{t_0^2 + t^2}$$

Set $t = t_0 \tan\theta$, then $dt = t_0 \sec^2\theta\, d\theta$. As for the limits, as $t \to \infty$: $\theta \to \frac{\pi}{2}$ as $t \to -\infty$, $\theta \to -\frac{\pi}{2}$;

$$\begin{aligned} \vec{\mathbf{I}} &= F_0 t_0^2 \hat{\mathbf{i}} \int_{-\frac{\pi}{2}}^{\frac{\pi}{2}} \frac{t_0 \sec^2\theta}{t_0^2 + t_0^2 \tan^2\theta}\, d\theta \\ &= F_0 \hat{\mathbf{i}} \int_{-\frac{\pi}{2}}^{\frac{\pi}{2}} \frac{t_0 \sec^2\theta}{1 + \tan^2\theta}\, d\theta \end{aligned}$$

Since $1 + \tan^2\theta = \sec^2\theta$

$$\begin{aligned} \vec{\mathbf{I}} &= \hat{\mathbf{i}} F_0 t_0 \int_{-\frac{\pi}{2}}^{\frac{\pi}{2}} d\theta = \hat{\mathbf{i}} F_0 t_0 \theta \Big|_{-\frac{\pi}{2}}^{\frac{\pi}{2}} \\ &= \hat{\mathbf{i}} t_0 F_0 \left(\frac{\pi}{2} + \frac{\pi}{2}\right) \\ &= \pi F_0 t_0 \hat{\mathbf{i}} \end{aligned}$$

6.97 Just as the hourglass is turned, the entire weight of sand and structure is supported. Immediately, sand begins to fall, and the sand in transit is unsupported, so the normal force provided by the scale will decrease. If we assume that sand leaves the upper container at a constant rate

$$\frac{dm}{dt}\left(= 10.0\ \mathrm{kg}/\mathrm{h} = 2.78 \times 10^{-3}\ \mathrm{kg}/\mathrm{s}\right),$$

the necessary normal force ($\equiv$ scale reading for our ideal scale) is

$$n(t) = \left(M - \frac{dm}{dt}t\right) g$$

This situation holds for the time

$$\tau = \sqrt{\frac{2\ s}{g}} = \sqrt{\frac{0.400\ \text{m}}{9.81\ \text{m}/\text{s}^2}} = 0.202\ \text{s}$$

when the first grains of sand begin to strike the bottom of the hourglass. From this time the amount of falling sand ceases to increase; in fact the amount begins to decrease because the arriving sand piles up and reduces the distance the sand has to fall—that is, reduces the volume occupied by the falling sand. Now in addition to supporting the sand not falling, the scales have to deliver impulse to stop the sand hitting the bottom. In time dt the impulse required is

$$(dm \text{ that is stopped}) \begin{pmatrix} \text{speed of sand} \\ \text{just before it hits} \end{pmatrix}$$

But the speed of the sand is $g \times$ the time it has been falling. So, the impulse is

$$dI = \left(\frac{dm}{dt}dt + \begin{array}{c}\text{a little due to} \\ \text{rising sand level}\end{array}\right)(g\tau_{\text{fall}})$$

and the necessary normal force is

$$\delta N = \frac{dI}{dt} = g\tau_{\text{fall}}\frac{dm}{dt} + \begin{array}{c}\text{a little due to} \\ \text{rising sand level}\end{array}$$

The first term in δN equals the weight of the column of unsupported sand. So, if it weren't for the rising sand level, the scale reading would return to Mg exactly. Instead, the scale reading rises ever so slightly above Mg and remains above Mg until the end of the hour when the last sand grain hits the top of the pile. Because gravity delivers momentum to the system at the constant rate (Mg, downward), the *average* scale reading for the hour must equal Mg, so the system ends up with zero net momentum.

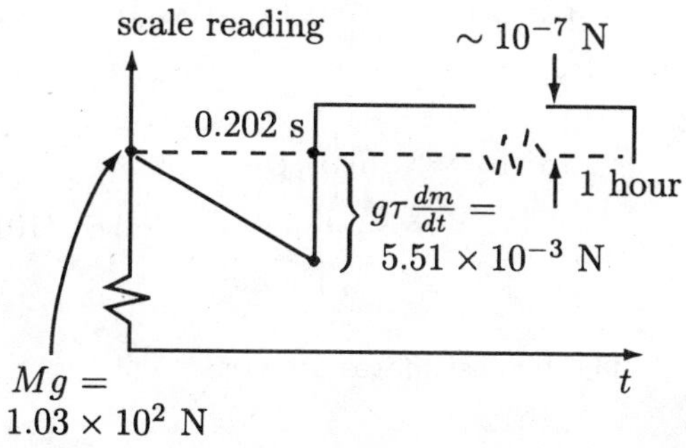

The average excess of the scale reading over Mg for the rest of the hour is such that

$$(1\ \text{hour})\,(\text{average excess}) = \begin{array}{l}\text{area of triangular} \\ \text{part of graph} \\ \text{during first } 0.202\ \text{s}\end{array}$$

So average excess:

$$\begin{aligned} &= \frac{dm}{dt}\frac{g\tau^2}{1\ \text{h}} \\ &= \frac{1}{2}\left(5.51 \times 10^{-3}\ \text{N}\right)\frac{(0.002\ \text{s})}{(3600\ \text{s})} \\ &= 1.5 \times 10^{-7}\ \text{N} \end{aligned}$$

You won't want to pay for a scale accurate enough to detect these effects!

Chapter 7

Work and Kinetic Energy

7.1

$$\begin{aligned} K_{\text{car1}} &= \frac{1}{2}(1000 \text{ kg})(12 \text{ m/s})^2 \\ &= 7.2 \times 10^4 \text{ J} \\ K_{\text{truck}} &= \frac{1}{2}(1.0 \times 10^4 \text{ kg})(12 \text{ m/s})^2 \\ &= 7.2 \times 10^5 \text{ J} \\ K_{\text{car2}} &= \frac{1}{2}(1000 \text{ kg})(35 \text{ m/s})^2 \\ &= 6.1 \times 10^5 \text{ J} \end{aligned}$$

Since the first car and truck have the same speed the ratio of their energies is the ratio of their masses

$$\frac{K_{\text{car1}}}{K_{\text{truck}}} = 10^{-1}$$

The mass of the first and second car are equal so the ratio of their energy is the ratio of the square of their speeds.

$$\frac{K_{\text{car1}}}{K_{\text{car2}}} = 0.12$$

7.3 The initial K of the car is:

$$K_i = \frac{1}{2}(1200 \text{ kg})(28 \text{ m/s})^2 = 4.7 \times 10^5 \text{ J}$$

Since the car is finally at rest, its final K is 0 J. Thus the change in the car's K is

$$K = K_f - K_i = -4.7 \times 10^5 \text{ J}$$

By the W-E-T, this change of K must equal the work done on the car (in this case the work done on the car by the haystack).

$$W_{\text{haystack on car}} = -4.7 \times 10^5 \text{ J}$$

Although its not obvious, the car does $+4.7 \times 10^5$ J of work on the hay. The hay stops the car by exerting a force on the car. *N3L* tells us that the car exerts a reaction force which is equal in magnitude but opposite in direction on the hay.

7.5 The vectors can be written out explicitly. Let the horizontal axis be the x-axis and the vertical axis be the y-axis.

$$\begin{aligned} \vec{\mathbf{a}} &= (6.4 - 3)\hat{\mathbf{i}} + (4.3 - 2)\hat{\mathbf{j}} = 3.4\hat{\mathbf{i}} + 2.3\hat{\mathbf{j}} \\ \vec{\mathbf{b}} &= (0.2 - 3)\hat{\mathbf{i}} + (4.8 - 2)\hat{\mathbf{j}} = -2.8\hat{\mathbf{i}} + 2.8\hat{\mathbf{j}} \end{aligned}$$

Following the rules for evaluating a dot product by components,

$$\begin{aligned} \vec{\mathbf{a}} \cdot \vec{\mathbf{b}} &= a_x b_x + a_y b_y \\ &= (3.4)(-2.8) + (2.3)(2.8) \\ &= -3.1 \end{aligned}$$

Now we'll estimate the dot product by using $\vec{\mathbf{a}} \cdot \vec{\mathbf{b}} = |\vec{\mathbf{a}}| \left|\vec{\mathbf{b}}\right| \cos\theta$. We measured the angle between the vectors $\vec{\mathbf{a}}$ and $\vec{\mathbf{b}}$ at 101°. Next, using a ruler we measured the magnitudes (lengths) of the vectors: $|\vec{\mathbf{a}}| = 4$ units and $\left|\vec{\mathbf{b}}\right| = 4$ units. Thus,

$$\begin{aligned} \vec{\mathbf{a}} \cdot \vec{\mathbf{b}} &= |\vec{\mathbf{a}}| \left|\vec{\mathbf{b}}\right| \cos\theta \\ &= (4 \text{ units})(4 \text{ units}) \cos(101°) \\ &= -3.1 \text{ unit}^2 \end{aligned}$$

The two estimates are consistent.

7.7 We are told that the rocket does not move during the $\frac{1}{2}$ s time interval. Remember how we defined work:

$$W = \int \vec{\mathbf{F}}\, ds$$

for non-constant force along s, or

$$W = \vec{\mathbf{F}} \cdot \vec{s}$$

for constant force along s. The point is, a displacement which is not $\perp$ to the applied force is necessary for work to be done. Has the rocket been displaced (moved) during the $\frac{1}{2}$ sec time interval? *No.* Thus, has work been done? The startling answer is *no*. In physics we often use terms which are commonplace: *force*, *velocity*, *energy*, *work*, etc. However the demands of physics require us to be more specific about what we say, and this problem shows the pitfalls that are present when we're not careful. Given any two people chosen at random you'll probably get two very different answers to the question "what is work?" Find two physicists and the answers you get will probably be very much the same. Answer: power = zero.

7.9 At maximum speed, power lost to (friction + drag) = power provided by engine

$$\begin{aligned} P &= Fv \\ 257\,\frac{\text{km}}{\text{h}} &= \frac{257 \times 10^3 \text{ m}}{3600 \text{ s}} \\ &= 71.4 \text{ m/s} \end{aligned}$$

so

$$\begin{aligned} F &= \frac{P}{v} \\ &= (270 \text{ hp})\left(\frac{746 \text{ W}}{1 \text{ hp}}\right)\left(\frac{1}{71.4 \text{ m/s}}\right) \\ &= 2.8 \text{ kN} \end{aligned}$$

7.13 To make the hammer spin around him in circular motion, the athlete must accelerate the hammer centripetally. By now we know that the force needed to produce centripetal acceleration always points in a radial line towards the centre of the circular motion. However the velocity of the hammer is always tangential to its path, as shown. The work that the athlete does is given by

$$W = \int \vec{\mathbf{F}} \cdot d\vec{s}$$

where $d\vec{s}$ is a small displacement in the direction of the hammer's velocity. We are not given $\vec{\mathbf{F}}$, but since we know the hammer's speed increases from 0 to 10 m/s, its kinetic energy increases from 0 to $\frac{1}{2}mv^2$

$$= \frac{1}{2}(2 \text{ kg})(10 \text{ m/s})^2 = 100 \text{ J}$$

Thus from the work/energy theorem, the work done is 100 J.

7.17 If an object is already in motion, and the force acting on it is directed opposite its velocity (in other words the force is decelerating the object), then $\vec{\mathbf{F}}$ and $\vec{s}$ are in opposite directions.

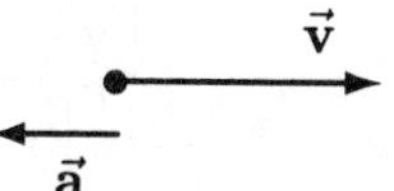

The product $\vec{\mathbf{F}} \cdot d\vec{s}$ will be a non-zero negative number. Thus, choices **a)** and **e)** are eliminated. If the force acts in the same direction as the object's velocity then it also acts in the same direction as the object's displacement. In this case $\vec{\mathbf{F}} \cdot d\vec{s}$ will yield a positive non-zero number. Thus, choice **b)** is eliminated. If the force produces centripetal acceleration, then the object's displacement is always tangential to its circular path. The force is always $\perp$ to this path and points radially towards the center of the circular path.

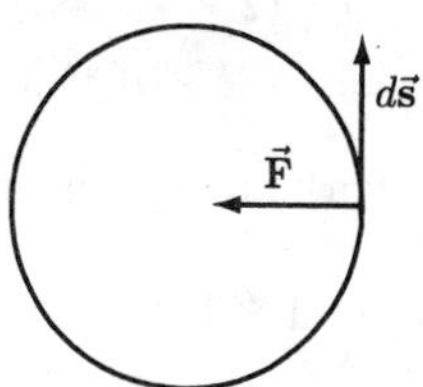

Thus $d\vec{s}$ and $\vec{\mathbf{F}}$ are $\perp$ to each other and the product $\vec{\mathbf{F}} \cdot d\vec{s}$ will be identically 0. Thus, choice **c)** is eliminated. Thus choice **d)** is the correct choice.

7.21 $K = \frac{1}{2}(250 \times 10^{-3} \text{ kg})(28 \text{ m/s})^2 = 98 \text{ J}$

7.25 The particle's speed is given by

$$\begin{aligned} v^2 &= (2.2 \text{ m/s})^2 + (4.9 \text{ m/s})^2 + (2.7 \text{ m/s})^2 \\ &= 36.14(\text{m/s})^2 \end{aligned}$$

Thus,

$$\begin{aligned} K &= \frac{1}{2}(5 \text{ kg})\left[36.14(\text{m/s})^2\right] \\ &= 9.0 \times 10^1 \text{ J} \end{aligned}$$

7.29 $\vec{\mathbf{a}} \cdot \vec{\mathbf{b}} = (1,2) \cdot (2,-1) = 2 + (-2) = 0.$

7.33 $(3,4,5)\cdot(1,2,3) = 3+8+15 = 26$

7.37 We could ask another question: "what angle does the vector $(1,2,2)$ make with the vector $\hat{\mathbf{i}} = (1,0,0)$ which lies on the x-axis?" From what we know of dot products,

$$(1,2,2)\cdot(1,0,0) = |(1,2,2)|\,|(1,0,0)|\cos\theta$$

so,

$$\begin{aligned}\theta &= \cos^{-1}\left[\frac{(1,2,2)\cdot(1,0,0)}{|(1,2,2)|\,|(1,0,0)|}\right]\\ &= \cos^{-1}\left[\frac{1}{\sqrt{1+4+4}\sqrt{1}}\right]\\ &= \cos^{-1}\left[\frac{1}{3}\right] = 70^\circ\end{aligned}$$

7.41 a) Since constant forces are at work here,

$$W = \vec{\mathbf{F}}\cdot\vec{\mathbf{s}}$$

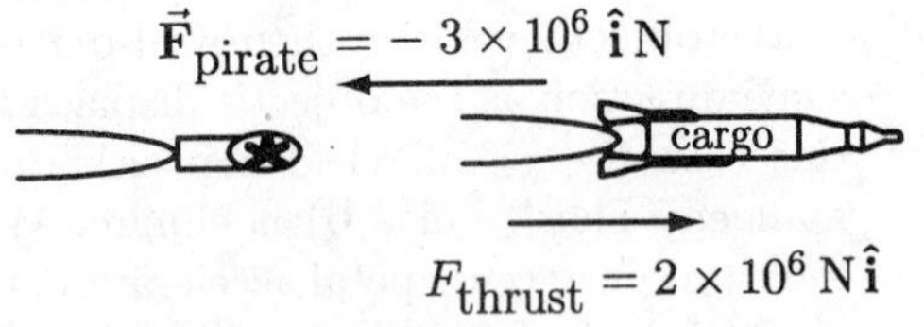

$\vec{\mathbf{F}}_{\text{pirate}} = -3\times10^6\,\hat{\mathbf{i}}\,\text{N}$

cargo

$F_{\text{thrust}} = 2\times10^6\,\text{N}\,\hat{\mathbf{i}}$

$$\begin{aligned}W_{\text{pirate}} &= (-3\times10^6\ \text{N}\hat{\mathbf{i}})(-100\times10^3\ \text{m}\hat{\mathbf{i}})\\ &= 3\times10^{11}\ \text{J}\\ W_{\text{captain}} &= (2\times10^6\ \text{N}\hat{\mathbf{i}})(-100\times10^3\ \text{m}\hat{\mathbf{i}})\\ &= -2\times10^{11}\ \text{J}\end{aligned}$$

From the work energy theorem

$$\frac{1}{2}mv_f^2 = 1\times10^{11}\ \text{J} \Rightarrow v_f = 450\ \text{m/s}$$

b) We will assume that the carrier is saved if it accelerates away from the pirates. We know that the carrier will accelerate in the direction of the net force. The vertical forces sum to 0:

$$2\times10^6\ \text{N} + (2\times10^6\ \text{N})\sin30^\circ - 3\times10^6\ \text{N} = 0$$

There is a net force of

$$(2\times10^6\ \text{N})\cos30^\circ = \sqrt{3}\times10^6\ \text{N}$$

to the left.

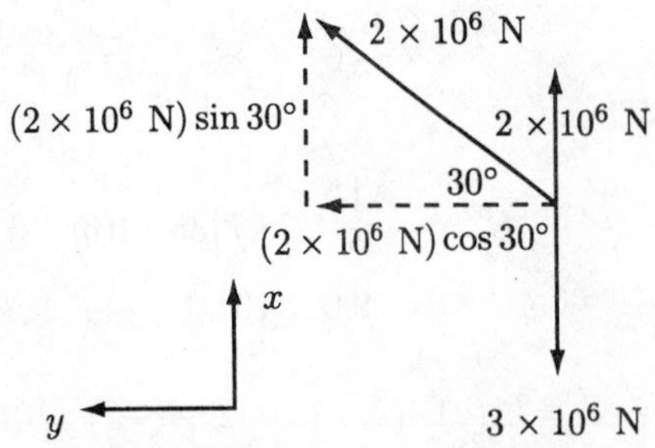

2 × 10⁶ N

(2 × 10⁶ N) sin 30°

2 × 10⁶ N

30°

(2 × 10⁶ N) cos 30°

x

y

3 × 10⁶ N

Since the carrier is now accelerating away from the pirates they are saved. The carrier accelerates to the left at

$$\frac{\sqrt{3}\times10^6\ \text{N}}{10^6\ \text{kg}} = \sqrt{3}\ \text{m/s}^2$$

Thus in 1 min it travels left a total distance of

$$\Delta s = \frac{1}{2}\left(\sqrt{3}\ \text{m/s}^2\right)(60\ \text{s})^2 = 3000\ \text{m}$$

The carrier's velocity component toward the pirates remains constant at 450 m/s. In 1 min the carrier moves

$$(450\ \text{m/s})(60\ \text{s}) = 27\ \text{km}$$

Thus the work done by the cruiser is

$$\begin{aligned}W &= \vec{\mathbf{F}}\cdot\vec{\mathbf{s}}\\ &= (2\times10^6\ \text{N})\left(\frac{\sqrt{3}}{2}\hat{\mathbf{j}}+\frac{1}{2}\hat{\mathbf{i}}\right)\\ &\quad\cdot\left(3\ \text{km}\hat{\mathbf{j}} - 27\ \text{km}\hat{\mathbf{i}}\right)\\ &= (2\times10^6\ \text{N})\left(\frac{3\sqrt{3}}{2}\ \text{km} - \frac{27}{2}\ \text{km}\right)\\ &= -2\times10^{10}\ \text{N}\end{aligned}$$

7.43 a) The space craft's thrusters are said to fire for 1 s. At the given speed it travels a distance of

$$s = v_it(1.0\times10^3\ \text{km/s})(1\ \text{s}) = 1.0\times10^3\ \text{km}$$

while the thrusters are fired for that one second. (We are told to ignore any change in speed during the thrust). Thus, in thrusting, the space craft did work equal to:

$$W = Fs(1.0\times10^9\ \text{N})(1.0\times10^6\ \text{m}) = 1.0\times10^{15}\ \text{J}$$

But according to the WET this also equal to the change in the spacecraft's kinetic energy, so

$$\Delta K = K_f - K_i = 1.0\times10^{15}\ \text{J}$$

The initial K is

$$\frac{1}{2}(1.0\times10^4\ \text{kg})(1.0\times10^3\ \text{km/s})^2 = 5.0\times10^{15}\ \text{J}$$

Thus the final K, after the thrusting is done, is:

$$\begin{aligned}K_f &= \frac{1}{2}mv_a^2 = \frac{1}{2}mv_i^2 + Fs\\ &= 1\times10^{15}\ \text{J} + 5\times10^{15}\ \text{J}\\ &= 6.0\times10^{15}\ \text{J}\end{aligned}$$

And so, the final speed v_a is

$$v_a = \sqrt{\frac{2(6\times10^{15}\ \text{J})}{1.0\times10^4\ \text{kg}}} = 1.1\times10^6\ \text{m/s}$$

b) Actually the spacecraft's speed changes during the 1 s thrust:

$$\begin{aligned} v &= v_i + at \\ s &= v_i t + \frac{1}{2}at^2 \end{aligned}$$

Thus the work done is

$$Fs = Fv_i t + \frac{1}{2}at^2 F = \frac{1}{2}m\left(v_f^2 - v_i^2\right)$$

and the final speed is given by:

$$\begin{aligned} v_f^2 &= v_i^2 + \frac{2}{m}Fv_i t + \left(\frac{F}{m}t\right)^2 \\ &= v_a^2 + \left(\frac{Ft}{m}\right)^2 \end{aligned}$$

$$\begin{aligned} &\approx \frac{1}{2}\left(\frac{Ft}{mv_a}\right)^2 \\ &= \frac{1}{2}\left[\frac{(10^9\ \text{N})(1\ \text{s})}{(10^4\ \text{kg})(1.1\times10^6\ \text{m/s})}\right]^2 \\ &= 4\times10^{-3}, \end{aligned}$$

and is negligible.

7.49 If each furrow is 1 m apart then there are

$$(1\times10^3\ \text{m})(1\ \text{furrow/m}) = 1\times10^3\ \text{furrows}$$

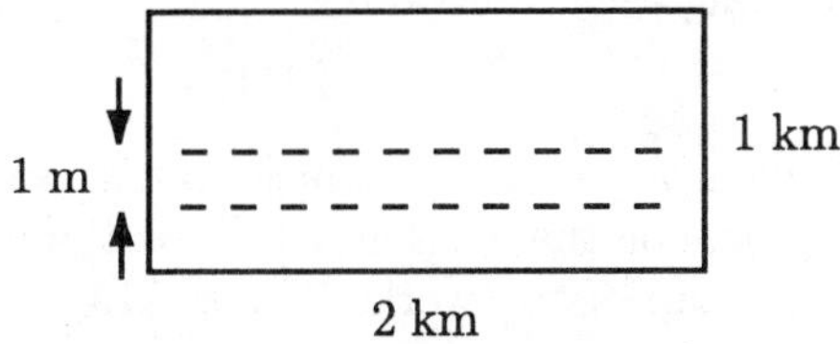

Now each furrow is 2×10^3 m long so the tractor needs to plow

$$(1\times10^3\ \text{furrows})(2\times10^3\ \text{m/furrow}) = 2\times10^6\ \text{m}$$

worth of furrows. If the tractor exerts 1×10^3N on average, then the amount of work it does to plow the whole field is:

$$\begin{aligned} W_{\text{tractor}} &= Fs = (1\times10^3\ \text{N})(2\times10^6\ \text{m}) \\ &= 2\times10^9\ \text{J} \end{aligned}$$

If there is a slope, the tractor must do work to compensate for the work done by the component of weight along the slope. However, as it turns around to plow the next furrow it will be plowing down the slope. Gravity will help it plow the field by doing work on the tractor (pulling it down the slope). If there's an even number of furrows, (which there is), each upslope—where the tractor does more work—has a corresponding downslope where it does less work. The net amount of work that the tractor does is the same as the field with no slope.

7.53 a) We can obtain the components of any vector $\vec{v}$ by taking dot products:

$$\begin{aligned} v_x &= \vec{v}\cdot\hat{\mathbf{i}} \\ v_y &= \vec{v}\cdot\hat{\mathbf{j}} \end{aligned}$$

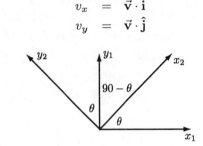

Now suppose that $\vec{v} = \hat{\mathbf{i}}_2$. Then:

$$\hat{\mathbf{i}}_2 = \left(\hat{\mathbf{i}}_2\cdot\hat{\mathbf{i}}_1\right)\hat{\mathbf{i}}_1 + \left(\hat{\mathbf{i}}_2\cdot\hat{\mathbf{j}}_1\right)\hat{\mathbf{j}}_1$$

Since the dot product of two unit vectors is the product of their magnitudes (1 times 1) times the cosine of the angle between them we have:

$$\begin{aligned} \hat{\mathbf{i}}_2 &= \cos\theta\hat{\mathbf{i}}_1 + \cos(90-\theta)\hat{\mathbf{j}}_1 \\ &= \cos\theta\hat{\mathbf{i}}_1 + \sin\theta\hat{\mathbf{j}}_1 \end{aligned}$$

It might be good to convince yourself that $\left|\hat{\mathbf{i}}_2\right| = 1$, as it should be. Now,

$$\begin{aligned} \hat{\mathbf{j}}_2 &= (\hat{\mathbf{j}}_2\cdot\hat{\mathbf{i}}_1)\hat{\mathbf{i}}_1 + (\hat{\mathbf{j}}_2\cdot\hat{\mathbf{j}}_1)\hat{\mathbf{j}}_1 \\ &= \cos(90+\theta)\hat{\mathbf{i}}_1 + \cos\theta\hat{\mathbf{j}}_1 \\ &= -\sin\theta\hat{\mathbf{i}}_1 + \cos\theta\hat{\mathbf{j}}_1 \end{aligned}$$

b) Given:

$$\vec{\mathbf{A}} = A_{x,2}\hat{\mathbf{i}}_2 + A_{y,2}\hat{\mathbf{j}}_2$$

Using what we obtained from Part a,

$$\begin{aligned} \vec{\mathbf{A}} &= A_{x,2}\left[\cos\theta\hat{\mathbf{i}}_1 + \sin\theta\hat{\mathbf{j}}_1\right] \\ &\quad + A_{y,2}\left[-\sin\theta\hat{\mathbf{i}}_1 + \cos\theta\hat{\mathbf{j}}_1\right] \\ &= \hat{\mathbf{i}}_1\left[A_{x,2}\cos\theta - A_{y,2}\sin\theta\right] \\ &\quad + \hat{\mathbf{j}}_1\left[A_{x,2}\sin\theta + A_{y,2}\cos\theta\right] \end{aligned}$$

c) In the second coordinate system it's simple enough:

$$\begin{aligned} \vec{\mathbf{A}}\cdot\vec{\mathbf{B}} &= (A_{x,2}\hat{\mathbf{i}}_2 + A_{y,2}\hat{\mathbf{j}}_2)\cdot(B_{x,2}\hat{\mathbf{i}}_2 + B_{y,2}\hat{\mathbf{j}}_2) \\ &= A_{x,2}B_{x,2} + A_{y,2}B_{y,2} \end{aligned}$$

We can use the result of part b) to compute the dot product in the first coordinate system.

$$\vec{\mathbf{A}}\cdot\vec{\mathbf{B}} = \begin{pmatrix} \hat{\mathbf{i}}_1\,[A_{x,2}\cos\theta - A_{y,2}\sin\theta] \\ +\hat{\mathbf{j}}_1\,[A_{x,2}\sin\theta + A_{y,2}\cos\theta] \end{pmatrix} \cdot \begin{pmatrix} \hat{\mathbf{i}}_1\,[B_{x,2}\cos\theta - B_{y,2}\sin\theta] \\ +\hat{\mathbf{j}}_1\,[B_{x,2}\sin\theta + B_{y,2}\cos\theta] \end{pmatrix}$$

$$\begin{aligned} &= [A_{x,2}\cos\theta - A_{y,2}\sin\theta]\,[B_{x,2}\cos\theta - B_{y,2}\sin\theta] \\ &\quad + [A_{y,2}\cos\theta + A_{x,2}\sin\theta]\,[B_{y,2}\cos\theta + B_{x,2}\sin\theta] \\ &= A_{x,2}B_{x,2}\cos^2\theta - A_{x,2}B_{y,2}\cos\theta\sin\theta \\ &\quad - A_{y,2}B_{x,2}\cos\theta\sin\theta + A_{y,2}B_{y,2}\sin^2\theta \\ &\quad + A_{x,2}B_{x,2}\sin^2\theta + A_{x,2}B_{y,2}\sin\theta\cos\theta \\ &\quad - A_{y,2}B_{x,2}\sin\theta\cos\theta + A_{y,2}B_{y,2}\cos^2\theta \\ &= A_{x,2}B_{x,2} + A_{y,2}B_{y,2} \end{aligned}$$

7.55 a) If there were no friction the tension force would have to point towards the center of the circle (O) to provide the necessary force to produce a_c.

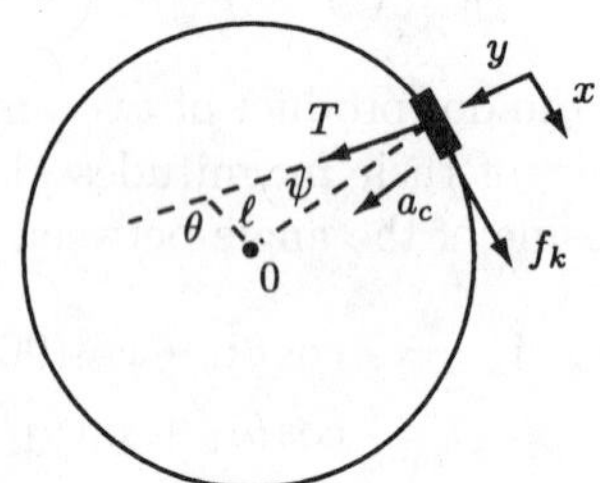

However, in this case the tension force must also counteract f_k.

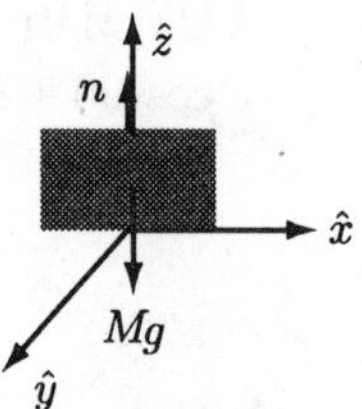

As can be seen in the figure, the tension force must be carefully maintained so that its y component produces the proper centripetal force while its $-x$ component exactly balances f_k, the frictional force. Presumably the point at which the string is held must also rotate about O in order to preserve the proper angle θ. It is this angle that dictates the proper ratio of T_x and T_y to accelerate the block. Using N2L along each direction:

$$\begin{aligned} T\cos\psi &= M\omega^2 R &(1) \\ n - mg &= 0 &(2) \\ f_k - T\sin\psi &= 0 &(3) \end{aligned}$$

$$f_k = \mu n = \mu M g \quad (4)$$

The forces acting in this system are constantly changing direction. To get a detailed view of what's going on we need to look at a differential time interval.

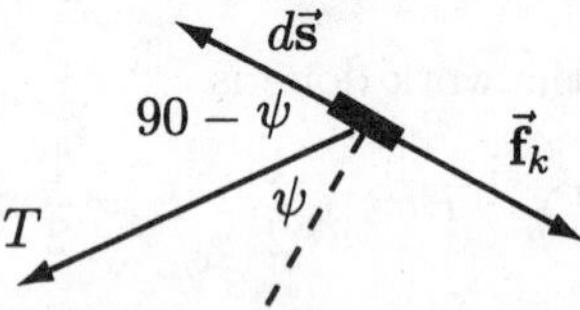

During a small displacement $d\vec{s}$, the tension force does an amount of work

$$\begin{aligned} dW_T &= \vec{\mathbf{T}}\cdot d\vec{s} = \left|\vec{\mathbf{T}}\right| |d\vec{s}| \cos(90^\circ - \phi) \\ &= T\,ds\sin\psi \end{aligned}$$

In the same displacement, friction does an amount of work

$$\begin{aligned} dW_f &= \vec{\mathbf{f}}_k\cdot d\vec{s} = \left|\vec{\mathbf{f}}_k\right| |d\vec{s}| \cos(180^\circ) \\ &= -\vec{\mathbf{f}}_k\,ds \\ &= -T\sin\psi\,ds \end{aligned}$$

by Eq. (3). As mentioned, θ (and therefore ψ) must be preserved throughout a whole revolution and so the magnitudes of all the forces must be the same even if their directions aren't. Thus, we can see that since

$$dW_T = -dW_f$$

during each interval ds,

$$W_T = -W_f$$

during a complete revolution. The work done by a force on the mass in one revolution is equal to the sum of the works it does in each displacement.

$$\begin{aligned} W_T &= T\sin\psi \sum_{\text{all } ds} ds = T\sin\psi 2\pi R \\ &\stackrel{(3)}{=} (f_k)2\pi R \\ &\stackrel{(4)}{=} \mu M g 2\pi R \end{aligned}$$

b) The hand needs to go around a circle about the origin with the same angular velocity as the mass.

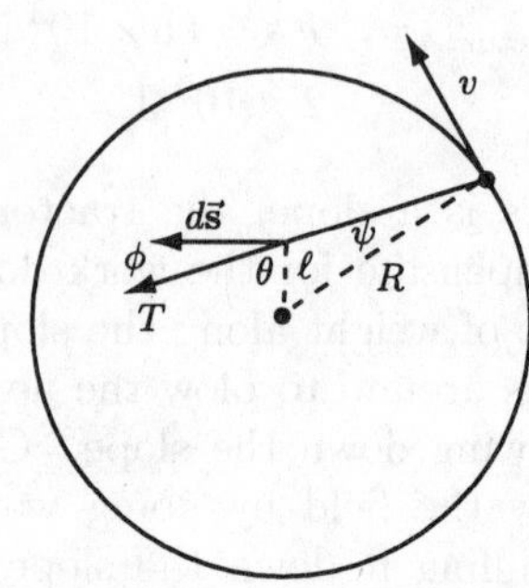

The path that the hand follows is along a circle of radius ℓ. A displacement on that path is given by

$$d\vec{s} = \ell\, d\phi\, \hat{\phi}$$

We can find ℓ using the sine law.

$$\frac{R}{\sin(180^\circ - \theta)} = \frac{\ell}{\sin\psi} \Rightarrow \ell = \frac{R\sin\psi}{\sin(\theta)}$$

where we used the identity $\sin(180^\circ - \theta) = \sin\theta$. A differential quantity of work is given by (using Eq. (3)

$$dW = \vec{\mathbf{T}} \cdot d\vec{s} = \frac{\mu_k Mg}{\sin\psi}\left(\frac{R\sin\psi}{\sin\theta}\right)\cos\phi\, d\phi$$

However, $\phi = 90^\circ - \theta$ and $\cos(90^\circ - \theta) = \sin\theta$ so,

$$dW = \frac{\mu_k MgR}{\sin\theta}\sin\theta\, d\phi = \mu_k RMg\, d\phi$$

The work done in one revolution is

$$W = 2\pi\mu_k MgR$$

Now we'll compute the work done by friction.

$$\vec{\mathbf{f}}_s = +\mu_k Mg\hat{\mathbf{i}}$$

(opposite the direction of velocity). The path along which the mass travels is a circle of radius R. A displacement along this path is $d\vec{s} = -ds\,\hat{\mathbf{i}}$.

$$\begin{aligned} dW &= \vec{\mathbf{f}}_s \cdot d\vec{s} = -\mu_k Mg\, ds \\ W &= \int_{\text{path}} -\mu_k Mg\, ds = -\mu_k 2\pi MgR \end{aligned}$$

The negative of the work done by the hand.

7.59 The whole point of a rocket is to increase its kinetic energy *wrt* the center of the Earth to escape the Earth's gravitational field. It can do this by maximizing its velocity which in turn is accomplished by pointing its thrust $\parallel$ to its velocity.

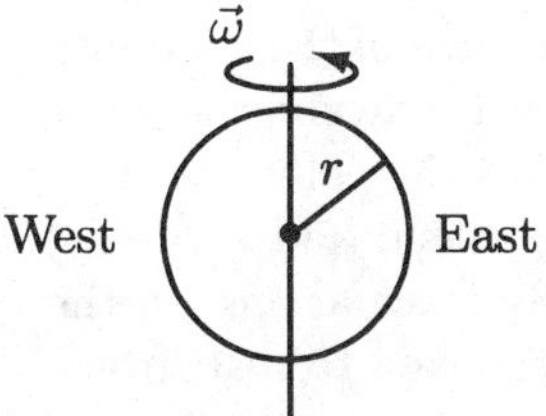

The Earth spins from west to east—so everything on its surface has a tremendous speed in the same direction (which we don't feel because were at rest *wrt* the rest of the Earth even if we are moving *wrt* the universe), recall that $V = \omega_{\text{earth}} r_{\text{earth}}$. By turning East, the rocket is thrusting in the same direction that it's already moving in—increasing its K as efficiently as possible. Note that if safety permitted, a rocket would be launched horizontally and facing the East to start with! The fact that it goes straight up and then to the East is to get some distance between itself and the ground; else a gust of wind might prove disastrous to the rocket if it were only half a kilometer away from the surface of the Earth.

7.63 For an ideal machine, work out = work in, on, $F_O s_O = F_I s_I$. This means:

$$MA = \frac{F_O}{F_I} = \frac{s_I}{s_O}$$

For this machine,

$$\begin{aligned} MA &= \frac{s_I}{s_O} = \frac{10 \text{ m}}{1 \text{ m}} = 10 \\ F_I &= \frac{F_O}{MA} = \frac{(50 \text{ kg})(9.8 \text{ m/s}^2)}{10} = 50 \text{ N} \end{aligned}$$

7.67 Providing machine X with an input force of 10 N enables it to lift (and therefore exert an output force of):

$$(5 \text{ kg})(9.81 \text{ m/s}^2) = 50 \text{ N}$$

By the definition of MA,

$$MA = \frac{F_O}{F_I} = \frac{50 \text{ N}}{10 \text{ N}} = 5$$

By the definition of efficiency,

$$e = \frac{W_O}{W_I} = \frac{(50 \text{ N})(1 \text{ m})}{(10 \text{ N})(6 \text{ m})} = \frac{5}{6}$$

7.71 To keep itself going at constant speed, the force that the train must exert has to balance friction, as shown below.

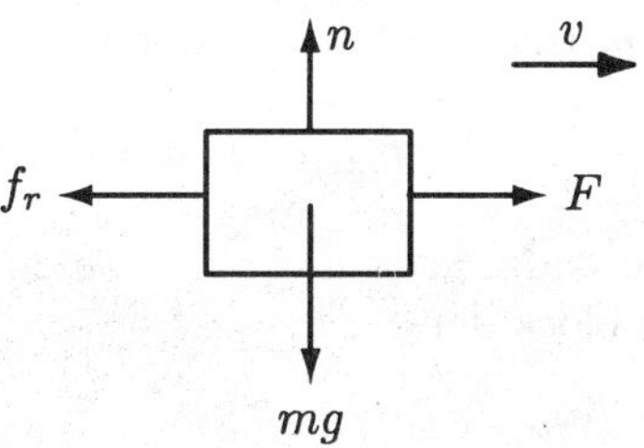

$$F = fr = \mu_r n = \mu_r mg$$

Now friction acts opposite the train's velocity, but this force acts in the direction of the velocity—thus,

$$\begin{aligned} P &= \vec{\mathbf{F}} \cdot \vec{\mathbf{v}} = fv = \mu_r mgv \\ (45\ \mathrm{km/h}) &= (45\ \mathrm{km/h}) \left(\frac{10^3\ \mathrm{m}}{1\ \mathrm{km}}\right) \left(\frac{1\ \mathrm{h}}{3600}\right) \\ &= 12.5\ \mathrm{m/s} \end{aligned}$$

So

$$\begin{aligned} P &= \left(5.0 \times 10^{-3}\right) \left(1.5 \times 10^6\ \mathrm{kg}\right) \\ &\quad \cdot \left(9.81\ \mathrm{m/s^2}\right) (12.5\ \mathrm{m/s}) \end{aligned}$$

Thus

$$P = 9.2 \times 10^5\ \mathrm{W}$$

7.75 We can use N2L to find what value of n is required to balance F (the wedge is not to accelerate vertically). As a forethought, we hope that this setup increases n since that will, in turn, increase the frictional force available to stop the trolley.

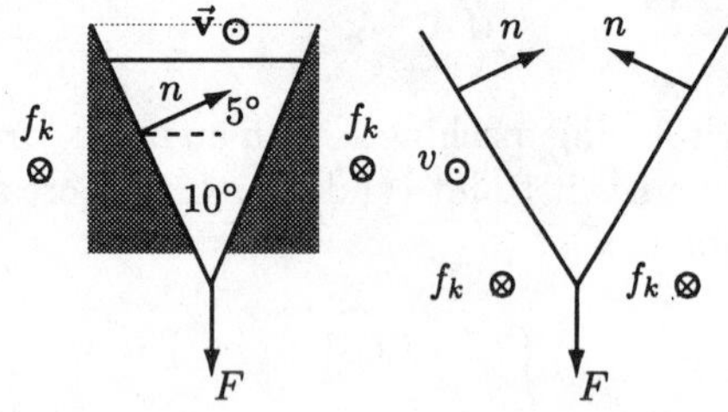

$$2N \sin 5° = F \Rightarrow n = \frac{F}{2 \sin 5°}$$

Friction is produced on both sides of the wedge.

$$f_k = \mu_k n = \frac{\mu_k F}{2 \sin 5°}$$

The total amount of friction acting on the wedge is:

$$f_{k,\mathrm{tot}} = 2 f_k = \frac{\mu_k F}{\sin 5°} \approx 11 \mu_k F$$

For the case of pushing down on the block we know that the total friction is

$$f_{k,\mathrm{tot}} = \mu_k n = \mu_k F$$

The wedge gives us 11 times more friction than the block does!

7.79 We were given the input and output force:

$$\begin{aligned} F_I &= (10.0\ \mathrm{g})(9.8\ \mathrm{m/s^2}) = 98\ \mathrm{N} \\ F_O &= (35\ \mathrm{kg})(9.8\ \mathrm{m/s^2}) = 343\ \mathrm{N} \end{aligned}$$

To compute the efficiency we need to find how s_I and s_O are related. In the diagram we can imagine pulling the rope down a distance s. Then each of the 4 ropes around the middle pulley must get shorter by $\frac{1}{4}s$, so $4s_{\mathrm{out}} = s_{\mathrm{in}}$

$$\begin{aligned} e &= \frac{W_O}{W_I} = \frac{F_O s_O}{F_I s_I} = \frac{343\ \mathrm{N} \cdot \frac{s_{\mathrm{in}}}{4}}{98\ \mathrm{N} \cdot s_{\mathrm{in}}} = .875 \\ &= 0.88 \text{ to 2 sig. fig.} \end{aligned}$$

We know that $W_O = .875 W_I$ for this system. The lost work goes into thermal energy. If the 10 kg object falls 4.0 m,

$$\begin{aligned} W_I &= F_I s_I = (10\ \mathrm{kg})(9.8\ \mathrm{m/s^2})(4.0\ \mathrm{m}) \\ &= 392\ \mathrm{J} \\ W_O &= e W_I = (0.875)(390\ \mathrm{J}) \\ &= 343\ \mathrm{J} \end{aligned}$$

The lost work went to thermal energy

$$W_I - W_O = 50\ \mathrm{J}$$

50 J went to thermal energy.

7.83 The object's net change in kinetic energy is 0—it starts at rest and ends at rest. Therefore, any work that gravity does must be "undone" by the string. Intuitively the string is the reason why the object eventually stops. The object has some K just before the string goes taut, so by the WET the string must do negative work to reduce K to 0. Note that this implies that the string can't be completely ideal—the string only does work if it exerts a force over a distance. If the string didn't stretch it would exert a force only at a point and would therefore do no work!

7.87

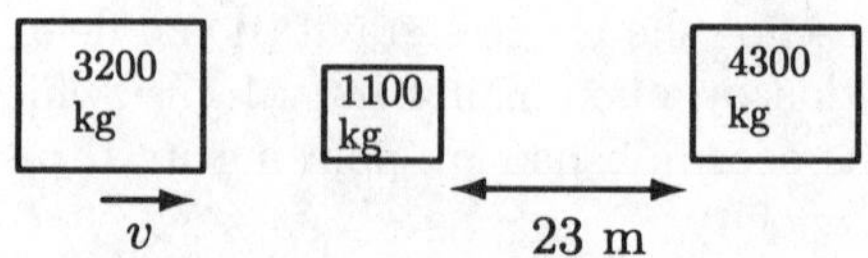

a) The mass of the combined wreckage is 4300 kg and it was traveling at some unknown speed v_c just after the collision. The wreckage, right after the crash, had a K of $\frac{1}{2} M_{\mathrm{tot}} v_c^2$. Friction is the only force acting on the system that could bring the cars to rest; from N2L

$$f_k = \mu_k n = \mu_k mg$$

In exerting this force on the wreckage over a distance of 23 m it must do an amount of work

$-\frac{1}{2}mv_c^2$ on the wreckage. When this happens, the wreckage has zero K and comes to rest. Thus,

$$\mu_k mgs \overset{\text{set}}{\equiv} \frac{1}{2}mv_c^2$$

Solving for v_c, we get

$$\begin{aligned} v_c &= \sqrt{2\mu_k g(s)} \\ &= \sqrt{2(0.58)(9.8\ \text{m/s}^2)(23\ \text{m})} \\ &= 16\ \text{m/s} \end{aligned}$$

b) From part a) we know the momentum of the system just after the crash is

$$p_f = (4300\ \text{kg})(16.2\ \text{m/s}) = 6.97 \times 10^4\ \text{N}\cdot\text{s}$$

If we consider the collision to be instantaneous then no net forces have a chance to do work on the system (in particular, friction). thus, momentum of the system, just after the crash must equal the momentum of the system just before the crash. If the Mercedes has some (as of yet unknown) speed v_m just before the crash, we can set $p_i = p_f$

$$\begin{aligned} (3200\ \text{kg})v_m &= 6.97 \times 10^4\ \text{N}\cdot\text{s} \\ \Rightarrow v_m &= 21.8\ \text{m/s} \\ &= 78\ \text{km/h} \end{aligned}$$

(the speed of the VW just before the crash is 0 m/s.)

c) In other words, the VW had 0.5 s to accelerate up to speed of 16.2 m/s. We've seen this in kinematics many times before.

$$16.2\ \text{m/s} = a(0.5\ \text{s}) \Rightarrow a = 32\ \text{m/s}^2$$

The VW probably does not accelerate uniformly; this is an average acceleration.

d) In reality momentum was not constant between just before and just after the collision. Now we will find out whether our approximation was good. We calculate the impulse due to friction.

$$I = f\Delta t$$

The momentum of the system which was taken away by friction during the collision is

$$\begin{aligned} p_f &= \mu_k mg\left(\frac{1}{2}s\right) \\ &= 0.58(4300\ \text{kg})(9.81\ \text{m/s})\frac{1}{2} \\ &= 1.2 \times 10^4\ \text{N}\cdot\text{s} \end{aligned}$$

The total momentum of the system before the collision was

$$p_{\text{sys}} = (3200\ \text{kg})(21.7\ \text{m/s}) = 6.9 \times 10^4\ \text{N s}$$

The ratio between them is:

$$\frac{p_f}{p_{\text{sys}}} = \frac{1.2\ \text{N}\cdot\text{s}}{6.9\ \text{N}\cdot\text{s}} = 0.17$$

In other words, friction reduced the system's momentum by 17% during the collision. Our approximation was perhaps a marginal one—fine for backs of envelopes but not OK for a detailed accident report.

e) We'll calculate energies before and after the collision from the calculated speeds.

$$\begin{aligned} K_{\text{before}} &= \frac{1}{2}(3200\ \text{kg})(21.8\ \text{m/s})^2 \\ &= 7.6 \times 10^5\ \text{J} \end{aligned}$$

$$\begin{aligned} K_{\text{after}} &= \frac{1}{2}(3200\ \text{kg} + 1100\ \text{kg})(16.2\ \text{m/s})^2 \\ &= 5.6 \times 10^5\ \text{J} \end{aligned}$$

$$\Delta K = 2.0 \times 10^5\ \text{J}$$

Where did the kinetic energy go? Some of it went into friction (as discussed in part c), but most of it went into the crumpling of metal. Friction might have been larger than we expected but the energy it uses up doesn't hold a candle to the amount of energy it takes to "total" 2 cars!

7.91 Considering forces acting on pulley 1, the two strings must make equal angles with the horizontal.

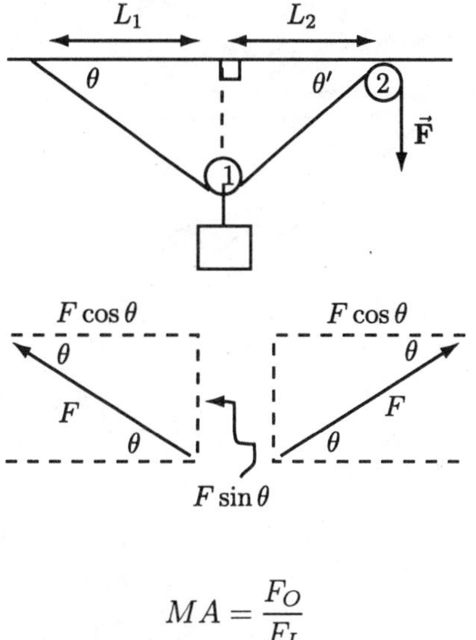

$$MA = \frac{F_O}{F_I}$$

F_I is the force you supply to the machine which well call F. Ideal rope implies that the tension you set up in the rope by exerting F is uniform throughout the whole string. Only the component that acts vertically on the pulley does work. By supplying an input force $F_I = F$ the machine exerts an output force of $F_O = 2F\sin\theta$, but to raise the block without acceleration we require

$$F_O = 2F\sin\theta = Mg$$

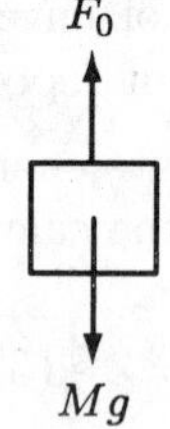

Thus,

$$\begin{aligned} F_O &= Mg \\ F_I &= \frac{Mg}{2\sin\theta} \end{aligned}$$

and the mechanical advantage is

$$MA = \frac{F_O}{F_I} = \frac{Mg}{\frac{Mg}{2\sin\theta}} = 2\sin\theta$$

The machine exerts a constant F_O of Mg on the block. If the block is moved up by a height h, then the work done on the block is

$$W = \vec{\mathbf{F}} \cdot \vec{s} = Fs = Mgh$$

The input work done is

$$F_I\, dy = \frac{mg}{2\sin\theta} dy$$

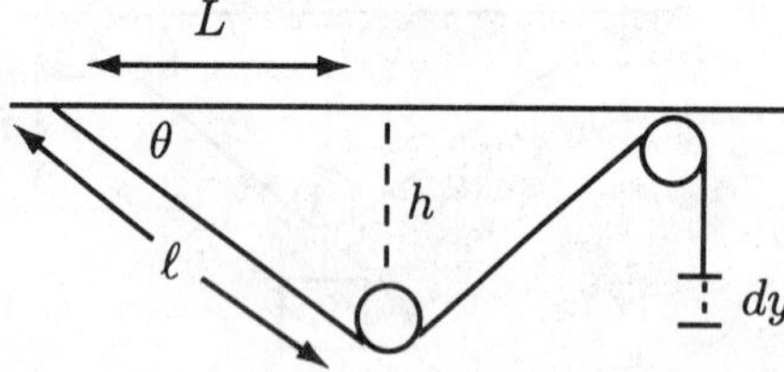

If the string shortens by dy; then ℓ shortens by

$$d\ell = \frac{dy}{2}$$

But $\ell\sin\theta = h$

$$(\ell - d\ell)(\sin(\theta + d\theta)) = h - dh \qquad (5)$$

Also $\ell\cos\theta = L =$ constant

$$\begin{aligned} (\ell - d\ell)\cos(\theta + d\theta) &= L \\ (\ell - d\ell)(\cos\theta\cos d\theta - \sin\theta d\theta) &= L \\ \cos d\theta &\approx 1, \end{aligned}$$

so, neglecting the term in $d\ell \times d\theta$

$$\ell\cos\theta - d\ell\cos\theta - \ell\sin\theta\, d\theta = L$$

$$d\theta = -\frac{d\ell\cos\theta}{\ell\sin\theta}$$

Substituting into (5)

$$\begin{aligned} (\ell - d\ell)(\sin\theta\cos d\theta + \cos\theta d\theta) &= h - dh \\ \ell\sin\theta - d\ell\sin\theta + \ell\cos\theta\, d\theta &= h - dh \end{aligned}$$

$(*) =$ (term in $d\ell \times d\theta$).

$$\begin{aligned} -d\ell\sin\theta + \ell\cos\theta\left[\frac{-d\ell\cos\theta}{\ell\sin\theta}\right] &= -dh \\ -d\ell\left[\frac{\sin^2\theta + \cos^2\theta}{\sin\theta}\right] &= -dh \end{aligned}$$

$$F_I \cdot dy = \frac{mg}{2\sin\theta} 2dh\sin\theta = mgdh$$

So total work done =

$$\begin{aligned} \int F_I\, dy &= \int mgdh = mgh \\ &= \text{output work} \end{aligned}$$

7.95

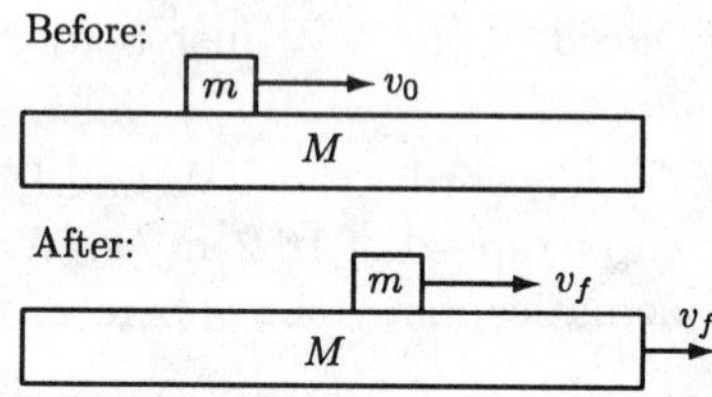

a) Free body diagrams

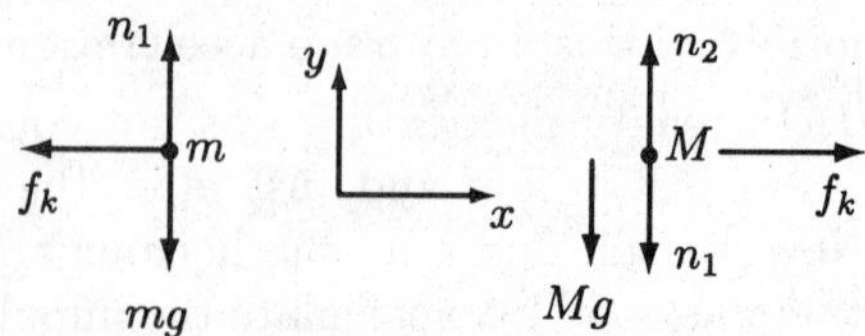

Small block:

$$\begin{aligned} n_1 &= mg \\ f_k &= \mu n_1 = \mu mg \\ -f_k &= m a_{1x} \\ a_{1x} &= -\mu g \end{aligned}$$

Long block:

$$\begin{aligned} f_k &= M a_{2x} \\ a_{2x} &= \frac{\mu mg}{M} = +\mu\alpha g \quad \left(\alpha = \frac{m}{M}\right) \end{aligned}$$

Kinematics

m:

$$\begin{aligned} v &= v_0 + a_1 xt \\ &= v_0 - \mu gt \\ s_1 &= \frac{v_f^2 - v_i^2}{2a} \\ &= \frac{v_f^2 - v_0^2}{-2\mu g} \end{aligned}$$

M:

$$\begin{aligned} v &= 0 + a_{2x}t = \alpha\mu gt \\ s_2 &= \frac{v_f^2 - v_i^2}{2a} = \frac{v_f^2}{2\alpha\mu g} \end{aligned}$$

Time when speeds are equal:

$$v_0 - \mu gt = \alpha\mu gt \Rightarrow t = \frac{v_0}{\mu g}\frac{1}{(1+\alpha)}$$

Final speed:

$$v_f = \alpha\mu g\left[\frac{v_0}{\mu g}\frac{1}{(1+\alpha)}\right] = \left(\frac{\alpha}{1+\alpha}\right)v_0$$

Displacements:

$$\begin{aligned} s_1 &= \frac{v_0^2\left[\left(\frac{\alpha}{1+\alpha}\right)^2 - 1\right]}{-2\mu g} \\ &= \frac{v_0^2}{2\mu g}\left(1 - \left(\frac{\alpha}{1+\alpha}\right)^2\right) \end{aligned}$$

$$\begin{aligned} s_2 &= v_0^2\left(\frac{\alpha}{1+\alpha}\right)^2\frac{1}{2\alpha\mu g} \\ &= \frac{v_0^2}{2\mu g}\frac{\alpha}{(1+\alpha)^2} \end{aligned}$$

Work done:

$$\begin{aligned} W &= -f_k s_1 + f_k s_2 \\ &= \mu_k mg\left[\frac{v_0^2}{2\mu g}\right]\left[\left(\frac{\alpha}{1+\alpha}\right)^2 - 1 + \frac{\alpha}{(1+\alpha)^2}\right] \\ &= \frac{1}{2}mv_0^2\left[\frac{\alpha^2 - (1+2\alpha+\alpha^2)+\alpha}{(1+\alpha)^2}\right] \\ &= \frac{1}{2}mv_0^2\left[\frac{-1-\alpha}{(1+\alpha)^2}\right] = -\frac{1}{2}\frac{mv_0^2}{(1+\alpha)} \end{aligned}$$

Kinetic energy:

Before:

$$\frac{1}{2}mv_0^2$$

After:

$$\begin{aligned} \frac{1}{2}(m+M)v_f^2 &= \frac{1}{2}m\left(1+\frac{1}{\alpha}\right)v_0^2\left(\frac{\alpha}{1+\alpha}\right)^2 \\ &= \frac{1}{2}mv_0^2\left(\frac{\alpha}{1+\alpha}\right) \end{aligned}$$

Change in KE:

$$\begin{aligned} \frac{1}{2}(m+M)v_f^2 - \frac{1}{2}mv_0^2 &= \frac{1}{2}\,\mathrm{m}\,v_0^2\left[\frac{\alpha}{1+\alpha} - 1\right] \\ &= -\frac{1}{2}\frac{mv_0^2}{(1+\alpha)} \\ i &= \text{work done} \end{aligned}$$

Momentum conservation

before: mv_0

after: $(m+M)v_f$.

Set them equal:

$$v_f = \frac{m}{m+M}v_0 = \frac{\alpha}{1+\alpha}v_0,$$

as above.

b) We need $s_1 - s_2 < L$

$$\begin{aligned} \frac{v_0^2}{2\mu g}\left[1 - \left(\frac{\alpha}{1+\alpha}\right)^2 - \frac{\alpha}{(1+\alpha)^2}\right] &< L \\ \frac{1}{(1+\alpha)^2}\left[1+2\alpha+\alpha^2-\alpha^2-\alpha\right] &< \frac{2\mu gL}{v_0^2} \end{aligned}$$

$$\begin{aligned} \frac{1}{1+\alpha} &< \frac{2\mu gL}{v_0^2} \\ v_0 &< \sqrt{2\mu gL(1+\alpha)} = v_{\max} \end{aligned}$$

For $v_0 > v_{\max}$, the after state is

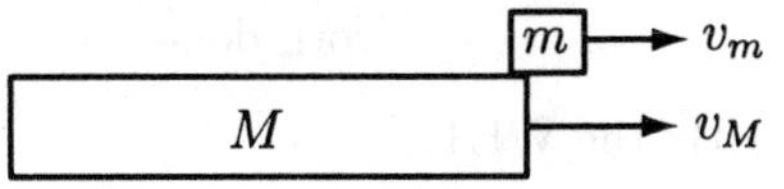

Writing s_1 and s_2 in terms of t:

$$\begin{aligned} s_1 &= v_0 t - \frac{1}{2}\mu gt^2 \\ s_2 &= \frac{1}{2}\alpha\mu gt^2 \end{aligned}$$

$$s_1 - s_2 = v_0 t + \frac{1}{2}\mu gt^2(-1-\alpha) \overset{\text{set}}{\equiv} L$$

$$\frac{\mu gt^2}{2}(1+\alpha) - v_0 t + L = 0$$

$$t = \frac{v_0 \pm \sqrt{v_0^2 - 2\mu gL(1+\alpha)}}{\mu g(1+\alpha)}$$

Now as $L \to 0$ we need $t \to 0$, so take $-$ sign

$$t = \frac{v_0 - \sqrt{v_0^2 - 2\mu gL(1+\alpha)}}{\mu g(1+\alpha)}$$

Then $v_m = v_0 - \mu gt$

$$\begin{aligned} v_m &= v_0 - \mu g\left[\frac{v_0 - \sqrt{v_0^2 - 2\mu gL(1+\alpha)}}{\mu g(1+\alpha)}\right] \\ &= v_0\left[1 - \frac{1}{1+\alpha}\right] + \frac{\sqrt{v_0^2 - 2\mu gL(1+\alpha)}}{1+\alpha} \\ &= v_0\frac{\alpha}{1+\alpha} + \frac{\sqrt{v_0^2 - 2\mu gL(1+\alpha)}}{1+\alpha} \end{aligned}$$

$$\begin{aligned} v_M &= \alpha\mu g\frac{v_0 - \sqrt{v_0^2 - 2\mu gL(1+\alpha)}}{\mu g\,(1+\alpha)} \\ &= \frac{\alpha}{1+\alpha}v_0 - \frac{\alpha}{1+\alpha}\sqrt{v_0^2 - 2\mu gL(1+\alpha)} \end{aligned}$$

Work done:

$$\begin{aligned} W &= -\mu_k mgs_1 + \mu_k mgs_2 \\ &= -\mu_k mg(s_1 - s_2) = -\mu_k mgL \end{aligned}$$

Final KE =

$$\frac{1}{2}mv_m^2 + \frac{1}{2}Mv_M^2$$

$$\begin{aligned} &= \frac{1}{2}m\left[v_0\frac{\alpha}{1+\alpha} + \frac{\sqrt{\ }}{1+\alpha}\right]^2 \\ &\quad + \frac{1}{2}\frac{m}{\alpha}\left[\frac{\alpha}{1+\alpha}v_0 - \frac{\alpha}{1+\alpha}\sqrt{\ }\right]^2 \\ &= \frac{1}{2}\frac{m}{(1+\alpha)^2}\left[v_0^2\left[\alpha^2 + 1 + 2\alpha\right] - 2\mu gL\,(1+\alpha)^2\right] \\ &= \frac{1}{2}mv_0^2 - \mu gmL \end{aligned}$$

$$\begin{aligned} \Delta KE &= \text{Final } KE - \text{initial } KE \\ &= -\mu gmL \\ &= \text{work done} \end{aligned}$$

7.97 a) By the WET:

$$W_{\text{avg}} = F_{\text{avg}}\Delta s = \frac{1}{2}mv_f^2 - \frac{1}{2}mv_i^2$$

With $v_f = 0$,

$$\begin{aligned} F_{\text{avg}} &= \frac{mv_i^2}{2\Delta s} = \frac{-(.025\text{ kg})(350\text{ m/s})^2}{2(.23\text{ m})} \\ &= 6.7\times 10^3\text{ N} \end{aligned}$$

By impulse/momentum:

$$\begin{aligned} F_{\text{avg}} &= \frac{mv}{t} = \frac{(.025\text{ kg})(-350\text{ m/s})}{1.5\times 10^{-3}\text{ sec}} \\ &= 5.8\times 10^3\text{ N} \end{aligned}$$

The estimates seem to be off by a factor of .87.

b) We have the following equations for WET and momentum-impulse theorem respectively: For a constant force:

$$Fd = \frac{1}{2}mv^2 \tag{6}$$

$$Ft = mv \tag{7}$$

For constant force, $d = \frac{1}{2}at^2$ and $v = at$. Combining the two expressions and solving for v gives

$$v = \frac{2d}{t} \tag{8}$$

Using Eq. (8) in Eq. (6) gives:

$$F\cdot d = \frac{1}{2}mv\left(\frac{2d}{t}\right) \Rightarrow tF = mv$$

which is Eq. (7). Thus for constant force the two methods produce the same result.

c) Let T be the time in which the bullet comes to rest and d be the distance which it travels during T. Then estimates for the force using both WET and momentum is:

$$\begin{aligned} F_{\text{est,WET}} &= \frac{1}{d}\int F(x)\,dx \\ &= \frac{1}{d}\int F(x(t))\frac{dx}{dt}\,dt \\ &= \frac{1}{d}\int F(t)\,v(t)\,dt \\ F_{\text{est},p} &= \frac{1}{T}\int F(t)\,dt \end{aligned}$$

We can consider the ratio

$$\begin{aligned} F_R &= \frac{F_{\text{WET,est}}}{F_{\text{est},p}} = \frac{T\int Fv dt}{d\int F dt} \\ &= \frac{\int F\left(\frac{v}{\frac{d}{T}}\right)dt}{\int F\,dt} = \frac{\int F\left(\frac{v}{v_{avg}}\right)dt}{\int F dt} \end{aligned}$$

The above ratio is > or < 1 depending on the ratio $\frac{v}{v_{\text{avg}}}$. The graph below shows v vs. t.

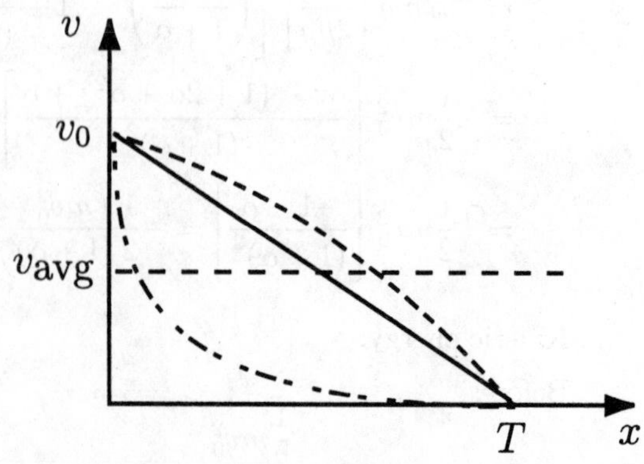

For constant force v is linearly dependent on t ($v = at$) so the solid line shows constant F. If F is increasing the bullet's acceleration is small at first but increases. The upper dashed line shows such an F. If F is decreasing the acceleration starts out large but then tapers off. The lower dashed line shows such an F.

For *increasing* F, $\frac{v}{v_{\text{avg}}}$ is less than the constant force value at every x, so $F_R < 1$.

For *decreasing* F, $\frac{v}{v_{\text{avg}}}$ is greater than the constant force value and $F_R > 1$. From the given data our F_{WET} was a larger estimate; our F_R would be greater than 1. Therefore, F must have been decreasing.

Chapter 8

Conservation of Energy

8.1 The energy stored in a spring is $U = \frac{1}{2}ks^2$ where s is the amount of stretch or compression of the spring.

$$\frac{U_1}{U_2} = \frac{\frac{1}{2}k\,(1\,\mathrm{cm})^2}{\frac{1}{2}k\,(2\,\mathrm{cm})^2} = \frac{1}{4}$$

8.3 Energy stored in the spring's compression gets converted to the ball's kinetic energy.

$$\frac{1}{2}ks^2 \overset{\text{set}}{\equiv} \frac{1}{2}mv^2$$

$$\begin{aligned} \Rightarrow s &= v\sqrt{\frac{m}{k}} \\ &= (2\,\mathrm{m/s})\sqrt{\frac{(.1\,\mathrm{kg})}{90\,\mathrm{N/m}}} \\ &= 7\times 10^{-2}\,\mathrm{m} \end{aligned}$$

This is the compression of the spring. The spring must be compressed to a length of

$$\ell = 10\ \mathrm{cm} - 7\ \mathrm{cm} = 3\ \mathrm{cm}$$

8.5 Reference frame: floor

$$\begin{aligned} U_i &= (45\ \mathrm{kg})(9.81\ \mathrm{m/s^2})(3.2\ \mathrm{m}) = 1400\ \mathrm{J} \\ U_f &= (45\ \mathrm{kg})(9.81\ \mathrm{m/s^2})\,(0\ \mathrm{m}) = 0 \end{aligned}$$

$$\Delta U = U_f - U_i = 0 - 1400\ \mathrm{J} = -1400\ \mathrm{J}$$

Reference frame: initial position

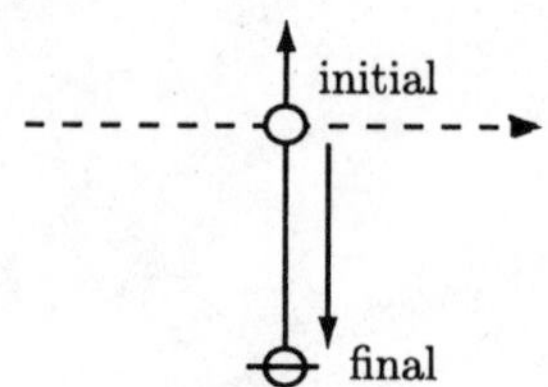

$$\begin{aligned} U_i &= (45\ \mathrm{kg})(9.81\ \mathrm{m/s^2})\,(0\ \mathrm{m}) \\ &= 0\ \mathrm{J} \\ U_f &= (45\ \mathrm{kg})(9.81\ \mathrm{m/s^2})(-3.2\ \mathrm{m}) \\ &= -1400\ \mathrm{J} \end{aligned}$$

$$\Delta U = U_f - U_i = -1400\ \mathrm{J}$$

Reference frame: highest shelf

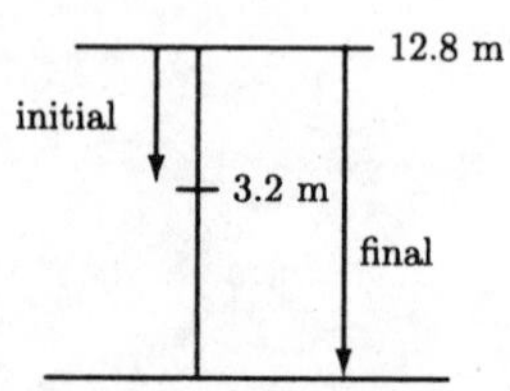

$$\begin{aligned} U_i &= (45\ \mathrm{kg})(9.81\ \mathrm{m/s^2})(-9.6\ \mathrm{m}) \\ &= -4200\ \mathrm{J} \\ U_f &= (45\ \mathrm{kg})(9.81\ \mathrm{m/s^2})(-12.8\ \mathrm{m}) \\ &= -5600\ \mathrm{J} \end{aligned}$$

$$\Delta U = U_f - U_i = -1400\ \mathrm{J}$$

The loss of 1400 J goes into the drums K.

$$1400 \text{ J} = \frac{1}{2}(45 \text{ kg})v^2$$

Solving for v yields $v = 7.9 \text{ m/s}$.

8.7 The wrench falls and has velocity; it has positive K. The value of its potential energy depends on the chosen reference level. If we take $U = 0$ on the surface then the total energy is not zero so choice c) is false. In the absence of non-conservative forces the TME of a system is conserved—that is, unchanged. Therefore, TME neither increases nor decreases but remains constant—choice d) is correct.

8.9 At the rocket's max height the payload has potential energy

$$\begin{aligned} U &= mgh \\ &= (500 \text{ kg})(9.81 \text{ m/s}^2)(3 \times 10^3 \text{ m}) \\ &= 1 \times 10^7 \text{ J} \end{aligned}$$

taking the ground as reference level. The rocket obtained this energy directly from burning fuel. It was transformed from fuel chemical energy to the payload's gravitational energy.

8.11 Friction is the force which is converting mechanical to thermal energy. The power at which mechanical energy is converted to thermal energy is equal in magnitude but opposite in sign to the rate at which friction does work on the block. This power is equal to:

$$P = -\left[\vec{\mathbf{f}}_k \cdot \vec{\mathbf{v}}\right] = -\left[-\mu mg\hat{\mathbf{i}} \cdot v\hat{\mathbf{i}}\right] = \mu mgv$$
$$= (.7)(100 \text{ kg})(9.81 \text{ m/s}^2)(1 \text{ m/s}) = 700 \text{ W}$$

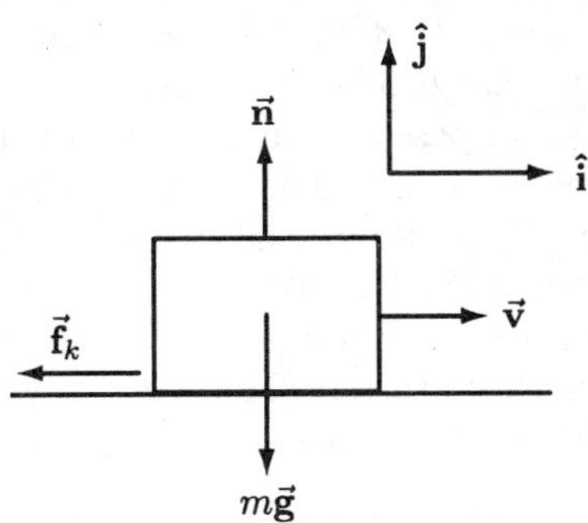

8.13 The spring is compressed by an amount Δs where:

$$k\Delta s = mg \Rightarrow \Delta s = \frac{mg}{k}$$

The energy stored in the spring is:

$$U_{\text{spring}} = \frac{1}{2}k\Delta s^2 = \frac{1}{2}k\left(\frac{mg}{k}\right)^2 = \frac{(mg)^2}{2k}$$

If we double the spring constant $k' = 2k$ the spring stretches by an amount

$$\Delta s' = \frac{mg}{k'} = \frac{mg}{2k}$$

and the spring potential energy becomes:

$$U'_{\text{spring}} = \frac{1}{2}k'\Delta s^{2'} = \frac{1}{2}(2k)\left(\frac{mg}{k}\right)^2 = \frac{(mg)^2}{4k}$$

The spring energy decreases by a factor of 2.

8.17 The question suggests that we use N2L to analyze the block.

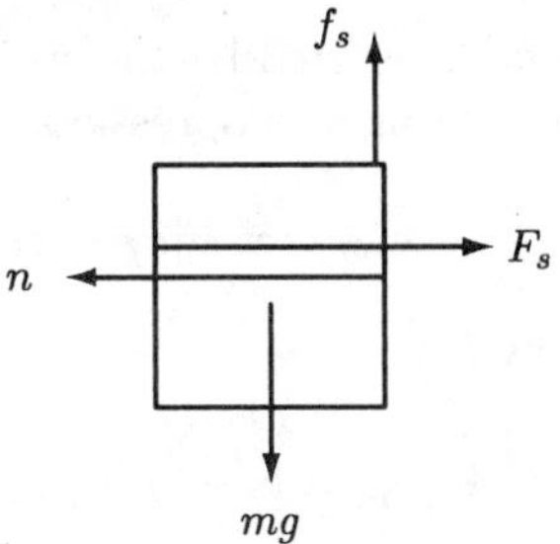

The spring is compressed by an amount $\Delta s = \ell - L$

$$F_s = k\Delta s = k(\ell - L) = n$$

The maximum value of f_s is

$$f_s = \mu n = mg$$

From the two equations we get

$$\mu[k(\ell - L)] = mg$$

Solving for the minimum spring constant,

$$k = \frac{mg}{\mu(\ell - L)}$$

The spring potential energy stored is:

$$\begin{aligned} U &= \frac{1}{2}k(\ell - L)^2 = \frac{1}{2}\frac{mg}{\mu(\ell - L)}(\ell - L)^2 \\ &= \frac{mg(\ell - L)}{2\mu} \end{aligned}$$

8.21 We will try to mimic Example 8.2 as closely as possible—changes will have to be made but the overall plan will be the same. The cars will not travel equal distances—it takes more force to accelerate a larger mass. We can find the relationship between the cars' displacements by considering their momentum. They start out with zero momentum. Since the only net force that acts on them is internal (i.e. the spring) the momentum must always remain zero.

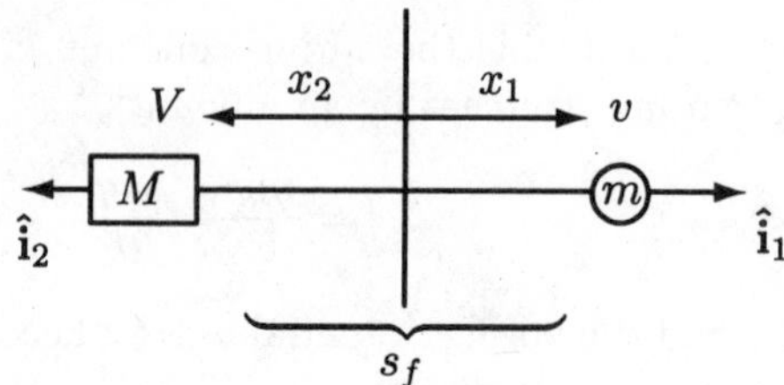

Conservation of momentum requires

$$M\,|V| = m\,|v| \tag{1}$$

Thus

$$\frac{x_2}{x_1} = \frac{m}{M}$$

The force exerted by the spring is proportional to the spring's compression

$$\vec{\mathbf{F}}_{\text{on } m} = k\,[s_f - x_1 - x_2]\,\hat{\mathbf{i}}_1 \tag{2}$$

and by (1)

$$\vec{\mathbf{F}}_{\text{on } m} = k\left[s_f - x_1\left(1 + \frac{m}{M}\right)\right]\hat{\mathbf{i}}_1$$

Likewise

$$\vec{\mathbf{F}}_{\text{on } M} = k\left[s_f - x_2\left(1 + \frac{M}{m}\right)\right]\hat{\mathbf{i}}_2 \tag{3}$$

Each mass goes through a series of displacements

$$\begin{aligned} d\vec{\mathbf{s}}_M &= dx_2\hat{\mathbf{i}}_2 \\ d\vec{\mathbf{s}}_m &= dx_1\hat{\mathbf{i}}_1 \end{aligned}$$

The work done by the spring during each displacement is:

$$\begin{aligned} dW_M &= \vec{\mathbf{F}}_{\text{on } M}\cdot d\vec{\mathbf{s}}_M = k\left[s_f - x_2\left(1+\frac{M}{m}\right)\right]dx_2 \\ dW_m &= \vec{\mathbf{F}}_{\text{on } m}\cdot d\vec{\mathbf{s}}_m = k\left[s_f - x_1\left(1+\frac{m}{M}\right)\right]dx_1 \end{aligned}$$

Limits: The spring starts with a compression of s_f so by (2) and (3), we can see that the lower limit is

$$x_1 = x_2 = 0$$

We can obtain an upper limit by noting that the force on the blocks at the upper limit is 0. This happens when

$$x_1 + x_2 = s_f \tag{4}$$

Using Eq. (1) with (4) gives us the following upper limits:

$$x_1 = \frac{Ms_f}{M+m},\; x_2 = \frac{ms_f}{M+m}$$

We are now ready to do the integrals.

$$\begin{aligned} W_M &= k\int_0^{\frac{ms_f}{M+m}} \left[s_f - x_2\left(1+\tfrac{M}{m}\right)\right]dx_2 \\ &= k\left[s_f\left(\tfrac{ms_f}{M+m}\right) - \frac{1}{2}\left(\tfrac{ms_f}{M+m}\right)^2\left(1+\tfrac{M}{m}\right)\right] \\ &= ks_f^2\tfrac{m}{M+m}\left(1 - \frac{1}{2}\left(\tfrac{m}{M+m}\right)\left(\tfrac{M+m}{m}\right)\right) \\ &= \frac{1}{2}ks_f^2\left(\tfrac{m}{M+m}\right) \end{aligned}$$

$$\begin{aligned} W_m &= k\int_0^{\frac{Ms_f}{M+m}} \left[s_f - x_1\left(1+\tfrac{m}{M}\right)\right]dx_1 \\ &= k\left[s_f\left(\tfrac{Ms_f}{M+m}\right) - \frac{1}{2}\left(\tfrac{Ms_f}{M+m}\right)^2\left(1+\tfrac{m}{M}\right)\right] \\ &= \frac{ks_f^2}{2}\left(\tfrac{M}{M+m}\right)\left(2 - \tfrac{M}{M+m}\left(\tfrac{M+m}{M}\right)\right) \end{aligned}$$

$$\begin{aligned} W_M + W_m &= \frac{1}{2}ks_f^2\frac{1}{(M+m)}(m+M) \\ &= \frac{1}{2}ks_f^2 \end{aligned}$$

This is precisely the energy stored in the spring. In the limit as $M \gg m$ the two cars still must satisfy the momentum requirement $MV = mv$, that is, they share momentum equally. In this limit, energy is not shared equally. $K = \frac{p^2}{2m}$ shows that the smaller mass will have most of the energy.

8.23 The bead with the greatest U_{grav} will have the greatest speed at the bottom. U_{grav} results from a conservative force so are not interested in bumps or curves in the bead. The bead which is highest will have the greatest U_{grav} and therefore be the fastest at the end of the wire. It appears that the beads on wires **b** and **d** are of equal height. However, the bead on wire **d** cannot pass the hump without being given more energy. It would need a greater U_{grav} to get over the hump than it has where it starts out. Conservation of energy says that a bead released from rest cannot get higher than the height at which it started. The bead on wire **b** will have the greatest speed at the end of the wire.

8.25 The exact formula comes from computing work done by gravity using Newton's Law of Gravity. The gravitational force law $F = \frac{GM_1M_2}{r^2}$ applies to all planets, therefore its potential function, our exact formula, applies to all planets, whether Earth or Romulus. The practical formula is derived from the exact formula so it also applies to all planets as long as you use the correct value for the acceleration due to gravity.

8.27 No dissipative forces act on the bead so TME is conserved. We take the reference level for GPE at point E. Then the bead starts out with a $U_{\text{grav}} = mgh$ and a zero K (it starts from rest) at point A and ends with a $U_{\text{grav}} = 0$ and some kinetic energy $\frac{1}{2}mv_E^2$ at point E. Since TME is conserved,

$$\begin{aligned} E_{\text{initial}} &= E_{\text{final}} \\ \Rightarrow mgh &= \frac{1}{2}mv_E^2 \\ \Rightarrow v_E &= \sqrt{2gh} \end{aligned}$$

8.29 a) The ski lift needs to lift

$$(700)(70 \text{ kg}) = 50\,000 \text{ kg}$$

a distance 300 m vertically per hour. The power required to do this is:

$$\begin{aligned} P &= \frac{mgh}{\Delta t} \\ &= \frac{(5\times 10^4 \text{ kg})(9.81 \text{ m/s}^2)(300 \text{ m})}{3600 \text{ sec}} \\ &= 4\times 10^4 \text{ W} \end{aligned}$$

b) The KE given to the people requires an additional power

$$\begin{aligned} P_k &= \frac{1}{2}\frac{mv^2}{\Delta t} = \frac{\frac{1}{2}(5\times 10^4 \text{ kg})(1 \text{ m/s})^2}{3600 \text{ sec}} \\ &= 7 \text{ W} \end{aligned}$$

negligible compared with the power required to lift them.

8.33 If we assume that the celestial stuff falls from infinity, its change in U per second is:

$$\frac{\Delta U}{1 \text{ sec}} = \frac{GM_{\text{star}}M_{\text{stuff}}}{R_{\text{star}}(1 \text{ sec})} \overset{\text{set}}{\equiv} P = 10^{30} \text{ W}$$

where we assume that the U lost by the stuff goes directly into emitting x-rays. Solving for $\frac{M_{stuff}}{1 \text{ sec}}$,

$$\begin{aligned} \frac{M_{\text{stuff}}}{1 \text{ sec}} &= \frac{(P)R_{\text{star}}}{GM_{\text{star}}} \\ &= \frac{(10^{30} \text{ W})(10\times 10^3 \text{ m})}{\left(6.7\times 10^{-11} \frac{\text{N}\cdot\text{m}^2}{\text{kg}^2}\right)(10^{30} \text{ kg})} \\ &= 10^{14} \text{ kg/s} \end{aligned}$$

8.37 No dissipative forces are mentioned so TME is conserved.

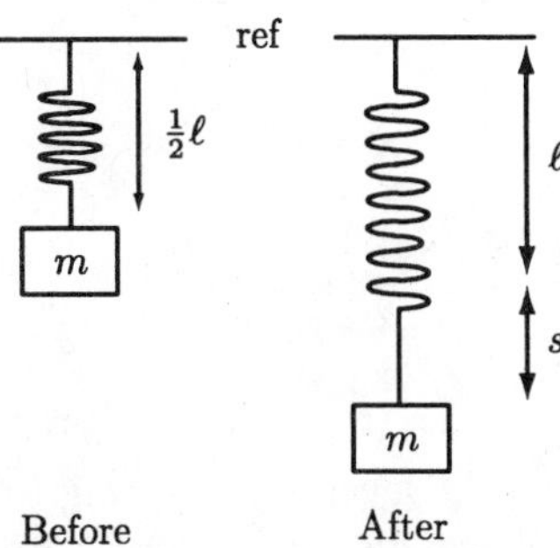

Let the GPE be zero when the box is touching the roof. Then the initial TME is:

$$\begin{aligned} E_i &= \frac{1}{2}k\left(\frac{1}{2}\ell\right)^2 + mg\left(-\frac{1}{2}\ell\right) \\ &= \frac{1}{8}k\ell^2 - \frac{1}{2}mg\ell \end{aligned}$$

Suppose the block is first brought to rest a distance $\ell + s$ below the roof. The TME of the system is

$$E_f = \frac{1}{2}ks^2 - mg(\ell + s)$$

Conservation of TME implies

$$\begin{aligned} E_i &= E_f \\ \Rightarrow \frac{1}{8}k\ell^2 - \frac{1}{2}mg\ell &= \frac{1}{2}ks^2 - mg(\ell + s) \end{aligned}$$

Which is a quadratic expression in s.

$$s^2 - \frac{2mg}{k}s - \ell\left[\frac{1}{4}\ell + \frac{mg}{k}\right]$$

Solving for s (which needs to be positive) gives:

$$\begin{aligned} s &= +\frac{mg}{k} + \frac{1}{2}\sqrt{\left(\frac{2mg}{k}\right)^2 + 4\ell\left[\frac{1}{4}\ell + \frac{mg}{k}\right]} \\ &= \frac{mg}{k} + \frac{1}{2}\left(\frac{2mg}{k} + \ell\right) = \frac{2mg}{k} + \frac{\ell}{2} \end{aligned}$$

Thus the box is a distance

$$\ell + s = \frac{3\ell}{2} + \frac{2Mg}{k}$$

below the roof.

8.41

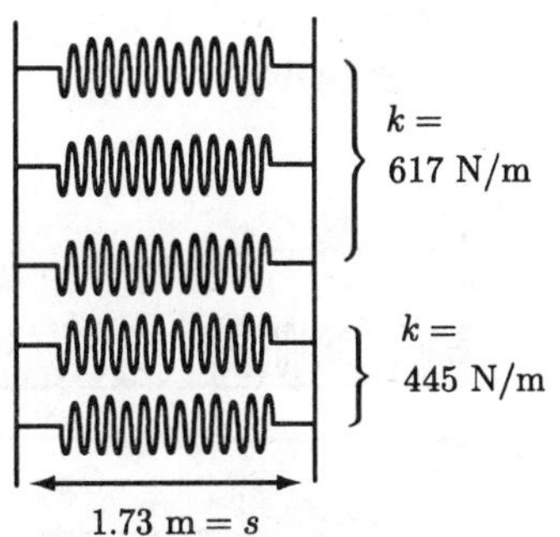

$$\begin{aligned} U &= \sum \frac{k}{2} s^2 \\ &= \frac{1}{2}(1.73\ \text{m})^2 \left(\begin{array}{c} 3 \times (617\ \text{N/m}) \\ +2 \times (445\ \text{N/m}) \end{array} \right) \\ &= 4.1\ \text{kJ} \end{aligned}$$

8.45 Take the level of the spring to be the reference level. No dissipative forces are present so TME is conserved. Initially the car has a TME of $E_i = M_1 gh$. It starts from rest, and eventually reaches the reference level. When it does all the potential energy will have been converted to K. Finally, as the car compresses the spring this K is converted to spring U. When the spring is fully compressed by an amount s all the K will have been converted to spring U.

$$E_f = \frac{1}{2} k s^2$$

$$\begin{aligned} E_i &= E_f \\ \Rightarrow M_1 gh &= \frac{1}{2} k s^2 \\ \Rightarrow k &= \frac{2 M_1 gh}{s^2} \\ k &= \frac{2(1.0 \times 10^4\ \text{kg})(9.81\ \text{m/s})(1.0 \times 10^1\ \text{m})}{(1.0\ \text{m})^2} \\ &= 2.0 \times 10^6\ \text{N/m} \end{aligned}$$

Now the unloaded car starts back up the track. Now the initial TME is $\frac{1}{2}ks^2$ and the final TME will be $m_2 gh + \frac{1}{2} m_2 v_f^2$.

$$\begin{aligned} E_i &= E_f \Rightarrow \frac{1}{2} k s^2 = m_2 gh + \frac{1}{2} m_2 v_f^2 \\ v_f &= \sqrt{\frac{ks^2}{m_2} - 2gh} \\ &= \sqrt{\frac{(2.0 \times 10^6\ \text{N/m})(1.0\ \text{m})^2}{1.0 \times 10^3\ \text{kg}} - 2\,(9.81\ \text{m/s}^2)\,(10\ \text{m})} \\ &= 42\ \text{m/s} \end{aligned}$$

8.49 Note that if both springs have a relaxed length of $\frac{1}{2}\ell$ they are both stretched (unless one is stretched by more than $\frac{3}{2}\ell$.)

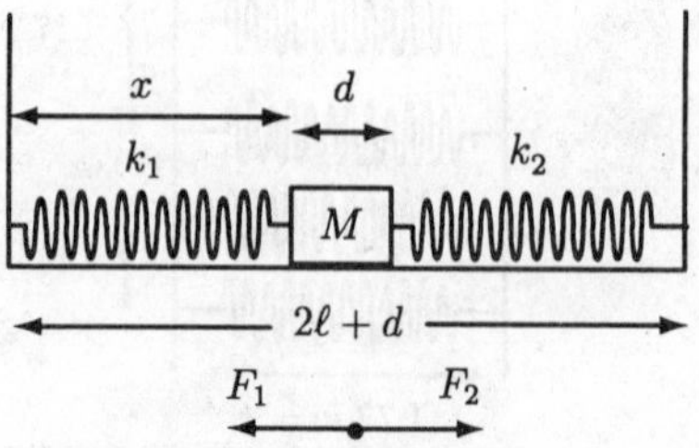

If $x > \frac{1}{2}\ell$ the first spring is stretched by an amount $x - \frac{1}{2}\ell$ and the other spring is stretched by an amount

$$(2\ell - x) - \frac{1}{2}\ell = \frac{3}{2}\ell - x$$

The potential energy of the system is

$$U_{\text{spring}} = \frac{1}{2} k_1 \left(x - \frac{1}{2}\ell \right)^2 + \frac{1}{2} k_2' \left(\frac{3}{2}\ell - x \right)^2$$

Because we are squaring the term it doesn't matter whether the springs are stretched or compressed. Now we'll find $P_{e,\min}$

$$\frac{dU}{dx} = k_1 \left(x - \frac{1}{2}\ell \right) + k_2 \left(x - \frac{3}{2}\ell \right) \overset{\text{set}}{\equiv} 0$$

Solving for x,

$$x = \frac{\ell\,[k_1 + 3k_2]}{2\,(k_1 + k_2)}$$

If the system is in equilibrium the forces acting on the mass must be equal, that is:

$$F_{\text{net}} = k_1 \left[x - \frac{1}{2}\ell \right] - k_2 \left[\frac{3}{2}\ell - x \right] = 0$$

When

$$x = \frac{\ell\,[k_1 + 3k_2]}{2\,(k_1 + k_2)}$$

$$\begin{aligned} F_{\text{net}} &= k \left[\frac{\ell(k_1+3k_2)}{2(k_1+k_2)} - \frac{\ell(k_1+k_2)}{2(k_1+k_2)} \right] \\ &\quad -k_2 \left[\frac{3\ell(k_1+k_2)}{2(k_1+k_2)} - \frac{\ell(k_1+3k_2)}{2(k_1+k_2)} \right] \\ &= \frac{k_1}{2(k_1+k_2)}\,[2\ell k_2] - \frac{k_2}{2(k_1+k_2)}\,[2\ell k_1] \\ &= 0 \end{aligned}$$

8.53 If gravity were the only force acting on the car we know from Chapter 5 that the car would accelerate down the slope of angle θ at $g \sin\theta$. The problem states that the car goes down the slope at constant velocity so there must be another force at work here; the most likely one is friction. As the car goes down the hill its K remains constant but its U decreases. The car's TME $= K + U$ decreases. We know that friction is a non-conservative (dissipative) force. The decrease of TME is due to friction forces within the car's brakes dissipating energy as heat.

8.57 The force that opposes the weight force is the normal force that the ground exerts on the foot, the net force on your body accelerates you.

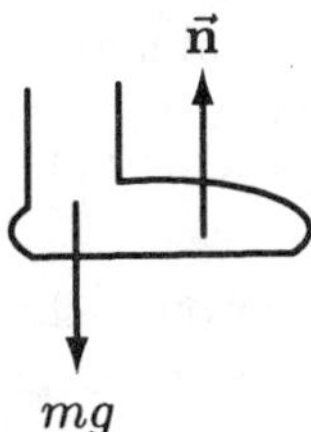

Since the point at which the normal force acts does not move, the normal force does no work. The forces that actually cause your body to rise occur in the muscles in your legs. They use chemical energy stored in the body to transform the energy needed to increase the body's potential energy.

8.61 The normal and weight forces are in balance and don't displace the truck—thus they don't do any work on the truck.

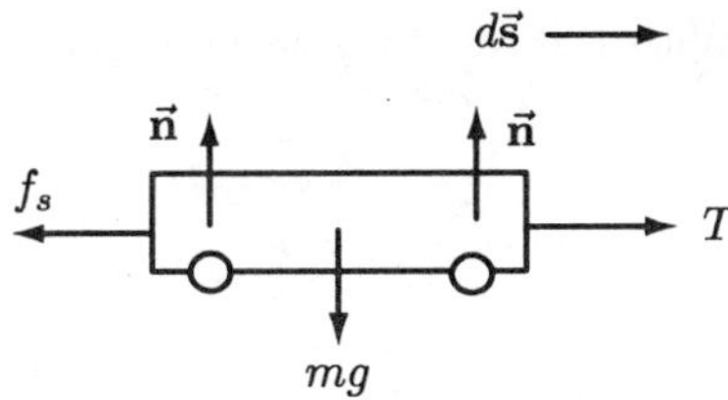

We are told that the truck moves with constant velocity. Another force—friction must be acting on the truck so that the car doesn't accelerate horizontally. Since the truck is moving in the direction of the tension force, the work done by T,

$$W_T = \int \vec{\mathbf{T}} \cdot d\vec{s}$$

is positive (non-zero) and the work done by friction

$$W_{f_k} = \int \vec{\mathbf{f}}_k \cdot d\vec{s}$$

is negative (non-zero). In fact,

$$W_{\text{total}} = W_T + W_{f_k} = 0 \Rightarrow W_T = -W_{f_s}$$

The total work done on the truck is zero, that's why the truck's K is constant. The work Isaac does ends up as thermal energy in the truck and the floor.

8.65 With the spring stretched 6.3 cm the cart has a TME consisting of spring U only.

$$E_i = \frac{1}{2}(1200\text{ N/ m})(.063\text{ m})^2 = 2.4\text{ J}$$

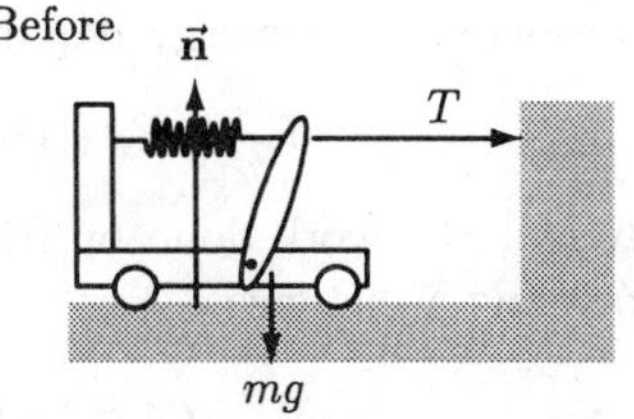

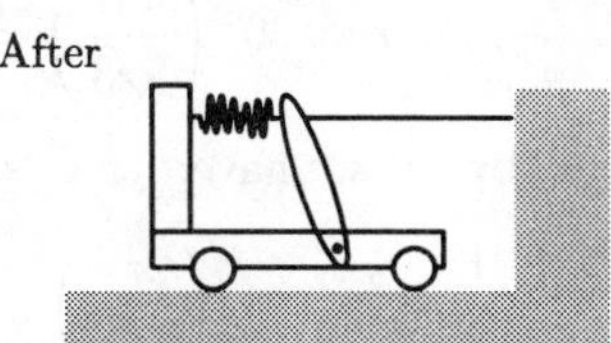

When the spring reaches its unstretched length all of this stored U will have been converted to K.

$$\frac{1}{2}mv^2 = 2.4\text{ J} = K$$

The forces acting on the system are the normal + weight force which balance each other and do no work. The tension force acts horizontally but since the string does not move, the tension also does no work. Spring force acting at the *left* end of the cart does do work. This is an example of work done by internal forces.

8.69 We'll assume the bannister is straight first. Then the friction force acting on the child is

$$f_s = \mu_k n = \mu_k mg\cos 30° \tag{5}$$

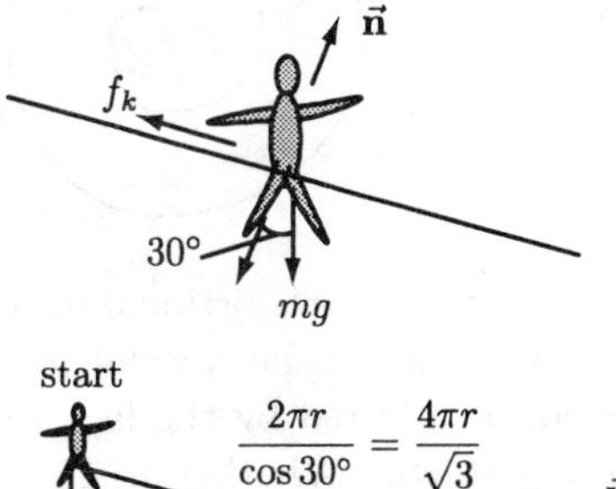

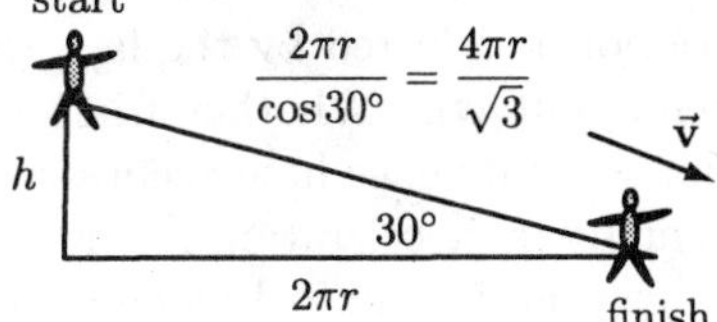

The child starts from rest and has an initial TME (with reference level at the bottom of the bannister)

$$E_i = U_{\text{grav}} = mgh$$

and

$$h = 2\pi r \tan 30° = \frac{2\pi r}{\sqrt{3}}$$

At the bottom of the banister the child will have a TME of

$$E_f = K = \frac{1}{2}mv_f^2$$

Conservation of energy requires:

$$E_f = E_i + W_{f_k} \tag{6}$$

By (5), the work done by friction in the child traveling a distance $\frac{4\pi r}{\sqrt{3}}$ is

$$-\mu_k mg \cos 30^\circ \left(\frac{4\pi r}{\sqrt{3}}\right) = -\mu_k mg 2\pi r$$

Thus, by (6) we have:

$$\frac{1}{2}mv_f^2 = mg\left(\frac{2\pi r}{\sqrt{3}}\right) - 2\pi r \mu_k mg$$

Solving for v_f,

$$\begin{aligned} v_f &= \sqrt{2g\pi r\left(\frac{1}{\sqrt{3}} - \mu_k\right)} \\ &= \sqrt{(9.8 \text{ m/s}^2)(10\pi \text{ m})\left[\frac{1}{\sqrt{3}} - (0.5)\right]} \\ &= 5 \text{ m/s} \end{aligned}$$

Now we will consider the curvature of the banister by first asking the question "what is different?" The child must now have centripetal acceleration to achieve the helical path.

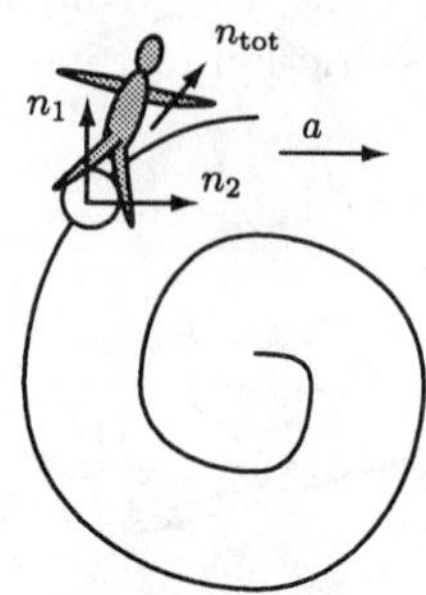

Since friction is proportional to the normal force we can compare the normal forces to see how friction is affected by the fact that the bannister is curved. Whether friction is affected or not will influence how accurate 5 m/s is. This centripetal acceleration is achieved by a normal force pointing inward along a radial line to the center line of the spiral staircase (shown as n_2 in the figure). The child must lean inward to get this component of normal force. The radius of her quasi-circular path is $R = 5$ m so the normal force needed to supply the centripetal acceleration is $n_2 = \frac{mv^2}{R}$. Thus:

$$\frac{n_2}{n_1} = \frac{\frac{mv^2}{R}}{mg\cos\theta} = \frac{v^2}{Rg\cos\theta}$$

with $v = 5$ m/s,

$$\frac{n_2}{n_1} = 0.6$$

The centripetal normal force is certainly significant and will make friction increase a great deal. The 5 m/s is probably a large overestimate.

8.73 The pole converts the jumpers K due to the running start into U_{grav} which he needs to clear the pole. Let's find the speed he needs to clear a 9 m pole:

$$\begin{aligned} E_i &= E_f \Rightarrow \frac{1}{2}mv^2 = mgh \\ \Rightarrow v &= \sqrt{2gh} = \sqrt{2(9.8 \text{ m/s}^2)(9 \text{ m})} \\ &= 13 \text{ m/s} \end{aligned}$$

Considering the record 100 m run was $v_{\text{avg}} = 10$ m/s it is unlikely the jumper can muster the needed initial K to clear a 9 m pole. [*Note:* In a 100 m run all the acceleration occurs in the first few meters, so $v_{\text{av}} \approx v_{\text{max}}$]

8.77 With the reference level as shown,

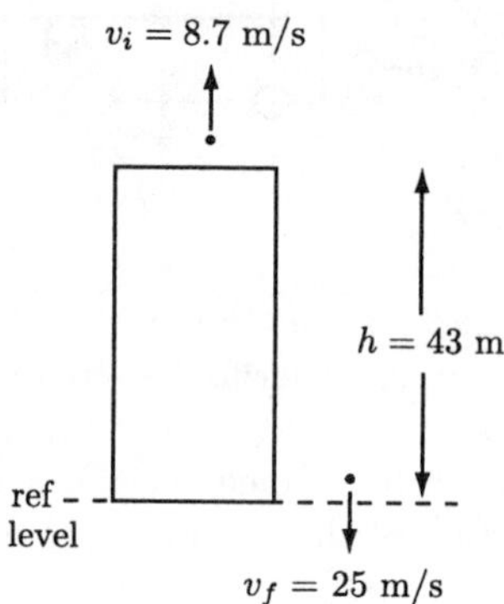

the initial TME is given by

$$\begin{aligned} E_i &= mgh + \frac{1}{2}mv_i^2 \\ &= (.25 \text{ kg})\left[(9.8 \text{ m/s}^2)(43 \text{ m}) + \frac{1}{2}(8.7 \text{ m/s})^2\right] \\ &= 115 \text{ J} \end{aligned}$$

The final TME of the ball is:

$$\begin{aligned} E_f &= \frac{1}{2}mv_f^2 \\ &= \frac{1}{2}(.25 \text{ kg})(25 \text{ m/s})^2 = 78 \text{ J} \\ \Delta E &= -40 \text{ J} \end{aligned}$$

(The final result has only one sig. fig.) The loss in TME is due to the (negative) work done on the ball by air resistance, a non-conservative, dissipative force.

8.81 We need to look at the total mechanical energy

	Before	After
PE:	mgh	0
KE:	$\frac{1}{2}mv_i^2$	$\frac{1}{2}mv_f^2$
Total:	$mgh+\frac{1}{2}mv_i^2$	$\frac{1}{2}mv_f^2$

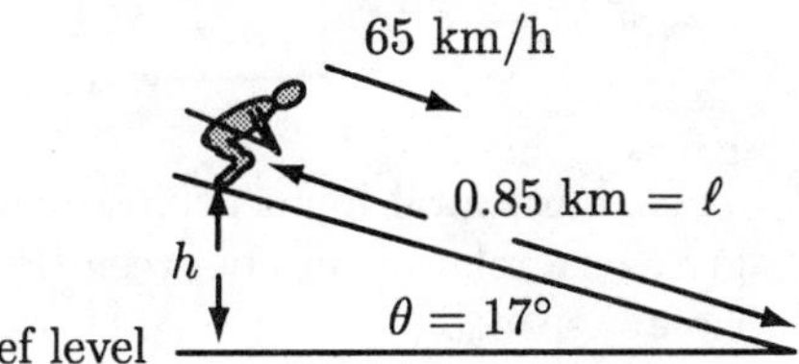

The change in the total energy equals the energy dissipated by friction and wind drag.

$$\begin{aligned} E_f - E_i &= \Delta E \\ &= -m\left[gh+\frac{1}{2}v_i^2-\frac{1}{2}v_f^2\right] \end{aligned}$$

where $h=\ell\sin\theta$ and

$$1\,\frac{\text{km}}{\text{h}}=\frac{10^3\text{ m}}{3600\text{ s}}=0.278\text{ m/s}$$

So

$$\begin{aligned} \Delta E &= -(87\text{ kg})\left[(9.8\text{ m/s}^2)(850\text{ m})\sin 17^\circ\right. \\ &\quad \left.+\frac{1}{2}(0.278\text{ m/s})^2\left(65^2-84^2\right)\right] \\ &= -2.0\times10^5\text{ J} \\ \Delta E &= \vec{\mathbf{f}}\cdot\vec{\mathbf{s}}=-f_{\text{av}}\ell \\ \Rightarrow f_{\text{av}} &= \frac{2.0\times10^5\text{ J}}{850\text{ m}}=238\text{ N} \end{aligned}$$

$$f_{\text{av}}(\text{friction + drag})=240\text{ N}$$

8.85 We can model the system as two point masses a distance of 3 km apart. The 'x' marks where you're standing.

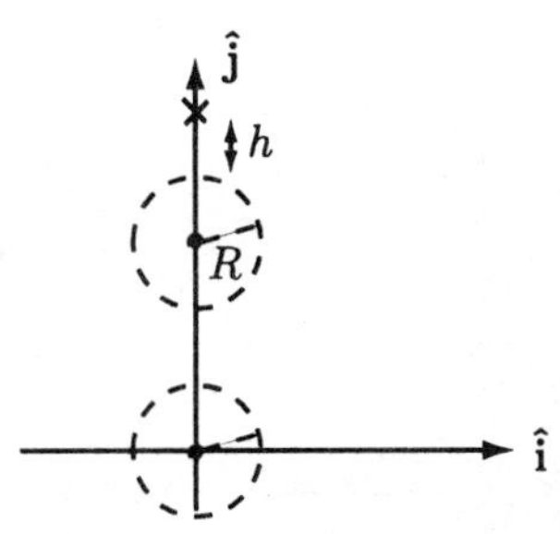

According to the exact formula,

$$\begin{aligned} U &= -GMm\left[\frac{1}{R+h}+\frac{1}{4R+h}\right] \\ &= -GMm\left[\frac{1}{R}\left(\frac{1}{1+\frac{h}{R}}\right)+\frac{1}{4R}\left(\frac{1}{1+\frac{h}{4R}}\right)\right] \end{aligned}$$

Using the binomial theorem,

$$(1+x)^\alpha=1+\alpha x+\frac{\alpha(\alpha-1)x^2}{2!}+\cdots$$

we can write the U as:

$$\begin{aligned} U &= -GMm\left[\frac{1}{R}\left(1-\frac{h}{R}+\frac{(-1)(-2)\left(\frac{h}{R}\right)^2}{2!}+\cdots\right)\right. \\ &\quad \left.+\frac{1}{4R}\left(1-\frac{h}{4R}+\frac{(-1)(-2)\left(\frac{h}{4R}\right)^2}{2!}+\cdots\right)\right] \end{aligned}$$

Neglecting second order terms + higher we can put U into the form

$$U=-\frac{GMm}{R}\left[\frac{5}{4}-\frac{17h}{16R}\right]$$

which is in the form of mgh + constant with $g=\frac{17GM}{16R^2}$. The approximation becomes not so good when the second order terms which we dropped, $\frac{h^2}{R^2}$, become significant compared to the first order term $\frac{h}{R}$. This happens when $h\approx R$ so that's the height at which the approximation starts to lose accuracy.

8.89 The initial energy of the system is $E_i=\frac{1}{2}ks_0^2$.

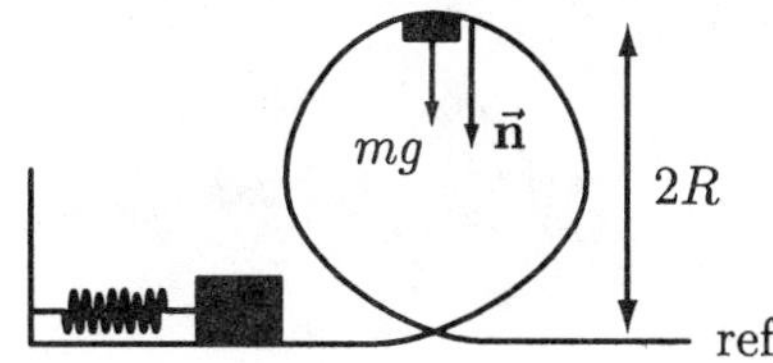

The TME of the system when the mass reaches the top of the loop is:

$$E_t=\frac{1}{2}mv_t^2+mg(2R)$$

We are told that the track is frictionless so we'll assume that TME is conserved

$$\begin{aligned} E_i &= E_t\Rightarrow\frac{1}{2}ks_0^2 \\ &= \frac{1}{2}mv_t^2+mg(2R) \\ \Rightarrow v_t &= \sqrt{\frac{ks_0^2}{m}-4gR} \end{aligned}$$

N2L places some restrictions on the mass's speed at the top of the loop. We have:

$$n + mg = \frac{mv_t^2}{R}$$

If we want to find the minimum speed, this is the speed at which the mass just barely stays on the track. In this situation, $n \to 0$ to give:

$$mg = \frac{mv_{t,\min}^2}{R} \Rightarrow v_{t,\min} = \sqrt{Rg}$$

Thus the minimum speed at the top of the loop which will let the mass stay on the track is $v_{t,\min} = \sqrt{Rg}$. To find the compression needed we can use this result in our expression for $v(s_0)$ and solve for s_0.

$$\sqrt{Rg} = \sqrt{\frac{ks_0^2}{m} - 4gR} \Rightarrow s_0 = \sqrt{\frac{5mRg}{k}}$$

8.93 According to Hooke's Law

$$\begin{aligned} F &= ks \\ U &= \frac{1}{2}ks^2 \end{aligned}$$

This bungee does not behave like a Hooke's Law spring—the more it is stretched the more it deviates (see appendix, Figure A). We can estimate k from F or U

i) k is the slope of the F vs. s curve

ii) ks is the slope of the U vs. s curve.

Using a spreadsheet, we may calculate these values

$$k_{\text{est1}}(N) = \frac{F(N+1) - F(N-1)}{S(N+1) - S(N-1)}$$

and

$$k_{\text{est2}}(N) = \frac{U(N+1) - U(N-1)}{S(N)(S(N+1) - S(N-1))}$$

The results are shown in Figure B in the appendix. The two estimates agree well at $k = 1140$ N/m when the cord's behavior is well represented by Hooke's Law (stretch ≤ 0.9 m). For larger stretches, the two estimates differ increasingly with $k_1 > k_2$.

8.97 The system starts with zero momentum; the only net external force that might be acting is the normal force acting on M exerted by the ground (we will not take the ground to be part of the system).

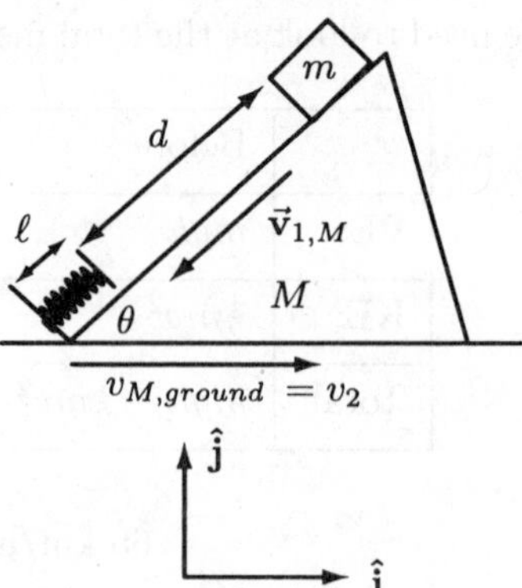

Thus, horizontal momentum is conserved. This gives us a relationship between the object's horizontal speeds.

$$\begin{aligned} p_i = p_f \quad &\Rightarrow \quad 0 = m\vec{v}_{1x} + M\vec{v}_{2x} \\ &\Rightarrow \quad m\vec{v}_{1x} = -M\vec{v}_{2x} \end{aligned} \tag{7}$$

If we take the ground to be the reference level the system starts out with a TME of $E_i = mgd\sin\theta$. Just before the block m reaches the spring, the system has a TME of

$$E_f = mg\ell\sin\theta + \frac{1}{2}mv_1^2 + Mv_2^2$$

Eq (7) suggests that we might want to work with components, so:

$$\begin{aligned} E_f = mg\ell + \sin\theta \\ \tfrac{1}{2}m\left(v_{1x}^2 + v_{1y}^2\right) + \tfrac{1}{2}Mv_{2x}^2 \end{aligned} \tag{8}$$

We have 2 equations and 3 unknowns. Our third equation comes from the fact that m is confined to slide on the slope of M. Relative motion from Chapter 3 allows us to express $\vec{v}_1$ as a combination of velocity *wrt* the M reference frame and velocity *wrt* the ground reference frame.

$$\vec{v}_1 = \vec{v}_{1,M} + \vec{v}_2$$

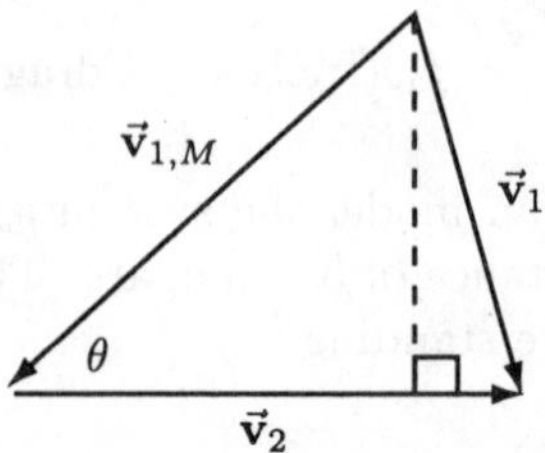

$$\begin{aligned} v_{1x} - v_{2x} &= v_{1x}, \text{wrt wedge} \\ v_{1y} - v_{2y} &= v_{1y} = v_{1y} \text{ wrt wedge} \\ \frac{v_x \text{ wrt wedge}}{v_y \text{ wrt wedge}} &= \cot\theta \end{aligned}$$

$$\frac{v_{1y}}{v_{1x} - v_{2x}} = \tan\theta = \frac{v_{1y}}{u + \frac{m}{M}u}$$

where $v_{1x} = u$ and $\frac{m}{M} = \alpha$; then:

$$mg\ell \sin\theta + \tfrac{1}{2}m\left(u^2 + u^2(1+\alpha)^2\tan^2\theta\right) + \tfrac{1}{2}M(\alpha u)^2 = mgd\sin\theta$$

$$\begin{aligned} mg(d-\ell)\sin\theta &= \frac{1}{2}mu^2\left[1+(1+\alpha)^2\tan^2\theta\right] \\ &\quad +\frac{1}{2}m\left(\frac{M}{m}\right)\alpha^2u^2 \\ &= \frac{1}{2}mu^2\left[1+(1+\alpha)^2\tan^2\theta+\alpha\right] \end{aligned}$$

So

$$u^2 = \frac{2g(d-\ell)\sin\theta}{1+(1+\alpha)^2\tan^2\theta+\alpha}$$

Then:

$$v^2 = v_{1x}^2 + v_{1y}^2 = u^2\left(1+(1+\alpha)^2\tan^2\theta\right)$$

So

$$\begin{aligned} v^2 &= \frac{2g(d-\ell)\sin\theta\left[1+(1+\alpha)^2\tan^2\theta\right]}{(1+\alpha)\left[1+(1+\alpha)\tan^2\theta\right]} \\ v &= \sqrt{\frac{2g(d-\ell)\sin\theta}{(1+\alpha)}\frac{\left[1+(1+\alpha)^2\tan^2\theta\right]}{\left[1+(1+\alpha)\tan^2\theta\right]}} \end{aligned}$$

Chapter 9

Angular Momentum

9.1 a) $L_O = rp\sin\phi$ where $\phi = \frac{\pi}{2}$ for a circular path,

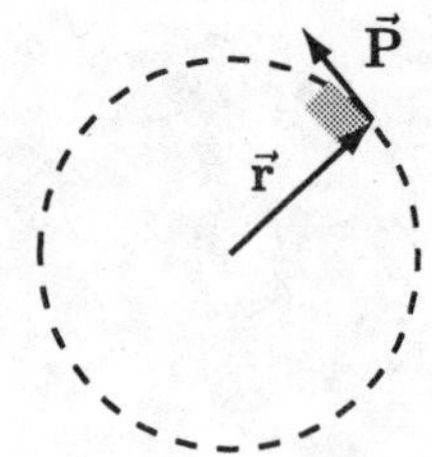

so,

$$\begin{aligned} L_{O,\text{car}} &= \left(1.0\times10^3\ \text{m}\right)(1000\ \text{kg})\left(55\ \frac{\text{km}}{\text{h}}\right) \\ &\cdot\sin\left(\frac{\pi}{2}\right)\left(\frac{10^3\ \text{m}}{1\ \text{km}}\right)\left(\frac{1\ \text{h}}{3600\ \text{s}}\right) \\ &= 1.5\times10^7\ \text{J}\cdot\text{s} \end{aligned}$$

(don't forget to convert to SI units)

b)

$$\begin{aligned} \frac{L_{O,\text{car}}}{L_{O,\text{truck}}} &= \frac{rp_{\text{car}}\sin\phi}{rp_{\text{truck}}\sin\phi} \\ &= \frac{rm_{\text{car}}v\sin\phi}{rm_{\text{truck}}v\sin\phi} \\ &= \frac{1000\ \text{kg}}{5000\ \text{kg}} = .20 \end{aligned}$$

Since the car and truck round the same turn at the same speed, the ratio of their angular momenta is the ratio of their masses. The direction of $\vec{\mathbf{L}}$ is the same for both.

9.3

$$\begin{aligned} \vec{\mathbf{a}}\times\vec{\mathbf{b}} &= \left[(5.0\ \text{m})\,\hat{\mathbf{i}}\right]\times\left[(4.3\ \text{m})\,\hat{\mathbf{k}}\right] \\ &= \left(22\ \text{m}^2\right)\left(\hat{\mathbf{i}}\times\hat{\mathbf{j}}\right) \\ &= -\left(22\ \text{m}^2\right)\hat{\mathbf{j}} \end{aligned}$$

9.5

$$\begin{aligned} \vec{\tau} &= \vec{\mathbf{r}}\times\vec{\mathbf{F}} = \left(\ell\hat{\mathbf{i}}\right)\times\left(-Mg\hat{\mathbf{j}}\right) \\ &= -Mg\ell\left(\hat{\mathbf{i}}\times\hat{\mathbf{j}}\right) = -Mg\ell\hat{\mathbf{k}} \end{aligned}$$

$\hat{\mathbf{j}}$

$\vec{\mathbf{r}} = 2m\hat{\mathbf{i}}$ $\hat{\mathbf{i}}$

$\vec{\mathbf{F}} = m - Mg\hat{\mathbf{j}}$

Using the given values,

$$\begin{aligned} \vec{\tau} &= -(10\ \text{kg})\left(9.81\ \text{m}/\text{s}^2\right)(2\ \text{m})\,\hat{\mathbf{k}} \\ &= -(200\ \text{N}\cdot\text{m})\,\hat{\mathbf{k}} \end{aligned}$$

9.7 After one of the people gets up the remaining person's weight force will exert a torque on the bench that will cause the bench to rotate. The rotation will be in the direction which dumps the remaining person to the ground; think of the bench as a see-saw.

9.9 From experience it is much easier to balance something from above. From the figures it can be seen that for the case where we balance from above, gravitation force produces a torque that places the center of mass back underneath the support. For the case of balancing from below the same torque serves to spin the object so that its center of mass is further from the support.

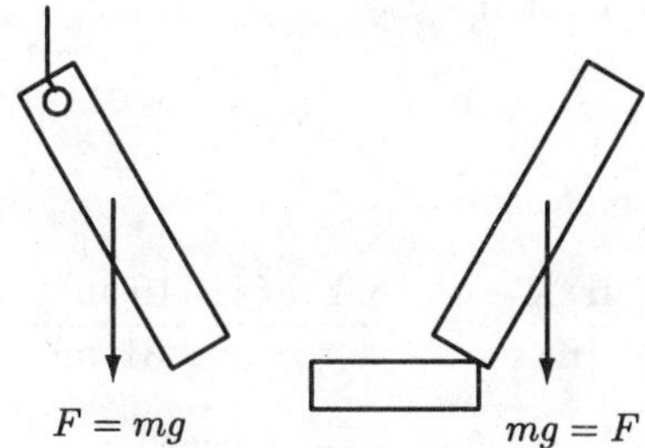

9.11 Using Eq. (9.16),

10	20	30	40	x (cm)
10 gm	20 gm	30 gm	40 gm	

$$\begin{aligned}\vec{\mathbf{r}}_{cm} &= \frac{(10\text{ g})(10\text{ cm})\hat{\mathbf{i}} + (20\text{ g})(20\text{ cm})\hat{\mathbf{i}} + (30\text{ g})(30\text{ cm})\hat{\mathbf{i}} + (40\text{ g})(40\text{ cm})\hat{\mathbf{i}}}{100\text{ g}} \\ &= (30\text{ cm})\hat{\mathbf{i}}\end{aligned}$$

9.13 Let the particle on the left be number 1 and the one on the right be number 2. Then

$$\begin{aligned}\vec{L}_{O,\text{sys}} &= \vec{\mathbf{L}}_1 + \vec{\mathbf{L}}_2 \\ &= \vec{\mathbf{r}}_1 \times \vec{\mathbf{p}}_1 + \vec{\mathbf{r}}_2 \times \vec{\mathbf{p}}_2 \\ &= \left[(-2.0\text{ m})\hat{\mathbf{i}}\right] \times \left[M(2.0\text{ m/s})\hat{\mathbf{j}}\right] \\ &\quad + \left[(1.0\text{ m})\hat{\mathbf{i}}\right] \times \left[M(2.0\text{ m/s})\hat{\mathbf{j}}\right] \\ &= M\left(-2.0\text{ m}^2/\text{s}\right)\left(\hat{\mathbf{i}} \times \hat{\mathbf{j}}\right) \\ &= (-1.0\text{ J}\cdot\text{s})\,\hat{\mathbf{k}}\end{aligned}$$

9.15 Let $L_{O,1}$ and $L_{O,2}$ be the angular momentum for the cars. If the cars are the same the masses are equal. If the turns are the same then $r_1 = r_2 = r$. Thus we can write:

$$\frac{L_{O,1}}{L_{O,2}} = \frac{rMv_1 \sin\theta}{rMv_2 \sin\theta} = \frac{v_1}{v_2} = \frac{55}{75} = 0.73$$

9.19 **a)** $\vec{\mathbf{L}}_O$ will be zero if $\vec{\mathbf{r}} \parallel \vec{\mathbf{v}}$, that is for any point 0 on a line through the particle's position parallel to its velocity.

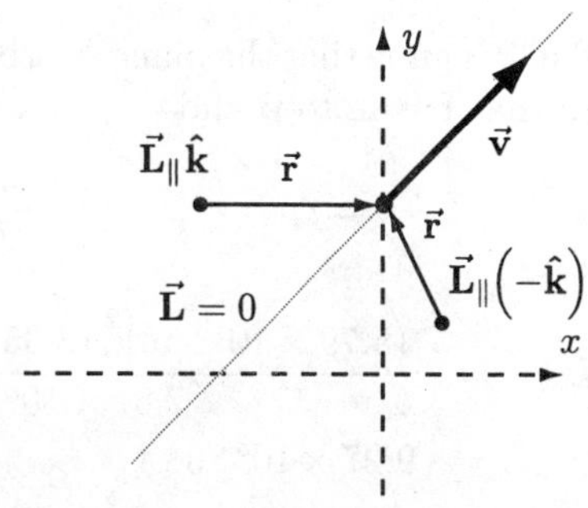

b) To have $\vec{\mathbf{r}} \times (m\vec{\mathbf{v}}) \parallel \hat{\mathbf{k}}$, $\vec{\mathbf{r}}$ must need a counter-clockwise rotation to become parallel to $\vec{\mathbf{v}}$ (right-hand rule). This is true for points 0 in the x-y plane above the $\vec{\mathbf{L}} = 0$ line.

c) For $\vec{\mathbf{r}} \times (m\vec{\mathbf{v}}) \parallel \left(-\hat{\mathbf{k}}\right)$, $\vec{\mathbf{r}}$ must rotate clockwise to become $\parallel \vec{\mathbf{v}}$. So 0 must be in the x-y plane below the $\vec{\mathbf{L}} = 0$ line.

d) For $\vec{\mathbf{L}}$ in the x-y plane, $\vec{\mathbf{r}}$ must be $\perp$ to the x-y plane. So 0 has to be on the line parallel to the z-axis through the particle's position.

e) If the particle follows a straight path at constant velocity, its angular momentum about any point is a constant.

9.25

$$\begin{aligned}\vec{\mathbf{A}} \times \vec{\mathbf{B}} &= (2.2\text{ m})\hat{\mathbf{j}} + (4.7\text{ m})\hat{\mathbf{k}} \\ &\quad \times \left[(1.3\text{ N}\cdot\text{s})\hat{\mathbf{i}} + (0.71\text{ N}\cdot\text{s})\hat{\mathbf{k}}\right] \\ &= (2.2\text{ m})(1.3\text{ N}\cdot\text{s})\left(\hat{\mathbf{j}} \times \hat{\mathbf{i}}\right) \\ &\quad + (2.2\text{ m})(0.71\text{ N}\cdot\text{s})\left(\hat{\mathbf{j}} \times \hat{\mathbf{k}}\right) \\ &\quad + (4.7\text{ m})(1.3\text{ N}\cdot\text{s})\left(\hat{\mathbf{k}} \times \hat{\mathbf{i}}\right) + 0 \\ &= \left[(2.9)\left(-\hat{\mathbf{k}}\right) + (1.6)\hat{\mathbf{i}} + (6.1)\hat{\mathbf{j}}\right]\text{J}\cdot\text{s}\end{aligned}$$

$\vec{\mathbf{C}}$ describes an angular momentum vector.

9.29

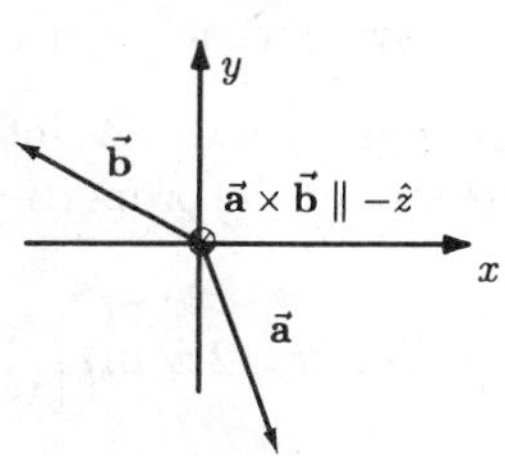

$$\begin{aligned}\vec{\mathbf{a}} \times \vec{\mathbf{b}} &= \left(\hat{\mathbf{i}} - 2\hat{\mathbf{j}}\right) \times \left(\hat{\mathbf{j}} - 2\hat{\mathbf{i}}\right) \\ &= \hat{\mathbf{i}} \times \hat{\mathbf{j}} - 0 - 0 + 4\hat{\mathbf{j}} \times \hat{\mathbf{i}} \\ &= 3\hat{\mathbf{j}} \times \hat{\mathbf{i}} = -3\hat{\mathbf{k}}\end{aligned}$$

9.33 We'll assume that the planet's orbits are circular so that $\vec{\mathbf{r}}$ is $\perp$ to $\vec{\mathbf{p}}$ and

$$L = rp = r^2 m w = \frac{r^2 m 2\pi}{T}$$

$$\begin{aligned}
L_M &= \frac{\left(5.79 \times 10^{10}\ \text{m}\right)^2 \left(3.35 \times 10^{23}\ \text{kg}\right) 2\pi}{7.61 \times 10^6\ \text{s}} \\
&= 9.27 \times 10^{38}\ \text{J} \cdot \text{s} \\
L_V &= \frac{\left(1.08 \times 10^{11}\ \text{m}\right)^2 \left(4.87 \times 10^{24}\ \text{kg}\right) 2\pi}{1.94 \times 10^7\ \text{s}} \\
&= 1.84 \times 10^{40}\ \text{J} \cdot \text{s} \\
L_E &= \frac{\left(1.50 \times 10^{11}\ \text{m}\right)^2 \left(5.98 \times 10^{24}\ \text{kg}\right) 2\pi}{3.16 \times 10^7\ \text{s}} \\
&= 2.67 \times 10^{40}\ \text{J} \cdot \text{s} \\
L_M &= \frac{\left(2.28 \times 10^{11}\ \text{m}\right)^2 \left(6.40 \times 10^{23}\ \text{kg}\right) 2\pi}{5.93 \times 10^7\ \text{s}} \\
&= 3.52 \times 10^{39}\ \text{J} \cdot \text{s} \\
L_J &= \frac{\left(7.78 \times 10^{11}\ \text{m}\right)^2 \left(1.90 \times 10^{27}\ \text{kg}\right) 2\pi}{3.74 \times 10^8\ \text{s}} \\
&= 1.93 \times 10^{43}\ \text{J} \cdot \text{s} \\
L_S &= \frac{\left(1.43 \times 10^{12}\ \text{m}\right)^2 \left(5.69 \times 10^{26}\ \text{kg}\right) 2\pi}{9.30 \times 10^8\ \text{s}} \\
&= 7.86 \times 10^{42}\ \text{J} \cdot \text{s} \\
L_U &= \frac{\left(2.87 \times 10^{12}\ \text{m}\right)^2 \left(8.67 \times 10^{25}\ \text{kg}\right) 2\pi}{2.65 \times 10^9\ \text{s}} \\
&= 1.69 \times 10^{42}\ \text{J} \cdot \text{s} \\
L_N &= \frac{\left(4.50 \times 10^{12}\ \text{m}\right)^2 \left(1.02 \times 10^{26}\ \text{kg}\right) 2\pi}{5.20 \times 10^9\ \text{s}} \\
&= 2.49 \times 10^{42}\ \text{J} \cdot \text{s} \\
L_P &= \frac{\left(5.9 \times 10^{12}\ \text{m}\right)^2 \left(1.2 \times 10^{22}\ \text{kg}\right) 2\pi}{7.83 \times 10^9\ \text{s}} \\
&= 3.9 \times 10^{38}\ \text{J} \cdot \text{s}
\end{aligned}$$

$$\sum_{i=\text{mercury}}^{\text{pluto}} L_i = 3.14 \times 10^{43}\ \text{J} \cdot \text{s}$$

Clearly, Jupiter contributes the most (61 %) to the solar system's angular momentum.

9.37 The cross product can be interpreted as an area. See the section on properties of the cross product for details.

$$\vec{\mathbf{a}} \times \vec{\mathbf{b}} = (3.2\ \text{m})\,(2.6\ \text{m}) \left[\left(\hat{\mathbf{i}} + 2\hat{\mathbf{j}} - \hat{\mathbf{k}}\right) \times \left(\hat{\mathbf{i}} - \hat{\mathbf{j}} + \hat{\mathbf{k}}\right)\right]$$

$$\begin{aligned}
&= \left(8.3\ \text{m}^2\right) \begin{vmatrix} \hat{\mathbf{i}} & \hat{\mathbf{j}} & \hat{\mathbf{k}} \\ 1 & 2 & -1 \\ 1 & -1 & 1 \end{vmatrix} \\
&= \left(8.3\ \text{m}^2\right) \left[\hat{\mathbf{i}}\,(2-1) - \hat{\mathbf{j}}\,(1+1) + \hat{\mathbf{k}}\,(-1-2)\right] \\
&= \left(8.3\ \text{m}^2\right) \left(+\hat{\mathbf{i}} - 2\hat{\mathbf{j}} - 3\hat{\mathbf{k}}\right)
\end{aligned}$$

The area of the parallelogram is given by $\left|\vec{\mathbf{a}} \times \vec{\mathbf{b}}\right|$.

$$\left|\vec{\mathbf{a}} \times \vec{\mathbf{b}}\right| = \left(8.3\ \text{m}^2\right) \sqrt{(1+4+9)} = 31\ \text{m}^2$$

We can use the dot product to find θ, the angle between the normal vector $\vec{\mathbf{n}}$, and the $\hat{\mathbf{k}}$-direction.

$$\vec{\mathbf{n}} \cdot \hat{\mathbf{k}} = |\vec{\mathbf{n}}| \left|\hat{\mathbf{k}}\right| \cos\theta = |\vec{\mathbf{n}}| \cos\theta$$

so

$$\begin{aligned}
\frac{\vec{\mathbf{n}} \cdot \hat{\mathbf{k}}}{|\vec{\mathbf{n}}|} &= \frac{8.3\ \text{m}^2\,\hat{\mathbf{i}} - 16\ \text{m}^2\,\hat{\mathbf{j}} - 25\ \text{m}^2\,\hat{\mathbf{k}}}{31\ \text{m}^2} \cdot \hat{\mathbf{k}} \\
&= -\frac{25}{31} = \cos\theta
\end{aligned}$$

so

$$\theta = 90^\circ + \cos^{-1}\left(\frac{25}{31}\right) = 126^\circ$$

9.41 We will begin by plugging in the expression for $\vec{\mathbf{r}}$ into $\vec{\mathbf{L}}_O = \vec{\mathbf{r}} \times \vec{\mathbf{p}}$

$$\vec{\mathbf{L}}_O = \left(\vec{\mathbf{r}}_0 \times \vec{\mathbf{v}}_0 t\right) \times m\vec{\mathbf{v}}_0 = m\left(\vec{\mathbf{r}}_0 \times \vec{\mathbf{v}}_0\right)$$

Since $\vec{\mathbf{v}}_0$ is $\parallel$ to itself, $\vec{\mathbf{v}}_0 \times \vec{\mathbf{v}}_0 = 0$. Note that time dropped out of the expression for $\vec{\mathbf{L}}_O$. This explicitly shows that in this case $\vec{\mathbf{L}}_O$ is constant in time. When $\vec{\mathbf{r}}_0$ and $\vec{\mathbf{v}}_0$ are $\perp$ to each other,

$$\left|\vec{\mathbf{r}}_0 \times \vec{\mathbf{v}}_0\right| = r_0 v_0 \sin\left(\frac{\pi}{2}\right) = r_0 v_0$$

and

$$\left|\vec{\mathbf{L}}_O\right| = m r_0 v_0$$

The figure below shows a particle following an arbitrary path given by $\vec{\mathbf{r}} = \vec{\mathbf{r}}_0 + \vec{\mathbf{v}}_0 t$.

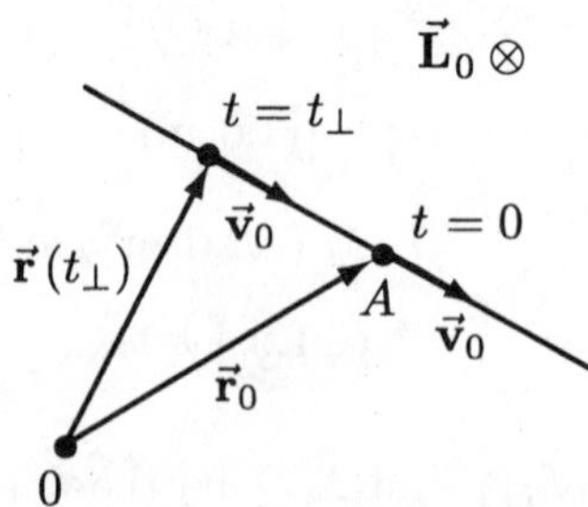

At $t = 0$ the particle is at point A, at position $\vec{\mathbf{r}} = \vec{\mathbf{r}}_0$. The vector $\vec{\mathbf{r}}\,(t_\perp)$ is $\perp$ to $\vec{\mathbf{v}}_0$. At this point, $t = t_\perp$. Since $\vec{\mathbf{r}}\,(t_\perp)$ is $\perp$ to $\vec{\mathbf{v}}_0$ at this point their dot product is 0.

$$\begin{gathered}
\left(\vec{\mathbf{r}}_0 + t_\perp \vec{\mathbf{v}}_0\right) \cdot \vec{\mathbf{v}}_0 = 0 \\
\vec{\mathbf{r}}_0 \cdot \vec{\mathbf{v}}_0 + v_0^2 t_\perp = 0 \\
t_\perp = -\frac{\vec{\mathbf{r}}_0 \cdot \vec{\mathbf{v}}_0}{v_0^2} = -\frac{\vec{\mathbf{r}}_0 \cdot \hat{\mathbf{v}}_0}{v_0}
\end{gathered}$$

Then, from the definition of the cross product,

$$\left|\vec{\mathbf{L}}_O\right| = mr\,(t_\perp)\,v_0$$

The angular momentum of a particle following a straight line at constant speed is not only constant in time but seems to depend only on the $\perp$ distance between the origin and the path. This is the distance between the particle and origin when the particle is at the closest point to the origin We typically call this distance $r_\perp$.

9.45

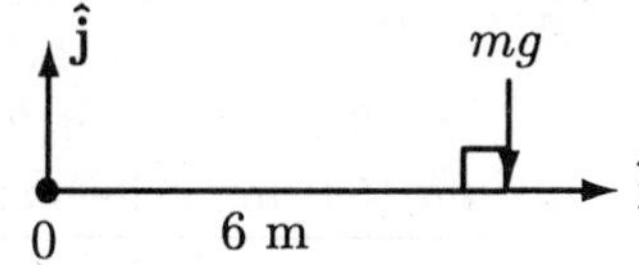

$$\begin{aligned}\vec{\tau} &= \vec{\mathbf{r}}\times\vec{\mathbf{F}}\\ &= (6\text{ m})\,\hat{\mathbf{i}}\times(80\text{ kg})\,(9.81\,\text{m}/\text{s}^2)\left(-\hat{\mathbf{j}}\right)\\ &= 4.7\times10^3\text{ J}\cdot\text{s}\left(\hat{\mathbf{i}}\times-\hat{\mathbf{j}}\right)\\ &= -\left(5\times10^3\text{ J}\cdot\text{s}\right)\hat{\mathbf{k}}\end{aligned}$$

9.49 The torque exerted on the food is due to the power output of the mixer, not the power input to the machine. We have a convenient relationship between P_I and P_O, efficiency.

$$\begin{aligned}P_O &= eP_I = .95\,(270\text{ W}) = 256\text{ W}\\ \tau &= \frac{P_O}{\omega} = \frac{256\text{ W}}{2.0\text{ cycles}/\text{s}}\left(\frac{1\text{ cycles}}{2\pi}\right)\\ &= 20\text{ N}\cdot\text{m}\end{aligned}$$

9.53 To lift the block of mass m you must apply a torque τ_1 to the wheel which is equal in magnitude but opposite in direction to the torque, τ_2, which is produced by the force that gravity exerts on the block.

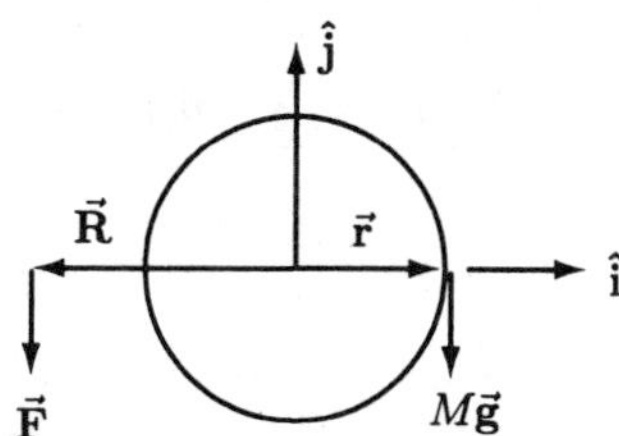

Assuming $\vec{\mathbf{F}}$ is always $\perp$ to $\vec{\mathbf{R}}$,

$$\begin{aligned}\vec{\tau}_1 &= R\hat{\mathbf{r}}\times F\hat{\phi} = RF\hat{\mathbf{k}}\\ \vec{\tau}_2 &= r\hat{\mathbf{i}}\times\left(-Mg\hat{\mathbf{j}}\right) = -rMg\hat{\mathbf{k}}\end{aligned}$$

Lifting the block without acceleration requires

$$\vec{\tau}_1 + \vec{\tau}_2 = 0$$

$$RF\hat{\mathbf{k}} - rMg\hat{\mathbf{k}} = 0 \Rightarrow F = \frac{rMg}{R}$$

The power required is

$$P = \tau w = rMgw$$

Gravity exerts a force Mg in the direction opposite to the block's motion over some distance Δs over a time Δt.

$$P_{\text{grav}} = -\frac{Mg\Delta s}{\Delta t} = -\frac{Mgr\Delta\theta}{\Delta t} = -Mgrw$$

9.57 The initial angular momentum for the system of two masses in circular motion is

$$\vec{\mathbf{L}}_i = 2\left(\ell^2 m\omega_0\hat{\mathbf{k}}\right),$$

since each contributes the same amount to the total. The frictional force acting on one of the masses is $\vec{\mathbf{f}}_k = -\mu mg\hat{\phi}$, opposite in direction to $\vec{\mathbf{v}} = v_0\hat{\phi}$. The torque produced by the frictional force on one of the masses is

$$\vec{\tau} = \vec{\mathbf{r}}\times\vec{\mathbf{f}}_k = \ell\hat{\mathbf{r}}\times(-\mu mg)\,\hat{\phi} = -\ell\mu mg\hat{\mathbf{k}}$$

The net torque on the system (due to friction acting on both particles is

$$\vec{\tau}_{\text{net}} = -2\ell\mu mg\hat{\mathbf{k}}$$

This constant torque produces a constant rate of change in the angular momentum, so

$$\vec{\mathbf{L}} = \vec{\mathbf{L}}_i - \vec{\tau}\Delta t$$

which is zero when

$$2\ell^2 m\omega_0\hat{\mathbf{k}} - 2\ell\mu mg\Delta t\hat{\mathbf{k}} = 0$$

Solving for Δt yields

$$\Delta t = \frac{\ell\omega_0}{\mu g}$$

After this time the dumbbell has no angular momentum; it is stationary.

b) The energy of the dumbbell at $t = 0$ is

$$K_i = \frac{1}{2}\,(2m)\,\omega_0^2\ell^2 = m\omega_0^2\ell^2$$

The power dissipated by friction is

$$|\vec{\tau}|\,\omega = 2\mu_k mg\ell\omega$$

Careful—this power is not constant! Friction is indeed constant but ω is not; the system has an angular acceleration. We can get an expression for $\alpha = \frac{d\omega}{dt}$ by noting that

$$\begin{aligned} \frac{dE_{sys}}{dt} &= -P_{f_k} \\ \Rightarrow \frac{d}{dt}\left(m\omega^2\ell^2\right) &= -2\mu mg\ell\omega \\ \Rightarrow \frac{d}{dt}\omega^2 &= -\frac{2\mu_k g\omega}{\ell} \\ \Rightarrow 2\omega\frac{d\omega}{dt} &= -\frac{2\mu_k\omega g}{\ell} \\ \alpha &= \frac{d\omega}{dt} = -\frac{\mu_k g}{\ell} \end{aligned}$$

If the system starts at ω_0,

$$\begin{aligned} \omega(t) &= \omega_0 + \alpha t \\ &= \omega_0 - \frac{\mu_k g}{\ell}t \stackrel{\text{set}}{\equiv} 0 \\ \Rightarrow t &= \frac{\omega_0\ell}{\mu_k g} \end{aligned}$$

The same time we calculated earlier, as expected.

9.61 The motion could be due to a large "planet" orbiting the star. The star-planet system would be isolated, with the objects interacting by gravitational force. The observed motion could result from the star along with a massive companion traveling through space and orbiting about the system CM as they go—in a very similar fashion as do the skaters in study problem #8 in the text.

9.65 We would support the stick underneath its CM. We can safely assume that the meter stick has a uniform mass distribution so its CM without the weights is at its geometric center at $\frac{1}{2}$ m.

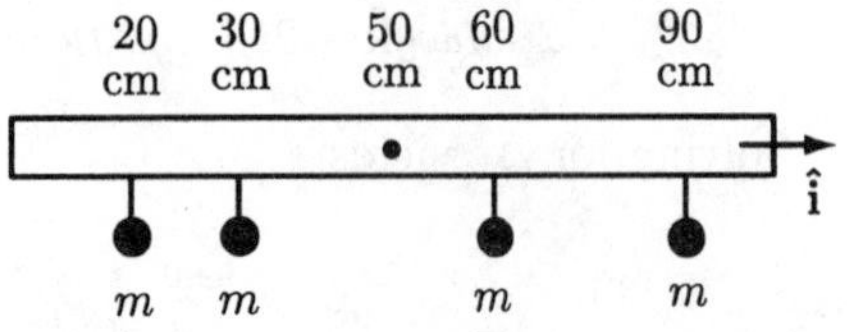

20 cm, 30 cm, 50 cm, 60 cm, 90 cm; m m m m; $\hat{\mathbf{i}}$

We will use the point $x = 0$ to be an origin with which to compute CM. Note that if we placed an origin on one of the masses we would have one less calculation to do (since the mass would be located at $x = 0$) but it's easier to see what's going on this way.

$$\text{CM} = \frac{(20\text{ cm})\,m + (30\text{ cm})\,m + (50\text{ cm})\,m + (60\text{ cm})\,m + (90\text{ cm})\,m}{5m} = 50\text{ cm}$$

the CM lies at the geometric center of the ruler. Thus you should support the stick at its center.

9.69 The first step is to convert to the CM reference frame.

$$\begin{aligned} \vec{\mathbf{r}}_{\text{CM}} &= \frac{\sum m_i\vec{\mathbf{r}}_i}{\sum m_i} = \frac{1}{4}\left(\vec{\mathbf{r}}_1 + \vec{\mathbf{r}}_2 + \vec{\mathbf{r}}_3 + \vec{\mathbf{r}}_4\right) \\ &= \frac{\left[\begin{array}{c}\left(3\hat{\mathbf{i}} + 3\hat{\mathbf{j}}\right) + \left(3\hat{\mathbf{i}} + \frac{1}{2}\hat{\mathbf{j}}\right) \\ + \left(\frac{1}{2}\hat{\mathbf{i}} + \frac{1}{2}\hat{\mathbf{j}}\right) + \left(\frac{1}{2}\hat{\mathbf{i}} + 3\hat{\mathbf{j}}\right)\end{array}\right](1\text{ m})}{4} \\ &= \left(\frac{7}{4}\text{ m}\right)\hat{\mathbf{i}} + \left(\frac{7}{4}\text{ m}\right)\hat{\mathbf{j}} \end{aligned}$$

$$\begin{aligned} \vec{\mathbf{v}}_{\text{CM}} &= \frac{\begin{array}{c} m\,(3\text{ m/s})\,\hat{\mathbf{j}} - m\,(2.0\text{ m/s})\,\hat{\mathbf{j}} \\ +m\,(5.0\text{ m/s})\,\hat{\mathbf{i}} - m\,(3.0\text{ m/s})\,\hat{\mathbf{i}}\end{array}}{4m} \\ &= \left(\frac{1}{2}\text{ m/s}\right)\hat{\mathbf{i}} + \left(\frac{1}{4}\text{ m/s}\right)\hat{\mathbf{j}} \end{aligned}$$

The total kinetic energy is

$$\begin{aligned} K &= \sum K_i \\ &= \frac{1}{2}(1.0\text{ kg})\cdot\Big[(3.0\text{ m/s})^2 + (5.0\text{ m/s})^2 \\ &\quad + (3.0\text{ m/s})^2 + (2.0\text{ m/s})^2\Big] \\ &= 23.5\text{ J} \end{aligned}$$

The CM kinetic energy is

$$\begin{aligned} K_{\text{CM}} &= \frac{1}{2}Mv_{\text{CM}}^2 \\ &= \frac{1}{2}(4.0\text{ kg})\left(\frac{1}{4} + \frac{1}{16}\right)(\text{m/s})^2 \\ &= 0.625\text{ J} \end{aligned}$$

The internal kinetic energy is

$$K_{\text{int}} = K - K_{\text{CM}} = 22.9\text{ J}$$

The total angular momentum is

$$\begin{aligned} \vec{\mathbf{L}} &= \sum\vec{\mathbf{L}}_i = \sum m_i\vec{\mathbf{r}}_i \times \vec{\mathbf{v}}_i \\ &= (1.0\text{ kg})\left[\begin{array}{l} (3.0\text{ m})\,(3.0\text{ m/s})\,\hat{\mathbf{k}} \\ +\left(\frac{1}{2}\text{ m}\right)(5.0\text{ m/s})\left(-\hat{\mathbf{k}}\right) \\ +\left(\frac{1}{2}\text{ m}\right)(3.0\text{ m/s})\,\hat{\mathbf{k}} \\ +\left(\frac{1}{2}\text{ m}\right)(2.0\text{ m/s})\left(-\hat{\mathbf{k}}\right)\end{array}\right] \\ &= (7.0\text{ J}\cdot\text{s})\,\hat{\mathbf{k}} \end{aligned}$$

The CM angular momentum is

$$\vec{\mathbf{L}}_{\text{CM}} = M\vec{\mathbf{r}}_{\text{CM}} \times v_{\text{CM}}$$

$$= (4.0\text{ kg})\left[\left(\frac{7}{4}\text{ m}\right)\left(\hat{\mathbf{i}}+\hat{\mathbf{j}}\right)\right] \times\left[\left(\frac{1}{4}\text{ m/s}\right)\left(2\hat{\mathbf{i}}+\hat{\mathbf{j}}\right)\right]$$
$$= \left(\frac{7}{4}\text{ J}\cdot\text{s}\right)[-2+1]\left(\hat{\mathbf{k}}\right)$$
$$= -\left(\frac{7}{4}\text{ J}\cdot\text{s}\right)\hat{\mathbf{k}}$$

The internal angular momentum is

$$\vec{\mathbf{L}}_{\text{int}} = \vec{\mathbf{L}} - \vec{\mathbf{L}}_{\text{CM}} = (8.8\text{ J}\cdot\text{s})\,\hat{\mathbf{k}}$$

9.73 The uniform wire of length L is bent to form three sides of a square. Each side is of length $\frac{1}{3}L$.

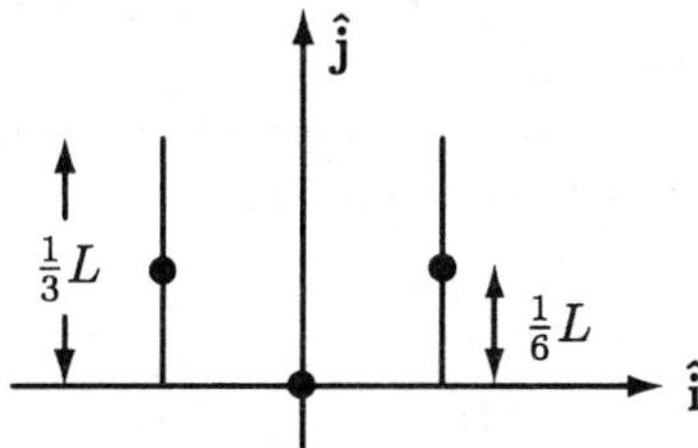

The C of each individual side must lie at the geometric center of the individual side and represents $\frac{1}{3}$ of the wire's total mass, the C of the points (and thus the $\frac{3}{4}$ square) lies at the geometric center of the three points. Since the two point masses are at equal distances from the y-axis their contributions towards x_{CM} cancel; the point mass on the x-axis is at $x = 0$ so $x_{\text{CM}} = 0$ m. We will now compute y_{CM}.

$$y_{\text{CM}} = \frac{\left(\frac{1}{6}L\right)\left(\frac{1}{3}M\right) + \left(\frac{1}{6}L\right)\left(\frac{1}{3}M\right) + (0L)\left(\frac{1}{3}M\right)}{M} = \frac{1}{9}L$$

Which is a distance

$$\frac{1}{6}L - \frac{1}{9}L = \frac{1}{18}L$$

below the square's geometric center.

9.77 The first thing we need to do is break down the velocities into components.

$$\vec{\mathbf{v}}_A = -(100\text{ m/s})\,\hat{\mathbf{i}} - (2.5\text{ m/s})\,\hat{\mathbf{k}}$$
$$\vec{\mathbf{v}}_B = (100\text{ m/s})\,\hat{\mathbf{j}} + (2.5\text{ m/s})\,\hat{\mathbf{k}}$$

The velocity of the C is:

$$\vec{\mathbf{v}}_{\text{CM}} = \frac{\sum m_i\vec{\mathbf{v}}_i}{\sum m_i}$$
$$= \frac{(3000\text{ kg})(-100\text{ m/s}\,\hat{\mathbf{i}} - 2.5\text{ m/s}\,\hat{\mathbf{k}}) + (4000\text{ kg})\left(100\text{ m/s}\,\hat{\mathbf{j}} + 2.5\text{ m/s}\,\hat{\mathbf{k}}\right)}{3000\text{ kg} + 4000\text{ kg}}$$
$$= -40\text{ m/s}\,\hat{\mathbf{i}} + 60\text{ m/s}\,\hat{\mathbf{j}} + 0.4\text{ m/s}\,\hat{\mathbf{k}}$$

9.81 The satellite begins to rotate. By Newton's third law, the satellite exerts an equal and opposite torque on the astronaut, who rotates in the opposite direction.

9.85 The velocity of the skaters is given by:

$$\vec{\mathbf{v}}_{\text{tot}} = \vec{\mathbf{v}}_{\text{icm}} + \vec{\mathbf{v}}_{\text{CM}}$$
$$\vec{\mathbf{v}}_{58,\text{tot}} = \omega' r_{58}\hat{\mathbf{i}} + \vec{\mathbf{v}}_{\text{CM}}$$
$$\vec{\mathbf{v}}_{82,\text{tot}} = -\omega' r_{82}\hat{\mathbf{i}} + \vec{\mathbf{v}}_{\text{CM}}$$

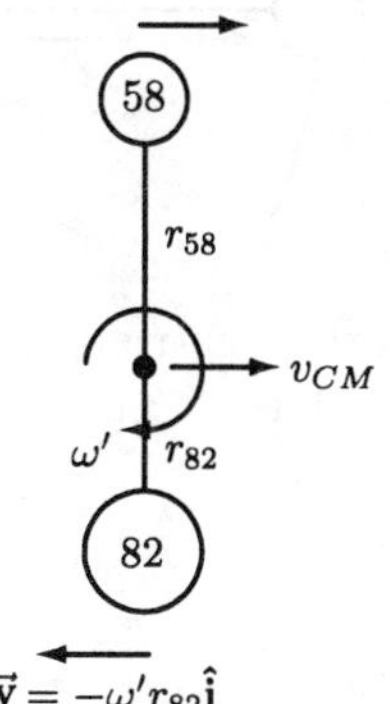

Using the given numbers,

$$\vec{\mathbf{v}}_{58,\text{tot}} = (8.0\text{ /s})(0.59\text{ m})\,\hat{\mathbf{i}} + (0.17\text{ m/s})\,\hat{\mathbf{i}}$$
$$= (4.9\text{ m/s})\,\hat{\mathbf{i}}$$
$$\vec{\mathbf{v}}_{82,\text{tot}} = -(8.0\text{ /s})(0.41\text{ m})\,\hat{\mathbf{i}} + (0.17\text{ m/s})\,\hat{\mathbf{i}}$$
$$= -(3.1\text{ m/s})\,\hat{\mathbf{i}}$$

9.89 The explosion occurs rapidly, and the satellite's momentum is conserved. Since half the satellite is brought to rest with respect to Earth, the other half ends up with twice its initial velocity. In a circular orbit, the satellite's speed before the explosion is found from Newton's Second Law:

$$\frac{GMm_{\text{sat}}}{r^2} = \frac{m_{\text{sat}}v^2}{r} \Rightarrow v^2 = \frac{GM}{r}$$

The speed of the non-stationary satellite half, after the explosion is

$$v_{\text{after}} = 2v_{\text{before}} = 2\sqrt{\frac{GM}{r}}$$

So, its total mechanical energy is

$$E_{\text{tot}} = K + U_{\text{grav}}$$
$$= \frac{1}{2}m_{\text{sat}}\left[2\sqrt{\frac{GM}{r}}\right]^2 + \frac{-GMm_{\text{sat}}}{r}$$
$$= \frac{+GMm_{\text{sat}}}{r} > 0$$

With positive total energy, the satellite half escapes from the Earth.

9.93 a)

$$M = 3.6\times10^5 \text{ kg}$$
$$\ell = 1.0 \text{ km}$$
$$r = \frac{\ell}{2}$$

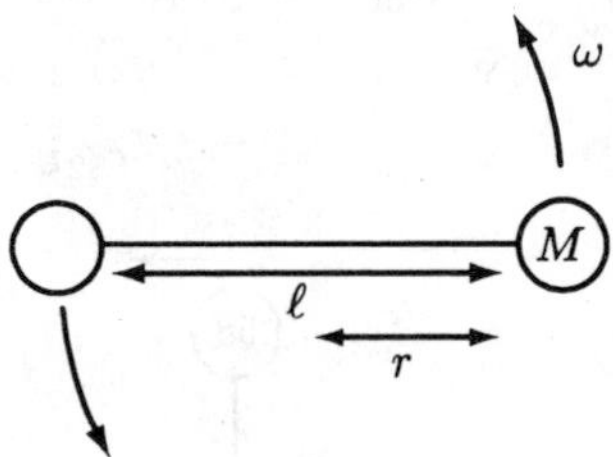

We need the centripetal acceleration $\omega^2 r$ to equal 1 g. Thus

$$\omega^2 = \frac{g}{r}$$
$$\omega = \sqrt{\frac{g}{r}} = \sqrt{\frac{9.8 \text{ m/s}^2}{\frac{(10^3 \text{ m})}{2}}}$$
$$= 0.14 \text{ rad/s}$$

The system's angular momentum is

$$L = 2m\omega r^2 = 2m\omega\left(\frac{\ell}{2}\right)^2$$
$$= 2\left(3.6\times10^5 \text{ kg}\right)(0.14 \text{ rad/s})\left(\frac{1.0\times10^3 \text{ m}}{2}\right)^2$$
$$= 2.5\times10^{10} \text{ J}\cdot\text{s}$$

b) The remaining spacecraft now has a total mass $\frac{3M}{2}$.

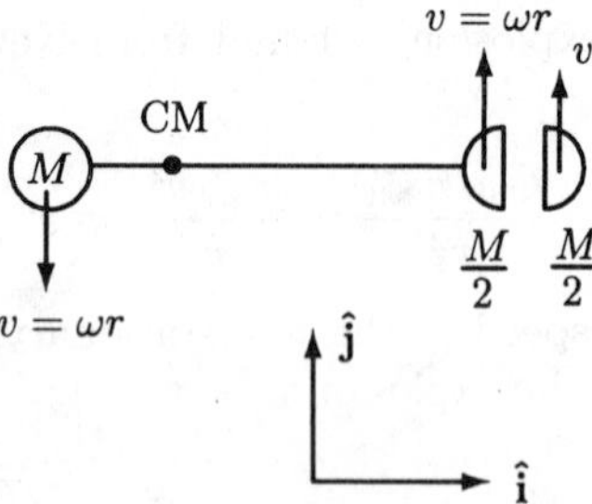

Its C is at a distance from the intact module of

$$x_{\text{CM}} = \frac{M(0) + \ell\left(\frac{M}{2}\right)}{3\frac{M}{2}} = \frac{\ell}{3}$$

The velocity of the C is

$$\vec{\mathbf{v}}_{\text{CM}} = \frac{-\frac{\omega\ell}{2}M\hat{\mathbf{j}} + \frac{M}{2}\frac{\omega\ell}{2}\hat{\mathbf{j}}}{3\frac{M}{2}} = -\frac{\omega\ell}{6}\hat{\mathbf{j}}$$

The rotational speed about the C is found by computing the velocity of each module in the C frame

$$\vec{\mathbf{v}}_{M,\text{CM}} = -\frac{\omega\ell}{2}\hat{\mathbf{j}} - \left(-\frac{\omega\ell}{6}\hat{\mathbf{j}}\right) = -\frac{\omega\ell}{3}\hat{\mathbf{j}}$$
$$\vec{\mathbf{v}}_{\frac{M}{2},\text{CM}} = \frac{\omega\ell}{2}\hat{\mathbf{j}} - \left(-\frac{\omega\ell}{6}\hat{\mathbf{j}}\right) = \frac{2}{3}\omega\ell\hat{\mathbf{j}}$$

Thus

$$\omega = \frac{v_{M,\text{CM}}}{d_{M,\text{CM}}} = \frac{\frac{\omega\ell}{3}}{\frac{\ell}{3}} = \omega \qquad (1)$$

and

$$\omega = \frac{v_{\frac{M}{2},\text{CM}}}{d_{\frac{M}{2},\text{CM}}} = \frac{\frac{2\omega\ell}{3}}{\frac{2\ell}{3}} = \omega \qquad (2)$$

The results of (1) and (2) are consistent. The new system continues to rotate with angular speed ω about its C . The system's C moves at speed $\frac{\omega\ell}{6}$.

The new angular momentum *about the old origin* has several components.

1. Separated piece

$$\vec{\mathbf{L}} = \frac{M}{2}\cdot\frac{\ell}{2}\cdot\frac{\vec{\omega}\ell}{2} = \frac{M\omega\ell^2}{8}\hat{\mathbf{k}}$$

2. Internal angular momentum of the remaining spacecraft

$$\vec{\mathbf{L}} = M\vec{\omega}\left(\frac{\ell}{3}\right)^2 + \frac{M}{2}\vec{\omega}\left(\frac{2\ell}{3}\right)^2$$
$$= \frac{M\omega\ell^2}{3}\hat{\mathbf{k}}$$

3. Angular momentum of spacecraft's C about original origin:

$$\vec{\mathbf{L}} = \frac{3M}{2}\left(\frac{\ell}{2} - \frac{\ell}{3}\right)\left(\frac{\vec{\omega}\ell}{6}\right)$$
$$= \frac{M\omega\ell^2}{24}\hat{\mathbf{k}}$$

The total is

$$\vec{\mathbf{L}} = M\vec{\omega}\ell^2\left(\frac{1}{8} + \frac{1}{3} + \frac{1}{24}\right)$$
$$= \frac{M\omega\ell^2}{2}\hat{\mathbf{k}}$$

the same as before the separation.

9.103 By definition

$$\vec{\tau} = \int \vec{\mathbf{r}} \otimes dm\vec{\mathbf{g}}$$

Suppose we were talking about a line so that $\vec{\mathbf{r}} = r\hat{\ell}$ and $\vec{\mathbf{g}} = g(h)\,\hat{g}$. Then,

$$\vec{\tau} = \int \vec{\mathbf{r}} \otimes dm\vec{\mathbf{g}} = \left(\hat{\ell}\otimes\hat{g}\right)\int r\,dmg \qquad (3)$$

$\vec{\mathbf{r}}_{\mathrm{CG}}$ should be the point where, if you put a support under it, there would be no net torque acting on the body. So we would like

$$\vec{\tau} = \vec{\mathbf{r}}_{\mathrm{CG}} \otimes \int dm\vec{\mathbf{g}} = r_{\mathrm{CG}}\left(\hat{\ell} \otimes \hat{g}\right) \int dmg \quad (4)$$

Relating (3) and (4),

$$\vec{\mathbf{r}}_{\mathrm{CG}} = \frac{\int r\vec{\mathbf{g}}\,(h)\,dm}{\int dm\,g} \qquad (5)$$

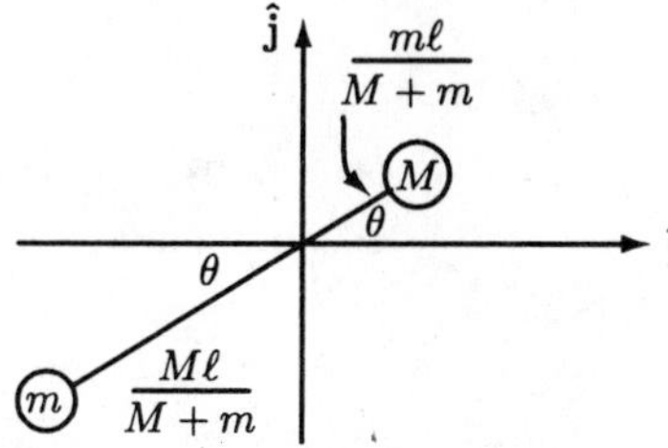

Place the origin of the coordinate system at the C . We will use Eq. (5) to calculate $\vec{\mathbf{r}}_{\mathrm{CG}}$. Note that in $g(h)$, the quantity $h - h_0$ is the height above the C .

$$\vec{\mathbf{r}}_{\mathrm{CG}} = \frac{\begin{array}{l} M\left[g_0 - \alpha\left(\frac{m\ell}{M+m}\sin\theta\right)\right]\frac{m\ell}{M+m} \\ \cdot\left(\cos\theta\hat{\mathbf{i}} + \sin\theta\hat{\mathbf{j}}\right) \\ +m\left[g_0 + \alpha\left(\frac{M\ell}{M+m}\sin\theta\right)\right] \\ -\frac{M\ell}{M+m}\left(\cos\theta\hat{\mathbf{i}} + \sin\theta\hat{\mathbf{j}}\right) \end{array}}{\begin{array}{l} M\left[g_0 - \alpha\left(\frac{m\ell}{M+m}\sin\theta\right)\right] \\ +m\left[g_0 + \alpha\left(\frac{M\ell}{M+m}\sin\theta\right)\right] \end{array}}$$

Where $\dfrac{m\ell}{M+m}\sin\theta$ is the height of M above the origin and $\dfrac{M\ell}{M+m}\sin\theta$ is the height of m below the origin. Using the given numerical values,

$$\vec{\mathbf{r}}_{\mathrm{CG}} = -\frac{\sqrt{2}\alpha m^2}{36 g_0}$$

Since we took the C to be the origin, distance between CG and C is $\frac{\sqrt{2}\alpha m^2}{36 g_0}$. Note that it clearly lies on the rod.

Chapter 10

Collisions

10.3 This type of collision was analyzed in the text. The result is given. Let

$$\begin{aligned} v_1 &= 1.0 \text{ m/s} \\ u_1 &= .1 \text{ m/s} \\ m &= m_1 = m_2 \end{aligned}$$

We need to solve for v_2. Using the expression for u_1 in Eq. (10.1), and solving for v_2,

$$v_2 = \frac{u_1(m+m) - (m-m)v_m}{2m}$$

so

$$v_2 = u_1 = 0.1 \text{ m/s}$$

The initial velocity of the target ball is 0.1 m/s in the direction of the incident ball.

10.5 From the definition of v_{CM},

$$\begin{aligned} \vec{\mathbf{v}}_{\text{CM}} &= \frac{\sum m_i \vec{\mathbf{v}}_i}{\sum m_i} \\ &= \frac{(200 \text{ kg})(6 \text{ m/s})\hat{\mathbf{i}}}{200 \text{ kg} + 1000 \text{ kg}} \\ &= (1 \text{ m/s})\,\hat{\mathbf{i}} \end{aligned}$$

200 kg 6 m/s 1000 kg $\hat{\mathbf{j}}$ $\hat{\mathbf{i}}$

$\vec{\mathbf{v}}_{int}$ $\vec{\mathbf{v}}_{CM}$

From $\vec{\mathbf{v}}_{\text{tot}} = \vec{\mathbf{v}}_{\text{int}} + \vec{\mathbf{v}}_{\text{CM}}$,

$$\begin{aligned} \vec{\mathbf{v}}_{\text{int}} &= \vec{\mathbf{v}}_{\text{tot}} - \vec{\mathbf{v}}_{\text{CM}} \\ \vec{\mathbf{v}}_{\text{astr}} &= 6 \text{ m/s}\,\hat{\mathbf{i}} - 1 \text{ m/s}\,\hat{\mathbf{i}} \\ &= 5 \text{ m/s}\,\hat{\mathbf{i}} \\ \vec{\mathbf{v}}_{\text{sat}} &= 0 \text{ m/s}\,\hat{\mathbf{i}} - 1 \text{ m/s}\,\hat{\mathbf{i}} \\ &= -1 \text{ m/s}\,\hat{\mathbf{i}} \end{aligned}$$

The astronaut catches on to the satellite and doesn't let go—this means her collision with the satellite was a perfectly inelastic collision. Both have zero velocity in the CM frame.

10.7 Initially the spring is relaxed and does not exert a force on the atoms during the collision. From Eq. (10.3) in the text the leading atom in the molecule (#2) is at rest immediately after the collision.

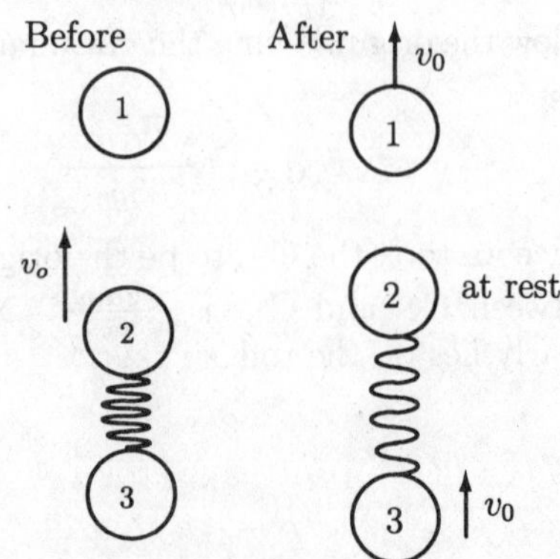

The following atom (#3) travels up, compressing the spring. Atom #1 moves upward but atoms 2 and 3 begin to oscillate. Some of the initial K is used to set the spring oscillating. Therefore we can say that some K of the system gets converted to E_{int} (represented as spring PE) by the collision. Thus, the collision is inelastic. After the collision the molecule has CM velocity

$$\frac{M(0)+M(v_0)}{2M}=\frac{v_0}{2}$$

so we have

	Before	After
KE	$\frac{1}{2}(2M)v_0^2$	$\frac{1}{2}Mv_0^2+\frac{1}{2}(2M)\left(\frac{v_0}{2}\right)^2$
	Mv_0^2	$\frac{3M}{4}v_0^2$

So the energy converted to internal energy is

$$Mv_0^2-\frac{3}{4}Mv_0^2=\frac{Mv_0^2}{4}$$

10.9 The collision must give the blue ball a non zero horizontal impulse and a zero vertical impulse (or at least a negligible vertical impulse). Even an off center shot cannot give the blue ball a zero vertical momentum while conserving both energy and momentum. A direct shot to the center right pocket is impossible.

10.11 Compare the KE before and after the collision (assume the collision is elastic so K is conserved). Note that conservation of momentum says that if m_2 ends up with m_1's initial momentum then m_1 must end up with m_2's initial momentum.

$$\begin{aligned} K_i &= \frac{P_1^2}{2m_1}+\frac{P_2^2}{2m_2} \\ K_f &= \frac{P_2^2}{2m_1}+\frac{P_1^2}{2m_2} \end{aligned}$$

K_i can equal K_f only if:

$$\begin{aligned} \frac{P_1^2}{m_1}+\frac{P_2^2}{m_2} &= \frac{P_2^2}{m_1}+\frac{P_1^2}{m_2} \\ &\Rightarrow P_1^2\left(\frac{1}{m_1}-\frac{1}{m_2}\right) \\ &= P_2^2\left(\frac{1}{m_1}-\frac{1}{m_2}\right) \\ &\Rightarrow m_1=m_2 \end{aligned}$$

Thus the switching of momentum between particles in an elastic collision can happen only if the particles have equal mass.

10.15 We are given the hint that the collision is elastic by the description of the bumper. Use the general solution for 1-dimensional elastic collisions (Eq. 10.1 and 10.2). Let

$$\begin{aligned} v_1 &= 6.5\ \text{m}/\text{s} \\ m_1 &= m=1100\ \text{kg} \\ m_2 &= M=1200\ \text{kg} \end{aligned}$$

u_1 the final speed of car 1 and u_2 the final speed of car 2.

$$\begin{aligned} u_1 &= v_1\frac{m_1-m_2}{m_1+m_2}=-.28\ \text{m}/\text{s} \\ u_2 &= v_1\frac{2m_1}{m_1+m_2}=6.2\ \text{m}/\text{s} \end{aligned}$$

The final speeds are 0.28 m / s and 6.2 m / s.

10.19 The collision is elastic and 1-dimensional. Apply the general solution given by Eqns. 10.1 and 10.2.

520 kg + x — v_m = 5.0 m/s → — 520 kg

520 kg + x — 520 kg → 6.2 m/s = u_M

Object 1 (m) is the loaded car and object 2 (M) is the empty car. The initial speed is $v_m=$ 5.0 m / s

$$\begin{aligned} m &= 520\ \text{kg}+x \\ M &= 520\ \text{kg} \\ u_M &= 6.2\ \text{m}/\text{s} \end{aligned}$$

The equation for u_M gives:

$$6.2\ \text{m}/\text{s}=(5.0\ \text{m}/\text{s})\frac{2\,[520\ \text{kg}+x]}{1040\ \text{kg}+x}$$

Solving for x gives:

$$x=330\ \text{kg}$$

10.23 The slope is frictionless. TME is conserved and from our work in Chapter 9, $E_i=E_f\Rightarrow$

$$mgh=\frac{1}{2}mv_{1i}^2\Rightarrow\vec{\mathbf{v}}_{1i}=\sqrt{2gh}\left(+\hat{\mathbf{i}}\right)$$

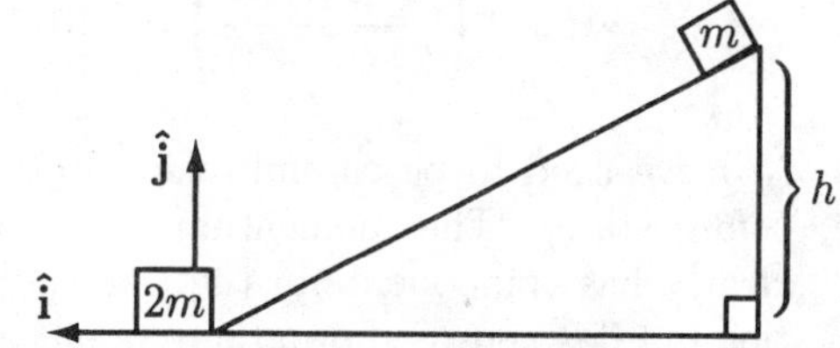

Note that since these masses are particles, the collision itself occurs on level ground—not on the slope. This 1-dimensional elastic collision follows the general solution given by Eqns. 10.1. and 10.2. Let $m = m$, $M = 2m$, $v_m = \sqrt{2gh}$ and u_m is the speed of the mass m at the bottom of the slope. Then,

$$u_m = \sqrt{2gh}\left(\frac{m-2m}{m+2m}\right) = -\frac{1}{3}\sqrt{2gh}$$

The minus sign means that the mass m rebounds back with a speed $\frac{1}{3}\sqrt{2gh}$ up the slope. After the collision, the mass m has a K of

$$\frac{1}{2}m\left[\frac{1}{3}\sqrt{2gh}\right]^2 = \frac{1}{9}mgh$$

and will go up a height of $\frac{1}{9}h$ before it starts sliding back down.

10.27 We will begin by using momentum conservation.

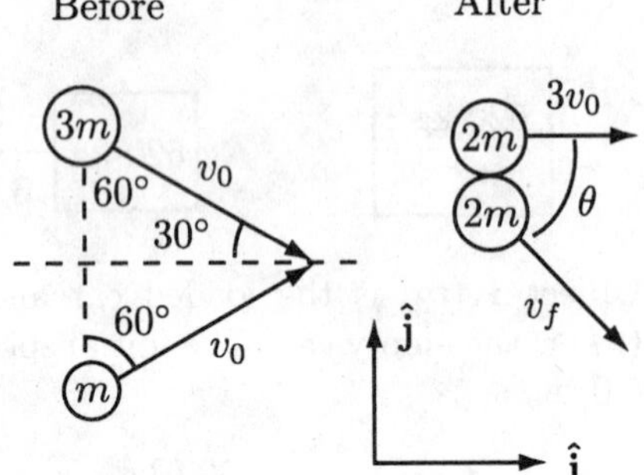

P_x:

$$\begin{aligned}(3m)v_0\sin 60° + mv_0\sin 60° \\ = (2m)3v_0 + (2m)v_f\cos\theta \quad (1) \\ \Rightarrow (\sqrt{3}-3)\,v_0 = v_f\cos\theta\end{aligned}$$

P_y:

$$\begin{aligned}mv_0\cos 60° - (3m)\,v_0\cos 60° \\ = -(2m)v_f\sin\theta \quad (2) \\ \Rightarrow v_0 = 2v_f\sin\theta\end{aligned}$$

From (2) we have $v_f = \frac{v_0}{2\sin\theta}$. Using this to eliminate v_f in Eq. (1) and solving for θ,

$$\theta = \tan^{-1}\left(\frac{1}{2\,(\sqrt{3}-3)}\right) = -21.5°$$

Now we need to be careful. First look at the *before* state. The momentum of the system clearly has components in the $\hat{\mathbf{i}}$ and $-\hat{\mathbf{j}}$ direction but if θ were as shown there would be more p_x in the after state than there is in the before state. This cannot be, since momentum is conserved, so θ must be so that v_f has a negative x-component rather than a positive one, as shown below:

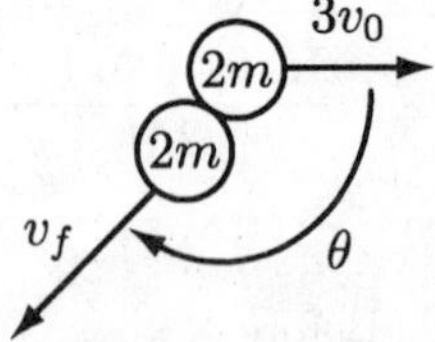

Furthermore, in Eq. (2) we have defined

$$\sin\theta > 0 \Rightarrow 0 < \theta < 180°$$

since magnitudes must be positive. We need to question $\theta = -22°\ldots$ it cannot be correct, but it can lead us to the proper angle. We get the same tangent value for $180° - 22°$.

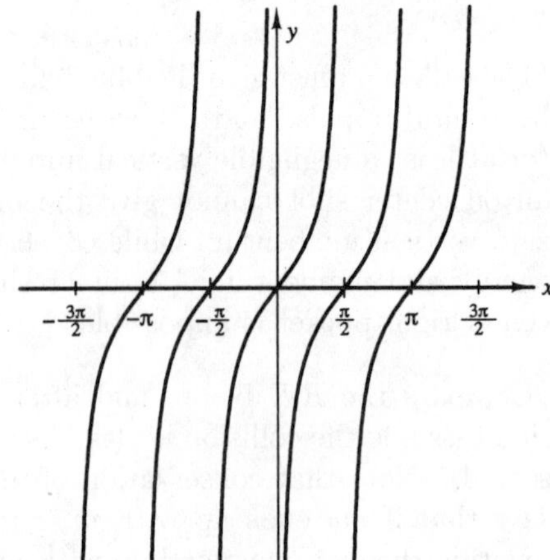

From the reasoning about p_x we expect θ to be a quadrant 2 angle. Thus, $180° - 22° = 158°$ is the angle we're looking for

$$\theta = 158°$$

From (2)

$$\begin{aligned}\vec{\mathbf{v}}_f &= \frac{v_0}{2\sin\theta}\left[\cos\theta\hat{\mathbf{i}} - \sin\theta\hat{\mathbf{j}}\right] \\ &= -\frac{v_0}{2}\left[\cot\theta\hat{\mathbf{i}} - \hat{\mathbf{j}}\right] \\ &= -\frac{v_0}{2}\left[2.5\hat{\mathbf{i}} + \hat{\mathbf{j}}\right]\end{aligned}$$

(The change in the K should equal the amount of energy converted from E_{int}. $|v_f| = 1.33v_0$.)

$$\begin{aligned}\Delta K &= \frac{1}{2}(2m)(1.33\,v_0)^2 + \frac{1}{2}(2m)(3v_0)^2 \\ &\quad -\frac{1}{2}(m)v_0^2 - \frac{1}{2}\,(3m)\,v_0^2 \\ &= 8.8\,mv_0^2\end{aligned}$$

The incoming particles release $8.8\,mv_0^2$ of internal energy.

10.31 We will transform this scenario into the CM frame $\vec{\mathbf{v}}_{CM} = \frac{1}{2}v_0\hat{\mathbf{i}}$.

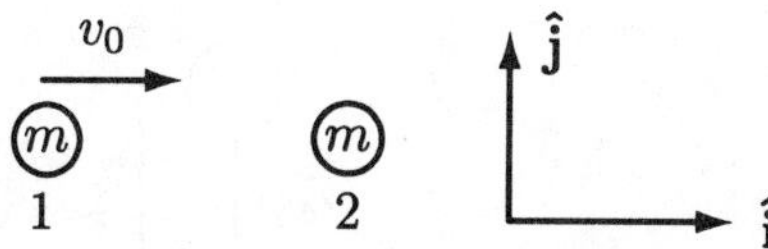

$$\vec{\mathbf{v}}_{1,CM} = v_0\hat{\mathbf{i}} - \frac{1}{2}v_0\hat{\mathbf{i}} = \frac{1}{2}v_0\hat{\mathbf{i}}$$
$$\vec{\mathbf{v}}_{2,CM} = 0 - \frac{1}{2}v_0\hat{\mathbf{i}} = -\frac{1}{2}v_0\hat{\mathbf{i}}$$

Before

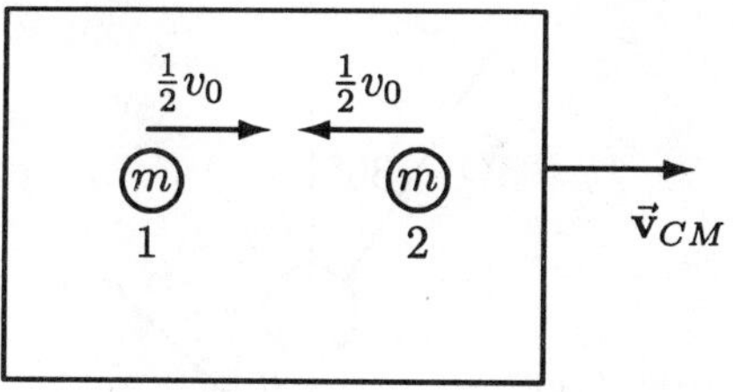

The energy of each particle before collision (in the CM frame) is:

$$E_{CM,i} = \frac{1}{2}m\left(\frac{1}{2}v_0\right)^2 = \frac{1}{8}mv_0^2$$

The energy of each ball after the collision (in the CM frame) is:

$$\begin{aligned} E_{CM,f} &= E_{CM,i}(1-f) \\ &= \frac{1}{8}mv_0^2(1-f) \\ &= \frac{1}{2}m\left[\frac{1}{2}v_0\sqrt{1-f}\right]^2 \end{aligned}$$

So the new speed of each ball is $\frac{1}{2}v_0\sqrt{1-f}$. The CM frame after the collision looks like:

After

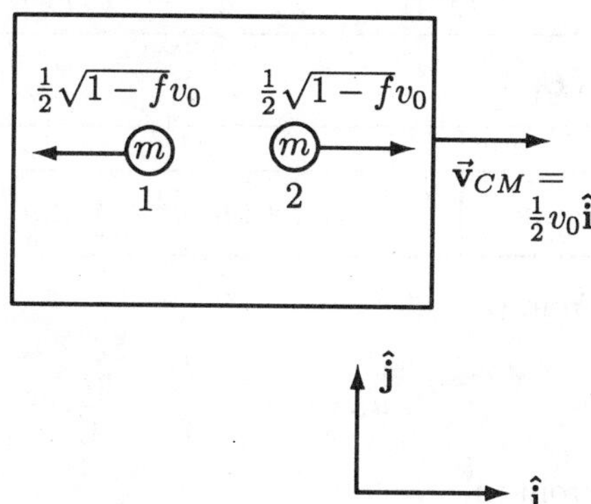

Converting to the lab frame:

lab, after

$\frac{1}{2}v_0\left[1-\sqrt{1-f}\right]$ $\frac{1}{2}v_0\left(1+\sqrt{1-f}\right)$

m 1 m 2

In the lab frame, the second ball gained energy while the first ball lost energy. A fraction such as f would not be the same for the two balls even though they are identical. A more sensible approach would be to define f_{lab} such that:

$$\begin{aligned} f_{lab} &= 1 - \frac{E_{\text{tot after}}}{E_{\text{tot before}}} \\ &= 1 - \left(\frac{\frac{1}{2}m\left[\frac{1}{2}v_0\left(1-\sqrt{1-f}\right)\right]^2 + \frac{1}{2}m\left[\frac{1}{2}v_0\left(1+\sqrt{1-f}\right)\right]^2}{\frac{1}{2}mv_0^2}\right) \\ &= \frac{1}{2}f \end{aligned}$$

In Example 10.6, $f = .3$. From the equation above we have $f_{lab} = \frac{1}{2}(.3) = .15$ which indeed is the energy lost to thermal energy.

10.35 Since no net external forces act on the system, v_{CM} remains constant during and after the collision. Furthermore, the bullet penetrating the block is a totally inelastic collision. We know that the outgoing particle of a totally inelastic collision has come to rest in the CM frame. We do know that after the bullet strikes—after some of the bullet's K is used to puncture a hole in the block–energy is conserved.

$$v_{CM} = \frac{mv}{M+m}$$

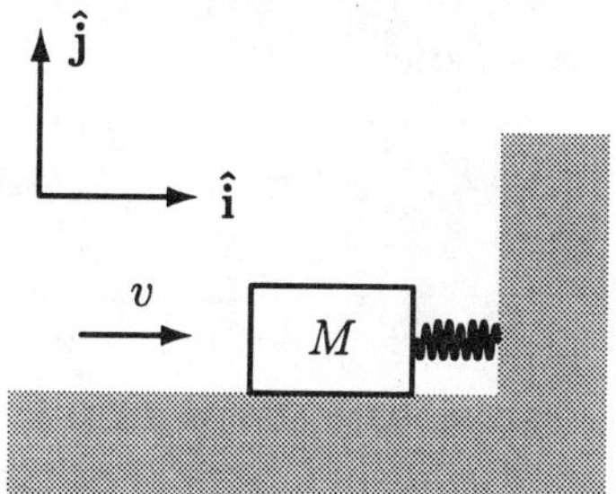

Let the initial state be right after the bullet strikes the block and the final state be when the combined mass $M+m$ comes to rest in the lab frame (at the spring's maximum compression).

$$E_i = \frac{1}{2}(M+m)v_{CM}^2 = \frac{(mv)^2}{2(M+m)}$$

$$E_f = \frac{1}{2}ks^2$$
$$E_i = E_f$$
$$\Rightarrow v = \frac{s}{m}\sqrt{k(M+m)}$$
$$= \frac{0.30m}{0.030\ \text{kg}}\sqrt{(1.0\times10^3\ \text{N/m})(2.03\ \text{kg})}$$
$$v = 450\ \text{m/s}$$

10.39 Since no net forces act on the system of two masses, momentum is conserved.
intial:

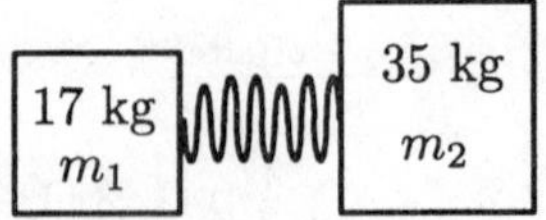

final:

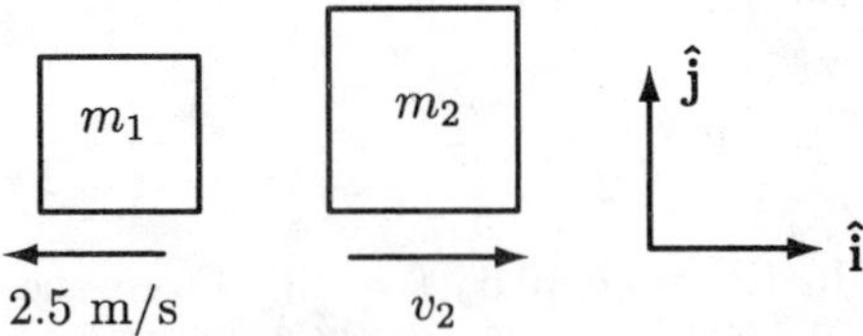

$$\vec{\mathbf{p}}_i = 0$$
$$\vec{\mathbf{p}}_f = m_2\vec{\mathbf{v}}_2 - m_1v_1\hat{\mathbf{i}}$$
$$\vec{\mathbf{p}}_i = \vec{\mathbf{p}}_f$$
$$\vec{\mathbf{p}}_{35} = m_2\vec{\mathbf{v}}_2 = m_1v_1\hat{\mathbf{i}}$$
$$= (17\ \text{kg})(2.5\ \text{m/s})\,\hat{\mathbf{i}}$$
$$= (43\ \text{kg}\cdot\text{m/s})\,\hat{\mathbf{i}}$$

This momentum is unchanged when the astronaut catches the piece. Thus

$$\vec{\mathbf{p}} = (43\ \text{kg}\cdot\text{m/s})\,\hat{\mathbf{i}}$$

The TME of the system in the final state is entirely kinetic:

$$E_f = \frac{1}{2}m_1v_1^2 + \frac{1}{2}m_2v_2^2$$
$$= \frac{p^2}{2m_1} + \frac{p^2}{2m_2}$$

In the initial state all the TME is potential energy stored in the compressed spring. No net external forces act, so

$$E_i = E_f$$

Energy stored in spring:

$$E_f = \frac{1}{2}(425\ \text{kg}\cdot\text{m/s})^2\left[\frac{1}{17\ \text{kg}} + \frac{1}{35\ \text{kg}}\right]$$
$$= 79\ \text{J}$$

10.43 Before:

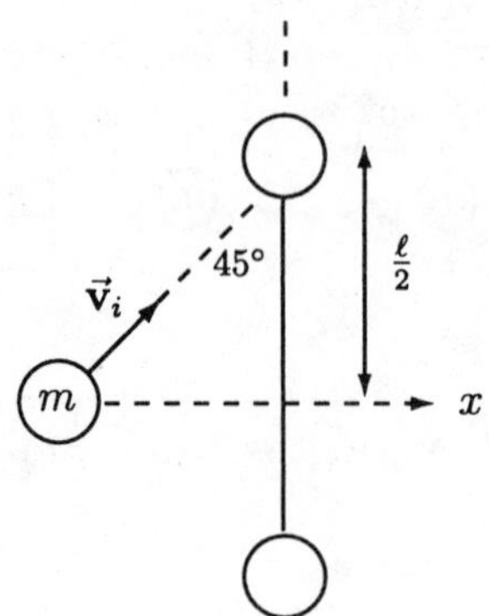

After:

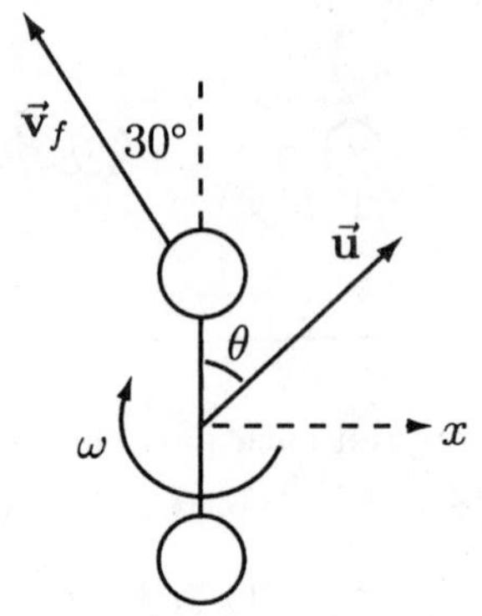

P_x:

Before	After	
$mv_i\frac{\sqrt{2}}{2}$	$\frac{-mv_f}{2} + 2mu\sin\theta$	(i)

P_y:

Before	After	
$mv_i\frac{\sqrt{2}}{2}$	$mv_f\frac{\sqrt{3}}{2} + 2mu\cos\theta$	(ii)

L_z:

Before	After	
$-mv_i\frac{\ell}{2}\frac{1}{\sqrt{2}}$	$mv_f\frac{\ell}{2}\cdot\frac{1}{2} - 2m\left(\frac{\ell}{2}\right)^2\omega$	(iii)

KE:

Before	After	
$\frac{1}{2}mv_i^2$	$\frac{1}{2}mv_f^2 + \frac{1}{2}(2m)u^2 + \frac{1}{2}(2m)\left(\frac{\omega\ell}{2}\right)^2$	(iv)

From (i):

$$u\sin\theta = v_i\frac{\sqrt{2}}{4} + \frac{v_f}{4}$$

From (ii):

$$u\cos\theta = v_i\frac{\sqrt{2}}{4} - \frac{v_f\sqrt{3}}{4}$$

Thus

$$\begin{aligned} u^2 &= \tfrac{v_i^2}{8} + \tfrac{v_i v_f \sqrt{2}}{8} + \tfrac{v_f^2}{16} \\ &\quad + \tfrac{v_i^2}{8} - v_i v_f \tfrac{\sqrt{6}}{8} + v_f^2 \tfrac{3}{16} \\ &= \tfrac{v_i^2}{4} + v_i v_f \tfrac{\sqrt{2}}{8}\left(1-\sqrt{3}\right) + \tfrac{v_f^2}{4} \quad ((\text{v})) \end{aligned}$$

From (iii)

$$\omega \ell = \frac{v_f}{2} + v_i \frac{\sqrt{2}}{2} \qquad ((\text{vi}))$$

From (iv)

$$v_i^2 - v_f^2 = 2u^2 + \frac{(\omega \ell)^2}{2}$$

Substitute in from (v) and (vi)

$$\begin{aligned} v_i^2 - v_f^2 &= 2\left(\frac{v_i^2 - v_f^2}{4} + v_i v_f \frac{\sqrt{2}\,(1-\sqrt{3})}{8}\right) \\ &\quad + \left(\frac{v_f^2}{4} + \frac{v_i^2}{2} + v_i v_f \frac{\sqrt{2}}{2}\right)\frac{1}{2} \\ 0 &= \frac{13}{8}v_f^2 - \frac{v_i^2}{4} + v_i v_f \frac{\sqrt{2}}{4}\left(2-\sqrt{3}\right) \end{aligned}$$

Solve for v_f:

$$\begin{aligned} \frac{v_f}{v_i} &= \frac{-\sqrt{2}\,(2-\sqrt{3}) \pm \sqrt{2\,(4+3-4\sqrt{3}) + 2 \times 13}}{13} \\ &= \frac{-\sqrt{2}\,(2-\sqrt{3}) + \sqrt{2}\sqrt{20-4\sqrt{3}}}{13} \\ &= \frac{2\sqrt{2}}{13}\left(\sqrt{5-\sqrt{3}} - 1 + \frac{\sqrt{3}}{2}\right) \end{aligned}$$

We took the + root since v_f, being a speed, must be positive. Thus

$$v_f = 0.364 v_i$$

Then

$$\omega \ell = 0.182 v_i + 0.707 v_i = 0.889 v_i$$

and

$$\begin{aligned} u &= \frac{v_i}{2}\sqrt{1 + (0.364)^2 + \frac{\sqrt{2}}{2}\left(1-\sqrt{3}\right)(0.364)} \\ &= 0.486 v_i \end{aligned}$$

and

$$\begin{aligned} \sin\theta &= \frac{1}{4}\frac{(\sqrt{2}+0.364)}{0.486} = 0.915 \\ \Rightarrow \theta &= 66^\circ \end{aligned}$$

So we have:

$$\begin{aligned} v_f &= 0.36 v_i \\ \vec{\mathbf{u}} &= (0.49 v_i \text{ at } 66^\circ \text{ to original axis of molecule}) \\ \omega &= 0.89 \frac{v_i}{\ell} \text{ where } \ell = \text{length of molecule} \end{aligned}$$

ω is clockwise in Fig. 10.31.

10.47 a) The impulse gives the whole system a momentum which we usually call $\vec{\mathbf{p}}_{\text{CM}}$ (or equivalently a velocity $\vec{\mathbf{v}}_{\text{CM}}$). This means the whole system translates at a velocity $\vec{\mathbf{v}}_{\text{CM}} = \frac{\Delta \vec{\mathbf{p}}}{2M}$. In addition the barbell will rotate about its CM at some angular speed

$$\omega = \frac{\Delta \vec{\mathbf{p}}}{2M} \cdot \frac{2}{\ell} = \frac{\Delta p}{M\ell}$$

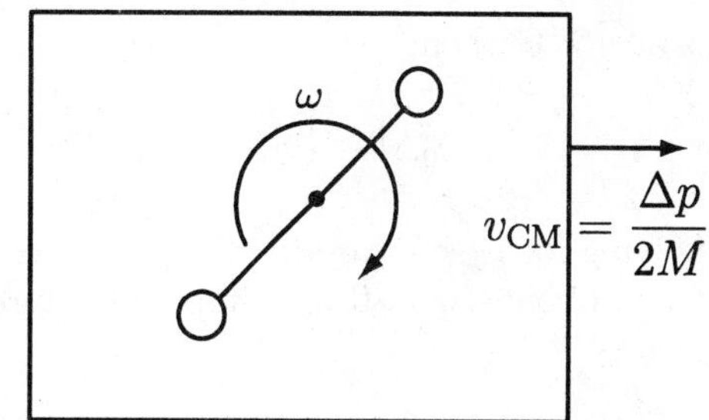

b) In this case the dumbbell travels along a straight path without rotating. Its CM velocity is the same as in a).

c) The first dumbbell gets an energy due to translation of its CM (remember there are 2 masses) which is:

$$E_{\text{CM}} = \frac{(\Delta p)^2}{4m}$$

It also has an energy due to its rotation. Its change in angular momentum about its CM is

$$\Delta L = \frac{1}{2}\ell \Delta p$$

This also is equal to

$$\begin{aligned} L &= 2\left(\left[\frac{1}{2}\ell\right]^2 M\omega\right) = \frac{1}{2}\ell^2 M\omega \\ \frac{1}{2}\ell \Delta p &= \frac{1}{2}\ell^2 M\omega \Rightarrow \omega = \frac{\Delta p}{M\ell} \end{aligned}$$

$$\begin{aligned} E_{\text{rot}} &= 2\left[\frac{1}{2}M v_r^2\right] = M(\omega r)^2 \\ &= \frac{M(\Delta p)^2}{M^2 \ell^2}\left(\frac{1}{4}\ell^2\right) = \frac{(\Delta p)^2}{4M} \end{aligned}$$

$$\begin{aligned} E_{\text{tot}} &= E_{\text{CM}} + E_{\text{rot}} \\ &= \frac{\Delta p^2}{4m} + \frac{\Delta p^2}{4m} = \frac{\Delta p^2}{2m} \end{aligned}$$

The other dumbbell gets only the linear E_{CM} term.

$$E_{\text{CM}} = \frac{(\Delta p)^2}{4m}$$

10.51 The block and bullet combination

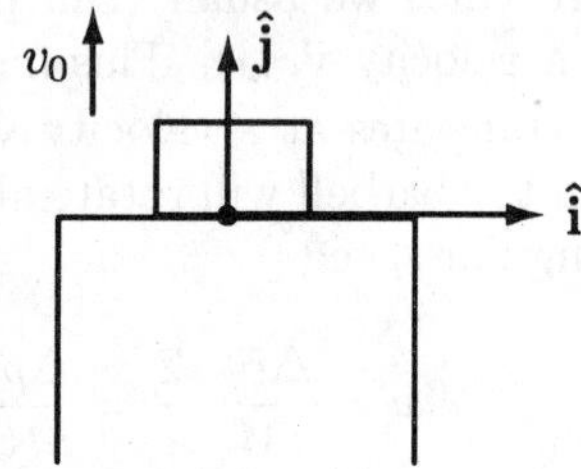

is at $y = 0$ when:

$$v_0 \Delta t - \frac{1}{2} g \Delta t^2 = 0$$

Solving for v_0 (we can divide by Δt because we know about $\Delta t = 0$ and it is not interesting) gives

$$\begin{aligned} v_0 &= \frac{1}{2} g \Delta t \\ &= \frac{1}{2} \left(9.8 \text{ m}/\text{s}^2\right)(0.86 \text{ s}) \\ &= 4.2 \text{ m}/\text{s} \end{aligned}$$

The block-bullet combination has a momentum of :

$$p_{\text{block+bullet}} = (2.015 \text{ kg})(4.2 \text{ m}/\text{s}) = 8.5 \text{ N} \cdot \text{s}$$

During the collision external forces on the block-bullet system are neglible, so momentum is constant. The block-bullet combination got its momentum from the bullet's initial momentum. Before the collision with the block,

$$\begin{aligned} p_{\text{bullet}} &= 8.5 \text{ N} \cdot \text{s} \\ \Rightarrow v_{\text{initial bullet}} &= \frac{p_{\text{bullet}}}{m_{\text{bullet}}} = 570 \text{ m}/\text{s} \end{aligned}$$

$$\begin{aligned} E_{\text{initial}} &= \frac{1}{2}(.015 \text{ kg})(570 \text{ m}/\text{s})^2 = 2400 \text{ J} \\ E_{\text{final}} &= \frac{1}{2}(2.01 \text{ kg})(4.2 \text{ m}/\text{s})^2 = 18 \text{ J} \end{aligned}$$

The change in energy,

$$|\Delta E| = |E_f - E_{\text{initial}}| = 2400 \text{ J}$$

went to thermal energy.

10.55 Since no net external forces act on the stars, momentum, total energy and angular momentum are all conserved. We'll begin with momentum: (we assume that the second star is initially stationary).

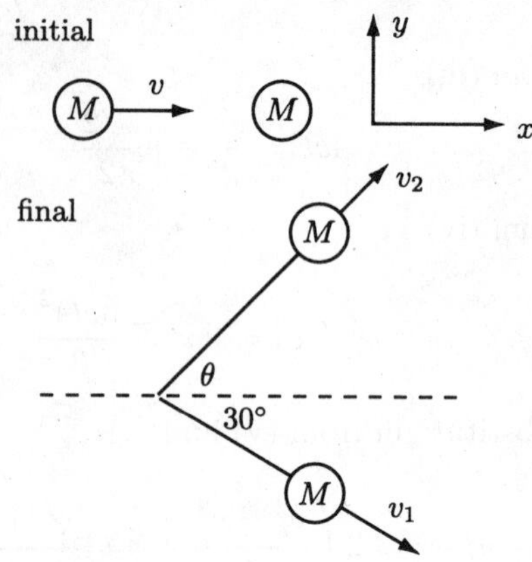

$$\begin{aligned} P_i &= P_f \\ P_x : M v_0 &= M v_1 \cos 30° + M v_2 \cos \theta \quad (3) \\ P_y : 0 &= -M v_1 \sin 30° + M v_2 \sin \theta \, (4) \end{aligned}$$

From (4) we have $v_2 = \dfrac{v_1}{2 \sin \theta}$. Putting this into (3) and solving for θ gives

$$\tan \theta = \frac{v_1}{2 v_0 - \sqrt{3} v_1}$$

Now we may use the result of Example 10.3: the outgoing velocities are perpendicular, so $\theta = 60°$. Then

$$\begin{aligned} \tan \theta &= \tan 60° = \sqrt{3} \\ \sqrt{3}\left(2 v_0 - \sqrt{3} v_1\right) &= v_1 \\ 2\sqrt{3} v_0 &= v_1 + 3 v_1 = 4 v_1 \end{aligned}$$

so

$$v_1 = \frac{\sqrt{3}}{2} v_0$$

then

$$v_2 = \frac{\sqrt{3}}{2} \frac{v_0}{2} \frac{2}{\sqrt{3}} = \frac{v_0}{2}$$

So

$$\begin{aligned} \vec{\mathbf{v}}_1 &= \left(\frac{\sqrt{3}}{2} v_0, \text{ at } 30°\right) \\ \vec{\mathbf{v}}_2 &= \left(\frac{v_0}{2}, \text{ at } 60°\right) \end{aligned}$$

10.59 The first thing we need to do is conserve momentum.

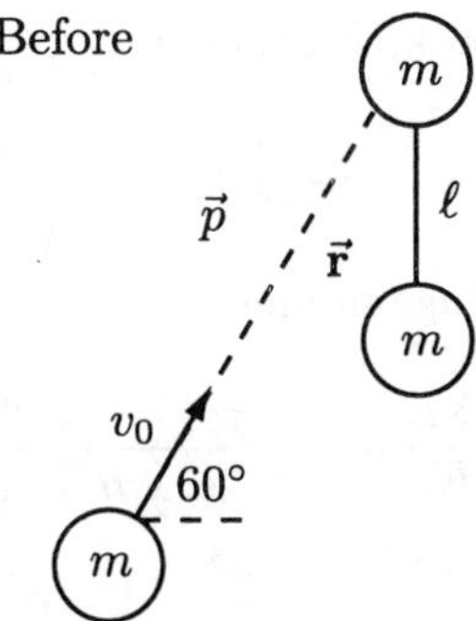

We can compare the magnitudes of v_0 and v_f because momentum in the CM frame is zero, thus v_0 and v_f must be in the same direction.

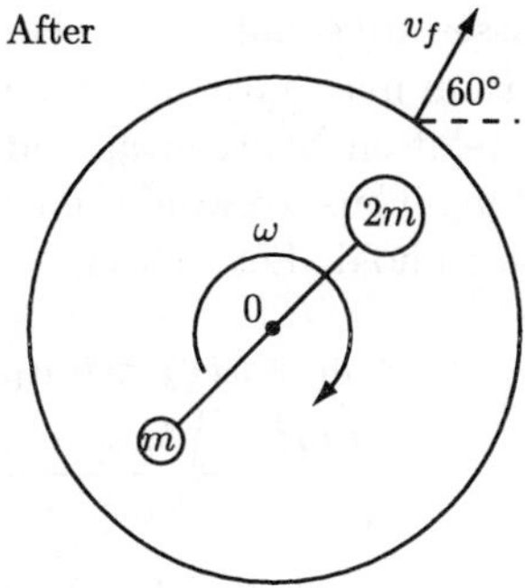

$$mv_0 = 3mv_f \Rightarrow \vec{\mathbf{v}}_f = \frac{1}{3}\vec{\mathbf{v}}_0$$

The collision is totally inelastic, and the objects rotate—this suggests that we should look to conservation of angular momentum rather than energy. We will compute L about the system's CM just as the collision takes place. The CM is located $\frac{2}{3}$ closer to the $2m$ mass than the $1m$ mass.

$$\begin{aligned} r_\perp &= \frac{1}{3}\ell \sin 30^\circ = \frac{1}{2}\cdot\frac{1}{3}\ell = \frac{1}{6}\ell \\ L_i &= (\ell m v_0)\frac{1}{6} \end{aligned}$$

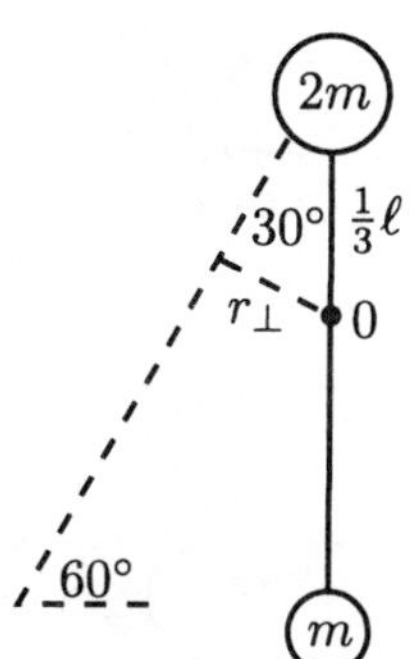

v_{CM} travels on a straight line from point 0, thus:

$$\vec{\mathbf{r}} \times m\vec{\mathbf{v}}_{\text{CM}} = 0$$

There is no CM angular momentum, only L_{int} due to the rotation of the dumbbell.

$$L_f = \omega \sum r^2 m = \omega\left[\left(\frac{1}{3}\ell\right)^2(2m) + \left(\frac{2}{3}\ell\right)^2 m\right]$$

No net external torques means that the system's angular momentum is conserved.

$$\begin{aligned} L_i &= L_f \\ \Rightarrow \frac{1}{6}(\ell m v_0) &= \omega\left[\frac{2}{9}\ell^2 m + \frac{4}{9}\ell^2 m\right] \\ &= \omega\frac{2}{3}\ell^2 m \\ \Rightarrow \vec{\omega} &= \left(\frac{v_0}{4\ell}, \text{ clockwise}\right) \end{aligned}$$

Now we'll compute how much energy was dissipated as heat.

$$\begin{aligned} E_{\text{therm}} &= -\Delta K_{\text{sys}} = K_i - K_f \\ &= K_i - (K_{\text{CM}} + K_{\text{rot}}) \\ &= \frac{1}{2}mv_0^2 - \left(\begin{array}{l} \frac{1}{2}(3m)\left(\frac{1}{3}v_0\right)^2 \\ +\frac{1}{2}(2m)\left(\frac{v_0}{4\ell}\cdot\frac{1}{3}\ell\right)^2 \\ +\frac{1}{2}m\left(\frac{v_0}{4\ell}\cdot\frac{2}{3}\ell\right)^2 \end{array}\right) \\ &= \frac{1}{2}mv_0^2 - \frac{3}{16}mv_0^2 = \frac{5}{16}mv_0^2 \end{aligned}$$

10.63 a) Before:

v_n → ○ ○
n $n+1$

After:

u_n → ○ ○ → u_{n+1}

Momentum:

$$mv_n = mu_n + fmu_{n+1}$$

Energy:

$$\frac{1}{2}mv_n^2 = \frac{1}{2}mu_n^2 + \frac{1}{2}fmu_{n+1}^2$$

From the first equation:

$$u_{n+1} = \frac{1}{f}(v_n - u_n) \quad (5)$$

Using this result in the second equation:

$$v_n^2 - u_n^2 = f\left[\frac{v_n - u_n}{f}\right]^2$$

So

$$f(v_n + u_n) = v_n - u_n$$

or

$$u_n = v_n \frac{(1-f)}{(1+f)}$$

as required. Then from equation (5)

$$u_{n+1} = \frac{1}{f} v_n \left[1 - \frac{(1-f)}{(1+f)}\right] = \frac{2v_n}{1+f}$$

also as required. Using equations (10.3) gives the same results.

b) Between collisions, friction acts on each cart.

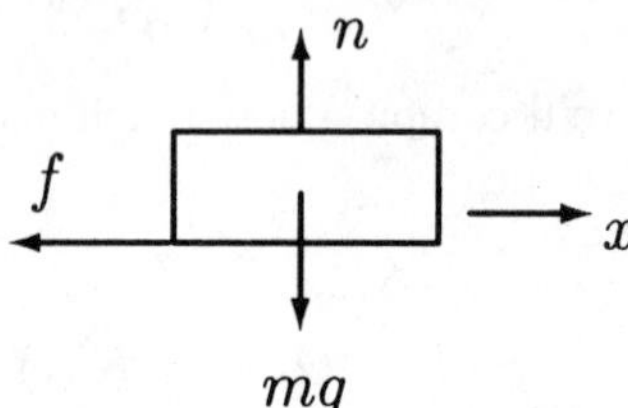

$$f = \mu_k n = \mu_k mg = -ma_x \Rightarrow a_x = -\mu_k g$$

Work done by friction as the cart travels a distance d is

$$W = -\mu_k mgd = \Delta K$$

So if v_{n+1} is the speed with which the $(n+1)$st cart hits the $(n+2)$nd:

$$\begin{aligned} \frac{1}{2}m(v_{n+1})^2 - \frac{1}{2}mv_n^2 &= -\mu_k mgd \\ \Rightarrow v_{n+1} &= \sqrt{v_n^2 - \mu_k gd} \end{aligned}$$

10.67 The coefficient of restitution is defined by the relation

$$\epsilon = \frac{|u_1 - u_2|}{|v_1 - v_2|}$$

In the CM frame, the total momentum is zero so

$$\begin{aligned} mu_1 + Mu_2 &= 0 \\ u_2 &= -\frac{m}{M}u_1 \end{aligned}$$

Thus

$$\epsilon = \frac{\left(1 + \frac{m}{M}\right)u_1}{\left(1 + \frac{m}{M}\right)v_1} = \frac{u_1}{v_1}$$

The initial kinetic energy is

$$\begin{aligned} \frac{1}{2}mv_1^2 + \frac{1}{2}Mv_2^2 &= \frac{1}{2}mv_1^2 + \frac{1}{2}M\left(\frac{m}{M}v_1\right)^2 \\ &= \frac{1}{2}mv_1^2\left(1 + \frac{m}{M}\right) \end{aligned}$$

The final kinetic energy

$$= \frac{1}{2}mu_1^2\left(1 + \frac{m}{M}\right)$$

The change in kinetic energy is

$$\begin{aligned} \frac{\Delta E}{K_i} &= \frac{\frac{1}{2}m\left(u_1^2 - v_1^2\right)\left(1 + \frac{m}{M}\right)}{\frac{1}{2}mv_1^2\left(1 + \frac{M}{m}\right)} \\ &= \frac{u_1^2}{v_1^2} - 1 = \epsilon^2 - 1 \end{aligned}$$

or

$$\frac{|\Delta E|}{K_i} = 1 - \epsilon^2$$

as required. The relation holds whether or not the masses are equal. In the lab frame the total momentum may have any value and there is no simple relation between v_1 and v_2, or between u_1 and u_2. Thus we would not expect the same relation to hold. Let's check.

$$\begin{aligned} p &= mv_1 + Mv_2 = mu_1 + Mu_2 \\ \frac{\Delta E}{K_i} &= \frac{\frac{1}{2}mu_1^2 + \frac{1}{2}Mu_2^2 - \frac{1}{2}mv_1^2 - \frac{1}{2}Mv_2^2}{\frac{1}{2}mu_1^2 + \frac{1}{2}Mu_2^2} \\ &= 1 - \left(\frac{mv_1^2 + Mv_2^2}{mu_1^2 + Mu_2^2}\right) \end{aligned}$$

if the masses are equal

$$\begin{aligned} \frac{\Delta E}{K_i} &= 1 - \left(\frac{v_1^2 + v_2^2}{u_1^2 + u_2^2}\right) = 1 - \left(\frac{p^2 - 2v_1v_2}{p^2 - 2u_1u_2}\right) \\ \epsilon^2 &= \frac{u_1^2 + u_2^2 - 2u_1u_2}{v_1^2 + v_2^2 - 2v_1v_2} = \frac{p^2 - 4u_1u_2}{p^2 - 4v_1v_2} \end{aligned}$$

Thus

$$\frac{|\Delta E|}{K_i} = 1 - \epsilon^2$$

if, and only if, $p \equiv 0$, i.e, in the CM frame.

Chapter 11

Rigid Bodies in Equilibrium

11.1 We will assume that each figure has constant density and use the result of Ex. 11.2: To be in equilibrium, the CM must be over the base of support.

a) The CM of this object is located at its geometric center, easily found at point O. A line dropped from O shows that the CM is *not* over the base of support. *This object is not in equilibrium.*

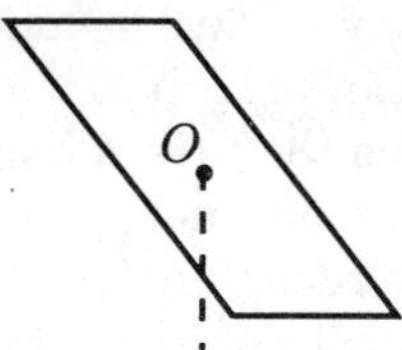

b) Although we cannot make a precise determination of where the CM is for this figure, we can see that the parts A and B have equal area, and therefore equal mass.

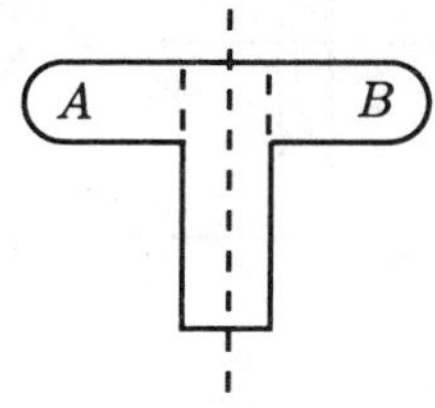

The CM must therefore lie somewhere on the dashed line. Regardless of the position of the CM on the dashed line, it will always be over the base of support. *This structure is in equilibrium.*

c) The rightmost three pieces have a CM as shown (labeled C_1).

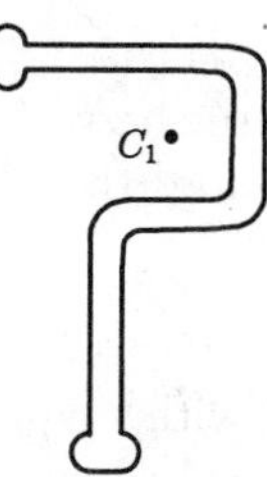

C_1 falls almost on the base. The other two masses look like they would bring the CM over the base.

d) We can break this object up into two parts as shown.

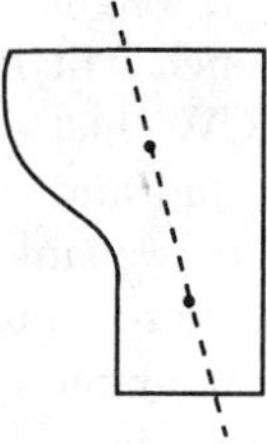

The CM of the system will be somewhere on the dashed line which connects the two CM's of the smaller halves. Regardless of where the CM is on the dashed line, it will always be over the base of support so the *system is in equilibrium.*

e) The skinny part of this object represents only a fraction of the circle's mass.

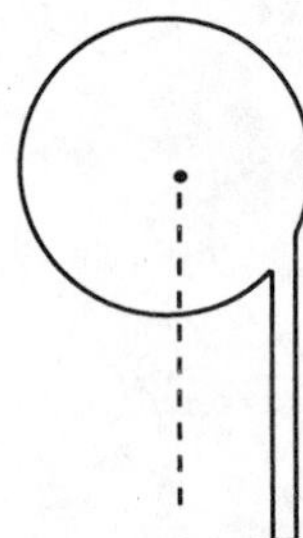

Therefore the CM of this system falls a little to the right and below the geometric center of the circle. The CM is certainly not over the tiny base so the *system is not in equilibrium.*

f) The circle's CM is at its geometric center;

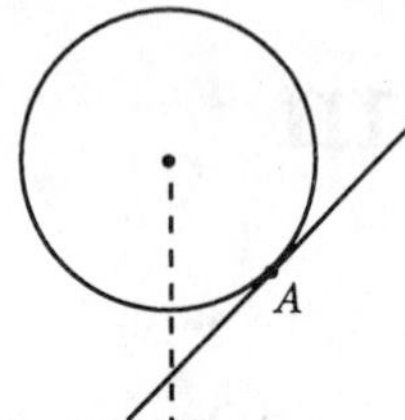

it is not over the base of support (which is point A) and therefore the object is not in equilibrium.

11.3 Without the hint given in Exercise 11.1 you would be left with 2 equations and 3 unknowns. If this happened to you, you forgot to consider what happens as the bench just begins to topple.

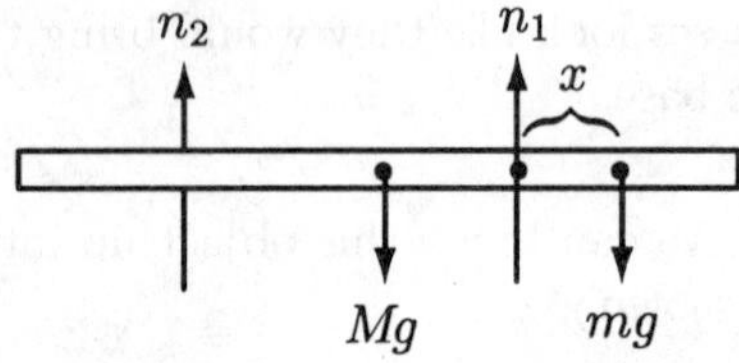

If the person of weight mg sits to the right, as shown, the bench will have a tendency to rotate (topple) CW. Just as it rotates, the left leg leaves the ground and $n_2 \to 0$. That reduces the problem to two unknowns, n_1 and x, and two equations, force and torque balance. We can eliminate an unknown, n_1, from our torque-balance equation by taking point 0 to be the origin. Thus, let x denote the maximum distance that the person can sit from the right leg. Torque balance requires:

$$\begin{aligned} Mg(.25 \text{ m}) &= mg(x) \\ \Rightarrow x &= \frac{M(.25 \text{ m})}{m} = 0.04 \text{ m} \end{aligned}$$

measured from 0. Measured from the end of the bench we have

$$0.25 \text{ m} - 0.04 \text{ m} = 0.21 \text{ m}$$

Another method would be to realize that we want the CM of the system to be beneath the part of the bench that is supported.

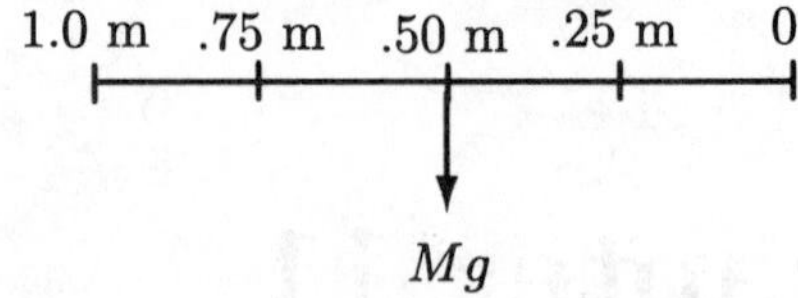

In other words, where can we place the mass m so that the CM of M and m falls between 0.75 m and 0.25 m?

$$\frac{M(.50 \text{ m}) + m(x)}{M+m} \overset{\text{set}}{\equiv} 0.25 \text{ m},\ 0.75 \text{ m}$$

Doing the above calculation gives $x = 0.21$ m, 0.79 m, the same result as above.

11.5 Frictional force is what accelerates vehicles centripetally around a turn. It also exerts a torque on the vehicle about its CM.

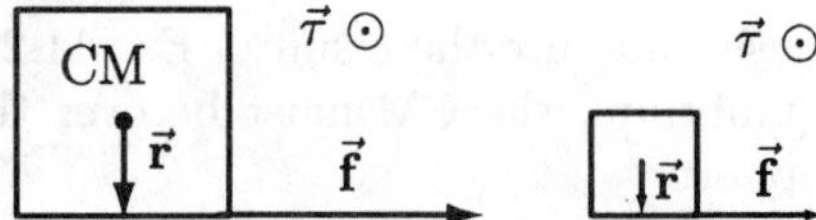

The bus, being of greater mass will require and get a larger friction force to turn than the car. In addition, its CM is farther away from the friction forces point of application. The net result is that $\vec{\tau} = \vec{\mathbf{r}} \times \vec{\mathbf{F}}$ will be much larger for the bus than the car. The balancing torque from normal force on the tires is limited by the width of the bus, not that much different than the car's width. In order to reduce the torque acting on it the bus will slow down for the turn to reduce the needed friction force $f = \frac{mv^2}{r}$.

11.7 Focusing on the block of mass M,

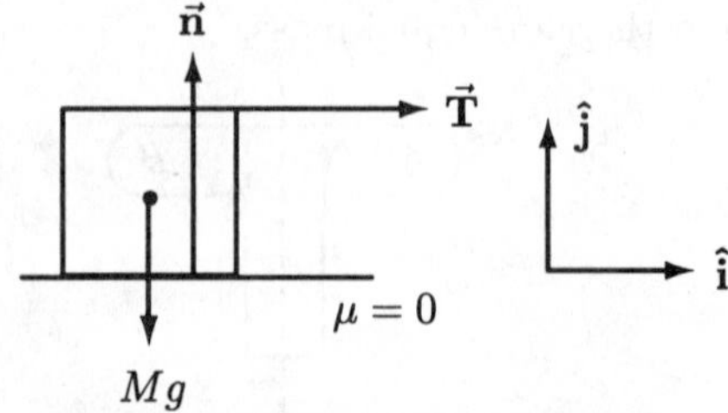

we can see that regardless of the point of application of $\vec{\mathbf{n}}$ ($\vec{\mathbf{n}}$ is the only force which can conceivably change its point of application here), the horizontal tension force is always unbalanced without another horizontal force acting towards $-\hat{\mathbf{i}}$, like friction. Now consider $\mu = 1$ and $T = Mg$. Note that both f_s and T produce torques into the page.

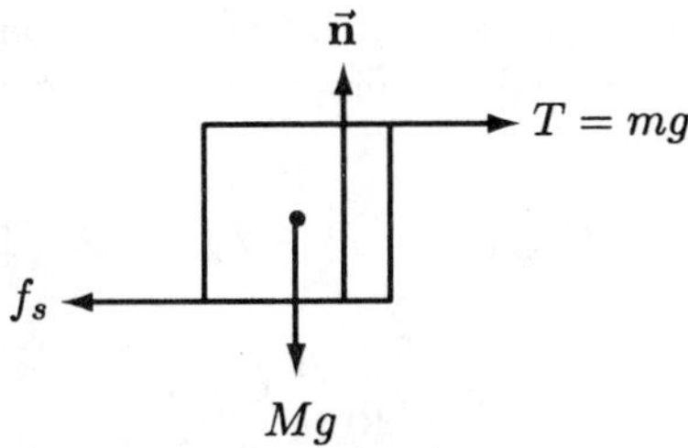

To keep the block from rotating $\vec{n}$'s point of application must be to the right of the CM, as shown. ($\vec{n}$ to the left of CM would produce even more torque into the page). Force balance

$$\begin{aligned} n &= Mg \quad (1) \\ f_s &= mg \quad (2) \\ f_s &\leq \mu n = \mu Mg \quad (3) \end{aligned}$$

If $\mu = 0$, $f_s = 0$ and equilibrium is not possible for any value of m (except the trivial case $m = 0$). If $m = M$,

$$F_{\max} = \mu Mg = Mg$$

for $\mu = 1$, so force balance can be achieved. The block may still rotate. Well compute the net torque about the block's CM. Take the block's side to be a length of 2ℓ, and let $\vec{n}$ act at the edge of the block. Then

$$\begin{aligned} \vec{\tau} &= \ell n \odot + \ell f_s \otimes + \ell T \otimes \\ &= \ell Mg \odot + \ell Mg \otimes + \ell Mg \otimes \\ &= \ell Mg \otimes \end{aligned}$$

The net torque makes the block rotate CW. *Equilibrium is not possible.*

11.9 One of the first things we should do in investigating equilibrium is draw a FBD.

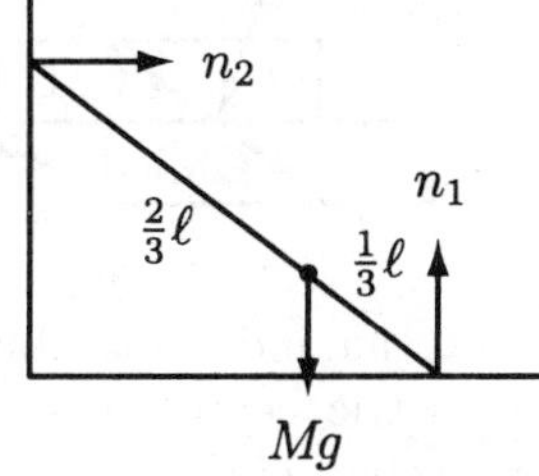

Note that n_2 is necessarily an unbalanced force. The ladder must, if nothing else, accelerate to the right. *The sign cannot be in equilibrium.*

11.11 The "T" has been divided into two rectangles of equal area and therefore equal mass since we are told that the sheet is uniform.

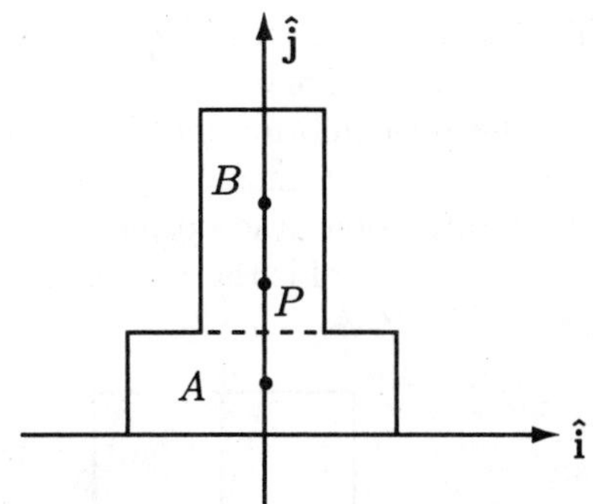

The CM of each individual rectangle is at the geometric center of the respective rectangle. Since each CM represents an equal amount of mass, the CM of the whole system = the CM of the two pieces, which is midway between A and B, indicated as point P. The y-coordinate of P is

$$y_P = \frac{\frac{\ell}{2} + \left[\ell + \frac{3\ell}{2}\right]}{2} = \frac{3\ell}{2}$$

Thus P is at $\left(0, \frac{3}{2}\ell\right)$.

11.13 The person sits down at point O.

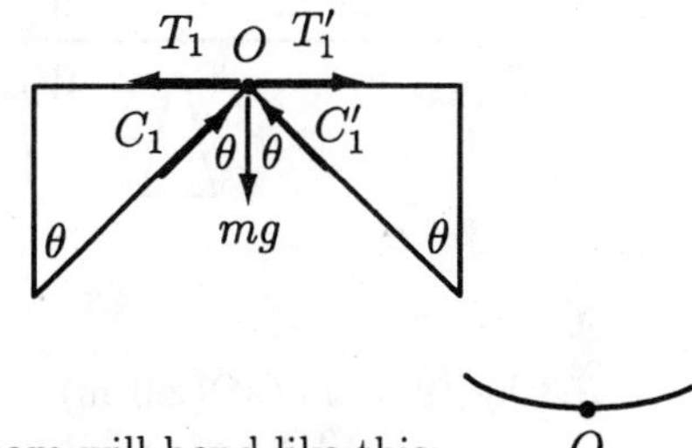

The top beam will bend like this: O
This will cause the 2 diagonal beams to be squeezed, or compressed. Because the bench does not accelerate horizontally, the horizontal components of $\vec{\mathbf{C}}_1$ and $\vec{\mathbf{C}}_1'$ are equal. Because the angles between $\vec{\mathbf{C}}_1$ and $\vec{\mathbf{C}}_1'$ are equal $\left|\vec{\mathbf{C}}_1\right| = \left|\vec{\mathbf{C}}_1'\right|$. Balancing vertical forces:

$$2C_1 \cos 45^\circ = mg \Rightarrow C_1 = \frac{mg}{\sqrt{2}} = 590 \text{ N}$$

11.15 The figure shows the crate about to topple.

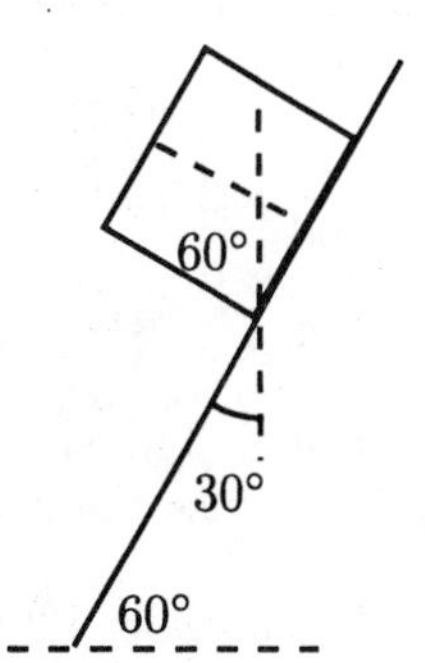

This means that the CM lies just over the limit of the base of support, over the dashed line. From the given information about how the crate balances, the CM must also lie in the middle of the crate, along the dotted line. The exact location of the CM is the intersection of the two lines.

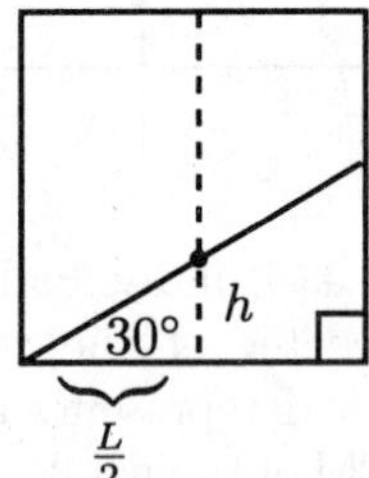

From some geometry, the CM is located a distance

$$h = \frac{1}{2}L\tan 30° = \frac{1}{2\sqrt{3}}L$$

from the bottom of the crate.

11.19 Torque balance about O:

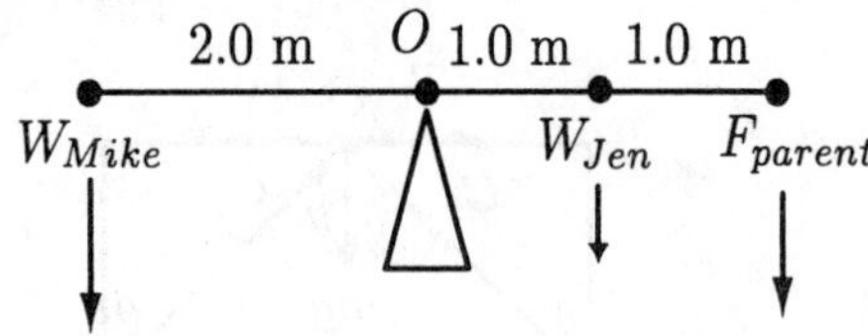

$$\begin{aligned}&(53\text{ kg})(9.81\text{ m}/\text{s}^2)(2.0\text{ m})\\ &= (25\text{ kg})(9.81\text{ m}/\text{s}^2)(1.0\text{ m}) + F_{\text{parent}}(2.0\text{ m})\\ &\Rightarrow F_{\text{parent}} = 4.0\times 10^2\text{ N}\end{aligned}$$

The children can sit anywhere as long as the following holds true:

$$\begin{aligned}M_{\text{Mike}}\ell_{\text{Mike}} &= M_{\text{Jen}}\ell_{\text{Jen}}\\ \Rightarrow \ell_{\text{Mike}} &= \frac{M_{\text{Jen}}}{M_{\text{Mike}}}\ell_{\text{Jen}}\end{aligned}$$

Where ℓ_{Mike} and ℓ_{Jen} are the distances between Mike, Jen and the center of the seesaw. Since the maximum distance Jenny can sit from the center is 2.0 m, then Mike must sit at a maximum distance

$$\frac{25}{53}\times 2.0\text{ m} = 0.94\text{ m}$$

from the center.

11.23 Force balance: The shaft exerts a force on the pulley to balance the normal force

$$F_{\text{shaft}} = n = 1.2\times 10^5\text{ N}$$

Any force exerted by the shaft on the pulley must also be exerted by the bearing on the shaft:

$$F_{\text{bearings}} = F_1 + F_2 = 1.2\times 10^5\text{ N} \qquad (4)$$

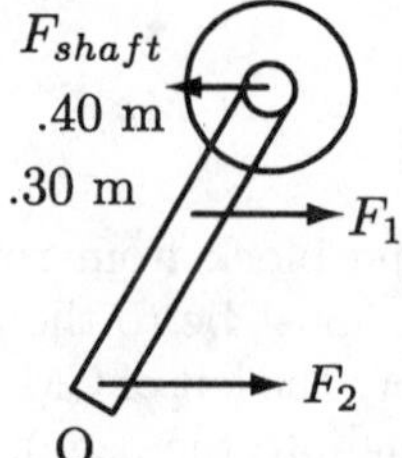

The belt exerts a torque of magnitude

$$\tau = F_{\text{shaft}}\,(0.10\text{ m}) = 1.2\times 10^4\text{ N}\cdot\text{m}$$

about the closest bearing. The bearings must also exert a compensating example of this magnitude.

Torque balance (take O as origin)

$$F_1(.30\text{ m}) - F_{\text{shaft}}(.40\text{ m}) = 0$$

$$\begin{aligned}\Rightarrow F_1 &= \frac{(1.2\times 10^5\text{ N})\,(0.40\text{ m})}{0.30\text{ m}}\\ &= 1.6\times 10^5\text{ N}\end{aligned}$$

By Eq. (4),

$$\begin{aligned}F_2 &= 1.2\times 10^5\text{ N} - 1.6\times 10^5\text{ N}\\ &= -0.4\times 10^5\text{ N}\end{aligned}$$

Note that we get a negative answer for F_2, indicating that F_2 is actually in the opposite direction from that shown—the same direction as F_{shaft}.

11.27 Picture two stacked blocks.

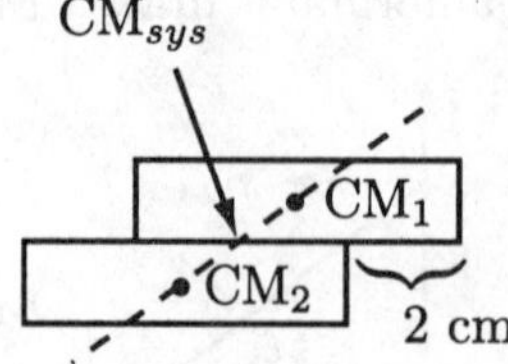

The CM of the individual blocks is located at the geometric center of each block. The CM of the system is the directly between CM_1 and CM_2 since they represent equal amounts of mass.

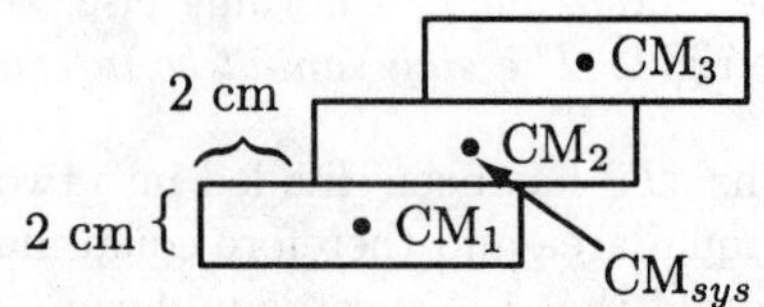

For the same reason the CM of three stacked blocks is located at the CM of the second block. We can generalize that each block added moves the CM to the right by 1 cm and up by 1 cm. The system becomes unstable when CM_{sys} moves more than 8 cm away from CM_1 since that would make CM_{sys} not lie over the base of support. Thus, if nine blocks or more were stacked the system would become unstable.

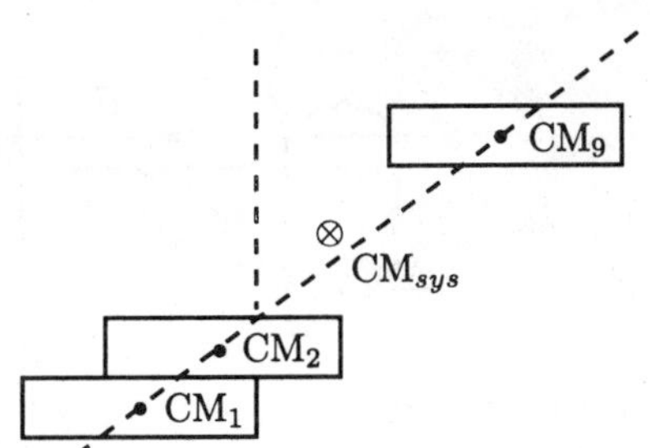

11.33 The screws must serve two functions. They must balance the weight of the amplifier and the torque it produces. We will model this as the screws exerting two forces each: n and f as shown below:

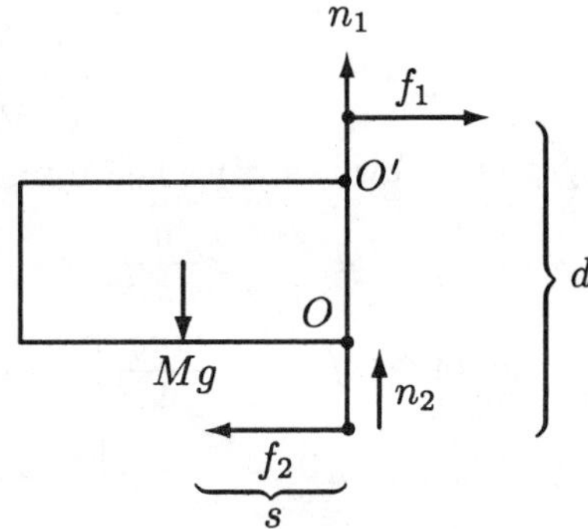

Mg produces a torque about O which is out of the page. For the purpose of balancing torque we chose f_1 to point to the right and f_2 to point to the left. This produces a torque about O which is into the page and balances $\vec{\tau}_{Mg}$. Since the amplifier doesn't accelerate horizontally $f_1 = f_2$; they form a couple which exerts a torque of

$$\tau_f = f_1 d = Mgs = \tau_{Mg}$$

since they balance the torque due to the amp's weight.

$$\begin{aligned} f_1 &= f_2 = \frac{Mgs}{d} \\ &= \frac{(9.7 \text{ kg})(9.81 \text{ m/s}^2)(0.09 \text{ cm})}{(0.08 \text{ cm})} \\ &= 110 \text{ N} \end{aligned}$$

which is each screw's horizontal force. Force balance gives:

$$n_1 + n_2 = Mg = 95 \text{ N}$$

which is the total vertical force. Put bottom screws in first. Normal force can contribute f_1, but not f_2.

11.37 a) Clearly a tension force exerted up and to the right is needed to balance the climber's weight and normal force.

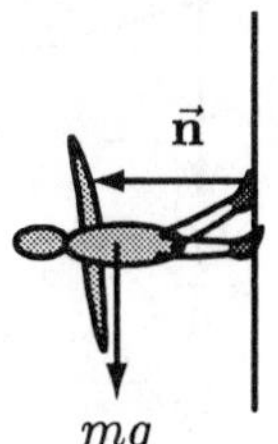

The climber wants torque balance as well force balance; rotation is (almost) as undesirable as acceleration when rock climbing. Note that the weight and normal force both point into the climber's CM and therefore exert no torque. For torque balance we want the rope's tension force to exert zero torque as well. This means the climbing rope needs to be attached at her CM.

b) The posture does not change the fact that the tension force needs to be exerted at the climber's CM.

c) Torques about CM:

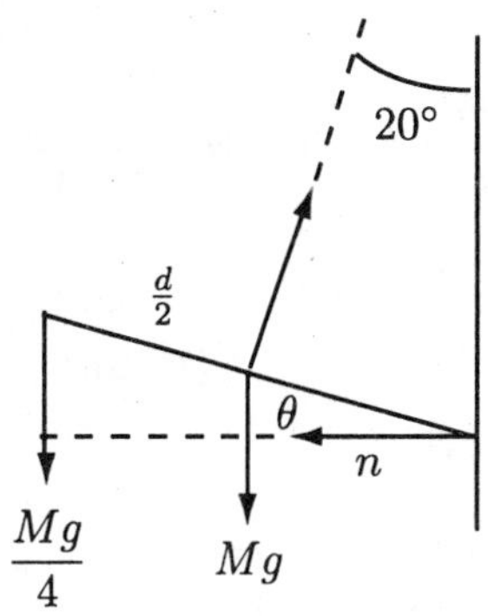

$$\begin{aligned} nd\sin\theta - \frac{Mg}{4}\frac{d}{2}\cos\theta &= 0 \\ \Rightarrow \tan\theta &= \frac{Mg}{8n} \end{aligned}$$

Force balance:

$$\begin{aligned} n &= T\sin 20^\circ \\ \frac{5Mg}{4} &= T\cos 20^\circ \\ n &= \frac{5Mg}{4}\tan 20^\circ \end{aligned}$$

$$\tan\theta = \frac{Mg}{(8)\frac{5Mg}{4}\tan 20^\circ} = \frac{0.10}{\tan 20^\circ}$$
$$\Rightarrow \theta = 15^\circ$$

d) Force balance:

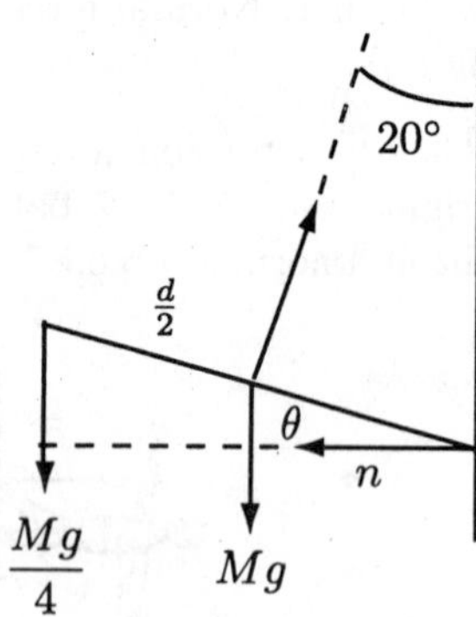

$$T\cos\phi = Mg + \frac{Mg}{4}$$
$$T\sin\phi = n$$

Torque balance about CM

$$\frac{Mg}{4}\frac{d}{2}\cos(45^\circ+\theta) = n\,(0.9d\sin\theta + 0.1d\sin(\theta+45^\circ))$$

Eliminate T

$$\tan\phi = \frac{n}{\frac{5Mg}{4}}$$
$$n = \frac{5Mg}{4}\tan 20^\circ$$

$$\frac{Mgd}{8}\cos(45^\circ+\theta) = \frac{5Mgd}{4}\tan 20^\circ\,(0.9\sin\theta + 0.1\sin(\theta+45^\circ))$$
$$\frac{Mgd}{8}[\cos 45^\circ\cos\theta - \sin 45^\circ\sin\theta] = \frac{5Mgd}{4}\tan 20^\circ\,(0.9\sin\theta + 0.1\,(\sin\theta\cos 45^\circ + \cos\theta\sin 45^\circ))$$

$$\cos\theta\left[\frac{\sqrt{2}}{4} - 5\tan 20^\circ \times 0.1\frac{\sqrt{2}}{2}\right] = 5\sin\theta\left[0.9\tan 20^\circ + 0.1\frac{\sqrt{2}}{2}\tan 20^\circ + 0.1\frac{\sqrt{2}}{2}\right]$$

$$\tan\theta = \frac{0.225}{2.12} = 0.106$$
$$\theta = 6^\circ$$

e)

$$T\cos\phi + f = \frac{5Mg}{4}$$
$$T\sin\phi = n$$

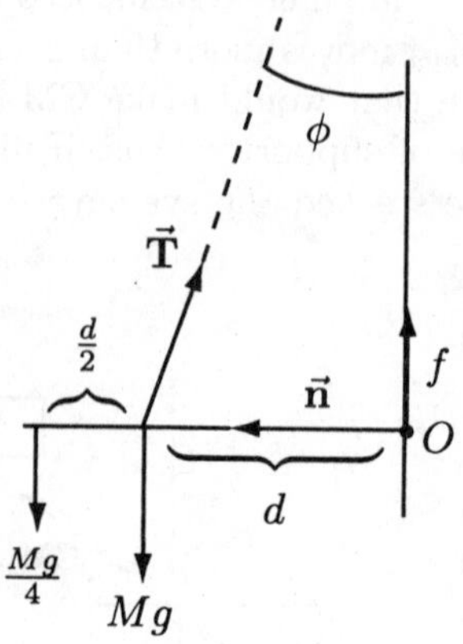

Torque about O

$$\frac{Mg}{4}\left(\frac{3d}{2}\right) + Mgd - T(\cos\phi)\,d = 0$$
$$Mgd\left[\frac{3}{8}+1\right] - d\left[\frac{5Mg}{4} - f\right] = 0$$
$$Mgd\left[\frac{3}{8}+1-\frac{5}{4}\right] + fd = 0$$
$$Mg\left(\frac{1}{8}\right) = -f$$

(f points downward!) Then from force balance

$$T = \frac{n}{\sin\phi}$$
$$n\cot\phi + f = \frac{5mg}{4}$$
$$n = \tan\phi\left[\frac{5mg}{4} + \frac{Mg}{8}\right] = \tan\phi\frac{11Mg}{8}$$
$$\frac{f}{n}\,(\leq \mu_s) = \frac{\frac{Mg}{8}}{\tan\phi\frac{11Mg}{8}}$$

So

$$\mu_{s,\min} = \frac{1}{11\tan\phi} = 0.25$$

11.41 The handle of the scissors rotate at the same rate as the blades. Whatever torque is applied to the handle about scissor's screw is the same torque that the blades can deliver to whatever is being cut. The metal shears must cut metal; which is considerably harder to cut than the cloth that the tailor cuts. We want the force delivered by the shears to be large; a requirement which is not so important for the tailor's scissors. The product rF must be the same for the handle and blade.

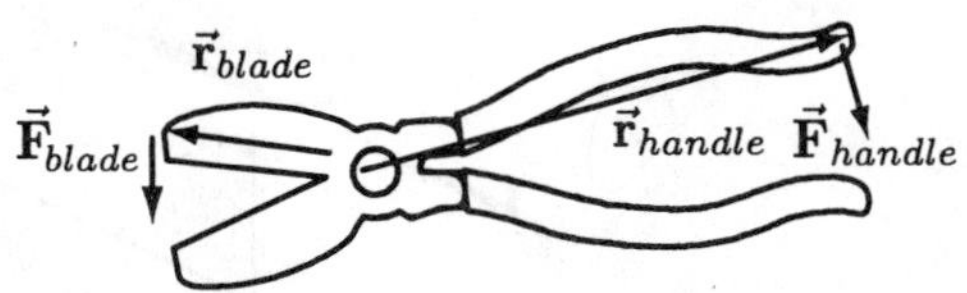

Thus:

$$F_{\text{blade}} = \frac{r_{\text{handle}}}{r_{\text{blade}}} F_{\text{handle}}$$

The force exerted by the blade is much larger if $r_{\text{handle}} > r_{\text{blade}}$ for any given force exerted on the handle. The shears have a very large $\frac{r_{\text{handle}}}{r_{\text{blade}}}$ ratio compared with the tailor's scissors. This reflects the fact that we want a larger force to be exerted by the shear's blades than the tailor's blades.

11.45 If the water is to be pulled up at constant velocity the pulley needs to rotate at constant angular speed.

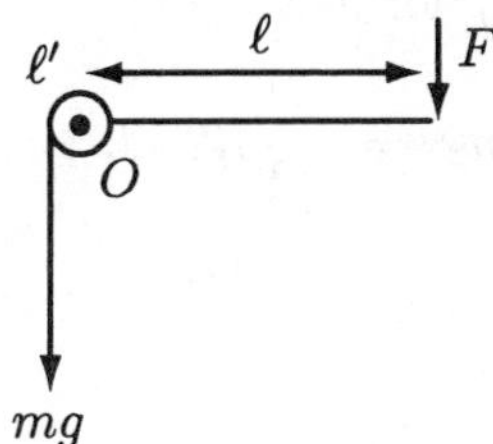

This implies that $\vec{\tau}_{net,0} = 0$. Taking torques about O,

$$\vec{\tau}_{net} = \ell F \hat{\otimes} + \ell' mg \hat{\odot} = 0$$

Solving for F gives:

$$\begin{aligned} F &= \frac{\ell' mg}{\ell} \\ &= \frac{(0.02\ \text{m})(30\ \text{kg})(9.81\,\text{m}/\text{s}^2)}{(0.3\ \text{m})} \\ &= 20\ \text{N} \end{aligned}$$

11.49 We've chosen the box's acceleration to be towards the left.

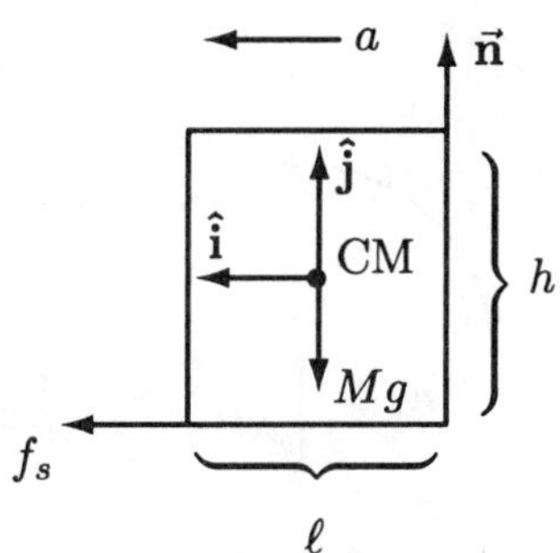

An application of the right hand rule shows that the torque f_s exerts about the CM is $\otimes$, so the normal force must be producing a torque in the $\odot$ direction, so n must be acting through a line to the right of the CM. At the point where the box is just about to topple, n acts at the very edge of the box, as shown.

Force balance:

$$\sum F_y = n - Mg = 0 \tag{5}$$

$$\sum F_x = f_s = Ma \tag{6}$$

Torque balance:

$$\sum \tau = \frac{1}{2}\ell n \odot + \frac{1}{2} h f_s \otimes = 0 \tag{7}$$

From (7),

$$\ell n = h f_s$$

Using (5) + (6),

$$\begin{aligned} \ell\,[Mg] &= h\,[Ma] \\ \Rightarrow a_{\max} &= \frac{\ell g}{h} \\ &= \frac{0.20\ \text{m}}{0.25\ \text{m}} g = 0.80g \end{aligned}$$

If the acceleration of the truck were any greater, a larger friction force would be needed to accelerate the box with the truck. τ_{f_s} would be greater so τ_n would have to increase. The only way for τ_n to increase is for it to move further to the right of CM, which it can't do since it's already acting at $\frac{1}{2}\ell$ (right).

11.53 We will apply force and torque balance to analyze what conditions, if any, are necessary for equilibrium. Let the boxes have a side length of 2ℓ

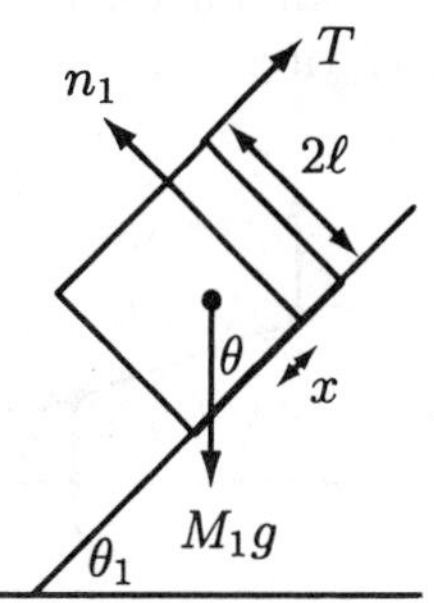

Force Balance:

$$\begin{aligned} n_1 - M_1 g \cos\theta_1 &= 0 && (8) \\ T - M_1 g \sin\theta_1 &= 0 && (9) \\ n_2 - M_2 g \cos\theta_2 &= 0 && (10) \\ T - M_2 g \sin\theta_2 &= 0 && (11) \end{aligned}$$

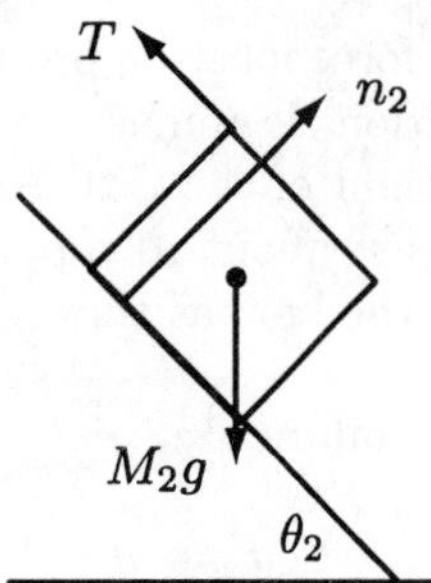

Solving Eq. (9) and (11) for T and setting the equations equal, (we are assuming that tension is constant along the length of string), gives us a necessary condition for equilibrium:

$$M_1 \sin\theta_1 = M_2 \sin\theta_2$$

Now well turn to the question of rotation. We'll compute torque about its CM:

$$\vec{\tau}_{net} = x n_1 \odot + \ell T \otimes = 0 \Rightarrow n_1 = T\left(\frac{\ell}{x}\right)$$

but from (8) and (9) this means

$$\begin{aligned} M_1 \cos\theta_1 &= \left(\frac{\ell}{x}\right) M_1 \sin\theta_1 \\ \tan\theta_1 &= \frac{x}{\ell} \end{aligned}$$

If the box of mass M_1 is just about to topple, the normal force will be right at the edge, $x = \ell$. Thus $\tan\theta_1 \leq 1$, and θ_1 can be no larger than 45°. A similar calculation shows that θ_2 can't be greater than 45° either. Thus equilibrium is not possible unless θ_1 and $\theta_2 \leq 45°$ and $M_1 \sin\theta_1 = M_2 \sin\theta_2$.

11.57 To find the tension in the cable we'll apply torque balance to the car. Since the normal force is unknown we'll place the origin at the car's back tire.

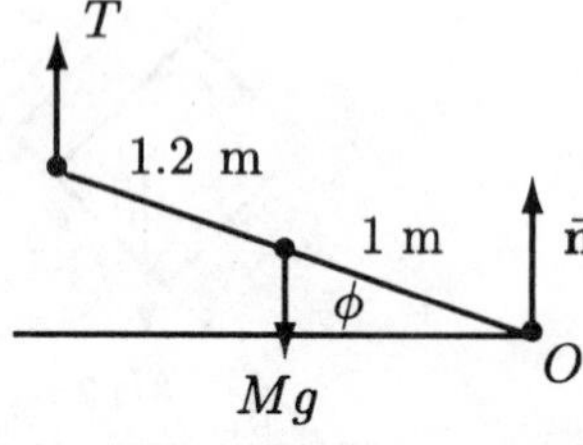

$$\vec{\tau}_{\rm net} = (2.2 \text{ m}) \cos\phi T \otimes + (1 \text{ m}) \cos\phi M g \odot = 0$$

$$\Rightarrow T = \frac{Mg\,(1.0 \text{ m})}{(2.2 \text{ m})} = 4500 \text{ N}$$

We'll model the truck as a rod parallel to the boom which passes through the CM.

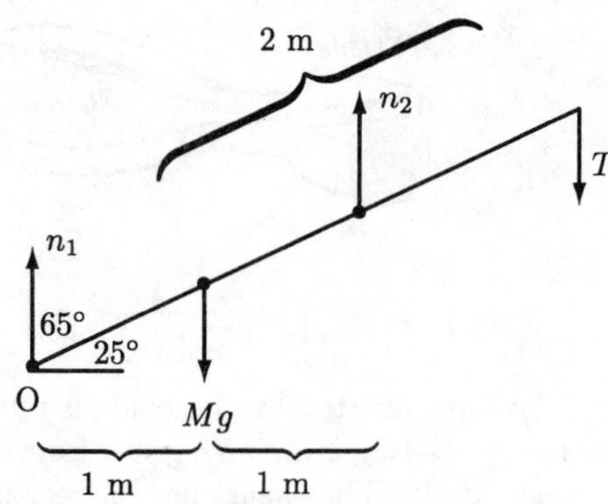

Since the truck doesn't rotate, $\vec{\tau}_{\rm net}$ about any point is 0.

$$\begin{aligned} \vec{\tau}_{{\rm net},O} &= (1 \text{ m})\, Mg \otimes + (2 \text{ m})\, n_2 \odot \\ &\quad + [(2 \text{ m}) \cos 25° + (1 \text{ m})]\, T \otimes \\ &= 0 \end{aligned}$$

Solving for n_2,

$$\begin{aligned} n_2 &= \frac{1}{2}\Big[(2000 \text{ kg})\,(9.81\,\text{m}/\text{s}^2) \\ &\quad + [2\cos 25° + 1]\,(4500 \text{ N})\Big] \\ &= 1.6 \times 10^4 \text{ N} \end{aligned}$$

It is easier to use force balance to find n_1 than to compute more torques.

Force balance:

$$n_1 + n_2 - Mg - T = 0$$

$$n_1 = Mg + T - n_2 = 8.1 \times 10^3 \text{ N}$$

So

Force on each front tire $= 4.0 \times 10^3$ N

Force on each back tire $= 8.0 \times 10^3$ N

11.61 Top barrel force balance:

$$\begin{aligned} Mg - n_1 \sin 45° &= 0 \Rightarrow n_1 = \sqrt{2} Mg \\ n_2 \cos 45° - n_2 &= 0 \Rightarrow n_2 = \frac{1}{\sqrt{2}} n_1 = Mg \end{aligned}$$

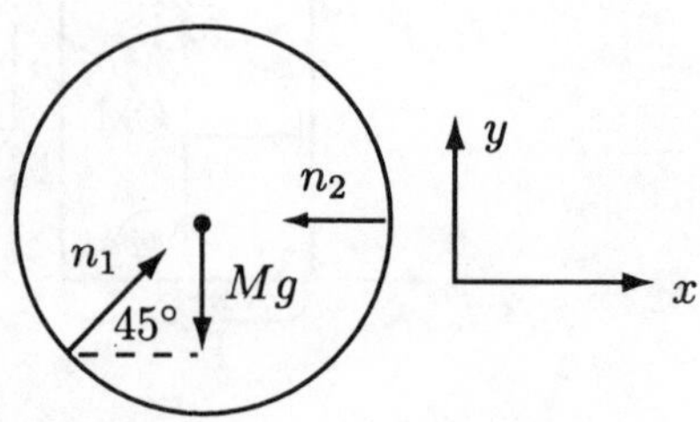

Middle barrel force balance:

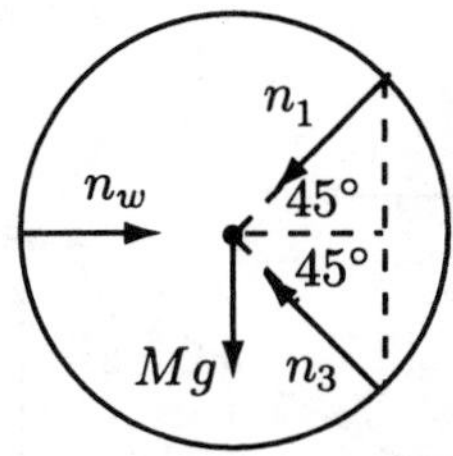

Note that n_1 is known beforehand via Newton's Third Law

$$n_3 \cos 45° - n_1 \cos 45° - Mg = 0$$

$$n_w = (n_1 + n_3)\cos 45° = (n_1 + n_3)\frac{\sqrt{2}}{2}$$

$$n_3 = n_1 + Mg\sqrt{2} = 2Mg\sqrt{2}$$

So

$$n_w = 3Mg$$

Bottom barrel force balance:

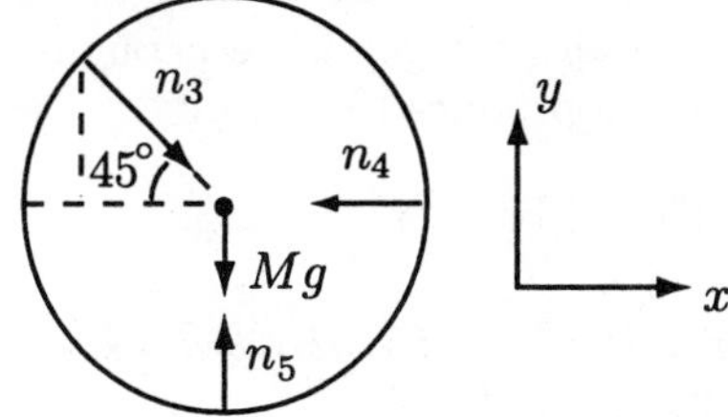

$$n_5 - Mg - n_3 \cos 45° = 0$$

$$\begin{aligned} \Rightarrow n_5 &= Mg + n_3 \cdot \frac{1}{\sqrt{2}} \\ &= Mg + 2\sqrt{2}Mg\left(\frac{1}{\sqrt{2}}\right) \\ &= 3Mg \end{aligned}$$

$$\begin{aligned} n_3 \sin 45° - n_4 &= 0 \\ \Rightarrow n_4 &= \frac{1}{\sqrt{2}} n_3 = 2Mg \end{aligned}$$

The barrels remain in force balance without the need for friction. Each force acting on any barrel passes through that barrel's CM, so $\vec{\tau}_{\text{net,CM}} = 0$. Friction is not needed for torque balance either, thus the value of μ_s is irrelevant.

11.65 The wheel and log system:

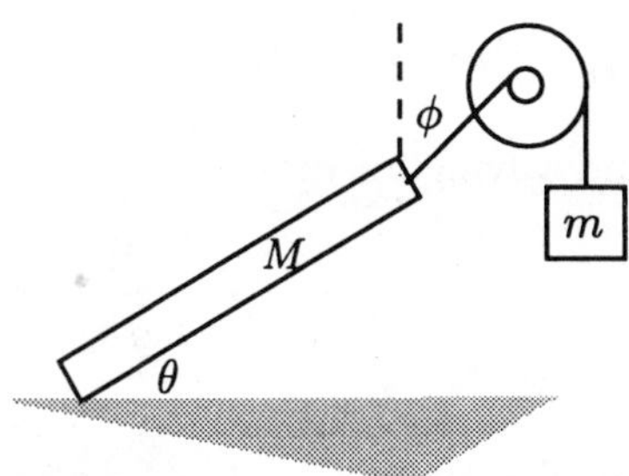

wheel:

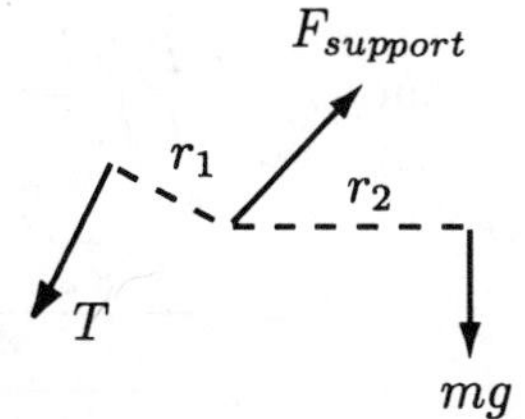

log:

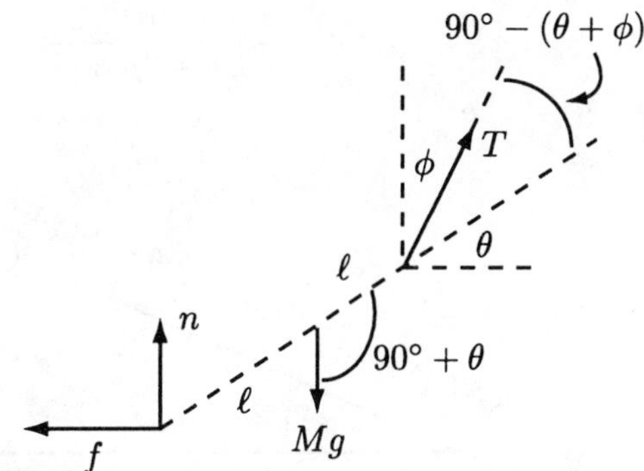

Torque balance on the wheel yields:

$$Tr_1 = r_2 mg$$

The equations for the log are a bit more complex:

vertical forces:

$$n + T\cos\phi = Mg$$

horizontal forces:

$$-f + T\sin\phi = 0$$

Torques about the contact point: (+ equals out-of-page.)

$$2\ell T\cos(\theta + \phi) - Mg\ell\cos\theta = 0$$

To find a limiting valve of m, we shall set $f = \mu n$.

Solution:

$$\begin{aligned} Mg\cos\theta &= 2T\cos(\theta + \phi) \\ &= 2T[\cos\theta\cos\phi - \sin\theta\sin\phi] \\ &= 2\{\cos\theta[Mg - n] - \sin\theta\mu n\} \end{aligned}$$

So

$$-Mg\cos\theta = -2n[\cos\theta + \mu\sin\theta]$$

Thus,

$$n = \frac{Mg}{2}\frac{\cos\theta}{\cos\theta + \mu\sin\theta}$$

Then,

$$\begin{aligned}\tan\phi &= \frac{T\sin\phi}{T\cos\phi} = \frac{\mu n}{Mg-n} \\ &= \frac{\frac{\frac{\mu}{2}\cos\theta}{[\cos\theta+\mu\sin\theta]}}{1-\frac{\frac{1}{2}\cos\theta}{[\cos\theta+\mu\sin\theta]}}\end{aligned}$$

So

$$\tan\phi = \frac{\mu\cos\theta}{\cos\theta+2\mu\sin\theta}$$

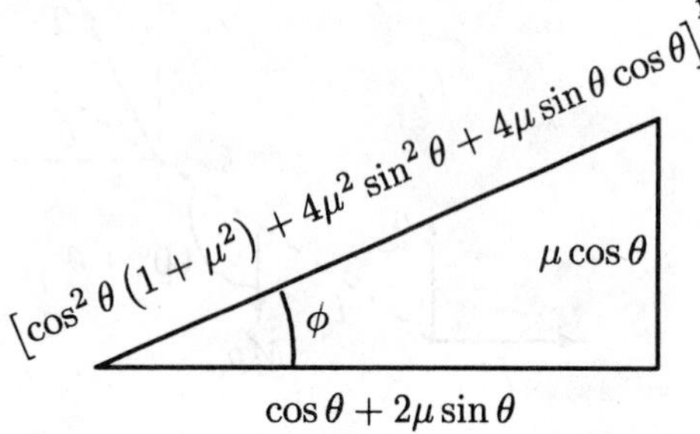

So

$$\begin{aligned}T &= \frac{\mu n}{\sin\phi} \\ &= \mu\frac{Mg}{2}\cdot\frac{\cos\theta}{\cos\theta+\mu\sin\theta} \\ &\cdot\frac{\left[\cos^2\theta\left(1+\mu^2\right)+4\mu^2\sin^2\theta+4\mu\sin\theta\cos\theta\right]^{\frac{1}{2}}}{\mu\cos\theta}\end{aligned}$$

So

$$T = \frac{Mg}{2}\frac{\left[\cos^2\theta\left(1+\mu^2\right)+4\mu^2\sin^2\theta+4\mu\sin\theta\cos\theta\right]^{\frac{1}{2}}}{\cos\theta+2\mu\sin\theta}$$

So, the maximum mass m for which the system can be in equilibrium is

$$m_{\text{max}}(\theta) = \frac{M}{2}\frac{r_1}{r_2}\frac{\left[\cos^2\theta\left(1+\mu^2\right)+4\mu^2\sin^2\theta+4\mu\sin\theta\cos\theta\right]^{\frac{1}{2}}}{\cos\theta+2\mu\sin\theta}$$

The minimum mass m corresponds to maximum friction acting to the right instead of to the left. In this case the cable will be on the other side of the vertical. The result for $m_{\text{min}}(\theta)$ can be found by replacing μ by $-\mu$ in the above formula.

11.69 Force balance:

$$\begin{aligned}-n_2\sin\theta+f_s &= 0 \quad (12) \\ f_{s,\text{max}} &= \mu_s n_1 \quad (13) \\ n_1+n_2\cos\theta-g\,(M+m) &= 0 \quad (14)\end{aligned}$$

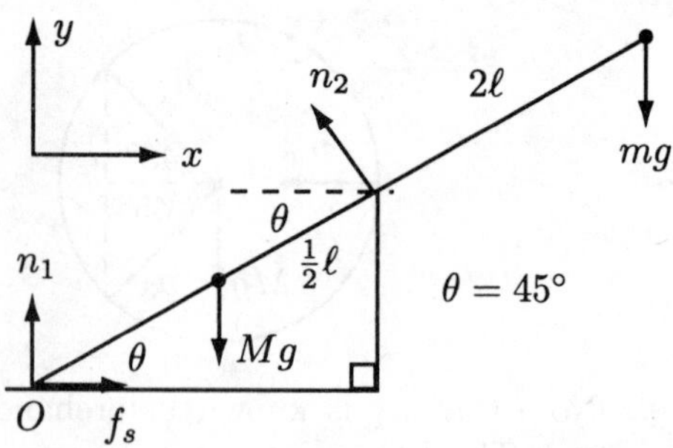

We wiil assume a massless prize. As the person goes up the ladder and pole, the person's weight exerts a larger torque about O. Only two other forces exert a torque about O: Mg and n_2; since Mg is fixed, n_2 must get larger to balance torque. However, the horizontal component of n_2 is limited by Eq. (12) since f_s is limited by Eq. (13). Thus, there is some maximum mass of a person who can win the prize. If the person were more massive, the torque produced by n_2 could not get large enough to balance the torque produced by mg.

Torque balance:

$$\begin{aligned}\vec{\tau}_{\text{net},O} &= \frac{1}{2}\ell Mg\cos\theta\otimes+\ell n_2\odot+3\ell mg\cos\theta\otimes \\ &= 0 \quad (15)\end{aligned}$$

From (15) with $\theta = 45°$, we get

$$n_2 = \frac{g}{\sqrt{2}}\left[\frac{1}{2}M+3m\right]$$

and using this in (12) will give us f_s,

$$\begin{aligned}f_s &= n_2\sin\theta = \frac{1}{2}g\left[\frac{1}{2}M+3m\right] \\ &= \frac{1}{4}g\,[M+6m]\end{aligned}$$

and in (14) will give us n_1.

$$\begin{aligned}n_1 &= g(M+m)-\frac{1}{4}g\,[M+6m] \\ &= \frac{1}{4}g\,[3M-2m]\end{aligned}$$

Now we'll use the condition of static friction:

$$f_s \le \mu_s n_1 \Rightarrow \frac{1}{4}g\,[M+6m] \le \mu_s\frac{1}{4}g\,[3M-2m]$$

and solving for m:

$$m \le \frac{M\,[3\mu_s-1]}{2\,(3+\mu_s)} \doteq 56 \text{ kg}$$

11.73 We can look at any two pieces of the rod equally distant from the rod's geometric center.

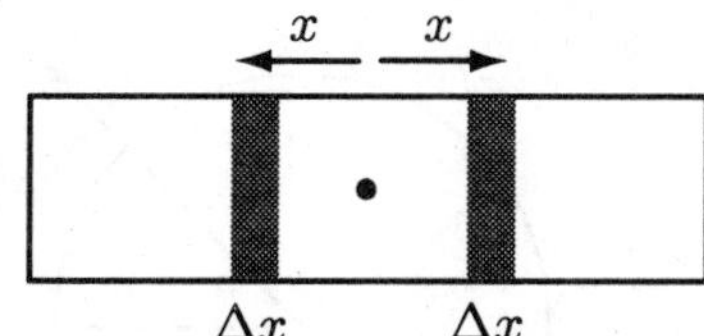

The pieces have equal densities and thus equal mass (we are talking about equal elements of volume). They are of equal mass and equal distance from the origin so their CM is at the geometric center of the rod. The rod is made up of many such pairs, so its CM is at its geometric center.

11.77 There are equal amounts of mass on either side of the y-axis, and the mass is distributed symmetrically *wrt* the y-axis.

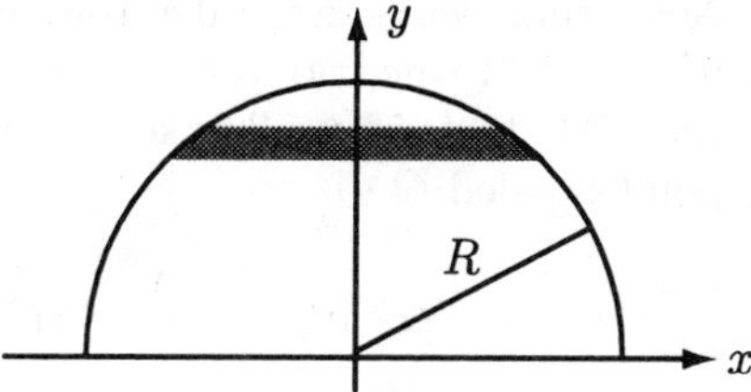

For every dm at $x = x_0$ there is a dm at $x = -x_0$; thus

$$x_{CM} = \frac{1}{m}\int x\,dm = 0$$

The only question is what is y_{CM}? Since $y_{CM} = \int y\,dm$ we will model the semicircle as a collection of strips of width dy, and length $2x$. The semicircle has an equation of $x = \sqrt{R^2 - y^2}$ (obtained from the equation of a circle knowing that, since we are integrating *wrt* y, we need to express the curve as a function of y).

$$\overbrace{\qquad\qquad}^{x}$$
$$\boxed{dm = \sigma\ dA}\ \}\ dy$$

The mass of the semicircle is:

$$M = \frac{1}{2}\pi R^2\sigma$$

where σ is the mass per unit area. The mass of one differential strip is:

$$dm = \sigma\ dA = \sigma 2x\,dy = 2\sigma\sqrt{R^2 - y^2}\,dy$$

y varies from 0 to R.

$$\begin{aligned} y_{CM} &= \frac{1}{M}\int y\ dm = \frac{2}{\pi R^2\sigma}\int_0^R 2\sigma y\sqrt{R^2 - y^2}\,dy \\ &= \frac{4}{\pi R^2}\int_0^R y\sqrt{R^2 - y^2}\,dy \\ &= \frac{2}{\pi R^2}\int_0^{R^2} u^{1/2}\,du = \frac{4R}{3\pi} \end{aligned}$$

$(u = R^2 - y^2,\ du = -2y\,dy)$

11.81 Consider the end of the boom as a massless particle:

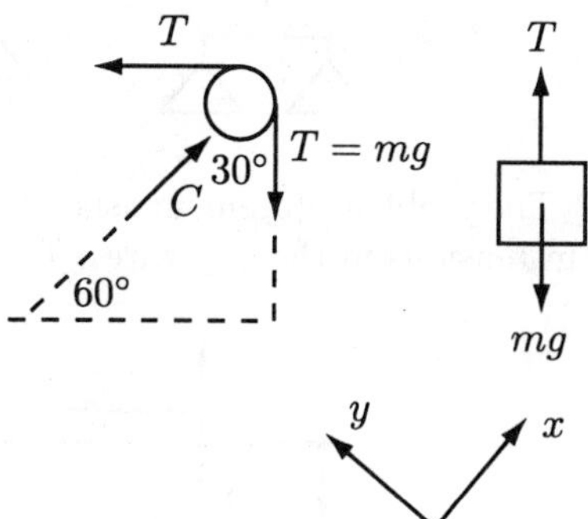

The rope around the boom has constant tension, so tension is constant along its length. The rope supports the mass m of sand, so $T = mg$. The compression of the boom must equal the component of both tension forces which are $\parallel$ to the boom: the x-component.

$$\begin{aligned} C - 2T\cos 30° &= 0 \\ \Rightarrow C &= 2mg\cos 30° \\ &= 8.5 \times 10^4\ \text{N} \end{aligned}$$

11.85 We can start off with

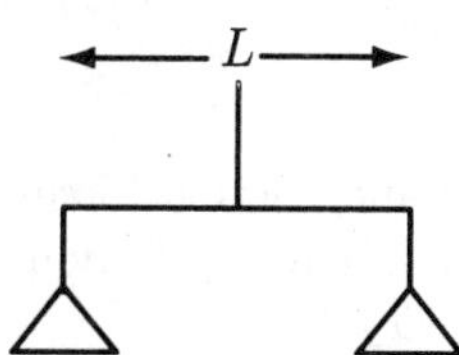

but we need to place another prism. According to torque balance, if one side is twice as heavy the other side needs to have twice the moment arm for torque to balance.

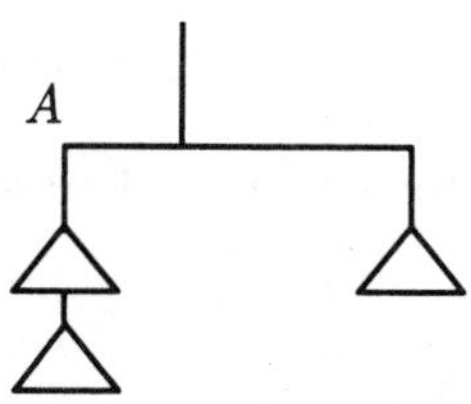

We have another stick to deal with. We can hang the two prisms on it just as long the system of stick plus two prisms hang from point A to insure

$$\tau_{\text{left}} = \tau_{\text{right}} \Rightarrow \ell\,(2mg) = (2\ell)mg$$
$$3\ell = L \Rightarrow \ell = \frac{L}{3}$$

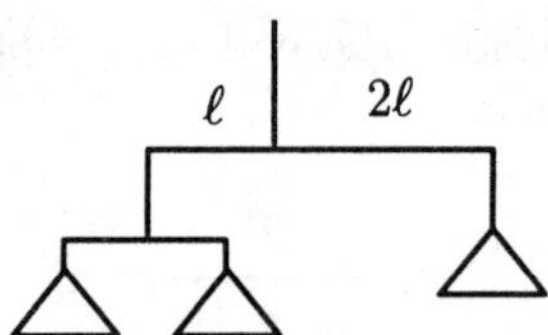

11.87 The problem statement asked for the difference in tensions in the guy wires, $T_1 - T_2$.

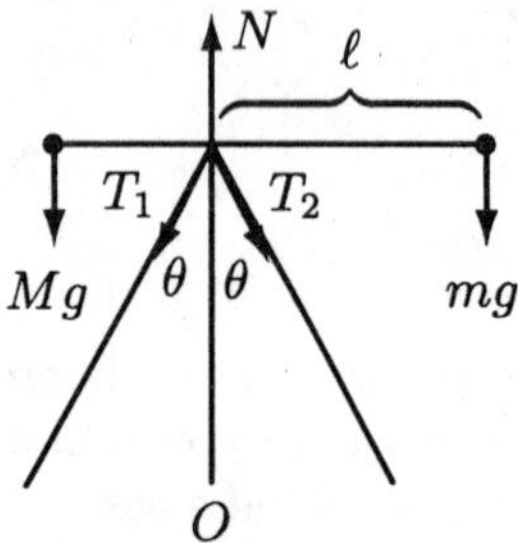

We'll compute torque about O. Note

$$\theta = \tan^{-1}\left(\frac{10 \text{ m}}{100 \text{ m}}\right) = 5.7°$$

$$\begin{aligned}\vec{\tau}_{\text{net},O} &= \ell mg\hat{\otimes} + (5 \text{ m})\, Mg\hat{\odot} \\ &\quad + (100 \text{ m})\sin(5.7°)\, T_2 \otimes \\ &\quad + (100 \text{ m})\sin(5.7°)\, T_1 \odot \\ &= 0\end{aligned}$$

Normal and friction force at O do not produce a torque about O. Solving the above equation for $T_1 - T_2$,

$$T_1 - T_2 = \frac{g\,[\ell m - (5 \text{ m})\, M]}{(100 \text{ m})\sin(5.7°)}$$

When the load is all the way out to the end, $\ell = 30$ m and

$$T_1 - T_2 = 2.5 \times 10^3 \text{ N}$$

When the load is next to the tower, $\ell = 0$ m.

$$T_1 - T_2 = -1.2 \times 10^4 \text{ N}$$

meaning that $T_2 > T_1$.

11.91 The moment F becomes large enough to lift the wheel over the step, $n_2 \to 0$.

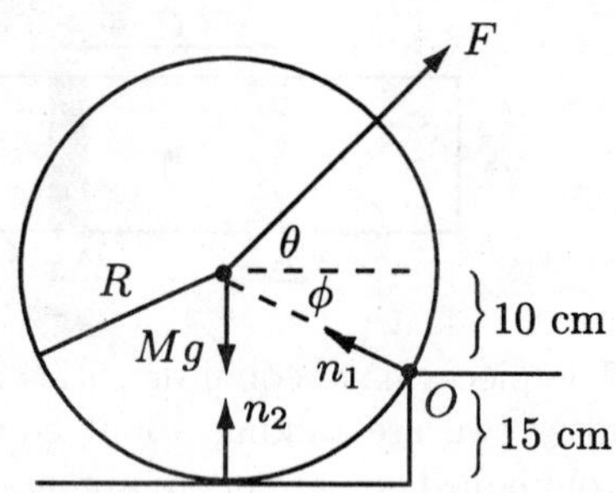

Also, note that n_1 being a force normal to the wheel's surface acts through the wheel's CM. Thus,

$$\phi = \sin^{-1}\left(\frac{10 \text{ cm}}{25 \text{ cm}}\right) = 23.6°$$

Note that we can't take torques about the wheel's CM since no force has a moment about the CM. Instead we'll take torques about the point labeled O.

$$\begin{aligned}\sum \vec{\tau}_O &= RF\sin(\theta + \phi) \otimes + RMg\cos\phi \odot = 0 \\ \Rightarrow F &= \frac{Mg\cos\phi}{\sin(\theta + \phi)}\end{aligned}$$

Minimum F occurs for max $\sin(\theta + \phi) \Rightarrow \theta = 90° - \phi$. So

$$\begin{aligned}F_{\min} &= (75 \text{ kg})\left(9.81 \text{ m/s}^2\right)\cos 23.6° \\ &= 670 \text{ N about } 66° \text{ from horizontal}\end{aligned}$$

11.95 We'll picture building a stack of blocks by placing the lowest block so that the CM of the combined system falls on the edge of the lowest block.

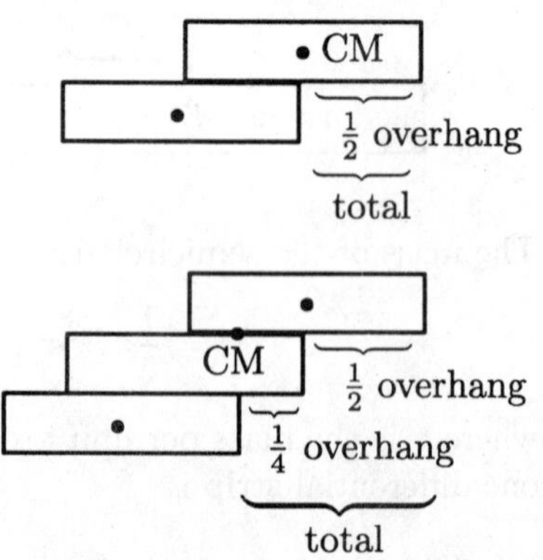

We need to imagine where the CM of the upper $N - 1$ blocks is and support it by making sure that it falls directly above the edge of the lowest block.

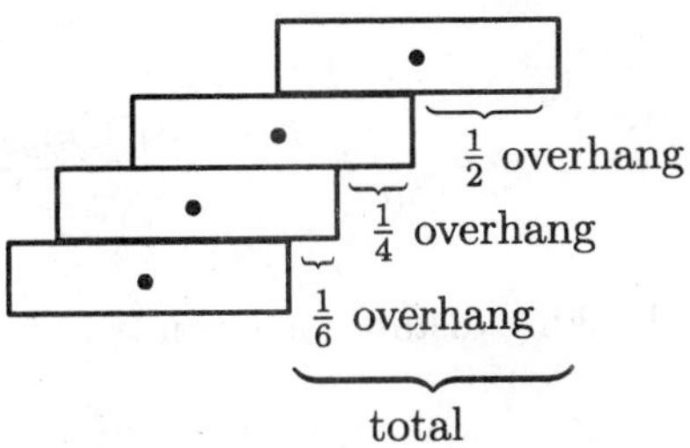

If the blocks have a length of ℓ there is a pattern that emerges:

$$\text{total overhang} = \frac{1}{2}\sum_{i=1}^{N}\left(\frac{1}{i}\right)\ell$$

when there are $N+1$ blocks. For 10 blocks, overhang:

$$\begin{aligned} &= \frac{1}{2}\left(1+\frac{1}{2}+\frac{1}{3}+\frac{1}{4}+\frac{1}{5}+\frac{1}{6}+\frac{1}{7}+\frac{1}{8}+\frac{1}{9}\right)\ell \\ &= 1.41\ell = 11.3 \text{ cm} \end{aligned}$$

Note that the overhang increases as the harmonic series; an infinite amount of blocks could yield an infinite amount of overhang.

11.99 a) There are two types of small triangles:

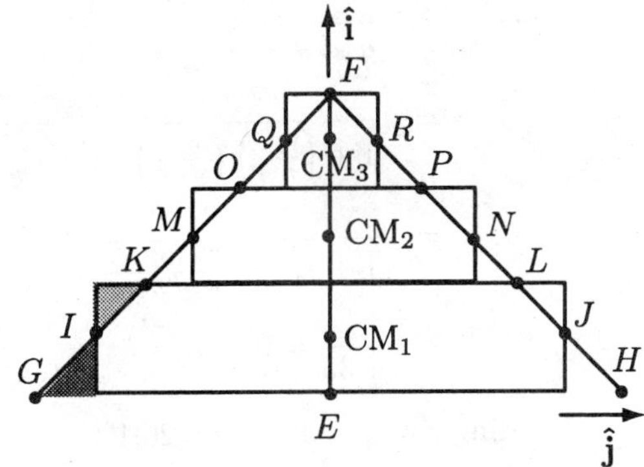

the triangle formed by the corner of a rectangle (lighter) and the part of the large triangle that is not covered by the rectangle (darker). These two types of triangles come in pairs; if we can show that their areas are equal without using facts concerning how many rectangles there are, $A_{\text{triangle}} = A_{\text{rectangle}}$ for any number, N, of rectangles.

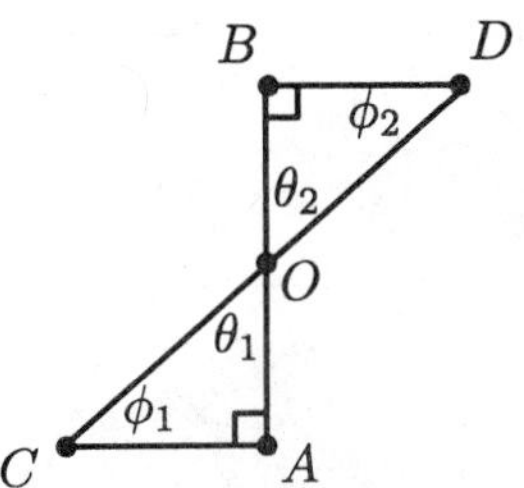

Note that θ_1 and θ_2 are opposite angles and must therefore be equal. Then since the triangles both have right angles in them, $\phi_1 = \phi_2$. Since all the angles are equal, the two triangles must be at least similar. We are told that the side of the large triangle intersects the side of the rectangle at its midpoint. AB represents the side of one of the rectangles, so $AO = AB$. Thus $\Delta AOC \equiv \Delta BOD$, and the total area of the rectangles equals the area of the large triangle.

b) If you're like us, the first question you asked is: "How am I supposed to know the length of the rectangles?" The answer lies in the fact that the midpoint of any rectangle's side is centered on the triangle's side and that the triangles' sides are straight lines. Take the $N = 3$ case. At the bottom rectangle's midpoint level (at y_1), the length GH decreased by $\frac{1}{6}$, that is, $IJ = \frac{5}{6}GH$. Similarly, $KL = \frac{4}{6}GH$, $MN = \frac{3}{6}GH$, $OP = \frac{2}{6}GH$ and $QR = \frac{1}{6}GH$. This will give us the length of any rectangle. Since the rectangles are uniform, the CM of each is at its geometric center, which will always be on FE. Thus the CM of all the rectangles must lie on that line too. If $GH = L$ and $EF = H$.

$N = 3$: Each rectangle has a height of $\frac{H}{3}$. Let y_n be the y-coordinate of the CM of the nth rectangle, counted from the bottom and M_n be the mass of that rectangle.

$$\begin{aligned} y_1 &= \frac{H}{6} \\ y_2 &= \frac{H}{2} \\ y_3 &= \frac{5H}{6} \\ M_1 &\propto \left(\frac{H}{3}\right)\left(\frac{5}{6}L\right) = \frac{5}{18}HL \\ M_2 &\propto \left(\frac{H}{3}\right)\left(\frac{L}{2}\right) = \frac{HL}{6} \\ M_3 &\propto \left(\frac{H}{3}\right)\left(\frac{L}{6}\right) = \frac{HL}{18} \end{aligned}$$

$$h_{\text{CM all rectangles}} = \frac{\sum m_i y_i}{\sum m_i}:$$

$$\begin{aligned} &= \frac{\left(\frac{5}{18}HL\right)\left(\frac{H}{6}\right) + \left(\frac{HL}{6}\right)\left(\frac{H}{2}\right) + \left(\frac{HL}{18}\right)\left(\frac{5H}{6}\right)}{\left(\frac{5}{18}+\frac{1}{6}+\frac{1}{18}\right)HL} \\ &= \frac{1}{6}H\frac{(5+9+5)}{(5+3+1)} \\ &= \frac{19H}{54} = 0.35H = 0.53 \text{ m} \end{aligned}$$

Remember,

$$m_\Delta = m_{\text{all rectangles}}$$

We may use a spreadsheet to perform the calculations for $n > 3$. The results are shown in the graph in the appendix.

11.103 Let's draw a force-torque diagram for the aircraft:

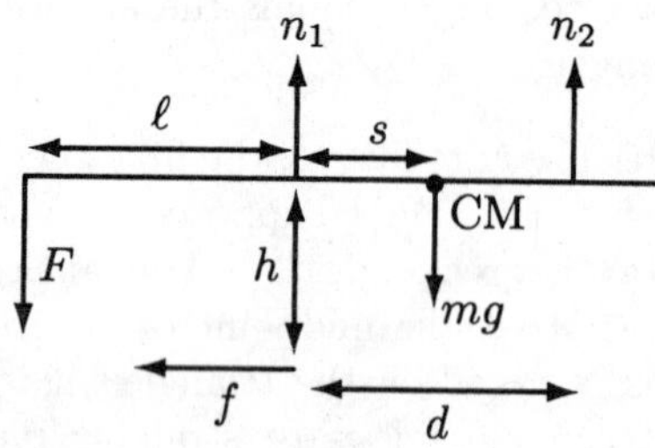

$$\begin{aligned} \ell &= 7.0 \text{ m} \\ s &= 0.70 \text{ m} \\ d &= 2.0 \text{ m} \end{aligned}$$

Force balance:

$$n_1 + n_2 = mg + F \tag{16}$$

Newton's Second Law

$$f = ma \tag{17}$$

friction (the wheels skid)

$$f = \mu_k n_1 \tag{18}$$

Torque about the CM ($\vec{\mathbf{L}}$ about CM remains constant as the aircraft slows.)

$$F(\ell + s) + n_2(d - s) = n_1 s + fh \tag{19}$$

From (16)

$$n_2 = mg + F - n_1$$

From (18)

$$n_1 = \frac{f}{\mu}$$

Then from (19)

$$F(\ell + s) + \left(mg + F - \frac{f}{\mu}\right)(d - s) = \frac{f}{\mu}s + fh$$

Solve for f

$$\begin{aligned} f\left[\frac{s}{\mu} + h + \frac{(d-s)}{\mu}\right] &= F(\ell + s) + F(d - s) \\ &\quad + mg\,(d - s) \\ f\left(h + \frac{d}{\mu}\right) &= F(\ell + d) + mg\,(d - s) \end{aligned}$$

Thus

$$a = \frac{f}{m} = \frac{F(\ell + d) + mg\,(d - s)}{m\left(\frac{h+d}{\mu}\right)}$$

Braking distance is given by:

$$0 - v_i^2 = -\frac{2f}{m} \cdot x$$

Thus

$$x = \frac{mv_i^2}{2f} = \frac{v_i^2}{2} \cdot \frac{m\left(\frac{h+d}{\mu}\right)}{[F(\ell + d) + mg\,(d - s)]}$$

Without F

$$x = \frac{v_i^2}{2} \frac{\not{m}\left(\frac{h+d}{\mu}\right)}{\not{m}g\,(d - s)}$$

With F

$$x_F = \frac{v_i^2}{2} \frac{\left(\frac{h+d}{\mu}\right)}{g\,(d - s) + \frac{F}{m}(\ell + d)}$$

Comparing:

$$\begin{aligned} \frac{x_F}{x} &= \frac{g\,(d - s)}{g\,(d - s) + \frac{F}{m}(\ell + d)} \\ &= \frac{1}{1 + \left(\frac{F}{mg}\right)\left(\frac{\ell+d}{d-s}\right)} \\ &= \frac{1}{1 + \frac{(1.0\times10^3 \text{ N})}{(2\times10^3 \text{ kg})(9.8 \text{ m/s}^2)}\left(\frac{9.0 \text{ m}}{1.3 \text{ m}}\right)} \\ &= \frac{1}{1.35} = 0.739 \end{aligned}$$

Landing roll is reduced by 26.1%.

Chapter 12

Dynamics of Rigid Bodies

12.1 We can place a coordinate system centered on the docking module so that the whole station rotates on the $\hat{\mathbf{k}}$ axis.

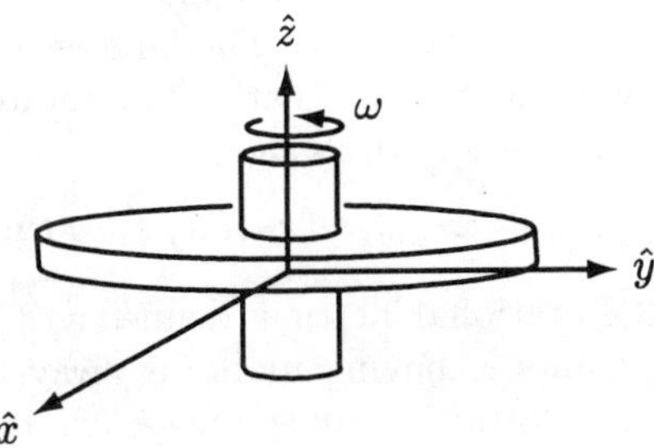

The advantage is two-fold. The station is radially symmetric *wrt* the $\hat{\mathbf{k}}$ axis with the angular velocity parallel or anti-parallel to the axis. Furthermore, any point on the station we wish to look at will be at a constant radial distance from the origin, undergoing uniform circular motion.

12.3 From $w_f^2 - w_i^2 = 2\alpha\Delta\theta$,

$$\begin{aligned} \Delta\theta &= \frac{w_f^2}{2\alpha} = \frac{\left(1.0 \times 10^2 \text{ rad}/\text{s}\right)^2}{2\,(1.0 \times 10^{-3} \text{ rad}/\text{s}^2)} \\ &= 5.0 \times 10^6 \text{ rad} \left(\frac{1 \text{ rev}}{2\pi \text{ rad}}\right) \\ &= 8.0 \times 10^5 \text{ rev} \end{aligned}$$

12.5 Assuming $\vec{\mathbf{L}}$ is $\|$ to $\vec{\omega}$,

$$\vec{\mathbf{L}} = I\vec{\omega} = (12 \text{ J} \cdot \text{s}^2)(5.0 \,/\, \text{s})\hat{\omega} = (60 \text{ J} \cdot \text{s})\,\hat{\omega}$$

12.7 We will find the angular acceleration of the flywheel and relate it to the torque that the tension force exerts on the flywheel by $\tau = I\alpha$. From τ_T we can find T. The flywheel starts from rest, so:

$$\theta = \frac{1}{2}\alpha t^2 \Rightarrow \alpha = \frac{2\Delta\theta}{t^2}$$

Since the flywheel is symmetric about the axis of rotation, I is $\|$ to $\vec{\omega}$ and:

$$\begin{aligned} \tau &= I\alpha = \frac{1}{2}MR^2\alpha \\ |\vec{\tau}| &= \left|\vec{\mathbf{R}} \times \vec{\mathbf{T}}\right| = TR \end{aligned}$$

since $\vec{\mathbf{T}}$ and $\vec{\mathbf{R}}$ are $\perp$

$$I\alpha \stackrel{\text{set}}{\equiv} TR$$

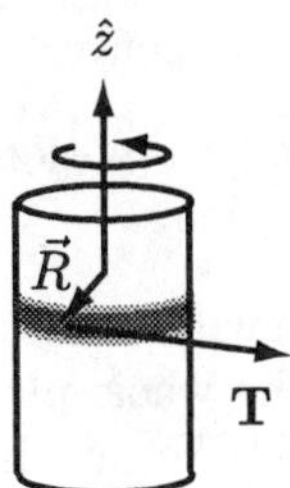

$$\begin{aligned} \Rightarrow T &= \frac{I\alpha}{R} = \frac{1}{2}MR\alpha = \frac{MR\Delta\theta}{t^2} \\ &= \frac{(220 \text{ kg})\,(2.0 \text{ m})\,(250 \text{ rad})}{(120 \text{ s})^2} \\ &= 7.6 \text{ N} \end{aligned}$$

12.9 While in the air, the diver is free of external torques and has constant angular momentum about her center of mass. So,

$$\begin{aligned} L_{\text{before}} &= I_i\omega_i = I_f\omega_f = L_{\text{tucked,ie.after}} \\ \omega_f &= \frac{I_i}{I_f}\omega_i = \frac{10.3\ \text{kg}\cdot\text{m}^2}{3.61\ \text{kg}\cdot\text{m}^2}(3.14\ \text{rad/s}) \\ &= 8.96\ \text{rad/s} \end{aligned}$$

"Extra kinetic energy" $= K_f - K_i$

$$\begin{aligned} &= \frac{1}{2}I_f\omega_f^2 - \frac{1}{2}I_i\omega_i^2 = \frac{1}{2}\left[I_f\left(\frac{I_i}{I_f}\omega_i\right)^2 - I_i\omega_i^2\right] \\ &= \frac{1}{2}I_i\omega_i^2\left(\frac{I_i}{I_f} - 1\right) \\ &= \frac{1}{2}(10.3\ \text{kg}\cdot\text{m}^2)(3.14\ \text{rad/s})^2\left(\frac{10.3}{3.61} - 1\right) \\ &= 94.1\ \text{J} \end{aligned}$$

12.11 The gyroscope's angular momentum is given by

$$L = I_{\text{disk}}\omega = \frac{1}{2}mr^2\omega$$

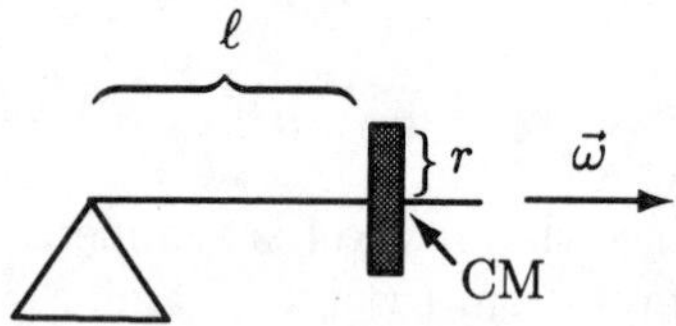

Its torque is $\tau = mg\ell$, so the gyroscope's precession is:

$$\begin{aligned} \Omega &= \frac{\tau}{L} = \frac{2g\ell}{r^2\omega} \\ &= \frac{2(9.8\ \text{m/s}^2)(2.0\times10^{-3}\ \text{m})}{(2.0\times10^{-2}\ \text{m})^2(3.0\times10^3\ \text{rpm})(2\pi)\left(\frac{1\ \text{min}}{60\ \text{s}}\right)} \\ &= 0.31\ \text{rad/s} \end{aligned}$$

Dont forget to convert ω to SI units!

12.17 The propeller's motion is a combination of two types of motion:

1. Motion with CM: even if the propeller stopped rotating, the whole plane would still move *wrt* the ground. We say that this is the motion of the plane's CM. Since the propeller is fixed *wrt* the CM, $v_{\text{CM}} = v_{\text{prop}}$ (if the prop stopped spinning).

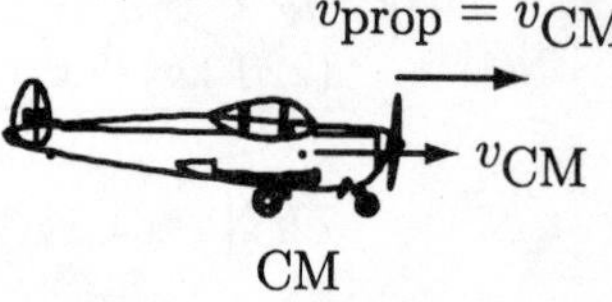

2. Motion *wrt* the CM: The propeller does spin, and this motion is non-zero *wrt* the plane's CM; see below.

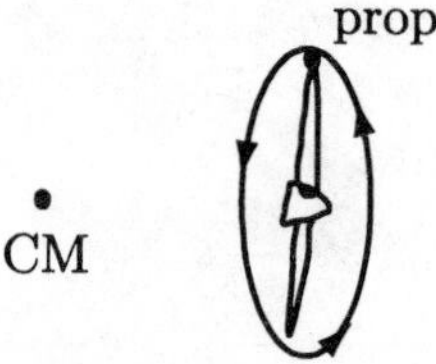

Each type of motion by itself could be less than the speed of sound, but the propeller's total velocity, which is the vector sum of both velocities, may have a magnitude greater than the speed of sound.

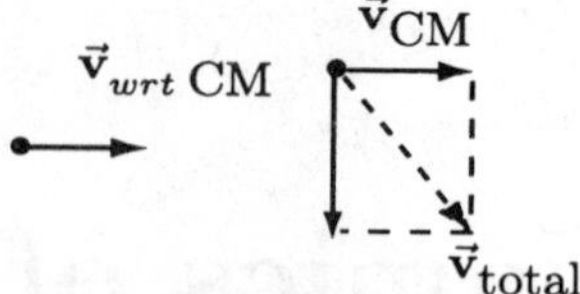

Even if $|\vec{v}_{\text{CM}}| < |\vec{v}_{\text{sound}}|$ and $|\vec{v}_{\text{wrt CM}}| < |\vec{v}_{\text{sound}}|$, $|\vec{v}_{\text{CM}} + \vec{v}_{\text{wrt CM}}|$ may be greater than $|\vec{v}_{\text{sound}}|$.

12.19 Since gears are not allowed to skip teeth, the 5-tooth gear must go around twice for every time the 10-tooth gear goes around once. Thus the 5-tooth gear has a angular speed double that of the 10-tooth gear. The large gear is stationary wrt the 5-tooth gear. They rotate at the same angular speed, .8 rad / s.

$$v_{\text{rim}} = \omega r_{\text{rim}} = (.8\ \text{rad/s})(.40\ \text{m}) = .3\ \text{m/s}$$

12.23 Rotational inertia is defined as $\int r_\perp^2\,dm$. What counts is having mass far away from the axis of rotation. Since they each have the same mass the question becomes which has most of its mass located far away from the axis of rotation; the farther away this mass is distributed, the greater the object's rotational inertia.

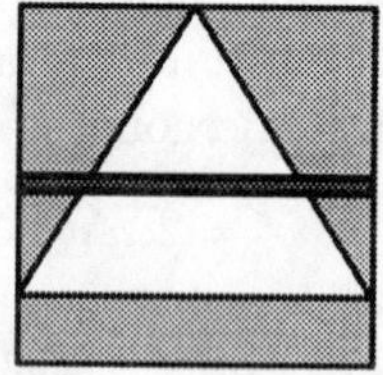

We superimposed the rod and the triangle on the square. Clearly the square has more mass located farther away (represented by the shaded area), so from this diagram,

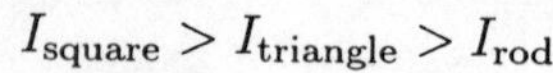

$$I_{\text{square}} > I_{\text{triangle}} > I_{\text{rod}}$$

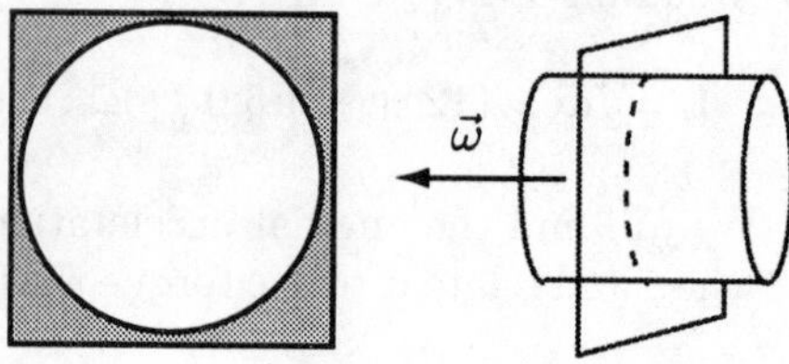

Now we superimposed the square and cylinder. Remember that its the mass far away from the axis of rotation that counts—all of the mass of the cylinder is closer to this axis than the shaded portion of the square. Thus, we have

$$I_{\text{square}} > I_{\text{cylinder}}$$

By the same reasoning, the next picture shows $I_{\text{cylinder}} > I_{\text{sphere}}$.

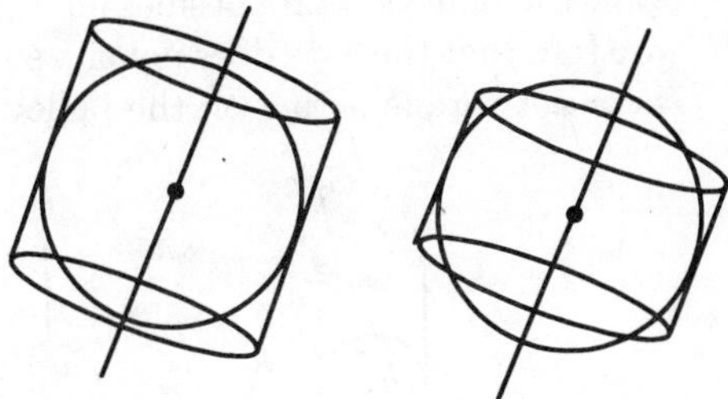

The height of the cylinder is irrelevant since the sphere's sides drop away immediately after its equator and the cylinder's do not. Thus:

$$I_{\text{square}} > I_{\text{cylinder}} > I_{\text{sphere}}$$

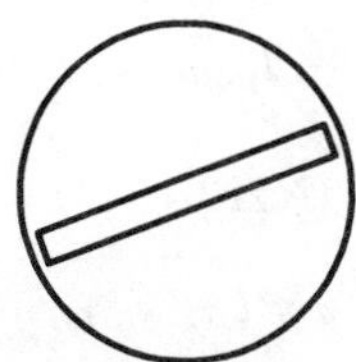

The square has the greatest rotational inertia. Clearly, $I_{\text{sphere}} > I_{\text{rod}}$ so the rod has the smallest.

12.27 The total energy of the bicycle system is due to the speed of the bicycle's CM and rotation of its two tires.

$$\begin{aligned} E_{\text{tot}} &= \frac{1}{2}m_{\text{tot}}v^2 + 2\left(\frac{1}{2}I_{\text{wheel}}\omega^2\right) \\ &= \frac{1}{2}m_{\text{tot}}v^2 + I_{\text{wheel}}\left(\frac{v}{r_{\text{wheel}}}\right)^2 \end{aligned}$$

$I_{\text{wheel}} = mr^2$, so

$$\begin{aligned} E_{\text{tot}} &= \frac{1}{2}m_{\text{tot}}v^2 + m_{\text{wheel}}v^2 \\ &= \frac{1}{2}v^2\left[m_{\text{tot}} + 2m_{\text{wheel}}\right] \end{aligned}$$

The power delivered to the bike is the rate of change of the bike's energy.

$$\begin{aligned} P &= \frac{dE_{\text{tot}}}{dt} = \frac{1}{2}\left[m_{\text{tot}} + 2m_{\text{wheel}}\right]\frac{d}{dt}v^2 \\ &= \left[m_{\text{tot}} + 2m_{\text{wheel}}\right]v \cdot \frac{dv}{dt} \\ &= \left[15\text{ kg} + 75\text{ kg} + 2\,(4.0\text{ kg})\right] \\ &\quad \times (5.0\text{ m/s})\left(1.0\text{ m/s}^2\right) \\ &= 490\text{ W} \end{aligned}$$

12.31 We are to assume that the frictional forces, and by extension, frictional torques remain constant. Then, the wheel undergoes uniform angular acceleration when the motor is turned off. We find the required force from the wheel's observed deceleration. The required power from the motor is given by the product $\tau\omega$ when the motor is running. The torque τ is the same as that found from observation when the motor is off.

$$|\vec{\tau}| = I\,|\vec{\alpha}| = \left|\vec{\mathbf{f}}\right| r_{\text{shaft}}$$

So,

$$\begin{aligned} \left|\vec{\mathbf{f}}\right| &= \frac{I\alpha}{r_{\text{shaft}}} \\ &= \frac{\frac{1}{2}m_{\text{wheel}}(r_{\text{wheel}})^2(350\text{ rpm})\left(\frac{2\pi\text{ rad}}{\text{rev}}\right)\left(\frac{60\text{ s}}{1\text{ min}}\right)^{-1}}{(2.5\times10^{-2}\text{ m})\;45\text{ s}} \\ &= \frac{1}{2}(4.2\text{ kg})\,(0.28\text{ m})^2\,\frac{(0.814\text{ rad/s}^2)}{2.5\times10^{-2}\text{ m}} \\ &= 5.4\text{ N} \end{aligned}$$

Then (motor on) the required power is

$$\begin{aligned} P &= \tau\omega \\ &= \left|\vec{\mathbf{f}}\right| r_{\text{shaft}}\,(350\text{ rpm})\left(\frac{2\pi\text{ rad}}{\text{rev}}\right)\left(\frac{1\text{ min}}{60\text{ s}}\right) \\ &= (5.4\text{ N})(2.5\times10^{-2}\text{ m})(36.7\text{ rad/s}) \\ &= 4.9\text{ W} \end{aligned}$$

12.35 Steps 1, 2, 3:

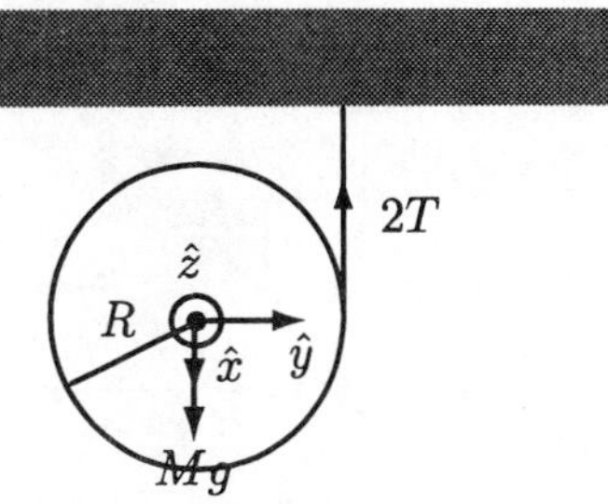

The cylinder will unwind CCW, so $\vec{\omega}$ and $\vec{\alpha}$ will both point out of the page. We chose a coordinate system with $\hat{\mathbf{k}}$ also pointing out of the page.

$$I_{\text{cylinder}} = \frac{1}{2}MR^2 \tag{1}$$

Step 4a: (N2L)

$$\sum F_x = Mg - 2T = Ma_x \tag{2}$$

Step 4b: (torque about CM)

$$\sum \tau_{net,z} = 2TR = I\alpha_z \qquad (3)$$

Step 5: (relations and constraints) With x positive downward, counterclockwise rotation ($\omega_z > 0$) corresponds to $v_x > 0$. The cylinder rolls along the string, so

$$v_x = \omega_z R \qquad (4)$$

Differentiated, this gives

$$a_x = \alpha_z R$$

Step 6: We are certainly going to need the cylinder's linear acceleration, so that is a good unknown to solve for first. From (3) and (4),

$$TR = \frac{Ia_x}{2R} \Rightarrow T = \frac{Ia_x}{2R^2} \overset{Eq.(1)}{\Rightarrow} T = \frac{1}{4}Ma_x$$

Using this result in (2) gives:

$$a_x = \frac{2}{3}g$$

It will take the cylinder a time Δt to roll a distance L. Once it rolls this distance the string will have completely unwound. Think of a spool of thread rolling with a length L of thread:

$$L \equiv \frac{1}{2}a_x\Delta t^2 \Rightarrow \Delta t = \sqrt{\frac{3L}{g}}$$

At $t = \Delta t$ the cylinder's CM velocity is:

$$\vec{\mathbf{v}} = \vec{\mathbf{a}}\Delta t = \frac{2}{3}g\sqrt{\frac{3L}{g}}\hat{\mathbf{i}} = \sqrt{\frac{4gL}{3}}\hat{\mathbf{i}}$$

Its angular velocity is

$$\begin{aligned}
\vec{\omega} &= \frac{v_x}{R}\hat{\mathbf{k}} = \frac{1}{R}\sqrt{\frac{4gL}{3}}\hat{\mathbf{k}} \\
K_{\text{CM}} &= \frac{1}{2}Mv_{\text{CM}}^2 = \frac{2}{3}MgL \\
K_{\text{rot}} &= \frac{1}{2}I\omega^2 = \frac{1}{2}\left(\frac{1}{2}MR^2\right)\left[\frac{1}{R}\sqrt{\frac{4gL}{3}}\right]^2 \\
&= \frac{1}{3}MgL \\
K_{\text{tot}} &= K_{e,\text{CM}} + K_{e,\text{rot}} = MgL
\end{aligned}$$

Which is the potential energy at the top.

12.39 Steps 1, 2, 3:

Now the pulleys rotate besides having mass. We are told that the rope does not slip so there must be a net torque acting on the pulley.

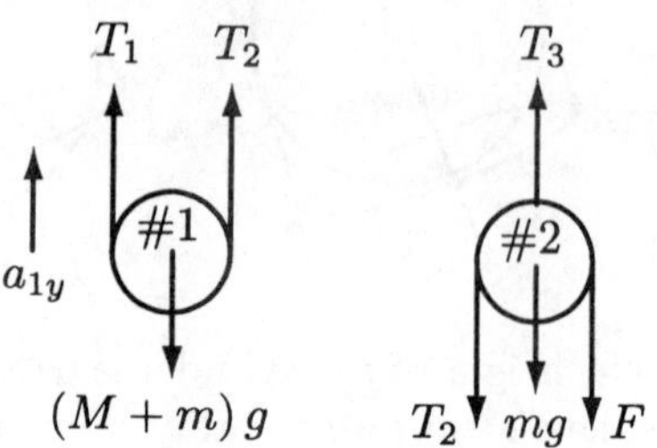

For the net torque on the moving pulley to be out of the page, T_2 must be greater than T_1.

$$I_{\text{pulley}} = I_{\text{disk}} = \frac{1}{2}Mr^2$$

Step 4a: (N2L)

$$T_1 + T_2 - (M+m)\,g = (M+m)\,a_{y1} \qquad (5)$$

Step 4b: (torque)

$$rT_2 - rT_1 = I\alpha_{z1} \qquad (6)$$

Applying torque balance to the second pulley,

$$rT_2 - rF = I\alpha_{z2} \qquad (7)$$

Step 5: For rolling without slipping,

$$a_{y1} = r\alpha_{z1} \qquad (8)$$

The string segment between the two pulleys has the same speed at each point along its length. In particular, its speed is the same at the points where it rolls off the lower pulley and where it rolls onto the upper pulley.

$$v_{y1} = r\omega_{z1} = -r\omega_{z2}$$

or

$$a_{y1} + r\alpha_{z1} = -r\alpha_{z2} \qquad (9)$$

From (8) $\rightarrow$ (6),

$$T_2 - T_1 = \frac{Ia_{y1}}{r^2} = \frac{1}{2}ma_{y1}$$

From (5)

$$T_2 + T_1 = (M+m)(g + a_{y1})$$

Adding this to the previous equation, we eliminate T_1:

$$T_2 = \frac{1}{4} m a_{y1} + \frac{1}{2}(M+m)(g + a_{y1}) \qquad (10)$$

Now, combining (7), (8), and (9)

$$\begin{aligned} T_2 - F &= \frac{I}{r}\alpha_{z2} = -\frac{I}{r}\left(\alpha_{z1} + \frac{\alpha_{y1}}{r}\right) \\ &= -\frac{I}{r}\left(2\frac{a_{y1}}{r}\right) = -m a_{y1} \end{aligned}$$

Finally, from (10) and the above expression,

$$F = \frac{5}{4} m a_{1y} + \frac{1}{2}(M+m)(a_{1y} + g) = 57.7 \text{ N}$$

If the pulleys were frictionless the rope would slip over their "highly polished wooden surface." Without the torque due to friction the pulleys would not rotate. No force would be necessary for angular acceleration of the pulleys. Also, without this friction force, the tension along the length of the rope will be constant at F. If the pulleys don't rotate, there is no net torque acting on them. Then, in the frictionless case,

$$F = \frac{1}{2}(M+m)(a_{1y} + g) = 57.0 \text{ N}$$

12.43 Suppose the barrel's angular velocity is into the page and the ball's position is at an angle θ from the bottom, as shown in the figure.

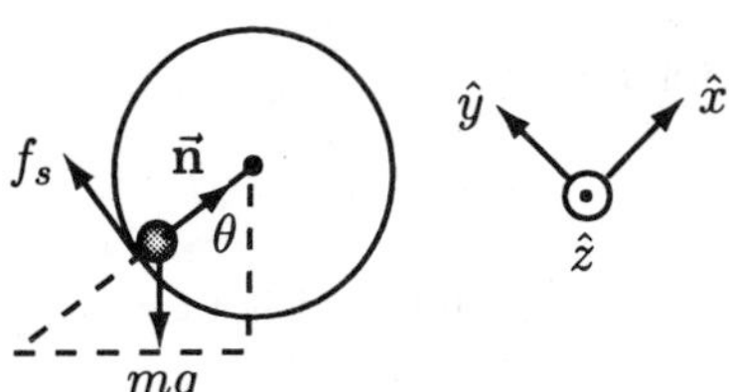

The forces that act on the ball are its weight, normal force, and possibly friction. For these forces to balance, friction must be directed upward, as shown in the figure. However, friction is the only force producing torque about the ball's CM; unbalanced, this torque would produce an angular acceleration of the ball. But—the ball cannot have both an angular acceleration and roll without slipping on the barrel moving at constant angular speed. Only at the bottom of the barrel ($\theta = 0$) can forces on the ball balance without friction. Then, with torques also in balance, the ball can roll at constant angular speed. Now, suppose the barrel has an angular acceleration (assumed also into the page). Then, to continue rolling without slipping, the ball must also accelerate. Friction produces the necessary torque, and the ball will be at an angle $\theta > 0$. Apply Newton's laws to find the value of θ:

Step 4a:

$$n - mg\cos\theta = 0 \qquad (11)$$

$$f_s - mg\sin\theta = 0 \qquad (12)$$

Step 4b:

$$-r f_s = I_{\text{ball}}\alpha_{\text{ball},z} \qquad (13)$$

Step 5: Rolling without slipping means the points of contact on the ball and the barrel must have the same velocity:

$$\Omega R = \omega r$$

Differentiating this relation gives

$$R\alpha_{\text{barrel},z} = \alpha_{\text{ball},z} r \qquad (14)$$

Start with (13) and use (12) to eliminate f_s:

$$-rmg\sin\theta = \frac{2}{5} m r^2 \alpha_{\text{ball},z}$$

Since $\alpha_{\text{barrel},\alpha}$ was given, use (14) to eliminate $\alpha_{\text{ball},z}$. Then solve for θ:

$$\theta = \sin^{-1}\left[-\frac{2\alpha_{\text{barrel},z} R}{5g}\right] \qquad (15)$$

(Recall: $\alpha_{\text{barrel},z} < 0$.)
Now, the largest possible $\alpha_{\text{barrel},z}$ for which the ball does not slip, occurs when the maximum value of friction is just large enough to accelerate the ball.

$$f_{s,\max} = \mu n = \mu m g\cos\theta \qquad (16)$$

Using (16) in (13) we find

$$\begin{aligned} -\mu g\cos\theta &= \frac{2}{5} r\alpha_{\text{ball}} = \frac{2}{5} R\alpha_{\text{barrel}} \\ \left(-\frac{2}{5} R\alpha_{\text{barrel}}\right) &= \mu g\sqrt{1 - \sin^2\theta} \end{aligned}$$

so

$$\begin{aligned} \left(-\frac{2}{5} R\alpha_{\text{barrel}}\right)^2 &= (\mu g)^2\left[1 - \left(\frac{-\frac{2}{5}R\alpha_{\text{barrel}}}{g}\right)^2\right] \\ \left(-\frac{2}{5} R\alpha_{\text{barrel}}\right) &= \frac{\mu g}{\sqrt{1+\mu^2}} \end{aligned}$$

so

$$\alpha_{\text{barrel}} = -\frac{5\mu g}{2R\sqrt{1+\mu^2}}$$

12.49 In traveling 1 m down the ramp the ball drops a height

$$\Delta h = -(\sin 32^\circ)(1.0 \text{ m})$$

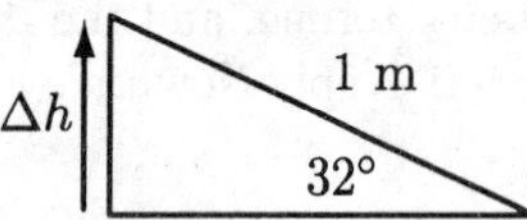

In losing height the ball loses PE—whatever is lost in PE gets converted to kinetic energy, that is:

$$-\Delta PE = \Delta K$$

There are two terms in the kinetic energy: K due to velocity of the ball's CM and K due to the ball's rotation.

$$\begin{aligned} -mg\Delta h &= \frac{1}{2}mv_{\rm CM}^2 + \frac{1}{2}I_{\rm ball}\omega^2 \\ -g\Delta h &= \frac{1}{2}v_{\rm CM}^2 + \frac{1}{5}r^2\left(\frac{v_{\rm CM}}{r}\right)^2 \end{aligned}$$

So,

$$\begin{aligned} v_{\rm CM} &= \sqrt{-\frac{10}{7}g\Delta h} \\ &= \sqrt{\frac{10}{7}\sin 32^\circ\,(1.0 \text{ m})\,(9.8 \text{ m}/\text{s}^2)} \\ &= 2.7 \text{ m}/\text{s} \end{aligned}$$

12.53 In order for the ball to stay on the track, according to Newton's Second Law, its acceleration must be greater than that due to gravity alone.

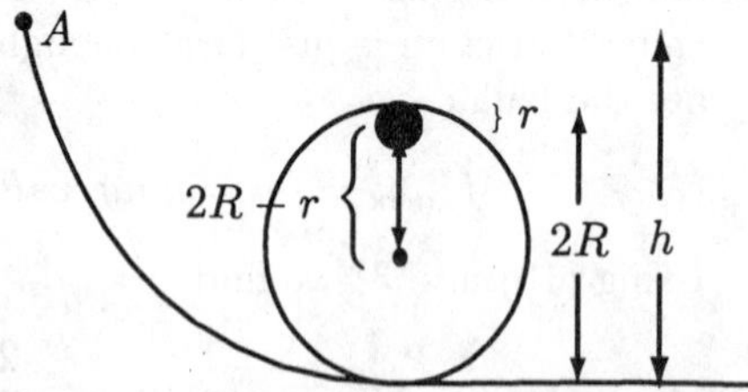

We will now consider the ball to be an extended body.

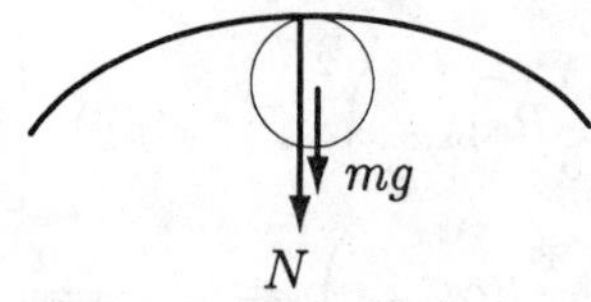

Applying N2L:

$$N + mg = \frac{mv_{\rm CM}^2}{(2R-r)} \qquad (17)$$

At the minimum $v_{\rm CM}$, the ball is just barely on the track and $N = 0$. When this happens, Eqn. (17) becomes

$$g = \frac{v_{\rm CM,min}^2}{(R-r)} \Rightarrow v_{\rm CM,min} = \sqrt{g(R-r)}$$

The linear kinetic and potential energies of the ball at the top of the loop are:

$$\begin{aligned} K &= \frac{1}{2}mv_{\rm CM,min}^2 = \frac{1}{2}mg(R-r) \\ PE &= mgh = mg\,(2R-r) \end{aligned}$$

taking the ground to be the reference level from which PE is computed. Now we are to consider rotational kinetic energy:

$$\begin{aligned} K_{\rm rot} &= \frac{1}{2}I_{\rm sphere}\omega^2 \\ &= \frac{1}{2}\left(\frac{2}{5}mr^2\right)\left(\frac{v_{\rm CM,min}}{r}\right)^2 \\ &= \frac{1}{5}mv_{\rm CM,min}^2 = \frac{1}{5}mg\,(R-r) \end{aligned}$$

This is the idea: at the top of the track (at point A) the ball has only PE. At the top of the loop it has linear kinetic and potential (and if we consider its rolling, its rotational kinetic) energies. Since no non-conservative forces act on the ball, its TME is conserved:

$$E_{\text{at } A} = E_{\text{at top of loop}}$$

If the ball rolls without slipping,

$$mgh = \frac{1}{2}mg(R-r) + mg(2R-r) + \frac{1}{5}mg(R-r)$$

Solving for h,

$$h = 2.7R - 1.7r$$

If the ball slides without rolling,

$$\begin{aligned} mgh &= \frac{1}{2}mg(R-r) + mg(2R-r) \\ \Rightarrow h &= \frac{1}{2}\,[5R - 3r] \\ &= 2.5R - 1.5r \end{aligned}$$

12.57 The problem states neither you nor the capsule rotates, therefore, if we declare you and the capsule to be a system, the system has no angular momentum. Since you are integral to the system, a part of it, there is nothing you can do to change the angular momentum of the system. Suppose you start spinning around so that you have an angular momentum of $\vec{\mathbf{L}}$.

The capsule would start turning in the opposite sense. It would turn at a speed that would give it an angular momentum of $-\vec{\mathbf{L}}$. Thus, the system would have a net angular momentum of $\vec{\mathbf{L}}+(-\vec{\mathbf{L}})=0$. This is the idea behind changing the direction of your window. Conservation of angular momentum requires $\vec{\mathbf{L}}_{\text{you}} = -\vec{\mathbf{L}}_{\text{capsule}}$, or:

$$I_y\omega_y = -I_c\omega_c \Rightarrow I_y\theta_y = -I_c\theta_c \qquad (18)$$

There is an important requirement: the point is to be able to look out the window. In order to be looking out the window, the relative angle that you turn with respect to the capsule must equal $2\pi n$ (see figure)

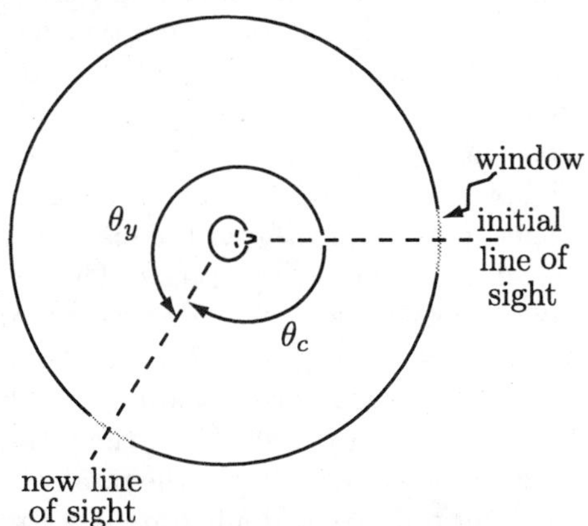

for some integer n

$$\theta_y - \theta_c = 2\pi n \qquad (19)$$

Solving (18) and (19) for θ_c, we find:

$$\theta_c = \frac{2\pi n}{1+\frac{I_c}{I_y}}$$

If the ratio $\frac{I_c}{I_y}$ is a rational number then there are only a finite number of directions in which the window can point. Take, for example $\frac{I_c}{I_y}=1$. Then there are only two angles which you can view through the window: $\theta_c = 0$ and $\theta_c = \pi$. Note that one more turn will get you to $\theta = 2\pi$, back where you started. As it turns out as long as $\frac{I_c}{I_y}$ is rational you will have a finite number of viewing angles. If $\frac{I_c}{I_y}$ is irrational then, given an infinite amount of time, you will be able to point the window in any direction.

12.61 Rotation about the k' axis clearly has the angular velocity pointing in the $\hat{\mathbf{k}}$ direction. The object has neither rotational symmetry about the z' axis nor mirror symmetry about any plane perpendicular to the z' axis. So $\vec{\mathbf{L}}$ cannot be completely in the $\hat{\mathbf{k}}$ direction, and $\vec{\mathbf{L}}$ and $\vec{\omega}$ are not $\|$.

When the object rotates about the k-axis we have rotational symmetry, so $\vec{\mathbf{L}}$ and $\vec{\omega}$ are $\|$. The angular momentum of the object rotating about z' is not $\|$ to $\vec{\omega}$ so we cannot make use of the relation $\vec{\mathbf{L}} = I\vec{\omega}$ to find $\vec{\mathbf{L}}$ given $\vec{\omega}$. However, $|\vec{\omega}|$ will give the correct value of rotational kinetic energy $E=\frac{1}{2}I\omega^2$. Since the parallel axis theorem is derived from an argument about kinetic energy the results for angular momentum neither confirm nor deny it. The theorem simply isn't useful for computing angular momentum about points on the z'-axis.

12.65 We suppose that the rotational inertia of a plane square sheet of side ℓ about an axis through its center and perpendicular to the sheet is $I_{sq} = \alpha M\ell^2$ where M is the mass of the sheet and α is an unknown coefficient that depends solely on the square shape of the sheet.

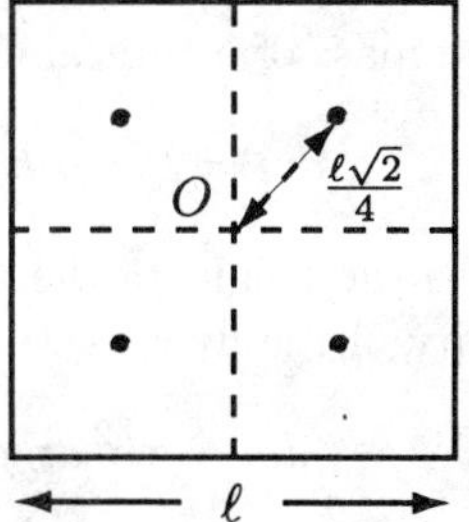

It turns out we can determine α by finding a second expression for I_{sq}—by viewing the whole square as a composite of the four smaller squares shown with dashed borders in the figure.

$$I_{sq} = 4\left[I_{\text{subsquare}} + M_{\text{subsquare}}\left(\frac{\ell\sqrt{2}}{4}\right)^2\right]$$

(parallel axis theorem)

$$\begin{aligned}\alpha M\ell^2 &= 4\left[\alpha\frac{M}{4}\left(\frac{\ell}{4}\right)^2 + \frac{M}{4}\frac{\ell^2}{8}\right]\\ &= M\ell^2\left[\frac{\alpha}{4}+\frac{1}{8}\right]\end{aligned}$$

So

$$\alpha = \frac{\alpha}{4}+\frac{1}{8} \Rightarrow \alpha = \frac{1}{6}$$

Therefore $I_{sq} = \frac{1}{6}M\ell^2$.

12.69 Step 1: *Warning.* This is a 3-dimensional object.

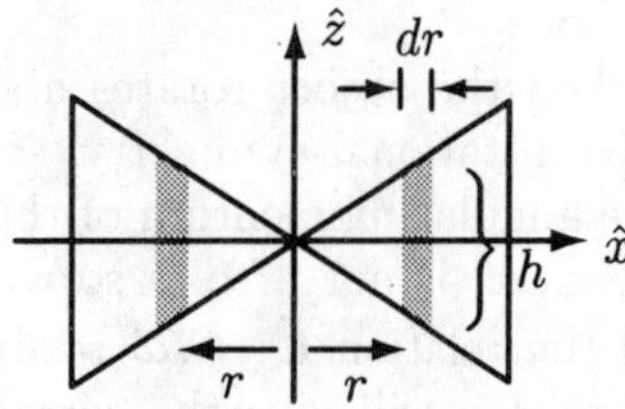

We can model the flywheel as a collection of differential cylindrical shells of height h and some differential thickness. The geometry of the problem suggests cylindrical coordinates.

Step 2: The cylindrical shell will have a radius r and have some height h. We can express h in terms of the distance between the axis of rotation and the differential element by noticing that our differential element defines similar triangles. Thus,

$$\frac{\frac{1}{2}R}{R} = \frac{h}{r} \Rightarrow h = \frac{1}{2}r$$

Thus, the mass of a typical element is:

$$dm = 2\pi r\, dr\, h\, \rho_m$$

where h is the height of the element and ρ_m is the density of the flywheel.

$$dm = \pi r^2\, dr\, \rho_m$$

Step 3

$$dI = r_\perp^2\, dm = \pi r^4\, dr\, \rho_m$$

Step 4 r varies from 0 to R. ($dr > 0$ is a thickness.)

Step 5

$$I = \int_0^R \pi r^4\, \rho_m\, dr = \pi \rho_m \int_0^R r^4\, dr = \frac{\pi \rho_m}{5} R^5$$

To express I in standard form we need to find the flywheel's volume to calculate ρ_m. We can use what we have previously done since $dm = \rho_m\, dv$

$$v = \int dv = \int_0^R \pi r^2\, dr = \frac{1}{3}\pi R^3$$

so

$$\rho_m = \frac{M}{\frac{1}{3}\pi R^3} = \frac{3M}{\pi R^3}$$

and

$$I = \frac{\pi R^5}{5}\left(\frac{3M}{\pi R^3}\right) = \frac{3MR^2}{2}$$

12.73 The object drawn below is rotating about the z-axis.

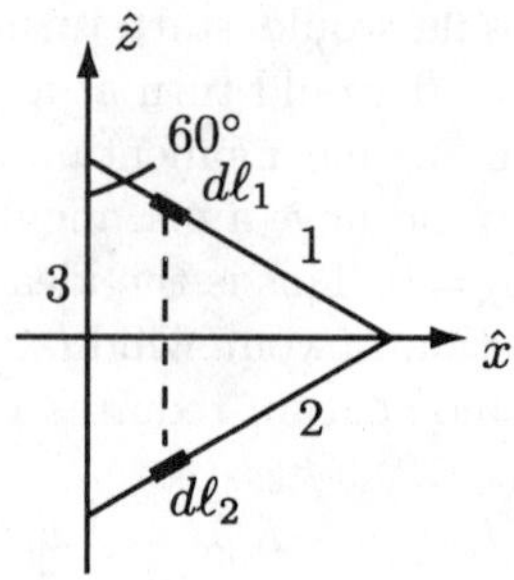

Note that two pieces cannot be treated as a rotating rod since they are at skewed angles *wrt* the axis of rotation. Thus, Table 13.1 in conjunction with the $\parallel$ axis theorem cannot help us. We must use integration to find I_{sys}. Note that rod 3 has a zero rotational inertia since there is no $\perp$ distance between it and the axis of rotation, so $r_\perp$ for rod 3 is zero and $I_{\text{rod3}} = 0$. Now consider rods 1 and 2. Take a typical differential piece $d\ell_1$ in rod 1. There is a corresponding element in rod 2, $d\ell_2$ with equal mass and an equal $\perp$ distance to the z-axis. Thus each piece contributes equally to the system's rotational inertia. Now we generalize—*every* element in rod 1 has such a corresponding element in rod 2 with an equal $r_\perp dm$. Thus, the rotational inertia of the system is twice the rotational inertia of either rod. We will integrate rod 1 to find its rotational inertia and multiply the result by 2 to find I_{sys}. The rod can be modeled as a collection of differential elements of length $d\ell$ and mass $dm = \lambda d\ell$ where λ is the linear density of the rod. The element has a $\perp$ distance of x to the axis of rotation. Thus, for a typical differential element we have:

$$dI = r_\perp^2 dm = x^2 \lambda\, d\ell \tag{20}$$

Now $r_\perp$ is a function of x but our differential element is $d\ell$. What we need to do is find the relationship between x and ℓ.

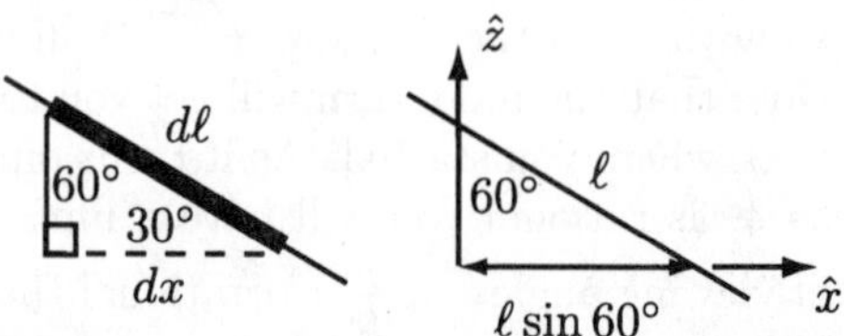

The triangle is equilateral, so by the figure, $dx = d\ell \cos 30°$. Solving for $d\ell$ and plugging into (20) gives:

$$dI = \frac{2}{\sqrt{3}} x^2 \lambda\, dx$$

and we can see from the last figure that x will

vary from 0 to $\ell \sin 60° = \frac{\sqrt{3}}{2}\ell$.

$$I_{\text{rod1}} = \int dI = \int_0^{\frac{\sqrt{3}}{2}\ell} \frac{2}{\sqrt{3}} x^2 \lambda\, dx$$
$$= \frac{2}{3\sqrt{3}} \lambda x^3 \Big|_0^{\frac{\sqrt{3}}{2}\ell} = \frac{1}{4}\lambda \ell^3$$

The convention is to express I in terms of the total mass. The linear density can be rewritten as $\lambda = \frac{M}{\ell}$ so,

$$I_{\text{rod1}} = \frac{1}{4} M\ell^2$$

and as discussed before,

$$I_{\text{sys}} = 2I_{\text{rod1}} = \frac{1}{2}M\ell^2 = \frac{1}{6}M_{\text{sys}}\ell^3$$

since $M_{\text{sys}} = 3M_{\text{rod}}$. Since the rods are uniform, the CM of each is at its geometric center. The system is symmetric, so its CM lies on the x-axis at coordinate

$$x_{\text{CM}} = \frac{\sum_i m_i x_i}{\sum m_i}$$
$$= \frac{M(0) + M\left(\frac{\sqrt{3}}{4}\ell\right) + M\left(\frac{\sqrt{3}}{4}\ell\right)}{3M}$$
$$= \frac{1}{2\sqrt{3}}\ell$$

Now we'll use the previous result along with the $\parallel$ axis theorem to find the new I_{sys}. Note that we're calling the previous axis of rotation the "A" axis.

$$I_A = I_{\text{CM}} + M_{\text{sys}} r_\perp^2$$

where $r_\perp$ is the $\perp$ distance between the A and CM axes, $r_\perp = \frac{1}{2\sqrt{3}}\ell$

$$I_{\text{CM}} = I_A - M_{\text{sys}} r_\perp^2$$
$$= \frac{1}{2}M\ell^2 - 3M\left(\frac{1}{2\sqrt{3}}\ell\right)^2$$
$$= \frac{1}{4}M\ell^2$$

12.77 The analysis of the Earth's response to torque is no different from that for a spinning gyroscope—it is the scale and the cause of the torque that are different. So, the necessary torque (assumed perpendicular to Earth's angular momentum) is

$$\tau \simeq \Omega L = \Omega \frac{2}{5} M R^2 \omega$$
$$= \frac{2\pi}{(26,000 \text{ y})(\pi \times 10^7 \frac{\text{s}}{\text{y}})}$$
$$\times \frac{2}{5}(6.0 \times 10^{24} \text{ kg})(6.4 \times 10^6 \text{ m})^2 \frac{2\pi}{86,400 \text{ s}}$$
$$= 5.5 \times 10^{22} \text{ N} \cdot \text{m}$$

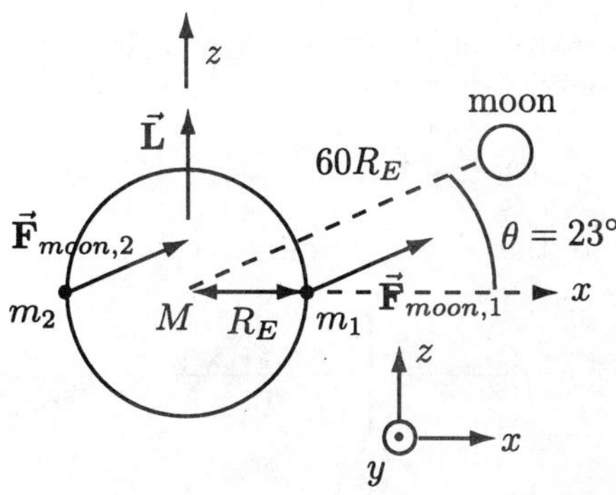

Let the Moon be 23° above the plane of the equator, the Earth's angular momentum define the z-direction, and the Moon be in the x-z-plane, as shown. To compute the torque the Moon exerts on the two bulges of mass m, we need some vector trickery. The Moon's position is

$$\vec{\mathbf{r}}_{\text{E-M}} = 60R_E(\hat{\mathbf{k}} \sin\theta + \hat{\mathbf{i}} \cos\theta)$$

(origin at Earth's center) Since the Earth rotates, bulges 1 and 2 have variable position vectors.

$$\vec{\mathbf{r}}_1 = -\vec{\mathbf{r}}_2 = R_E(\hat{\mathbf{i}} \cos\omega t + \hat{\mathbf{j}} \sin\omega t)$$

Then the torque exerted by the Moon about Earth's center is

$$\vec{\tau} = \vec{\mathbf{r}}_1 \times \vec{\mathbf{F}}_{\text{moon,1}} + \vec{\mathbf{r}}_2 \times \vec{\mathbf{F}}_{\text{moon,2}}$$

The gravitational force exerted by the Moon on mass 1 is

$$\vec{\mathbf{F}}_{\text{moon,1}} = \frac{GM_{\text{moon}}m}{(\text{distance from Moon to 1})^2}\hat{\mathbf{r}}_{1\text{M}}$$

where $\hat{\mathbf{r}}_{1\text{M}}$ unit vector from 1 to moon. Now

$$\vec{\mathbf{r}}_{\text{from 1 to moon}} = \vec{\mathbf{r}}_{\text{E,M}} - \vec{\mathbf{r}}_1$$

So

$$\vec{\mathbf{F}}_{\text{moon,1}} = \frac{GM_{\text{moon}}m}{|\vec{\mathbf{r}}_{\text{E,M}} - \vec{\mathbf{r}}_1|^2} \cdot \frac{\vec{\mathbf{r}}_{\text{E,M}} - \vec{\mathbf{r}}_1}{|\vec{\mathbf{r}}_{\text{E,M}} - \vec{\mathbf{r}}_1|}$$
$$= \frac{GM_{\text{moon}}m(\vec{\mathbf{r}}_{\text{E,M}} - \vec{\mathbf{r}}_1)}{|\vec{\mathbf{r}}_{\text{E,M}} - \vec{\mathbf{r}}_1|^3}$$

A similar expression gives $\vec{\mathbf{F}}_{\text{moon,2}}$. So

$$\vec{\tau} = \vec{\mathbf{r}}_1 \times \frac{GM_{\text{moon}}m(\vec{\mathbf{r}}_{\text{E,M}} - \vec{\mathbf{r}}_1)}{|\vec{\mathbf{r}}_{\text{E,M}} - \vec{\mathbf{r}}_1|^3} + \vec{\mathbf{r}}_2 \frac{GM_{\text{moon}}m(\vec{\mathbf{r}}_{\text{E,M}} - \vec{\mathbf{r}}_2)}{|\vec{\mathbf{r}}_{\text{E,M}} - \vec{\mathbf{r}}_2|^3}$$
$$= GM_{\text{moon}}m \left\{ \frac{\vec{\mathbf{r}}_1 \times \vec{\mathbf{r}}_{\text{E,M}}}{|\vec{\mathbf{r}}_{\text{E,M}} - \vec{\mathbf{r}}_1|^3} + \frac{(-\vec{\mathbf{r}}_1) \times (\vec{\mathbf{r}}_{\text{E,M}})}{|\vec{\mathbf{r}}_{\text{E,M}} + \vec{\mathbf{r}}_1|^3} \right\}$$

(Note: $\vec{\mathbf{r}}_1 \times \vec{\mathbf{r}}_1 = \vec{\mathbf{r}}_2 \times \vec{\mathbf{r}}_2 = 0$ and $\vec{\mathbf{r}}_2 = -\vec{\mathbf{r}}_1$) Now

$$|\vec{\mathbf{r}}_{\text{E,M}} \mp \vec{\mathbf{r}}_1|^2 = r_{\text{E,M}}^2 + r_1^2 \mp 2\vec{\mathbf{r}}_1 \cdot \vec{\mathbf{r}}_{\text{E,M}}$$
$$= (60R_E)^2 + (R_E)^2 \mp 120R_E^2 \hat{\mathbf{r}}_1 \cdot \hat{\mathbf{r}}_{\text{E,M}}$$

Neglecting 1 compared with 60^2, we have approximately

$$|\vec{\mathbf{r}}_{E,M} \mp \vec{\mathbf{r}}_1|^2 = (60R_E)^2\left[1 \mp 2\hat{\mathbf{r}}_1 \cdot \hat{\mathbf{r}}_{E,M}\right]$$

So, the torque simplifies to

$$\vec{\tau} = \frac{GM_{moon}m}{(60R_E)^3}\left\{\frac{60R_E^2\hat{\mathbf{r}}_1\times\hat{\mathbf{r}}_{E,M}}{\left(1-\frac{2\hat{\mathbf{r}}_1\cdot\hat{\mathbf{r}}_{E,M}}{60}\right)^{3/2}} - \frac{60R_E^2\hat{\mathbf{r}}_1\times\hat{\mathbf{r}}_{E,M}}{\left(1+\frac{2\hat{\mathbf{r}}_1\cdot\hat{\mathbf{r}}_{E,M}}{60}\right)^{3/2}}\right\}$$

The binomial theorem allows us to simplify the expression in braces:

$$\begin{aligned}\left(1 \mp \frac{2\hat{\mathbf{r}}_1\cdot\hat{\mathbf{r}}_{E,M}}{60}\right)^{3/2} &\simeq 1 \pm \left(\frac{3}{2}\right)\frac{2\hat{\mathbf{r}}_1\cdot\hat{\mathbf{r}}_{E,M}}{60} \\ &= 1 \pm \frac{\hat{\mathbf{r}}_1\cdot\hat{\mathbf{r}}_{E,M}}{20}\end{aligned}$$

So

$$\vec{\tau} = \frac{GM_{moon}m}{3600R_E}\left(\hat{\mathbf{r}}_1\times\hat{\mathbf{r}}_{E,M}\right)\left\{\frac{\hat{\mathbf{r}}_1\cdot\hat{\mathbf{r}}_{E,M}}{10}\right\}$$

Recall,

$$\begin{aligned}\hat{\mathbf{r}}_1 &= \left(\hat{\mathbf{i}}\cos\omega t + \hat{\mathbf{j}}\sin\omega t\right) \\ \hat{\mathbf{r}}_{E,M} &= \left(\hat{\mathbf{k}}\sin\theta + \hat{\mathbf{i}}\cos\theta\right)\end{aligned}$$

So

$$\hat{\mathbf{r}}_1\cdot\hat{\mathbf{r}}_{E,M} = \cos\theta\cos\omega t$$

and

$$\hat{\mathbf{r}}_1\times\hat{\mathbf{r}}_{E,M} = \hat{\mathbf{i}}\sin\omega t\sin\theta - \hat{\mathbf{j}}\cos\omega t\sin\theta - \hat{\mathbf{k}}\cos\theta\sin\omega t$$

So,

$$\begin{aligned}\vec{\tau} &= \frac{GM_{moon}m}{(3.6\times10^4)R_E}\cos\theta\cos\omega t \\ &\quad\cdot\left\{\hat{\mathbf{i}}\sin\omega t\sin\theta - \hat{\mathbf{j}}\cos\omega t\sin\theta - \hat{\mathbf{k}}\cos\theta\sin\omega t\right\}\end{aligned}$$

During one rotation of the Earth ωt runs from 0 to 2π, so we average $\vec{\tau}$ over that interval

$$\begin{aligned}\langle\vec{\tau}\rangle &= \frac{GM_{moon}m}{(3.6\times10^4)R_E} \\ &\quad\cdot\cos\theta\left\{\begin{array}{r}\hat{\mathbf{i}}\sin\theta\left\langle\sin\omega t\cos\omega t\right\rangle \\ -\hat{\mathbf{j}}\sin\theta\left\langle\cos^2\omega t\right\rangle \\ -\hat{\mathbf{k}}\cos\theta\left\langle\cos\omega t\sin\omega t\right\rangle\end{array}\right\} \\ &= \frac{GM_{moon}m}{(3.6\times10^4)R_E}\frac{\cos\theta\sin\theta}{2}\left(-\hat{\mathbf{j}}\right)\end{aligned}$$

Professional as astronomers will note that we got the direction right. We could worry about averaging over the Moon's orbit, the effect of the Sun..., but then we should do a better model of the Earth's shape. Let's set this expression equal to the estimated torque required and get an estimate of m;

$$\begin{aligned}m &\simeq \frac{|\langle\vec{\tau}\rangle|\left(3.6\times10^4\right)R_E}{GM_{moon}}\frac{4}{\sin 2\theta} \\ &= \frac{\left(5.5\times10^{22}\ \text{N}\cdot\text{m}\right)\left(3.6\times10^4\right)\left(6.4\times10^6\ \text{m}\right)4}{\left(6.7\times10^{-11}\frac{\text{m}^3}{\text{kg}\cdot\text{s}^2}\right)\frac{1}{81}\left(6.0\times10^{24}\ \text{kg}\right)\sin 46^\circ} \\ &= 1.4\times10^{22}\ \text{kg}\end{aligned}$$

As a fraction of Earth's mass,

$$\frac{2M}{M_E} = \frac{2\left(1.4\times10^{22}\ \text{kg}\right)}{\left(6.0\times10^{24}\ \text{kg}\right)} \simeq 5\times10^{-3}$$

Fact: The Earth's equatorial radius is 0.3% greater than its polar radius—about right.

12.79 If the flywheel axis were mounted horizontally, each time the car turned it would need to exert a torque on the flywheel to change the flywheel's direction of angular momentum.

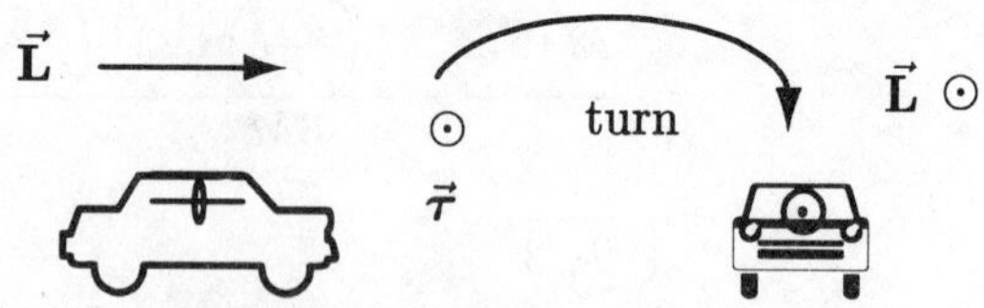

The figure shows that the car needs to exert a ⊙ torque on the flywheel in order to make the right turn. This is an adverse effect on the car's ability to corner. Now consider when the flywheel axis is vertical so that $\vec{\mathbf{L}}$ points upward. No matter how the car turns, $\vec{\mathbf{L}}$ remains upward and no torque is needed since the direction of the flywheel's angular momentum doesn't change. When the flywheel is mounted vertically and the car goes over a steep hill,

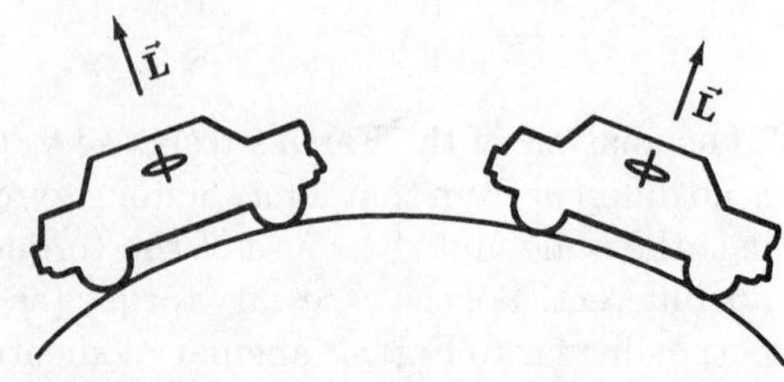

normal force provides the necessary torque. In this configuration the flywheel makes it harder to roll the car in a turn.

12.83 a) We'll choose the coordinate system so that the axis of rotation is the z-axis and $\vec{\omega} = \omega\hat{\mathbf{k}}$ as shown.

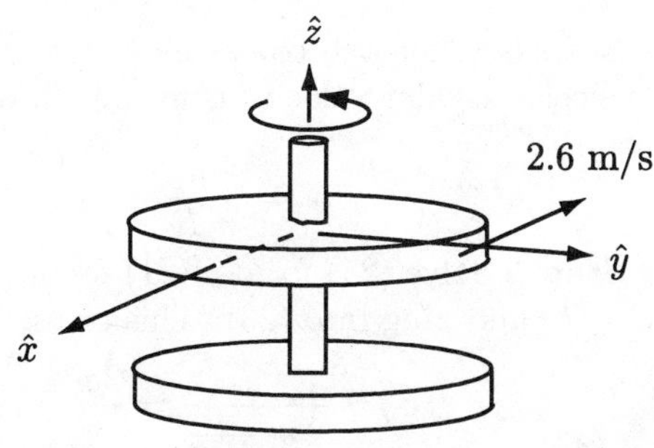

$$\begin{aligned}\omega &= \frac{v_{\text{rim}}}{r_{\text{rim}}} = \frac{2.6 \text{ m/s}}{.50 \text{ m}} = 5.2 \text{ rad/s} \\ \vec{\omega} &= (5.2 \text{ rad/s})\,\hat{\mathbf{k}}\end{aligned}$$

b) $\vec{\omega}_f - \vec{\omega}_0 = \vec{\alpha}t$ since $\omega_f = 0$, $\vec{\omega}_0$ and $\vec{\alpha}$ are in opposite directions. We can write out the components:

$$\begin{aligned}\omega_0 &= -\alpha_z t \Rightarrow \alpha_z = -\frac{\omega_z}{t} \\ &= -\frac{5.2 \text{ rad/s}}{45 \text{ s}} = -0.12 \text{ rad/s}^2\end{aligned}$$

so

$$\vec{\alpha} = a_z\hat{\mathbf{k}} = -\left(0.12 \text{ rad/s}^2\right)\hat{\mathbf{k}}$$

c) With the axis of rotation $\perp$ to the disk, $I = \frac{1}{2}MR^2$, so

$$\begin{aligned}K &= \frac{1}{2}I\omega^2 = \frac{1}{2}\left(\frac{1}{2}MR^2\right)\omega^2 \\ &= \frac{1}{4}(250 \text{ kg})(0.50 \text{ m})^2(5.2 \text{ rad/s})^2 \\ &= 420 \text{ J}\end{aligned}$$

Since the disk is rotationally symmetric *wrt* the $\hat{\mathbf{k}}$-axis and angular velocity is in the $\hat{\mathbf{k}}$ direction, $\vec{\mathbf{L}} = I\vec{\omega}$

$$\begin{aligned}L_k &= \frac{1}{2}MR^2\omega_k \\ &= \frac{1}{2}(250 \text{ kg})(0.50 \text{ m})^2(5.2 \text{ rad/s}) \\ &= (160 \text{ J}\cdot\text{s})\,\hat{\mathbf{k}}\end{aligned}$$

d) The torque necessary to provide the disk's angular acceleration is:

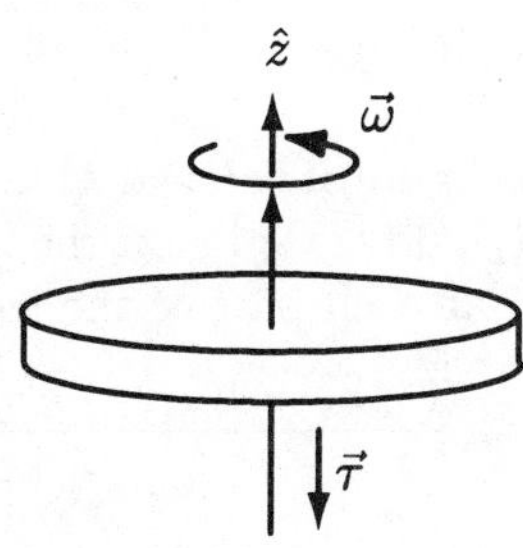

$$\begin{aligned}\vec{\tau} &= I\vec{\alpha} = \frac{1}{2}MR^2\left[\left(-0.12 \text{ rad/s}^2\right)\hat{\mathbf{k}}\right] \\ &= \frac{1}{2}(250 \text{ kg})(0.50 \text{ m})^2\left[\left(-0.12 \text{ rad/s}^2\right)\hat{\mathbf{k}}\right] \\ &= -(3.8 \text{ N}\cdot\text{m})\,\hat{\mathbf{k}}\end{aligned}$$

where we used the moment of inertia for a uniform disk about an axis $\perp$ to its plane. Now,

$$\begin{aligned}\Delta\theta &= \omega_0 t + \frac{1}{2}\alpha t^2 \\ &= (5.2\,/\text{s})(45 \text{ s}) + \frac{1}{2}(0.12\,/\text{s})(45 \text{ s})^2 \\ &= 110 \text{ rad} = 18 \text{ rotations}\end{aligned}$$

The work done is negative since the torque is opposite in direction from the wheel's angular velocity

$$W = -(3.8 \text{ N}\cdot\text{m})(110 \text{ rad}) = -420 \text{ N}\cdot\text{m}$$

e) Since no net external torques are exerted on the system of two disks, their angular momentum is constant.

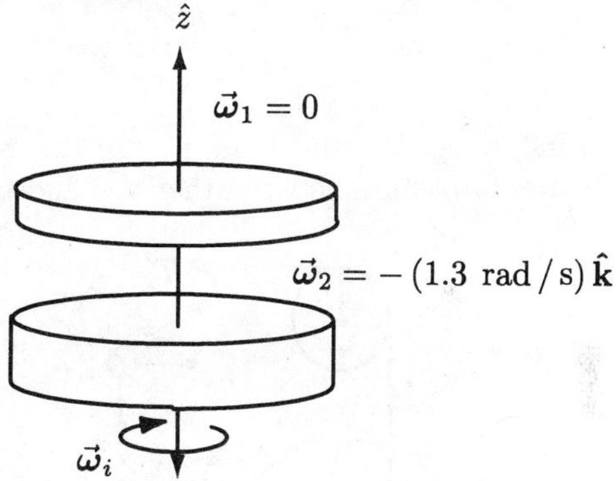

Note that rotational symmetry about the $\hat{\mathbf{k}}$ axis allows us to use $\vec{\mathbf{L}} = I\vec{\omega}$.

$$\begin{aligned}\vec{\mathbf{L}}_1 &= 0 \\ \vec{\mathbf{L}}_2 &= -(2I)\omega_2\hat{\mathbf{k}} \\ \vec{\mathbf{L}}_i &= \vec{\mathbf{L}}_1 + \vec{\mathbf{L}}_2 = I\hat{\mathbf{k}}\,[-2\omega_2]\end{aligned}$$

where I for a disk rotating about its center is

$$\frac{1}{2}MR^2 = (250 \text{ kg})(.50 \text{ m})^2 = 31 \text{ J}\cdot\text{s}^2$$

$$\begin{aligned}\vec{\mathbf{L}}_i &= 2\left(31 \text{ J}\cdot\text{s}^2\right)[-1.3 \text{ rad/s}] \\ &= -(81 \text{ J}\cdot\text{s})\,\hat{\mathbf{k}} \\ \vec{\mathbf{L}}_f &= 3I\vec{\omega}_f = 3\left(31 \text{ J}\cdot\text{s}^2\right)\vec{\omega}_f \\ &= \left(93 \text{ J}\cdot\text{s}^2\right)\vec{\omega}_f \\ \vec{\mathbf{L}}_i &= \vec{\mathbf{L}}_f = -81 \text{ J}\cdot\text{s}\,\hat{\mathbf{k}} \\ &= \left(93 \text{ J}\cdot\text{s}^2\right)\vec{\omega}_f \\ \Rightarrow \vec{\omega}_f &= (-0.87 \text{ rad/s})\,\hat{\mathbf{k}}\end{aligned}$$

Now we'll turn to kinetic energy.

$$\begin{aligned} K_i &= \frac{1}{2}I\omega_1^2 + \frac{1}{2}(2I)\omega_2^2 \\ &= I\left[\frac{1}{2}\omega_1^2 + \omega_2^2\right] = 53 \text{ J} \\ K_f &= \frac{1}{2}(3I)\omega_f^2 = 35 \text{ J} \\ \Delta K &= K_f - K_i = -18 \text{ J} \end{aligned}$$

This is energy that's lost to the system; the thermal energy gained by the rest of the universe is 18 J.

12.87 Steps 1, 2 and 3:

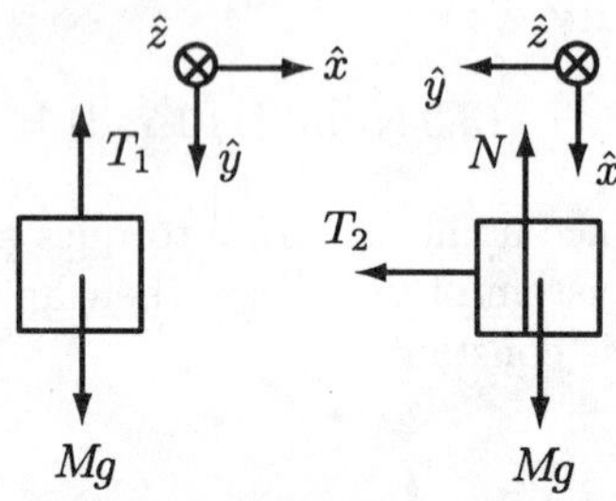

First note that the tensions are different because of friction between the rope and pulley.

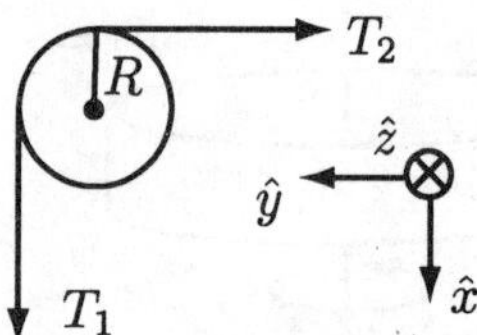

There are two ways in which you could have guessed this. In Step 4a you would have found an inconsistency in N2L if $T_1 = T_2$. The second way is that if $T_1 = T_2$, τ_{net} on the pulley is zero so the pulley couldn't have an angular acceleration. Since the masses accelerate linearly, the rope would have to slip on the pulley, but we are assuming that it doesn't slip. Thus, $T_1 \neq T_2$.

Step 4a

$$\begin{aligned} Mg - T_1 &= Ma_y && (21) \\ T_2 &= Ma_y && (22) \end{aligned}$$

Note: conservation of string was used here to justify the same name for the two accelerations.

Step 4b

$$\sum_i \tau_{i,z} = -RT_1 + RT_2 = I\alpha_z \qquad (23)$$

Step 5 When the blocks move towards $\hat{\mathbf{j}}$, the wheel's angular velocity is in the $-\hat{\mathbf{k}}$ direction. So,

$$\alpha_z = -\frac{a_y}{R} \qquad (24)$$

Step 6 Solve (24) for α_z, (21) for T_1 and (22) for T_2 and plug into (23). This gives:

$$a_y = \frac{g}{2 + \frac{I}{MR^2}} = \frac{2g}{5}$$

12.91 We are looking at the bola from above.

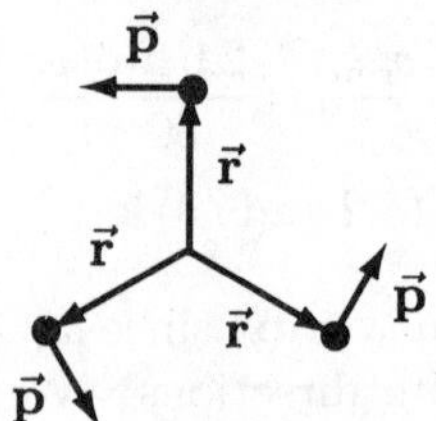

$\vec{\omega}$ is pointing straight out of the page. The balls have to be in the same plane as the strings that hold them together. Otherwise unbalanced tension forces perpendicular to that plane accelerate the balls back toward it. Also, any angles other than 120° between the strings produce unequal forces on the balls that accelerate the balls at different angular frequencies and return them to the 120° configuration. We assume that any oscillations produced by these effects have died out and left the bola in a stable configuration. By the right hand rule, any of the $\vec{\mathbf{r}} \times \vec{\mathbf{p}}$'s yield a vector straight out of the page, thus angular momentum is $\parallel$ to $\vec{\omega}$. To calculate $\vec{\mathbf{L}}$, we'll assume that the balls are point masses.

$$\begin{aligned} \vec{\mathbf{L}} &= I\vec{\omega} = 3\left(m\ell^2\right)\omega\hat{\odot} \\ &= 3m\ell^2\omega\hat{\odot} \\ K_{\text{rot}} &= \frac{1}{2}I\omega^2 = \frac{1}{2}\left(3m\ell^2\right)\omega^2 \\ &= \frac{3}{2}m\ell^2\omega^2 \end{aligned}$$

12.95 First we'll find out where the CM is:

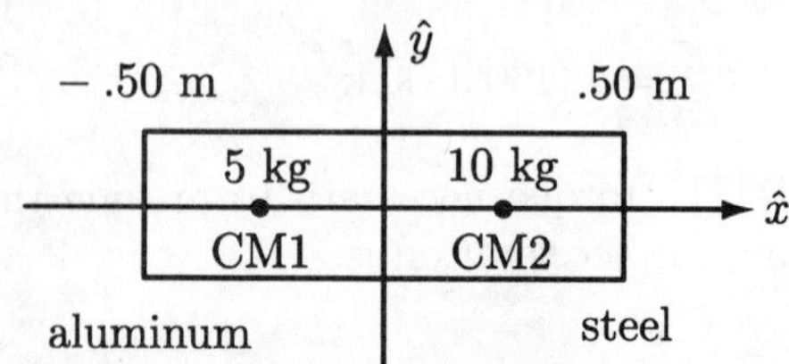

From the figure it is clear that $y_{\text{CM}} = 0$ since both CM_1 and CM_2 lie on the x-axis (we're assuming that each part of the rod is uniform).

$$\begin{aligned} \bar{x}_{\text{CM}} &= \frac{(-.25 \text{ m})(5 \text{ kg}) + (.25 \text{ m})(10 \text{ kg})}{15 \text{ kg}} \\ &= 8.3 \text{ cm} \end{aligned}$$

To obtain I about the system's CM, we may apply the parallel-axis theorem to each half rod and add the results:

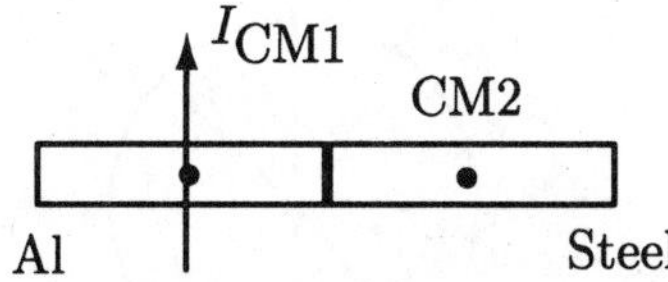

$$
\begin{aligned}
I_{CM} &= \frac{1}{12}(5.0\text{ kg})(0.25\text{ m})^2 \\
&\quad +(5.0\text{ kg})(0.33\text{ m})^2 \\
&\quad +\frac{1}{12}(10.0\text{ kg})(0.25\text{ m})^2 \\
&\quad +(10.0\text{ kg})(0.17\text{ m})^2 \\
&= 0.91\text{ kg}\cdot\text{m}^2
\end{aligned}
$$

To obtain the rotational inertia about axes through the two ends, we use the parallel-axis theorem again.

$$
\begin{aligned}
I_{\text{tip of steel}} &= I_{CM}+M\left(R^2\right) \\
&= 0.91\text{ kg}\cdot\text{m}^2 \\
&\quad +(15\text{ kg})(0.50\text{ m}-0.083\text{ m})^2 \\
&= 3.5\text{ kg}\cdot\text{m}^2 \\
I_{Al} &= 0.91\text{ kg}\cdot\text{m}^2 \\
&\quad +(15\text{ kg})(0.50\text{ m}+0.083\text{ m})^2 \\
&= 6.0\text{ kg}\cdot\text{m}^2
\end{aligned}
$$

12.99 The impulse delivered acts on a line through the ball's CM.

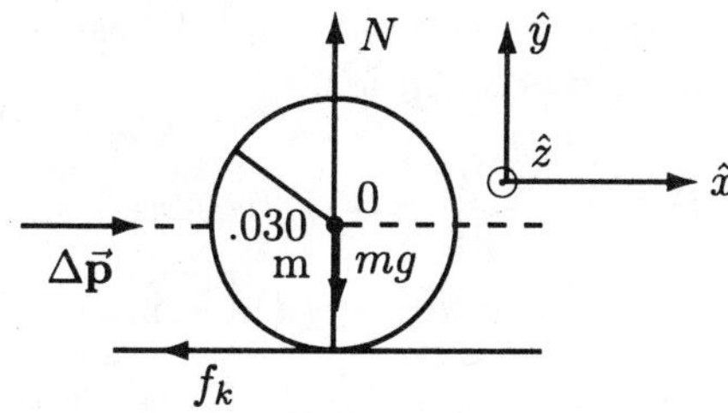

Thus, the impulsive force exerts no torque about the CM. The ball starts sliding without rolling. Kinetic friction acts on the ball a distance of .03 m away from the ball's CM, and thus exerts a torque, causing angular acceleration in the $-z$-direction. We'll start the problem by finding a_x and α_z

$$
\begin{aligned}
\sum F_x &= -f_k = ma_x \\
\sum F_y &= N - Mg = 0
\end{aligned}
$$

so,

$$a_x = -\frac{f_k}{m} = -\frac{\mu_k mg}{M} = -\mu_k g = -4.9\text{ m/s}^2$$

as expected, $a_x < 0$; the ball slows down. Now we'll use torque to find α_z.

$$\sum \vec{\tau}_{int} = rf_k\hat{\otimes} = I\alpha_z\hat{\mathbf{k}} \Rightarrow \alpha_z = -\frac{rf_k}{I}$$

or,

$$\alpha_z = -\frac{r\mu_k mg}{\frac{2}{5}mr^2} = -\frac{5\mu_k g}{2r} = -410\text{ rad/s}^2$$

Remember, initially the ball is sliding without rolling. Then it both rolls and slides. At some point its rotation will "catch up" to its speed *wrt* the table. When this happens, the ball will no longer slide. Then

$$v_x = -\omega_z r$$

Since we have a_x we should be able to write $v(t)$. Since we have α_z we should be able to write $\omega_z(t)$. Setting $v_x \equiv -\omega_z r$ will give us the time at which the ball ceases sliding. From the time we can, of course, find the distance.

$$
\begin{aligned}
v_x(t) &= v_0 + a_x t = \frac{\Delta p}{m} + a_x t \\
&= 2.5\text{ m/s} - \left(4.9\text{ m/s}^2\right)t \\
\omega_z(t) &= \left(\omega_0 + \alpha_z t\right) \\
&= \left(-410\text{ rad/s}^2\right)t
\end{aligned}
$$

Using the rolling without slipping constraint,

$$
\begin{aligned}
v_x(t) &\equiv -\omega_z(t)\,r \\
&\Rightarrow 2.5\text{ m/s} - \left(4.9\text{ m/s}^2\right)t \\
&\equiv \left(410\text{ rad/s}^2\right)t\,(.030\text{ m})
\end{aligned}
$$

solving for t gives $t = .145$ s. Finally the distance traveled by the ball in this time is:

$$
\begin{aligned}
\Delta x &= v_0 t + \frac{1}{2}at^2 \\
&= (2.5\text{ m/s})(.145\text{ s}) \\
&\quad -\frac{1}{2}\left(4.9\text{ rad/s}^2\right)(.145\text{ s}) \\
&= .31\text{ m}
\end{aligned}
$$

After the ball travels .31 m; it rolls without slipping.

12.103 With 149 stripes painted, there are 150 intervals between stripes, each corresponding to an angle $\Delta\theta = \frac{2\pi}{150}$ rad.

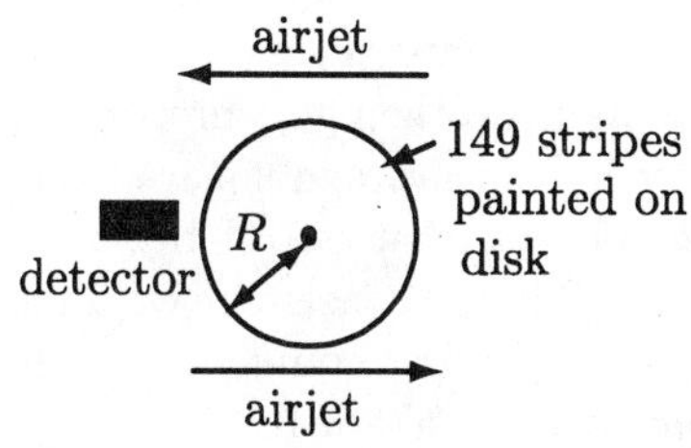

Thus, we can directly convert stripes counted in 1 s into an average angular speed during that 1 s interval.

$$\omega_{av} = \frac{N\Delta\theta}{(1.000\ \text{s})}$$

where N is the number of stripes counted. Since 1.000 s is 10^{-3} of the interval between counts and is 1 unit in the last significant figure of that interval, we can take measured angular speeds as being instantaneous values:

$$\omega = \frac{2\pi N}{(150.0\ \text{s})}$$

Thus we predict for the ith measurement

$$N_i = (23.87\ \text{s})\,\omega\,(t_i)$$

First theory: No friction, angular acceleration is constant, $\omega = \alpha t$. Then, the number of counts should be proportional to the time. This conclusion has already failed miserably by the second measurement: twice the first count, 504, differs from 496 by much more than the precision of the data.

For the approximate and exact formulae including friction, we have the spreadsheet predict the counts for each theory for each measurement and then compute sums of squares of the differences, actual counts minus predicted counts. One can then vary the parameter b in the theories to find the value which minimizes the sums of squared errors. Following the hint, estimate a first guess for b from the fifth data point:

$$\begin{aligned} 1173 \simeq\ & (23.87)\left(1.0783\times10^{-2}\ \text{rad/s}\right) \\ & \times\left(5.000\times10^{3}\ \text{s}\right) \\ & \times\left(1-\frac{b}{2}\left(5.000\times10^{3}\ \text{s}\right)+\text{neglect}\right) \end{aligned}$$

Then

$$\begin{aligned} 0.9114 &\simeq 1-\frac{b}{2}\left(5\times10^{3}\ \text{s}\right) \\ \Rightarrow b &= 3.54\times10^{-5}\ \text{rad/s} \end{aligned}$$

See the appendix for construction of the spreadsheet and the best fit found by the authors. The approximate theory fits OK for the first five points and then gets pretty bad. The exact theory has an average squared error of 0.2 for all 20 points, that is, off by ≈ 0.4 counts on average. That is consistent with fluctuations in how soon the first count occurs after the measurement interval starts.

12.107 a) Let the space station have a radius of R_O.

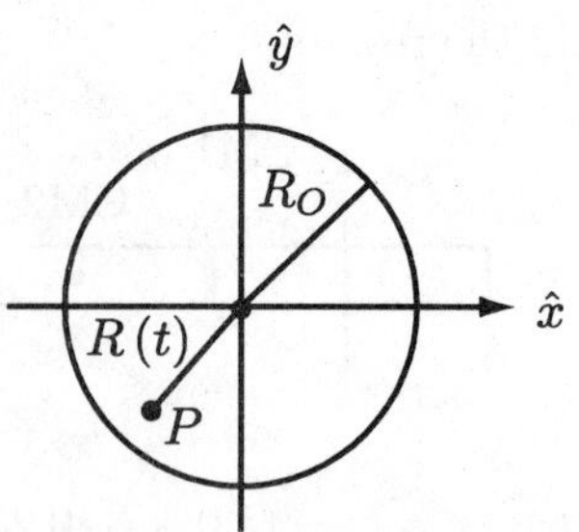

We know how to parameterize the position of an object in UCM. Suppose the astronauts are at some point P, a distance $R(t)$ from O, the axis of rotation. Let's take $R(t)$ to be a constant, R, then we know that the astronaut's position will be given by:

$$\vec{\mathbf{r}}(t) = R\left[\cos(\omega t)\hat{\mathbf{i}} + \sin(\omega t)\hat{\mathbf{j}}\right]$$

However, the astronauts' radial position also changes with time. They crawl towards O at constant velocity, $\vec{\mathbf{v}} = -v_O\hat{\mathbf{r}}$. Thus, their radial position from O as a function of time is

$$R(t) = R_O - v_O t$$

so,

$$\begin{aligned} \vec{\mathbf{r}}(t) &= R(t)\left[\cos\omega t\hat{\mathbf{i}} + \sin\omega t\hat{\mathbf{j}}\right] \\ &= (R_O - v_O t)\left[\cos\omega t\hat{\mathbf{i}} + \sin\omega t\hat{\mathbf{j}}\right] \end{aligned}$$

Now we differentiate to find the astronauts' velocity and acceleration.

$$\begin{aligned} \vec{\mathbf{v}}(t) &= \frac{d\vec{\mathbf{r}}(t)}{dt} = -v_O\left[\cos\omega t\hat{\mathbf{i}} + \sin\omega t\hat{\mathbf{j}}\right] \\ &\quad +\omega(R_O - v_O t)\left[-\sin\omega t\hat{\mathbf{i}} + \cos\omega t\hat{\mathbf{j}}\right] \\ &= \left[\omega\sin\omega t\,(v_O t - R_O) - v_O\cos\omega t\right]\hat{\mathbf{i}} \\ &\quad +\left[\omega\cos\omega t\,(R_O - v_O t) - v_O\sin\omega t\right]\hat{\mathbf{j}} \end{aligned}$$

$$\begin{aligned} \vec{\mathbf{a}}(t) &= \frac{d\vec{\mathbf{v}}(t)}{dt} \\ &= \begin{bmatrix} \omega^2\cos\omega t\,(v_O t - R_O) \\ +\omega v_O\sin\omega t + \omega v_O\sin\omega t \end{bmatrix}\hat{\mathbf{i}} \\ &\quad + \begin{bmatrix} -\omega^2\sin\omega t\,(R_O - v_O t) \\ -\omega v_O\cos\omega t - \omega v_O\cos\omega t \end{bmatrix}\hat{\mathbf{j}} \\ &= \left[\omega^2\cos\omega t\,(v_O t - R_O) + 2\omega v_O\sin\omega t\right]\hat{\mathbf{i}} \\ &\quad - \left[\omega^2\sin\omega t\,(R_O - v_O t) + 2\omega v_O\cos\omega t\right]\hat{\mathbf{j}} \end{aligned}$$

b) The astronauts crawl towards the center. The figure shows a view from above.

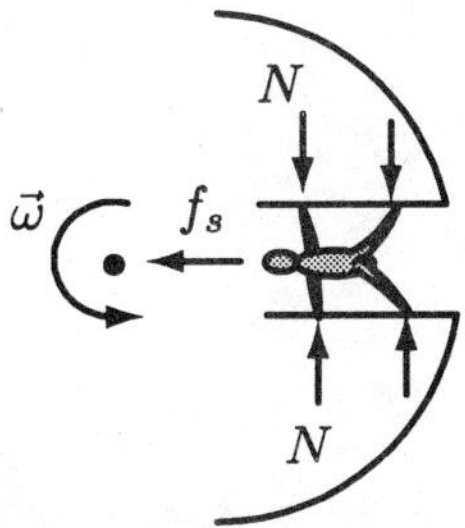

The station is spinning around; the normal forces are exerted by the walls on the astronauts and keep them rotating along with the rest of the station. The friction force provides the necessary centripetal force to keep them at their current radius from the station's center. The normal forces act in the direction of their velocity tangential to the station and therefore do work on them. Although friction does act in the direction of their radial velocity, their hands do not slip on the floor, so friction does no work on them. We can ask "what is the displacement of the point of contact of the force?", for this determines whether work was done or not. In the case of normal forces, the point of contact between their hands and walls displaces through some arc length ds which is in the direction of the normal force. Hence the normal force does work on the astronauts. The point of contact between the astronauts hands and floor does not move radially (they don't slip), so the point of contact of friction on the astronauts does not go through a displacement; friction does no work.

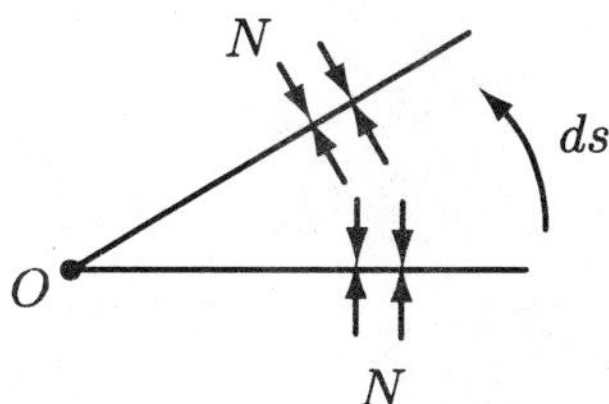

To keep themselves at their current radial position, the astronauts need to have an acceleration of $\vec{\mathbf{a}} = -\omega^2 r\hat{\mathbf{r}}$ so they need a force

$$\vec{\mathbf{F}} = m\vec{\mathbf{a}} = -m\omega^2 r\hat{\mathbf{r}}$$

In going from the outer radius to some inner radius, the work that they do is equal to

$$\begin{aligned} W &= \int_{R_O}^{r} \vec{\mathbf{F}} \cdot d\vec{s} = \int_{R_O}^{r} -m\omega^2 r\,(\hat{\mathbf{r}} \cdot \hat{\mathbf{r}}\,dr) \\ &= m\omega^2 \int_{r}^{R_O} r\,dr = \frac{1}{2} m\omega^2 \left[R_O^2 - r^2\right] \end{aligned}$$

They have to expend energy to accomplish this task. Where does this energy come from? It comes from chemical bonds from within their body to give their arm muscles the fuel necessary to apply the normal force. Where does the energy go to? In crawling closer to the axis of rotation, the astronauts decrease the rotational inertia of the station. Since no net external torques act on the station and astronauts, $\vec{\mathbf{L}}_{\text{sys}} = I\vec{\omega}$ remains constant. That means that ω (the station's angular speed) increases. The rotational K of the station, which is proportion to I and ω^2 must increase because of the angular speed squared term. Energy flows from the astronauts to the station. One last comment: we are told that the station is gigantic so $\Delta I_{\text{station}}$ will be very small. Thus we can also expect that the change in angular speed, $\Delta\omega$, will also be very small, and we shall neglect it.

c) The work that the astronauts did in getting to the radius r is independent of path or time. The only important thing was the end points of their journey: R_O and r_O. Recall, independence of path was one of our properties of potential energy. Also note that the work done by the astronauts, $\frac{1}{2}M\omega^2(R^2 - r)$ looks like the *difference* between two energies, another property of potential energy, only changes (the difference) in PE are important. Lastly, note the similarity between the derivation of gravitational PE and our expression for the work done by the astronauts. As the astronauts close in on the axis of rotation we would expect this PE to increase. Note in the expression shown, as $r \to 0$ the PE gets larger by becoming less and less negative.

d) Since friction can't keep the astronauts at a radial position r anymore, they slide to the outer end of the tube. The astronauts will have a radial and tangential component of velocity.

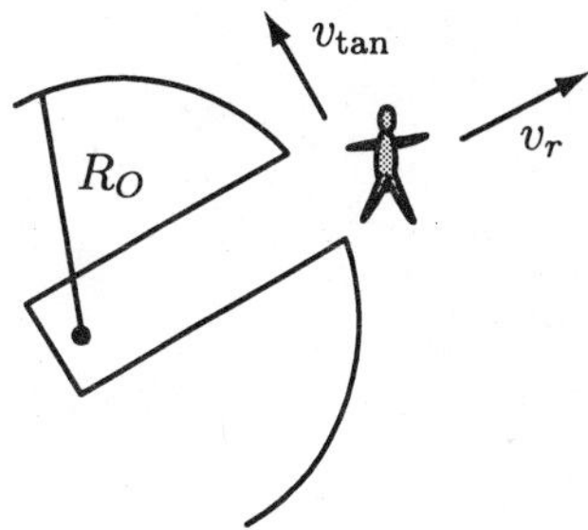

The tangential component will be equal to the tangential speed of the outer end of the station.

$$v_{\text{tan}} = \omega R_O$$

To finish the problem up we will use conservation of energy to get the radial velocity. What we are about to do is similar (in fact exactly the same) as setting $\Delta U = \frac{1}{2}Mv^2$ in Chapter 9.

The only difference is now we're using the effective radial PE instead of a gravitational PE. Let i denote initial state, when they're at $r = \frac{1}{2}R_O$ and f be the final state when they're at the station's outer edge at $r = R_O$

$$\begin{aligned} PE_i &= -\frac{1}{2}M\omega^2\left(\frac{1}{2}R_O\right)^2 = -\frac{1}{8}M\omega^2 R_O^2 \\ PE_f &= -\frac{1}{2}M\omega^2 R_O^2 \\ \Delta PE &= P_{e,f} - P_{e,i} = -\frac{1}{2}M\omega^2 R_O^2\left(1-\frac{1}{4}\right) \\ &= -\frac{3}{8}M\omega^2 R_O^2 \end{aligned}$$

Since friction does no work TME is constant, so $\Delta PE = -\Delta K$

$$-\frac{3}{8}M\omega^2 R_O^2 = -\frac{1}{2}Mv_r^2 \Rightarrow v_r = \frac{\sqrt{3}}{2}\omega R_O$$

So the speed is

$$\omega R_O\sqrt{1+\frac{3}{4}} = \sqrt{7}\frac{\omega R_O}{2}$$

Chapter 13

Fluids

13.1 If an ideal fluid didn't need support at every point of its lower surface, we would be able to support a horizontal column of water by supporting only one end:

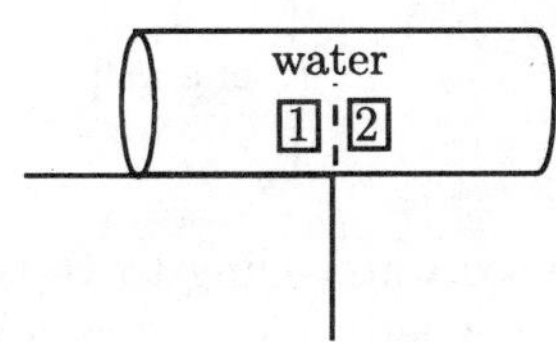

In order for the volume "2" not to fall, volume "1" (which is being supported from below) would have to exert a shear force upward on it. However, we know that an ideal fluid cannot support a shear force. Therefore, volume 2 will fall. This argument can be repeated for every little volume within the water column. From this we can conclude that an ideal liquid must be supported at every point along its base. When a plank or a rope extends across a gap, each element of the system requires mechanical support to balance its weight. A rope, which can only exert tension forces along its own direction accomplishes this by hanging in a curve so that the tension forces on an element can have vertical components. A board can exert shear forces perpendicular to its direction, so that it need not be greatly curved in order to support its parts.

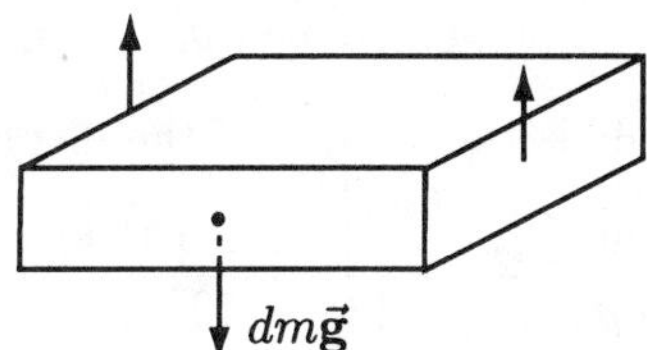

In an object we would call a "rigid beam" there is negligible curvature and the beam is primarily supported by shear.

13.3 $3.7 \times 10^6\,\text{Pa}\left(\frac{1\text{ atm}}{1.01\times10^5\text{ Pa}}\right) = 37\,\text{atm}$

$$\begin{aligned} F &= PA = (3.7 \times 10^6 \text{ Pa})(1.7 \times 10 \text{ m}^2) \\ &= 6.3 \times 10^8 \text{ N} \end{aligned}$$

13.5 Fluid pressure is approximately constant throughout a high pressure hydraulic system, so

$$P = \frac{F_1}{A_1} = \frac{F_2}{A_2}$$

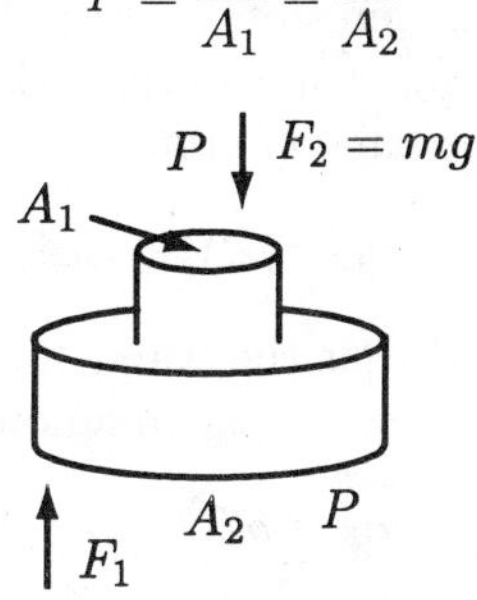

To produce a larger force than we put in we would want the ratio of areas to be a number larger than 1; thus, if $A_1 < A_2$,

$$F_2 = F_1\frac{A_2}{A_1}$$

and $F_2 > F_1$ so we would put the load on the larger piston. With

$$F_2 = M_{\text{load}}g = 1080\text{ N},$$

we would have to supply an input force of

$$F_1 = (1080\text{ N}) \cdot \frac{10\text{ m}^2}{1.0\text{ m}^2} = 110\text{ N}$$

13.7 $P = P_0 + \rho gh$ where P_0 is the pressure on the lake due to the atmosphere and ρgh is the pressure due to the water. h is measured from the lake's surface

$$\begin{aligned} P &= 1\text{ atm} + (103\text{ kg/m}^3)(9.81\text{ m/s}^2)(20\text{ m}) \\ &= 3 \times 10^5\text{ Pa} = 3\text{ atm} \end{aligned}$$

13.9 By Archimedes' Principle, the buoyant force which balances the cruiser's weight equals the weight of the displaced fluid.

$$m_{\text{cruiser}}g = \rho_{\text{water}}Vg$$

where V is the volume of the displaced fluid

$$V = \frac{m_{\text{cruiser}}}{\rho_{\text{water}}} = \frac{750 \times 10^3\text{ kg}}{1.00 \times 10^3\text{ kg/m}^3} = 750\text{ m}^3$$

13.11 From the continuity equation, the product of AV must be constant along any stream tube, but we need to be a bit careful here.

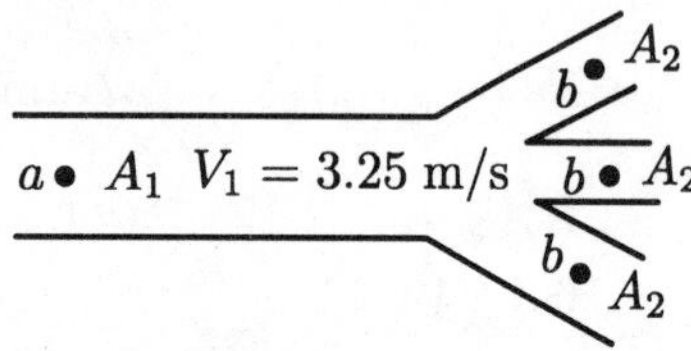

Recall that continuity is first and foremost a statement of conservation of mass. We will briefly reconstruct the argument. If you don't understand, reread the section on continuity in the text and come back. Mass flowing past point a in a time interval dt is:

$$dm_a = \rho dV_a = \rho A_1 V_1\, dt$$

We are told that the water flows through each of the exit tubes in equal amounts so

$$dm_b = \rho dV_b = \rho A_2 V_2\, dt$$

Since the water in the pipe entering the device gets divided into 3 equal parts,

$$\begin{aligned} dm_a &= 3dm_b \\ \Rightarrow \rho A_1 V_1\, dt &= 3\rho A_2 V_2\, dt \\ \Rightarrow V_2 &= \frac{A_1 V_1}{3A_2} = 5.85\text{ m/s} \end{aligned}$$

If your answer was 17.5 m/s you fell into the same trap we did the first time around solving this problem. That's the danger in plugging numbers into formulae.

13.13 From Example 13.11 this distance is given by

$$\begin{aligned} h_{\text{max}} &= \frac{P_{\text{atm}}}{\rho_{\text{gas}}g} \\ &= \frac{1.0 \times 10^5\text{ N/m}^2}{(6.8 \times 10^2\text{ kg/m}^2)(9.8\text{ m/s}^2)} \\ &= 14\text{ m} \end{aligned}$$

13.17 No, the mechanism by which water drops cling to the wall is intermolecular attraction. We know that there are no intermolecular forces between molecules of an ideal fluid. Furthermore, suppose that some molecules were able to cling to the wall somehow. To support the rest of the drop, it would have to exert a shear force ($F_{1,2}$ in the diagram). However, ideal fluids cannot support a shear force, so an ideal fluid cannot stick to a vertical surface.

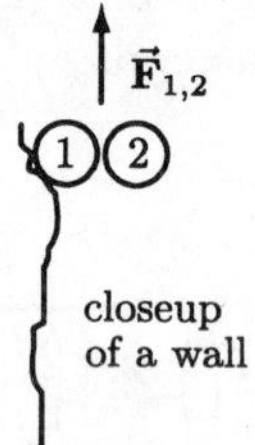

Since we see water cling to vertical surfaces in day to day experience, water must not be a completely ideal fluid.

13.21

$$\begin{aligned} m &= \rho V \\ &= (2000\text{ kg/m}^3)(.1\text{ m})(.05\text{ m})(.02\text{ m}) \\ &= .2\text{ kg} \end{aligned}$$

13.25 The hydraulic system must supply a force, $F_1 = PA$ to balance the weight force, mg acting on the mass.

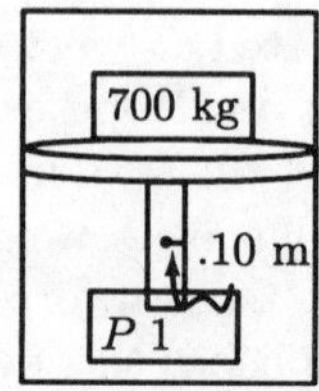

$$\begin{aligned} mg &= PA \\ \Rightarrow P &= \frac{mg}{A} = \frac{mg}{\pi(.1\text{ m})^2} \\ &= 2 \times 10^5\text{ Pa} \end{aligned}$$

However, this was assuming a vacuum. There is 1 atm of pressure above the piston, acting down on the piston. The result $P = 2 \times 10^5$ Pa means that region 1 must be 2×10^5 Pa greater than region 2 since $P_2 = 1$ atm, the pressure in region 1 must be

$$P = 2 \times 10^5 \text{ Pa} + 1 \text{ atm} = 3 \text{ atm}$$

13.29 From the given information we can compute the volume of 10.0 g of gold.

$$V = \frac{m}{\rho} = \frac{10.0 \times 10^{-3} \text{ kg}}{1.93 \times 10^4 \text{ kg/m}^3} = 5.18 \times 10^{-7} \text{ m}^3$$

The artisan can work this volume of gold into a sheet $d = 25.0\mu$m thick. If the whole volume was pounded to a uniform thickness d, its surface area would be

$$A = \frac{V}{d} = \frac{5.18 \times 10^{-7} \text{ m}^3}{25.0 \times 10^{-6} \text{ m}} = .0207 \text{ m}^2$$

If you have trouble seeing this, imagine the volume being shaped into a parallelepiped first and consider $V_{box} = (\text{area})(\text{height})$.

13.33 Let's estimate that the CM of the bicycle and rider falls somewhere midway between the tires so that each tire must support exactly half the weight of bicycle plus rider. Let A be the area of contact between the tire and road.

$$\begin{aligned} A &= \frac{\frac{1}{2}(M_{\text{bike}} + M_{\text{rider}})g}{P_{\text{tire}}} \\ &= \frac{\frac{1}{2}(100 \text{ kg})(9.8 \text{ m/s}^2)}{5 \times 10^5 \text{ Pa}} \\ &= 1 \times 10^{-3} \text{ m}^2 \end{aligned}$$

Which would be the area of a square of side 3 cm.

13.37 Although the new lake has 10× the surface area, we know that pressure in a fluid varies with height only, so the pressure on the dams are equal. Since the lakes are of equal height, the surface areas of the dams are equal. Therefore, the forces exerted on the dams are equal. No changes need to be made.

13.41 We want the barometer to be able to measure pressures that are about 1 atm, atmospheric pressure. From Section 3.3, Chapter 13:

$$\begin{aligned} L &= \frac{P_0}{\rho g} = \frac{1.0 \times 10^5 \text{ Pa}}{(8.0 \times 10^2 \text{ kg/m}^3)(9.81 \text{ m/s}^2)} \\ &= 13 \text{ m} \end{aligned}$$

13.45 From Eq. 13.11 in the text,

$$h_0 = \frac{1.01 \times 10^5 \text{ Pa}}{(1.3 \text{ kg/m}^3)(9.81 \text{ m/s}^2)} = 7920 \text{ m}$$

The given information allows us to write:

$$\begin{aligned} P &= P_0 e^{-z/h_0} \\ \Rightarrow 25 \text{ in. Hg} &= (30 \text{ in. Hg})\, e^{-z/(7920\,\text{m})} \end{aligned}$$

Solving for z,

$$z = 7920 \ln\left(\frac{6}{5}\right) = 1400 \text{ m}$$

13.49 We need only to consider the depth of the cork below the surface of the water. Since the top is "closed" we assume that the container acts like a barometer with a near vacuum at its top surface.

$$\begin{aligned} P &= +\rho_w g h \\ &= (1000 \text{ kg/m}^3)(9.8 \text{ m/s}^2)(.15 \text{ m}) \\ &= 1.5 \times 10^3 \text{ Pa} \end{aligned}$$

Therefore, the force on the cork is

$$F = PA = (1.5 \times 10^3 \text{ Pa})\pi(.01 \text{ m})^2 = 0.5 \text{ N}$$

13.53 Archimedes' principle makes no provisions about where it's true—it is equally as true in a bathtub as it is in a pond. The sailboat will displace an amount of water whose weight equals the buoyant force acting on the boat whether it's in a bathtub or a pond. Since the weight of the boat doesn't change, the buoyant force needed to float it doesn't change. Therefore the boat will displace equal amounts of water in the bathtub or pond.

13.59 The fraction below the surface is the ratio of the densities (see Example 13.6). Therefore, the fraction above the surface is:

$$1 - \frac{\rho_{\text{iron}}}{\rho_{\text{Hg}}} = 1 - \frac{7.96 \times 10^3 \text{ kg/m}^3}{1.36 \times 10^4 \text{ kg/m}^3} = 41.5\%$$

13.63 The captain has a mass m.

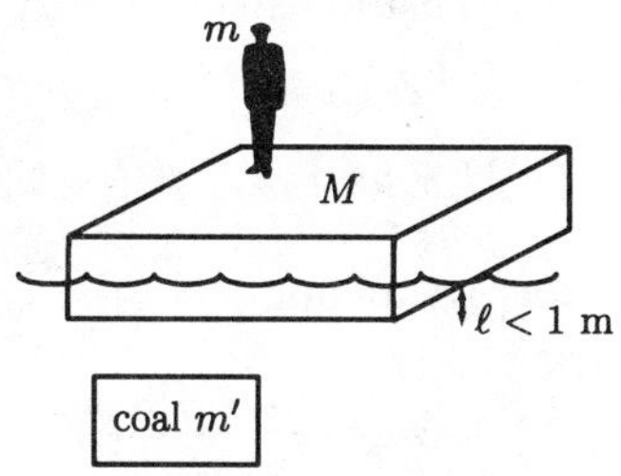

The barge has a mass M and cross-sectional area

$$A = (3.00 \text{ m})(10.5 \text{ m}) = 31.5 \text{ m}^2$$

Of course we want the barge to float but we also want it to float while not reaching a depth of 1.00 m. Let n be the number of bags of coal such that the barge does not float with its bottom reaching 1.00 m below the waters surface. The total mass on the barge is $m+M+nm'$. By Archimedes' principle, this mass must equal the mass of the displaced water when the barge's bottom is just above a depth of 1.00 m below the water's surface.

$$m + M + nm' = A\ell\rho_{\text{water}}$$

Solving for n,

$$\begin{aligned} n &= \frac{A\ell\rho_{\text{water}} - m - M}{m'} \\ &= \frac{(31.5 \text{ m}^2)(1.00 \text{ m})(1000 \text{ kg/m}^3) - 80 \text{ kg} - 9020 \text{ kg}}{50.0 \text{ kg}} \\ &= 448 \text{ bags} \end{aligned}$$

13.67 We are interested in ρ_{glass}, so the volume of the glass is of interest to us—not the volume V of the beaker which was given as $V = 1.00 \times 10^{-4} \text{ m}^3$, but the volume of the actual glass itself, V'.

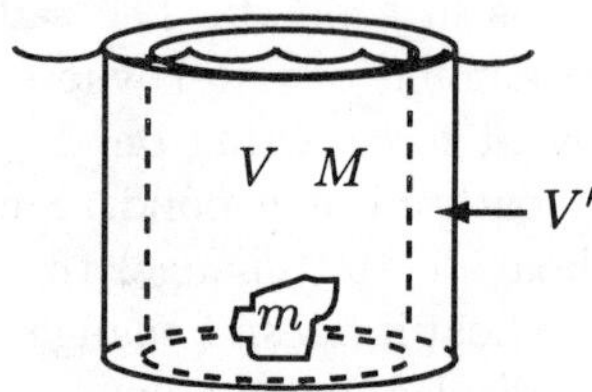

When the beaker is just about to sink, the water is right at the very top of the beaker. The volume of displaced water is not just equal to the volume V of the beaker, but is equal to the volume of the (inside) of the beaker plus the volume of the glass itself— $V_{\text{displaced water}} = V + V'$. Thus, by Archimedes' principle, the beaker will be in equilibrium when the water is right at the top if the total mass, $M + m$ is equal to the total mass of displaced water, $\rho_{\text{water}}(V + V')$.

$$M + m = \rho_{\text{water}}(V + V')$$

Solving for the volume V' of the glass,

$$V' = \frac{M + m}{\rho_{\text{water}}} - V$$

and the density of the glass is:

$$\begin{aligned} \rho_{\text{glass}} &= \frac{M}{V'} = \frac{M\rho_{\text{water}}}{M + m - \rho_{\text{water}}V} \\ &= 1.8 \times 10^3 \text{ kg/m}^3 \end{aligned}$$

13.73 Outside the window, winds travel at extremely high speeds. We know that rapidly flowing fluids (like air) are regions of low pressure. If the wind's speed is very great outside the house, the pressure on the outside of a window would be much less than the pressure on the inside of the window, which would be, more or less, normal atmospheric pressure. There will be a net outward force on the window.

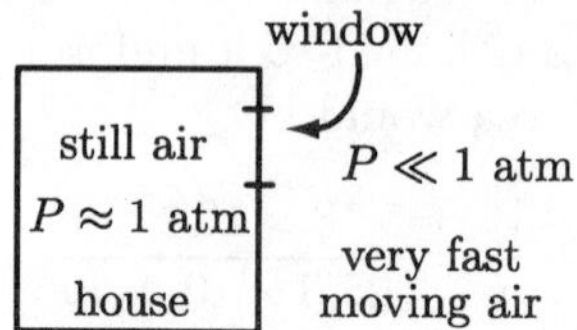

If the winds go fast enough so that the outside pressure drops far enough below the pressure inside the house, the pressure inside the house will push the window out of its frame. That's why in hurricanes you sometimes hear that the windows should be left open halfway, to equalize the pressure inside the house with the pressure outside. Note that we talk about forces that arise from pressure differences. There is no such thing as a "suction force."

13.77 In the discussion on siphons, it was found that

$$\frac{P_0}{\rho g} > D + H$$

(refer to Figure 13.33.) H is the height of the siphon *wrt* the water line. Recall that the location of point S is not important; just as long as it's located below the water's surface. From the above expression we see that a smaller acceleration due to gravity produces a larger maximum height for the siphon. Since $g_{\text{Moon}} < g_{\text{Earth}}$, Moon residents would find that their siphons would carry water over a greater height than Terran siphons can.

13.81 Applying Bernoulli's Law to points A and B,

$$\begin{aligned} P_A + \rho g z + \frac{1}{2}\rho v_A^2 &= P_B + \rho g z + \frac{1}{2}\rho v_B^2 \\ \rho g z &= P_A - P_{\text{atm}} + \frac{1}{2}\rho\left(v_A^2 - v_B^2\right) \end{aligned}$$

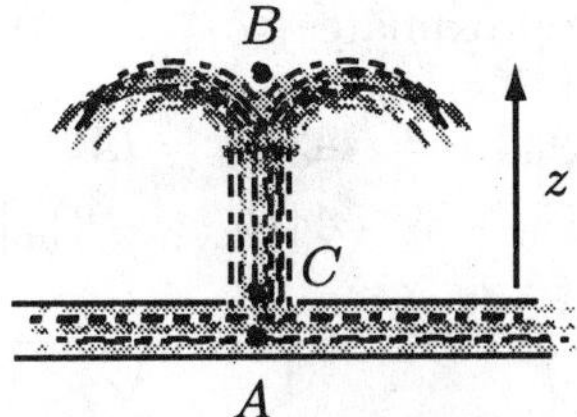

At the maximum height, the water is momentarily at rest before it starts to fall back down, so $v_B = 0$.

$$z = \frac{P_A - P_{atm}}{\rho g} + \frac{\frac{1}{2}v_A^2}{g}$$

Now consider point C which is a negligible distance above point A. If we assume a small break in the pipe then $r_{hole} \ll r_{pipe}$. The area of the pipe (hole) is proportional to the radius of the pipe (hole) squared:

$$\frac{A_{pipe}}{A_{hole}} = \left(\frac{r_{hole}}{r_{pipe}}\right)^2$$

But from continuity we know that $\rho A v$ = constant

$$\begin{aligned} \frac{v_{hole}}{v_{pipe}} &= \left(\frac{r_{pipe}}{r_{hole}}\right)^2 \\ \Rightarrow v_{pipe}^2 &= v_A^2 = v_{hole}^2\left(\frac{r_{hole}}{r_{pipe}}\right)^4 \end{aligned}$$

If $r_{hole} < r_{pipe}$ then $r_{hole}^4 \lll r_{pipe}^4$, so v_A^2 is very small compared to v_c. By approximating $v_A = 0$ (the same approximation was made in Example 13.10 in the text).

$$z = \frac{6.9 \times 10^5 \text{ Pa} - 1.01 \times 10^5 \text{ Pa}}{(1.00 \times 10^3 \text{ kg/m}^3)(9.81 \text{ m/s}^2)} = 60 \text{ m}$$

13.85 From continuity, $\rho A_A V_A = \rho A_B V_B$. Since $A_A = A_B$ (pipe is of constant radius), $V_A = V_B$. Take the height of point A to be the reference level. Applying Bernoulli's equation to points A and B,

$$P_A + \frac{1}{2}\rho V_A^2 = P_B + \rho g(100 \text{ m}) + \frac{1}{2}\rho V_B^2$$

since $V_A = V_B$,

$$P_A = P_B + \rho g(100 \text{ m})$$

As in a siphon, if the water is just to get over the hill, the pressure at point B is zero. Gravity will pull the water back down the hill.

$$P_A = \rho g(100 \text{ m}) = 9.8 \times 10^5 \text{ Pa} = 9.7 \text{ atm}$$

(We were given 2 sig. figs. for the height of the hill.) The power output of the motor equals the force it exerts on the water multiplied by the speed of the water leaving the pump.

Power = $F \cdot v$ = (Pressure)(Area)(speed) = Pressure × volume rate

$$\begin{aligned} &= (9.8 \times 10^5 \text{ Pa})(1.0 \text{ m}^3/\text{s}) = 9.8 \times 10^5 \text{ W} \\ &= 0.98 \text{ MW} \end{aligned}$$

13.89 We'll begin by applying Bernoulli's equation at either end of the channel.

$$P_A + \rho g h_A + \frac{1}{2}\rho V_A^2 = P_B + \rho g h_B + \frac{1}{2}\rho V_B^2$$

The channel is open to the atmosphere, so $P_A = P_B = P_{atm}$. Setting the reference height level to h_B, we get:

$$\rho g h_A + \frac{1}{2}\rho V_A^2 = \frac{1}{2}\rho V_B^2$$

or

$$V_B = \sqrt{2gh_A + V_A^2}$$

The area of the water flow at either end of the channel is the channel width times the fluid depth:

$$\begin{aligned} A_A &= (.60 \text{ m})(.15 \text{ m}) = .090 \text{ m}^2 \\ A_B &= (1.0 \text{ m})d \end{aligned}$$

And from continuity, $(\rho_A = \rho_B)$

$$\begin{aligned} A_A V_A &= A_B V_B \Rightarrow A_B = A_A\left(\frac{V_A}{V_B}\right) \\ A_B &= (.090 \text{ m}^2)\left[\frac{V_A}{\sqrt{2gh_A + V_A^2}}\right] \\ d &= (.090 \text{ m})\left[\frac{2.0 \text{ m/s}}{\sqrt{2gh_A + (2.0 \text{ m/s})^2}}\right] \end{aligned}$$

Now h is the height of point A *wrt* point B:

$$\begin{aligned} h &= (.10 \times 10^3 \text{ m})\sin(2°) = 3.5 \text{ m} \\ d &= (0.90 \text{ m})\left[\frac{2.0 \text{ m/s}}{\sqrt{2(9.81 \text{ m/s}^2)(3.5 \text{ m}) + (2.0 \text{ m/s})^2}}\right] \\ &= 2.1 \text{ cm} \end{aligned}$$

13.91 If we look at the flow lines below h, they pass under the wing uninterrupted and unbent.

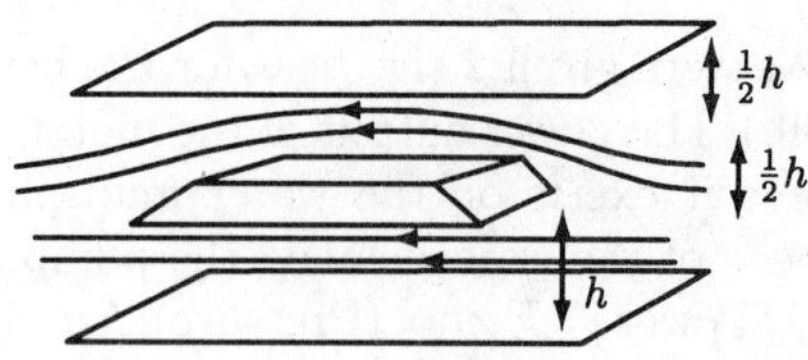

The area of flow in this region remains constant at hW as it passes under the wing. Therefore, by continuity, the speed of fluid flow under the wing remains constant at v_0. Streamlines of fluid above h are closer together above the wing. Before the wing, the area of this fluid flow is Wh and its speed is v_0. At the wing, the stream tube area decreases by half: $\frac{1}{2}Wh$. Therefore, by continuity, the flow speed must double to $2v_0$.

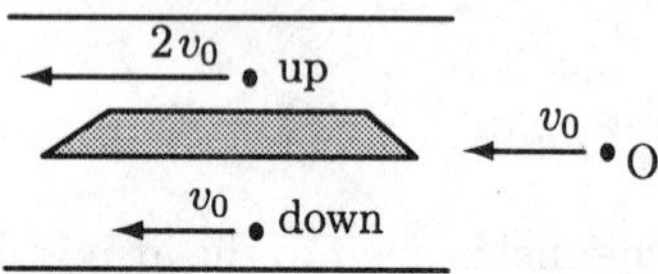

By Bernoulli's equation, if we neglect the height difference between the points up, down and 0, as instructed:

$$\begin{aligned} P_0 + \frac{1}{2}\rho v_0^2 &= P_{\text{up}} + \frac{1}{2}\rho\, v_{\text{up}}^2 && (1) \\ &= P_{\text{down}} + \frac{1}{2}\rho\, v_{\text{down}}^2 && (2) \end{aligned}$$

From (1),

$$\begin{aligned} P_{\text{up}} &= P_0 + \frac{1}{2}\rho\left[v_0^2 - v_{\text{up}}^2\right] \\ &= P_0 + \frac{1}{2}\rho\left[v_0^2 - (2v_0)^2\right] \\ &= P_0 - \frac{3}{2}\rho v_0^2 \end{aligned}$$

$$\begin{aligned} P_{\text{down}} &= P_0 + \frac{1}{2}\rho\left[v_0^2 - v_{\text{down}}^2\right] \\ &= P_0 + \frac{1}{2}\rho\left[v_0^2 - v_0^2\right] \\ &= P_0 \end{aligned}$$

The lift force on the wing results from the difference between P_{up} and P_{down}

P_{up} F_{up}

P_{down} F_{down}

$$F_{\text{lift}} = F_{\text{down}} - F_{\text{up}} = P_{\text{down}}A_{\text{down}} - P_{\text{up}}A_{\text{up}}$$

If we approximate

$$\begin{aligned} A_{\text{up}} &= A_{\text{down}} = LW \\ F_{\text{lift}} &= A\left[P_{\text{down}} - P_{\text{up}}\right] \\ &= A\left[P_0 - \left(P_0 - \frac{3}{2}\rho v_0^2\right)\right] \\ &= \frac{3}{2}\rho v_0^2 LW \end{aligned}$$

13.95 From (13.15), the lift force is

$$F_L = \frac{1}{2}C_L \rho v^2 A$$

and we can see that, with all things being equal, the lift force is smaller in regions where atmospheric air density is smaller. Therefore, it is safe to conclude that the region of greater altitude (less dense atmospheric pressure) would require a greater take off speed, v. Therefore take-off speed would be greater in Leadville, CO than Oakland, CA. The airspeed indicator compares ram pressure, $\frac{1}{2}\rho v^2$, with atmospheric pressure. The plane will not fly unless F_L is large enough. This means that a certain value of ρv^2 is required to lift off regardless of whether you're in CO or CA. However, ρv^2 is exactly what the airspeed indicator measures. Since ρv_{min}^2 is the same for all lift-offs everywhere for the same plane, the measured airspeed will be the same for all lift-offs everywhere as well.

13.97 From Eq. 13.16 and Figure 13.38 (here, $P_A = P_{\text{atm}} = 0.99 \times 10^5$ Pa)

$$\begin{aligned} P_B &= P_A + \frac{1}{2}\rho v^2 \\ &= (0.99 \times 10^5 \text{ Pa}) \\ &\quad + \frac{1}{2}(1.2 \text{ kg}/\text{m}^3)(68.0 \text{ m}/\text{s})^2 \\ &= 0.99 \times 10^5 \text{ Pa} + 0.028 \times 10^5 \text{ Pa} \\ &= 1.02 \times 10^5 \text{ Pa} \end{aligned}$$

13.99 The problem statement practically gives the answer away. We learned about three quantities which are absolutely conserved in the absence of external forces and torques. Which one involves swirling (rotational) quantities? Angular momentum of course! For an object moving in a fluid, we can place an origin on the path of the object, in which case its angular momentum is zero, and zero must the angular momentum stay. Any vortices have $\vec{\mathbf{L}}$ *wrt* the origin; the sum of all these angular momenta must be zero. That's why they must swirl in opposite directions.

13.103 The pitot tube actually measures ram pressure, which will always be greater than atmospheric pressure if the plane is moving *wrt* the atmosphere. We can trust the altimeter for a reading of atmospheric pressure. From Figure 13.17, the plane is at an altitude of 9.3 km.

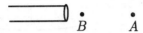

Now we'll apply Bernoulli's Law to two points at equal height; one point, B will be by the opening of the pitot tube. In the plane's reference frame, wind is travelling towards the plane at some speed v. (Incidently, a person standing on the Earth will see the plane travel towards point A at the same speed v). In the plane's reference frame, air is brought to rest at point B and $P_A = P_{\text{atm}}$ as measured by the altimeter.

$$P_A + \frac{1}{2}\rho V_A^2 = P_B \qquad (3)$$

The density of air at 9300 m is much less than ρ at sea level! From the discussion about the atmosphere, our isothermal atmosphere model has ρ proportional to P.

$$\begin{aligned} \rho_{\text{at 12 km}} &= \rho_{\text{sea level}} \frac{P_A}{P_{\text{sea level}}} \\ &= \left(1.3\ \frac{\text{kg}}{\text{m}^3}\right) \frac{2.96}{10.1} \\ &= 0.38\ \text{kg}/\text{m}^3 \end{aligned}$$

Solving (3) for V_A,

$$\begin{aligned} V_A &= \sqrt{\frac{(P_B - P_A)}{\frac{1}{2}\rho}} \\ &= \sqrt{\frac{(3.33 \times 10^4\ \text{Pa} - 2.96 \times 10^4\ \text{Pa})}{\frac{1}{2}(.38\ \text{kg}/\text{m}^3)}} \\ &= 140\ \text{m}/\text{s} \end{aligned}$$

13.105 Let O be the origin of coordinates.

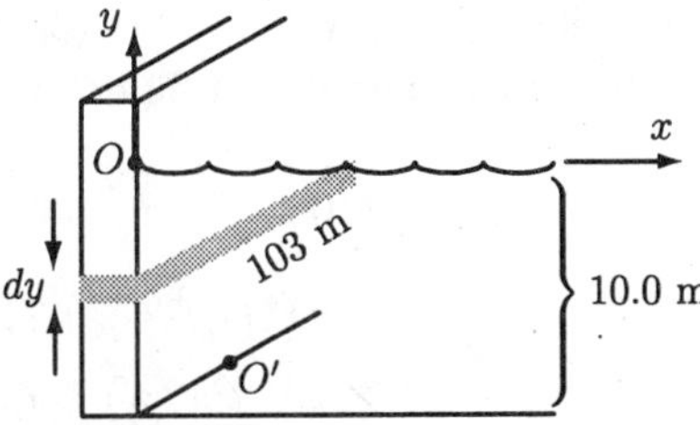

At coordinate y, below the water's surface, the pressure on the dam is:

$$P(y) = P_{\text{atm}} - \rho g y$$

Note that the minus sign arises because y is negative for all points below the water's surface. Now let's talk about the net force on the dam. At y, the water exerts a pressure $P_{\text{atm}} - \rho g y$ to the left on the dam. However, there is atmosphere to the right at the dam exerting a pressure P_{atm} to the right on the dam. The *net* pressure on the dam at a depth y is $-\rho g y$, to the left. The differential element, we are told, is a differential strip of height dy and length $W = 103$ m.

$$dA = (W)\,dy$$

So the net force on a differential strip of the dam at depth y is:

$$dF = P(y)\,dA = -\rho g y (W)\,dy$$

We will integrate over the entire face of the dam, from $y = -10.0$ m to $y = 0$ m.

$$\begin{aligned} F &= -\int_{-10.0\text{ m}}^{0\text{ m}} \rho g y (W)\,dy = -\rho g (W) \int_{-10.0\text{ m}}^{0\text{ m}} y\,dy \\ &= \frac{(1.00\times 10^3\ \text{kg}/\text{m}^3)(9.81\ \text{m}/\text{s}^2)(103\ \text{m})}{2} y^2 \Big|_{-10.0\text{ m}}^{0\text{ m}} \\ &= (5.00 \times 10^5\ \text{kg}/\text{m}\cdot\text{s}^2)\, y^2 \Big|_{0\text{ m}}^{-10.0\text{ m}} \\ &= 5.05 \times 10^7\ \text{N} \end{aligned}$$

The torque at y about O' is:

$$\begin{aligned} d\tau &= r dF \odot \\ &= (d + y)(-\rho g y\,[W])\,dy \odot \end{aligned}$$

remember: $y < 0$!

$$d\tau = -\rho g y\,(d + y)\,(W)\,dy \odot$$

We integrate from $y = -10.0$ m to $y = 0$ m to get the total torque.

$$\begin{aligned} \tau &= \int d\tau = -\int_{-10.0\text{ m}}^{0\text{ m}} \rho g y\,(d + y)\,(W)\,dy \odot \\ &= \rho g\,(W) \int_{0\text{ m}}^{-10.0\text{ m}} (yd + y^2)\,dy \\ &= \rho g\,(W) \left[\frac{d}{2} y^2 + \frac{1}{3} y^3\right]_{0\text{ m}}^{-10.0\text{ m}} \\ &= (1.00 \times 10^3\ \text{kg}/\text{m}^3)\,(9.81\ \text{m}/\text{s}^2)\,(103\ \text{m}) \\ &\quad \times \left[(5.00\ \text{m})\,(-10.0\ \text{m})^2 - \frac{1}{3}\,(10.0\ \text{m})^3\right] \\ &= 1.68 \times 10^8\ \text{N}\cdot\text{m} \end{aligned}$$

We can ask ourselves: "where do we have to place the net force due to pressure in order to calculate the net torque which we calculated above?" The answer is rather simple once you see it. Since $\tau_{net} = rF_{net}$,

$$r = \frac{\tau_{net}}{F_{net}} = \frac{1.68 \times 10^8 \text{ N} \cdot \text{m}}{5.05 \times 10^7 \text{ N}} = 3.33 \text{ m}$$

Since torque was calculated about the bottom of the dam, the center of pressure is 3.33 m from the bottom, or 6.67 m below the surface of the water.

13.109 Bernoulli's equation says:

$$P + \frac{1}{2}\rho v^2 + \rho g h = \text{constant}$$

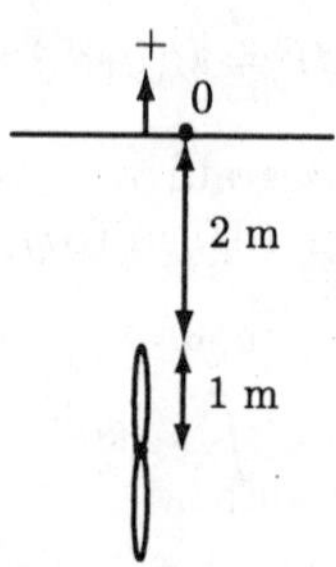

Let's assume that at least as an order of magnitude estimate, the water travels with the propeller, so $v_{water} \approx v_{propeller}$. What happens when v becomes really large? The only term that can decrease as $\frac{1}{2}\rho v^2$ increases is P. Of course, $\frac{1}{2}\rho v^2$ cannot become so large that P must be negative, since pressures cannot be negative. We expect a limit on v that makes $P \to 0$.

$$\frac{1}{2}\rho v_{max}^2 = -\rho g h \Rightarrow v_{max} = \sqrt{2g(-h)}$$

When $v \to v_{max}$, the pressure at the top of the propeller goes to zero. Here, we are told that the propeller is 3 m below the water's surface so the tip of the upper propeller is at a depth of 2 m

$$v_{max} = \sqrt{2(9.81 \text{ m/s}^2)(2 \text{ m})} = 6.26 \text{ m/s}$$

From rotational kinematics,

$$\begin{aligned} v_{max} &= \omega_{max} L = \omega_{max}(1 \text{ m}) \\ &= 6.26 \text{ m/s} \\ \Rightarrow \omega_{max} &= 6.26 \text{ rad/s} \end{aligned}$$

What happens when the pressure of the water $\to 0$? It boils! A chunk of water evaporates leaving a cavity in the water's surface. The top of the propeller is momentarily in air and then slams into the unevaporated water. When this happens, the propeller can be damaged.

13.113 a) Consider a differential volume of fluid in UCM.

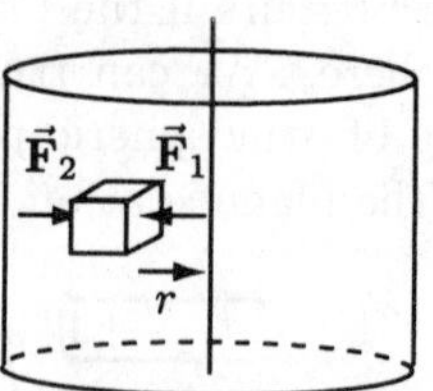

To remain in UCM, we know that

$$\sum F_r = F_1 - F_2 = -dm\omega^2 r \qquad (4)$$

The difference between F_1 and F_2 is caused by a difference of pressure between the two radial sides of the differential element. The volume accelerates inwards. Rewriting (4) in terms of pressure,

$$P(r) - P(r + dr) = -\frac{dm\omega^2 r}{dA}$$

From the definition of the derivative

$$\frac{df}{dx} = \lim_{dx \to 0} \frac{f(x + dx) - f(x)}{dx}$$

and $dm = \rho dA\, dr$.

$$\frac{P(r + dr) - P(r)}{dr} = -\rho\omega^2 r$$

or,

$$dP = \rho\omega^2 r\, dr$$

Integrating both sides yields

$$P(r) = \frac{1}{2}\rho\omega^2 r^2 + P_0$$

where P_0 is a constant of integration. Note that when $r = 0$, $P(r)$ must equal the pressure at the center of the cylinder.

$$P(r) = \frac{1}{2}\rho\omega^2 r^2 + P_{center} \qquad (5)$$

Now we'll compute $P(r = .175 \text{ m}) - P(r = 0)$, the difference in pressure between the side and center of the cylinder of fluid.

$$\begin{aligned} \Delta P &= \frac{1}{2}(1000 \text{ kg/m}^3)(1550 \text{ rpm})^2 \\ &\quad \times (2\pi)^2 \left(\tfrac{1 \text{ min}}{60 \text{ s}}\right)^2 (.175 \text{ m})^2 \\ &= 4 \times 10^5 \text{ Pa} \approx 4 \text{ atm} \end{aligned}$$

b) As we learned earlier, pressure increases linearly with depth from the fluid's surface.

$$P(z) = P_{atm} + \rho g z$$

here z is measured from the fluid's surface and increases with depth below the surface. Pressure is dependent on the square of the radial distance from the center and linearly dependent on the depth below the surface. Let's picture lines of constant pressure. If you move outward radially within the cylinder of fluid, pressure increases with the square of the new radius. To keep pressure constant, if we move radially outward, we need to move upward a lot since pressure is only linearly dependent on height. If we take a plane passing through the cylinder,

$$P \propto y$$
$$P \propto x^2$$

which is a parametric form of a parabola.

Chapter 14

Oscillatory Motion

14.3 We can describe the block's position *wrt* to the system's equilibrium position (where the spring is not compressed or stretched) by SHM.

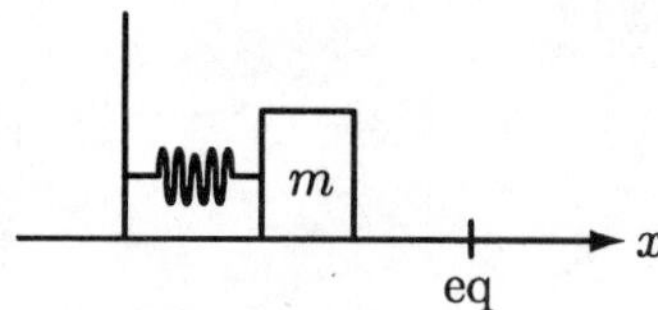

We know the position function will have the form:

$$x(t) = A\cos(\omega t + \phi_0)$$

At $t = 0$ s the spring is compressed by 7.5 cm so the mass is at $x = -7.5$ cm. This 7.5 cm also represents the maximum distance that the mass can be from the equilibrium position, also known as the mass's amplitude. Thus, $A = +7.5$ cm and $\phi_0 = \pi$ since we want $A\cos(\phi_0) = -7.5$ cm. The angular frequency is given by the text's equation (14.2)

$$\omega = \sqrt{\frac{k}{m}} = \sqrt{\frac{0.50\ \mathrm{N/m}}{0.075\ \mathrm{kg}}} = 2.6\ \mathrm{rad/s}$$

Thus,

$$\begin{aligned} x(t) &= +(7.5\ \mathrm{cm})\cos\left[(2.6\ \mathrm{rad/s})t + \pi\right] \\ &= -(7.5\ \mathrm{cm})\cos\left[(2.6\ \mathrm{rad/s})t\right] \end{aligned}$$

and

$$\begin{aligned} v(t) &= \frac{dx}{dt} \\ &= \left(2.6\ \frac{\mathrm{rad}}{\mathrm{s}}\right)(7.5\ \mathrm{cm})\sin\left[\left(2.6\ \frac{\mathrm{rad}}{\mathrm{s}}\right)t\right] \\ &= \left(19.5\ \frac{\mathrm{cm}}{\mathrm{s}}\right)\sin\left[\left(2.6\ \frac{\mathrm{rad}}{\mathrm{s}}\right)t\right] \end{aligned}$$

14.5 The pendulum's period is given by:

$$T = \frac{2\pi}{\omega} = 2\pi\sqrt{\frac{\ell}{g}}$$

The period is proportional to the square root of the rod's length. Thus,

$$\frac{T_A}{T_B} = \frac{\sqrt{\ell_A}}{\sqrt{\ell_B}} = \sqrt{\frac{2\ell}{\ell}} = \sqrt{2}$$

The period of pendulum A is $\sqrt{2}$ times the period of pendulum B.

14.7 From 14.2.1:

$$\theta(t) = [\theta_{\mathrm{max}}]\cos\left(\sqrt{\frac{g}{\ell}}t + \phi_0\right)$$

We want $\theta(0\ \mathrm{s}) = 3.0^\circ$ and this is the amplitude of the pendulum, so $\theta_{\mathrm{max}} = 3^\circ$ and $\phi_0 = 0$. Thus,

$$\begin{aligned} \theta(t) &= 3^\circ\cos\left(\sqrt{\frac{9.81\ \mathrm{m/s^2}}{.75\ \mathrm{m}}}t\right) \\ &= 3.0^\circ\cos\left([3.6\ \mathrm{rad/s}]\,t\right) \end{aligned}$$

14.9 At the equilibrium point all the pendulum's energy is in the kinetic form.

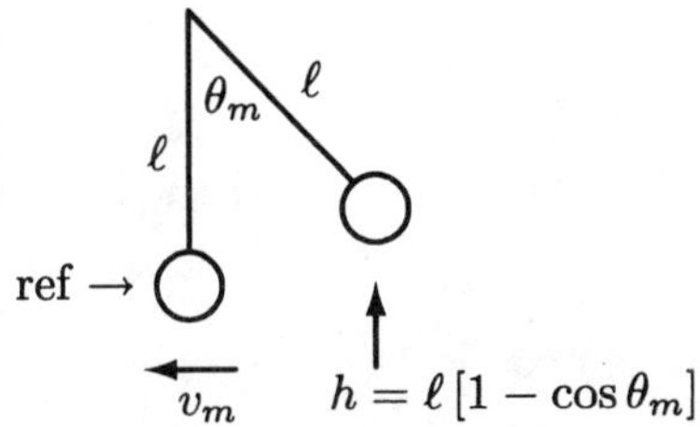

$$E = K = \frac{1}{2}mv_m^2$$

At the max height the pendulum makes an angle of θ_m with the vertical. All its energy is in the potential form and is equal to

$$mgh = mg\,[\ell(1 - \cos\theta_m)]$$

where we took the reference level from which h is measured to be at the equilibrium level where the bob's height is a distance ℓ below the pivot point. Since

$$\begin{aligned} E_{\rm tot} &= K_{\rm max} = P_{\rm max} \\ \frac{1}{2}mv_m^2 &= mg\ell\,[1 - \cos\theta_m] \\ \Rightarrow \theta_m &= \cos^{-1}\left[1 - \frac{v^2}{2g\ell}\right] \\ &= \cos^{-1}\left[1 - \frac{(0.37\ {\rm m/s})^2}{2(9.81\ {\rm m/s^2})(0.60\ {\rm m})}\right] \\ &= 8.7^\circ \end{aligned}$$

14.11 The amplitude of a forced oscillation was given in the text in Section 14.4. The natural angular frequency of the spring/mass system is $\omega_0 = \sqrt{\frac{k}{m}} = 7$ rad / s and the harmonic force has an angular frequency of 15 rad / s.

$$\begin{aligned} B &= \frac{F_0}{M(\omega_0^2 - \omega^2)} \\ &= \frac{10\ {\rm N}}{(0.2\ {\rm kg})\,[(7\ {\rm rad/s})^2 - (15\ {\rm rad/s})^2]} \\ &= -0.3\ {\rm m} \end{aligned}$$

The minus sign reflects the fact that the resulting oscillations are out of phase with the external force. The amplitude of the system's oscillation is the absolute value of B, 0.3 m.

14.13 The angular frequency, ω, is $\sqrt{\frac{k}{m}}$ for a mass oscillating on a spring. Thus:

$$\omega = \frac{2\pi}{T} = \sqrt{\frac{k}{m}} \Rightarrow T = 2\pi\sqrt{\frac{m}{k}}$$

The period of the oscillating system is proportional to the square root of the mass. Thus, if we double the mass the period increases by a factor of $\sqrt{2}$. Thus

$$T = \sqrt{2}\left(\frac{1}{2}\ {\rm s}\right) = 0.7\ {\rm s}$$

14.17 The position and speed of the mass will be in the form of:

$$\begin{aligned} x(t) &= A\cos(\omega t + \phi_0) \qquad (1) \\ v_x(t) &= -A\omega\sin(\omega t + \phi_0) \qquad (2) \end{aligned}$$

The angular frequency for a mass on spring system is

$$\omega = \sqrt{\frac{k}{m}} = \sqrt{\frac{.50\ {\rm N/m}}{.075\ {\rm kg}}} = 2.6\ {\rm rad/s}$$

The coordinate system is centered on the mass when the spring is relaxed, so

$$\begin{aligned} y(0) &= 0\ {\rm m} = A\cos(\omega t + \phi_0) \\ \Rightarrow \phi_0 &= \frac{\pi}{2}\ {\rm or}\ \frac{3\pi}{2} \end{aligned}$$

The impulse, Δp, gives the mass an initial speed of

$$v_{x0} = \frac{\Delta p}{m} = \frac{0.22\ {\rm N\cdot s}}{0.075\ {\rm kg}} = 2.9\ {\rm m/s}$$

Using the initial speed condition,

$$v_x(0) = 2.9\ {\rm m/s} = -A\omega\sin(\omega t + \phi_0)$$

Because A is positive by definition, we choose $\phi_0 = \frac{3\pi}{2}$ so that $\sin\phi_0 = -1$. Then

$$A = \frac{+2.9\ {\rm m/s}}{(2.6\ {\rm rad/s})(1)} = +1.1\ {\rm m}$$

Then Eq. (1) becomes:

$$x(t) = (+1.1\ {\rm m})\cos\left([2.6\ /{\rm s}]\,t + \frac{3\pi}{2}\right)$$

14.21 The puck leaves the center of the ice surface with a speed of 1.5 m / s and hits the right spring with the same speed, thereby compressing it. The spring exerts a force on the puck directed towards the center of the ice surface, thereby accelerating the puck to the left. Assuming an ideal system, conservation of energy requires that the puck must leave the spring at the same speed with which it first struck the spring. The puck returns to the center at a speed of 1.5 m / s, and travels at that speed to the left spring. The process is repeated and the puck leaves the left spring at 1.5 m / s back towards the center. The puck continues to do this indefinitely. The puck undergoes oscillatory motion. It takes a time

$$t_1 = \frac{x}{v} = \frac{.50\ \text{m}}{1.5\ \text{m}/\text{s}} = \frac{1}{3}\ \text{s}$$

for the puck to reach the right spring. Once it reaches the spring we have a mass on a spring system with an angular frequency of

$$\omega = \sqrt{\frac{k}{m}} = 2.0\ \text{rad}/\text{s}$$

In compressing the spring, stopping and returning to the unstretched spring length the puck (mass on a spring system) completes $\frac{1}{2}$ period,

$$t_2 = \frac{1}{2}T = \frac{1}{2}\frac{2\pi}{\omega} = \frac{1}{2}\pi\ \text{s}$$

After reaching the unstretched spring length, the puck needs to cover 1.0 m at 1.5 m / s which takes a time

$$t_3 = \frac{1.0\ \text{m}}{1.5\ \text{m}/\text{s}} = \frac{2}{3}\ \text{s}$$

The puck then reaches the left spring at 1.5 m / s. Since the left spring has the same spring constant as the first it takes the same time for the puck to compress and reach the springs unstretched length as it did for the right spring,

$$t_4 = \frac{1}{2}\pi\ \text{s}$$

Finally the puck has to travel .50 m at 1.5 m / s to reach its initial position,

$$t_5 = t_1 = \frac{1}{3}\ \text{s}$$

Total time to make one oscillation is:

$$\sum_{i=1}^{5} t_i = \left(\frac{4}{3} + \pi\right)\ \text{s} = 4.5\ \text{s}$$

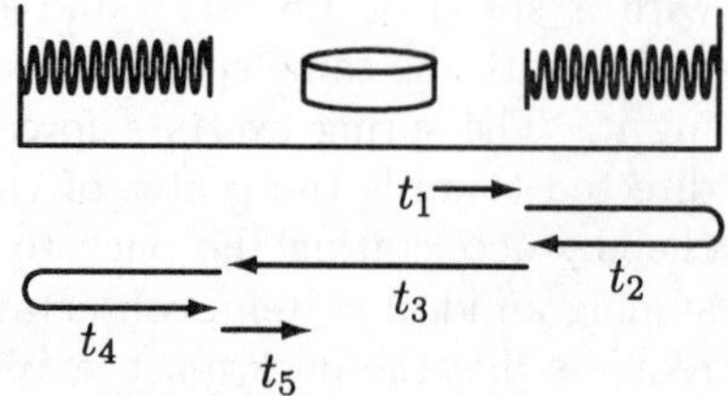

14.25 The angular frequency of the mass on spring system is

$$\omega = \sqrt{\frac{k}{m}} = \sqrt{15}\ \text{rad}/\text{s}$$

The position and speed of the particle will have the forms of

$$x(t) = A\cos[\omega t + \phi_0] \quad (3)$$
$$v(t) = -A\omega\sin[\omega t + \phi_0] \quad (4)$$

The two pieces of data given are

$$x(0.50\ \text{s}) = 0.33\ \text{m}$$

and

$$v(0.50\ \text{s}) = 1.7\ \text{m}/\text{s}$$

From (4) and the given speed data we can write an expression for A:

$$A = -\frac{1.7\,\text{m}/\text{s}}{(\sqrt{15}\ \text{rad}/\text{s})\,(\sin[1.9 + \phi_0])} \quad (5)$$

Using (3), (5) and the given position data,

$$\begin{aligned} 0.33\ \text{m} &= -\left(\frac{0.44\ \text{m}}{\sin[1.9 + \phi_0]}\right)\cos(1.9 + \phi_0) \\ &= -(0.44\ \text{m})\cot[1.9 + \phi_0] \end{aligned}$$

Rewriting the above gives

$$\tan[1.9 + \phi_0] = -\frac{4}{3} \Rightarrow 1.9 + \phi_0 = -0.93\ \text{rad}$$

Thus $\phi_0 = -2.83$ rad,

$$A = \frac{0.33\ \text{m}}{\cos[1.9 - 2.83]} = +0.55\ \text{m}$$

The amplitude of the oscillation is $A = 0.55$ m.

14.29 In Section 14.2 we found that the period of a pendulum was $T = 2\pi\sqrt{\frac{\ell}{g}}$ where ℓ is the distance between the pivot point and mass. Thus the period is proportional to $\sqrt{\ell}$. To decrease the period (and make the clock run faster) we need to decrease ℓ by moving the disk towards the support.

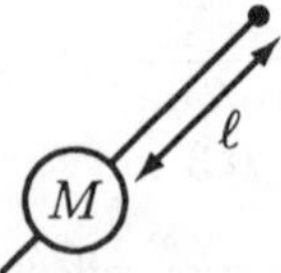

(Note that increasing the acceleration due to gravity by flying the clock close to Jupiter would work as well.)

14.33 From Example 14.4 in the text we found that for a simple pendulum $T = 2\pi\sqrt{\frac{\ell}{g}}$. So $T \propto \frac{1}{\sqrt{g}}$, and at the equator,

$$\begin{aligned} T_{\text{eq}} &= T_{\text{np}}\sqrt{\frac{g_{\text{np}}}{g_{\text{eq}}}} = (2.000\ \text{s})\sqrt{\frac{9.832}{9.780}} \\ &= 2.005\ \text{s} \end{aligned}$$

14.37 The pendulum's initial speed is

$$v = \frac{p}{m} = \frac{0.12\ \mathrm{N\cdot s}}{0.30\ \mathrm{kg}} = 0.40\ \mathrm{m/s}$$

and its initial angular speed, $\frac{d\theta}{dt}$, is

$$\frac{v}{r} = 0.16\ \mathrm{rad/s}$$

The pendulum will have an angle and angular speed of the form

$$\begin{aligned}\theta &= \theta_{\max}\sin\left(\sqrt{\frac{g}{\ell}}t\right)\\ \frac{d\theta}{dt} &= \theta_{\max}\sqrt{\frac{g}{\ell}}\cos\left(\sqrt{\frac{g}{\ell}}t\right)\end{aligned}$$

Notice that $\theta(0\ \mathrm{s}) = 0°$, thus the choice of using $\sin(\omega t)$ for $\theta(t)$ instead of the conventional $\cos(\omega t + \phi_0)$ is prudent because we don't need to find the phase angle, ϕ_0. Here

$$\omega = \sqrt{\frac{g}{\ell}} = \sqrt{\frac{9.81\ \mathrm{m/s^2}}{2.50\ \mathrm{m}}} = 1.98\ \mathrm{rad/s}$$

We know that:

$$\begin{aligned}\frac{d\theta}{dt} &= 0.16\ \mathrm{rad/s}\\ 0.16\ \mathrm{rad/s} &= \sqrt{\frac{g}{\ell}}\theta_{\max}\\ \Rightarrow \theta_{\max} &= \frac{0.16\ \mathrm{rad/s}}{1.98\ \mathrm{rad/s}} = 0.081\ \mathrm{rad}\\ \theta(t) &= (0.081\ \mathrm{rad})\sin\left[(1.98\ \mathrm{rad/s})t\right]\end{aligned}$$

14.41 The hoop is a physical pendulum.

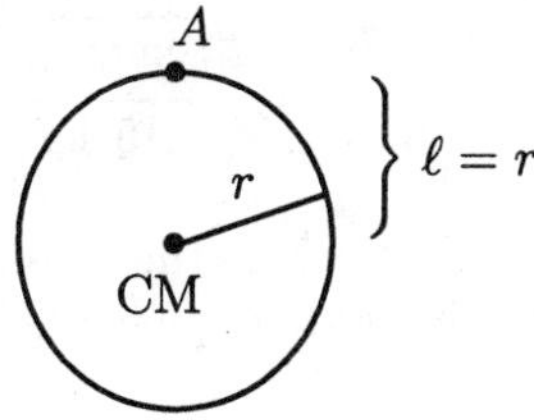

The CM of the hoop is at the hoop's geometric center, so the distance between the CM and point of suspension is the hoop's radius, 0.30 m. The calculation of I_{hoop} is particularly easy. All the hoop's mass is at a $\perp$ distance of r from the hoop's CM, that is, $r_\perp$ remains constant at r in the integral for rotational inertia.

$$I_{\text{hoop,CM}} = \int r_\perp^2\, dm = r^2 \int dm = Mr^2$$

So, by the parallel axis theorem,

$$I_A = Mr^2 + Mr^2 = 2Mr^2$$

The angular frequency of the hoop is given by:

$$\omega = \sqrt{\frac{Mgr}{2Mr^2}} = \sqrt{\frac{g}{2r}}$$

So the frequency of the hoop's oscillation is

$$f = \frac{\omega}{2\pi} = \frac{1}{2\pi}\sqrt{\frac{9.8\ \mathrm{m/s^2}}{2\,(0.30\ \mathrm{m})}} = 0.64\ \mathrm{Hz}$$

14.45 For half the swing the pendulum has a string length of ℓ and for the other half it has a string length of $\frac{1}{2}\ell$. We know that the period should change as the length of string changes. The period of this pendulum is thus:

$$\begin{aligned}T &= \frac{1}{2}T_{\text{“}\ell\text{”}=\ell} + \frac{1}{2}T_{\text{“}\ell\text{”}=\frac{1}{2}\ell}\\ \frac{1}{2}T_{\text{“}\ell\text{”}=\ell} &= \frac{\pi}{\sqrt{\frac{g}{\text{“}\ell\text{”}}}} = \pi\sqrt{\frac{\ell}{g}}\\ \frac{1}{2}T_{\text{“}\ell\text{”}=\frac{1}{2}\ell} &= \frac{\pi}{\sqrt{\frac{g}{\frac{1}{2}\ell}}} = \frac{\pi\sqrt{2}}{2\sqrt{\left(\frac{g}{\ell}\right)}}\end{aligned}$$

thus,

$$\begin{aligned}T &= \pi\sqrt{\frac{\ell}{g}}\left(1+\frac{1}{\sqrt{2}}\right)\\ &= \pi\sqrt{\frac{1.2\ \mathrm{m}}{9.8\ \mathrm{m/s^2}}}\left(1+\frac{1}{\sqrt{2}}\right) = 1.9\ \mathrm{s}\end{aligned}$$

14.49 We can compare the maximum PE of each pendulum, when they are at their maximum angle. Let the pendulum have rods of length ℓ and take the reference level for PE computation to a distance ℓ below their pivot point.

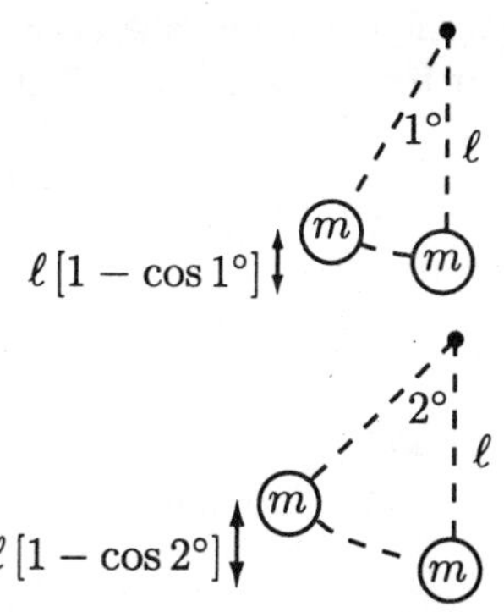

As the figures show,

$$\begin{aligned}E_{1°} &= PE_{\max,1°} = mg\ell\,[1-\cos 1°]\\ E_{2°} &= PE_{\max,2°} = mg\ell\,[1-\cos 2°]\end{aligned}$$

Since $\cos 1° > \cos 2°$, $E_{2°} > E_{1°}$ by a factor of

$$\frac{E_{2°}}{E_{1°}} = \frac{1 - \cos 2°}{1 - \cos 1°} = 4$$

Equivalently, we may use Equation 14.11

$$E = \frac{1}{2} mg\ell\theta^2_{\text{max}}$$

So

$$\frac{E_{2°}}{E_{1°}} = \left(\frac{2°}{1°}\right)^2 = 4$$

14.53 We will assume that the cylinder is uniform so its CM is at its geometric center. We need to consider it as a physical pendulum.

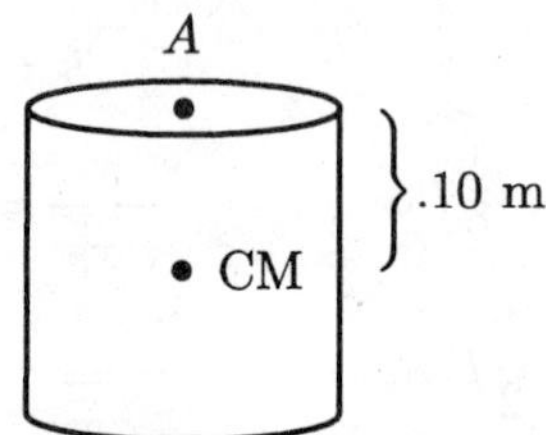

The rotational inertia about an axis along a diameter through the CM is given in Table 12.3:

$$I = \frac{M}{4}\left(r^2 + \frac{h^2}{3}\right)$$

Using Eq. 14.9, and the parallel axis theorem:

$$\begin{aligned} \omega &= \sqrt{\frac{Mg(0.10 \text{ m})}{I_{\text{cylinder CM}} + M(0.10 \text{ m})^2}} \\ &= \sqrt{\frac{g(0.10 \text{ m})}{\frac{1}{4}\left([9.0\times10^{-2} \text{ m}]^2 + \frac{[0.20 \text{ m}]^2}{3}\right) + (0.10 \text{ m})^2}} \\ &= 8.0 \text{ rad/s} \end{aligned}$$

Since the pendulum's energy remains constant: PE at top of swing $= KE$ at bottom of swing

$$\frac{1}{2} mg\ell\theta^2_{\text{max}} = \frac{1}{2} I \left(\frac{d\theta}{dt}_{\text{max}}\right)^2$$

where $\ell =$ distance to CM $= \frac{h}{2}$. Thus

$$\left.\frac{d\theta}{dt}\right|_{\text{max}} = \sqrt{g\ell\frac{M}{I}}\theta_{\text{max}} = \omega\theta_{\text{max}}$$

Remember that θ_{max} is in *radians*, and ℓ is distance to CM. Thus

$$\begin{aligned} \left.\frac{d\theta}{dt}\right|_{\text{max}} &= (7.988 \text{ rad/s})\left(\frac{2°}{180} \times \pi \text{ rad}\right) \\ &= 0.279 \text{ rad/s} \end{aligned}$$

and

$$\begin{aligned} v_{\text{bot}} &= h\left.\frac{d\theta}{dt}\right|_{\text{max}} \\ &= (0.20 \text{ m})(0.279 \text{ rad/s}) \\ &= 0.056 \text{ m/s} \end{aligned}$$

14.57 Resonance means that $\omega_0 = \omega_{\text{force}}$; for a mass on spring system, $\omega_0 = \sqrt{\frac{k}{m}}$.

$$\sqrt{\frac{k}{m}} \overset{\text{set}}{\equiv} (2.5 \text{ rad/s})$$

$$\begin{aligned} \Rightarrow k &= (2.5 \text{ rad/s})^2 (520 \text{ kg}) \\ &= 3.3 \times 10^3 \text{ N/m} \end{aligned}$$

14.61 From the N2L chapters, the effective spring constant for springs in tandem (parallel) is sum of the individual spring constants.

$$k_{\text{eff}} = 88 \text{ N/m}$$

The angular frequency of the carriage (which is essentially a mass-on-spring system) is:

$$\omega_0 = \sqrt{\frac{k}{m}} = \sqrt{\frac{88 \text{ N/m}}{15 \text{ kg}}} = 2.4 \text{ rad/s}$$

The baby supplies the external driving force with an angular frequency $\omega = (6 \text{ rad/s})$ and an amplitude of

$$m_{\text{baby}}g = 51 \text{ N}$$

The response of the carriage is given by (Eq. 14.12) and M is the mass of the thing being accelerated. In this case, although the baby supplies the driving force, the baby's mass must be included with the mass of the accelerating object since the baby is also accelerating:

$$\begin{aligned} M &= M_{\text{car}} + M_{\text{baby}} \\ Z_{\text{res}} &= \frac{(51 \text{ N})\cos[(6 \text{ rad/s})t]}{(20.2 \text{ kg})[(2.4 \text{ rad/s})^2 - (6 \text{ rad/s})^2]} \\ &= -0.08 \text{ m} \cos[(6 \text{ rad/s})t] \end{aligned}$$

14.65 The thing to focus on here is that the torque is proportional to and opposite the angle through which the fiber is twisted. This should remind you of the common theme that restoring force is proportional to and has direction opposite displacement. In this case we can write

$$\tau = -C\theta$$

where θ is the angle by which the fiber is rotated and C is the constant of proportionality, given by the problem statement. Using the N2L equivalent for rotational motion

$$\tau = -C\theta = I\alpha = I\frac{d^2\theta}{dt^2}$$

or

$$\frac{d^2\theta}{dt^2} = -\frac{C}{I}\theta$$

which is the SHM differential equation. Relating this to what we already know,

$$\omega = \sqrt{\frac{C}{I}}$$

The period of the torsional pendulum is

$$T = \frac{2\pi}{\omega} = 2\pi\sqrt{\frac{I}{C}}$$

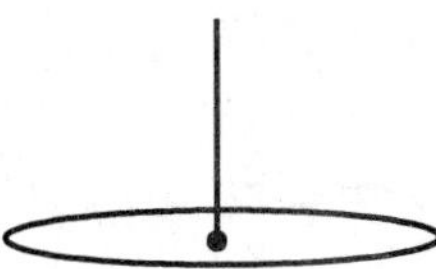

From Table 12.3,

$$I_{\text{disk}} = \frac{1}{2}MR^2$$

so

$$\begin{aligned} T &= 2\pi\sqrt{\frac{MR^2}{2C}} \\ &= 2\pi\sqrt{\frac{(0.50\ \text{kg})(5\times10^{-2}\ \text{m})^2}{2\,(10^{-3}\ \text{N}\cdot\text{m}/\text{rad})}} = 5.0\ \text{s} \end{aligned}$$

At time $t = 0$

$$\theta = \theta_0 = 5^\circ$$

so we can write the angle of the pendulum as a function of time

$$\theta(t) = \theta_{\text{max}}\cos(\omega t)$$

Differentiating both sides

$$\frac{d\theta(t)}{dt} = -\theta_{\text{max}}\omega\sin(\omega t)$$

which has its maximum value when $\omega t = \frac{\pi}{2}$, so

$$\begin{aligned} \left.\frac{d\theta}{dt}\right|_{\text{max}} &= \theta_{\text{max}}\omega = 5\left(\frac{\pi}{180^\circ}\right)\left(\frac{2\pi}{5.0\ \text{s}}\right) \\ &= 0.11\ \text{rad}/\text{s} \end{aligned}$$

14.69 The figure shows the block in equilibrium where the springs are at their unstretched length.

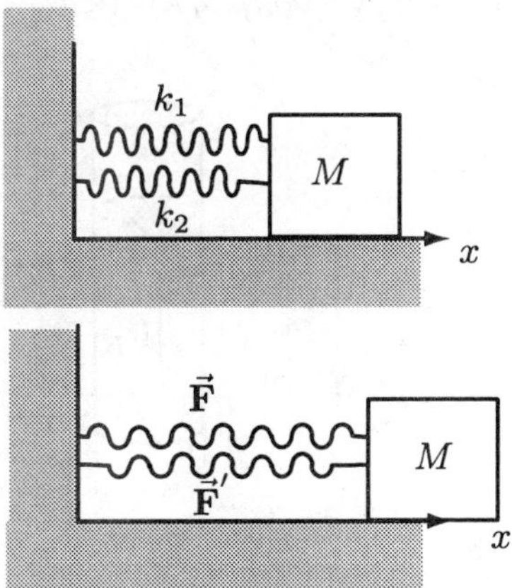

Suppose we displace the block by a distance x so that the springs exert a net force; N2L gives:

$$\begin{aligned} \sum F_x &= -k_1x - k_2x = m\frac{d^2x}{dt^2} \\ \Rightarrow \frac{d^2x}{dt^2} &= -\frac{(k_1+k_2)}{m}x \end{aligned}$$

which is the SHM equation for which $\omega = \sqrt{\frac{k_1+k_2}{m}}$ and

$$T = \frac{2\pi}{\omega} = 2\pi\sqrt{\frac{m}{k_1+k_2}}$$

14.73 We can use the text's result to find v_{max} of the log. The text gave the log's position function as

$$z(t) = A\cos\left[(2.27\ \text{rad}/\text{s})t\right]$$

where A, the amplitude, is the distance the log was displaced. The problem statement gives $A = 1.0$ cm so,

$$\begin{aligned} v(t) &= \frac{dz(t)}{dt} \\ &= (1.0\ \text{cm})(2.27\ \text{rad}/\text{s})\cos\left[(2.27\ \text{rad}/\text{s})t\right] \end{aligned}$$

At maximum speed the cos term equals one so

$$\begin{aligned} v_{\text{max}} &= (1.0\ \text{cm})(2.27\ \text{rad}/\text{s}) \\ &= .023\ \text{m}/\text{s}. \end{aligned}$$

Let's find the work done by the buoyant force. From the text, the net force acting on the log after it's displaced any distance z from its equilibrium position is

$$\vec{\mathbf{F}}_{\text{net}} = -z\pi r^2 \rho_\omega g \hat{\mathbf{k}}$$

If we imagine releasing the log from its displaced position at z_0 the buoyant force will do an amount of work

$$W = \int_{z_0}^{0} (-z\pi r^2 \rho_\omega g\,\hat{\mathbf{k}}) \cdot (\hat{\mathbf{k}}\,dz) = +\frac{1}{2}\pi r^2 \rho_\omega g z_0^2$$

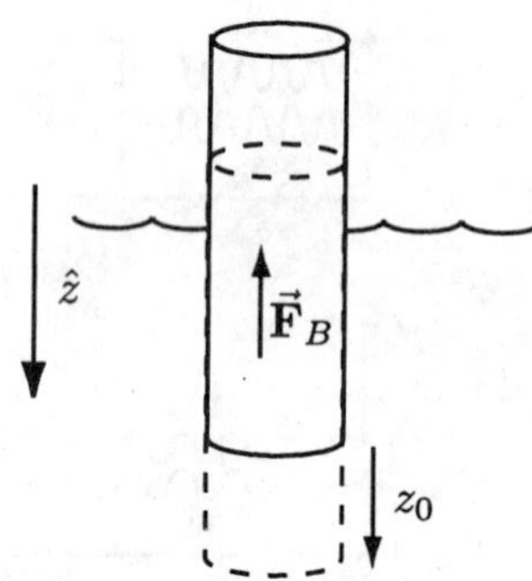

We are going to ascribe the work done by the buoyant force to our potential. Note that if the buoyant force does work on the log z_0 becomes smaller, so our expression for W, our potential energy function, decreases. Thus, positive net work done by the buoyant force corresponds to the decrease of our PE function. Secondly, the limits of integration depend only on the initial and final positions of the log and the value of W is dependant only on the limits, not on how the log gets from the lower limit to upper limit. W is independent of path. Finally, we measured z *wrt* the log's equilibrium position. We could have just as easily measured it from an arbitrary depth, a. The force itself wouldn't change and both limits would contain an additive constant a in expressing the log's position *wrt* a. These terms would cancel in calculating W. Thus, it makes no difference where we take the origin to be in the calculation of W. The important thing is the change in the log's position.

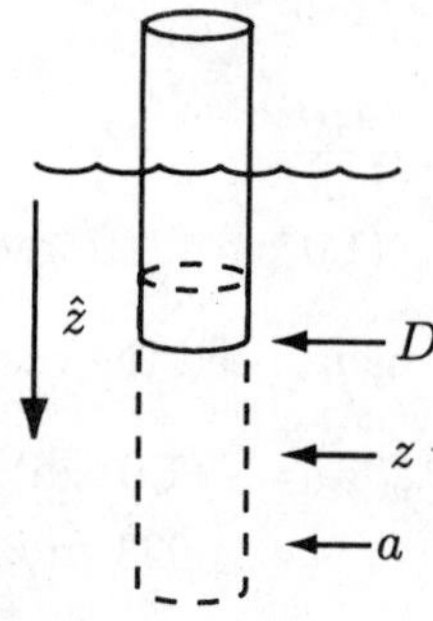

Thus this PE does posess properties 1–4.

14.77

$$\begin{aligned}
\theta(t+\Delta t) &= \theta(t) + \frac{d\theta}{dt}\Delta t + \frac{1}{2}\frac{d^2\theta}{dt^2}(\Delta t)^2 \\
\frac{d\theta}{dt}(t+\Delta t) &= \frac{d\theta}{dt}(t) + \Delta t\frac{d^2\theta}{dt^2} + \frac{1}{2}\frac{d^3\theta}{dt^3}(\Delta t)^2
\end{aligned}$$

From the differential equation

$$\begin{aligned}
\frac{d^2\theta}{dt^2} &= -\left(\frac{g}{\ell}\right)\sin\theta \\
\frac{d^3\theta}{dt^2} &= -\left(\frac{g}{\ell}\right)\cos\theta\frac{d\theta}{dt}
\end{aligned}$$

Substituting:

$$\begin{aligned}
\theta(t+\Delta t) &= \theta(t) + \frac{d\theta}{dt}\Delta t - \left(\frac{g}{2\ell}\right)\sin\theta(\Delta t)^2 \\
&= \theta(t) + \frac{d\theta}{dt}\Delta t - \frac{\omega_0^2}{2}\sin\theta(\Delta t)^2
\end{aligned}$$

and

$$\begin{aligned}
\frac{d\theta}{dt}(t+\Delta t) &= \frac{d\theta}{dt}(t) + \Delta t\left(-\frac{g}{\ell}\sin\theta\right) \\
&\quad + \frac{1}{2}(\Delta t)^2\left(-\frac{g}{\ell}\cos\theta\frac{d\theta}{dt}\right) \\
&= \frac{d\theta}{dt}(t)\left[1 - \frac{(\omega_0\Delta t)^2}{2}\cos\theta\right] \\
&\quad -\omega_0^2\Delta t\sin\theta
\end{aligned}$$

The period of the simple pendulum is

$$T = \frac{2\pi}{\omega_0} = 2\pi\sqrt{\frac{\ell}{g}} = 2.006 \text{ s}$$

Thus the time step will be 2.006×10^{-2} s. In the spreadsheet shown in the appendix,

$$v' = \frac{d\theta}{dt}\left(t + \frac{\Delta t}{2}\right)$$

is used to calculate $d\theta = \frac{d\theta}{dt}dt$. The computed period is 2.10 s, 5% larger than for the simple pendulum.

14.81 First we find the CM of the block:

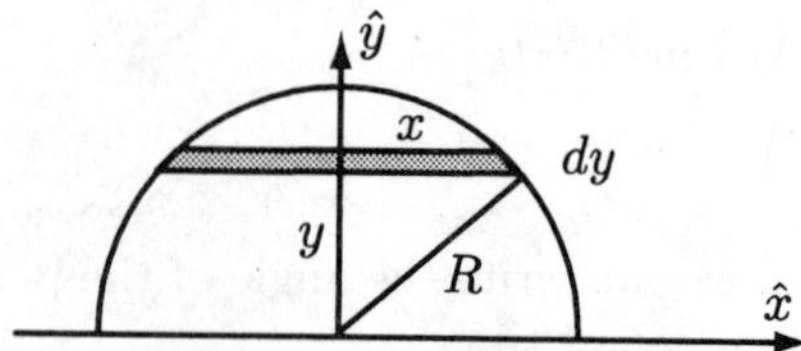

The block will be modeled as a collection of differential strips of mass dm and length $2x$. Any strip will have a height y above the x-axis. Its mass is

$$dm = \sigma\,dA = \sigma 2x\,dy$$

By the Pythagorean identity we have $x = \sqrt{R^2 - y^2}$ so,

$$dm = 2\sigma\sqrt{R^2 - y^2}dy$$

and so

$$dy_{\text{CM}} = \frac{y\,dm}{m} = \frac{2\sigma y\sqrt{R^2 - y^2}\,dy}{m}$$

The y-coordinate will vary from 0 to R, the radius of the semicircle.

$$\begin{aligned} y_{\text{CM}} &= \frac{1}{M}\int_0^R 2\sigma y\sqrt{R^2 - y^2}dy \\ &= \frac{2\sigma}{M}\int_{R^2}^0 yu^{\frac{1}{2}}\left(\frac{du}{-2y}\right) \end{aligned}$$

where

$$\begin{aligned} u &= R^2 - y^2, du = -2y\,dy \\ y_{\text{CM}} &= \frac{\sigma}{m}\int_0^{R^2} u^{\frac{1}{2}}\,du = \frac{2}{3M}\sigma u^{3/2}\bigg|_0^{R^2} \\ &= \frac{2}{3}\left(\frac{R^3}{\frac{1}{2}\pi R^2}\right) = \frac{4R}{3\pi} \end{aligned}$$

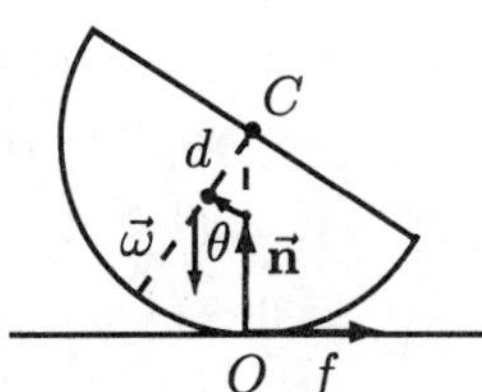

When displaced, the weight of the block exerts an unbalanced torque about the contact point 0 which causes the oscillation. By taking torques about 0 we eliminate the forces n and f from consideration.

$$\tau = mgd\sin\theta = I\alpha = -I\frac{d^2\theta}{dt^2}$$

To find I, we may note that the rotational inertia of the block about the center of the circle is the same as for a disk, $I = \frac{mr^2}{2}$, and use the parallel axis theorem twice.

$$\begin{aligned} I_C &= I_{\text{CM}} + mr^2 \\ \Rightarrow I_{\text{CM}} &= I_C - md^2 \\ &= \frac{mr^2}{2} - md^2 \end{aligned}$$

Then

$$\begin{aligned} I_O &= I_{\text{CM}} + mr_\perp^2 \\ &= I_{\text{CM}} + m(r^2 + d^2 - 2rd\cos\theta) \end{aligned}$$

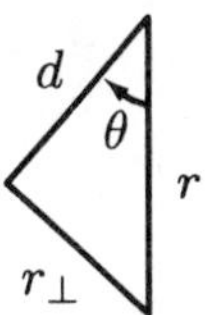

If θ is small,

$$\cos\theta \simeq 1 - \frac{\theta^2}{2} \simeq 1$$

Thus

$$\begin{aligned} I_O &= \frac{mr^2}{2} - md^2 + mr^2 + md^2 - 2mrd \\ &= \frac{3mr^2}{2} - 2mr\left(\frac{4r}{3\pi}\right) = mr^2\left[\frac{3}{2} - \frac{8}{3\pi}\right] \end{aligned}$$

Then the equation of motion becomes, for small θ,

$$\frac{d^2\theta}{dt^2} = -\frac{mgd\theta}{I}$$

So we have SHM with frequency

$$\begin{aligned} \omega &= \sqrt{\frac{mgd}{I}} = \sqrt{\frac{mg\frac{4R}{3\pi}}{mr^2\left(\frac{3}{2} - \frac{8}{3\pi}\right)}} \\ &= \sqrt{\frac{g}{r}}\sqrt{\frac{4}{\frac{9\pi}{2} - 8}} = \sqrt{\frac{g}{r}}\sqrt{\frac{8}{9\pi - 16}} \\ &= 0.81\sqrt{\frac{g}{r}} \end{aligned}$$

and period $\frac{2\pi}{\omega} = 7.8\sqrt{\frac{r}{g}}$.

Chapter 15

Introduction to Wave Motion

15.1 Two waves on identical strings, tensions differ.

$$T_1 = 3T_2$$

Wave speed is directly proportional to square root of tension

$$v \propto \sqrt{T}$$

$\left\{\text{since } v = \sqrt{\frac{T}{\mu}}\right\}$

$$\frac{v_1}{v_2} = \sqrt{\frac{T_1}{T_2}} = \sqrt{\frac{3T_2}{T_2}} = \sqrt{3}$$

15.3 $v = 330$ m/s; $\lambda = 0.75$ m. Frequency is related to wave speed and wavelength by

$$f = \frac{v}{\lambda} \qquad \text{(Eqn. 15.2)}$$

$$f = \frac{330 \text{ m/s}}{0.75 \text{ m}} = 440 \text{ Hz}$$

15.5 Flower on water surface:

equilibrium position:

maximum height 0.12 s later:

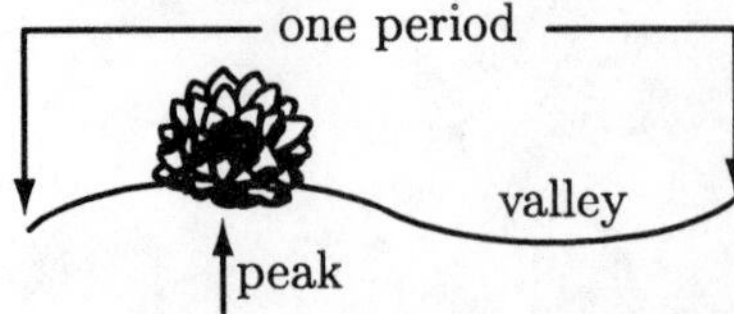

The time for the flower to rise from equilibrium to peak is $\frac{1}{4}$ of a period. In one full period the flower would rise from equilibrium position to a peak then fall through equilibrium position to a valley then back up to equilibrium.

$$T = 4\,(0.12 \text{ s}) = 0.48 \text{ s}$$

15.7 Wave equation for sound wave in air:

$$\frac{\partial^2 y}{\partial t^2} = \frac{7}{5}\frac{P}{\rho}\frac{\partial^2 y}{\partial x^2}$$

$$P = \text{air pressure}, \rho = \text{density}$$

Compare to the general wave equation:

$$\frac{\partial^2 y}{\partial t^2} = v^2 \frac{\partial^2 y}{\partial x^2} \qquad \text{(Eqn. 15.10)}$$

$$v = \sqrt{\frac{7}{5}\frac{P}{\rho}}$$

Supposing that

$$P = 1 \text{ atm} = 1.01 \times 10^5 \text{ Pa}; \rho = 1.3 \text{ kg/m}^3$$

then

$$v = \sqrt{\frac{7\,(1.01 \times 10^5 \text{ Pa})}{5\,(1.3 \text{ kg/m}^3)}} = 330 \text{ m/s at } 0^\circ \text{ C}$$

15.9 Two identical strings. Tensions differ:

$$T_1 = 4T_2$$

Waves of equal frequency and amplitude propagate. Wave speed is:

$$v = \frac{\omega}{k} = \sqrt{\frac{T}{\mu}}$$

$$\begin{aligned} P &\propto Tk\omega = T\frac{k}{\omega}\omega^2 \\ &= T\sqrt{\frac{\mu}{T}}\omega^2 = \sqrt{T\mu}\omega^2 \end{aligned}$$

$P \propto \sqrt{T}$ given that μ, ω are constant.

$$\frac{P_1}{P_2} = \sqrt{\frac{T_1}{T_2}} = \sqrt{\frac{4T_2}{T_2}} = 2$$

15.11 Average power transmitted

$$\begin{aligned} \langle|P|\rangle &= \frac{1}{2}\mu v_w \omega^2 A^2 (\text{Eqn. } 15.12) \\ \mu &= 4.0 \text{ g} / \text{m} \\ v_w &= 44 \text{ m} / \text{s} \\ \omega &= 220 \text{ rad} / \text{s} \\ A &= 0.100 \text{ m} \end{aligned}$$

$$\begin{aligned} \langle|P|\rangle &= \frac{1}{2}\left(4.0 \times 10^{-3} \text{ kg} / \text{m}\right)(44 \text{ m} / \text{s}) \\ &\quad \times (220 \text{ rad} / \text{s})(0.100 \text{ m})^2 \\ &= 43 \text{ W} \end{aligned}$$

15.13 Wave y_1 is left-moving. At a fixed boundary a reflected harmonic wave is inverted (i.e. differs in phase from the incident wave by π.) The reflected wave is right-moving.

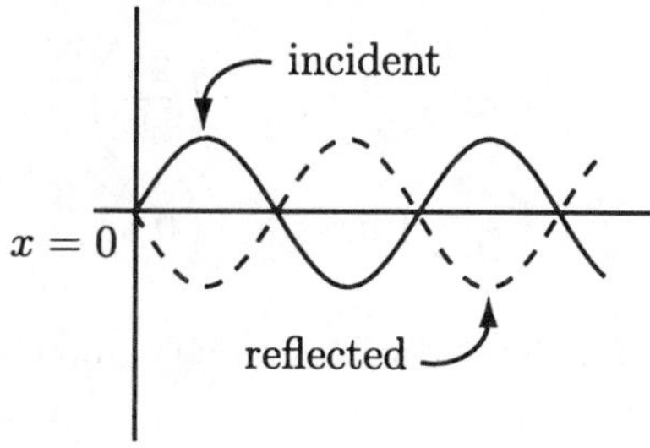

Reflected wave function:

$$\begin{aligned} y_1'(x,t) = (5.0 \text{ cm}) \cos [&(3.0 \text{ rad} / \text{m})\, x \\ &- (5.0 \text{ rad} / \text{s})\, t + \pi] \end{aligned}$$

For a free boundary, the phase difference would be zero rather than π.

15.15 Standing wave on a string wavelength:

$$\lambda = 1.0 \text{ m}$$

fundamental frequency:

$$f_1 = 110 \text{ Hz}$$

tension

$$T = 120 \text{ N}$$

Wave speed is

$$v = \lambda f = \sqrt{\frac{T}{\mu}}$$

Mass per length:

$$\begin{aligned} \mu &= (\lambda f)^{-2} T \\ &= [(1.0 \text{ m})(110 \text{ Hz})]^{-2}\, 120 \text{ N} \\ &= 9.9 \times 10^{-3} \text{ kg} / \text{m} = 9.9 \text{ g} / \text{m} \end{aligned}$$

15.19 No. The speed of a wave on a string depends only on physical characteristics of the string such as mass per unit length and tension in the string.

15.23 The frequency of the wavetrain corresponds to the frequency of air puffs. There are 11 puffs for each rotation and 40 rotations per second. Number of puffs per second:

$$\begin{aligned} f &= (11 \text{ puffs/rotation})(40 \text{ rotations/second}) \\ &= 440 \text{ Hz} \end{aligned}$$

15.27 Harmonic wave on a string:

$$y(x,t) = (1.00 \text{ m}) \sin [(62.8 / \text{m})\, x + (314 / \text{s})\, t]$$

Compare to general form:

$$y(x,t) = A \sin\left[\frac{2\pi x}{\lambda} \pm \frac{2\pi t}{T}\right]$$

"+" indicates wave is left-moving

wavelength:

$$\frac{2\pi}{\lambda} = 62.8 / \text{m} \Rightarrow \lambda = \frac{2\pi}{62.8 / \text{m}} = 0.100 \text{ m}$$

period:

$$\frac{2\pi}{T} = 314 / \text{s} \Rightarrow T = \frac{2\pi}{314 / \text{s}} = 0.0200 \text{ s}$$

frequency:

$$f = \frac{1}{T} = \frac{1}{.0200 \text{ s}} = 50.0 \text{ Hz}$$

Maximum displacement of any string segment, is the amplitude.

$$\begin{aligned} A &= 1.00 \text{ mm} \\ v &= \lambda f = 5.00 \text{ m} / \text{s} \end{aligned}$$

15.31 Harmonic wave:

right moving	"−"
frequency	$f = 50$ Hz
speed	$v = 10.0$ m / s
amplitude	$A = 0.750$ mm

at $t = 0$, $x = 3.25$ m; $y = 0$ and $\frac{\partial y}{\partial t} < 0$.

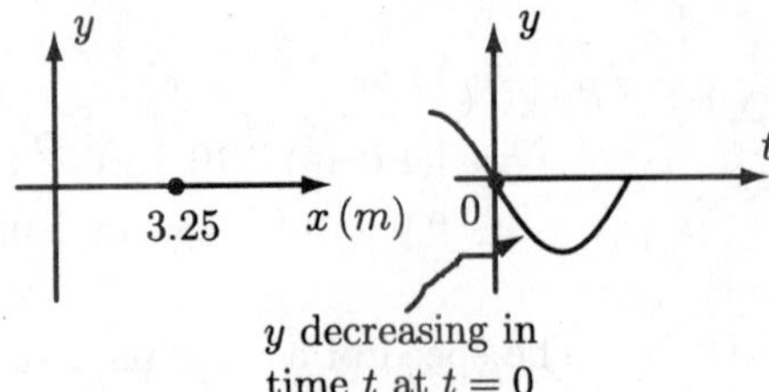

This is a right moving harmonic wave and so has the form:

$$\begin{aligned} y(x,t) &= A\cos(kx - \omega t + \phi_0) \\ \omega &= 2\pi f = 2\pi\,(50\text{ Hz}) = 314\,/\,\text{s} \\ k &= \frac{\omega}{v} = (314\,/\,\text{s})(10.0\text{ m}\,/\,\text{s}) = 31.4\,/\,\text{m} \\ y\,(x,t) &= (0.750\text{ mm})\cos[(31.4\text{ rad}\,/\,\text{m})\,x \\ &\quad - (314\text{ rad}\,/\,\text{s})t + \phi_0] \end{aligned}$$

Use the initial conditions to find ϕ_0: Expect a cosine curve to be zero and decreasing with time at $\phi = \frac{3\pi}{2}$.

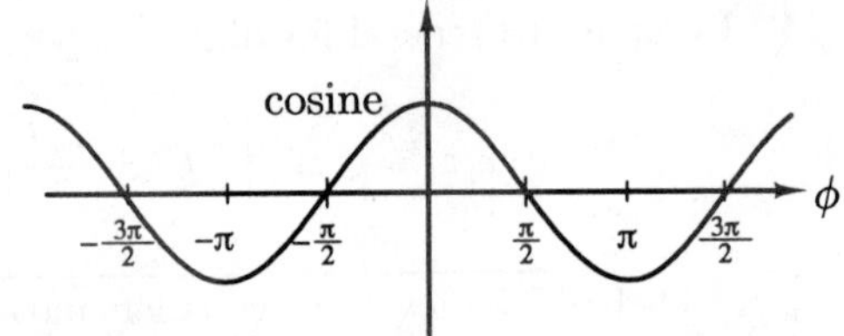

At $t = 0$, $x = 3.25$ m, the wave phase is

$$\left(31.4\,\frac{\text{rad}}{\text{m}}\right)(3.25\text{ m}) - (0) + \phi_0 = \frac{3\pi}{2}$$

Thus

$$\begin{aligned} \phi_0 &= \left[\frac{3\pi}{2} - (10\pi)(3.25)\right]\text{rad} \\ &= \frac{3\pi}{2} - 32\pi - \frac{1}{2}\pi \\ &= -31\pi \end{aligned}$$

or equivalently, π rad.

$$\begin{aligned} y\,(x,t) &= (0.750\text{ mm})\cos[(31.4\text{ rad}\,/\,\text{s})\,x \\ &\quad - (314\text{ rad}\,/\,\text{s})\,t + \pi] \end{aligned}$$

15.35 Wave on a guitar string:

amplitude:

$$A = 5.0\,\mu\text{m}$$

wavelength:

$$\lambda = 0.33\text{ m}$$

mass per unit length:

$$\mu = 6.31\,\frac{\text{g}}{\text{m}} = 0.00631\,\frac{\text{kg}}{\text{m}}$$

tension:

$$T = 290\text{ N}$$

The weight of a very small segment of the string is equal to the mass of the segment $\mu\,dx$ times the acceleration due to gravity $g = 9.8$ m / s^2

$$\begin{aligned} \mu g\,dx &= \left(0.00631\,\frac{\text{kg}}{\text{m}}\right)(9.8\text{ m}\,/\,\text{s}^2)\,dx \\ &= \left(0.0618\,\frac{\text{N}}{\text{m}}\right)dx \end{aligned}$$

The maximum net y component of tension $T\left(\frac{\partial^2 y}{\partial x^2}\right)dx$ can be found by differentiating the function

$$y = A\cos(kx - \omega t + \phi)$$

twice.

$$\frac{\partial^2 y}{\partial x^2} = -k^2 y$$

Thus

$$\begin{aligned} T\left.\frac{\partial^2 y}{\partial x^2}\right|_{\text{max}} &= \left(\frac{2\pi}{\lambda}\right)^2 AT \\ T\left(\frac{\partial^2 y}{\partial x^2}\right)dx &\cong \left(0.53\,\frac{\text{N}}{\text{m}}\right)dx \gg \mu g\,dx \end{aligned}$$

for the two values to be comparable we require that:

$$\mu g \cong T\left(\frac{(2\pi)^2 A}{\lambda^2}\right)$$

or:

$$\begin{aligned} A &\cong \frac{\mu g \lambda^2}{(2\pi)^2 T} \\ &= \frac{\left(0.00631\,\frac{\text{kg}}{\text{m}}\right)(9.8\text{ m}\,/\,\text{s}^2)\,(0.33\text{ m})^2}{(2\pi)^2\,(290\text{ N})} \\ &\cong 6\times 10^{-7}\text{ m} = 0.6\,\mu\text{m} \end{aligned}$$

15.39 At a given time the instantaneous power is a function of position x.

$$P = \pm Tk\omega A^2 \sin^2(kx \pm c)$$

where c is some constant. *No, the power is not the same at every point.* The average power is not a function of position

$$\langle |P| \rangle = \frac{1}{2}\mu v_w \omega^2 A^2$$

Yes, the average power is the same at every point.

15.43 Waves on a string:

Frequency:

$$f = 440 \text{ Hz or } \omega = 2\pi f$$

Average power output:

$$\langle |P| \rangle = 120 \text{ W}$$

Tension:

$$T = 950 \text{ N}$$

Mass per unit length:

$$\mu = 25 \frac{\text{g}}{\text{m}} = 0.025 \frac{\text{kg}}{\text{m}}$$

Average power:

$$\langle |P| \rangle = \frac{1}{2}Tk\omega A^2$$

To eliminate k, note that

$$\begin{aligned} k &= \frac{\omega}{v_w} \\ v_w^2 &= \frac{T}{\mu} \end{aligned}$$

Thus

$$\begin{aligned} \langle |P| \rangle &= \frac{1}{2}T\left(\frac{\omega}{v_w}\right)\omega A^2 \\ &= \frac{\frac{1}{2}T\omega^2 A^2}{\sqrt{\frac{T}{\mu}}} = \frac{\sqrt{T\mu}\,(\omega A)^2}{2} \end{aligned}$$

Convert frequency to angular frequency $\omega = 2\pi f$

$$\langle |P| \rangle = \frac{\sqrt{T\mu}\,(2\pi f A)^2}{2} = 2\sqrt{T\mu}\,(\pi f A)^2$$

Solve for amplitude;

$$\begin{aligned} A^2 &= \frac{\langle |P| \rangle}{2\sqrt{T\mu}\,(\pi f)^2} \\ &= \frac{120 \text{ W}}{2\sqrt{(950 \text{ N})\left(.025 \frac{\text{kg}}{\text{m}}\right)}\left(\pi \left[440 \text{ s}^{-1}\right]\right)^2} \\ &= 6.44 \times 10^{-6} \text{ m}^2 \\ A &= 2.5 \times 10^{-3} \text{ m} = 2.5 \text{ mm} \end{aligned}$$

15.47 If the wave numbers are different, then either the frequencies or the speeds of the two waves also differ. Thus the wave displacements will not always be equal and opposite at *any* point. There are no nodes and no standing waves.

15.51 a) $\frac{1}{2}$ wavelength:

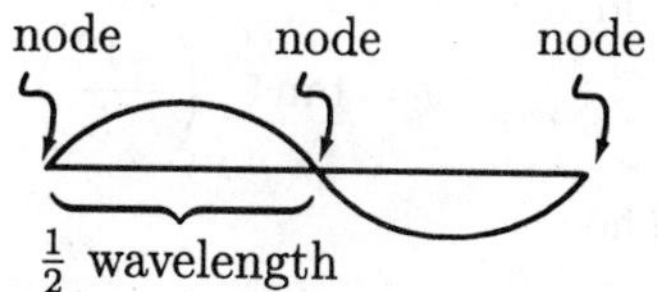

b) $\frac{1}{2}$ wavelength:

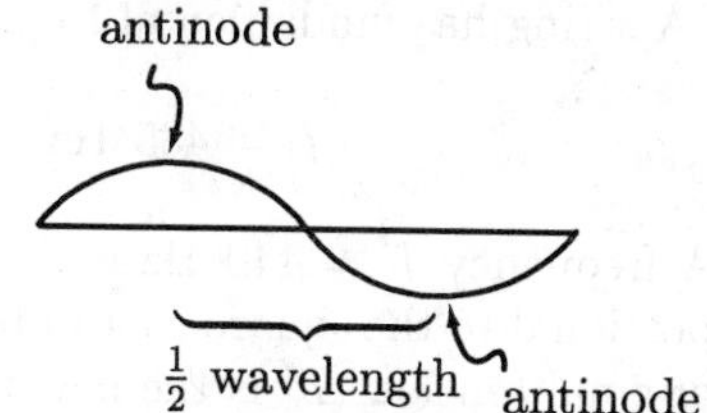

c) $\frac{1}{4}$ wavelength:

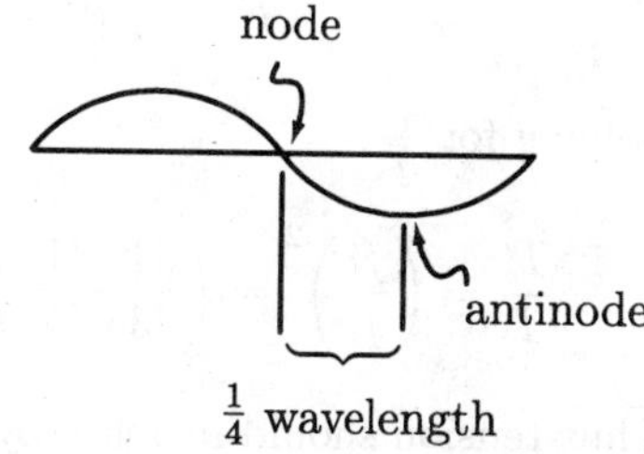

15.55 The figure shows the two waves and their sum.

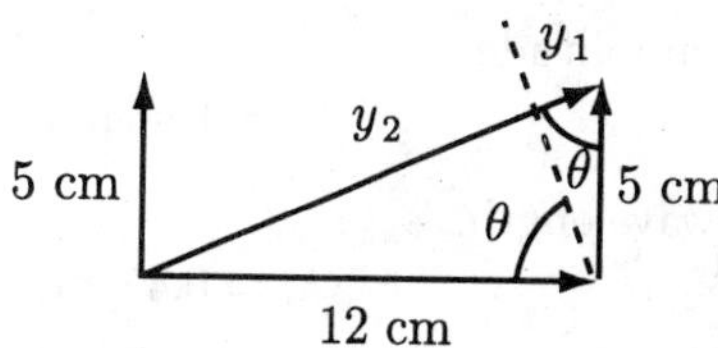

With

$$\theta \equiv \left[\left(2.0 \frac{\text{rad}}{\text{m}}\right)x + \left(4.0 \frac{\text{rad}}{\text{s}}\right)t\right]$$

We have

$$y_1 = (5.0 \text{ cm})\cos\theta$$

and

$$y_2 = (12 \text{ cm})\sin\theta$$

We can write the sum in the form

$$\begin{aligned} y_1 + y_2 &= A\cos(\theta + \phi) \\ &= A(\cos\theta\cos\phi - \sin\theta\sin\phi) \end{aligned}$$

Then

$$A\cos\phi = 5.0 \text{ cm}$$

and

$$A\sin\phi = -12 \text{ cm}$$

So

$$A = \sqrt{(5.0 \text{ cm})^2 + (-12 \text{ cm})^2} = 13 \text{ cm}$$

and

$$\phi = \tan^{-1}\left(-\frac{12}{5}\right) = -67^\circ$$

Then

$$y_1 + y_2 = (13 \text{ cm})\cos(\theta - 67^\circ)$$

15.59 A string has fundamental frequency

$$f_1 = 435 \text{ Hz}$$

A frequency $f_1' = 440$ Hz is desired. f_1 is proportional to the square root of tension. T is the original tension. T' is the new tension.

$$\frac{f_1}{f_1'} = \frac{\sqrt{T}}{\sqrt{T'}}$$

Solving for $\frac{T'}{T}$:

$$\frac{T'}{T} = \left(\frac{f_1'}{f_1}\right)^2 = \left(\frac{440 \text{ Hz}}{435 \text{ Hz}}\right)^2 = 1.023$$

Thus tension should increase by 2.3%.

15.63 Incident wave:

amplitude:

$$A_i = 0.35 \text{ mm}$$

wavelength:

$$\lambda_i = 0.45 \text{ m}$$

mass per unit length:

$$\mu_1 = 6.5 \frac{\text{g}}{\text{m}}$$

Transmitted wave:

amplitude:

$$A_t = 0.29 \text{ mm}$$

Wave speed relates μ and λ:

$$v = \lambda f = \sqrt{\frac{T}{\mu}}$$

Amplitudes are related by:

$$A_t = \frac{2v_2}{v_1 + v_2} A_i \qquad \text{(Eqn. 15.18)}$$

Relating wave speed to μ using:

$$v = \sqrt{\frac{T}{\mu}}$$

$$A_t = \frac{2\sqrt{\frac{T}{\mu_2}}A_i}{\sqrt{\frac{T}{\mu_1}} + \sqrt{\frac{T}{\mu_2}}} = \frac{2A_i}{\sqrt{\frac{\mu_2}{\mu_1}} + 1}$$

Tension is the same in both strings, so a factor of $\sqrt{T}$ cancels out. Solving for μ_2:

$$\sqrt{\mu_2} = \sqrt{\mu_1}\left(\frac{2A_i}{A_t} - 1\right)$$

$$\mu_2 = \mu_1\left(\frac{2A_i}{A_t} - 1\right)^2$$

$$= \left(6.5 \frac{\text{g}}{\text{m}}\right)\left(\frac{2(.35 \text{ m})}{.29 \text{ mm}} - 1\right)^2$$

$$\mu_2 = 13 \frac{\text{g}}{\text{m}}$$

Amplitude of the reflected wave:

$$A_r = \frac{v_2 - v_1}{v_2 + v_1} A_i \qquad \text{(Eqn. 15.18)}$$

$$= \frac{\sqrt{\frac{T}{\mu_2}} - \sqrt{\frac{T}{\mu_1}}}{\sqrt{\frac{T}{\mu_1}} + \sqrt{\frac{T}{\mu_2}}} A_i$$

$$= \frac{\sqrt{\mu_1} - \sqrt{\mu_2}}{\sqrt{\mu_1} + \sqrt{\mu_2}} A_i$$

$$= \frac{\sqrt{6.5 \frac{\text{g}}{\text{m}}} - \sqrt{13 \frac{\text{g}}{\text{m}}}}{\sqrt{6.5 \frac{\text{g}}{\text{m}}} + \sqrt{13 \frac{\text{g}}{\text{m}}}} 0.35 \text{ mm}$$

$$= \frac{1 - \sqrt{2}}{1 + \sqrt{2}} 0.35 \text{ mm} = -0.060 \text{ mm}$$

Negative sign indicates that the wave is inverted.

15.67 Fluid elements are pushed by surrounding elements, with forces due to pressure. Each element has weight. The path of an element is elliptical (see text figure 15.10) net force on the particle is not vertical.

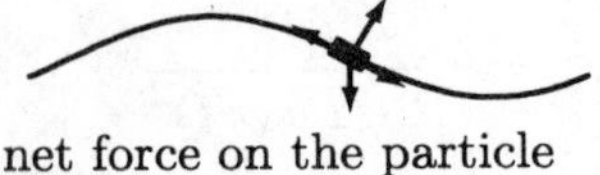

net force on the particle is not vertical

15.69 If there is a node at the junction, then ℓ = one half wavelength on string of mass/length μ and 2ℓ = one half wavelength on string with mass/length $\frac{\mu}{4}$.

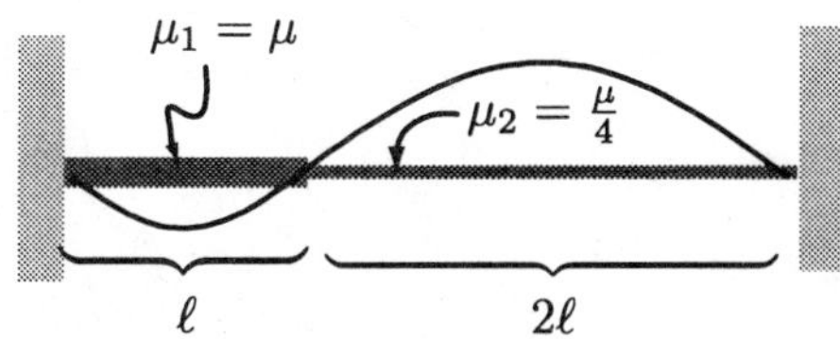

The tension is the same on both pieces of string, so $v_1 = \sqrt{\frac{T}{\mu}}$ and

$$v_2 = \sqrt{\frac{4T}{\mu}} = 2v_1$$

Thus at frequency f,

$$\lambda_1 = \frac{v_1}{f}$$

and

$$\lambda_2 = \frac{v_2}{f} = \frac{2v_1}{f} = 2\lambda_1$$

as required. Thus we need

$$\ell = \frac{\lambda_1}{2} = \frac{v_1}{2f} = \sqrt{\frac{T}{\mu}}\frac{1}{2f}$$

Any number of half wavelengths will fit, so we can also have

$$\ell_1 = m\left(\frac{\lambda}{2}\right) \Rightarrow f = \frac{m}{2\ell}\sqrt{\frac{T}{\mu}},\quad m = \text{ integer}$$

15.73 Net inward force on chain element is

$$\begin{aligned} dF &= 2T\left(\frac{d\theta}{2}\right) = ma = m\omega^2 r \\ &= \mu r\, d\theta \omega^2 r \end{aligned}$$

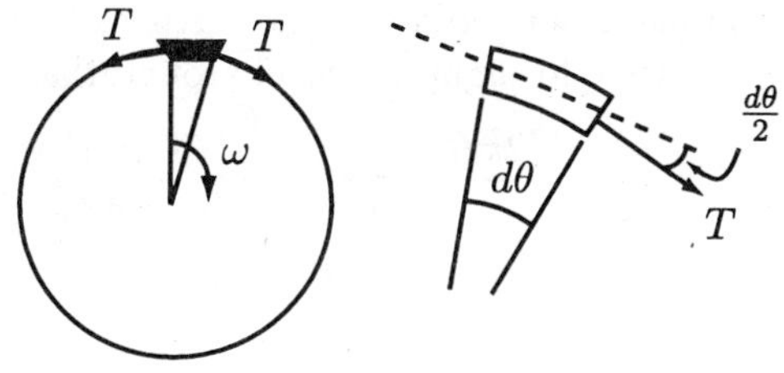

Thus

$$T = \mu r^2 \omega^2$$

The speed of the waves is

$$v = \sqrt{\frac{T}{\mu}} = \sqrt{\omega^2 r^2} = \omega r = v_{\text{chain}}$$

One pulse will appear to stand still and the other will race around at $2v$.

15.75 No. This is not a standing wave. The points on the string where $y = 0$ are not at fixed places, but travel along the string. Thus there are no nodes.

15.77 The superposition is a pulse that travels $\approx$ 40 m in 0.12 s, i.e., its speed is 330 m / s, the speed of each individual wave.

15.83

$$\begin{aligned} y(x,t) &= (1.2 \text{ mm}) \sin\left[\left(\pi\frac{\text{rad}}{\text{m}}\right)x\right] \\ &\quad \times \cos\left[\left(200\pi\frac{\text{rad}}{\text{s}}\right)t\right] \\ &= A \sin kx \cos \omega t \end{aligned}$$

$$\begin{aligned} P &= -T\frac{\partial y}{\partial x}\frac{\partial y}{\partial t} \\ &= -T(k\cos kx \cos\omega t)(-\omega \sin\omega t \sin kx)A^2 \\ &= A^2 T k\omega \cos kx \sin kx \sin\omega t \cos\omega t \\ &= \frac{A^2 T k\omega}{4}\sin 2kx \cos 2\omega t \end{aligned}$$

$$\begin{aligned} u &= \frac{1}{2}T\left(\frac{\partial y}{\partial x}\right)^2 \\ &= \cos^2(kx)\cos^2(\omega t) \end{aligned}$$

The kinetic energy density is

$$\text{k.e.} = \frac{1}{2}\mu\left(\frac{\partial y}{\partial x}\right)^2 = \frac{1}{2}\mu A^2\omega^2 \sin^2 kx \sin^2 \omega t$$

Since

$$v^2 = \frac{T}{\mu},\ \mu = \frac{T}{v^2} = T\left(\frac{k}{\omega}\right)^2$$

$$\begin{aligned} \text{k.e.} &= \frac{1}{2}TA^2\frac{k^2}{\omega^2}\omega^2 \sin^2 kx \sin^2 \omega t \\ &= \frac{1}{2}TA^2k^2 \sin^2 kx \sin^2 \omega t \end{aligned}$$

ωt	$\frac{P}{A^2Tk\omega}$	$\frac{u}{A^2Tk^2}$	$\frac{ke}{A^2Tk^2}$
0	$\frac{\sin 2kx}{4}$	$\frac{\cos^2 kx}{2}$	0
$\frac{\pi}{4}$	0	$\frac{\cos^2 kx}{4}$	$\frac{\sin^2 kx}{4}$
$\frac{\pi}{2}$	$-\frac{\sin 2kx}{4}$	0	$\frac{\sin^2 kx}{2}$
$\frac{3\pi}{4}$	0	$\frac{\cos^2 kx}{4}$	$\frac{\sin^2 kx}{4}$
π	$\frac{\sin kx}{4}$	$\frac{\cos^2 kx}{2}$	0

Chapter 16

Sound and Light Waves

16.1 Wavelength and frequency are inversely related:

$$\lambda = \frac{v_s}{f}$$

where v_s is the speed of sound. From Table 16.1 $v_s = 330\,\mathrm{m/s}$

$$\lambda = \frac{330\ \mathrm{m/s}}{440\ \mathrm{Hz}} = 0.75\ \mathrm{m}$$

Wavelength is inversely proportional to frequency when the speed of sound is constant:

$$\lambda \propto \frac{1}{f}$$

If f is quadrupled, then

$$\lambda_{\mathrm{new}} = \frac{1}{4 f_{\mathrm{old}}} = \frac{1}{4}\lambda_{\mathrm{old}} = \frac{.75\ \mathrm{m}}{4} = .19\ \mathrm{m}$$

16.3 From the figure there are 6 displacement nodes

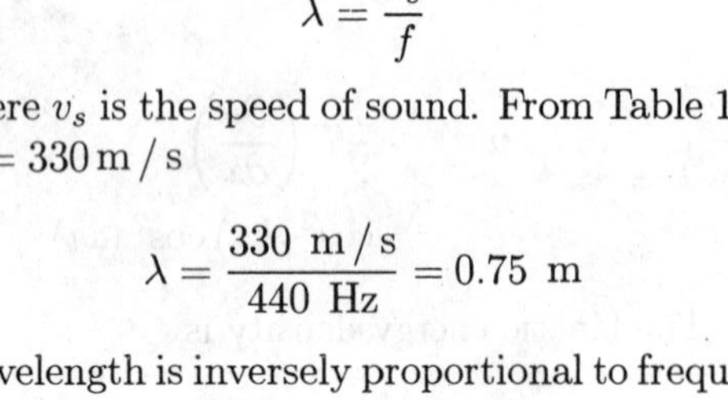

open pipe

DISP.

1.0 m

PRESSURE

There are 7 pressure nodes. Now the wavelength is

$$\lambda_n = \frac{2L}{n}$$

So $f_n = \frac{v_s}{\lambda_n} = \frac{n v_s}{2L}$:

$$f_6 = \frac{3(340\,\mathrm{m/s})}{1.0\ \mathrm{m}} \cong 1.0\ \mathrm{kHz}$$

16.5 All electromagnetic waves travel at $c = 3.0 \times 10^8\ \mathrm{m/s}$ in vacuum.

16.7 **a)** At constant speed, the speed is equal to distance traveled over time of travel:

$$v = \frac{d}{t}$$

where $v = c$, the speed of light. So time of travel is:

$$t = \frac{d}{c} = \frac{10\ \mathrm{m}}{3.0 \times 10^8\,\mathrm{m/s}} = 33\ \mathrm{ns}$$

b) The radius of the Earth is 6370 km $= R_E$. So the circumference of the earth is $C_E = 2\pi R_E$ The time for light to travel around the earth is:

$$t = \frac{C_E}{c} = \frac{2\pi R_E}{c} = \frac{2\pi(6370\ \mathrm{km})}{3.0 \times 10^5\ \mathrm{km/s}} = .13\ \mathrm{s}$$

c)

$$t = \frac{d}{c} = \frac{4 \times 10^{16}\ \mathrm{m}}{3 \times 10^8\,\mathrm{m/s}} = 1.3 \times 10^8\,\mathrm{s}$$

$$\frac{1.3 \times 10^8\,\mathrm{s}}{(3600\ \frac{\mathrm{s}}{\mathrm{h}})(24\ \frac{\mathrm{h}}{\mathrm{d}})(365\ \frac{\mathrm{d}}{\mathrm{y}})} = 4.2\ \mathrm{y} \sim 4\ \mathrm{y}$$

16.9 From the definition of intensity

$$I = \frac{P}{A} = \frac{P_0}{4\pi r^2} I$$

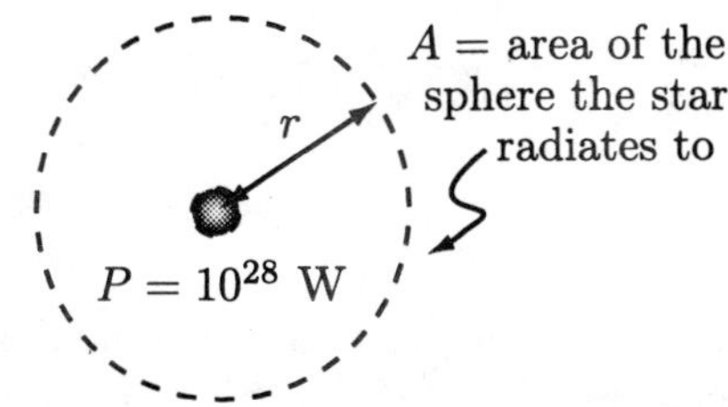

$$= \frac{10^{28}\text{ W}}{4\pi(5\times10^{17}\text{ m})^2} = \frac{10^{28}}{100\pi\times10^{34}}\frac{\text{W}}{\text{m}^2}$$
$$= 3\times10^{-9}\frac{\text{W}}{\text{m}^2}$$

16.11 Neither the observer nor the source are moving relative to one another. Rather, the air is moving ∴ there is no frequency shift. $f = 440$ Hz.

16.13 $v_e = 21$ km/s $= 2.1\times10^4$ m/s (conversion 1 m/s = 3.6 km/h) Wave fronts are farther apart so the wavelength gets longer. Since $\Delta\lambda = \lambda - \lambda_0$, the red shift is:

$$\frac{\Delta\lambda}{\lambda_0} = \frac{\lambda-\lambda_0}{\lambda_0} = \frac{\lambda}{\lambda_0} - 1$$

For light which is an EM wave $\lambda = \frac{c}{f}$, so

$$\frac{\Delta\lambda}{\lambda_0} = \frac{\frac{c}{f}}{\frac{c}{f_0}} - 1 = \frac{f_0}{f} - 1$$

for a receding source

$$f = f_0\sqrt{\frac{1-\frac{v_e}{c}}{1+\frac{v_e}{c}}}$$

Thus,

$$\frac{\Delta\lambda}{\lambda_0} = \sqrt{\frac{1+\frac{v_e}{c}}{1-\frac{v_e}{c}}} - 1$$

Where v_e is the source speed,

$$\frac{\Delta\lambda}{\lambda_0} = \sqrt{\frac{1+\frac{2.1\times10^4\text{ m/s}}{3.0\times10^8\text{ m/s}}}{1-\frac{2.1\times10^4\text{ m/s}}{3.0\times10^8\text{ m/s}}}} - 1$$
$$= \sqrt{\frac{1+7\times10^{-5}}{1-7\times10^{5}}} - 1$$
$$= 7\times10^{-5}$$

16.15 The beam is reflected at a 35° angle since the angle of reflection is equal to the angle of incidence.

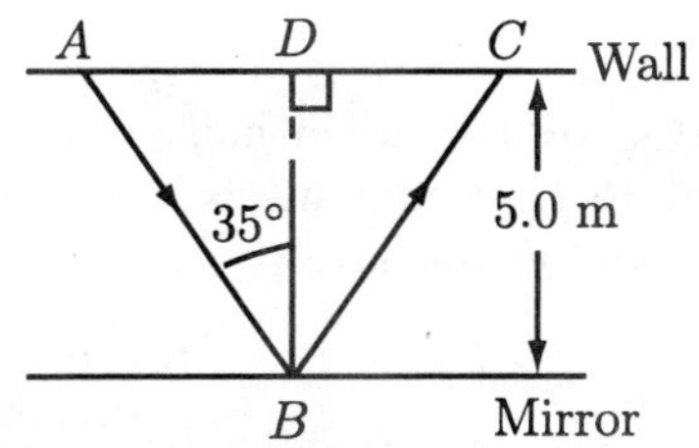

Consider the isosceles triangle ABC. Triangle ABC is composed of two equivalent right triangles, BCD and ABD. AC is the distance between the hole and the place where the reflected beam strikes the wall.

$$AC = AD + DC$$

Since triangle BCD and triangle ABD are congruent $AD = DC$, so $AC = 2AD$. Examining the right triangle ABD:

$$\tan 35° = \frac{AD}{BD}$$

Thus

$$AC = 2(BD\tan35°) = 2(5\text{ m})\tan35° = 7.0\text{ m}$$

16.17 The critical angle θ_c is the incident angle corresponding to a 90° angle of refraction.

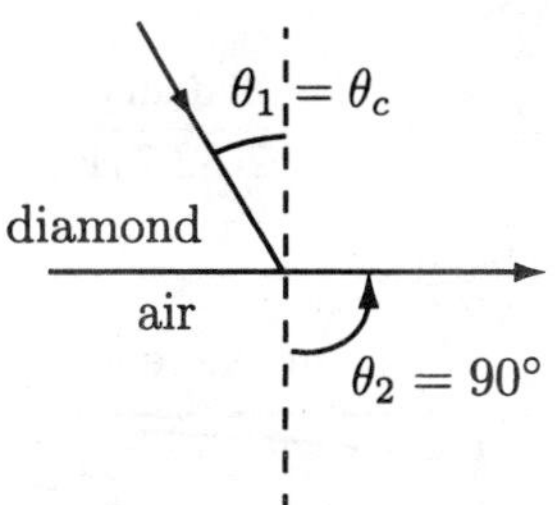

Snell's law for a light wave is:

$$n_1\sin\theta_1 = n_2\sin\theta_2$$

For this problem: $n_{\text{diamond}} = 2.418$ and $n_{\text{air}} = 1$

$$n_{\text{diamond}}\sin\theta c = n_{\text{air}}\sin 90° = 1$$
$$\theta_c = \arcsin\left(\frac{1}{2.418}\right)$$
$$= 24.4°$$

16.19 The wave corresponding to the fundamental frequency is sketched in each tube.

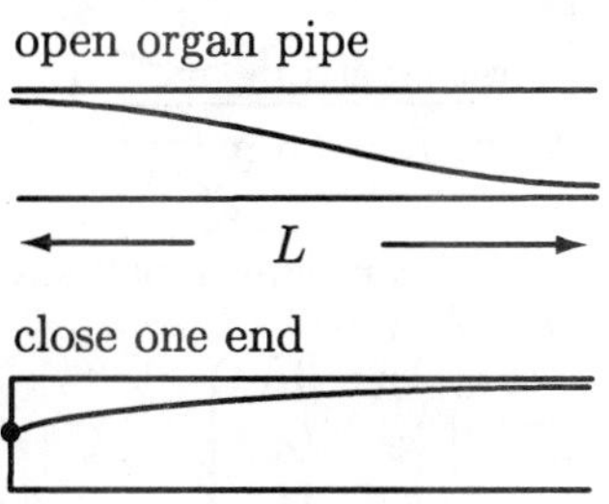

In the open pipe case:

$$\lambda_1 = 2L$$

closed pipe case:

$$\lambda_1 = 4L$$

So, since the wavelength is inversely proportional to frequency, it follows that the frequency is cut in half when the pipe is closed.

$$f_1' = \frac{1}{2} f_1 = \frac{262 \text{ Hz}}{2} = 131 \text{ Hz}$$

16.23 a) Wavelength is $\lambda_1 = 2L$.

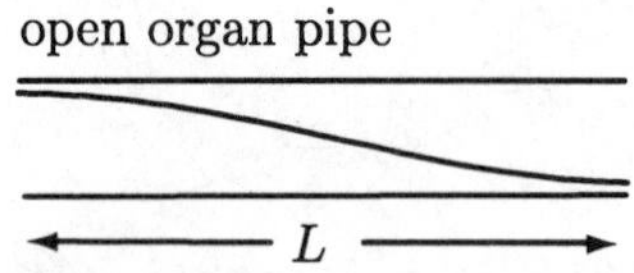

Frequency:

$$f_1 = \frac{v_w}{\lambda_1} = \frac{v_w}{2L}$$

Thus L is:

$$L = \frac{v_w}{2f_1} = \frac{340 \, \text{m/s}}{2(32.7 \, \text{Hz})} = 5.2 \text{ m}$$

b) Wavelength:

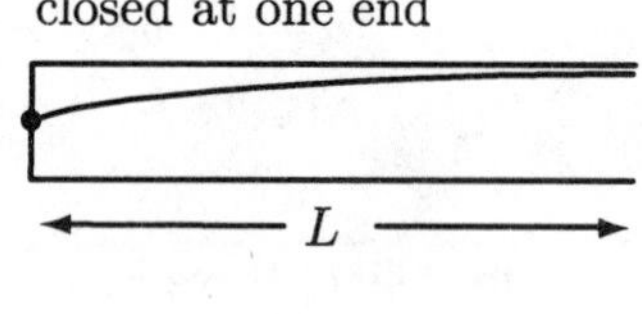

$$\lambda_1 = 4L$$

Frequency:

$$f_1 = \frac{v_w}{\lambda_1} = \frac{v_w}{4L} \; \therefore \; L = \frac{v_w}{4f_1}$$

This is half the answer of part a): $L = 2.6$ m

16.27 The chamber must have a displacement node at each end.

Vacuum chamber cylinder

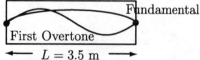

Fundamental

First Overtone

$L = 3.5$ m

In general the wavelength of a standing wave is:

$$\left.\begin{array}{l} \lambda_1 = 2L \\ \lambda_2 = L \\ \lambda_3 = \frac{2L}{3} \\ \lambda_4 = \frac{L}{2} \\ \vdots \end{array}\right\} \lambda_n = \frac{2L}{n}$$

Frequency is related to wavelength and the speed of sound v_s by:

$$\lambda_n f_n = v_s \rightarrow f_n = \frac{v_s}{\lambda_n} = \frac{n v_s}{2L}$$

if $v_s = 340$ m/s, then the fundamental frequency is

$$f_1 = \frac{340 \text{ m/s}}{2(3.5 \text{ m})} = 49 \text{ Hz}$$

The frequency of the first overtone is

$$f_2 = 2f_1 = 97 \text{ Hz}$$

16.31 The average earth-sun distance is $d = 1.5 \times 10^{11}$ m. Time for light to travel distance d:

$$\begin{aligned} t &= \frac{d}{c} = \frac{1.5 \times 10^{11} \text{ m}}{3.00 \times 10^8 \text{ m/s}} \\ &= 500 \text{ s} = 8.3 \text{ min}. \end{aligned}$$

16.35 No, the inverse square law does not hold exactly in practice. In the real world obstacles can scatter and absorb the energy of the wave. In the case of the street lamp, light has lower intensity than predicted by the inverse square law, since fog droplets scatter and absorb light rays.

16.39 Since $I \propto P_*^2$,

$$\begin{aligned} \Delta SIL &= SIL_2 - SIL_1 \\ &= 10 \text{ dB} \left[\log_{10} \left(\frac{P_{*2}}{P_{*1}} \right)^2 \right] \end{aligned}$$

Solving for $\frac{P_{*2}}{P_{*1}}$:

$$\begin{aligned} \log_{10} \left(\frac{P_{*2}}{P_{*1}} \right) &= \frac{\Delta SIL}{20 \text{ dB}} \\ \frac{P_{*2}}{P_{*1}} &= 10^{\frac{\Delta SIL}{20 \text{ dB}}} \\ &= 10^{\frac{10}{20}} = \sqrt{10} = 3.2 \end{aligned}$$

16.43 The explosion releases 1.0×10^7 J in 1 second, so the power of the explosion is:

$$P_{\text{Total}} = \frac{1.0 \times 10^7 \text{ J}}{1 \text{ s}} = 1.0 \times 10^7 \text{ W}$$

Half of the energy is energy is converted to sound waves (the other half presumably is converted to light and mechanical deformation). The sound waves have power:

$$\begin{aligned} P_{\text{sound}} &= (.5)(1.0 \times 10^7 \text{ W}) \\ &= 5.0 \times 10^6 \text{ W} \end{aligned}$$

Assuming spherical wave fronts, the intensity is

$$I = \frac{P_{\text{sound}}}{4\pi r^2}$$

where r is the distance at which we wish to find the intensity. $r = 110$ m

$$\begin{aligned} I &= \frac{5.0 \times 10^6 \text{ W}}{4\pi (110 \text{ m})^2} \\ &= 33 \text{ W/m}^2 \end{aligned}$$

Sound intensity in decibels is the SIL:

$$\begin{aligned} SIL &= 10 \text{ dB} \log_{10}\left(\frac{I}{10^{-12} \text{ W/m}^2}\right) \\ &= 10 \text{ dB} \log_{10}\left(\frac{32.88 \text{ W/m}^2}{10^{-12} \text{ W/m}^2}\right) \\ &= 140 \text{ dB} \end{aligned}$$

16.49 The speed of light is the same in all frames

$$c = 3 \times 10^8 \text{ m/s}$$

16.53 Andromeda nebula approaches Earth at 275 km/s

$$\begin{aligned} \frac{\Delta\lambda}{\lambda_0} &= \frac{\lambda - \lambda_0}{\lambda_0} = \frac{\lambda}{\lambda_0} - 1 \\ &= \frac{f_0}{f} - 1 \\ &= \sqrt{\frac{1 + \frac{(-v_e)}{c}}{1 - \frac{(-v_e)}{c}}} - 1 \text{(Eqn. 16.24)} \\ &= \sqrt{\frac{1 - \frac{2.75 \times 10^5}{3 \times 10^8}}{1 + \frac{2.75 \times 10^5}{3 \times 10^8}}} - 1 \\ &= -9.17 \times 10^{-4} \end{aligned}$$

Galaxy is approaching and so is blue-shifted.

16.57 The source is moving, so we use Eqn. 16.20. ($\vec{\mathbf{v}}_e$ and $\vec{\mathbf{v}}_w$ are parallel)

$$\begin{aligned} f &= \frac{f_0}{1 - \frac{v_e}{v_w}} \\ &= \frac{850 \text{ Hz}}{1 - \frac{20.0 \text{ m/s}}{340 \text{ m/s}}} = 903 \text{ Hz} \\ &= 9.0 \times 10^2 \text{ Hz} \end{aligned}$$

16.65 When the beam enters the water, it is refracted so that $\theta_2 < \theta_1$

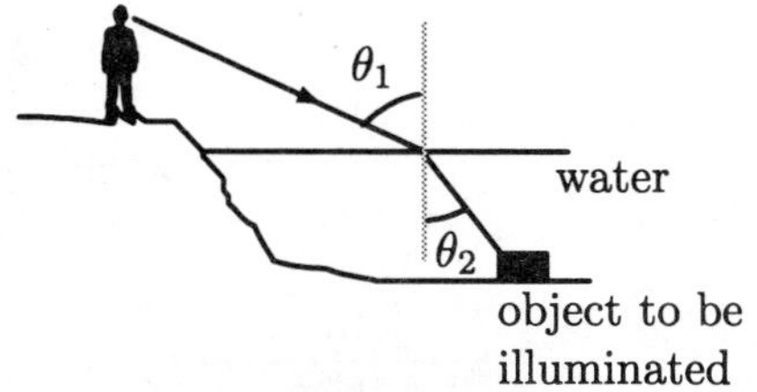

Of course light from the object that enters your eye is bent in the same way. You should point the laser at the apparent position of the object. The laser beam is pointed above the actual position of the object.

16.69 Flint Glass Prism

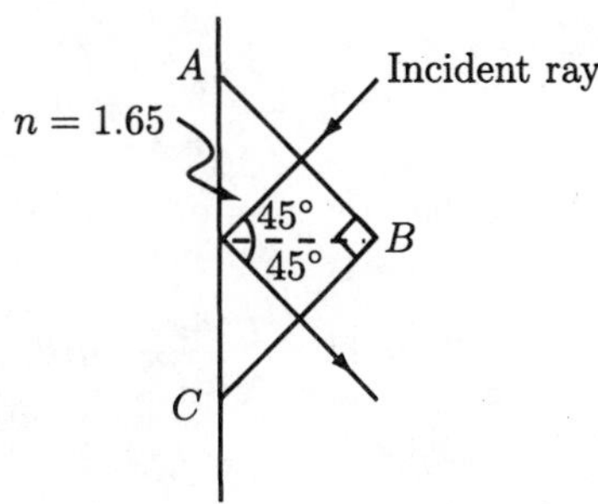

The incident angle is $\theta_1 = 0°$. By Snell's law

$$n_1 \sin\theta_1 = n_2 \sin\theta_2$$

Thus

$$n_2 \sin\theta_2 = 0$$

and $\theta_2 = 0$. The ray passes through side AB. Applying Snell's law to side AC:

$$\sin\theta_2 = \frac{n_1}{n_2}\sin\theta_1 = \frac{1.65}{1.00}\sin 45° > 1$$

Since $\sin\theta_2 > 1$, beam reflects internally at side AC. The incident angle for side BC is 0°, ∴ the ray emerges at 0° to the normal of side BC.

16.73 Mirror:

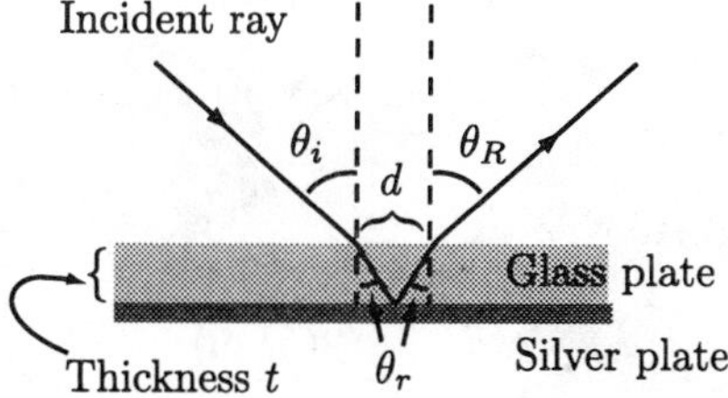

Angle at which light beam leaves mirror: From the symmetry of geometry of light rays: (see figure). The angle of reflection θ_R equals the angle of incidence θ_i

$$\theta_R = \theta_i$$

Where light beam leaves mirror: The beam emerges a distance d from where the incident beam hits the mirror.

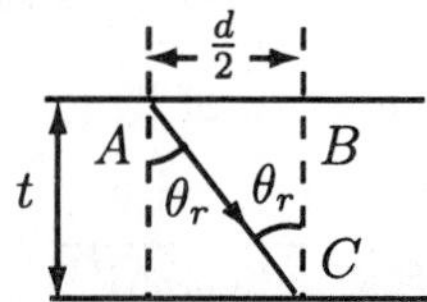

Examining the geometry of triangle ABC:

$$\tan\theta_r = \frac{\frac{d}{2}}{t} \rightarrow d = 2t\tan\theta_r$$

Snell's law:

$$n_{\text{air}}\sin\theta_i = n_{\text{glass}}\sin\theta_r$$

Thus,

$$\sin\theta_r = \frac{n_{\text{air}}}{n_{\text{glass}}}\sin\theta_i$$

And using the Trig. identity $\sin^2 A + \cos^2 B = 1$:

$$\cos\theta_r = \left[1 - \left(\frac{n_{\text{air}}}{n_{\text{glass}}}\sin\theta_i\right)^2\right]^{\frac{1}{2}}$$

Thus,

$$\begin{aligned} d &= 2t\tan\theta_r = \frac{2t\sin\theta_r}{\cos\theta_r} \\ &= \frac{2t\left(\frac{n_{\text{air}}}{n_{\text{glass}}}\right)\sin\theta_i}{\left[1 - \left(\frac{n_{\text{air}}}{n_{\text{glass}}}\sin\theta_i\right)^2\right]^{\frac{1}{2}}} \\ &= \frac{2t}{\sqrt{\left(\frac{n_{\text{glass}}}{n_{\text{air}}}\csc\theta_i\right)^2 - 1}} \end{aligned}$$

16.79 a)

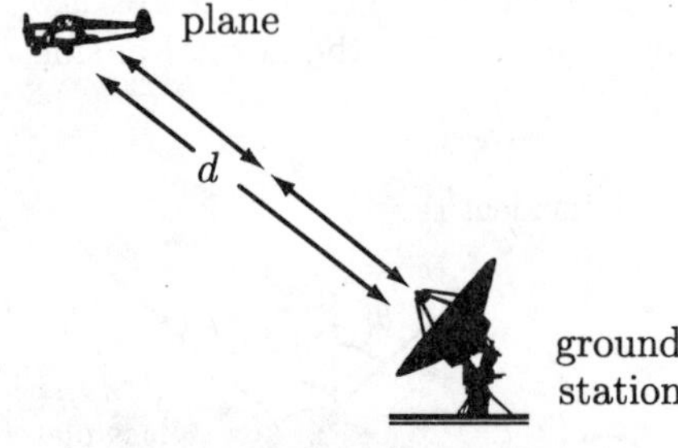

$$t = 2\frac{d}{c} = \frac{2(5\,\text{km})}{3\times10^8\,\text{m/s}} = 33\mu\text{s}$$

b) $t = \dfrac{2(200\,\text{km})}{3\times10^8\,\text{m/s}} = 1.3$ ms. During this time the aircraft travels a distance $v_{\text{plane}}t$ which is completely negligible so long as $v_{\text{plane}} \ll c$, i.e. always. The speed of the airplane does not cause a significant error. To compute speed, the DME compares the measured distance with the time of the measurement, and estimates v as $\frac{\Delta d}{\Delta t}$. Only the component of velocity toward or away from the station is measured.

16.83 $\lambda f = c$ (the speed of light). Thus

$$\lambda = \frac{c}{f} = \frac{3.00\times10^8\,\text{m/s}}{2450\,\text{MHz}} = 0.122\text{ m}$$

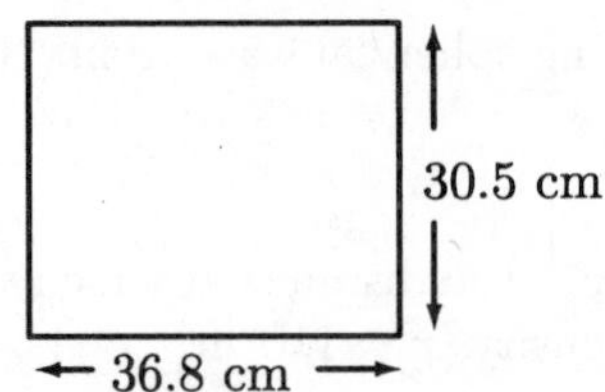

a) Along the cavity,

$$N = \frac{\ell}{\lambda} = \frac{36.8\text{ cm}}{12.2\text{ cm}} = 3$$

b) Across the cavity,

$$N = \frac{\ell}{\lambda} = \frac{30.5\text{ cm}}{12.2\text{ cm}} = 2.5$$

A whole number of half-wavelengths fit inside the cavity. This is consistent with a requirement that there be a node at each wall, i.e., e.g. zero displacement at the wall.

16.87 To find the orbital speed, we need the orbit radius.

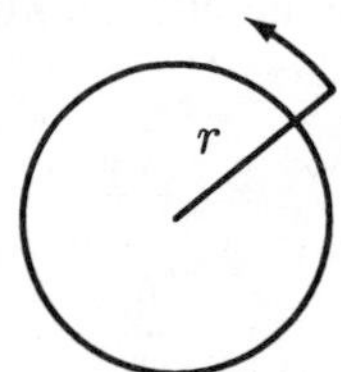

Since gravity provides the centripetal force:

$$\begin{aligned} \frac{GM_E m}{r^2} &= m\omega^2 r \\ r^3 &= \frac{GM_E}{\omega^2} = \frac{GM_E}{(2\pi)^2}T^2 \end{aligned}$$

and

$$\begin{aligned} v &= wr = \left(\frac{2\pi}{T}\right)\left(\frac{GM_E}{(2\pi)^2}T^2\right)^{\frac{1}{3}} \\ &= (GM_E)^{\frac{1}{3}}\left(\frac{T}{2\pi}\right)^{-\frac{1}{3}} \\ &= \left(6.7\times10^{-11}\frac{\text{m}^3}{\text{kg}\cdot\text{s}^2}\cdot 6\times10^{24}\,\text{kg}\right)^{\frac{1}{3}}\left(\frac{2\pi}{105\text{ min}}\right)^{\frac{1}{3}} \\ &= 7.4\times10^3\text{ m/s} \end{aligned}$$

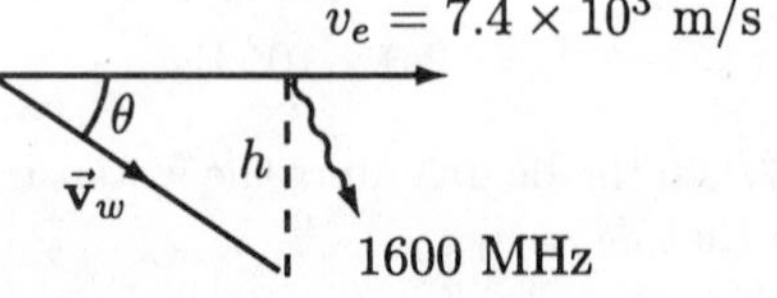

We use Eqn. (16.23).

$$\begin{aligned} f &= \frac{f_0\sqrt{1-\frac{v_e^2}{c^2}}}{1-\frac{\vec{v}_e\cdot\hat{v}_w}{c}} = \frac{f_0\sqrt{1-\frac{v_e^2}{c^2}}}{1-\frac{v_e\cos\theta}{c}} \\ \frac{v_e}{c} &= \frac{7.4\times10^3\,\text{m/s}}{3\times10^8\,\text{m/s}} = 2.5\times10^{-5} \end{aligned}$$

Thus we may approximate the square root in the numerator as 1.00. Then

$$\begin{aligned} f &= \frac{f_0}{1 - 2.5 \times 10^{-5} \cos\theta} \\ &\approx f_0(1 + 2.5 \times 10^{-5} \cos\theta) \end{aligned}$$

As the satellite passes directly overhead ($\cos\theta = 0$) , $f \to f_0$ and then f decreases as the satellite recedes. The maximum frequency shift is (with $\cos\theta = 1$)

$$\begin{aligned} f_0 \cdot 2.5 \times 10^{-5} &= (1600 \text{ MHz})\,(2.5 \times 10^{-5}) \\ &= 40 \text{ kHz} \gg 50 \text{ Hz} \end{aligned}$$

Yes. This effect does have to be taken into account. Data for graph

$$h = r - R_E = r - 6.4 \times 10^4 \text{ km}$$

and

$$\begin{aligned} r &= \frac{v}{\omega} = \left(\frac{7.4 \times 10^3 \text{ m/s}}{2\pi}\right)(105 \text{ min})\left(\frac{60 \text{ s}}{\text{min}}\right) \\ &= 7.4 \times 10^4 \text{ km} \end{aligned}$$

Thus

$$h = 1.0 \times 10^4 \text{ km}$$

And

$$\tan\theta = \frac{h}{|vt|} = \frac{1.0 \times 10^4 \text{ km}}{(1.6 \times 10^3 \text{ m/s})t}$$

where $\sec^2\theta = \dfrac{1}{\cos^2\theta} = 1 + \tan^2\theta$,

$$\begin{aligned} f &= f_0\left[1 + (2.5 \times 10^{-5})\sqrt{\frac{1}{1 + \left(\frac{h}{vt}\right)^2}}\right], \ (t < 0) \\ &= f_0\left[1 - (2.5 \times 10^{-5})\sqrt{\frac{1}{1 + \left(\frac{h}{vt}\right)^2}}\right], \ (t > 0) \end{aligned}$$

Where we have taken $t = 0$ when the satellite passes directly overhead.

16.91 Triangles OAB and OBC are isosceles, since in each case two of the sides equal the radius of the sphere.

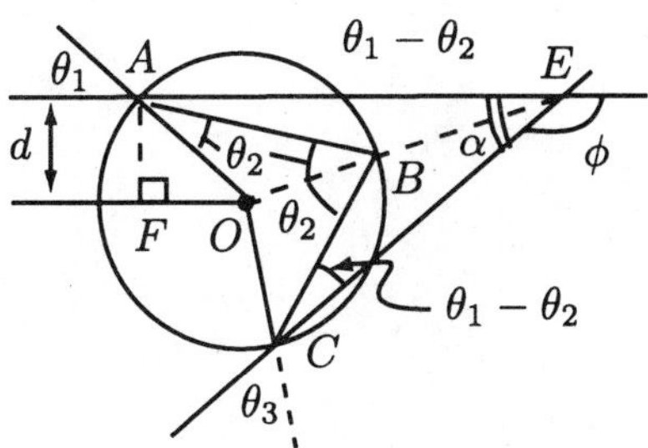

Internal reflection occurs at B. Thus the angles OAB, OBA, OBC and OCB are all equal. Call this angle θ_2. Applying Snell's law at A and at C:

$$n_1 \sin\theta_1 = n_2 \sin\theta_2 = n_1 \sin\theta_3 \Rightarrow \theta_1 = \theta_3$$

Now

$$\phi = 180^\circ - \alpha$$

Now the two triangles OAB and OBC are congruent (They have 2 equal sides and 2 equal angles) Thus $AB = BC$. Thus triangles ABE and CBE also have two equal sides and an equal angle. They are also congruent, and OBE is a straight line. Thus

$$\frac{\alpha}{2} = 180^\circ - (\theta_1 - \theta_2 + 180^\circ - \theta_2) = 2\theta_2 - \theta_1$$

and

$$\phi = 180^\circ - \alpha = 180^\circ - 4\theta_2 + 2\theta_1$$

as required. From ΔAOF,

$$\sin\theta_1 = \frac{d}{r}$$

And from Snell's law

$$\sin\theta_2 = \frac{d}{nr}$$

Thus for a given $\frac{d}{r}$ we may calculate

$$\theta_1 = \sin^{-1}\frac{d}{r}, \ \theta_2 = \sin^{-1}\left(\frac{d}{nr}\right)$$

and

$$\phi = 180^\circ - 4\theta_2 + 2\theta_1$$

The results show that maximum deflection $\phi \approx 138^\circ$ occurs for $\frac{d}{r} \sim 0.86$.

16.95 a) The reason the pavement looks blurry is that Joel is seeing light from the sky refracted upward toward his eyes instead of light reflected from the surface of the road.

90° − θ_J To Joel $\theta = 90°$ From sky   15 m   3 km

A light ray entering Joel's eye that appears to come from the edge of the blurry region is horizontal at the road surface and refracted upward to an angle from the horizontal of

$$\frac{\pi}{2} - \theta_J = \frac{15 \text{ m}}{3 \text{ km}}$$

(in radians!) Since $n \sin\theta$ is constant along the refracted ray, we have

$$\begin{aligned} n_{\text{road}} \sin 90^\circ &= n_{\text{Joel}} \sin(90^\circ - 5 \times 10^{-3} \text{ rad}) \\ n_{\text{road}} &= n_{\text{Joel}} \cos\left(5 \times 10^{-3} \text{ rad}\right) \\ &\simeq n_{\text{Joel}} \left(1 - \frac{1}{2}\left(5 \times 10^{-3}\right)^2\right) \end{aligned}$$

Thus,

$$\Delta n \equiv n_{\text{Joel}} - n_{\text{road}} = 1.3 \times 10^{-5}$$

Since $n_{air} \simeq 1.0003$ in usual atmospheric conditions, this is a 3% change in $n - 1$. (Correspondingly, the air above the road is 3% less dense than air at 15 m above the road—a huge difference that results in strong updrafts.)
To argue that $n \sin\theta$ is constant, apply Snell's law to adjacent differentially thick layers of air.

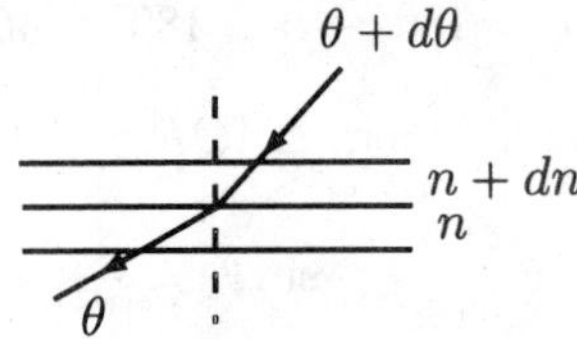

$$\begin{aligned} n \sin\theta &= (n + dn)\sin(\theta + d\theta) \\ &\equiv n\sin\theta + d(n\sin\theta) \Rightarrow d(n\sin\theta) = 0 \end{aligned}$$

b) If Joel can just see the mountain when refraction effects are negligible his line of sight is tangent to Earth's surface, as shown. Then, neglecting any difference between ℓ and the distance to the mountain along Earth's surface,

$$\begin{aligned} (R_E + h)^2 &= R_E^2 + \ell^2 \\ R_E^2 + 2R_E h + h^2 &= R_E^2 + \ell^2 \\ \ell^2 &= 2R_E h + h^2 \\ \Rightarrow \ell &= \sqrt{2R_E h + h^2} \\ &\simeq 2(6.4 \times 10^6 \text{ m})(400 \text{ m}) \\ &= 72 \text{ km} \end{aligned}$$

The mountain could be seen from as far away as 72 km. Now, if we include refraction, Joel's line of sight looking toward the horizon is actually 5×10^{-3} rad above the tangent to Earth's surface.

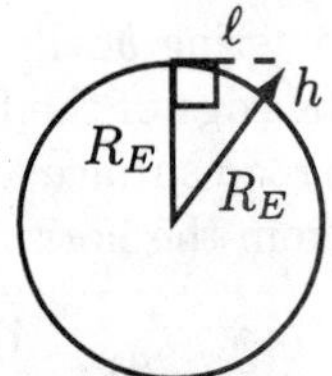

The height of a mountain that Joel can barely see is: necessary height = height to be seen at horizon + height ray of light goes above horizon line

$$\begin{aligned} &= h_1 + \ell\theta_J \\ &= \frac{\ell^2}{2R_E} + \ell \tan\left(5 \times 10^{-3} \text{ rad}\right) \\ &= \frac{(60.0 \text{ km})^2}{2(6.4 \times 10^3 \text{ km})} + (60.0 \text{ km})(5 \times 10^{-3}) \\ &= 0.58 \text{ km} \end{aligned}$$

Joel cannot see his destination.

Chapter 17

Interference and Diffraction

17.1 We use the method of Example 17.3.

$$\begin{aligned} \lambda &= 600.0\,\text{nm} = 600 \times 10^{-9} \text{ m} \\ &= 6.000 \times 10^{-7} \text{ m} \end{aligned}$$

Two narrow slits separated by

$$\begin{aligned} d &= 1.00 \text{ mm} \\ &= 1.00 \times 10^{-3} \text{ m} \end{aligned}$$

Screen is 2.00 m from the slits:

$$L = 2.00 \text{ m}$$

Solution: The slits are far away from the point of observation $\Rightarrow$ we can use for maxima

$$d \sin\theta = m\lambda$$

or

$$\sin\theta = m\frac{\lambda}{d}$$

For $L \gg \Delta x$ we can use

$$\sin\theta \approx \tan\theta$$

so

$$\begin{aligned} \tan\theta &\simeq \frac{m\lambda}{d} \simeq \theta \\ \frac{\Delta x_m}{L} &\approx \frac{m\lambda}{d} \end{aligned}$$

1st maximum: $m = 1$

$$\begin{aligned} \Delta x_1 &= \frac{\lambda L}{d} = \frac{(6.000 \times 10^{-7} \text{ m})(2.00 \text{ m})}{1.00 \times 10^{-3} \text{ m}} \\ &= 1.2 \times 10^{-3} \text{ m} = 1.20 \text{ mm} \end{aligned}$$

2nd maximum: $m = 2$

$$\begin{aligned} \Delta x_2 &= \frac{2\lambda L}{d} \\ &= 2(1.2 \times 10^{-3} \text{ m}) \\ &= 2.40 \times 10^{-3} \text{ m} \\ &= 2.40 \text{ mm} \end{aligned}$$

3rd maximum $m = 3$

$$\begin{aligned} \Delta x_3 &= 3(1.2 \times 10^{-3} \text{ m}) \\ &= 3.60 \times 10^{-3} \text{ m} = 3.60 \text{ mm} \end{aligned}$$

In our domain of $d \ll L$ ie $\theta \ll 1$

$$\begin{aligned} \Delta x_{m+1} - \Delta x_{m-1} &= \frac{(m+1)\lambda L}{d} - \frac{m\lambda L}{d} \\ &= \frac{\lambda L}{d} = 1.2 \times 10^{-3} \text{ m} \end{aligned}$$

$\Rightarrow$ Separation of maxima is

$$1.20 \times 10^{-3} \text{ m} = 1.20 \text{ mm}$$

Since we used the approximation $\sin\theta \approx \tan\theta$ for $\theta \ll 1$ the separation is not constant except near the center of the screen.

17.3 For destructive interference, the reflected ray on the exterior of the coating must exhibit a phase difference of an odd number of π from the ray reflected from the glass surface. Model normal incidence. The thickness of the coating is δt.

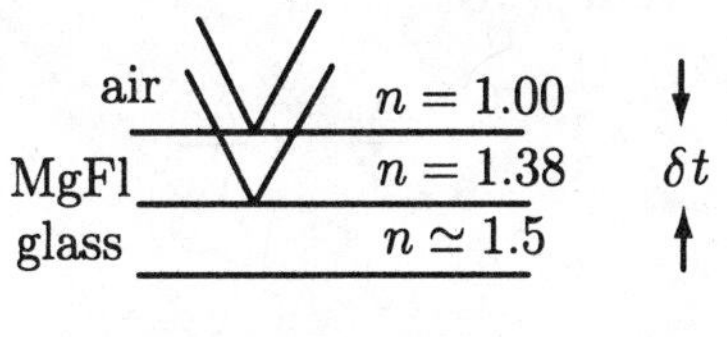

air $n = 1.00$

MgFl $n = 1.38$ δt

glass $n \simeq 1.5$

The phase difference through the coating is

$$\Delta\phi = 2(n\,\delta t)\cdot 2\frac{\pi}{\lambda}$$

There is a phase change of π at *each* reflection, so there is no net phase difference between the beams due to reflections. For destructive interference, we want

$$\Delta\phi = (2m+1)\pi \quad \left(\text{odd}\ \frac{\lambda}{2}\right)$$

pick the lowest order m ($\Rightarrow$ thinnest coating) then

$$\begin{aligned} \pi &= \frac{4\pi n\,\delta t}{\lambda} \\ \delta t &= \frac{\lambda}{4n} \\ &= \frac{550\ \text{nm}}{4\times 1.38} \\ &= 100\ \text{nm (2 sig. figs.)} \end{aligned}$$

17.5

$$\begin{aligned} \lambda_1 &= 488\ \text{nm} = 488\times 10^{-9}\ \text{m} \\ &= 4.88\times 10^{-7}\ \text{m} \end{aligned}$$

$$\lambda_2 = 514.5\ \text{nm} = 5.145\times 10^{-7}\ \text{m}$$

$$\text{slit } d = 3.00\ \mu\text{m} = 3.00\times 10^{-6}\ \text{m}$$

We consider Fraunhofer Diffraction.

The blue-green ($\lambda_1 = 488$ nm) light:

$$\sin\theta_1 = \frac{m_1\lambda_1}{a}$$

The green ($\lambda_2 = 514.5$ nm) light:

$$\sin\theta_2 = \frac{m_2\lambda_2}{a}$$

There are two diffraction patterns superimposed on the screen. Since $\frac{\lambda_1}{\lambda_2}$ is not a simple fraction, there will be no order m_1 or m_2 where the minima occur at $\theta_1 = \theta_2$. So there is a bright blue-green band at the center of the pattern, then alternating blue then blue-green bands. Since $\sin\theta \propto \lambda$ and for $\theta \ll 1$, $\ell \equiv$ distance to screen, then

$$\frac{\Delta x}{\ell} \approx \frac{m\lambda}{a}$$

or

$$\Delta x \propto \lambda$$

$\Rightarrow$ the green is wider. Consider the approximate width of the central maximum of each:

$$\begin{aligned} \Delta x_1 &= \frac{2\lambda_1\ell}{a} \\ \Delta x_2 &= \frac{2(\lambda_2\ell)}{a} \\ \Delta x_2 - \Delta x_1 &= 2(\lambda_2-\lambda_1)\frac{\ell}{a} \end{aligned}$$

$$\begin{aligned} &= \frac{2(5.145\times 10^{-7}\ \text{m} - 4.88\times 10^{-7}\ \text{m})(1.00\ \text{m})}{(3.00\times 10^{-6}\ \text{m})} \\ &= 2\left(8.83\times 10^{-3}\ \text{m}\right) = 17.7\ \text{mm} \end{aligned}$$

17.7

$$\begin{aligned} \lambda &= 600.0\ \text{nm} = 6\times 10^{-7}\ \text{m} \\ a &= 1.0\times 10^{-6}\ \text{m} \\ d &= 5.00\times 10^{-6}\ \text{m} \end{aligned}$$

a) Interference maxima occur when

$$d\sin\theta = m\lambda$$

or

$$\begin{aligned} \sin\theta &= \frac{m\lambda}{d} = \frac{6\times 10^{-7}}{5.0\times 10^{-6}}\times m \\ &= 0.120, 0.240, 0.360, 0.480, 0.600, \ldots 0.960 \end{aligned}$$

or

$$\theta = 6.89^\circ, 13.9^\circ, 21.1^\circ, 28.7^\circ, 36.9^\circ, \ldots, 73.7^\circ$$

b) Diffraction minima occur when

$$\sin\theta = \frac{m\lambda}{a} = m\frac{(6\times 10^{-7}\ \text{m})}{(1\times 10^{-6}\ \text{m})}$$

for $m = 1$:

$$\sin\theta = 0.600$$

No other minima are present.

$$\Rightarrow \theta = 36.9^\circ$$

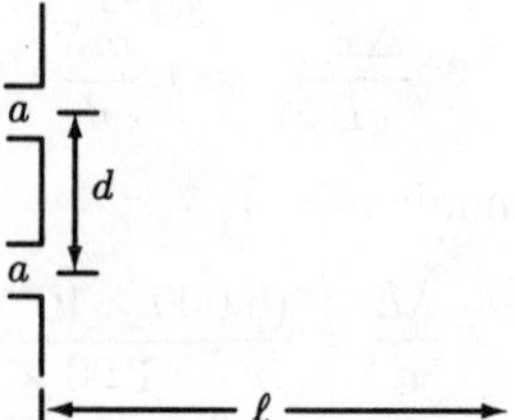

Every fifth interference maximum is missing.

17.9 Let

$$y_1 = (3\text{ mm})\cos\left[\left(3\frac{\text{rad}}{\text{m}}\right)x - \left(10^3\frac{\text{rad}}{\text{s}}\right)t\right]$$
$$y_2 = (1\text{ mm})\cos\left[\left(3\frac{\text{rad}}{\text{m}}\right)x - \left(10^3\frac{\text{rad}}{\text{s}}\right)t + \frac{\pi}{6}\right]$$

Put y_1 along the y-axis. y_2 makes an angle $\frac{\pi}{6}$ with the y-axis. Measuring off the diagram, y has amplitude 3.9 mm and relative phase 8°.

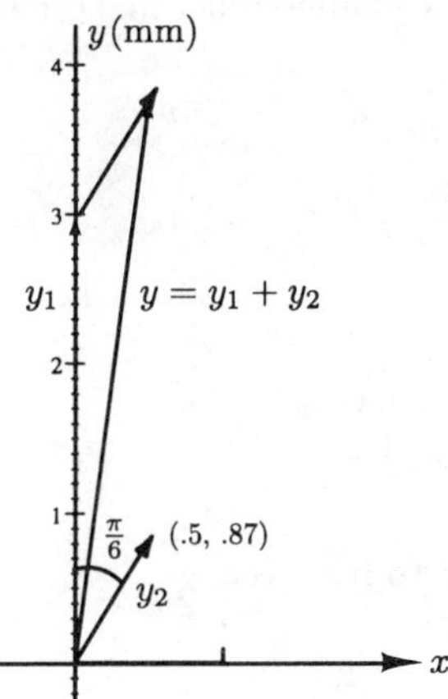

Thus

$$y = (3.9\,\text{mm})\cos\left[\left(3\frac{\text{rad}}{\text{m}}\right)x - \left(10^3\frac{\text{rad}}{\text{s}}\right)t + 0.14\text{ rad}\right]$$

17.11 For a minimum, the equal amplitude phasors must form a regular polygon, of 7 sides. The relative phase change between sources is $\frac{2\pi}{7}$.

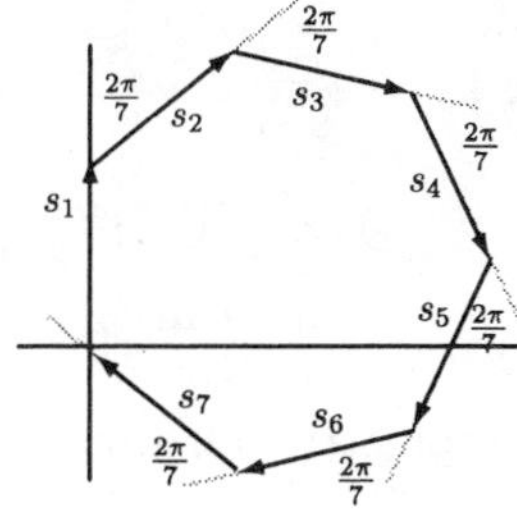

17.13

$$\theta_i = 20°$$
$$d = 0.314\text{ nm} = 3.14\times10^{-10}\text{ m}$$

for constructive interference:

$$2d\sin\theta = m\lambda$$

or

$$\lambda = \frac{2d\sin\theta}{m}$$

for $m = 1$,

$$\lambda = 2(0.314\text{ nm})(\sin(20°)) = 0.215\text{ nm}$$

$\Rightarrow$ wavelengths exhibiting strong diffraction are

$$\lambda = \frac{(0.215\text{ nm})}{m},\quad m = 1,2,3\ldots$$

17.15 Hold the soap film vertically; the film deforms into a *wedge* shape that is thicker at the bottom. The soap film has refractive index $n = 1.33$.

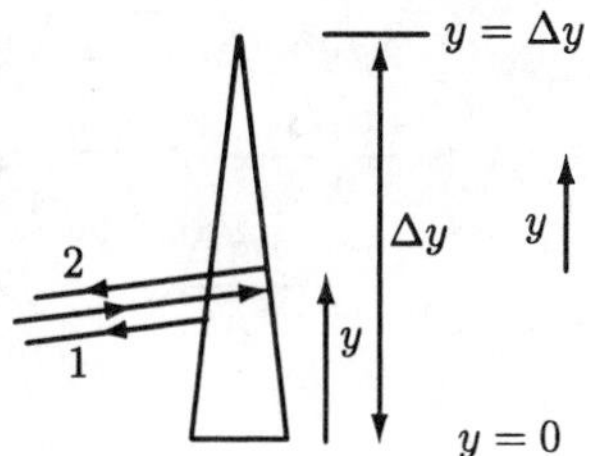

As the incident ray strikes the film at different values of y_1 the thickness is different (assume a linear change in thickness.)

$$t = \Delta t - y\frac{\Delta t}{\Delta y}$$
$$= \Delta t\left(1 - \frac{y}{\Delta y}\right)$$

In our figure, constructive interference occurs when

$$2t = \left(m + \frac{1}{2}\right)\frac{\lambda}{n}$$
$$2\Delta t\left(1 - \frac{y}{\Delta y}\right) = \left(m + \frac{1}{2}\right)\frac{\lambda}{n}$$

When light of a certain wavelength has constructive interference, the color of that wavelength dominates. Thus for a single order of m, the dominating color's wavelength is

$$\lambda = \frac{2\Delta t n}{m + \frac{1}{2}}\left(1 - \frac{y}{\Delta y}\right)$$

which leads to bands of color at constant y.

17.19 thickness = 250 nm, index of refraction= 1.20

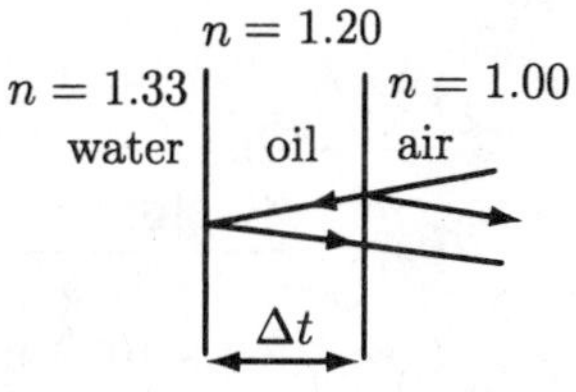

The dominant color is the one which undergoes constructive interference. Reflection at the air-oil interface gives a phase change of π:

$$\Delta\phi_1 = \pi$$

Reflection at the oil-water interface also gives a phase change of π so

$$\Delta\phi_2 = \pi + 2\pi \cdot \frac{2n\Delta t}{\lambda}$$

Constructive interference occurs for

$$\begin{aligned} \Delta\phi_2 - \Delta\phi_1 &= 2\pi m \\ \not\pi + \frac{4\pi n\Delta t}{\lambda} - \not\pi &= 2m\pi \\ \frac{4\pi n\Delta t}{\lambda} &= 2m\pi \end{aligned}$$

for $m = 1$

$$\begin{aligned} \frac{2n\Delta t}{\lambda} &= 1 \\ \lambda &= 2n\Delta t \\ &= 2(1.20)(250 \text{ nm})\lambda \\ &= 600 \text{ nm (orange)} \end{aligned}$$

17.23

$$\begin{aligned} n &= 1.55 \\ \alpha &= 3^\circ = \frac{3^\circ}{180^\circ} \times \pi = \frac{\pi}{60} \\ \lambda &= 510 \text{ nm} \\ \ell &= 1 \times 10^{-5} \text{ m} \end{aligned}$$

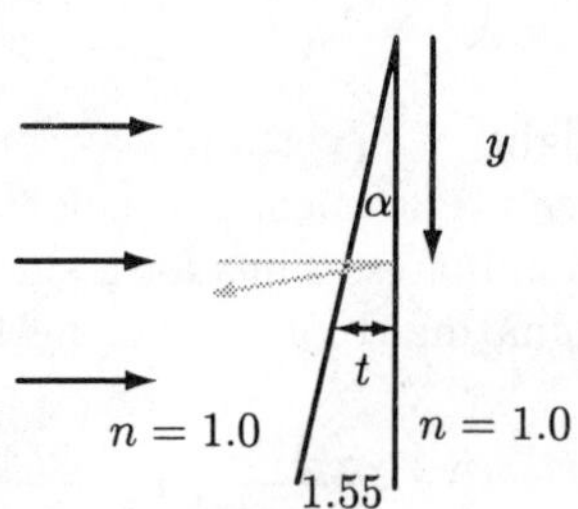

Interference occurs along the wedge where the path difference is less than the coherence length. So interference will occur for incident rays with $\ell > n2t$ in the glass, i.e., for

$$\begin{aligned} t &< \frac{\ell}{2n} \\ t &< \frac{1 \times 10^{-5} \text{ m}}{2 \times 1.55} \\ t &< 3.2 \times 10^{-6} \text{ m} \end{aligned}$$

Now $\alpha = \frac{\pi}{60}$. Constructive interference occurs for

$$\Delta\phi = \Delta\phi_2 - \Delta\phi_1 = 2m\pi$$

where

$$\Delta\phi_1 = \pi$$

due to reflection at glass ($n = 1.55$)

$$\Delta\phi_2 = 2\left(\frac{2\pi nt}{\lambda}\right)$$

(No phase change reflecting off air interface.) So

$$\begin{aligned} \frac{4\not\pi nt}{\lambda} - \not\pi &= 2m\not\pi \\ \frac{4nt}{\lambda} - 1 &= 2m \\ m &= \frac{2nt}{\lambda} - \frac{1}{2}, \quad m = 0, 1, 2, \ldots \\ \left(m + \frac{1}{2}\right) &= \frac{2nt}{\lambda} \end{aligned}$$

We need to have $t < \frac{\ell}{2n}$, so

$$\begin{aligned} \left(m + \frac{1}{2}\right) &< 2n \cdot \frac{\ell}{2n\lambda} \\ m + \frac{1}{2} &< \frac{2\ell}{2\lambda} \\ m &< \frac{\ell}{\lambda} - \frac{1}{2} \\ m &< \frac{1 \times 10^{-5} \text{ m}}{510 \times 10^{-9} \text{ m}} - \frac{1}{2} \\ m &< 19.6 - \frac{1}{2} \end{aligned}$$

$\Rightarrow$ have 19 fringes. This gives

$$\begin{aligned} \left(19 + \frac{1}{2}\right) &= \frac{2nt}{\lambda} \\ t &= \frac{\lambda}{2n}\left(19\tfrac{1}{2}\right) = 3.21 \times 10^{-6} \text{ m} \\ &= 3.2\,\mu\text{m} \end{aligned}$$

Since by geometry $\tan\alpha \approx \frac{t}{y}$ then the last fringe is at

$$\begin{aligned} \frac{\pi}{60} &= \frac{t}{y} \\ y &= t \cdot \frac{60}{\pi} = 3.2 \times 10^{-6} \text{ m} \times \frac{60}{\pi} \\ &= 6.1 \times 10^{-5} \text{ m} \end{aligned}$$

17.27

$$\begin{aligned} \lambda &= 630 \text{ nm} = 6.3 \times 10^{-7} \text{ m} \\ d &= 1.0 \text{ mm} = 1.0 \times 10^{-3} \text{ m} \\ L &= 2.0 \text{ m} \end{aligned}$$

Constructive interference occurs for

$$2\frac{\pi}{\lambda}d\sin\theta = 2m\pi$$

$$\sin\theta = m\frac{\lambda}{d}, \quad m = 0, \pm 1, \pm 2, \ldots$$

for small θ,

$$\sin\theta \approx \tan\theta = \frac{y}{L}$$

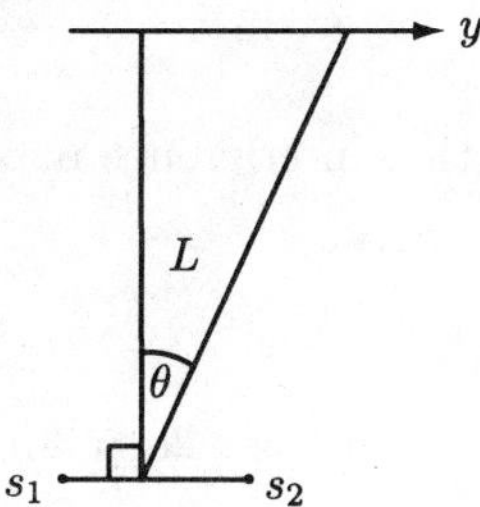

So

$$\begin{aligned} \frac{y}{L} &\approx \frac{m\lambda}{d} \\ y &= \frac{m\lambda L}{d} = \frac{(6.30 \times 10^{-7}\ \text{m})(2.0\ \text{m})}{(1 \times 10^{-3}\ \text{m})} \times m \\ &= m(1.26 \times 10^{-3}\ \text{m}) \end{aligned}$$

$\Rightarrow y = 0,\ \pm 1.3$ mm, ± 2.5 mm, ± 3.8 mm etc.

Put a glass slide of index $n+$ thickness t into one of the beams. The plate introduces a phase change

$$\Delta\phi = \frac{2\pi}{\lambda}(n-1)t$$

Constructive interference occurs for

$$\begin{aligned} \frac{\cancel{2}\cancel{\pi}}{\lambda}d\sin\theta_1 - \frac{\cancel{2}\cancel{\pi}}{\lambda}(n-1)t &= \cancel{2}m\cancel{\pi} \\ d\sin\theta_1 - (n-1)t &= m\lambda \end{aligned}$$

where, for the *central* bright fringe, $m = 0$. Thus

$$d\sin\theta_1 - (n-1)t = 0$$

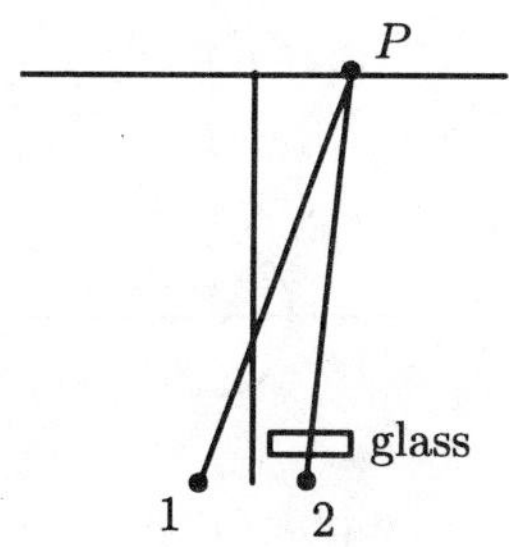

The extra path length from s_1 to P is compensated by the extra optical path length through the glass. Thus

$$\begin{aligned} n - 1 &= \frac{d}{t}\sin\theta_1 \\ n &= 1 + \frac{1.0 \times 10^{-3}\ \text{m}}{0.5 \times 10^{-3}\ \text{m}}\sin 15^\circ \\ &= 1 + 2(0.259) \\ n &= 1.52 \end{aligned}$$

The refractive index of the glass at 630 nm is 1.52.

17.31 Tilt one of the mirrors in the interferometer by angle α

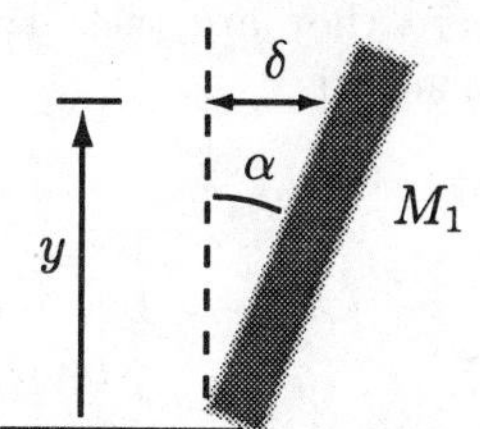

tilt introduces extra path length $2\delta = 2\alpha y$ since

$$\tan\alpha \approx \alpha = \frac{\delta}{y}$$

for

$$\alpha = 1\ \text{arc second} = \frac{1}{206265}\ \text{rad}$$

$$\Rightarrow \text{extra path length is } \frac{2y}{206265}$$

If L_1 is measured from the lower end of M_1 then

$$2(L_2 - L_1 + \delta) = m\lambda$$

Now set L_2 and L_1 so that at $y = 0$, $L_2 - L_1 = 0$ then $2\delta = m\lambda = \lambda$ for $m = 1$ which is the next bright fringe

$$\begin{aligned} \frac{2y}{206265} &= \lambda \\ y &= \frac{\lambda}{2} \times 206265 \end{aligned}$$

for $\lambda = 650$ nm $= 650 \times 10^{-9}$ m (red)

$$\begin{aligned} y &= \frac{650 \times 10^{-9}\ \text{m}}{2} \times 206265 \\ &= 67\ \text{mm} \end{aligned}$$

for $\lambda = 430$ nm (blue)

$$\begin{aligned} y &= \frac{430 \times 10^{-9}\ \text{m}}{2} \times 206265 \\ &= 44\ \text{mm} \end{aligned}$$

17.35 Assume there is no direct line of sight to the radio stations, and that the hillsides absorb 100% of the radio waves. The tunnel looks something like this:

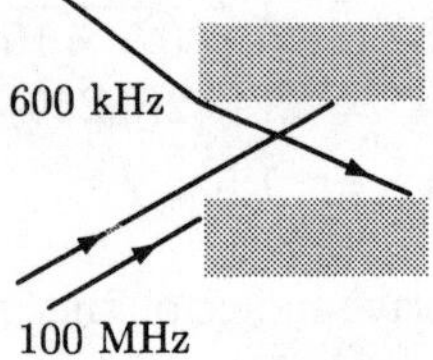

Longer wavelengths are diffracted more around obstacles. Thus the 600 kHz signals, with their longer wavelengths are bent further into the tunnel were they are picked up, than are the 100 MHz signals.

17.41

$$\begin{aligned}\lambda &= 694 \text{ nm}\\ a &= 2.00\,\mu\text{m}\\ D &= 2.00 \text{ m}\end{aligned}$$

Diffraction minima are given by

$$\sin\theta = m\frac{\lambda}{a},\quad m = 1, 2, 3, \ldots$$

for $m = 1$

$$\begin{aligned}\sin\theta &= \frac{\lambda}{a}\\ &= \frac{694 \text{ nm} \times 10^{-9} \text{ m/nm}}{2.00\,\mu\text{m} \times 10^{-6} \text{ m}/\mu\text{m}}\\ \sin\theta &= 0.347\\ \theta &= 20.3^\circ\end{aligned}$$

Since θ is not small, we cannot use the approximation $\sin\theta \approx \tan\theta = \frac{y}{D}$. Instead

$$\begin{aligned}y &= D\tan\theta\\ y &= (2.00 \text{ m})\tan 20.3^\circ\\ y &= 0.74 \text{ m}\end{aligned}$$

17.45 $D = 10$ m, $\lambda = 600$ nm $= 600 \times 10^{-9}$ m. Using the Rayleigh criterion:

$$\begin{aligned}\theta_{\min} &= 1.22\frac{\lambda}{D}\\ &= \frac{(1.22)(600 \times 10^{-9}\text{ m})}{10 \text{ m}}\\ &= 7.32 \times 10^{-8} \text{ rad} \times 206265\,\frac{\text{arcsec}}{\text{rad}}\\ \theta_{\min} &= 0.02''\end{aligned}$$

In reality, atmospheric distortions limit the angle to $\sim 0.5''$ on really good nights.

17.49 Want a two slit system, each slit 2.0 μm wide with every other fringe missing

$$\lambda = 488 \text{ nm}$$

Interference maxima are at

$$\sin\theta = m\frac{\lambda}{d}\ldots$$

diffraction minima are at

$$\sin\theta = m_1\frac{\lambda}{a}$$

If every other interference maximum is missing then

$$\begin{aligned}\frac{\not\lambda}{a} &= 2\frac{\not\lambda}{d}\\ d &= 2a = 4.0\,\mu\text{m}\end{aligned}$$

See appendix for plot of intensity vs. angle for two 2.0 μm slits.

17.53

$$\begin{aligned}f_0 &= 262 \text{ Hz}\\ f_1 &= 256 \text{ Hz}\\ f_B &= |f_1 - f_0| = 6 \text{ Hz}\end{aligned}$$

Phasor representing the fork is

$$\delta P_1 = A\cos[kx - 2\pi f_0 t]$$

The string has 4× the intensity ⇒ 2× amplitude

$$\begin{aligned}\delta P_2 &= 2A\cos[kx - 2\pi f_1 t]\\ &= 2A\cos[kx - 2\pi f_1 t + 2\pi f_0 t - 2\pi f_0 t]\\ \delta P_2 &= 2A\cos[kx - 2\pi f_0 t - 2\pi f_B t]\end{aligned}$$

Take both phasors at $x = 0$. At $t = 0$, the diagram is

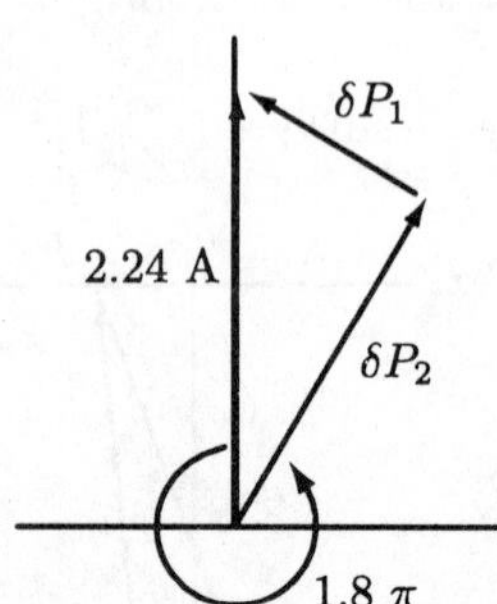

At $t = \dfrac{1}{4f_B}$ the phasors are

$$\delta P_1 = A\cos\left[\frac{-2\pi f_0}{4f_B}\right]$$

$$\begin{aligned} &= A\cos\left[\frac{-2\pi\cdot(262\text{ Hz})}{4\cdot(6\text{ Hz})}\right] \\ &= A\cos\left[-1.8\pi\right] \\ \delta P_2 &= 2A\cos\left[\frac{-2\pi f_0}{4f_B}-\frac{2\pi f_B}{4f_B}\right] \\ &= 2A\cos\left[\frac{-2\pi\,(262\text{ Hz})}{4\cdot(6\text{ Hz})}-\frac{\pi}{2}\right] \end{aligned}$$

The two phasors are at 90°

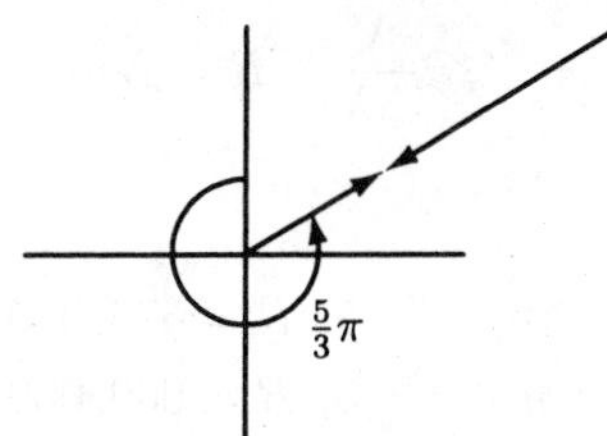

Resulting phasor is

$$\delta P = (2.24A)\cos\left[kx - 0.01\pi\right]$$

at $t = \dfrac{1}{2f_B}$ phasors are

$$\begin{aligned} \delta P_1 &= A\cos\left[\frac{-2\pi f_0}{2f_B}\right] \\ &= A\cos\left[\frac{-2\pi\cdot 262\text{ Hz}}{2\cdot 6\text{ Hz}}\right] \\ &= A\cos\left[\frac{-2\pi\cdot 262}{12}\right] \\ &= A\cos(-1.7\pi) \\ \delta P_2 &= 2A\cos\left[-2\pi\cdot\frac{262}{12}-2\pi\frac{f_B}{2f_B}\right] \\ &= 2A\cos\left[-2\pi\cdot\frac{262}{12}-\pi\right] \end{aligned}$$

The phasors are in opposite directions

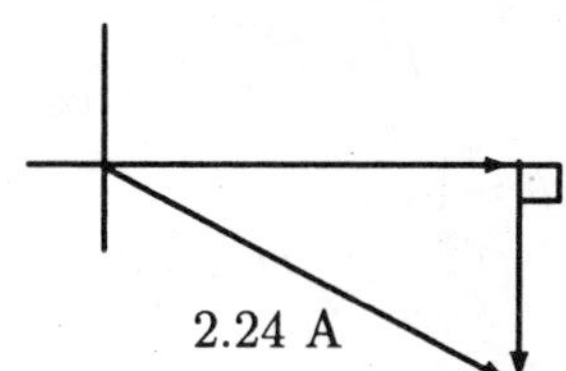

Resulting phasor is $\delta = A\cos\left[-\frac{5}{3}\pi\right]$.

At $t = \dfrac{3}{4f_B}$ phasors are

$$\begin{aligned} \delta P_1 &= A\cos\left[-2\pi\frac{f_0\cdot 3}{4f_B}\right] \\ &= A\cos\left[-2\pi\cdot\frac{262\cdot 3\text{ Hz}}{46\text{ Hz}}\right] \\ \delta P_1 &= A\cos\left[\frac{-3\pi}{2}\right] \\ \delta P_2 &= 2A\cos\left[\frac{-3\pi}{2}-2\pi\frac{f_B\cdot 3}{4f_B}\right] \end{aligned}$$

The phasors are again at right angles

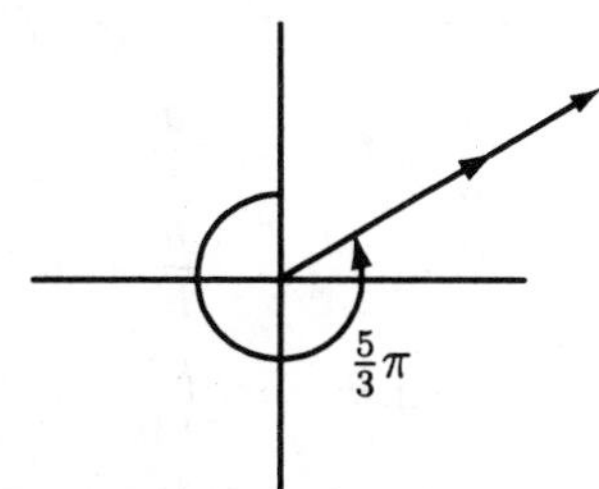

At $t = \dfrac{1}{f_B}$ phasors are

$$\begin{aligned} \delta P_1 &= A\cos\left[\frac{-2\pi f_0}{f_B}\right] \\ &= A\cos\left[\frac{-2\pi\cdot 262\text{ Hz}}{6\text{ Hz}}\right] \\ &= A\cos\left[-\frac{5}{3}\pi\right] \\ \delta P_2 &= 2A\cos\left[-\frac{5}{3}\pi-\frac{2\pi\cdot f_B}{f_B}\right] \end{aligned}$$

Resulting phasor

$$\delta P = 3A\cos\left[-\frac{5}{3}\pi\right]$$

The phasors are aligned Maximum intensity is $9A^2$ or 9 times that of the fork. Minimum intensity is A^2 or 1 times that of the fork.

17.57 For *six* slits spaced equally apart; $d = 3\lambda$. Assume very narrow fringes $\Rightarrow$ do not consider diffraction minima.

$3\lambda\ \{$: (six slits)

$N = 6$

$d = 3\lambda$

Minima occur at angles where the phasors sum to 0 i.e,

$$\Delta\phi = 2n\frac{\pi}{N} + 2m\pi$$

where

$$n < N,\ m = 0, 1, 2, \ldots$$

and

$$n = 1, 2, \ldots, N-1$$

but

$$kd\sin\theta = \Delta\phi$$

So

$$\begin{aligned}\frac{2\pi d\sin\theta}{\lambda} &= 2n\frac{\pi}{N}+2m\pi\\ \frac{d}{\lambda}\sin\theta &= \frac{n}{N}+m\\ 3\sin\theta &= \frac{n}{N}+m\\ \sin\theta &= \left(\frac{n}{6}+m\right)\frac{1}{3}\end{aligned}$$

$\theta \simeq 3.2°, 6.4°, 9.6°, 12.8°, 16.1°, 22.9°, 26.4°, 30°$, etc. Primary maxima occur when

$$\begin{aligned}kd\sin\theta &= 2m\pi\\ \frac{2\pi d}{\lambda}\sin\theta &= 2m\pi\\ 3\sin\theta &= m\\ \sin\theta &= \frac{m}{3},\ m=0,\pm1,\pm2,\ldots\\ \theta &= 0, 19.5°, 41.8°, 90°.\end{aligned}$$

Secondary maxima occur between minima, but a little closer to the central maximum, i.e., approximately at $4.5°, 8°, 11°, 14.5°$, etc.

17.61 a) Define

$$R \equiv \frac{\lambda}{\Delta\lambda} \text{ and } \lambda_1 \approx \lambda_2$$

Show $R = Nm$ where N = number of "slits," m = order. Maxima occur for $\sin\theta = m\frac{\lambda}{d}$. Take the differential

$$\Delta(\sin\theta) = \Delta\left(\frac{m\lambda}{d}\right)$$

With m and d constant,

$$\Delta\theta\cos\theta = m\frac{\Delta\lambda}{d} \qquad (1)$$

For small $\Delta\theta$,

$$\begin{aligned}\sin(\theta+\Delta\theta) &= \sin\theta\cos\Delta\theta+\cos\theta\sin\Delta\theta\\ &\approx \sin\theta+\Delta\theta\cos\theta\end{aligned}$$

since

$$\begin{aligned}\sin\Delta\theta &\approx \Delta\theta\\ \cos\Delta\theta &\doteq 1\end{aligned}$$

Rayleigh's Criterion says that two sources will be resolved if the maximum of one lies on the first minimum of the other. The phase difference between the two lines is

$$\begin{aligned}\Delta\phi &= \frac{2\pi}{N} = \left(\frac{2\pi d}{\lambda}\sin(\theta+\Delta\theta)-\frac{2\pi d}{\lambda}\sin\theta\right)\\ \Rightarrow \frac{1}{N} &= \frac{d}{\lambda}[\sin(\theta+\Delta\theta)-\sin\theta]\\ &= \frac{d}{\lambda}\Delta\theta\cos\theta\end{aligned}$$

Now using

$$\Delta\theta\cos\theta = \frac{m\Delta\lambda}{d}$$

then

$$\begin{aligned}\frac{1}{N} &= \frac{d}{\lambda}\frac{m\Delta\lambda}{d}\\ \frac{1}{Nm} &= \frac{\Delta\lambda}{\lambda}\end{aligned}$$

or

$$\frac{\lambda}{\Delta\lambda} \equiv R = Nm$$

b) For

$$\begin{aligned}m &= 5,\ R = 676{,}000\\ m &= 6,\ R = 660{,}000\\ m &= 11,\ R = 1{,}200{,}000\end{aligned}$$

Using

$$\begin{aligned}R &= mN\\ N &= \frac{R}{m}\end{aligned}$$

we get 3 values of N. Take average:

$$N = \left\{\begin{array}{l}676{,}000/5\\ 660{,}000/6\\ 1{,}200{,}000/11\end{array}\right\}$$

$$N_{\text{average}} = 1.2\times10^5$$

The slit separation is

$$\frac{L}{N} = \frac{0.20\ \text{m}}{1.2\times10^5} = 1.7\times10^{-6}\ \text{m}$$

From (1) the angular dispersion

$$D = \frac{\Delta\theta}{\Delta\lambda} = \frac{m}{d\cos\theta}$$

with $\cos\theta \approx 1$,

$$D = \frac{m}{d}$$

for $m = 2$,

$$\begin{aligned}D &= \frac{2}{1.7\times10^{-6}\ \text{m}}\\ \Rightarrow D &= 1.2\times10^6\ \text{m}^{-1}\end{aligned}$$

17.65

$$\left.\begin{array}{l}f_1 = 224\ \text{Hz}\\ f_2 = 220\ \text{Hz}\end{array}\right\}\ I_1 = 3I_2$$

the beat frequency is

$$\begin{aligned}f_B &= 224-220\ \text{Hz}\\ f_B &= 4\ \text{Hz}\end{aligned}$$

Since Intensity $\propto |\text{Amplitude}|^2$ then

$$A_1 = \sqrt{3}A_2$$

If the sound waves interfere constructively, then

$$A = A_1 + A_2 = \left(\sqrt{3}+1\right) A_2$$

So

$$I_{\text{max}} = \left(\sqrt{3}+1\right)^2 A_2^2 \simeq 7.5A_2^2 = 7.5I_2$$

$\Rightarrow$ Maximum intensity is ≈ 7.5 times that of the fork.

Destructive interference

$$\begin{aligned} |A| &= |A_1 - A_2| \\ |A| &= \left|\left(\sqrt{3}-1\right)\right| A_2 \end{aligned}$$

or

$$I_{\text{min}} = \left(\sqrt{3}-1\right)^2 A_2^2 \approx 0.54A_2^2 = 0.54I_2$$

$\therefore$ Minimum intensity is ≈ 0.54 that of the fork.

17.69 $d = 5.1$ nm. Constructive interference occurs for

$$\sin\theta = \frac{m\lambda}{2d}$$

for $m = 1$:

$$\sin\theta = \frac{\lambda}{2d}$$

so

$$\lambda = 2d\sin\theta$$

for $\theta = 15°$:

$$\begin{aligned} \lambda &= 2(5.1\text{ nm})(\sin 15°) \\ \Rightarrow \lambda &= 2.6\text{ nm} \end{aligned}$$

for $\theta = 30°$:

$$\begin{aligned} \lambda &= 2(5.1\text{ nm})(\sin 30°) \\ \Rightarrow \lambda &= 5.1\text{ nm} \end{aligned}$$

for $\theta = 45°$:

$$\begin{aligned} \lambda &= 2(5.1\text{ nm})(\sin 45°) \\ \Rightarrow \lambda &= 7.2\text{ nm} \end{aligned}$$

17.73 This is the standard configuration for Bragg scattering.

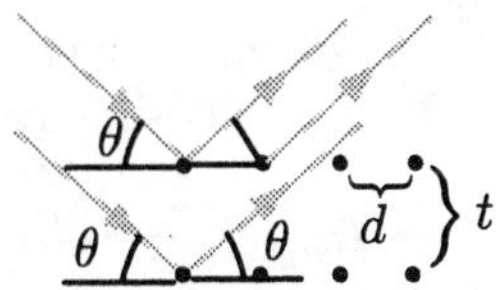

Constructive interference occurs for

$$\sin\theta = \frac{m\lambda}{2t}, \text{ i.e, } \lambda = \frac{2t\sin\theta}{m}$$

Now consider the other orientation. FIGURE The angle of incidence is $\theta+\phi$ and constructive interference occurs for

$$\sin(\theta+\phi) = \frac{m\lambda}{2s}$$

where the plane spacing

$$s = d\sin\phi$$

and

$$\tan\phi = \frac{t}{5d}$$

Thus

$$\begin{aligned} \lambda &= \frac{2s\sin(\theta+\phi)}{m} \\ &= \frac{2d}{m}\sin\phi\,(\sin\theta+\phi) \\ &= \frac{2d}{m}\sin\phi\,[\sin\theta\cos\phi + \cos\theta\sin\phi] \end{aligned}$$

Now,

$$\sin\phi = \frac{t}{\sqrt{t^2+25d^2}}$$

and

$$\cos\phi = \frac{5d}{\sqrt{t^2+25d^2}}$$

so,

$$\lambda = \frac{2dt}{m}\frac{5d\sin\theta + t\cos\theta}{t^2+25d^2}$$

The reflected beam emerges at angle $\theta + 2\phi$ to the surface i.e, $\theta + 2\tan^{-1}\left(\frac{t}{5d}\right)$.

17.79 Constructive interference occurs for a total phase difference of $2\pi m$ $(m = 0, 1, 2, \ldots)$

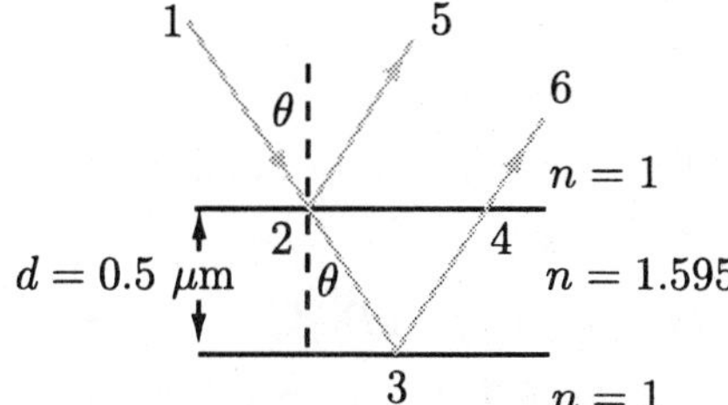

i.e,

$$\Delta\phi_{5,6} = 2m\pi$$

$$\begin{aligned} \Delta\phi_{5,6} &= \text{(phase change due to path length)} \\ &\quad -\text{(phase change due to reflection at 2)} \\ &= \frac{2\cdot 2\pi dn}{\lambda\cos\theta} - \pi \end{aligned}$$

$$2m \not{\pi} = \not{\pi}\left(\frac{4nd}{\lambda\cos\theta}-1\right)$$
$$2m+1 = \frac{4nd}{\lambda\cos\theta}$$
$$\lambda = \frac{4nd}{(2m+1)\cos\theta}$$

a) for $\theta = 30°$

$$\lambda = \frac{4nd}{(2m+1)\left(\frac{\sqrt{3}}{2}\right)} = \frac{3683}{(2m+1)}\text{ nm}$$

for

$m = 1, 2, \lambda =$ not visible
$m = 3, \lambda = 530$ nm green
$m = 4, \lambda = 410$ nm violet

b) for $\theta = 60°$

$$\lambda = \frac{4nd}{(2m+1)\left(\frac{1}{2}\right)} = \frac{6380}{2m+1}\text{ nm}$$

we get for

$m = 5, \lambda = 580$ nm yellow
$m = 6, \lambda = 490$ nm blue
$m = 7, \lambda = 425$ nm violet

c) for $\theta = 0$

$$\lambda = \frac{4nd}{(2m+1)} = \frac{3.19\times10^{-6}\text{ m}}{(2m+1)} = \frac{3,190\text{ nm}}{2m+1}$$

$m = 3$ gives $\lambda = 456$ nm (blue)
$m = 2$ gives $\lambda = 638$ nm (red)

17.83

Wave speed $= v_s$
Energy density $= u$
Intensity $= uv_s$

$$u = \frac{I}{v_s}$$

Two sources each of intensity I_0, pressure amplitude A_0

$$\begin{aligned}\delta P &= \delta P_1 + \delta P_2\\ &= A_0\cos(k_1x-\omega_1t) + A_0\cos(k_2x-\omega_2t)\\ &= 2A_0\cos\left(\tfrac{(k_1+k_2)x-(\omega_1+\omega_2)t}{2}\right)\cdot\\ &\quad\times\cos\left(\tfrac{(k_2-k_1)x-(\omega_2-\omega_1)t}{2}\right)\\ &= 2A_0\cos(kx-\omega t)\cos\left(\frac{\Delta kx-\Delta\omega t}{2}\right)\\ \delta P &= 2A_0\cos(kx-\omega t)\sqrt{\frac{1+\cos(\Delta kx-\Delta\omega t)}{2}}\end{aligned}$$

with

$$\begin{aligned}k &\equiv \frac{k_1+k_2}{2}\\ \Delta k &\equiv k_2-k_1\\ \omega &\equiv \frac{\omega_1+\omega_2}{2}\\ \Delta\omega &\equiv \omega_2-\omega_1 = v_s\Delta k\end{aligned}$$

The power per unit area transmitted by the wave (cf., § 16.3.2) is proportional to the square of the pressure perturbation δP, where

$$\begin{aligned}\delta P^2 &= 4A_0^2\cos^2(kx-\omega t)\\ &\quad\times\left[\frac{\cos((\Delta k)x-(\Delta\omega)t)+1}{2}\right]\\ &= 2A_0^2\cos^2(kx-\omega t)\left[1+\cos(\Delta kx-\Delta\omega t)\right]\end{aligned}$$

To find the average over short distances (a few wavelengths of each of the beating waves) we note that $x_{\text{short}}\Delta k \ll 1$ so that the change in the quantity $\cos[\Delta kx-\Delta\omega t]$ is small and may be neglected. However the average of $\cos^2(kx-\omega t)$ over a few wavelengths $\lambda = \frac{2\pi}{k}$ is $\frac{1}{2}$. Thus

$$\left\langle\delta P^2\right\rangle_{\text{short}} = A_0^2\left[1+\cos(\Delta kx-\Delta\omega t)\right]$$

Then the power per unit area is [see text, p. 538 and Eqn. 16.10]

$$\begin{aligned}\frac{\langle\text{Power}\rangle_{\text{short}}}{A} &= \frac{(\delta P)^2}{\rho_0 v_s}\\ &= \frac{A_0^2}{\rho_0 v_s}\left[1+\cos(\Delta kx-\Delta\omega t)\right]\\ &= 2I_0\left[1+\cos(\Delta kx-\Delta\omega t)\right]\end{aligned}$$

where we used Eqn. 16.15 to express the result in terms of I_0. Notice that this power/area varies from 0 to $4I_0$. Now we express the result in terms of u:

$$\langle u\rangle_{\text{short}} = \frac{2I_0}{v_s}\left[1+\cos(\Delta kx-\Delta\omega t)\right]$$

The distance between two maxima of $\langle u\rangle_{\text{short}}$ is the distance between two peaks of the cosine function, i.e,

$$\Delta kd = 2\pi \Rightarrow d = \frac{2\pi}{\Delta k}$$

These maxima pass an observer in time

$$\Delta t = \frac{d}{v_s} = \frac{2\pi}{v_s\Delta k} = \frac{2\pi}{\Delta\omega} = \frac{1}{f_b}$$

where $f_b = f_2 - f_1$ is the beat period. Now to get the long-distance average we average over many d. Then the cosine averages to zero, and

$$\langle u\rangle_{\text{short}} = \frac{2I_0}{v_s} = 2\langle u\rangle_1$$

where $\langle u \rangle_1$ is the average energy density due to one of the beating waves.

17.91 Like the standard two source problem (see figure)

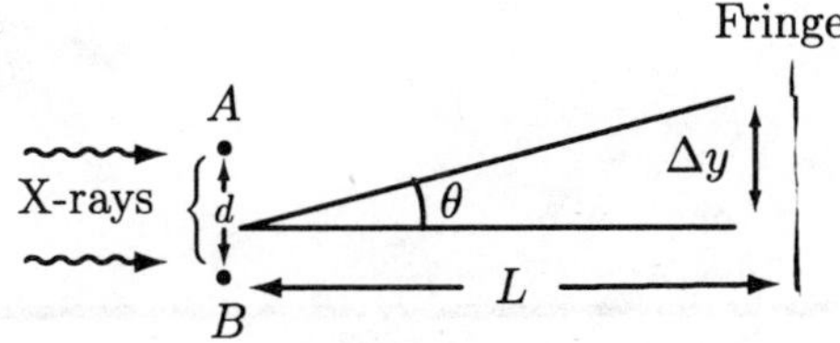

constructive interference occurs for

$$\begin{aligned} kd\sin\theta &= 2m\pi \\ \frac{2\pi}{\lambda}d\sin\theta &= 2\pi m \\ \sin\theta &= \frac{m\lambda}{d} \end{aligned}$$

$\Rightarrow \frac{\lambda}{d}$ for one fringe since $\ell \ll L$

$$\sin\theta \approx \theta \approx \tan\theta = \frac{\Delta y}{L}$$

For fringe spacing ℓ

$$\Delta y = \ell$$

then

$$\frac{\ell}{L} = \frac{\lambda}{d} \Rightarrow d = \frac{\lambda L}{\ell}$$

At a maximum

$$I = (A_A + A_B)^2$$

A_A

A_B

At a minimum

$$I = (A_A - A_B)^2$$

A_A A_B

We have

$$\begin{aligned} 4 &= \left(\frac{A_A + A_B}{A_A - A_B}\right)^2 \\ 4 &= \left(\frac{N_A + N_B}{N_A - N_B}\right)^2 \end{aligned}$$

since

$$\begin{aligned} A_A &\propto N_A \\ A_B &\propto N_B \end{aligned}$$

$$2 = \pm\frac{N_A + N_B}{N_A - N_B}$$

Taking the + sign:

$$\begin{aligned} 2(N_A - N_B) &= N_A + N_B \\ N_A &= 3N_B \\ \frac{N_A}{N_B} &= 3 \end{aligned}$$

Taking the − sign:

$$\begin{aligned} -2(N_A - N_B) &= N_A + N_B \\ N_B &= 3N_A \\ \frac{N_A}{N_B} &= \frac{1}{3} \end{aligned}$$

(Note that the choice of sign decides which label, A or B, is given to the larger atom.)

Chapter 18

Geometrical Optics

18.1

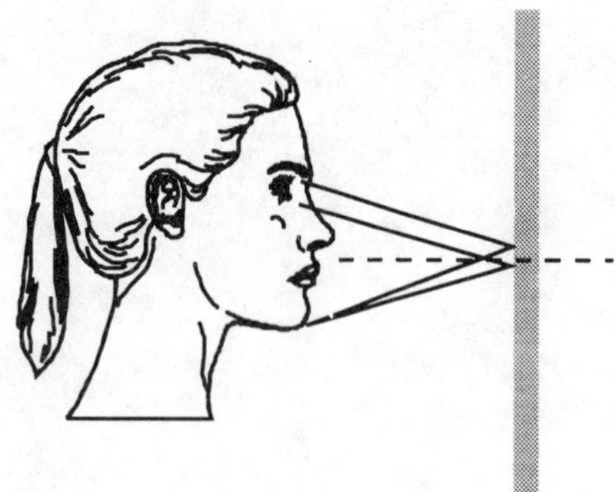

18.3 Mirror distance = 1 m, image distance

$$= 5 - 1 = 4 \text{ m}$$

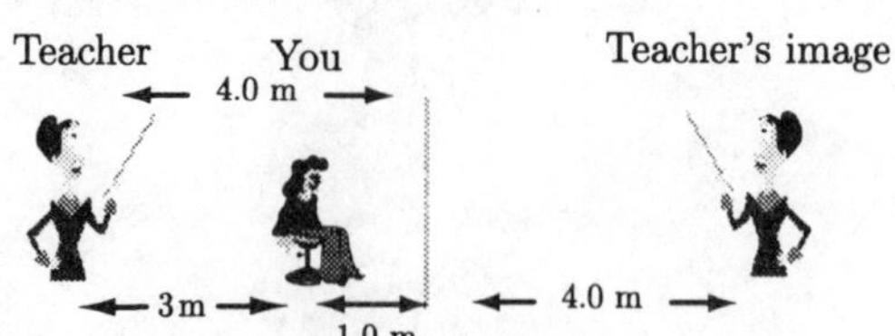

For mirror, object-mirror distance = image-mirror distance, so teacher is 4 m from mirror, or $4 - 1 = 3$ m behind student.

18.5 From Equations 18.3, 18.4 and 18.5

$$\frac{1}{x_O} + \frac{1}{x_i} = \frac{1}{f} \text{ with } f = \frac{r}{2}$$

$x_O = 10$ cm, then

$$\begin{aligned} x_i &= \left(\frac{2}{r} - \frac{1}{x_O}\right)^{-1} \\ &= \left(\frac{2}{8.0 \text{ cm}} - \frac{1}{10.0 \text{ cm}}\right)^{-1} \\ x_i &= +6.7 \text{ cm; real image} \end{aligned}$$

if $x_O = 3$ cm, then

$$x_i = \left(\frac{2}{8.0 \text{ cm}} - \frac{1}{3.0 \text{ cm}}\right)^{-1} = -12 \text{ cm}$$

The $-$ sign indicates that a virtual image exists behind the mirror.

18.7 There is one curved refracting surface, so use equation 18.6

$$\begin{aligned} \frac{n_1}{x_O} + \frac{n_2}{x_i} &= \frac{n_2 - n_1}{r} \\ n_1 &= \text{object medium} = 1.33 \\ n_2 &= \text{observer medium} = 1.00 \end{aligned}$$

The center of curvature is to the right of the surface, so $r = -5.0$ m (using the rule for images.) (cf Fig 18.18c)

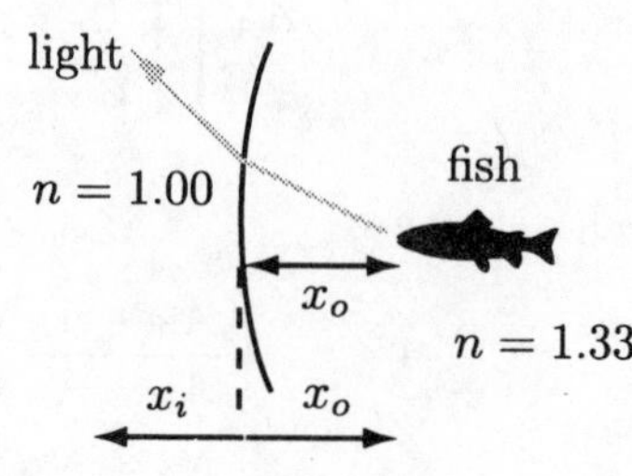

a) $x_0 = 3.0$ m

$$\begin{aligned}\frac{n_2}{x_i} &= \frac{n_2 - n_1}{r} - \frac{n_1}{x_0} \\ x_i &= \left[n_2\left(\frac{n_2-n_1}{r} - \frac{n_1}{x_0}\right)^{-1}\right] \\ &= \left[\frac{-0.33}{-5.0 \text{ m}} - \frac{1.33}{3.0 \text{ m}}\right]^{-1} = -2.7 \text{ m}\end{aligned}$$

This is a virtual image 2.7 m behind the glass.

b) $x_0 = 1.0$ m

$$x_i = \left\{\frac{-0.33}{-5.0} - \frac{1.33}{1}\right\}^{-1} = -0.79 \text{ m}$$

Virtual image 0.79 m behind glass. In both cases the fish appears closer than it really is.

18.9 Equation 18.10 relates f, n, and the curvature of the lens:

$$\frac{1}{f} = (n-1)\left(\frac{1}{r_1} - \frac{1}{r_2}\right)$$

For two lenses with the same shape, differences in f only depend on differences in n

$$\frac{f_A}{f_B} = \frac{n_B - 1}{n_A - 1} = \frac{2.0-1}{1.5-1} = 2.0.$$

18.11 Use Equation 18.4,

$$\frac{1}{x_0} + \frac{1}{x_i} = \frac{1}{f}$$

Here $x_0 = 10.0$ cm and $f = +25$ cm (lens is convex, used in air) so

$$\begin{aligned}x_i &= \left(\frac{1}{f} - \frac{1}{x_0}\right)^{-1} \\ &= +\left(\frac{1}{25 \text{ cm}} - \frac{1}{10.0 \text{ cm}}\right)^{-1} \\ &= -17 \text{ cm}\end{aligned}$$

A virtual image is formed 17 cm from the lens, on the same side as the object.

18.13 The image is formed at the focal point since the object is "distant" i.e., at ∞.

A ray through the center of the lens is undeflected. The figure shows rays from the foot and the top of the tree. The height of the image is given by

$$\begin{aligned}\frac{h_i}{f} &= \tan\theta \\ h_i &= (57.0 \text{ cm})\tan(10.0°) = 10.1 \text{ cm}\end{aligned}$$

18.15 Use Equation 18.4:

$$\frac{1}{x_0} + \frac{1}{x_i} = \frac{1}{f}$$

with $f = \frac{r}{2}$. Here r is positive (see figure.) Here $x_0 = 10$ cm and

$$\begin{aligned}f &= \frac{16 \text{ cm}}{2} = 8.0 \text{ cm} \\ x_i &= \left(\frac{1}{f} - \frac{1}{x_0}\right)^{-1} \\ &= \left(\frac{1}{8.0} - \frac{1}{10.0}\right)^{-1} \text{ cm} \\ &= +40 \text{ cm}\end{aligned}$$

Image is *real* since x_i is positive; located in front of mirror.

$$m = -\frac{x_i}{x_0} = -\frac{40 \text{ cm}}{10 \text{ cm}} = -4.0$$

Image is inverted.

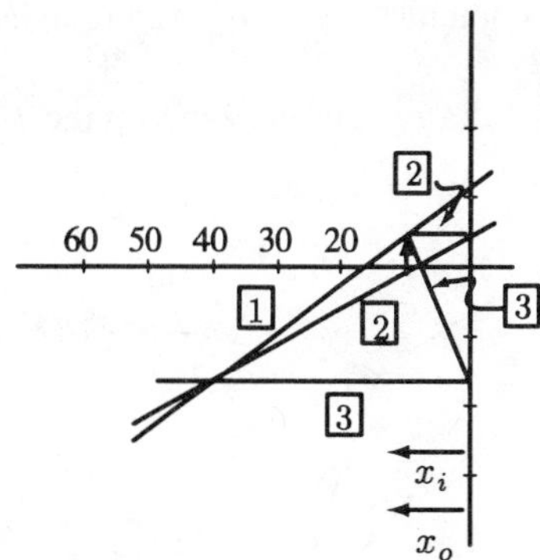

18.17 The magnification of the microscope is the product of the magnification of the objective and the angular magnification of the eyepiece. (Eqn. 18.13) where

$$m_0 = \frac{h_r}{h_0} = \frac{s}{f_0}$$

(see Figure 18.37) and

$$m_\theta = \frac{25.0 \text{ cm}}{f_e}$$

Thus

$$m = \frac{16 \text{ cm}}{3.1 \text{ cm}} \cdot \frac{25.0 \text{ cm}}{2.5 \text{ cm}} = 52.$$

18.19 Use the lensmaker's equation (18.10)

$$\frac{1}{f} = (n-1)\left(\frac{1}{r_1} - \frac{1}{r_2}\right)$$

a) $\lambda = 480$ nm:

$$\frac{1}{f} = (1.5296 - 1)\left(\frac{1}{6.500\text{ cm}} + \frac{1}{6.500\text{ cm}}\right)$$
$$f = 6.137\text{ cm @ }480.0\text{ nm}$$

b) $\lambda = 656.3$ nm:

$$\frac{1}{f} = (1.5203 - 1)\left(\frac{1}{6.500\text{ cm}} + \frac{1}{6.500\text{ cm}}\right)$$
$$f = 6.246\text{ cm @ }656.3\text{ nm}$$

18.21 There are 5 images:

I_1 is the image of O in mirror 1
I_2 is the image of O in mirror 2
I_{12} is the image of I_1 in mirror 2
I_{21} is the image of I_2 in mirror 1
I_{121} is the image of I_{12} in mirror 1;

it coincides with I_{212}, the image of I_{21} in mirror 2.

A ray used to view image I_{12} is shown.

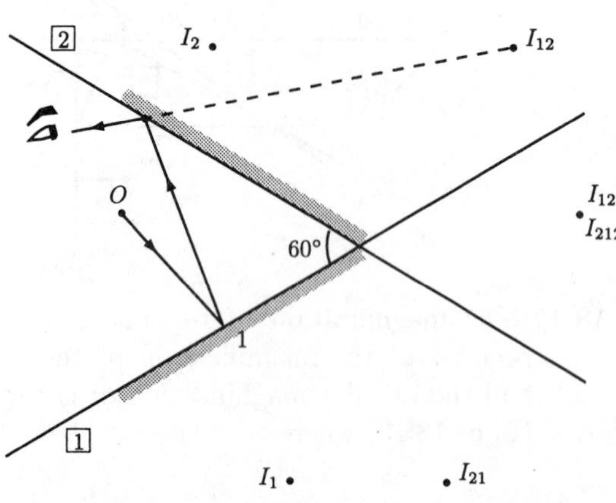

The other image may be viewed similarly. Rays used to view images I_{121} and I_{212} reflect 3 times before entering the eye.

18.25 By using 2 mirrors at an angle, you can see the second reflection of the back of your head:

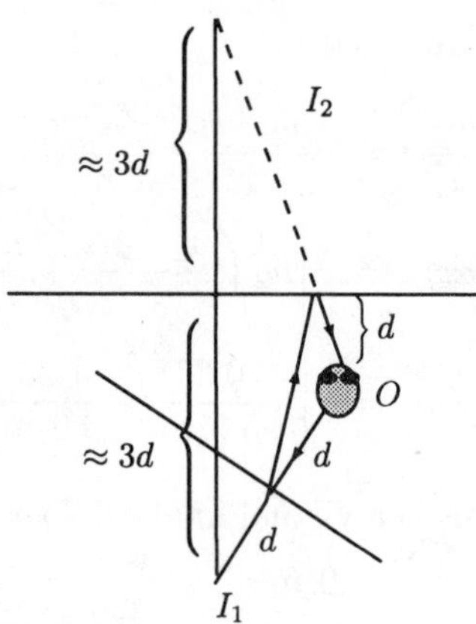

The image appears to be $\approx 4d$ away, where $2d$ is the separation of the 2 mirrors.

18.29 Use

$$\frac{x_0}{n_1} = \frac{-x_i}{n_2}$$

Here the glass thickness is $x_0 = 1.05$ cm, $n_1 = 1.53$. The image appears at coordinate

$$x_i = -\frac{1.05\text{ cm}}{1.53} = -0.686\text{ cm}$$

The image is virtual and is 0.686 cm below the top surface of the glass.

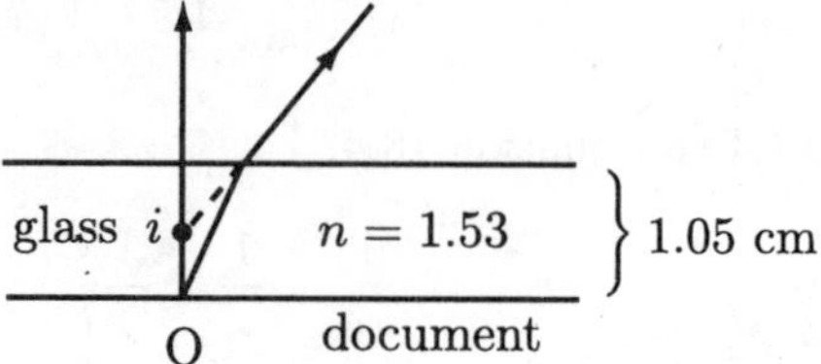

18.31 The angle of incidence ABD = the angle of reflection DBC.

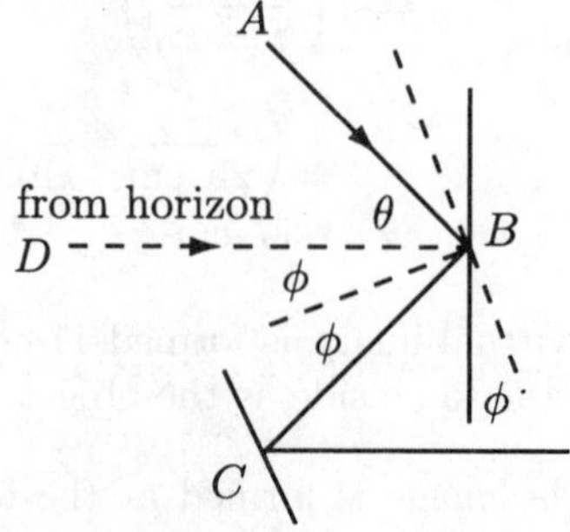

When viewing the horizon (mirror at dashed line) $\angle DBC = 2\phi$. Thus $\theta = 2\phi$.

18.35 No. A convex mirror always forms a virtual image and a concave mirror forms a virtual image if $x_0 < f$, as is the case here.

18.39

$$\frac{1}{x_0}+\frac{1}{x_i}=\frac{1}{f}=\frac{2}{r}$$

Here $x_0 = 1.0$ m, and $r = +0.50$ m

$$x_i=\left(\frac{2}{0.50\text{ m}}-\frac{1}{1.0\text{ m}}\right)^{-1}=0.33\text{ m}$$

A real, inverted image, reduced in size, is formed 0.33 m from the mirror.

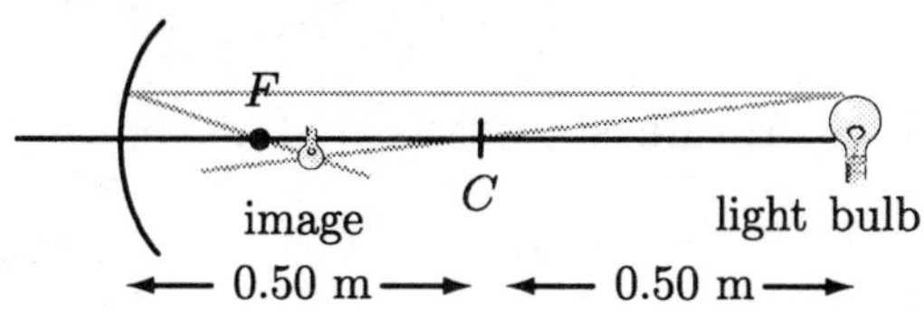

18.43 For a spherical refracting surface use Equation 18.6

$$\frac{n_1}{x_0}+\frac{n_2}{x_i}=\frac{n_2-n_1}{r}$$

Here $n_1 = 1.5, n_2 = 1.0, x_0 = 6.0$ cm and $r = -8.5$ cm. So the wick appears to be at

$$\begin{aligned} x_i &= n_2\left[\left(\frac{n_2-n_1}{r}-\frac{n_1}{x_0}\right)\right]^{-1} \\ &= 1.0\left[\frac{-0.5}{-8.5\text{ cm}}-\frac{1.5}{6.0\text{ cm}}\right]^{-1} \\ &= -5.2\text{ cm} \end{aligned}$$

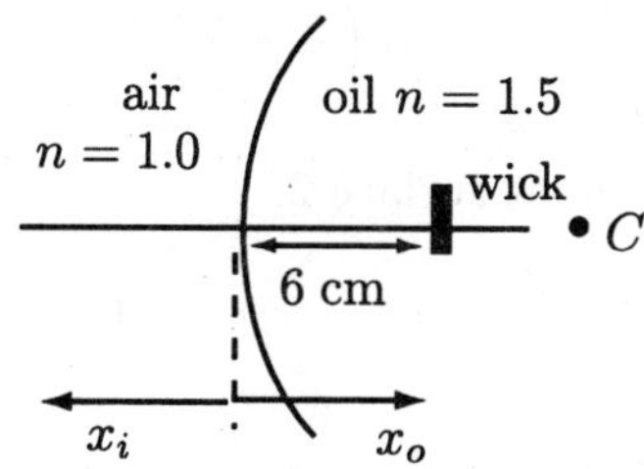

This is a virtual image, so the wick appears to be 5.2 cm behind the surface of the bottle.

18.47 $\frac{1}{x_0}+\frac{1}{x_i}=\frac{1}{f}$. Here $x_i = +2.00$ m and $f = +5.00$ cm, so

$$\begin{aligned} x_0 &= \left(\frac{1}{f}-\frac{1}{x_i}\right)^{-1} \\ &= \left(\frac{1}{5.00\text{ cm}}-\frac{1}{200\text{ cm}}\right)^{-1} \\ &= 5.13\text{ cm} \end{aligned}$$

Slide should be 5.13 cm behind the lens.

18.53 At the first surface the rays are bent to form the virtual image I_1 (Figure 18.20). The light rays are diverging as if from the image at I_1 when they meet the second surface. Thus, the *virtual* image at I_1 acts as a *real* object for the surface S_2. There is no physical thing at I_1, but the light rays behave as if there were. We use Equation 18.6 to describe how the rays bend at each surface. At the first surface $n_1 = 1.00$ (air), $n_2 = 1.50$ (glass) and the object distance $x_o = +0.50$ m. The radius of curvature is $r_1 = -1.0$ m. Equation 18.6 gives:

$$\begin{aligned} \frac{n_1}{x_o}+\frac{n_2}{x_i} &= \frac{n_2-n_1}{r} \\ &= \frac{1.00}{0.50\text{ m}}+\frac{1.50}{x_{i,1}} \\ &= \frac{(1.50-1.00)}{-1.0\text{ m}} \end{aligned}$$

$$\frac{1.50}{x_{i,1}}=-0.50\text{ /m}-2.0\text{ /m}=-2.5\text{ /m}$$

So,

$$x_{i,1}=-0.60\text{ m}.$$

The negative value of $x_{i,1}$ means that the image formed by the first surface is virtual and is 0.60 m from the surface along the optic axis. At the second surface $n_1 = 1.50$ (glass), $n_2 = 1.00$ (air), and the radius of curvature is $r_2 = -0.50$ m. The object position is

$$\begin{aligned} x_{o,2} &= \ell - x_{i,1} = 2.0\text{ cm}-(-60\text{ cm}) \\ &= +62\text{ cm} \end{aligned}$$

Using Equation 18.6 again, we have

$$\begin{aligned} \frac{1.50}{0.62\text{ m}}+\frac{1.00}{x_{i,2}} &= \frac{(1.00-1.50)}{-0.50\text{ m}} \\ \frac{1.00}{x_{i,2}} &= \frac{+1.0}{\text{m}}-\frac{1.50}{0.62\text{ m}} \\ x_{i,2} &= -0.70\text{ m} \end{aligned}$$

The image is virtual and is 0.70 m from the outside surface of the window. When you look through this window at the light bulb, it appears to be further away than it actually is. From the thin-lens approximation (Example 18.5) the image is 0.68 m from the center of the lens, 3% smaller.

18.55

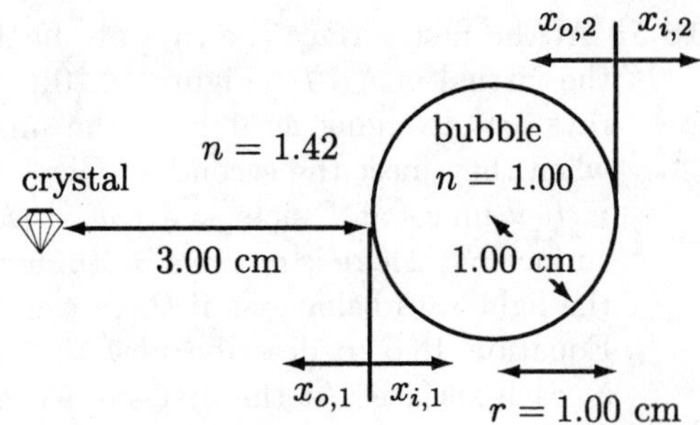

$$\frac{n_1}{x_{o,1}} + \frac{n_2}{x_{i,1}} = \frac{n_2 - n_1}{r_i}$$

with $r_i = +1.00$ cm

$$\frac{1.42}{3.00 \text{ cm}} + \frac{1.00}{x_{i,1}} = \frac{-0.42}{+1.00 \text{ cm}}$$

$$\begin{aligned} x_{i,1} &= (1.00)\left[\frac{-0.42}{1.00 \text{ cm}} - \frac{1.42}{3.00 \text{ cm}}\right]^{-1} \\ &= -1.12 \text{ cm} \end{aligned}$$

The first image is virtual. This images forms a real object for the second surface.

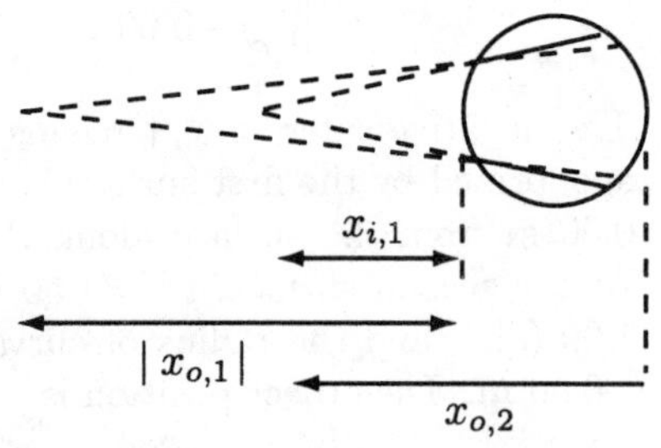

$$\begin{aligned} x_{o,2} &= |x_{i,1}| + 2r \\ &= 1.12 \text{ cm} + 2.00 \text{ cm} = 3.12 \text{ cm} \end{aligned}$$

At the second surface,

$$n_1 = 1.00, n_2 = 1.42, r_2 = -1.00 \text{ cm}$$

$$\begin{aligned} \frac{n_1}{x_{0,2}} + \frac{n_2}{x_{i,2}} &= \frac{n_2 - n_2}{r_2} \\ \Rightarrow \frac{1.00}{3.12 \text{ cm}} + \frac{1.42}{x_{i,2}} &= \frac{0.42}{-1.00 \text{ cm}} \end{aligned}$$

$$\begin{aligned} x_{i,2} &= -1.42\left[\frac{-0.42}{1.00 \text{ cm}} - \frac{1.00}{3.12 \text{ cm}}\right]^{-1} \\ &= -1.92 \text{ cm} \end{aligned}$$

The image is virtual and is located inside the bubble, 0.92 cm from the center, on the same side as the crystal. The bubble behaves like a diverging lens.

18.59 To find the focal points, consider an object at ∞.

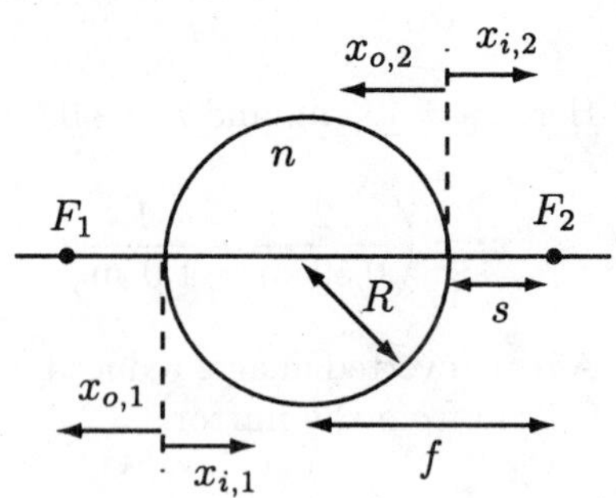

At surface 1,

$$x_{o,1} = \infty, \; n_1 = 1.00, \; n_2 = n, \; r = +R$$

$$\begin{aligned} \frac{n_1}{x_{o,1}} + \frac{n_2}{x_{i,1}} &= \frac{n_2 - n_1}{r} \\ 0 + \frac{n}{x_{i,1}} &= \frac{n-1}{R} \\ x_{i,1} &= R\left(\frac{n}{n-1}\right) > R \end{aligned}$$

Thus a real image is formed beyond the sphere if $n < 2$, as expected for glass. This image forms a virtual object for surface 2, with

$$\begin{aligned} x_{o,2} &= -(x_{i,1} - 2R) = R\left[\frac{2n - 2 - n}{n-1}\right] \\ &= R\left(\frac{n-2}{n-1}\right) \end{aligned}$$

Then, at surface 2,

$$n_1 = n, n_2 = 1, r = -R$$

$$\begin{aligned} \frac{n}{x_{o,2}} + \frac{1}{x_{i,2}} &= \frac{1-n}{-R} = \frac{n-1}{R} \\ \frac{n}{R}\left(\frac{n-1}{n-2}\right) + \frac{1}{x_{i,2}} &= \frac{n-1}{R} \\ \frac{1}{x_{i,2}} &= \left(\frac{n-1}{R}\right)\left(\frac{1-n}{n-2}\right) \\ &= \left(\frac{n-1}{R}\right)\left(\frac{-2}{n-2}\right) \\ x_{i,2} &= -\frac{R}{2}\left(\frac{n-2}{n-1}\right) \end{aligned}$$

if $1 < n < 2$

$$x_{i,2} = \frac{R}{2}\left(\frac{2-n}{n-1}\right) \equiv s$$

is > 0, and the image is real. The distance of F from the center of the sphere is

$$\begin{aligned} f &= R+s=R\left[1+\frac{1}{2}\frac{(2-n)}{(n-1)}\right] \\ &= \frac{R\,[2n-2+2-n]}{2(n-1)} = \frac{Rn}{2(n-1)} \end{aligned}$$

Because the sphere is symmetric, this result holds for both focal points.

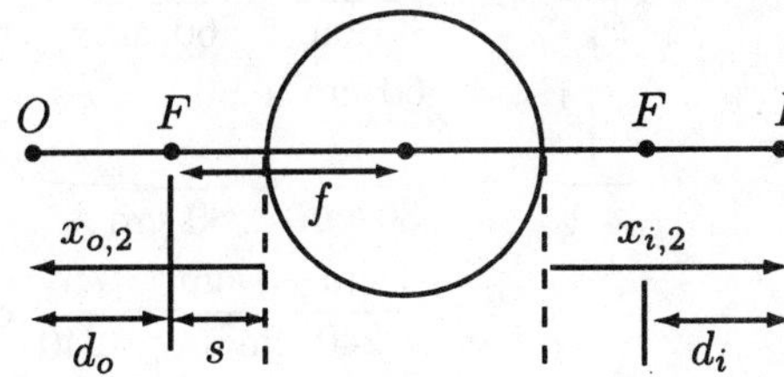

Repeating the calculation for an object at a finite distance from the sphere: At surface 1:

$$\begin{aligned} \frac{1}{x_{o,1}}+\frac{n}{x_{i,1}} &= \frac{n-1}{R} \\ \Rightarrow \frac{n}{x_{i,1}} &= \frac{n-1}{R}-\frac{1}{x_{o,1}} \\ &= \frac{(n-1)x_{o,1}-R}{Rx_{o,1}} \\ \frac{x_{i,1}}{n} &= \frac{Rx_{o,1}}{(n-1)x_{o,1}-R} \qquad (1) \end{aligned}$$

At surface 2

$$\frac{n}{x_{o,2}}+\frac{1}{x_{i,2}}=\frac{1-n}{-R}=\frac{n-1}{R}$$

The first image becomes the object for the second surface, with coordinate $x_{o,2}=2R-x_{i,1}$

$$\begin{aligned} \frac{n}{2R-x_{i,1}}+\frac{1}{x_{i,2}} &= \frac{n-1}{R} \\ \Rightarrow \frac{n}{2R-x_{i,1}} &= \frac{n-1}{R}-\frac{1}{x_{i,2}} \\ \frac{2R-x_{i,1}}{n} &= \frac{Rx_{i,2}}{(n-1)x_{i,2}-R} \qquad (2) \end{aligned}$$

Adding Eqns. (1) and (2), we find

$$\frac{2R}{n}=\frac{Rx_{o,1}}{(n-1)x_{o,1}-R}+\frac{Rx_{i,2}}{(n-1)x_{i,2}-R}$$

$$\begin{aligned} &\frac{2}{n}\left[\{(n-1)x_{o,1}-R\}\{(n-1)x_{i2}-R\}\right] \\ &= x_{o,1}\left[(n-1)x_{i,2}-R\right]+x_{i,2}\left[(n-1)x_{o,1}-R\right] \end{aligned}$$

$$\begin{aligned} &\frac{2}{n}\left[(n-1)^2x_{o,1}x_{i,2}-R(n-1)(x_{o,1}+x_{i2})+R^2\right] \\ &= 2(n-1)x_{o,1}x_{i,2}-R(x_{o,1}+x_{i,2}) \end{aligned}$$

$$\begin{aligned} &2(n-1)(x_{o,1}x_{i,2})\left[\tfrac{n-1}{n}-1\right] \\ &-\left[2R\tfrac{(n-1)}{n}-R\right](x_{o,1}+x_{i,2}) \\ &+\frac{2R^2}{n}=0 \end{aligned}$$

$$\begin{aligned} &2(n-1)(x_{o,1}x_{i,2})\left(-\tfrac{1}{n}\right) \\ &-R\left[\frac{2n-2-n}{n}\right](x_{o,1}+x_{i,2}) \\ &+\frac{2R^2}{n}=0 \end{aligned}$$

Now

$$x_{o,1}=d_o+s,\, x_{i2}=d_i+s$$

$$\begin{aligned} &-2\left(\tfrac{n-1}{n}\right)(d_o+s)(d_i+s) \\ &-R\left(\tfrac{n-2}{n}\right)(d_o+d_i+2s) \\ &+\tfrac{2R^2}{n}=0 \end{aligned}$$

$$\begin{aligned} &-2\left(\tfrac{n-1}{\not n}\right)(d_od_i+s(d_o+d_i)+s^2) \\ &-R\left(\tfrac{n-2}{\not n}\right)(d_o+d_i+2s) \\ &+\tfrac{2R^2}{\not n}=0 \end{aligned}$$

$$\begin{aligned} &d_od_i+(d_o+d_i)\left(s+\tfrac{R}{2}\left(\tfrac{n-2}{n-1}\right)\right) \\ &+s^2+\tfrac{sR(n-2)}{n-1}-\tfrac{R^2}{n-1}=0 \end{aligned}$$

$$\begin{aligned} d_od_i &= \frac{R^2}{n-1}-\frac{sR(n-2)}{n-1}-s^2 \\ &= \frac{R^2}{n-1}+\left(\frac{R}{n-1}\right)\frac{R}{2}\left(\frac{(2-n)^2}{(n-1)}\right) \\ &\quad -\left(\frac{R}{2}\right)^2\left(\frac{2-n}{n-1}\right)^2 \end{aligned}$$

$$\begin{aligned} &= \tfrac{R^2}{[2(n-1)]^2}\cdot\left[4(n-1)+2(2-n)^2-(4-4n+n^2)\right] \\ &= \tfrac{R^2}{[2(n-1)]^2}\cdot\left[4(n-1)+(4-4n+n^2)\right] \\ &= \tfrac{R^2n^2}{[2(n-1)]^2}=f^2 \end{aligned}$$

Thus $d_od_i=f^2$, as required.

18.63 For a convex mirror, $R<0$, so

$$\frac{1}{x_O}+\frac{1}{x_i}=\frac{2}{R}=-\frac{2}{|R|}$$

Solving for x_i:

$$\begin{aligned} \frac{1}{x_i} &= -\frac{2}{|R|}-\frac{1}{x_O} \\ &= \frac{-2x_o-|R|}{|R|\,x_o} \\ x_i &= \frac{-|R|\,x_o}{|R|+2x_o} \end{aligned}$$

Then the magnification is:

$$m = -\frac{x_i}{x_O} = \frac{|R|}{|R| + 2x_o} = \frac{1}{1 + \frac{2x_o}{|R|}}$$

which is always less then 1 for real objects $(x_o > 0)$.

18.67 For a lens in air, use Equation 18.10:

$$\frac{1}{f} = (n-1)\left(\frac{1}{r_1} - \frac{1}{r_2}\right)$$

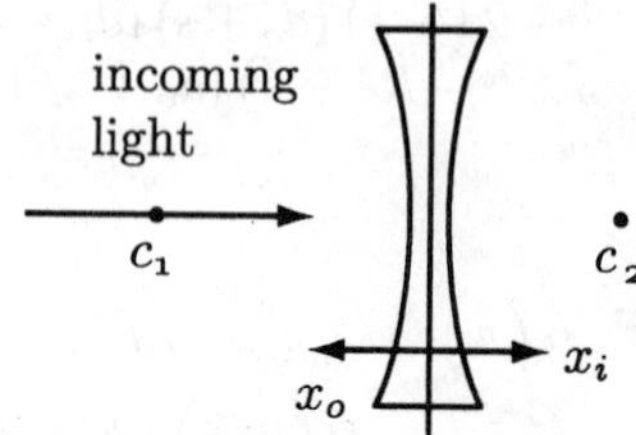

Here $n = 1.50$ and $r_1 = -8.0$ cm, $r_2 = +16$ cm (see figure). So

$$f = \left[0.50\left(-\frac{1}{8.0 \text{ cm}} - \frac{1}{16 \text{ cm}}\right)\right]^{-1} = \left(-\frac{1}{2}\left(\frac{3.0}{16 \text{ cm}}\right)\right)^{-1} = -\frac{32}{3} \text{ cm} = -11 \text{ cm}$$

For an object 16 cm from lens, $x_O = 16$ cm

$$x_i = \left(\frac{1}{f} - \frac{1}{x_O}\right)^{-1} = \left(-\frac{3.0}{32 \text{ cm}} - \frac{1}{16}\right)^{-1} = -\frac{32 \text{ cm}}{5} = -6.4 \text{ cm}$$

$$m = -\frac{h_i}{h_O} = -\frac{x_i}{x_O} = \frac{\frac{-32}{5} \text{ cm}}{16 \text{ cm}} = -\frac{2}{5} = -0.40$$

This is a virtual image, since $x_i < 0$.

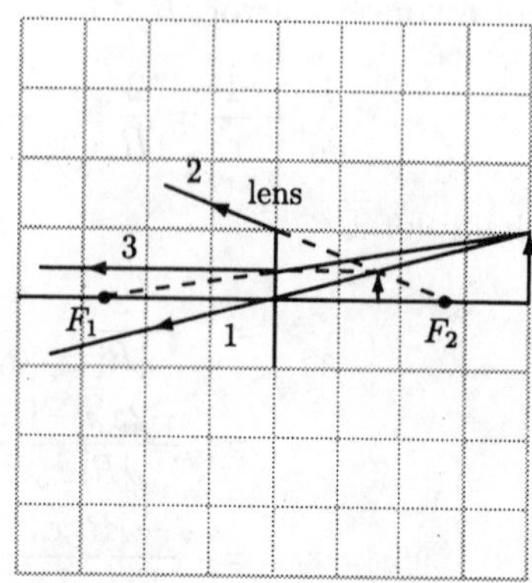

Horizontal scale: 1 square = 4 cm
Vertical scale: not needed.
Image location $x_i = -1.6$ squares $= -6.4$ cm
Image is virtual and 0.4 the size of the object.

18.71 a) object length = 3.3 cm
magnification $= \frac{2.3}{3.3} = 0.70$
image length = 2.3 cm. See figure in Appendix.

b)

$$\frac{1}{x_{i,1}} = \frac{1}{30 \text{ cm}} - \frac{1}{60 \text{ cm}} = \frac{1}{60 \text{ cm}}$$

$$x_{i,1} = 60 \text{ cm}$$

$$\frac{1}{x_{i,2}} = \frac{1}{30 \text{ cm}} - \frac{1}{80 \text{ cm}} = \frac{8 \text{ cm} - 3 \text{ cm}}{240 \text{ cm}^2} = \frac{5}{240} \text{ cm}^{-1}$$

$$\Rightarrow x_{i,2} = \frac{240 \text{ cm}}{5} = 48 \text{ cm}$$

$$m_{i,1} = \frac{-x_{i,1}}{x_{o,1}} = -1 \Rightarrow y_{i,1} = -10 \text{ cm}$$

$$m_{i,2} = \frac{-x_{i,2}}{x_o} = \frac{-240 \text{ cm}}{5(80 \text{ cm})} = -\frac{3}{5}$$

$$\Rightarrow y_{i,1} = -6 \text{ cm}$$

Thus

$$\ell_i^2 = (x_{i,1} - x_{i,2})^2 + (y_{i,1} - y_{i,2})^2 = (60 - 48)^2 + (10 - 6)^2 \text{ cm}^2 = (122 + 42) \text{ cm}^2 = (144 + 16) \text{ cm}^2 = 160 \text{ cm}^2$$

So

$$\ell_i = 12.65 \text{ cm}$$

Thus

$$m = \frac{\ell_i}{\ell_o} = \frac{12.65 \text{ cm}}{20 \text{ cm}} = 0.63$$

This is consistent with the figure, given the accuracy with which we can measure.

c) For an arrow on axis, the image length would be

$$\ell_i = x_{i,1} - x_{i,2} = 12 \text{ cm}$$

and the magnification would be $\frac{12 \text{ cm}}{20 \text{ cm}} = 0.60$ (slightly less).

18.73 Focal length has to be the same; f is not affected by flipping lens over. Each lens is converging, and so is the combination. The combination's focal length is half that of either lens alone.

18.77 1) Converging lens, first:
Since object is at focal point of converging lens;

$$x_i = \left(\frac{1}{f} - \frac{1}{x_o}\right)^{-1} \to \infty$$

Diverging lens then "sees" object at $x_o = \infty$. So

$$x_i = \left(-\frac{1}{5} - \frac{1}{\infty}\right)^{-1} = -5.0 \text{ cm}$$

virtual image at position of 1st lens

2) Diverging lens first:

$$x_i = \left(\frac{1}{f} - \frac{1}{x_o}\right)^{-1} = \left(-\frac{1}{5} - \frac{1}{5}\right)^{-1} = -\frac{5}{2} \text{ cm}$$

This is object viewed by converging lens so object distance is

$$\begin{aligned} x_o &= 5 \text{ cm} + \frac{5}{2} \text{ cm} = \frac{15}{2} \text{ cm} \\ x_i &= \left(\frac{1}{c} - \frac{1}{x_o}\right)^{-1} \\ &= \left(\frac{1}{+5} - \frac{2}{15}\right)^{-1} \text{ cm} \\ &= +15 \text{ cm} \end{aligned}$$

$\therefore$ real image 15 cm beyond second lens.

18.81 1) The *first lens* produces image at $x_i = \infty$ since object is at its focal point.

2) The *second lens* has $x_o = \infty$, so image is produced 5.0 cm from lens, at its focal point. The image is inverted, and has the same size as original image: $m = -1$.

3) At the third lens

$$x_o = 15 \text{ cm} - 5 \text{ cm} = 10 \text{ cm},$$

so

$$\begin{aligned} x_i &= \left(\frac{1}{f} - \frac{1}{x_o}\right)^{-1} \\ &= \left(\frac{1}{5.0 \text{ cm}} - \frac{1}{10 \text{ cm}}\right)^{-1} \\ &= 10 \text{ cm} \\ m &= -\frac{h_i}{h_O} = -\frac{x_i}{x_o} = \frac{-10}{10} = -1.0 \end{aligned}$$

So *net* magnification is $m_{net} = +1.0$

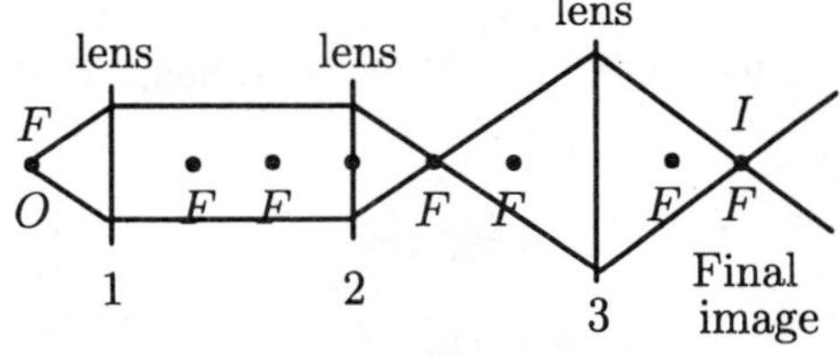

18.85 a) $x_o = \dfrac{3f}{4}$; $f > 0$ both lenses
1st lens:

$$x_i = \left(\frac{1}{f} - \frac{4}{3f}\right)^{-1} = -3.0f$$

Virtual image $-3f$ from lens 1
2nd lens:

$$x_o = 2f + 3f = 5f,$$

so

$$x_i = \left(\frac{1}{f} - \frac{1}{5f}\right)^{-1} = 1.25f = \frac{5f}{4}$$

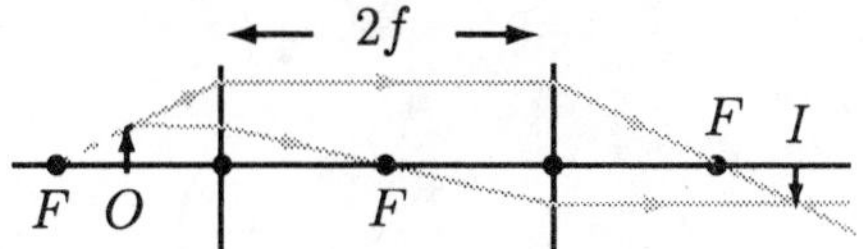

b) $x_o = f, f > 0$ both lenses$f$
lens 1:

$$x_i = \left(\frac{1}{f} - \frac{1}{f}\right)^{-1} = \infty$$

lens 2:

$$x_i = \left(\frac{1}{f} - \frac{1}{\infty}\right)^{-1} = f$$

c) $x_o = 1.5f,\ f > 0$ both lenses
lens 1:

$$x_i = \left(\frac{1}{f} - \frac{1}{x_o}\right)^{-1} = f\left(1 - \frac{1}{1.5}\right)^{-1} = 3f$$

lens 2:

$$x_o = 2f - 3f = -f,$$

so

$$x_i = \left(\frac{1}{f} - \frac{1}{-f}\right)^{-1} = +\frac{f}{2}$$

d) $x_o = 1.5f, f_1 = +f, f_2 = -f$
lens1:

$$x_i = \left(\frac{1}{f} - \frac{1}{1.5f}\right)^{-1} = +3.0f$$

lens 2:

$$x_o = -f,$$

so

$$x_i = \left(-\frac{1}{f} - \frac{1}{-f}\right)^{-1} = \infty$$

Image formed at ∞.

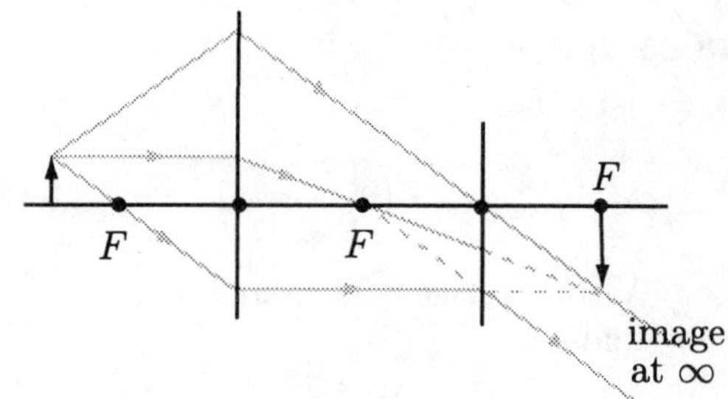

e) $x_o = f, f_1 = f, f_2 = -f$
lens 1:

$$x_i = \left(\frac{1}{f} - \frac{1}{f}\right)^{-1} = \infty$$

lens 2:

$$x_i = \left(-\frac{1}{f} - \frac{1}{\infty}\right)^{-1} = -f$$

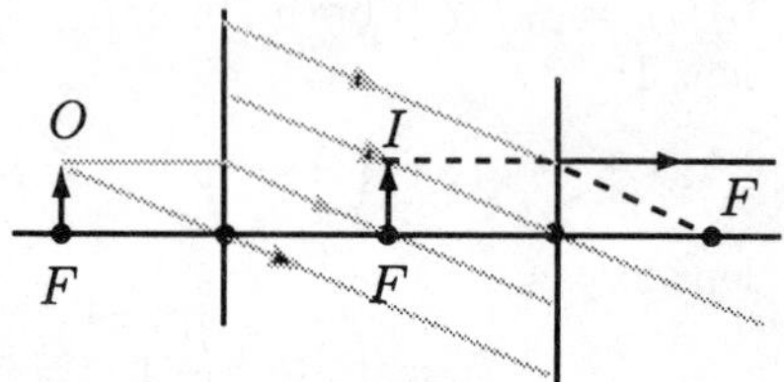

f) $x_o = \frac{3f}{4} f_1 = f, f_2 = -f$
lens 1:

$$x_i = \left(\frac{1}{f} - \frac{4}{3f}\right)^{-1} = -3.0f$$

lens 2:

$$x_o = 2f + 3f = 5f$$

$$x_i = \left(-\frac{1}{f} - \frac{1}{5f}\right)^{-1} = -\frac{5}{6}f$$

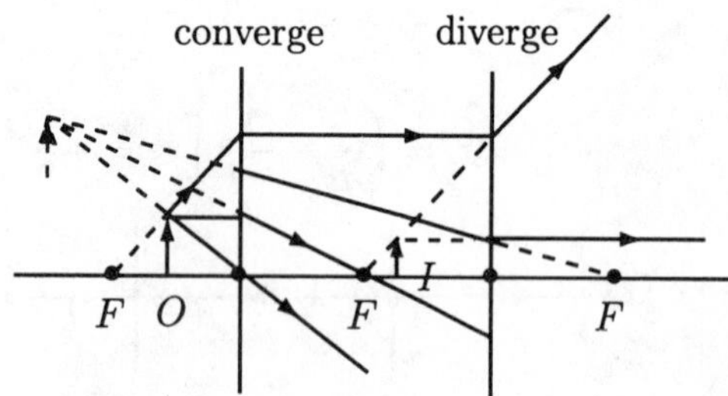

g) $x_o = 1.5f, f_1 = -f, f_2 = +f$
lens 1:

$$x_i = \left(\frac{1}{-f} - \frac{1}{1.5f}\right)^{-1} = -\frac{3}{5}f$$

lens 2:

$$x_o = 2f + \frac{3}{5}f = \frac{13}{5}f$$

$$x_i = \left(\frac{1}{f} - \frac{5}{13f}\right)^{-1} = \frac{13}{8}f$$

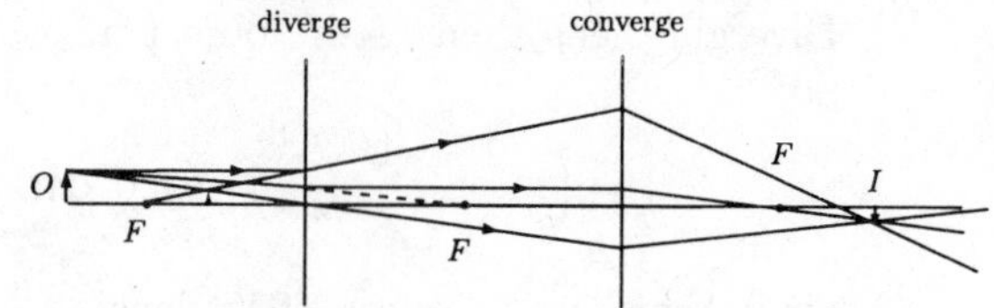

h) $x_o = f, f_1 = -f, f_2 = +f$
lens 1:

$$x_i = \left(-\frac{1}{f} - \frac{1}{f}\right)^{-1} = -\frac{f}{2}$$

lens 2:

$$x_o = 2f + \frac{f}{2} = \frac{5f}{2}$$

$$x_i = \left(\frac{1}{f} - \frac{2}{5f}\right)^{-1} = \frac{5}{3}f$$

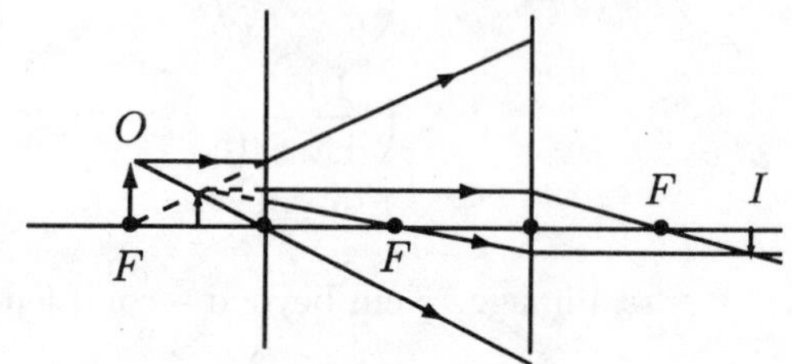

i) $x_o = \frac{3}{4}f, f_1 = -f, f_2 = +f$
lens 1:

$$x_i = \left(-\frac{1}{f} - \frac{4}{3f}\right)^{-1} = -\frac{3}{7}f$$

lens 2:

$$x_o = 2f + \frac{3}{7}f = \frac{17}{7}f$$

$$x_i = \left(\frac{1}{f} - \frac{7}{17f}\right)^{-1} = 1.7f$$

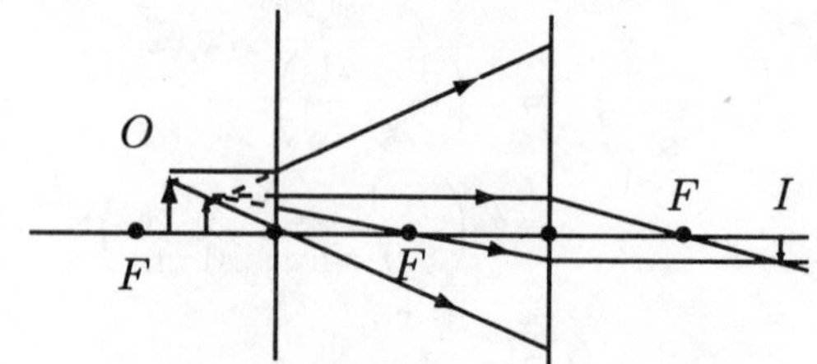

18.89 Use the lens makers equation,

$$\frac{1}{f} = (n-1)\left(\frac{1}{r_1} - \frac{1}{r_2}\right)$$

and find f for the two wavelengths.
500 nm:

$$\begin{aligned} f &= \left[(0.500)\left(\frac{1}{10.0 \text{ cm}} + \frac{1}{10.0 \text{ cm}}\right)\right]^{-1} \\ &= 10.0 \text{ cm} \end{aligned}$$

700 nm:

$$f = \left[(0.470) \left(\frac{2}{10.0 \text{ cm}} \right) \right]^{-1} = 10.6 \text{ cm}$$

Now use

$$\frac{1}{f} = \frac{1}{x_o} + \frac{1}{x_i}$$

to find x_i for these 2 focal lengths using $x_o = 3.0$ m

500 nm:

$$\begin{aligned} x_i &= \left(\frac{1}{f} - \frac{1}{x_o} \right)^{-1} \\ &= \left(\frac{1}{0.100 \text{ m}} - \frac{1}{3.0 \text{ m}} \right)^{-1} \\ &= 10.34 \text{ cm} \end{aligned}$$

700 nm:

$$\begin{aligned} x_i &= \left(\frac{1}{0.106 \text{ m}} - \frac{1}{3.0 \text{ m}} \right)^{-1} = 10.99 \text{ cm} \\ \Delta x_i &= 0.6 \text{ cm} \end{aligned}$$

18.93 Wave fronts propagate outward from the object, their curvature is altered as they propagate through the lens so that they converge onto the image, as sketched in the figure.

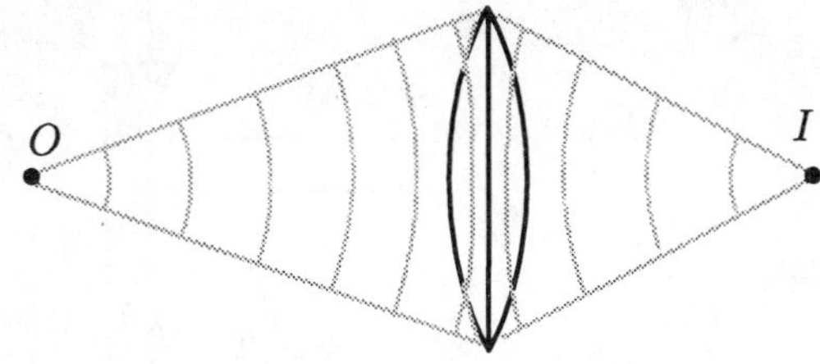

At any time there will be some number, N, of wavefronts between object and image. This number is fixed in time, as the frequency of the wave is constant and one wavefront arrives at I for every wave front emitted at O. Thus, any ray will have to cross this fixed number of wavefronts on its path from O to I through the lens. The travel time of light along any path through the lens from O to I is the same—the time for N new wavefronts to be emitted to replace the N that will pass through I as the light energy (moving at the same speed as the wave front it follows) moves from O to I. The wavefronts are closer together within the lens ($\lambda_{\text{lens}} = \frac{\lambda}{n}$, cf. Chapter 16) because they travel slower within the glass.

18.97 The mirror should be at 45°.

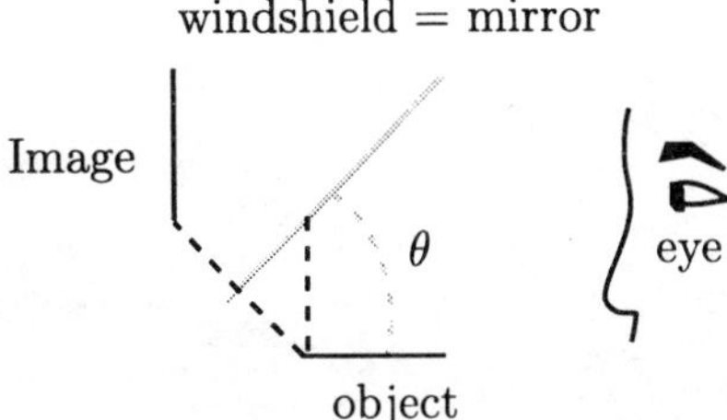

If the image is to be at ∞, a parallel light beam should be projected. This can be achieved by placing the instrument at the focal point of a convex lens.

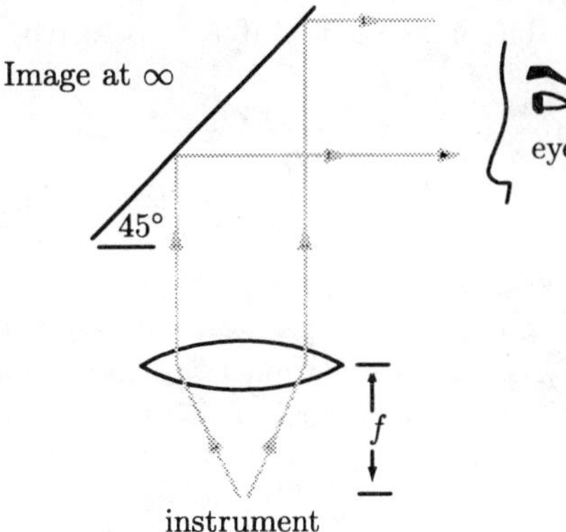

18.99 Use the lens-makers equation:

$$\frac{1}{f} = \left(\frac{n_\ell - n_m}{n_m} \right) \left(\frac{1}{r_1} - \frac{1}{r_2} \right)$$

$n_\ell = n$ of lens
$n_m = n$ of medium (air or water here)
air ($n_{\text{air}} = 1.00$):

$$\frac{1}{10.0 \text{ cm}} = \left(\frac{1}{r_1} - \frac{1}{r_2} \right) \left(\frac{n_\ell}{n_{\text{air}}} - 1 \right)$$

water ($n_{\text{water}} = 1.33$):

$$\frac{1}{f} = \left(\frac{1}{r_1} - \frac{1}{r_2} \right) \left(\frac{n_\ell}{n_{\text{water}}} - 1 \right)$$

$$\begin{aligned} \frac{f_{\text{water}}}{f_{\text{air}}} &= \frac{\left(\frac{n_\ell}{n_{\text{air}}} - 1 \right)}{\left(\frac{n_\ell}{n_{\text{water}}} - 1 \right)} = \frac{\frac{1.55}{1} - 1}{\frac{1.55}{1.33} - 1} = 3.33; \\ f_{\text{water}} &= 3.33\,(10 \text{ cm}) = 33 \text{ cm} \end{aligned}$$

The result is the same independent of the values of r_1 and r_2, and thus independent of the nature of the lens (converging or diverging).

18.103 The image is formed at x_i where

$$\frac{n_1}{x_o} + \frac{n_2}{x_i} = \frac{n_2 - n_1}{r}$$

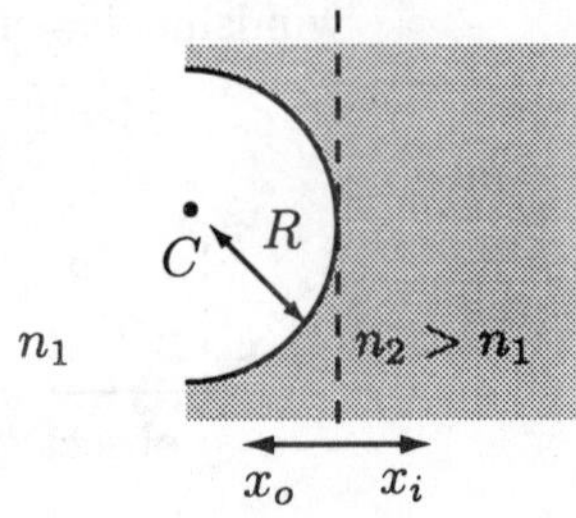

In this case, r is negative: $r = -|R|$

$$\frac{n_2}{x_i} = \frac{n_2 - n_1}{-|R|} - \frac{n_1}{x_o}$$

The image is real if x_i is positive, ie

$$\frac{n_1 - n_2}{|R|} - \frac{n_1}{x_o} > 0$$

$$\frac{n_1 - n_2}{|R|} > \frac{n_1}{x_o}$$

The LHS is negative, since $n_2 > n_1$. Thus the right-hand side must also be negative: the object must be virtual. Then we must also have

$$\left|\frac{n_1}{x_o}\right| > \left|\frac{n_1 - n_2}{R}\right|$$

or

$$|x_o| < \frac{Rn_1}{(n_2 - n_1)}$$

The object must be virtual and closer to the surface than $\dfrac{Rn_1}{(n_2 - n_1)}$.

18.111 From Example 18.12 we know that the radius of curvature of the secondary mirror is -18.3 m, in order to focus starlight 1.00 m behind the primary.

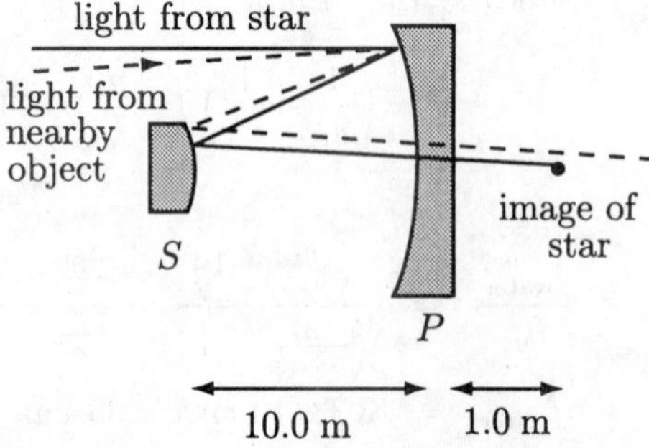

Light from a nearby object (dashed line) is focused more than 1 m behind the primary, suggesting that the secondary mirror needs to be moved farther from the primary. Assuming light from the nearby object is focused 1.00 m behind the primary when the secondary is moved backward its full 2 cm travel ($x_{i,s} = 11.02$ m) we can locate the object for the secondary (= image formed by the primary) and then locate the object. For the image formed by the secondary.

$$\frac{1}{x_{o,s}} + \frac{1}{x_{i},s} = \frac{1}{f_s}$$

or

$$x_{o,s} = \frac{1}{\frac{1}{f_s} - \frac{1}{x_{i,s}}}$$

$$f_s = -9.15 \text{ m}$$

So

$$\begin{aligned} x_{o,s} &= \frac{f_s}{1 - \frac{f_s}{x_{i,s}}} \\ &= \frac{-9.15 \text{ m}}{1 - \frac{-9.15 \text{ m}}{11.02 \text{ m}}} = -4.999 \text{ m} \end{aligned}$$

(Remember to keep an extra sig. figure in intermediate steps.) For the image formed by the primary.

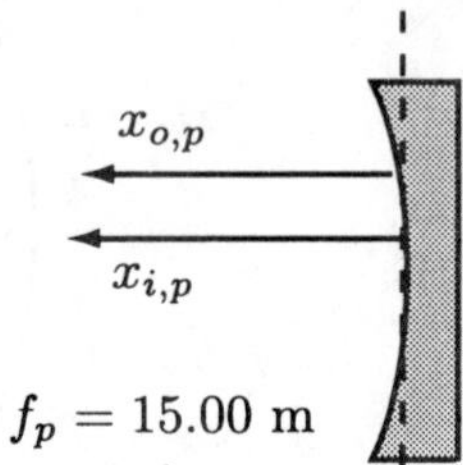

$$\begin{aligned} \frac{1}{x_{o,p}} + \frac{1}{x_{i,p}} &= \frac{1}{f_p} \\ \Rightarrow x_{o,p} &= \frac{f_p}{1 - \frac{f_p}{x_{i,p}}} \\ &= \frac{15.00 \text{ m}}{1 - \frac{15.00 \text{ m}}{(10.00 \text{ m} + 0.02 \text{ m} + 4.999 \text{ m})}} \\ &= \frac{15.00 \text{ m}}{1 - \frac{15.00 \text{ m}}{15.02 \text{ m}}} = 10 \text{ km} \end{aligned}$$

(Only one sig. figure survives the subtraction.)

Chapter 19

Temperature and Thermal Energy

19.1 Systems A and B are in thermal equilibrium by the Zero-th Law. The length of the mercury column correlates directly with temperature. Systems C & D are not in thermal equilibrium since they do not have the same temperature.

19.3 Ethyl boils at 78.5° C. Since K $= C + 273.15°$, Ethyl Alcohol boils at

$$T(\text{boiling}) = 78.5°C + 273.15° = 351.7 \text{ K}$$

Since

$$F = \frac{9}{5}C + 32$$

T(boiling) $= \frac{9}{5}(78.5° \text{ C}) + 32 = 173°$ F

19.5 $M = Nm = 2.5 \times 10^{-5}$ kg
$P = 1 \times 10^4$ Pa
$T = -80°$ C $= (273.15 - 80)°$ K $= 193.15$ K

$$V = 1.0 \text{ L}$$

a) Ideal Gas Law:

$$\begin{aligned} PV &= \mathcal{N}RT \\ \mathcal{N} &= \frac{PV}{RT} \\ &= \frac{(10^4 \text{ Pa})\,(1.0 \text{ L})}{(8.314 \text{ J/ mol}\cdot\text{K})(193.15° \text{ K})} \\ &= \frac{(10^4 \text{ N}\cdot\text{m}^2)(1.0 \text{ L})(10^{-3} \text{ m}^3/\text{L})}{(8.314 \text{ N})(193.15 \text{ K})} \\ \mathcal{N} &= 6.2 \times 10^{-3} \text{ mol} \end{aligned}$$

b)

$$\begin{aligned} GMW &= \frac{M}{\mathcal{N}} \\ &= \frac{2.5 \times 10^{-5} \text{ kg}}{6.2 \times 10^{-3} \text{ mol}} \times \frac{1000 \text{ g}}{\text{kg}} \\ &= 4 \text{ g} \end{aligned}$$

c) Since GMW $= 4$ g, the gas is (monatomic) helium.

19.7 Since the surroundings insulate, there is no heat flow into or out of the system, i.e,

$$Q = 0$$

So $\Delta U = Q - W$ (1st Law) becomes $\Delta U = -W$. The temperature of the nail indicates internal energy. So $U \uparrow$ and therefore work is done on the system as the nail is hammered.

19.9 **a)** 30 mol, 310 K, 1.0 atm

$$\begin{aligned} \frac{P_1V_1}{\mathcal{N}_1T_1} &= \frac{P_2V_2}{\mathcal{N}_2T_2} \\ PV &= \mathcal{N}RT \end{aligned}$$

when $R = 0.082 \frac{\text{L}\cdot\text{atm}}{\text{mol}\cdot\text{K}}$

$$\begin{aligned} V_a &= \frac{\mathcal{N}RT}{P} \\ V &= 76 \text{ L} \end{aligned}$$

b) $V_2 = 2.0 \times 10^{-3}$ m^3 $= 2.0$ L
$P_2 = 1.0$ atm

c) $V_2 = 1.0$ L
$\mathcal{N}_2 = 2.0$ mol
$T_2 = 310$ K

$$P_2 = \frac{\mathcal{N}RT}{V}$$

$$
\begin{aligned}
&= \frac{(0.082\ \frac{\mathrm{L\cdot atm}}{\mathrm{mol\cdot K}})}{(1.0\ \mathrm{L})} \times 2.0\ \mathrm{mol} \times 310\ \mathrm{K} \\
&= 51\ \mathrm{atm}
\end{aligned}
$$

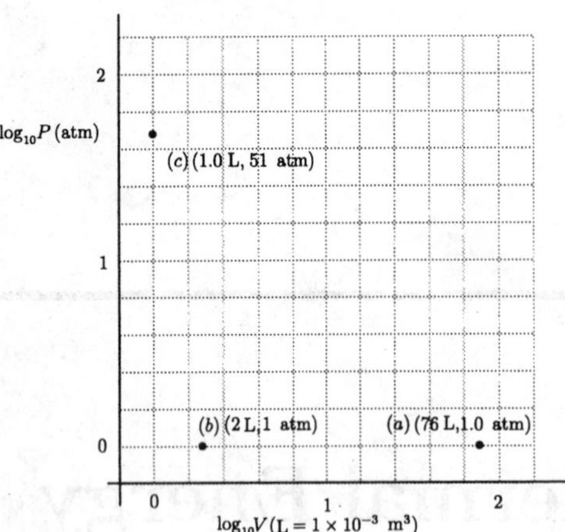

19.11 Use ideal gas model for neon (GMM = 20.2 g / mol) then

$$
\begin{aligned}
c_v &= \frac{3}{2}\frac{k}{m} \\
&= \frac{3}{2}\frac{(1.38 \times 10^{-23}\ \mathrm{J/K})}{(20.2 \times 1.66 \times 10^{-27}\ \mathrm{kg})} \\
c_v &= 617\ \mathrm{J/K \cdot kg} \\
c_p &= \frac{5}{2}\frac{k}{m} = \gamma c_v = \frac{5}{3} c_v \\
c_p &= 1030\ \mathrm{J/\,kg \cdot K}
\end{aligned}
$$

Measured value of 1046 J/kg · K agrees well, ≈ 1.3% error.

19.13 Process is adiabatic, ideal gas

$$
\begin{aligned}
P_1 V_1^\gamma &= P_2 V_2^\gamma \\
P_2 &= P_1 \left(\frac{V_1}{V_2}\right)^\gamma
\end{aligned}
$$

monatomic ideal gas $\Rightarrow \gamma = \frac{5}{3}$

$$
\begin{aligned}
P_1 &= 2.0\ \mathrm{atm} \\
V_1 &= 0.25\ \mathrm{m}^3 \\
V_2 &= 0.35\ \mathrm{m}^3 \\
P_2 &= (2.0\ \mathrm{atm}) \left(\frac{0.25\ \mathrm{m}^3}{0.35\ \mathrm{m}^3}\right)^{\frac{5}{3}} \\
&= 1.1\ \mathrm{atm}
\end{aligned}
$$

19.15 $\gamma = \frac{5}{3}$ at 75 K the gas is diatomic. But $\gamma = \frac{5}{3}$ corresponds to an ideal gas with only translational modes. At room temperature, we expect that the rotational modes (2) are excited. Equipartition of energy $\Rightarrow \frac{1}{2}R$ per mode. So

$$
\begin{aligned}
c_v' &= \frac{3}{2}R + \frac{R}{2} + \frac{R}{2} = \frac{5}{2}R \\
c_p' &= c_v + R = \frac{7}{2}R
\end{aligned}
$$

So

$$
\gamma = \frac{c_p'}{c_v'} = \frac{\frac{7}{2}}{\frac{5}{2}} = \frac{7}{5}
$$

19.17 Ideal, diatomic gas. Assume no vibrations are excited ⇒ 3 translational modes + 2 rotational modes = 5 modes

$$
\begin{aligned}
c_v &= \frac{5}{2}\frac{k}{m} \\
m &= 28 \times 1.67 \times 10^{-27}\ \mathrm{kg} \\
\Rightarrow c_v &= \frac{(5)(1.38 \times 10^{-23}\ \mathrm{J/K})}{2 \times 28 \times 1.67 \times 10^{-27}\ \mathrm{kg}} \\
&= 740\ \mathrm{J/\,kg \cdot K}
\end{aligned}
$$

19.19 a) Yes. Any heat entering the freezer is removed. The time scale is months to years.

b) Yes. Time scale is on the order of hours.

c) No. The body is in thermal contact with the hot sand and the cool air, and absorbs heat from the sun.

d) Yes. Minutes time scale. Air has low thermal conductivity.

19.23 Mercury freezes at −38.87° C.

$$
F = \frac{9}{5}C + 32^\circ\ \mathrm{F}
$$

$$
\begin{aligned}
T_{\mathrm{Freeze}} &= \left(\frac{9}{5}\right)(-38.87) + 32\ ^\circ\mathrm{F} \\
&= -37.97^\circ\ \mathrm{F} \\
&= {}^\circ\mathrm{C} + 273.15\ \mathrm{K} \\
&= -38.87 + 273.15\ \mathrm{K} \\
&= 234.28^\circ\ \mathrm{K}
\end{aligned}
$$

19.25 By definition of the Fahrenheit scale, freezing saturated brine is at 0° F. Melting pure ice is 32° F at atmospheric pressure. Boiling water is 212° F. Assume a linear variation in length with temperature

$$
\begin{aligned}
L &= 1.0\ \mathrm{cm} + \frac{10.0\ \mathrm{cm} - 1.0\ \mathrm{cm}}{212^\circ\ \mathrm{F} - 0^\circ\ \mathrm{F}} \times \mathrm{F} \\
&= 1.0\ \mathrm{cm} + 4.25 \times 10^{-2}\ \frac{\mathrm{cm}}{^\circ\mathrm{F}} \times \mathrm{F}
\end{aligned}
$$

For $F = 32$,

$$
\begin{aligned}
L &= 1.0\ \mathrm{cm} + 4.25 \times 10^{-2}\ \frac{\mathrm{cm}}{^\circ\mathrm{F}} \times 32^\circ\ \mathrm{F} \\
&= 2.4\ \mathrm{cm}
\end{aligned}
$$

19.27 Two gases of different molecular masses, same temperature. Model the gas as ideal. When mixed, the resulting mixture will still have the same temperature, say T. Let the masses be $m_1 + m_2$ with $m_1 > m_2$. Now

$$K_1 = \frac{1}{2}m_1\langle v_1^2\rangle = \frac{3}{2}kT$$
$$K_2 = \frac{1}{2}m_2\langle v_2^2\rangle = \frac{3}{2}kT$$

each molecule has the same kinetic energy but

$$\langle v_1^2\rangle = \frac{3kT}{m_1}$$
$$\langle v_2^2\rangle = \frac{3kT}{m_2}$$

Since $m_1 > m_2$ then $\langle v_2^2\rangle > \langle v_1^2\rangle$.

19.31 Two containers, same pressure

$$P_1 = P_2 = P$$

temperature:

$$T_1 = T_2 = T$$

but

$$V_1 = 2\text{ L}$$
$$V_2 = 1\text{ L}$$

We have that

$$PV = \frac{3}{2}NkT$$

and

$$U = \frac{3}{2}NkT$$

So $U = \frac{3}{2}PV$ then

$$\frac{U_1}{U_2} = \frac{\frac{3}{2}P_1V_1}{\frac{3}{2}P_2V_2} = \frac{V_1}{V_2} = 2$$

The 2 L container, since it contains more particles.

19.35 $T = 10^8$ K

a) $m_p = 1.67 \times 10^{-27}$ kg

$$v_{\text{rms}} = \left(\frac{3kT}{m}\right)^{\frac{1}{2}}$$
$$= \left(\frac{3 \times 1.38 \times 10^{-23}\ \text{J/K} \times 10^8\ \text{K}}{1.67 \times 10^{-27}\ \text{kg}}\right)^{\frac{1}{2}}$$
$$= 1.6 \times 10^6\ \text{m/s}$$

b) $m_e = 9.11 \times 10^{-31}$ kg

$$v_{\text{rms}} = \left(\frac{3 \times 1.38 \times 10^{-23}\ \text{J/K} \times 10^8\ \text{K}}{9.11 \times 10^{-31}\ \text{kg}}\right)^{1/2}$$
$$= 6.7 \times 10^7\ \text{m/s}$$

The speed of the electron is $\sim 23\%$ of the speed of light. The answer is therefore only approximate.

19.39 Isothermal expansion of 2.5 g He, $T = 290°$ K

$$V_1 = 11\ \text{m}^3$$
$$V_2 = 18\ \text{m}^3$$

$$W = P_1V_1\ln\left(\frac{V_2}{V_1}\right)$$
$$= \mathcal{N}RT_1\ln\left(\frac{V_2}{V_1}\right)$$

$$= (2.5\ \text{g})\left(\frac{1\ \text{mol}}{4.00\ \text{g}}\right)\left(8.314\ \frac{\text{J}}{\text{mol}\cdot\text{K}}\right)(290°\ \text{K})\ln\left(\frac{18}{11}\right)$$
$$W = 740\ \text{J}$$

19.41 Container has 3 moles of gas. Consider the gas to be the system. The block does work on the system; the heater adds or takes away heat.

a) From the 1st Law,

$$\underbrace{\Delta U}_{\text{Internal energy change}} = \underbrace{Q}_{\text{heat added to system}} - \underbrace{W}_{\text{work done by system}}$$

The block rises 2.4 m, so the work done by the system is

$$W = mg\Delta h$$
$$= (2.4\ \text{m})(9.8\ \text{m/s}^2)(1.0\ \text{kg})$$
$$= 24\ \text{J}$$

Since the temperature does not change,

$$\Delta U = O \Rightarrow Q = W = 24\ \text{J}$$

b) The block falls 1.2 m, so

$$W = -(1.2\ \text{m})(9.8\ \text{m/s}^2)(1.0\ \text{kg}) = -11.8\ \text{J}$$

(Work done on system.) Now

$$U = \frac{3}{2}\mathcal{N}RT$$

so

$$\Delta U = \frac{3}{2}\mathcal{N}R\Delta T$$

and

$$\begin{aligned}\Delta T &= +2.0\text{ K}\\ \Rightarrow \Delta U &= \frac{3}{2}(3.0\text{ mol})(8.314\text{ J/mol}\cdot\text{K})(+2.0\text{ K})\\ &= +74.8\text{ J}\end{aligned}$$

Now

$$\begin{aligned}\Delta U &= Q - W\\ Q &= \Delta U + W = 74.8\text{ J} - 11.8\text{ J}\\ &= +63\text{ J}\end{aligned}$$

c) $Q = 15$ J and $\Delta T = -1.0$ K

$$\begin{aligned}\Delta U &= \frac{3}{2}\mathcal{N}R\Delta T\\ &= \frac{3}{2}(3.0)(8.315\text{ J/mol}\cdot\text{K})(-1.0\text{ K})\\ \Delta U &= -37.42\text{ J}\end{aligned}$$

Now

$$\begin{aligned}\Delta U &= Q - W\\ W &= Q - \Delta U\\ &= 15\text{ J} - (-37.42)\text{ J}\\ &= +52.42\text{ J}\\ \Delta h &= \frac{W}{mg} = \frac{52.42\text{ J}}{(1.0\text{ kg})(9.8\text{ m/s}^2)}\\ &= 5.3\text{ m}\end{aligned}$$

The block rises since $W > 0$.

19.45

$$\begin{aligned}V &= 0.60\text{ m}^3\\ T &= 2.0\times10^2\text{ K}\\ P &= 0.050\text{ atm}\end{aligned}$$

a) Use

$$\begin{aligned}PV &= \mathcal{N}RT\\ \mathcal{N} &= \frac{PV}{RT}\\ &= \frac{(1.013\times10^5\ \frac{\text{Pa}}{\text{atm}}\times0.050\text{ atm})(0.60\text{ m}^3)}{(8.31\text{ J/mol}\cdot\text{K})(2.0\times10^2\text{ K})}\\ &= 1.8\text{ moles}\end{aligned}$$

b) now

$$\begin{aligned}V &= 0.50\text{ m}^3\\ \frac{P_1V_1}{T_1} &= \frac{P_2V_2}{T_2}\\ P_2 &= \frac{T_2}{T_1}\times\frac{P_1V_1}{V_2}\\ &= \frac{210\text{ K}}{200\text{ K}}\times\frac{0.6\text{ m}^3}{0.5\text{ m}^3}\times0.05\text{ atm}\\ &= 0.063\text{ atm}\end{aligned}$$

c) No we can't draw the actual process precisely. It was not isothermal, nor exactly adiabatic. (The product PV^γ changes by 8%.)

19.49 Constant volume process, *ideal* gas *monatomic*

$$\Rightarrow c_v' = \frac{3}{2}R$$

by 1st Law, if no work done then

$$\begin{aligned}\Delta U &= \mathcal{N}c_v'T = \frac{3}{2}\mathcal{N}R\Delta T\\ \Delta U &= 10.0\text{ J},\ \Delta T = 10.0\text{ K}\\ \mathcal{N} &= \frac{2}{3}\frac{\Delta U}{RT}\\ &= \frac{2}{3}\frac{10.0\text{ J}}{(8.314\text{ J/mol}\cdot\text{K})(10.0\text{ K})}\\ &= 0.080\text{ moles}\end{aligned}$$

The mass depends on the (unknown) molecular weight of the gas.

19.53 3.00 mol neon Ne, 2.00 mol H. The heat capacity of the neon is:

$$\begin{aligned}C_{\text{Ne}} &= \frac{3}{2}\mathcal{N}R\\ &= \frac{3}{2}(3.00\text{ mol})(8.3145\text{ J/mol}\cdot\text{K})\\ &= 37.42\text{ J/K}\end{aligned}$$

The heat capacity of the hydrogen is:

$$\begin{aligned}C_{\text{H}} &= \frac{3}{2}(2.00\text{ mol})(8.3145\text{ J/mol}\cdot\text{K})\\ &= 24.94\text{ J/K}\end{aligned}$$

$$\begin{aligned}c_v &= \frac{\text{total heat capacity}}{\text{total mass}}\\ &= \frac{C_{\text{H}} + C_{\text{Ne}}}{M_{\text{H}} + M_{\text{Ne}}}\\ &= \frac{37.42\text{ J/K}+24.94\text{ J/K}}{2.00(1.0\times10^{-3}\text{ kg})+3.00(20.0\times10^{-3}\text{ kg})}\\ &= 1010\text{ J/kg}\cdot\text{K}\end{aligned}$$

For c_p, the $\frac{3}{2}$ becomes $\frac{5}{2}$

$$\begin{aligned}\Rightarrow c_p &= \frac{2}{3}\times\frac{5}{2}\times1006\text{ J/kg}\cdot\text{K}\\ &= 1680\text{ J/kg}\cdot\text{K}\end{aligned}$$

Note $\gamma = 1.66$; no change in mixture.

19.57

$$v_s = \left(\frac{dP}{d\rho}\right)^{\frac{1}{2}}$$

now

$$PV^\gamma = \text{const}$$
$$\rho = \frac{M}{V}$$

So

$$V = \frac{M}{\rho}$$
$$V^\gamma = \left(\frac{M}{\rho}\right)^\gamma$$
$$\Rightarrow P\left(\frac{M}{\rho}\right)^\gamma = \text{const}$$

take $\frac{d}{d\rho}$:

$$\frac{dP}{d\rho}\left(\frac{M}{\rho}\right)^\gamma - \frac{\gamma M^\gamma}{\rho^{\gamma+1}} = 0$$
$$\frac{dP}{d\rho}\left(\frac{M}{\rho}\right)^\gamma - \gamma\left(\frac{M}{\rho}\right)^\gamma \frac{P}{\rho} = 0$$
$$\frac{dP}{d\rho} = +\frac{\gamma P}{\rho}$$

$$\Rightarrow v_s = \left(\frac{P\gamma}{\rho}\right)^{\frac{1}{2}}$$
$$= \left(\frac{(1.013\times10^5\ \text{Pa})(1.4)}{1.3\ \text{kg}/\text{m}^3}\right)^{\frac{1}{2}}$$
$$= 330\ \text{m}/\text{s}$$

Compare with

$$v_{\text{rms}} = \left(\frac{3kT}{m}\right)^{\frac{1}{2}}$$
$$= \left(\frac{3\times1.38\times10^{-23}\ \text{J}/\text{K}\times273\ \text{K}}{28\times1.67\times10^{-27}\ \text{kg}}\right)^{\frac{1}{2}}$$
$$= 490\ \text{m}/\text{s}$$

19.61 For one cylinder during the power stroke,

$$W = \frac{P_iV_i}{\gamma-1}\left(1-\left(\frac{V_i}{V_f}\right)^{1-\gamma}\right)$$

now

$$V_f - V_i = V_i\left(\frac{V_f}{V_i}-1\right)$$
$$V_i = \frac{V_f-V_i}{\left(\frac{V_f}{V_i}-1\right)}$$

So

$$W = \frac{P_i(V_f-V_i)}{\left(\frac{V_f}{V_i}-1\right)}\left(1-\left(\frac{V_i}{V_f}\right)^{1-\gamma}\right)\times\frac{1}{\gamma-1}$$

assume $\gamma = 1.4$, then

$$W = \frac{(4.1\times10^6\ \text{Pa})(397\ \text{cm}^3\times10^{-6}\ \text{m}^3/\text{cm}^3)}{(8-1)(1.4-1)}\left(1-\left(\tfrac{1}{8}\right)^{0.4}\right)$$
$$= 328\ \text{J}$$

For all cylinders multiply by 4:

$$W = 1300\ \text{J}$$

19.65 tire volume:

$$V = 5\times10^{-3}\ \text{m}^3$$
$$P = 1\ \text{atm}$$
$$T = 299\ \text{K}$$

a)

$$\mathcal{N} = \frac{PV}{RT}$$
$$= \frac{(5\times10^{-3}\ \text{m}^3)(1.0\ \text{atm})\left(1.013\times10^5\frac{\text{Pa}}{\text{atm}}\right)}{(8.314\ \text{J}/\text{mol}\cdot\text{K})(299\ \text{K})}$$
$$= 0.20\ \text{moles}$$

b)

$$V = 7\times10^{-4}\ \text{m}^3$$
$$= \frac{(0.20\ \text{mol})(7.0\times10^{-4}\ \text{m}^3)}{(5\times10^{-3}\ \text{m}^3)}$$
$$= 0.028\ \text{moles}$$

c)

$$V_i \equiv \text{volume of air at 1.0 atm}$$
$$= n(\text{volume of pump at 1.0 atm})$$
$$= nV_p$$

d)

$$\gamma = \frac{7}{5}$$
$$P_iV_i^\gamma = P_fV_f^\gamma$$
$$\left(\frac{P_i}{P_f}\right)^{1/\gamma} = \frac{V_f}{V_i}$$
$$V_f \equiv \text{volume of air at 5.0 atm}$$
$$= \text{volume of tire } V$$

So

$$\left(\frac{P_i}{P_f}\right)^{1/\gamma} = \frac{V_f}{nV_p}$$

and

$$n = \frac{V_f}{V_p}\left(\frac{P_f}{P_i}\right)^{1/\gamma}$$
$$= \frac{5\times10^{-3}\ \text{m}^3}{7\times10^{-4}\ \text{m}^3}\left(\frac{5.0\ \text{atm}}{1.0\ \text{atm}}\right)^{5/7}$$
$$= 22.5$$

$$\begin{aligned}\mathcal{N} &= \frac{P_i V_i}{RT} = \frac{P_i n V_p}{RT} \\ &= \frac{(1.0\ \text{atm})\left(1.013\times10^5\ \frac{\text{Pa}}{\text{atm}}\right)(22.5\ \text{strokes})\left(7\times10^{-4}\ \frac{\text{m}^3}{\text{stroke}}\right)}{(8.315\ \text{J/mol}\cdot\text{K})(299\ \text{K})} \\ &= 0.64\ \text{moles}\end{aligned}$$

e) final temp

$$\begin{aligned}T &= \frac{PV}{\mathcal{N}R} \\ &= \frac{(5.0\,\text{atm})\left(1.014\times10^5\frac{\text{Pa}}{\text{atm}}\right)(5\times10^{-3}\,\text{m}^3)}{(0.64\ \text{moles})(8.315\ \text{J/mol}\cdot\text{K})} \\ &= 475\ \text{K}\end{aligned}$$

f) Cool to 299 K then

$$\begin{aligned}P &= \frac{\mathcal{N}RT}{V} \\ &= \frac{(0.64\ \text{mol})(8.315\ \text{J/mol}\cdot\text{K})(299\ \text{K})}{5\times10^{-3}\ \text{m}^3} \\ &= 3.2\times10^5\ \text{Pa}\end{aligned}$$

19.69 For steam,

$$CMM = (2+16)\ \text{g/mol}$$

Use an ideal gas model with 3 rotational modes:

$$\begin{aligned}c_p &= \frac{3+3+2}{2}\frac{k}{m} \\ &= 4\frac{k}{m} = \frac{4R}{GMM} \\ &= \frac{4\times8.315\ \text{J/mol}\cdot\text{K}}{18\ \text{g/mol}}\left(\frac{1000\ \text{g}}{1\ \text{kg}}\right) \\ &= 1850\ \text{J/K}\cdot\text{kg}\end{aligned}$$

This is close to the experimental value (2009 J/K · kg) as long as all rotational modes are considered.

19.73

$$I = 2M\left(\frac{d}{2}\right)^2 = \frac{Md^2}{2}$$

So

$$\begin{aligned}E_{\text{min}} &= \frac{(1\times10^{-68}\ \text{J}^2\cdot\text{s}^2)\times2}{(14\times1.67\times10^{-27}\ \text{kg})(10^{-10}\ \text{m})^2} \\ &= 8\times10^{-23}\ \text{J}\end{aligned}$$

At 200 K

$$c_v' = \frac{5}{2}R \qquad (1)$$

This suggests 5 modes, so for a diatomic gas rotation is excited but vibration is not. Compare

$$\begin{aligned}\frac{1}{2}kT &= \frac{1}{2}(1.38\times10^{-23}\ \text{J/K})(200\ \text{K}) \\ &= 1.4\times10^{-21}\ \text{J} \gg E_{\text{min}}\end{aligned}$$

consistent with the excitation of rotational modes. At 80 K

$$\begin{aligned}\frac{1}{2}kT &= \frac{1}{2}(1.38\times10^{-23}\ \text{J/K})(80\ \text{K}) \\ &= 5.5\times10^{-22}\ \text{J}\end{aligned}$$

$\Rightarrow c_v$ would not be less than $\frac{5}{2}\frac{k}{m}$.

19.77 $v_{\text{rms}} = \left(\frac{3kT}{m}\right)^{\frac{1}{2}}$, i.e., $v_{\text{rms}} \propto \frac{1}{\sqrt{m}}$

$$\frac{m_e}{m_p} = \frac{1}{1830}$$

So

$$\frac{v_{\text{rms}}\ \text{electron}}{v_{\text{rms}}\ \text{ion}} = (23\times1830)^{\frac{1}{2}}$$

speed ratio = 205.

Since $PV = NkT$ is independent of the mass, the contributions to the pressure must be equal.

19.81 initial:

$$\left\{\begin{array}{lll}\text{radius} & 1.0\ \text{cm} & = r \\ \text{depth} & 20.0\ \text{m} & = d \\ \text{temp} & 4.0^\circ\ \text{C} & = T = 277\ \text{K}\end{array}\right.$$

final:

$$\left\{\begin{array}{lll}\text{radius} & ? & \\ \text{depth} & 0 & \\ \text{temp} & 27^\circ\ \text{C} & = 300\ \text{K}\end{array}\right.$$

Assume the air in the bubble is always in thermal equilibrium with the water. Use ideal gas law; $PV = \mathcal{N}RT$

$$\Rightarrow \frac{P_1V_1}{T_1} = \frac{P_2V_2}{T_2}$$

now

$$V = \frac{4}{3}r\pi^3$$

So

$$\frac{P_1r_1^3}{T_1} = \frac{P_2r_2^3}{T_2}$$

What is the pressure as function of depth, d? Water density is 1 kg/L (assume const)

$$\begin{aligned}P_1 &= P(d=0) + pgd \\ &= (1.0\ \text{atm}) \\ &\quad + \left(\frac{1.0\ \text{kg}}{1\ \text{L}}\right)\left(\frac{10^3\ \text{L}}{\text{m}^3}\right)\left(9.8\ \frac{\text{m}}{\text{s}^2}\right) \\ &\quad \times(20\ \text{m})\left(\frac{1\ \text{atm}}{1.013\times10^5\ \text{Pa}}\right) \\ &= 1.0\ \text{atm} + 1.93\ \text{atm} = 2.93\ \text{atm}\end{aligned}$$

So

$$P_1 = 2.93 \text{ atm}$$

Rearranging

$$\begin{aligned} r_2^3 &= \frac{T_2}{T_1}\frac{P_1}{P_2}r_1^3 \\ r_2 &= \left(\frac{T_2}{T_1}\frac{P_1}{P_2}r_1^3\right)^{1/3} \\ &= \left[\left(\frac{300}{277}\right)\left(\frac{2.93 \text{ atm}}{1.0 \text{ atm}}\right)(1.0 \text{ cm})^3\right]^{\frac{1}{3}} \\ &= 1.47 \text{ cm} \end{aligned}$$

19.83 The gas in cylinder one expands *adiabatically.* From the First Law,

$$\begin{aligned} \Delta U &= Q - W = -W \\ V_1 &= 0.700 \text{ m}^3 \text{ and } \gamma = \frac{5}{3} \\ V_2 &= 1.40 \text{ m}^3 \end{aligned}$$

$$\begin{aligned} W &= \frac{P_1 V_1}{\gamma - 1}\left[1 - \left(\frac{V_1}{V_2}\right)^{\gamma-1}\right] \\ &= \frac{\mathcal{N}RT_1}{\gamma - 1}\left[1 - \left(\frac{V_1}{V_2}\right)^{\gamma-1}\right] \end{aligned}$$

Now,

$$\begin{aligned} \mathcal{N}RT_1 &= (30.0 \text{ mol})(8.315 \text{ J/mol}\cdot\text{K})(300.0 \text{ K}) \\ &= 7.48 \times 10^4 \text{ J} \end{aligned}$$

So

$$\begin{aligned} W &= \frac{(7.48 \times 10^4 \text{ J})}{\frac{2}{3}}\left(1 - \left(\frac{0.700 \text{ m}^3}{1.40 \text{ m}^3}\right)^{\frac{2}{3}}\right) \\ &= 4.15 \times 10^4 \text{ J} \end{aligned}$$

Now

$$U = \frac{3}{2}\mathcal{N}RT$$

The system begins and ends at uniform temperature, and there are 60.0 moles in the system.

$$\begin{aligned} \frac{3}{2}\mathcal{N}R\Delta T &= \Delta U \\ \Rightarrow \frac{3}{2}\mathcal{N}R(T_2 - T_1) &= -W \end{aligned}$$

So,

$$\begin{aligned} T_2 &= T_1 - \frac{2W}{3\mathcal{N}R} \\ &= 300 \text{ K} - \frac{2}{3}\frac{(4.15 \times 10^4 \text{ J})}{(60.0 \text{ moles})(8.315 \text{ J/mol}\cdot\text{K})} \\ &= 245 \text{ K} \end{aligned}$$

19.87 The oven is *insulated.* No *heat* passes the walls of the oven. Turn the oven on, and the temperature does go up. The power company does work on the electrons in the heating element. So work is done on the system ($W < 0$). By the first law, $\Delta U = -W$ so the temperature goes up.

19.91 The atmosphere is in hydrostatic equilibrium:

$$\frac{dP}{dr} = -\rho g$$

assume ideal gas of monatomic H, then

$$\rho = \frac{Nm}{V}$$

and using

$$\begin{aligned} PV &= NkT \\ \rho &= \frac{mP}{kT} \\ \frac{dP}{dr} &= -\frac{mg}{kT}P \end{aligned}$$

Assume $g \equiv$ const. then integrating

$$\begin{aligned} \frac{dP}{P} &= -\frac{mg}{kT}dr \\ \ln\left(\frac{P}{P_0}\right) &= -\frac{mg}{kT}(r - r_0) \end{aligned}$$

let $r - r_0 \equiv h$ then

$$\begin{aligned} \frac{P}{P_0} &= e^{-\frac{hmg}{kT}} \\ \frac{P}{P_0} &= e^{-h/h_0} \end{aligned}$$

where $h_0 \equiv \frac{kT}{mg}$ is the scale height. Now $g = \frac{GM}{r_0^2}$ so

$$\begin{aligned} h_0 &= \frac{kTr_0^2}{GMm} \\ &= \frac{(1.38\times10^{-23} \text{ J/K})(10^4 \text{ K})(10^4 \text{ m})^2}{(6.67\times10^{-11} \text{ N}\cdot\text{m}^2\cdot\text{kg}^{-2})(1.67\times10^{-27} \text{ kg})(2\times10^{30} \text{ kg})} \\ &= 6.2 \times 10^{-5} \text{ m} \end{aligned}$$

Now, at $r = r_0$,

$$\begin{aligned} \frac{dg}{dr} &= -\frac{2GM}{r_0^3} = -\frac{2g}{r_0} \\ \Rightarrow \frac{\Delta g}{g} &= -\frac{2\Delta r}{r_0} \end{aligned}$$

Variation in g in this case is

$$2 \times \left(\frac{h_0}{r_0}\right) = 2 \times \frac{6.2 \times 10^{-5} \text{ m}}{10^4 \text{ m}} \ll 1$$

So, taking $g =$ const. was a good assumption.

Chapter 20

Thermodynamics of Real Substances

20.1 Equation of state

$$\left(P+\frac{a}{V_m^2}\right)(V_m-b)=RT \qquad (1)$$

for N_2

$$\begin{aligned} a &= 1.39\ \mathrm{L^2\cdot atm/mol^2} \\ b &= 0.039\ \mathrm{L/mol} \end{aligned}$$

Now

$$\left.\begin{array}{l} V=4.0\ \mathrm{L} \\ \#\ \text{moles}=1.2 \end{array}\right\} V_m = \frac{4.0\ \mathrm{L}}{1.2\ \mathrm{mol}} = \frac{10}{3}\frac{\mathrm{L}}{\mathrm{mol}}$$

$$T = 301\ \mathrm{K}$$

From (1) rearrange;

$$\begin{aligned} \left(P+\frac{a}{V_m^2}\right) &= \frac{RT}{V_m-b} \\ P &= \frac{RT}{V_m-b}-\frac{a}{V_m^2} \end{aligned}$$

$$\begin{aligned} P &= \frac{\left(8.314\ \frac{\mathrm{J}}{\mathrm{mol\cdot K}}\right)(301\mathrm{K})}{\left(\frac{10}{3}\frac{\mathrm{L}}{\mathrm{mol}}-0.039\ \frac{\mathrm{L}}{\mathrm{mol}}\right)\left(10^{-3}\ \frac{\mathrm{m^3}}{\mathrm{L}}\right)} \\ &\quad -\frac{\left(1.39\ \frac{\mathrm{L^2 atm}}{\mathrm{mol^2}}\right)\left(1.013\times10^5\ \frac{\mathrm{Pa}}{\mathrm{atm}}\right)}{\left(\frac{10}{3}\mathrm{L/mol}\right)^2} \\ &= 7.5\times10^5\ \mathrm{Pa}=7.4\ \mathrm{atm} \end{aligned}$$

20.3 Mass of iron is 10.0 kg
initial temp $T_1 = 1808$ K, and iron melts at 1808 K. If the iron starts out 100% solid then the amount of heat required is

$$\begin{aligned} 100\%\times10.0\ \mathrm{kg}\times289\ \mathrm{kJ/kg} &= 2.89\times10^3\ \mathrm{kJ} \\ &= 2.89\times10^6\ \mathrm{J} \end{aligned}$$

20.5

$$\left.\begin{array}{l} T_1=12^\circ\ \mathrm{C} \\ T_2=412^\circ\ \mathrm{C} \end{array}\right\}\Delta T=400\ \mathrm{K}$$

$$\begin{aligned} \mathrm{L}_1 &= 55.6\ \mathrm{cm} \\ \Delta\mathrm{L} &= 0.42\ \mathrm{cm} \end{aligned}$$

Now by definition

$$\frac{\Delta \mathrm{L}}{\mathrm{L}}=\alpha\Delta T$$

So

$$\begin{aligned} \alpha &= \frac{\Delta\mathrm{L}}{\mathrm{L}_1}\frac{1}{\Delta T} \\ &= \frac{0.42\ \mathrm{cm}}{55.6\ \mathrm{cm}}\times\frac{1}{400\ \mathrm{K}} \\ &= 1.9\times10^{-5}\ \mathrm{K^{-1}} \end{aligned}$$

20.7 The important factor is not the temperature difference alone but the total heat transferred in each case. Let's estimate.
Speck of ash:

$$\text{mass}\approx\frac{1}{10}g$$

specific heat (for carbon, Table 20.4)

$$\approx 700\ \mathrm{J/kg\cdot K}$$

Temperature difference

$$T_{\text{ash}} - T_{\text{body}} \simeq 700 \text{ K} - 300 \text{ K} \sim 400 \text{ K}$$

(cf Example 20.6)
Heat transferred =

$$\begin{aligned} mc\Delta T &= (10^{-4} \text{ kg})(700 \text{ J/kg} \cdot \text{K})(400 \text{ K}) \\ &\sim 28 \text{ J} \end{aligned}$$

Water
mass 1 g (small!)
Specific heat 4000 J/kg · K
Temperature difference

$$400 \text{ K} - 300 \text{ K} \sim 100 \text{ K}$$

(boiling water)
Heat transferred

$$\begin{aligned} = mc\Delta T &\sim (10^{-3} \text{ kg})(4000 \text{ J/kg} \cdot \text{K})(100 \text{ K}) \\ &\sim 400 \text{ J} \end{aligned}$$

Even with this very small assumed mass, the water transfers more than 10 times as much heat, and so would cause much more severe burns.

20.9 Ne is modeled as an ideal gas

$$\Rightarrow PV = \mathcal{N}RT$$

In the thermometer $V \equiv$ const. So

$$\frac{P}{T} = \frac{\mathcal{N}R}{V}$$

i.e.,

$$\begin{aligned} \frac{P_1}{T_1} &= \frac{P_2}{T_2} \\ T_2 &= \frac{P_2}{P_1}T_1 \\ &= \frac{1.10 \times 10^3 \text{ Pa}}{1.00 \times 10^3 \text{ Pa}} \times (19° \text{ C} + 273.15) \\ &= 321 \text{ K} = 48.2° \text{ C} \end{aligned}$$

20.13

$$\begin{aligned} V &= 5.50 \text{ L} \\ &= 5.50 \times 10^{-3} \text{ m}^3 \\ \mathcal{N} &= 2.50 \text{ mol} \\ &\Rightarrow V_m = \frac{5.50 \text{ L}}{2.50 \text{ mol}} \\ &= 2.20 \text{ L/mol} \\ \left(P + \frac{a}{V_m^2}\right)(V_m - b) &= RT \end{aligned}$$

$$\begin{aligned} \Rightarrow P &= \left(\frac{RT}{V_m - b}\right) - \frac{a}{V_m^2} \\ &= \frac{(8.315 \text{ J/mol}\cdot\text{K})(295 \text{ K})}{(2.20 \text{ L/mol} - 0.043 \text{ L/mol}) \times (10^{-3} \text{ m}^3\text{/L})} \\ &\quad - \frac{3.59 \text{ L}^2\cdot\text{atm/mol}^2}{(2.20 \text{ L/mol})^2} \times \frac{1.013 \times 10^5 \text{ Pa}}{1 \text{ atm}} \\ &= 1.06 \times 10^6 \text{ Pa} = 10.5 \text{ atm} \end{aligned}$$

20.17 Equation of state

$$\left(P + \frac{a}{V_m^2}\right)(V_m - b) = RT$$

$$\begin{aligned} V &= 2.30 \text{ L} \\ P &= 6.10 \text{ atm} \\ T &= 288 \text{ K} \end{aligned}$$

assume

$$P \gg \frac{a}{V_m^2}$$

then

$$\begin{aligned} P(V_m - b) &\cong RT \\ V_m - b &= \frac{RT}{P} \\ V_m &= \frac{RT}{P} + b \\ \frac{V}{\mathcal{N}} &= \frac{RT}{P} + b \\ \mathcal{N} &= \frac{V}{\frac{RT}{P} + b} \end{aligned}$$

$$\begin{aligned} \mathcal{N} &= \frac{2.30 \text{ L}}{\frac{8.315 \frac{\text{J}}{\text{mol}\cdot\text{K}} \times 288 \text{ K}\left(\times 10^3 \frac{\text{L}}{\text{m}^3}\right)}{(6.10 \text{ atm})\left(1.013 \times 10^5 \frac{\text{Pa}}{\text{atm}}\right)} + 0.032 \frac{\text{L}}{\text{mol}}} \\ &= 0.589 \text{ mol} \end{aligned}$$

So

$$V_m = \frac{2.30 \text{ L}}{0.589 \text{ mol}} = 3.91 \text{ L/mol}$$

from the table,

$$\begin{aligned} a &= 1.36 \text{ L}^2 \cdot \text{atm/mol}^2 \\ \Rightarrow \frac{a}{V_m^2} &= \frac{1.36}{(3.91)^2} = 0.089 \end{aligned}$$

this is 1% of the pressure of 6.10 atm $\Rightarrow$ the approximation was pretty good. Thus $\mathcal{N} =$ 0.59 mol.

20.21 One metric ton is 1000 kg. Latent heat of fusion is

$$L_f = 289 \text{ kJ/kg}$$

Assume the iron is at melting point

$$\begin{aligned} Q_f &= 289 \text{ kJ/kg} \times 1000 \text{ kg} \\ &= 2.89 \times 10^5 \text{ kJ} \end{aligned}$$

20.25 No. Both the bar and the upper and lower members of the frame have the same α, and the same length. Since the length and α are equal, then, over a given $\Delta\ T_1$, ΔL will be the same for both.

20.29 By definition,

$$\frac{\Delta V}{V} = \beta \Delta T$$

since mercury is isotropic. Now

$$\beta = 1.81 \times 10^{-4}\ \mathrm{K}^{-1}$$

Now $\rho = M/V$. Take the differential then

$$\Delta\rho = \frac{-M}{V^2}\Delta V$$

so

$$\begin{aligned} \frac{\Delta\rho}{\rho} &= \frac{-M\Delta V}{V^2}\frac{V}{M} \\ &= -\frac{\Delta V}{V} \\ \frac{\Delta\rho}{\rho} &= -\beta\Delta T \\ &= -(1.81 \times 10^{-4}\ \mathrm{K}^{-1})(10\ \mathrm{K}) \\ &= -2 \times 10^{-3} \end{aligned}$$

The density changes by $\sim 0.2\%$

20.33 T_{initial} is $5°\ \mathrm{C} = 278$ K
V_{initial} is 1.0 L
β for the carafe is $3.2 \times 10^{-5}\ \mathrm{K}^{-1}$. T increases to $25°\ \mathrm{C} = 298$ K then

$$\begin{aligned} \Delta T &= 20\ \mathrm{K} \\ \Delta V &= V\beta\Delta T \\ &= (1.0\ \mathrm{L})(3.2 \times 10^{-5}\ \mathrm{K}^{-1})(20\ \mathrm{K}) \\ &= 6.4 \times 10^{-4}\ \mathrm{L} \end{aligned}$$

$\Rightarrow$ Final volume of the carafe is 1.00064 L, 7 cc of wine overflowed; $(= 7\ \mathrm{mL} = 0.007\ \mathrm{L})$
final volume of wine is

$$\begin{aligned} V_f &= 1.00064\ \mathrm{L} + 0.007\ \mathrm{L} = 1.00764\ \mathrm{L} \\ \Rightarrow \Delta V &= V_f - V_i = V_i\beta\Delta T \end{aligned}$$

So

$$\begin{aligned} \beta &= \frac{V_f - V_i}{V_i\Delta T} = \frac{0.00764\ \mathrm{L}}{(1.0\ \mathrm{L})(20\ \mathrm{K})} \\ &= 3.8 \times 10^{-4}\ \mathrm{K}^{-1} \end{aligned}$$

20.37 Given C_A, C_B, T_A, T_B and an insulating enclosure $\Rightarrow$ heat lost from $A =$ heat gain in B.

$$\Rightarrow Q_A + Q_B = 0$$

The bodies come to equilibrium at final temperature T.

$$\left.\begin{aligned} Q_A = C_A(T_A - T) \\ Q_B = C_B(T_B - T) \end{aligned}\right\} \text{add}$$

$$\begin{aligned} 0 &= C_AT_A - C_AT + C_BT_B - C_BT \\ 0 &= C_AT_A + C_BT_B - T(C_A + C_B) \\ C_AT_A + C_BT_B &= T(C_A + C_B) \\ \therefore T &= \frac{C_AT_A + C_BT_B}{C_A + C_B} \end{aligned}$$

If heat is lost to the surroundings, the final temperature will be lower. If heat is gained from the surroundings, the final temperature will be higher.

20.41

$$\begin{aligned} m_x &= 0.300\ \mathrm{kg} \\ T_x &= 401°\ \mathrm{C} \\ m_w &= 0.100\ \mathrm{kg} \\ m_c &= 0.050\ \mathrm{kg} \\ T_w &= 20°\ \mathrm{C} \end{aligned}$$

Final Temp $= 60°$ C

$$m_xc_x(T_x - T) + m_wc_w(T_w - T) + m_cc_c(T_c - T) = 0$$

since the calorimeter is well insulated. Now

$$\begin{aligned} T_w &= T_c = 20\ °\mathrm{C} \\ T &= 60\ °\mathrm{C} \\ c_c &= c_w = 4186\ \frac{\mathrm{J}}{\mathrm{kg}\cdot\mathrm{K}} \end{aligned}$$

Now solve for c_x

$$\begin{aligned} c_x &= \frac{m_wc_w(T - T_w) + m_cc_c(T - T_c)}{m_x(T_x - T)} \\ &= \frac{(m_wc_w + m_cc_c)(T - T_c)}{m_x(T_x - T)} \\ &= \frac{c_w(m_w + m_c)(T - T_w)}{m_x(T_x - T)} \\ &= \frac{(4186\ \frac{\mathrm{J}}{\mathrm{kg}\cdot\mathrm{K}})(0.100\ \mathrm{kg}+0.050\ \mathrm{kg})(60-20)\mathrm{K}}{(0.300\ \mathrm{kg})(401-60)\mathrm{K}} \\ &= \frac{(4186)(0.150)(40)}{(0.300)(341)}\ \frac{\mathrm{J}}{\mathrm{kg}\cdot\mathrm{K}} \\ \therefore c_x &= 246\ \frac{\mathrm{J}}{\mathrm{kg}\cdot\mathrm{K}} \end{aligned}$$

20.43 Given $V = 0.523$ mV, $T_2 = ?$

$$\begin{aligned} V &= (0.0387\ \mathrm{mV/K})\, T_2\,(°\mathrm{C}) \\ \Rightarrow T_2\,(°\mathrm{C}) &= \frac{0.523\ \mathrm{mV}}{0.0387\ \mathrm{mV}} \\ &= 13.5\ °\mathrm{C} \end{aligned}$$

20.44 The rod is 6° from vertical.

20.45 For both mercury and glass

$$\Delta V = \beta V \Delta T$$

Assume the glass and mercury are at the same temperature. The *difference* in volume changes is

$$\begin{aligned}\delta V &= \beta_m V \Delta T - \beta_g V \Delta T \\ &= (\beta_m - \beta_g) V \Delta T\end{aligned}$$

Set $\delta V \equiv$ the amount of mercury that flows into the column.

$$\delta V = d \times \pi r^2$$

Also

$$\begin{aligned}V &= \frac{4}{3}\pi R^3 \\ \Rightarrow & \frac{4}{3}\pi R^3(\beta_m - \beta_g)\Delta T = \pi r^2 d \\ d &= \frac{4}{3}(\beta_m - \beta_g)\frac{R^3}{r^2}\Delta T\end{aligned}$$

Let

$$\begin{aligned}\Delta T &= 1^\circ \text{ C} \\ \Rightarrow d &= \frac{4}{3}(\beta_m - \beta_g)\frac{R^3}{r^2}(1^\circ \text{ C})\end{aligned}$$

20.47 Constant volume:

$$V = 20.0 \text{ cm}^3 = 2.00 \times 10^{-5} \text{ m}^3$$

Heat capacity of the bulb is due to the glass and gas together (assume ideal gas)

$$\begin{aligned}C_b &= M \times c_g + \frac{3}{2}\mathcal{N}R \\ &= (10.0 \times 10^{-3} \text{ kg})\left(5.00 \times 10^2 \frac{\text{J}}{\text{K} \cdot \text{kg}}\right) \\ &\quad + \frac{3}{2}(1.00 \times 10^{-3} \text{ mol})\left(8.315 \frac{\text{J}}{\text{mol} \cdot \text{K}}\right) \\ &= 5.012 \text{ J/K}\end{aligned}$$

Use

$$\begin{aligned}PV &= \mathcal{N}RT \\ P &= \frac{\mathcal{N}RT}{V} \\ &= \frac{(1.00 \times 10^{-3} \text{ mol})(8.315 \text{ J/mol} \cdot \text{K})(293 \text{ K})}{2.00 \times 10^{-5} \text{ m}^3} \\ &= 1.22 \times 10^5 \text{ Pa}\end{aligned}$$

At new pressure of 9.65×10^4 Pa $\equiv P_1$ the temperature is

$$\begin{aligned}T_1 &= \frac{P_1}{P}T \\ &= \left(\frac{9.65 \times 10^4 \text{ Pa}}{1.218 \times 10^5 \text{ Pa}}\right)(293 \text{ K}) \\ &= 232 \text{ K}\end{aligned}$$

Heat capacity of the system is

$$2.00 \times 10^2 \text{ J/K} \equiv C_s$$

Heat balance gives

$$\begin{aligned}&(T - T_1)C_s + (T_{\text{room}} - T_1)C_b = 0 \\ &\Delta T = (T - T_1) = (T_1 - T_{\text{room}})\frac{C_b}{C_s}\end{aligned}$$

$$\begin{aligned}\Delta T &= \frac{(232 - 293 \text{ K})(5.012 \text{ J/K})}{2.00 \times 10^2 \text{ J/K}} \\ &= -1.5 \text{ K}\end{aligned}$$

$\therefore$ the thermometer *lowered* the system temperature by 1.5 K (look at the definition of ΔT.)

20.51

$$\begin{aligned}m_{\text{ice}} &= 105 \text{ g} = 0.105 \text{ kg} \\ T_{\text{ice}} &= 0.00 \text{ C} \\ T_w &= 20.0 \text{ C}\end{aligned}$$

Final temperature $T = 5.0^\circ$ C. Set up heat balance

$$m_{\text{ice}}c_w(T - T_{\text{ice}}) + m_{\text{ice}}L_f + m_w c_w(T - T_w) = 0$$

$$m_w = \frac{-m_{\text{ice}}c_w(T - T_{\text{ice}}) - m_{\text{ice}}L_f}{c_w(T - T_w)}$$

$$= \frac{-(0.105 \text{ kg})\left(4186 \frac{\text{J}}{\text{K} \cdot \text{kg}}\right)(5^\circ \text{ C}) - (0.105 \text{ kg})\left(335 \times 10^3 \frac{\text{J}}{\text{K} \cdot \text{kg}}\right)}{\left(4186 \frac{\text{J}}{\text{K} \cdot \text{kg}}\right)(5 - 20) \text{ K}}$$

$$= 0.595 \text{ kg}$$

20.55 $C \equiv$ heat capacity of interior $= 8.4 \times 10^4$ J/K, $M \equiv$ mass of Freon 12. Apply a heat balance

$$C\Delta T + ML_v = 0$$

now

$$\Delta T = -15 \text{ K}$$

So

$$\begin{aligned}M &= -\frac{C\Delta T}{L_v} \\ &= -\frac{(8.4 \times 10^4 \text{ J/K})(-15 \text{ K})}{(165 \text{ kJ/kg})} \\ &= 7.6 \text{ kg}\end{aligned}$$

20.59

$$\begin{aligned}\text{drink volume} &= 118.3 \text{ mL} \\ \text{alcohol volume} &= 59.15 \text{ mL} = 59.15 \times 10^{-6} \text{ m}^3 \\ &= \text{water volume} \\ \text{mass of water} &= 59.15 \times 10^{-3}\text{kg} \\ \text{mass of alcohol} &= 789 \frac{\text{kg}}{\text{m}^3} \times 59.15 \times 10^{-6}\text{m}^3 \\ &= 46.67 \times 10^{-3}\text{kg}\end{aligned}$$

Set up a heat balance, neglect the glass and the environment:
initial temperature of drink $\equiv T_1 = 20°$ C
initial temperature of ice $\equiv T_2 = 0°$ C

$$m_w c_w (T - T_1) + m_{\text{alc}} c_{\text{alc}} (T - T_1) + m_{\text{ice}} c_w (T - 0) + m_{\text{ice}} L_f = 0$$

$$T(m_w c_w + m_{\text{alc}} c_{\text{alc}} + m_{\text{ice}} c_w) = T_1 (m_w c_w + m_{\text{alc}} c_{\text{alc}}) - m_{\text{ice}} L_f$$

$$T = \frac{T_1 (m_w c_w + m_{\text{alc}} c_{\text{alc}}) - m_{\text{ice}} L_f}{m_w c_w + m_{\text{alc}} c_{\text{alc}} + m_{\text{ice}} c_w}$$

$$= \frac{20° \, C \left(\begin{array}{c} (59.15 \times 10^{-3} \text{ kg}) \left(4186 \frac{\text{J}}{\text{kg·K}}\right) \\ +(46.67 \times 10^{-3} \text{ kg})(2450 \frac{\text{J}}{\text{K·kg}}) \end{array} \right) - (15 \times 10^{-3} \text{ kg})(334.7 \times 10^3 \frac{\text{J}}{\text{kg}})}{\left(\begin{array}{c} (59.15 \times 10^{-3} \text{ kg}) \left(4186 \frac{\text{J}}{\text{K·kg}}\right) \\ + (46.67 \times 10^{-3} \text{ kg}) \left(2450 \frac{\text{J}}{\text{K·kg}}\right) \\ + (15 \times 10^{-3} \text{ kg}) \left(4186 \frac{\text{J}}{\text{K·kg}}\right) \end{array} \right)}$$

$$T = 5.2° \, C$$

20.63 By Archimedes' principle, the ball will float since it displaces its mass in water

$$M_{\text{al}} = V_{water} \times \rho_{\text{water}} \text{ (@5° C)} \qquad (2)$$

Since the ball initially floats @ the diameter,

$$V_{\text{water}} = \left(\frac{1}{2} \times \frac{4}{3}\pi R^3\right) = \frac{2}{3}\pi R^3$$

Denote

$$\rho \equiv \text{density of water @ 5° C}$$
$$\rho' \equiv \text{density of water @ 80° C}$$

Since

$$\rho \propto \frac{1}{V}$$

we can write

$$\rho' = \rho \left(\frac{V_w}{V_w + \Delta V_w}\right) = \rho \left(\frac{1}{1 + \frac{\Delta V}{V_w}}\right)$$

but

$$\frac{\Delta V}{V} = \beta \Delta T$$

So

$$\rho' = \rho \left(\frac{1}{1 + \beta \Delta T}\right)$$

So now we know from (2) that the ball floats the mass of the ball is $\frac{2}{3}\pi R^3 \rho$ at 5° C. At 80° C, the volume is

$$V + \Delta V = \frac{2}{3}\pi R^3 + \Delta V$$

Again using Archimedes' Principle,

$$M = (V + \Delta V)\rho'$$

So

$$\overset{\text{cold}}{\frac{2}{3}\pi R^3 \rho} = \overset{\text{hot}}{\left(\frac{2}{3}\pi R^3 + \Delta V\right)} \rho \left(\frac{1}{1 + \beta \Delta T}\right)$$

$$\frac{2}{3}\pi R^3 (1 + \beta \Delta T) = \frac{2}{3}\pi R^3 + \Delta V$$

$$\Delta V = \frac{2}{3}\pi R^3 \beta \Delta T$$

Now the ball also expands, $R_{80} = R_5 + \Delta R$ where $\Delta R = \alpha R \Delta T$.

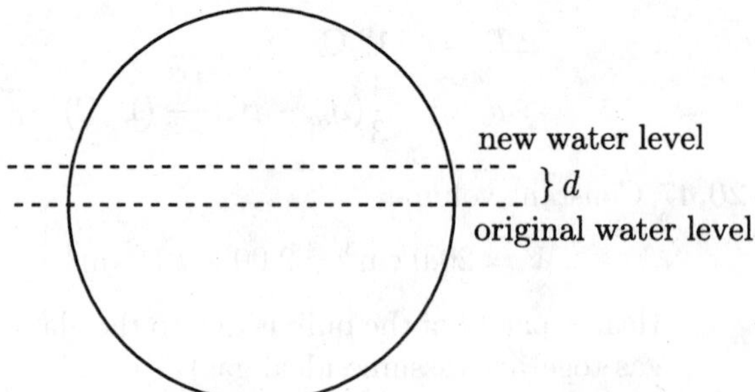

$$V_{\text{new}} = \frac{2}{3}\pi (R + \Delta R)^3 + \pi (R + \Delta R)^2 d$$
$$\simeq \frac{2}{3}\pi R^3 + \frac{2}{3}\pi \cdot 3R^2 \Delta R + \pi R^2 d$$

Thus

$$\Delta V = 2\pi R^2 \Delta R + \pi R^2 d = \pi R^2 (2\Delta R + d)$$

Setting the two expressions for ΔV equal:

$$\frac{2}{3} \not{\pi} R^3 \beta \Delta T = \not{\pi} R^2 (2\alpha R \Delta T + d)$$

Thus

$$d = 2\Delta T R \left(\frac{\beta}{3} - \alpha\right)$$

$$\frac{d}{R} = 2(75 \text{ K}) \left(\frac{2.6 \times 10^{-4}}{3} - 25 \times 10^{-6}\right) \text{K}^{-1}$$

$$= (150)(6.2 \times 10^{-5}) = 9.25 \times 10^{-3}$$

$\therefore$ water level is $9 \times 10^{-3} R$ *above* the equator.

Chapter 21

Heat Transfer

21.1 The rate of heat transfer by conduction is governed by Eqn. (21.1)

$$H = kA\left|\frac{dT}{dx}\right|$$

Assuming there is a uniform temperature gradient we can say,

$$\frac{dT}{dx} = \frac{65° \text{ F} - 35° \text{ F}}{7.0 \text{ cm}\left(\frac{1 \text{ m}}{100 \text{ cm}}\right)}\left(\frac{5 \text{ K}}{9° \text{ F}}\right) = 238 \text{ K/m}$$

The thermal conductivity of concrete is listed in Table 21.1. We choose a median value of $k = 1.1 \frac{\text{W}}{\text{m}\cdot\text{K}}$. Finally,

$$H = 1.1 \frac{\text{W}}{\text{m}\cdot\text{K}}\left(12 \text{ m}^2\right)(238 \text{ K/m}) = 3100 \text{ W}$$

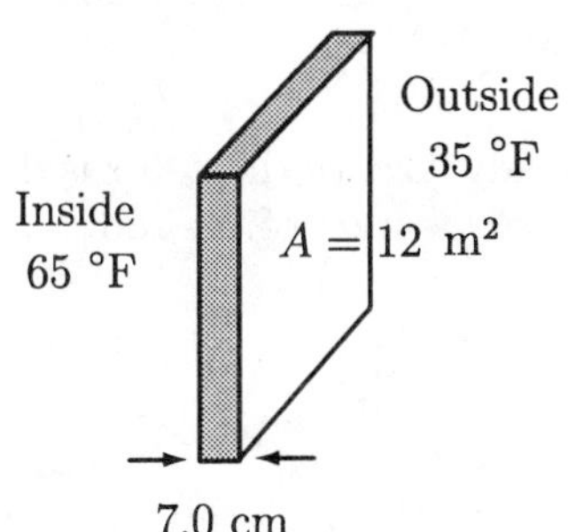

21.3 The rate of heat transfer by conduction is governed by Eqn. (21.1)

$$H = kA\left|\frac{dT}{dx}\right|$$

This can be rewritten as the heat loss per degree Kelvin per m^2

$$\frac{H}{A\cdot\Delta T} = \frac{k}{\Delta x}$$

From Table 21.1 we find the thermal conductivities of concrete and oak.

$$k_{\text{concrete}} \approx 1.1 \frac{\text{W}}{\text{m}\cdot\text{K}}; \; k_{\text{oak}} = 0.147 \frac{\text{W}}{\text{m}\cdot\text{K}}$$

Finally,

$$\begin{aligned}\left(\frac{H}{A\Delta T}\right)_{\text{concrete}} &= \frac{\left(1.1 \frac{\text{W}}{\text{m}\cdot\text{K}}\right)}{(15 \text{ cm})\left(\frac{1 \text{ m}}{100 \text{ cm}}\right)} \\ &= 7.3 \text{ W/m}^2 \\ \left(\frac{H}{A\Delta T}\right)_{\text{oak}} &= \frac{\left(0.147 \frac{\text{W}}{\text{m}\cdot\text{K}}\right)}{(3.0 \text{ cm})\left(\frac{1 \text{ m}}{100 \text{ cm}}\right)} \\ &= 4.9 \frac{\text{W}}{\text{m}^2\cdot\text{K}}\end{aligned}$$

Looking at these results, we can see that even though the oak is much thinner, it still has a smaller heat loss than the concrete. Therefore, the oak is more energy efficient.

21.5 **a)** microwave oven: primarily uses *radiation* to cook the food. Convection carries some energy away from the warm food.

b) electric oven: *radiation and convection*

c) room heated by a wood fire: *radiation and convection*

d) thermometer used to take a baby's temperature: *conduction.*

21.7 This is governed by the Wien displacement law, Eqn. (21.6).

$$\lambda_{\text{max}} = \frac{2.898 \times 10^{-3} \text{ K}\cdot\text{m}}{98° \text{ F}\,\frac{5° \text{ C}}{9° \text{ F}} + 273° \text{ C}} = 8.85 \; \mu\text{m}$$

21.9 Newton's law of cooling states that energy loss is proportional to the difference between the temperatures of object and environment.

$$H \propto (T - T_{\rm env})$$

The rate of change of the object's temperature is proportional to its rate of energy loss. So,

$$\frac{dT}{dt} \propto (T - T_{\rm env})$$

So, comparing the two temperature changes, we have

$$\frac{\frac{\Delta T_1}{\Delta t_1}}{\frac{\Delta T_2}{\Delta t_2}} \approx \frac{T_{\rm avg,1} - T_{\rm env}}{T_{\rm avg,2} - T_{\rm env}}$$

where Δt_2 is the unknown time interval.

$$\begin{aligned}\Delta t_2 &= \frac{T_{\rm avg,1} - T_{\rm env}}{T_{\rm avg,2} - T_{\rm env}} \frac{\Delta T_2}{\Delta T_1} \Delta t_1 \\ &= \frac{88^\circ\ \mathrm{C} - 20^\circ\ \mathrm{C}}{75^\circ\ \mathrm{C} - 20^\circ\ \mathrm{C}} \frac{10\ \mathrm{K}}{15\ \mathrm{K}} (10\ \mathrm{min}) \\ &= 8\ \mathrm{min}\end{aligned}$$

21.13 Metals have much larger conductivities than potatoes. Without a skewer, a potato will cook from the outside only by conduction of heat through the potato. With a metal skewer inserted through it, a potato will be partially cooked from the inside out. Because metal is a better conductor, it will be hotter than the surrounding potato and can act as an internal heat source.

21.17 The definition of the thermal resistance is given by Eqn. 21.5.

$$R = \frac{R_f}{A}$$

... where R_f can be found in Table 21.2.

$$\begin{aligned}R_{\rm plywood} &= \frac{0.109\ \mathrm{m^2 \cdot K/W}}{(10\ \mathrm{m} \times 3.5\ \mathrm{m}) - (1.2\ \mathrm{m} \times 2.3\ \mathrm{m})} \\ &= 3.4 \times 10^{-3}\ \mathrm{K/W} \\ R_{\rm window} &= \frac{0.159\ \mathrm{m^2 \cdot K/W}}{(1.2\ \mathrm{m} \times 2.3\mathrm{m})} = .058\ \mathrm{K/W}\end{aligned}$$

The relationship for the thermal resistance of insulators placed in parallel is given in Exercise 21.2

$$\begin{aligned}\frac{1}{R_{\rm wall}} &= \frac{1}{R_{\rm plywood}} + \frac{1}{R_{\rm window}} \\ R_{\rm wall} &= 3.2 \times 10^{-3}\ \mathrm{K/W}\end{aligned}$$

21.21 From Section 21.1.2, we know that the heat flow through any cross-sectional area normal to the temperature gradient is constant.

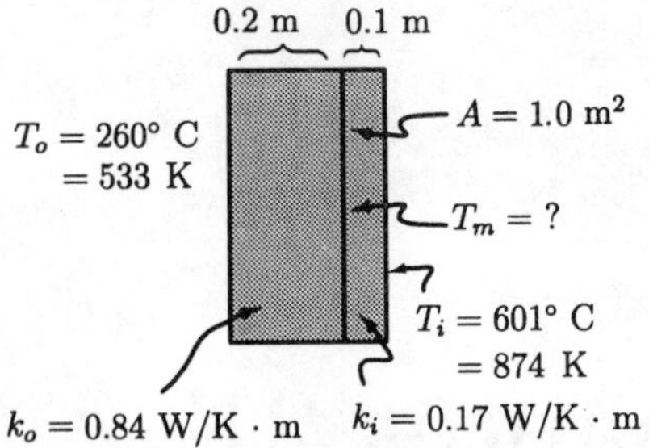

Therefore the heat flow through the outer material will equal the heat flow through the inner material.

$$\begin{aligned}H_o &= k_o A \left(\frac{T_m - T_o}{\Delta x_o}\right) \\ &= 0.84 \frac{\mathrm{W}}{\mathrm{m \cdot K}} \frac{T_m - 533\ \mathrm{K}}{0.20\ \mathrm{m}} 1.0\ \mathrm{m^2} \\ H_i &= k_i A \frac{(T_i - T_m)}{\Delta x_i} \\ &= 0.17 \frac{\mathrm{W}}{\mathrm{m \cdot K}} \frac{874\ \mathrm{K} - T_m}{0.10\ \mathrm{m}} 1.0\ \mathrm{m^2}\end{aligned}$$

Solve for the unknown temperature where the two materials meet.

$$H_o = H_i$$

$$\begin{aligned}4.2T_m - 2238.6 &= 1485.8 - 1.7T_m \\ T_m &= 631\ \mathrm{K} \\ &= 360\ ^\circ\mathrm{C}\end{aligned}$$

Now use this to determine the heat flux.

$$\begin{aligned}H &= 0.84 \frac{\mathrm{W}}{\mathrm{m \cdot K}} \frac{(631\ \mathrm{K} - 533\ \mathrm{K})}{0.20\ \mathrm{m}} (1.0\ \mathrm{m^2}) \\ &= 410\ \mathrm{W}\end{aligned}$$

21.25 In steady flow, the heat flux is the same through each cross section of the rod.

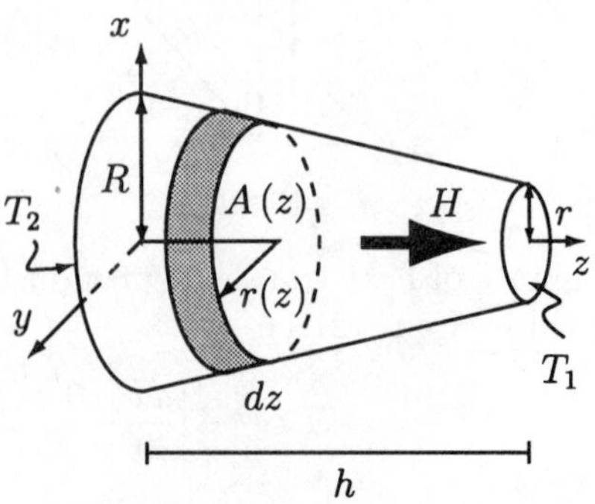

Thus

$$H = -kA(z)\frac{dT}{dz} = \text{Constant}$$

The cross section is $A(z) = \pi r(z)^2$, and the radius $r(z)$ is given by the linear function which has values R at $z = 0$ and r at $z = h$:

$$\begin{aligned} r(z) &= R - \left(\frac{z}{h}\right)(R - r) \\ &= \frac{R}{h}\left[h - z\left(1 - \frac{r}{R}\right)\right] \end{aligned}$$

So

$$\frac{dT}{dz} = -\frac{H}{k\pi R^2\left[1 - \frac{z}{h}\left(1 - \frac{r}{R}\right)\right]^2}$$

Thus

$$\begin{aligned} T_1 - T_2 &\equiv \int_0^h \frac{dT}{dz}\,dz \\ &= -\frac{H\,dz}{k\pi R^2\left[1 - \frac{z}{h}\left(1 - \frac{r}{R}\right)\right]^2} \\ &= -\frac{H}{k\pi R^2}\frac{-h}{\left(1-\frac{r}{R}\right)}\left[\frac{-1}{\left[1-\frac{z}{h}\left(1-\frac{r}{R}\right)\right]}\right]_0^h \\ &= \frac{hH}{k\pi R(R-r)}\left[\frac{-1}{\frac{r}{R}} + \frac{1}{1}\right] = \frac{-hH}{k\pi r R} \end{aligned}$$

So

$$H = \frac{k\pi r R}{h(T_2 - T_1)}$$

Reversing the rod, reverses the temperatures of its ends, and thus the sign of the heat flow with respect to the rod. The same quantity of heat flows through the rod in the same direction in space (i.e, from the higher temperature to the lower). The heat flow is unchanged.

21.29 A fire produces combustion products that are much hotter and less dense than the surrounding air. Buoyant force thus accelerates the hot gases upward. On a space station in free fall all parts of the station, including its air, accelerate together; and *no* forces—including buoyant forces—occur to balance gravity. Without buoyant forces to accelerate hot gases away from a burning object, all the nearby oxygen is quickly consumed and the object ceases to burn.

21.33 The "blue color" of the companion star implies that it has a relatively high surface temperature (according to Wien's law). Since power radiated per unit surface area increases as the fourth power of temperature, the star can be hot and faint only if it is very small. (Stars like Sirius B typically are the size of Earth and have surface temperature of 20,000° K, compared with 6,000° K for the Sun.)

21.37 A real body radiates less efficiently than a theoretical blackbody:

$$\text{Power Radiated} = \epsilon\sigma A T^4 \qquad (21.8)$$

$$\begin{aligned} A &= \frac{100\text{ W}}{0.9(5.67\times10^{-8}\text{ W/m}^2\cdot\text{K}^4)(2800\text{ K})^4} \\ &= 3\times10^{-5}\text{ m}^2 \end{aligned}$$

We can determine λ_{max} for a light bulb using the Wien displacement

$$\lambda_{\text{max}} = \frac{2.898\times10^{-3}\text{ K}\cdot\text{m}}{2800\text{ K}} = 1.04\times10^{-6}\text{ m}$$

This region where most of the radiation is emitted falls in the infrared. Since people cannot see IR light, this bulb is inefficient for our use.

21.39 The average value of $\frac{dT}{dt}$ is

$$\frac{\Delta T}{\Delta t} = \frac{10\text{ K}}{430\text{ s}} = 0.023\text{ K/s}$$

Newton's Law of Cooling says

$$\frac{dT}{dt} = -B\,(T - T_{\text{env}})$$

At 400 K:

$$0.023\text{ K/s} = -B\,(400 - 273)\text{ K}$$

so

$$B = -\frac{0.023\text{ K/s}}{127\text{ K}} = -1.8\times10^{-4}\text{ /s}$$

Then at 350 K,

$$\frac{\Delta T}{\Delta t} \simeq (1.8\times10^{-4}\text{ /s})\,(350 - 273)\text{ K}$$

so

$$\begin{aligned} \Delta t &\simeq \frac{(10\text{ K})}{0.0139\text{ K/s}} = 721\text{ s} \\ &\simeq 10\text{ min} \end{aligned}$$

21.43 An open fireplace heats a room primarily by radiation from the hot gases (i.e, flames) emitted and from the heated bricks from the fireplace structure. The convection of hot gas up the chimney and the cold outside air that replaces it result, some have claimed, in a net loss of thermal energy to the interior of the home especially in that drafty old castle. The racks that hold logs in a fireplace often hold them so that their inner surfaces form a cavity that burns the wood with a less violent motion of air and that radiates more efficiently into the room. Wood stoves are designed to control oxygen flow to the wood so as to burn it with minimum air flow. Exhaust gases are carried away by a vertical pipe that itself radiates to the room. Convective motion of air above the hot surface produces warm air in the room instead of up the chimney. Alas for the castle builders that good wood stoves weren't invented until 600 years later!

21.47 a) From Eqn. 21.9 we know,

$$\begin{aligned} H_{\rm rad} &= \epsilon\sigma A\left(T^4 - T^4_{\rm env}\right) \\ &= 0.7(5.67\times10^{-8}\ \mathrm{W/m^2\cdot K^4}) \\ &\quad\times\left(4\pi(.09\ \mathrm{m})^2\right)(368\ \mathrm{K}^4 - 285\ \mathrm{K}^4) \\ &= 47\ \mathrm{W} \end{aligned}$$

(Remember: You must work in absolute units of temperature like Kelvin or Rankine.)

b)

$$\begin{aligned} H_{\rm conv} = &\left(0.08\,\tfrac{\mathrm{W}}{\mathrm{m\cdot K}}\right)(0.18\ \mathrm{m})(368\ \mathrm{K} - 285\ \mathrm{K}) \\ &\times\left\{1.7 + 0.3\left[1 + \left(\tfrac{5.6\times10^4}{\mathrm{m^3\cdot K}}\right)\right.\right. \\ &\times\left.\left.(0.18\ \mathrm{m})^3(368\ \mathrm{K} - 285\ \mathrm{K})\right]^{1/4}\right\} \\ &= 7\ \mathrm{W} \end{aligned}$$

$$H_{\rm conv} < H_{\rm rad}$$

21.49 For steady state conditions, heat transfers out of the three reservoirs must sum to zero.

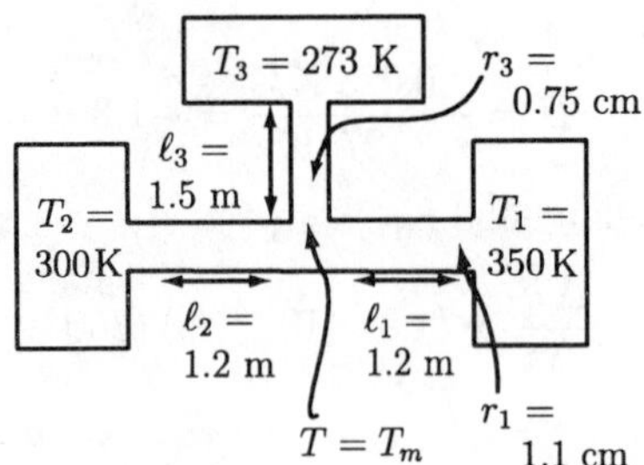

It is clear that energy will flow out of the highest temperature reservoir and that energy will flow into the lowest temperature reservoir. It isn't quite so clear whether the flow will be into or out of the intermediate temperature reservoir. We may proceed by computing the rate of heat transfer out of each reservoir in terms of the unknown temperature T_m of the junction of the rods. Demanding that the sum of the three heat flows be zero will determine T_m, and then we may find the separate rates of transfer.

$$\begin{aligned} H_1 &= k_{\rm cu}\pi r_1^2\frac{(T_1 - T_m)}{\ell_1}; \\ H_2 &= k_{\rm cu}\pi r_1^2\frac{(T_2 - T_m)}{\ell_1}, \\ H_3 &= k_{\rm cu}\pi r_3^2\frac{(T_3 - T_m)}{\ell_3} \end{aligned}$$

So

$$\frac{r_1^2}{\ell_1}(T_1 - T_m) + \frac{r_1^2}{\ell_1}(T_2 - T_m) + \frac{r_3^2}{\ell_3}(T_3 - T_m) \overset{\rm set}{\equiv} 0$$

And

$$\begin{aligned} T_m &= \left[\frac{r_1^2T_1}{\ell_1} + \frac{r_1^2T_2}{\ell_1} + \frac{r_3^2T_3}{\ell_3}\right] \div \left[\frac{2r_1^2}{\ell_1} + \frac{r_3^2}{\ell_3}\right] \\ &= \frac{\left[T_1 + T_2 + \frac{r_3^2}{r_1^2}\frac{\ell_1}{\ell_3}T_3\right]}{\left(2 + \frac{r_3^2}{r_1^2}\frac{\ell_1}{\ell_3}\right)} \\ &= \frac{650\ \mathrm{K} + \left(\frac{0.75}{1.1}\right)^2\left(\frac{1.2}{1.5}\right)(273\ \mathrm{K})}{2 + \left(\frac{0.75}{1.1}\right)^2\left(\frac{1.2}{1.5}\right)} = 316.8\ \mathrm{K} \end{aligned}$$

Then

$$\begin{aligned} H_1 &= \pi\left(400\,\frac{\mathrm{W}}{\mathrm{m\cdot K}}\right)\frac{(1.1\times10^{-2}\ \mathrm{m})^2}{1.2\ \mathrm{m}} \\ &\quad\cdot(350\ \mathrm{K} - 316.8\ \mathrm{K}) \\ &= 4.2\ \mathrm{W} \\ H_2 &= H_1\frac{(300\ \mathrm{K} - 316.8\ \mathrm{K})}{(350\ \mathrm{K} - 316.8\ \mathrm{K})} = -2.1\ \mathrm{W} \\ H_3 &= \pi\left(400\,\frac{\mathrm{W}}{\mathrm{m\cdot K}}\right)\frac{(0.75\times10^{-2}\ \mathrm{m})^2}{1.5\ \mathrm{m}} \\ &\quad\times(273\ \mathrm{K} - 316.8\ \mathrm{K}) \\ &= -2.1\ \mathrm{W} \end{aligned}$$

21.51

Each unit cell

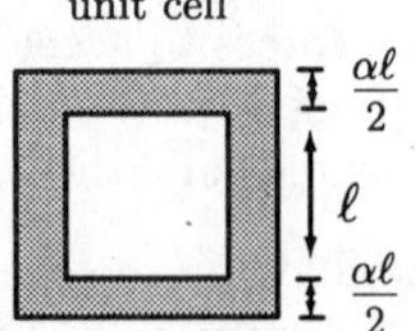

consists of an area ℓ^2, length ℓ of air **(1)**, in series with 2 faces each, an area ℓ^2, length $\frac{\alpha\ell}{2}$ of solid material **(2)**.

This is all in parallel with the side walls each an area $[\ell(1+\alpha)]^2 - \ell^2$ of solid, length $\ell(1+\alpha)$ **(3)**.

$$\begin{aligned} R_1 &= \frac{\Delta x}{kA} = \frac{\ell}{k_{\rm air}\ell^2} = \frac{1}{k_{\rm air}\ell} \\ R_2 &= \frac{\frac{\alpha\ell}{2}}{k_{\rm solid}\ell^2} = \frac{1}{2}\frac{\alpha}{k_{\rm solid}\ell} \end{aligned}$$

The total series resistance is

$$\begin{aligned} R_1 + 2R_2 &= \frac{1}{k_{\text{air}}\ell} + \frac{\alpha}{k_{\text{solid}}\ell} \\ R_3 &= \frac{\ell(1+\alpha)}{k_{\text{solid}}\ell^2(\alpha^2+2\alpha)} \\ &= \frac{1+\alpha}{k_{\text{solid}}\ell(\alpha+2)\alpha} \end{aligned}$$

In parallel, the total resistance of the unit cell is

$$\begin{aligned} \frac{1}{R} &= \frac{1}{R_{\text{series}}} + \frac{1}{R_3} \\ &= \ell\left\{\left[\frac{1}{k_{\text{air}}} + \frac{\alpha}{k_{\text{solid}}}\right]^{-1} + \frac{\alpha(\alpha+2)k_{\text{solid}}}{(1+\alpha)}\right\} \\ &= \ell k_{\text{solid}}\left\{\left[\frac{k_{\text{solid}}}{k_{\text{air}}} + \alpha\right]^{-1} + \frac{\alpha(\alpha+2)}{1+\alpha}\right\} \end{aligned}$$

So

$$R = \frac{1}{k_{\text{solid}}\ell}\left[\frac{k_{\text{air}}}{k_{\text{solid}} + \alpha k_{\text{air}}} + \frac{\alpha(\alpha+2)}{\alpha+1}\right]^{-1}$$

Now

$$R = \frac{\Delta x}{kA} = \frac{\ell(1+\alpha)}{k\left[\ell(1+\alpha)\right]^2} = \frac{1}{k\ell(1+\alpha)}$$

So

$$k = \frac{1}{R\ell(1+\alpha)}$$

Thus

$$\begin{aligned} k_{\text{wool}} &= \frac{\ell k_{\text{solid}}}{\ell(1+\alpha)}\left[\frac{k_{\text{air}}}{k_{\text{solid}} + \alpha k_{\text{air}}} + \frac{\alpha(\alpha+2)}{\alpha+1}\right] \\ (1+\alpha)^2 &= \left(\frac{k_{\text{solid}}}{k_{\text{wool}}}\right)\left(\frac{\alpha+1}{\frac{k_{\text{solid}}}{k_{\text{air}}} + \alpha} + \alpha(\alpha+2)\right) \end{aligned}$$

For glass wool,

$$\begin{aligned} k_{\text{wool}} &= 0.042 \\ k_{\text{solid}} &= 0.8 \\ k_{\text{air}} &= 0.025 \end{aligned}$$

So

$$\begin{aligned} \frac{k_{\text{solid}}}{k_{\text{wool}}} &= 19 \\ \frac{k_{\text{solid}}}{k_{\text{air}}} &= 32 \\ (1+\alpha)^2 &= (19)\left(\frac{\alpha+1}{\alpha+32} + \alpha(\alpha+2)\right) \\ (1+\alpha)^2(\alpha+32) &= 19\left[(\alpha+1) + (2\alpha+\alpha^2)(\alpha+32)\right] \end{aligned}$$

Now assume $\alpha \ll 32$, so

$$\alpha + 32 \approx 32$$

Then

$$\begin{aligned} (1+2\alpha+\alpha^2) &= \left(\frac{19}{32}\right)(1+\alpha+64\alpha+32\alpha^2) \\ (1+2\alpha+\alpha^2) &= \left(\frac{19}{32}\right)(1+65\alpha+32\alpha^2) \end{aligned}$$

$$\begin{aligned} 32\alpha^2(19-1) + \alpha\left[65\times 19 - 64\right] + 19 - 32 &= 0 \\ \alpha^2(32\times 18) + \alpha 1171 - 13 &= 0 \end{aligned}$$

$$\begin{aligned} \alpha &= \frac{-1171 \pm \sqrt{1171^2 + 4\cdot 13\cdot 32\cdot 18}}{2\cdot 32\cdot 18} \\ &= \frac{-1171 \pm 1184}{32\times 36} \end{aligned}$$

only the + sign makes sense, so $\alpha \approx 0.01$.

Chapter 22

Entropy and the Second Law of Thermodynamics

22.1 Three examples of irreversible processes are:

1) The melting of ice cubes in a glass of liquid at room temperature. As the ice absorbs heat from the surrounding liquid, it melts: passing from a more ordered state (solid) to one of less order (liquid). The entropy of the ice and liquid increases since an irreversible transfer of heat from liquid to ice takes place.

2) The free expansion of gas from a punctured tire. (See Example 22.4).

3) An inelastic collision, such as occurs in the ballistic pendulum experiment. Here, a large fraction of kinetic energy is lost to internal energy.

22.3 a)

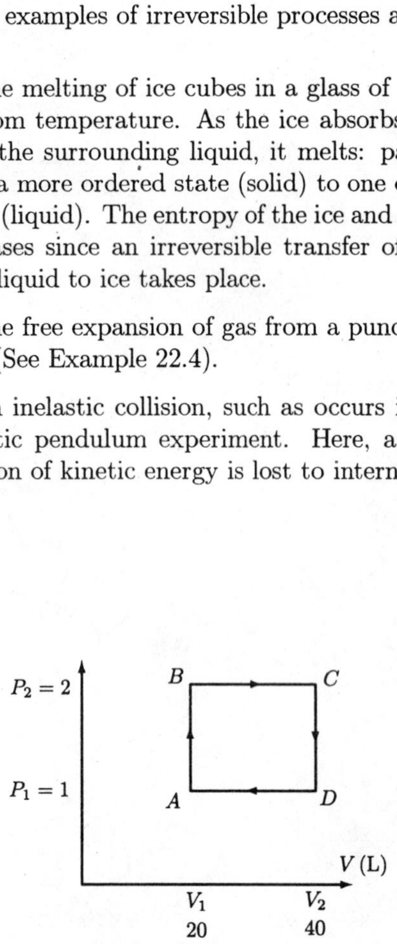

$P_2 = 2$ B C
$P_1 = 1$ A D
V (L)
V_1 20 V_2 40

Consider each process separately,

AB: isochoric ($\Delta V = 0$)

$$Q_{AB} = \mathcal{N} c_v'(T_B - T_A) = \mathcal{N}\left(\frac{3}{2}R\right)(T_B - T_A)$$

Since $\mathcal{N} = 1$,

$$\begin{aligned} Q_{AB} &= \frac{3}{2}V_1(P_2 - P_1) \\ &= \frac{3}{2}(20\text{ L})(2\text{ atm} - 1\text{ atm}) \\ &= 30\text{ L}\cdot\text{atm} \\ W_{AB} &= 0 \\ \Delta U_{AB} &= Q_{AB} - W_{AB} \\ &= Q_{AB} = 30\text{ L}\cdot\text{atm} \end{aligned}$$

where we have used the fact that, for a monatomic ideal gas, $c_v' = \frac{3}{2}R$ and $c_p' = \frac{5}{2}R$, and that $T = \frac{PV}{\mathcal{N}R}$.

BC: isobaric expansion ($\Delta P = 0$)

$$Q_{BC} = \mathcal{N} c_p'(T_C - T_B) = \frac{5}{2}\mathcal{N}R(T_C - T_B)$$

with $\mathcal{N} = 1$,

$$\begin{aligned} Q_{BC} &= \frac{5}{2}P_2(V_2 - V_1) = \frac{5}{2}(2\text{ atm})(40\text{ L} - 20\text{ L}) \\ &= 100\text{ L}\cdot\text{atm} \\ W_{BC} &= P_2(V_2 - V_1) = 2\text{ atm}(20\text{ L}) \\ &= 40\text{ L}\cdot\text{atm} \\ \Delta U_{BC} &= Q_{BC} - W_{BC} \end{aligned}$$

$$= \frac{3}{2}P_2(V_2 - V_1)$$
$$= \frac{3}{2}(2 \text{ atm})(20 \text{ L}) = 60 \text{ L} \cdot \text{atm}$$

CD: Isochoric

$$Q_{CD} = \mathcal{N}c'_v(T_D - T_C) = \frac{3}{2}V_2(P_1 - P_2)$$
$$= \frac{3}{2}(40 \text{ L})(-1 \text{ atm}) = -60 \text{ L} \cdot \text{atm}$$
$$W_{CD} = 0$$
$$\Delta U_{CD} = Q_{CD} - W_{CD}$$
$$= Q_{CD} = -60 \text{ L} \cdot \text{atm}$$

DA: isobaric compression

$$Q_{DA} = \frac{5}{2}P_1(V_1 - V_2) = \frac{5}{2}(1 \text{ atm})(20 \text{ L} - 40 \text{ L})$$
$$= -50 \text{ L} \cdot \text{atm}$$
$$W_{DA} = P_1(V_1 - V_2) = 1 \text{ atm}(-20 \text{ L})$$
$$= -20 \text{ L} \cdot \text{atm}$$
$$\Delta U_{DA} = \frac{3}{2}P_1(V_1 - V_2) = \frac{3}{2}(1 \text{ atm})(-20 \text{ L})$$
$$= -30 \text{ L} \cdot \text{atm}$$

Table: Since 1 L atm = 101 J, we have:

Process:	Q(J)	W(J)	ΔU(J)
AB	3030	0	3030
BC	10100	4040	6060
CD	−6060	0	−6060
DA	−5050	−2020	−3030

As a check, note that the sum of the third column gives:

$$\Delta U_{\text{net}} = 3030 \text{ J} + 6060 \text{ J} - 6060 \text{ J} - 3030 \text{ J} = 0$$

As required for a cyclic set of processes.

b) The efficiency is defined by:

$$e = \frac{\text{net work obtained}}{\text{heat input}} = \frac{\text{W}}{Q_{\text{in}}}$$

from the table in 3a),

$$W_{\text{net}} = 4040 \text{ J} - 2020 \text{ J} = +2020 \text{ J}$$
$$Q_{\text{in}} = 3030 \text{ J} + 10100 \text{ J} = 13130 \text{ J}$$

therefore,

$$e = \frac{W_{\text{net}}}{Q_{\text{in}}} = \frac{2020 \text{ J}}{13130 \text{ J}} = 0.154$$

22.5 Given:

$$T_C = 410 \text{ K}, T_H = 750 \text{ K}, Q_H = 3.0 \times 10^3 \text{ J}$$

Find: work, W and the heat rejected, Q_C

$$e_C = 1 - \frac{T_C}{T_H} = 1 - \frac{410}{750} = 0.453$$

The work done per cycle is:

$$W = e_C Q_H = (0.453)(3.0 \times 10^3 \text{ J})$$
$$= 1.4 \times 10^3 \text{ J}$$

The waste heat exhausted is:

$$|Q_C| = (1 - e_C)Q_H = (0.547)(3.0 \times 10^3 \text{ J})$$
$$= 1.6 \times 10^3 \text{ J}$$

As a check on the answers, note that

$$W + |Q_C| = 1.4 \times 10^3 \text{ J} + 1.6 \times 10^3 \text{ J}$$
$$= 3.0 \times 10^3 \text{ J} = Q_H$$

as required by the 1st Law of Thermodynamics.

22.7 No. For example, an irreversible processes carried out inside an insulated container involves no exchange of energy with the environment. Therefore the only entropy change possible is that of the system. When an ice cube melts, heat is transferred to the cube from the environment. The entropy change of the environment is negative, while the entropy change of the cube is positive and sufficiently large that the total entropy change (of cube plus environment) is also positive.

22.9 The Carnot cycle is the most efficient heat engine, and has the maximum efficiency possible, so

$$e < e_C = 1 - \frac{T_C}{T_H}$$

for $T_C = 300 \text{ K}, T_H = 600 \text{ K}$

$$e < 1 - \frac{300 \text{ K}}{600 \text{ K}} = 1 - (0.5) = 50\%$$

The engine can be no more than 50% efficient.

22.11

$$E_1 = -5.4 \times 10^{-19} \text{ J}; \ g_2 = 8$$
$$E_2 = -2.2 \times 10^{-18} \text{ J}; \ g_2 = 2$$
$$T = 10^4 \text{ K}$$

find:

$$\frac{n_2}{n_1}$$

We use Boltzmann's relation, Eq. (22.12), with

$$\begin{aligned} kT &= (1.38\times10^{-23}\ \mathrm{J\cdot K^{-1}})(10^4\ \mathrm{K}) \\ &= 1.38\times10^{-19}\ \mathrm{J} \end{aligned}$$

and

$$\begin{aligned} E_2-E_1 &= (-5.4\times10^{-19}\ \mathrm{J})-(-2.2\times10^{-18}\ \mathrm{J}) \\ &= 1.66\times10^{-18}\ \mathrm{J} \end{aligned}$$

then

$$\begin{aligned} e^{-\frac{E_2-E_1}{kT}} &= e^{-12.0}=6.0\times10^{-6} \\ \frac{n_2}{n_1} &= \frac{g_2}{g_1}e^{-\frac{E_2-E_1}{kT}}=\left(\frac{8}{2}\right)(6.0\times10^{-6}) \\ &= 2.4\times10^{-5} \end{aligned}$$

Thus there are only 2.4×10^{-5} as many atoms in the more energetic state.

22.13 Solar corona:

$$\begin{aligned} T &\approx 10^6\ \mathrm{K} \\ n &\approx 10^{14}\ \mathrm{m^{-3}} \\ \sigma &\approx 6.7\times10^{-29}\ \mathrm{m^2} \end{aligned}$$

$$\begin{aligned} \lambda &= \frac{1}{n\sigma}=\frac{1}{(10^{14}\ \mathrm{m^{-3}})(6.7\times10^{-29}\ \mathrm{m^2})} \\ &\approx 1\times10^{14}\ \mathrm{m} \end{aligned}$$

22.23 Diesel engine cycle (named after the German engineer, Dr. Rudolph Diesel, who first developed this cycle in 1895:

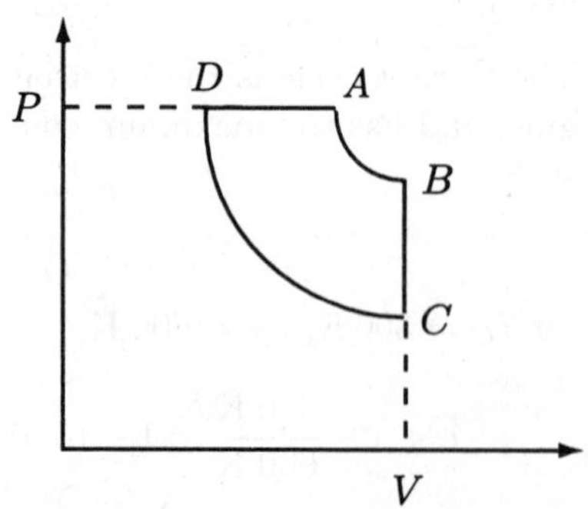

a) At point D the fuel is admitted and combustion at constant pressure takes place from $D \to A$. Thus, one has a constant pressure addition of heat in this part of the cycle. During $B \to C$, the exhaust opens and the burned gases escape releasing the pressure. This is a constant volume rejection of heat. There is no heat transfer during the adiabatic processes CD and AB.

b) Work is done *on* the system during the compression stroke $C \to D$. During this stroke, the gas is compressed adiabatically to a high pressure and temperature (above the ignition temperature of the fuel). Work is done by the system during the adiabatic expansion of the products of combustion ($A \to B$), and during the constant pressure process DA.

c) heat supplied:

$$Q_{\mathrm{in}}=Q_{DA}=c_{\mathrm{p}}(T_A-T_D)$$

heat rejected:

$$\begin{aligned} Q_{\mathrm{out}} &= Q_{BC}=c_{\mathrm{v}}(T_C-T_B) \\ e &= \frac{Q_{\mathrm{in}}+Q_{\mathrm{out}}}{Q_{\mathrm{in}}} \\ &= \frac{c_{\mathrm{p}}(T_A-T_D)+c_{\mathrm{v}}(T_C-T_B)}{c_{\mathrm{p}}(T_A-T_D)} \\ &= 1-\frac{c_{\mathrm{v}}}{c_{\mathrm{p}}}\left(\frac{T_B-T_C}{T_A-T_D}\right) \\ e &= 1-\frac{1}{\gamma}\left(\frac{T_B-T_C}{T_A-T_D}\right) \end{aligned}$$

d) Let

$$r_1=\frac{V_B}{V_A}=\frac{V_C}{V_A},\ r_2=\frac{V_C}{V_D}$$

Along the adiabats $C \to D$ and $A \to B$:

$$\begin{aligned} T_AV_A^{\gamma-1} &= T_BV_B^{\gamma-1} \\ T_DV_D^{\gamma-1} &= T_CV_C^{\gamma-1}=T_CV_B^{\gamma-1} \\ T_AV_A^{\gamma-1}-T_DV_D^{\gamma-1} &= (T_B-T_C)\,V_B^{\gamma-1} \end{aligned}$$

dividing through by $V_B^{\gamma-1}$ and recalling that $V_B=V_C$:

$$T_A\left(r_1^{-1}\right)^{\gamma-1}-T_D\left(r_2^{-1}\right)^{\gamma-1}=T_B-T_C$$

Now, along path $D \to A$ the *pressure* is *constant*:

$$\begin{aligned} \frac{T_D}{T_A} &= \frac{V_D}{V_A}=\frac{V_D}{V_B}\cdot\frac{V_B}{V_A} \\ &= \left(\frac{V_D}{V_C}\right)\left(\frac{V_B}{V_A}\right) \end{aligned}$$

therefore

$$\frac{T_D}{T_A}=r_2^{-1}r_1,$$

or

$$\frac{T_A}{T_D}=r_1^{-1}r_2$$

$$
\begin{aligned}
e &= 1-\frac{1}{\gamma}\left(\frac{T_B-T_C}{T_A-T_D}\right)\\
&= 1-\frac{1}{\gamma}\left(\frac{T_A\left(r_1^{-1}\right)^{\gamma-1}-T_D\left(r_2^{-1}\right)^{\gamma-1}}{T_A-T_D}\right)\\
&= 1-\frac{1}{\gamma}\left\{\frac{\left(\frac{T_A}{T_D}\right)\left(r_1^{-1}\right)^{\gamma-1}-\left(r_2^{-1}\right)^{\gamma-1}}{\left(\frac{T_A}{T_D}\right)-1}\right\}\\
&= 1-\frac{1}{\gamma}\left\{\frac{\left(r_1^{-1}r_2\right)\left(r_1^{-1}\right)^{\gamma-1}-\left(r_2^{-1}\right)^{\gamma-1}}{r_1^{-1}r_2-1}\right\}\\
&= 1-\frac{1}{\gamma}\left(\frac{r_1^{-\gamma}-r_2^{-\gamma}}{r_1^{-1}-r_2^{-1}}\right)
\end{aligned}
$$

from Figure 22.25,

$$
\begin{aligned}
r_1 &= \frac{8\text{ L}}{4\text{ L}}=2\\
r_2 &= \frac{8\text{ L}}{2\text{ L}}=4
\end{aligned}
$$

$$
\begin{aligned}
e_{\text{Diesel}} &= 1-\frac{1}{\gamma}\left(\frac{2^{-\gamma}-4^{-\gamma}}{2^{-1}-4^{-1}}\right)\\
e_{\text{Otto}} &= 1-\frac{1}{r^{\gamma-1}}=1-\frac{1}{4^{\gamma-1}}
\end{aligned}
$$

for $r=r_2$.

For diatomic gas, $\gamma=\frac{7}{5}$ and so $\gamma-1=\frac{2}{5}$. Then

$$
\begin{aligned}
e_{\text{Diesel}} &= 0.328\\
e_{\text{Otto}} &= 0.426
\end{aligned}
$$

(For the numbers in the Figure, $\gamma=\frac{7}{5}$.)

22.25 By increasing the hot temperature while keeping V_A constant, the entire cycle moves upward in the P-V diagram.

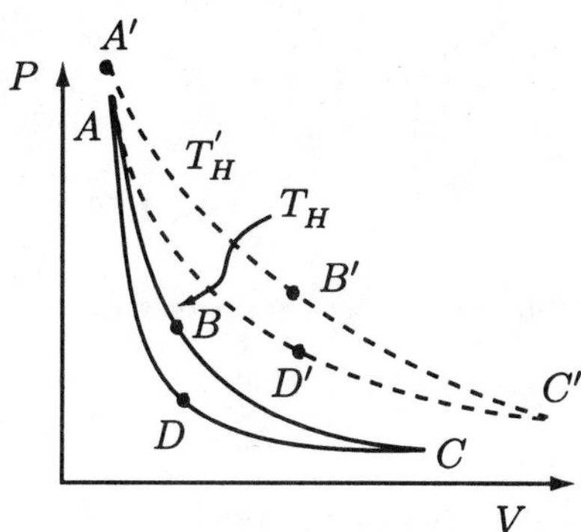

P_A increases, and since AD is an adiabat $P_DV_D^{\gamma}$ also increases. The product P_DV_D is fixed by the value of the temperature T_C of the reservoir, thus V_D must increase. The pressure and volume throughout the cycle increase. Similarly if T_C is decreased, pressure and volume throughout the cycle decrease.

22.29 According to Newton's law of cooling, the rate at which heat leaks into the freezer is proportional to the temperature difference between inside and outside.

$$H\propto T_H-T_C$$

The freezer uses net power P to remove this heat:

$$P=\frac{W_{\text{net}}\text{ per cycle}}{\Delta t\text{ per cycle}}$$

For the Carnot cycle (§22.3)

$$W_{\text{net}}=Nk(T_H-T_C)\ln\left(\frac{V_B}{V_A}\right)$$

to remove heat

$$Q_{\text{cold}}=NkT_C\ln\left(\frac{V_B}{V_A}\right)$$

each cycle. To remove heat faster, the refrigerator must cycle more rapidly.

$$\Delta t=\frac{Q_{\text{cold}}}{H}\propto\frac{T_C}{(T_H-T_C)}$$

Thus the power used is

$$P=\frac{W_{\text{net}}}{\Delta t}\propto\frac{(T_H-T_C)}{\frac{T_C}{(T_H-T_C)}}\propto(T_H-T_C)^2$$

since T_C remains fixed. In this problem,

$$T_H-T_C=55^\circ\text{ F}-25^\circ\text{ F}=30^\circ\text{ F}$$

requires

$$P=75\text{ W}$$

So

$$T_H-T_C=70^\circ\text{ F}-25^\circ\text{ F}=45^\circ\text{ F}$$

requires

$$P=\left(\frac{45}{30}\right)^2(75\text{ W})=\left(\frac{9}{4}\right)(75\text{ W})=170\text{ W}$$

So the required power increases by a factor of 2.25 or by 95 W.

22.37 The Sun decreases its entropy by radiating heat outwards (to the Earth, for example.) We choose an imaginary reversible process in which heat is conducted into a heat reservoir at the surface temperature of the Sun (6300 K, cf. Example 21.5)

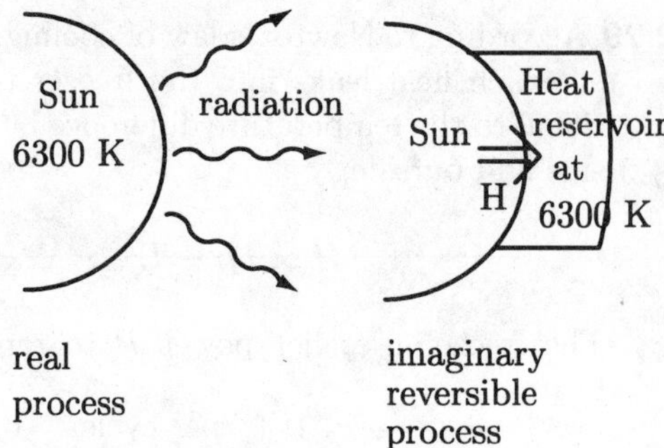

Then

$$\frac{dS}{dt} = \frac{\frac{dQ}{dt}}{T} = \frac{-3.9 \times 10^{26}\ \text{W}}{6300\ \text{K}}$$

(See front endpapers of text)

$$\frac{dS}{dt} = -6.2 \times 10^{22}\ \text{W/K}$$

At least as much entropy is generated elsewhere in the universe by the absorption of solar radiation.

22.39 Given

$$\begin{aligned} m &= 1\ \text{kg} \\ T_1 &= 15°\ \text{C} = 288\ \text{K} \\ T_2 &= 80°\ \text{C} = 353\ \text{K} \end{aligned}$$

The entropy increase of the water is

$$\begin{aligned} \Delta S &= \int \frac{dQ}{T} = \int_{T_1}^{T_2} \frac{mcdT}{T} = mc\ln (T)|_{T_1}^{T_2} \\ &= mc\ln\left(\frac{T_2}{T_1}\right) \\ &= (1.0 \times 10^3\ \text{g})(4.2\ \text{Jg}^{-1}\ °\text{C}^{-1}) \ln\left(\frac{353}{288}\right) \\ &= 850\ \text{J/K} \end{aligned}$$

22.43 The measured efficiency, e, may be expressed in terms of the Carnot efficiency, e_C:

$$e = (0.55)e_C = 0.17$$

therefore

$$e_C = 0.31$$

But we know $e_C = 1 - T_C/T_H$ with

$$T_C = (273 + 5)\ \text{K} = 278\ \text{K}$$

so

$$\begin{aligned} T_H &= \frac{T_C}{(1 - e_C)} = \frac{278\ \text{K}}{1 - (0.31)} \\ &= 402.9\ \text{K} = 4.0 \times 10^2\ \text{K} \end{aligned}$$

22.47 The Maxwellian velocity distribution (Eqn. 22.13)

$$f_m(v) = \left(\frac{m}{2\pi kT}\right)^{\frac{3}{2}} 4\pi v^2 \exp\left\{\frac{-mv^2}{2kT}\right\}$$

gives the fraction of molecules in a speed range v to $v + dv$. For a fixed value of v and dv, two things happen as T increases by a factor of 2:

i) The normalization factor $\left[\frac{m}{2\pi kT}\right]^{\frac{3}{2}}$ *decreases* by a factor $2^{\frac{3}{2}} = 2.83$ and

ii) The quantity $\frac{mv^2}{2}kT$ in the exponent decreases by a factor 2, so $e^{-\frac{mr^2}{2kT}}$ increases. Since

$$e^{\frac{x}{2}} = (e^x)^{\frac{1}{2}}$$

this change does depend on the value of v compared with $\sqrt{\frac{2kT}{m}} = v_{mp}$. However since $x < 0$,

$$0 < e^x < 1$$

and

$$\sqrt{e^x} > e^x$$

Thus this change at least partially offsets the decrease caused by the normalization factor. The total change is given by

$$\frac{f(2T)}{f(T)} = \frac{1}{2^{\frac{3}{2}}} \frac{e^{-\frac{mv^2}{4kTo}}}{e^{-\frac{mv^2}{2kTo}}} = \frac{1}{2^{\frac{3}{2}}} e^{\frac{mv^2}{2kTo}}$$

and is < 1 ($\frac{n}{v}$ decreases) if

$$e^{\frac{mv^2}{2kTo}} < 2^{\frac{3}{2}}$$

or

$$\left(\frac{v}{v_{mp}}\right)^2 < \ln 2^{\frac{3}{2}} = \frac{3}{2}\ln 2 = 1.04$$

or

$$\frac{v}{v_{mp}} < 1.02$$

Conversely, $n(v)$ *increases* for

$$\frac{v}{v_{mp}} > 1.02$$

Qualitatively, the peak of the curve decreases in value and the whole curve decreases in value and the whole curve shifts to the right, so $n(v)$ decreases for values to the left of the peak and increases for values to the right.

22.51

n_1 P_1 T_1		$n_2 = 2n_1$ $P_2 = P_1$ T_2

Mean free path $\lambda = \frac{1}{n\sigma}$. Thus

$$\lambda_2 = \frac{1}{2}\lambda_1$$

The collision time $\tau \sim \frac{1}{n\sigma v}$. Thus

$$\frac{\tau_1}{\tau_2} = \frac{n_2 v_2}{n_1 v_1} = \frac{n_2}{n_1}\sqrt{\frac{T_2}{T_1}}$$

We are given that the pressures are equal, Thus

$$n_1 kT_1 = n_2 kT_2 \Rightarrow \frac{T_2}{T_1} = \frac{n_1}{n_2}$$

so

$$\frac{\tau_1}{\tau_2} = \frac{n_2}{n_1}\sqrt{\frac{n_1}{n_2}} = \sqrt{\frac{n_2}{n_1}} = \sqrt{2}$$

22.55 Helium,

$$T = 20 \text{ K}, \; P = 0.5 \text{ atm}, \; d = 2 \times 10^{-10} \text{ m}$$

The mean free path is $\lambda = \frac{1}{n\sigma}$ with $\sigma = \pi d^2$ and n can be determined from the given pressure and temperature

$$P = nkT$$

Thus

$$\begin{aligned} n &= \frac{P}{kT} = \frac{(0.5 \text{ atm})\left(10^5 \text{ Pa/atm}\right)}{\left(1.4 \times 10^{-23} \frac{\text{J}}{\text{K}}\right)(20 \text{ K})} \\ &= 2 \times 10^{26} \text{ m}^{-3} \end{aligned}$$

Thus

$$\lambda = \frac{1}{\left(2 \times 10^{26} \text{ m}^{-3}\right)\pi\left(2 \times 10^{-10} \text{ m}\right)^2} = 40 \text{ nm}$$

22.59 The entropy of a macrostate is given by Boltzmann's relation:

$$S = k \ln \Omega$$

where $\Omega \equiv$ the number of microstates. For an adiabatic process, assume that the number of possible sets of positions and velocities of N molecules is given by:

$$\Omega = C\Theta(T)V^N$$

where C is a constant, and $\Theta(T)$ is a function of temperature T.

$$S = Nk \ln V + k \ln \Theta + k \ln C$$

for an adiabatic process,

$$\begin{aligned} \Delta S &= Nk \ln V_1 - Nk \ln V_2 + k \ln \Theta_1 - k \ln \Theta_2 \\ &= 0 \\ \ln \frac{\Theta_1}{\Theta_2} &= -N \ln \frac{V_1}{V_2} \end{aligned}$$

for an adiabatic process,

$$\begin{aligned} T_1 V_1^{\gamma-1} &= T_2 V_2^{\gamma-1} \\ \frac{V_1}{V_2} &= \left(\frac{T_2}{T_1}\right)^{\frac{1}{\gamma-1}} \end{aligned}$$

therefore,

$$\ln \frac{\Theta_1}{\Theta_2} = N \ln \left(\frac{T_1}{T_2}\right)^{\frac{1}{\gamma-1}} = N \ln \left(\frac{T_1}{T_2}\right)^{\frac{3}{2}}$$

since $\gamma = \frac{5}{3}$ for a monatomic gas. So,

$$\Theta(T) \propto T^{3N/2}$$

and the temperature dependence of the entropy is given by:

$$\frac{3}{2} Nk \ln T$$

Since

$$E = \frac{1}{2} m v_{\text{rms}}^2 = \frac{3}{2} kT$$

per monatomic atom, then for N atoms,

$$E = \frac{3}{2} NkT$$

and so the entropy may be written as

$$S = Nk \ln V + \frac{3}{2} Nk \ln E + k \ln C'$$

or,

$$S = k \ln \left(C' V^N E^{3N/2}\right)$$

where the number of atoms, N, is assumed to be fixed. Therefore the number of microstates, Ω, is proportional to the volume and the energy:

$$\Omega = CV^N E^{3N/2}$$

Since the energy is proportional to v_{rms}^2,

$$\Omega = (\text{const}) V^N \left(v_{\text{rms}}^3\right)^N$$

If we imagine the velocities (v_x, v_y, v_z) of the molecules as points in a velocity space, the volume of velocity space occupied is, roughly, $\frac{4}{3}\pi v_{\text{rms}}^3$. So,

$$\Omega = (\text{const}) \left(\begin{array}{c} \text{Volume} \\ \text{in space} \end{array}\right)^N \left(\begin{array}{c} \text{Volume in} \\ \text{velocity space} \end{array}\right)^N$$

22.63 Let $n =$ number of molecules $= 10^9$

a) The probability that all n molecules will be aimed toward the left after a single collision

$$= \underbrace{p \cdot p \cdot p \cdots p}_{n \text{ times}} = p^n$$

$$\text{Probability} = \left(\frac{1}{2}\right)^{10^9} = (.5)^{10^9} = 10^{-3\times 10^8}$$

b) If we suppose that the molecules' velocities are completely reordered once each collision time, then after N collisions, the probability that all n molecules will have been aimed toward the left at least once is given by

$$\begin{aligned} P &= 1 - \left(\begin{array}{c}\text{chance it has}\\ \text{never happened}\end{array}\right) \\ &= 1 - \left(\begin{array}{c}\text{chance it doesn't}\\ \text{happen on a given trial}\end{array}\right)^N \end{aligned}$$

So,

$$\begin{aligned} P &= 1-(1-p^n)^N = 1 - \left(1-(0.5)^{10^9}\right)^N \\ &\simeq N\left(10^{-3\times 10^8}\right) \end{aligned}$$

c) The time for N collisions to take place is $t = t_c N$ where

$$t_c = \text{the collision time} = 10^{-10}\text{ s}$$

We want the time necessary for P to be of order unity:

$$N \approx 10^{3\times 10^8} \overset{\text{set}}{\equiv} \frac{t}{t_c} = \frac{t}{\frac{1}{3}\times 10^{-17}\text{ y}}$$

So

$$t = \left(\frac{1}{3}\times 10^{-17}\text{ y}\right)\left(10^{3\times 10^8}\right) = 10^{3\times 10^8}\text{y}$$

Don't hold your breath!

22.67 When heat is added to a solid such as ice, the temperature of the solid rises by an amount that is directly proportional to the quantity of heat added *until* the melting point temperature of the solid is reached. At this point, there is no further increase in temperature as heat is added until *all* the solid is melted. The heat that must be added (per mass) to a solid to change it to a liquid (at the same pressure and temperature) is called the latent heat of fusion, L_F. For ice,

$$L_F = 334.7\times 10^3\text{ J/kg}$$

or

$$L_F = 80\text{ cal/g}$$

Therefore the amount of heat required to melt a block of ice, at 0° C, is the same as that required to raise the temperature of the resulting ice water from 0° C to 80° C! Since we have 10 g of ice at 0° C in 10 g of water at 80° C, and since

$$Q_{\text{lost by the water}} = Q_{\text{gained by the ice}}$$

If the final temperature is zero, and all the ice melts, then

$$\begin{aligned} Q_{\text{lost by the water}} &= 10\text{ g}\times 80°\text{ C}\times 1\text{ cal/g °C} \\ &= 800\text{ calories} \end{aligned}$$

and $Q_{\text{gained by the ice}} = 800$ calories, as required. Thus the final equilibrium state is 20 g of 'ice' water at $T_f = 0°$ C. For the ice:

$$\begin{aligned} Q &= m_{\text{ice}}\cdot L_F = (.01\text{ kg})\left(334.7\times 10^3\,\frac{\text{J}}{\text{kg}}\right) \\ &= 3347\text{ J} \end{aligned}$$

and

$$\Delta S_{\text{ice}} = \frac{Q}{T} = \frac{3347\text{ J}}{273\text{ K}} = 12.26\text{ J/K}$$

For the water

$$\begin{aligned} \Delta S_{\text{water}} &= \int_{T_i}^{T_f}\frac{dQ}{T} = \int_{T_i}^{T_f}\frac{mc\,dT}{T} \\ &= (0.0100\text{ kg})\left(4186\,\frac{\text{J}}{\text{kg}\cdot\text{K}}\right)\ln T\Big|_{353\text{ K}}^{273\text{ K}} \\ &= (41.86\text{ J/K})\ln\left(\frac{273}{353}\right) \\ &= -10.76\text{ J/K} \end{aligned}$$

Thus the net entropy change is

$$\begin{aligned} \Delta S_{\text{net}} &= \Delta S_{\text{ice}} + \Delta S_{\text{water}} \\ &= +12.26\text{ J/K} - 10.76\text{ J/K} \\ &= +1.5\text{ J/K} \end{aligned}$$

and is positive, as required.

22.71 Brayton or Joule cycle. From Figure 22.4,

$$\begin{aligned} Q_{\text{in}} &= \mathcal{N}c_p'(T_C - T_B) \\ Q_{\text{out}} &= \mathcal{N}c_p'(T_A - T_D) \end{aligned}$$

The net work in a cycle equals the total heat input, so the efficiency, e, is:

$$\begin{aligned} e_B &= \frac{W}{Q_{\text{in}}} = \frac{Q_{\text{in}} + Q_{\text{out}}}{Q_{\text{in}}} \\ &= 1 + \frac{(T_A - T_D)}{(T_C - T_B)} \end{aligned}$$

b) Since AB and CD are adiabats,

$$\begin{aligned} P_B V_B^\gamma &= P_A V_A^\gamma \\ P_C V_C^\gamma &= P_D V_D^\gamma \end{aligned}$$

But $P_B = P_C$ and $P_A = P_D$, so

$$\left(\frac{V_B}{V_C}\right)^\gamma = \left(\frac{V_A}{V_D}\right)^\gamma \Rightarrow \frac{V_B}{V_C} = \frac{V_A}{V_D}$$

Then, from the ideal gas law:

$$P_A V_A = \mathcal{N} R T_A$$

and

$$P_D V_D = \mathcal{N} R T_D$$

so

$$\frac{V_A}{V_D} = \frac{T_A}{T_D}$$

and similarly

$$\frac{V_B}{V_C} = \frac{T_B}{T_C}$$

so

$$\frac{T_D}{T_A} = \frac{T_C}{T_B}$$

therefore,

$$\frac{T_D - T_A}{T_C - T_B} = \frac{\left(\left(\frac{T_D}{T_A}\right) - 1\right)}{\left(\left(\frac{T_C}{T_B}\right) - 1\right)} \left(\frac{T_A}{T_B}\right) = \frac{T_A}{T_B}$$

So, the efficiency from part a) may be written as:

$$e_B = 1 - \frac{(T_D - T_A)}{(T_C - T_B)} = 1 - \frac{T_A}{T_B}$$

The efficiency of a Carnot cycle operating between the maximum and minimum temperatures of the Brayton cycle illustrated in Figure 22.4 would be:

$$e_C = 1 - \frac{T_A}{T_C}$$

Clearly $e_C > e_B$, since T_C is the maximum temperature of the cycle.

22.75 a)

$$\frac{N(v > v_{\rm rms})}{N} = 1 - \int_0^{v_{\rm rms}} f_M(v)\,dv$$

where

$$f_M(v) = \left(\frac{m}{2\pi kT}\right)^{\frac{3}{2}} 4\pi v^2 \exp\left\{-\frac{mv^2}{2kT}\right\}$$

Let $u = \alpha v$ where

$$\alpha = \left(\frac{m}{2kT}\right)^{\frac{1}{2}}$$

Then,

$$\begin{aligned} f_M(v) &= \left(\frac{m}{2kT}\right)^{\frac{1}{2}} 4u^2 e^{-u^2} \\ u_{\rm rms} &= \alpha v_{\rm rms} = \sqrt{\frac{3}{2}} \end{aligned}$$

and

$$\begin{aligned} f_M(v)\,dv &= \frac{4}{\sqrt{\pi}} u^2 e^{-u^2} du \\ \frac{N(v > v_{\rm rms})}{N} &= 1 - \frac{4}{\sqrt{\pi}} \int_0^{u_{\rm rms}} u^2 e^{-u^2} du \end{aligned}$$

Integrate by parts:

$$\frac{N(v > v_{\rm rms})}{N} = 1 + \frac{2}{\sqrt{\pi}}\left(ue^{-u^2}\Big|_0^{\sqrt{3/2}}\right) - \mathrm{erf}\left(\sqrt{\frac{3}{2}}\right)$$

where erf $(y) = \frac{2}{\pi}\int_0^y e^{-t^2}\,dt$. Looking up the values in a table, we find

$$\frac{N(V > v_{\rm rms})}{N} = 1 + 0.308 - (0.889) = 0.42$$

Therefore, 42% of the molecules in a Maxwellian have speeds greater than $v_{\rm rms}$.

b)

$$\frac{N(V > 3v_{\rm rms})}{N} = 1 - \int_0^{3v_{\rm rms}} f_M(v)\,dv$$

repeating the previous steps from a),

$$\begin{aligned} \frac{N(V > 3v_{\rm rms})}{N} &= 1 - \frac{4}{\sqrt{\pi}} \int_0^{3u_{\rm rms}} u^2 e^{-u^2} du \\ &= 1 + \frac{2}{\sqrt{\pi}} ue^{-u^2}\Big|_0^{3\left(\sqrt{3/2}\right)} \\ &\quad - \mathrm{erf}\left(3\sqrt{\frac{3}{2}}\right) \\ \frac{N(V > 3v_{\rm rms})}{N} &= 1 + 5.7 \times 10^{-6} - (\approx 1) \\ &= 6 \times 10^{-6} \end{aligned}$$

Therefore, only 6 millionths of the molecules in a Maxwellian distribution have speeds greater than $3v_{\rm rms}$.

22.79

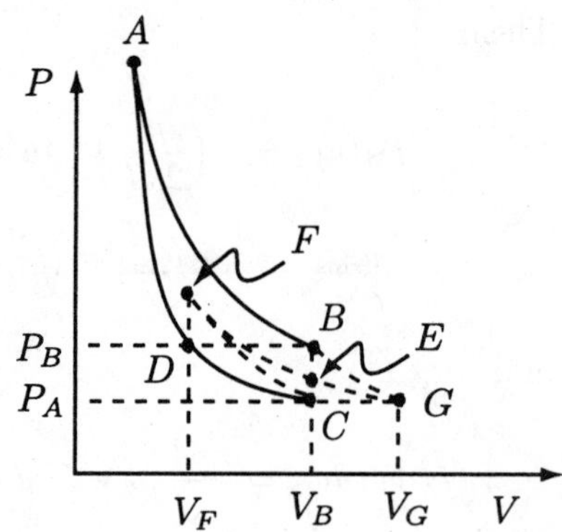

a) FG is an isotherm, Temperature $T_E = T_F = T_G$, CF and GB are adiabats. Thus

$$\begin{aligned}
P_E V_E &= P_F V_F = P_G V_G \\
\Rightarrow \frac{P_F}{P_G} &= \frac{V_G}{V_F} \\
P_F V_F^\gamma &= P_C V_C^\gamma \\
\Rightarrow V_C^\gamma &= \left(\frac{P_F}{P_C}\right) V_F^\gamma \\
P_B V_B^\gamma &= P_G V_G^\gamma
\end{aligned}$$

and $V_B = V_C$, so

$$P_B V_C^\gamma = P_G V_G^\gamma$$

so

$$\begin{aligned}
(P_B - P_C) V_C^\gamma &= P_G V_G^\gamma - P_F V_F^\gamma \\
(P_B - P_C) &= \frac{P_G V_G^\gamma - P_F V_F^\gamma}{V_F^\gamma}\left(\frac{P_C}{P_F}\right) \\
&= \frac{P_G P_C}{P_F}\left(\frac{V_G}{V_F}\right)^\gamma - P_C \\
&= P_C\left[\left(\frac{V_G}{V_F}\right)^{\gamma-1} - 1\right] \\
\frac{\Delta P}{P_C} &= \left(\frac{V_G}{V_F}\right)^{\gamma-1} - 1 \\
\frac{V_G}{V_F} &= \left[\frac{\Delta P}{P_C} + 1\right]^{\frac{1}{\gamma-1}}
\end{aligned}$$

b) The heat transfer along the isotherm is equal to the work done along the isotherm since $\Delta U = 0$

$$\begin{aligned}
Q_i &= W = \int_{V_F}^{V_G} P\, dV \\
&= \int_{V_F}^{V_G} \frac{P_E V_E}{V}\, dV \\
&= P_E V_E \ln\left[\frac{V_G}{V_F}\right]
\end{aligned}$$

Thus

$$\begin{aligned}
Q_i &= P_E V_E \ln\left\{\left[\frac{\Delta P}{P_C} + 1\right]^{\frac{1}{\gamma-1}}\right\} \\
&= \frac{P_E V_E}{\gamma - 1} \ln\left[\frac{\Delta P}{P_C} + 1\right]
\end{aligned}$$

c) Expanding the logarithm,

$$\ln(1+x) = x - \frac{x^2}{2} + \cdots$$

(see Appendix 1B);

$$Q_i = \frac{P_E V_E}{\gamma - 1}\left[\frac{\Delta P}{P_C}\right]$$

to first order in ΔP. Now

$$Q_{CV} = \mathcal{N} c_v' \Delta T$$

Also since $PV = \mathcal{N} RT$, in a constant volume process

$$(\Delta P) V = \mathcal{N} R \Delta T \Rightarrow \Delta T = (\Delta P)\frac{V}{\mathcal{N} R}$$

So

$$Q_{CV} = \left(\frac{c_v'}{R}\right) V \Delta P$$

Now,

$$P_E V_E = \mathcal{N} R T_E$$
$$P_C V_E = \mathcal{N} R T_C$$

So

$$\frac{P_E}{P_C} = \frac{T_E}{T_C} = 1 + \frac{(T_E - T_C)}{T_C}$$

But $T_E - T_C < \Delta T$; so to *first* order in small quantities,

$$Q_i = \frac{P_E}{P_C}\frac{V\Delta P}{(\gamma - 1)} = \frac{V \Delta P}{\gamma - 1}$$

Now

$$c_p' = c_v' + R$$

so

$$\frac{c_p'}{c_v'} = 1 + \frac{R}{c_v'} = \gamma$$

so

$$\frac{R}{c_v'} = \gamma - 1$$

Thus

$$Q_i = \frac{V\Delta P}{\gamma - 1} = \frac{V \Delta P}{R} c_v' = Q_{CV}$$

as required.

d) S is a state variable, and both Q and ΔS are zero for the two adiabats CF and GB. We have already shown that

$$\Delta S = \frac{Q_i}{T}$$

for an isotherm. Thus $\Delta S = \frac{Q_i}{T_E}$ for the segment CB of the constant volume process, and $Q_i = Q_{CV}$. Thus

$$dS = \frac{Q_{CV}}{T}$$

for the segment CB also.

22.81 a) Consider an imaginary cylinder, of height $v_z\,dt$, with the hole of area a as its base. Only molecules inside this cylinder (whose speeds lie in ranges v_x to $v_x + dv_x$, v_y to $v_y + dv_y$, and v_z to $v_z + dv_z$) can escape through the hole. Since the cylinder's volume is $a(v_z dt)$, the number of molecules inside the cylinder is:

$$dn_v = na(v_z dt)\, f(v_x)\, f(v_y)\, f(v_z)\, dv_x\, dv_y\, dv_z$$

where $n = \frac{N}{V}$ is the mean density of the gas in the container. Transforming from Cartesian coordinates to spherical polar coordinates (cf. Problem 61) yields:

$$dn_v = na(v\cos\theta)\, dt\; f(\vec{\mathbf{v}}) v^2 dv \sin\theta\, d\theta\, d\phi$$

The number of molecules, with speeds in the range v to $v+dv$, per unit solid angle ($\sin\theta\, d\theta\, d\phi$) which escape per second is:

$$\begin{aligned} d\nu &= \frac{dn_v}{dt} \\ &= na(v\cos\theta)\left(\frac{m}{2\pi kT}\right)^{\frac{3}{2}} v^2 e^{-\frac{mv^2}{2kT}}\, dv \sin\theta\, d\theta\, d\phi \end{aligned}$$

To find the total number of molecules which exit the hole in a unit time dt, integrate over all speeds and directions:

$$\begin{aligned} \nu &= na\left(\frac{m}{2\pi kT}\right)^{\frac{3}{2}} \int_{\theta=0}^{\pi/2} \cos\theta \sin\theta\, d\theta \int_0^{2\pi} d\phi \\ &\times \int_0^\infty v^3 e^{-mv^2/2kT}\, dv \end{aligned}$$

$$I_1 = \int_{\theta=0}^{\pi/2} \cos\theta\sin\theta\, d\theta$$

$$\text{Let } \begin{cases} \sin\theta = w \\ \cos\theta\, d\theta = dw \end{cases}$$

$$\begin{aligned} I_1 &= \int_0^1 w\, dw = \left.\frac{w^2}{2}\right|_0^1 = \frac{1}{2} \\ I_2 &= \int_0^{2\pi} d\phi = 2\pi \\ I_3 &= \int_0^\infty v^3 e^{-\frac{mv^2}{2kT}}\, dv \end{aligned}$$

$$\text{Let } \begin{cases} \frac{mv^2}{2kT} = x \\ \frac{mv}{kT}\, dv = dx \end{cases}$$

$$\begin{aligned} I_3 &= \int_0^\infty \left(\frac{2kT}{m}x\right) e^{-x} \left(\frac{kT}{m}dx\right) \\ &= 2\left(\frac{kT}{m}\right)^2 \int_0^\infty x e^{-x}\, dx \end{aligned}$$

Integrate by parts:

$$\begin{aligned} I_3 &= 2\left(\frac{kT}{m}\right)^2 \left[\left.\frac{xe^{-x}}{-1}\right|_0^\infty + \int_0^\infty e^{-x}\, dx\right] \\ &= 2\left(\frac{kT}{m}\right)^2 \left[0 - e^{-x}\Big|_0^\infty\right] = 2\left(\frac{kT}{m}\right)^2 \end{aligned}$$

So

$$\begin{aligned} \nu &= na\left(\frac{m}{2\pi kT}\right)^{\frac{3}{2}} \left(\frac{1}{2}\right)(2\pi)2\left(\frac{kT}{m}\right)^2 \\ &= na\left(\frac{kT}{2\pi m}\right)^{\frac{1}{2}} \end{aligned}$$

ν is the rate at which molecules leave the hole and

$$n = \frac{\mathcal{N} N_A}{V} = \frac{1}{4} na\langle v\rangle$$

where

$$\langle v\rangle = \left(\frac{8kT}{\pi m}\right)^{\frac{1}{2}}$$

is the mean speed.

b) When a molecule with speed v leaves it takes energy

$$E = \frac{1}{2}mv^2$$

with it, so the rate of energy change is

$$\begin{aligned} \frac{\Delta E}{\Delta t} &= E\, d\nu \\ &= \frac{1}{2}mv^2 nav\cos\theta \left(\frac{m}{2\pi kT}\right)^{\frac{3}{2}} \\ &\quad \times v^2 e^{-\frac{mv^2}{2kT}}\, dv \sin\theta\, d\theta\, d\varphi \end{aligned}$$

integrating over all speeds and directions,

$$\begin{aligned} \frac{dE}{dt} &= \frac{nam}{2}\left(\frac{m}{2\pi kT}\right)^{\frac{3}{2}} \left(\frac{1}{2}\right)(2\pi) \int_0^\infty v^5 e^{-\frac{mv^2}{2kT}}\, dv \\ &= \frac{nam}{2\sqrt{\pi}}\left(\frac{m}{2\pi kT}\right)^{\frac{3}{2}} \left(\frac{2kT}{m}\right)^3 \end{aligned}$$

(see the text's Appendix 1F for the integral)

$$\frac{dE}{dt} = \frac{nam}{2\sqrt{\pi}}\left(\frac{2kT}{m}\right)^{\frac{3}{2}}$$

Check dimensions:

$$\begin{aligned} kT &= \text{energy} \\ \sqrt{\frac{kT}{m}} &= \text{speed} \\ na &= \frac{1}{L} \end{aligned}$$

so we have

$$\frac{1}{L}\cdot\frac{L}{T}\,\text{Energy} = \frac{\text{Energy}}{\text{Time}}$$

$\therefore$ OK. The energy change per molecule escaping the hole is:

$$\epsilon_{esc} = \frac{dE/dt}{\nu} = \frac{\left(\frac{nam}{2\sqrt{\pi}}\right)\left(\frac{2kT}{m}\right)^{\frac{3}{2}}}{na\left(\frac{kT}{2\pi m}\right)^{\frac{1}{2}}} = 2kT$$

Note that the mean energy, ϵ_i, for a molecule inside the container is smaller than the mean escape energy, ϵ_{esc}:

$$\epsilon_i = \frac{3}{2}kT < \epsilon_{esc} = 2kT$$

or,

$$\epsilon_{esc} = \frac{4}{3}\epsilon_i$$

i.e., those molecules with speeds greater than the mean thermal speed (v_{rms}) are more likely to collide with the wall of the container, and so have a greater chance of exiting through the hole.

c) The rate at which the gas inside the container loses internal energy,

$$\frac{dE_i}{dt} = \frac{d}{dt}\left[\frac{3}{2}NKT\right] = \frac{3}{2}k(\dot{N}T + N\dot{T})$$

must equal the average energy change due to escaping molecules:

$$\frac{dE_{esc}}{dt} = -2\nu kT$$

Since $\dot{N} = \frac{dN}{dt} = -\nu$, this means that:

$$\begin{aligned} -\frac{3}{2}k\nu T + \frac{3}{2}kN\dot{T} &= -2\nu kT \\ \dot{T} &= -\frac{1}{3}\left(\frac{\nu}{N}\right)T \end{aligned}$$

since

$$\nu = \left(\frac{N}{V}\right)a\sqrt{\frac{k}{2\pi m}}T^{\frac{1}{2}}$$

$$\frac{dT}{dt} = -\frac{1}{3}\left(\frac{1}{V}\right)a\sqrt{\frac{k}{2\pi m}}T^{\frac{3}{2}}$$

Note:

$$\frac{dT}{dt} = -\frac{a}{3V}\langle v\rangle T$$

where $\langle v\rangle$ is the mean speed. Thus

$$\frac{dT}{dt} = \frac{-T}{\tau}$$

with

$$\tau = \frac{3V}{a}\frac{1}{\langle v\rangle}$$

We may integrate this expression to find $T(t)$:

$$\frac{dT}{dt} = -\frac{1}{3}KT^{\frac{3}{2}}$$

where

$$K = \frac{1}{V}a\sqrt{\frac{k}{2\pi m}}$$

$$\begin{aligned} \int_{T_0}^{T}\frac{dT}{T^{\frac{3}{2}}} &= \int_0^t -\frac{K\,dt}{3} = -\frac{Kt}{3} \\ -2\left(\frac{1}{\sqrt{T}} - \frac{1}{\sqrt{T_0}}\right) &= -\frac{Kt}{3} \\ \frac{1}{\sqrt{T}} &= \frac{1}{\sqrt{T_0}} + \frac{Kt}{6} \\ \sqrt{T} &= \frac{1}{\frac{1}{\sqrt{T_0}} + \frac{Kt}{6}} \\ &= \frac{\sqrt{T_0}}{1 + \frac{K\sqrt{T_0}t}{6}} \end{aligned}$$

$$T = \frac{T_0}{\left[1 + \frac{K\sqrt{T_0}t}{6}\right]^2} = \frac{T_0}{\left[1 + \frac{a}{6V}\sqrt{\frac{kT_0}{2\pi m}}t\right]^2}$$

22.83 a) In any cycle, the system's internal energy has the same value at the beginning and end of the cycle.

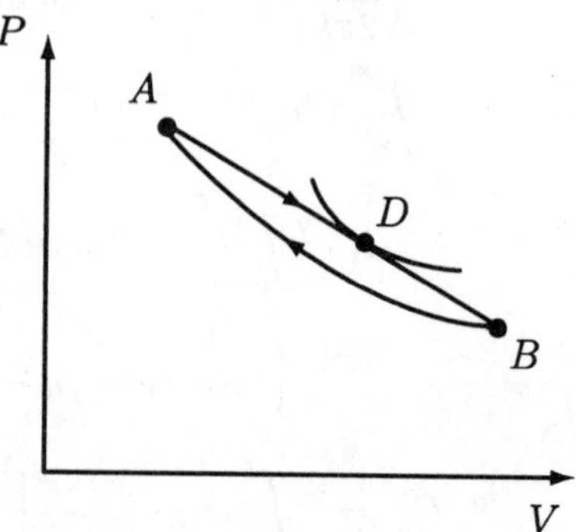

So, the first law requires $Q_{cycle} = W_{cycle}$. Since BA is an adiabat, $Q_{BA} = 0$ and $Q_{cycle} = Q_{AB} = W_{cycle}$, as required. Then, if Q_{AB} were entirely heat input (ie transfer into the system along the entire process) then the cycle's efficiency would be

$$e = \frac{W}{Q_{AB}} = 1$$

b) On an adiabat $\frac{dQ}{dV} \equiv 0$, so the same is true of the straight line process at point D. Between A and D, the adiabats intersecting line AB are steeper than the line, so thermal energy input is needed to account for greater work done and higher final temperature after an expansion dV along the line as compared with expansion along an adiabat. Between A and D, $\frac{dQ}{dV} > 0$ along the straight line. This argument is exactly reversed between D and A and $\frac{dQ}{dV} < 0$. To find point D, demand that the adiabat through D have the same slope as the straight line process:

adiabat:

$$P = \alpha V^{-\gamma}$$

and

$$\frac{dP}{dV} = -\alpha\gamma V^{-(\gamma+1)}$$

str. line:

$$\begin{aligned}\frac{P_A - P}{P_A - P_B} &= \frac{V_A - V}{V_A - V_B} \\ \Rightarrow P &= \frac{P_B V_A - P_A V_B}{V_A - V_B} + \frac{P_A - P_B}{V_A - V_B} V\end{aligned}$$

So we set slopes equal:

$$\frac{P_A - P_B}{V_A - V_B} = -\alpha\gamma V_D^{-(\gamma+1)}$$

and pressures equal

$$\begin{aligned}P_D &= \frac{P_B V_A - P_A V_B}{V_A - V_B} + \frac{P_A - P_B}{V_A - V_B} V_D \\ &= \alpha V_D^{-\gamma}\end{aligned}$$

at point D. So

$$\begin{aligned}\tfrac{P_B V_A - P_A V_B}{V_A - V_B} + \tfrac{P_A - P_B}{V_A - V_B} V_D &= -\tfrac{V_D}{\gamma}\left(-\alpha\gamma V_D^{-(\gamma+1)}\right) \\ &= -\tfrac{V_D}{\gamma}\left(\tfrac{P_A - P_B}{V_A - V_B}\right)\end{aligned}$$

Then

$$\begin{aligned}P_B V_A - P_A V_B &= (P_A - P_B)\left(-1 - \frac{1}{\gamma}\right) V_D \\ \Rightarrow V_D &= \frac{\gamma}{\gamma + 1} \frac{P_A V_B - P_B V_A}{P_A - P_B},\end{aligned}$$

as required.

c) The net work done by the system in a cycle is

$$\begin{aligned}W &= W_{AB} + W_{BA} \\ &= \quad (\text{Area beneath line } AB) \\ &\quad - (\text{Area beneath } BA) \\ &= \frac{P_A + P_B}{2}(V_B - V_B) + \frac{P_B V_B - P_A V_A}{\gamma - 1}\end{aligned}$$

(Equations 19.23 and 19.22 give the expression for W_{BA})

d) The heat input is

$$Q_{\text{in}} = Q_{A \to D} = W_{AD} + \Delta U_{AD}$$

$$\begin{aligned}&= \frac{P_A + P_D}{2}(V_D - V_A) + \mathcal{N} c_v'(T_D - T_A) \\ &= \frac{P_A + P_D}{2}(V_D - V_A) + \frac{(P_D V_D - P_A V_A)}{\gamma - 1}\end{aligned}$$

(Note: $c_v' = \dfrac{R}{\gamma - 1}$)

e) With $V_B = 3V_A$,

$$P_B = \frac{P_A V_A^\gamma}{V_B^\gamma} = \frac{P_A}{3}$$

With $\gamma = \frac{5}{3}$ then,

$$\begin{aligned}V_D &= \frac{\frac{5}{3} P_A V_A \left(3 - \frac{1}{3^\gamma}\right)}{\frac{8}{3} P_A \left(1 - \frac{1}{3^\gamma}\right)} \\ &= \frac{5V_A}{8} \frac{3^{\gamma+1} - 1}{3^\gamma - 1} = 2.114 V_A\end{aligned}$$

Now

$$\begin{aligned}P_D &= \frac{P_A V_A (3^{-\gamma} - 3)}{V_A (1 - 3)} + \frac{P_A (1 - 3^{-\gamma})}{V_A (1 - 3)} (2.114 V_A) \\ &= 0.532 P_A\end{aligned}$$

So

$$\begin{aligned}W &= \frac{P_A (1 + 3^{-\gamma})}{2} V V_A (3 - 1) \\ &\quad + \frac{P_A V_A \left(3^{1-\gamma} - 1\right)}{\frac{2}{3}} \\ &= 0.381 P_A V_A\end{aligned}$$

And

$$\begin{aligned}Q_{\text{in}} &= \frac{P_A (1 + 0.532)}{2} V_A (2.114 - 1) \\ &\quad + \frac{P_A V_A [(0.532)(2.114) - 1]}{\frac{2}{3}} \\ &= 1.040 P_A V_A\end{aligned}$$

So, the efficiency is

$$e \equiv \frac{W}{Q_{\text{in}}} = 0.366$$

f) Higher temperature isotherms lie to the right in the P–V diagram.

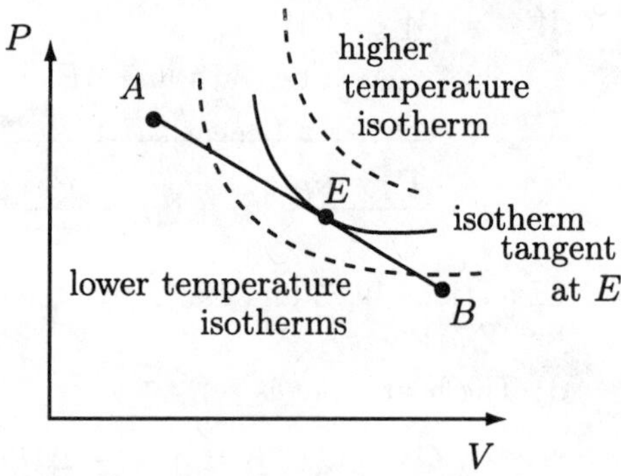

If an isotherm is tangent to the straight-line process at E, then any higher temperature isotherm will lie entirely to its right and not intersect the line AB at all. Thus, E is the maximum temperature point on the line AB.

Isotherm:

$$P = \frac{\mathcal{N}RT_E}{V}$$

Slope:

$$\frac{dP}{dV} = -\frac{\mathcal{N}RT_E}{V^2}.$$

Matching slope and pressure at point E,

$$P_E = \frac{P_BV_A - P_AV_B}{V_A - V_B} + \frac{P_A - P_B}{V_A - V_B}V_E \overset{\text{set}}{\equiv} \frac{\mathcal{N}RT_E}{V_E};$$

But

$$\frac{P_A - P_B}{V_A - V_B} = -\frac{\mathcal{N}RT_E}{V_E^2}$$

So

$$P_E = -\frac{P_A - P_B}{V_A - V_B} \cdot V_E$$

So

$$\begin{aligned} V_E &= \frac{P_AV_B - P_BV_A}{2(P_A - P_B)} \\ &= \frac{P_AV_A(3 - 3^{-\gamma})}{2P_A(1 - 3^{-\gamma})} \\ &= \frac{V_A}{2}\frac{(3^{\gamma+1} - 1)}{(3^{\gamma} - 1)} \\ &= 1.691V_A \end{aligned}$$

Then

$$\begin{aligned} P_E &= \frac{P_BV_A - P_AV_B}{V_A - V_B} + \frac{(P_A - P_B)}{(V_A - V_B)}\frac{(P_AV_B - P_BV_A)}{2(P_A - P_B)} \\ &= \frac{P_BV_A - P_AV_B}{2(V_A - V_B)} = \frac{P_AV_A(3^{-\gamma} - 3)}{2(1-3)V_A} \\ &= \frac{P_A}{4}(3 - 3^{-\gamma}) = 0.710P_A \end{aligned}$$

So

$$\begin{aligned} T_E &= \frac{P_EV_E}{\mathcal{N}R} = (1.691)(0.710)\frac{P_AV_A}{\mathcal{N}R} \\ &= (1.201)\frac{P_AV_A}{\mathcal{N}R} \end{aligned}$$

g) The efficiency of a Carnot cycle between the same maximum and minimum temperatures as the Sadly Cannot cycle is

$$\begin{aligned} e_c &\equiv 1 - \frac{T_C}{T_H} = 1 - \frac{\frac{P_BV_B}{(\mathcal{N}R)}}{T_E} \\ &= 1 - \frac{P_BV_B}{(1.201)P_AV_A} \\ &= 1 - \frac{3^{1-\gamma}}{1.201} = 0.5997 \end{aligned}$$

Chapter 23

Charge and the Electric Field

23.1 Conservation of charge requires that the charge on the anti neutrino equal the sum of the charge on the proton and electron, e and $-e$ respectively

$$Q_{\text{neutrino}} = +e - e = 0.$$

23.3 The triboelectric effect between the cotton T-shirt and wool sweater will leave equal amounts of +, − charge on the shirt and sweater. Looking at Table 24.1, wool is more likely to give up its electrons than cotton so a net positive charge will be left on the sweater and a net negative charge will be left on the cotton T-shirt and person.

23.5 At a distance 1.0 mm from the charge, $\vec{\mathbf{E}}$ will have a magnitude of

$$E = \frac{kQ}{r^2} = \frac{k(1.0\ \mu\text{C})}{(1.0 \times 10^{-3}\ \text{m})^2} = 9.0 \times 10^9\ \frac{\text{N}}{\text{C}}$$

Electric field lines go towards negative charge so the direction of $\vec{\mathbf{E}}$ at P is towards Q.

23.7 The electrostatic force on the test charge is zero wherever the electric field is zero. Since the charges are of equal sign and magnitude, the electric field is zero at points which are equidistant from the two charges and at which the electric field contributions from each charge have opposite directions. Choice a) is the correct answer.

23.9 Field lines cannot cross each other. Since all the charges are positive point charges, $\vec{\mathbf{E}}$ extends outward from each charge. Close to each charge, $\vec{\mathbf{E}}$ appears radial. Since the charges are equal, there will be an equal number of field lines extending from each charge. Since there's net charge in the region, the field lines are spherically symmetric when viewed from afar.

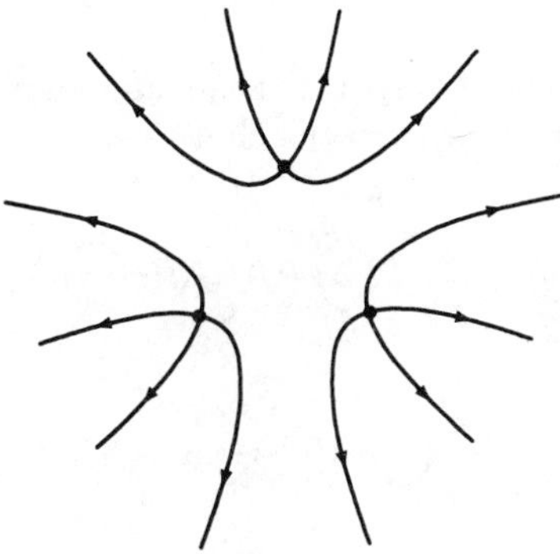

23.11 By Gauss' Law all we need to do to find the net flux through any volume is find the net charge within the volume and divide it by ϵ_0.

Volume A:

$$\begin{aligned} Q_{\text{net,inside}} &= 6Q + Q - Q = 6Q \\ \Phi_e &= \frac{1}{\epsilon_0} Q_{\text{net,inside}} \\ &= 6\frac{Q}{\epsilon_0} = \frac{1}{\epsilon_0} 6(4.425\ \mu\text{C}) \\ &= 3.00 \times 10^6\ \frac{\text{N}\cdot\text{m}^2}{\text{C}} \end{aligned}$$

Volume B:

$$Q_{\text{net,inside}} = 6Q - 4Q = 2Q$$

$$\begin{aligned}\Phi_e &= \frac{1}{\epsilon_0} Q_{\text{net,inside}} \\ &= 2\frac{Q}{\epsilon_0} = \frac{1}{\epsilon_0} 2(4.425\ \mu\text{C}) \\ &= 1.00 \times 10^6\ \frac{\text{N}\cdot\text{m}^2}{\text{C}}\end{aligned}$$

Volume C:

$$\begin{aligned}Q_{\text{net,inside}} &= Q - 3Q - 2Q = -4Q \\ \Phi_e &= \frac{1}{\epsilon_0} Q_{\text{net,inside}} \\ &= -4\frac{Q}{\epsilon_0} = \frac{1}{\epsilon_0}(-4)(4.425\ \mu\text{C}) \\ &= -2.00 \times 10^6\ \frac{\text{N}\cdot\text{m}^2}{\text{C}}\end{aligned}$$

23.15 From Coulomb's law,

$$\begin{aligned}F &= \frac{ke^2}{r^2} \\ &= \frac{\left(9.0 \times 10^9\ \frac{\text{N}\cdot\text{m}^2}{\text{C}^2}\right)(1.602 \times 10^{-19}\ \text{C})^2}{\left(10^{-10}\ \text{m}\right)^2} \\ &= 2 \times 10^{-8}\ \text{N}\end{aligned}$$

23.19 The charges are of the same sign so the Coulomb force between the charges is repulsive.

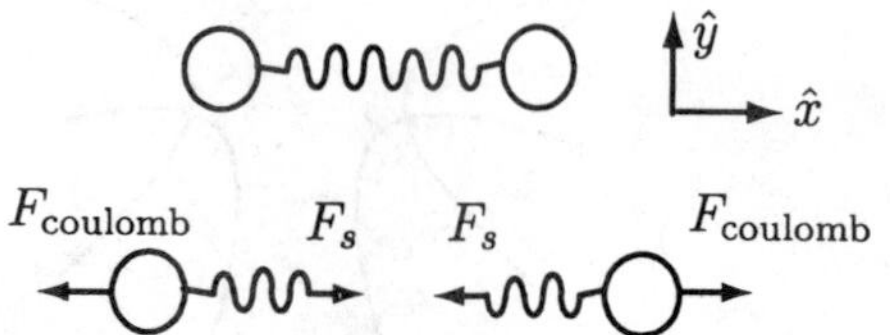

The spring force, F_s, keeps the objects from flying apart. From the FBD, the condition for equilibrium is

$$F_x = F_{\text{coulomb}} - F_s = 0$$

where

$$F_{\text{coulomb}} = \frac{kQ^2}{r^2}$$

$F_s = k_s r$ and r is the distance between the objects when they're in equilibrium.

$$\begin{aligned}\frac{kQ^2}{r^2} - k_s r &= 0 \\ \Rightarrow r^3 &= \frac{kQ^2}{k_s} \\ \Rightarrow r &= \left(\frac{kQ^2}{k_s}\right)^{\frac{1}{3}}\end{aligned}$$

23.23 From the equation for $\vec{\mathbf{E}}$ due to a point charge,

$$\begin{aligned}E &= \frac{kQ}{r^2} \\ Q &= \frac{Er^2}{k} \\ &= \frac{\left(56\ \frac{\text{N}}{\text{C}}\right)(0.17\ \text{m})^2}{9.0 \times 10^9\ \frac{\text{N}\cdot\text{m}^2}{\text{C}^2}} \\ &= 0.18\ \text{nC}\end{aligned}$$

23.27 The electric field vector points from + to −. The electron, being negatively charged wants to travel towards +, so we can conclude that the force on an electron in an electric field is opposite the direction of the electric field.

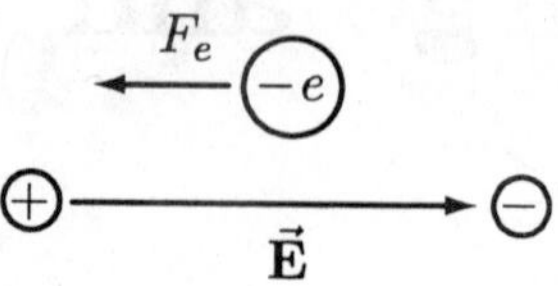

Since gravity pulls the electron downward, we want the coulomb force on the electron to act upward. Therefore $\vec{\mathbf{E}}$ must point downward and have a magnitude:

$$F_{\text{grav}} = F_e \Rightarrow m_e g = q_e E$$

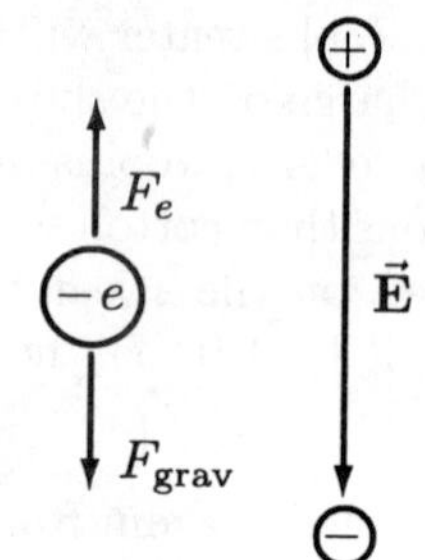

$$\begin{aligned}E &= \frac{m_e g}{q_e} \\ &= \frac{(9.109 \times 10^{-31}\ \text{kg})(9.81\ \text{m/s}^2)}{(1.602 \times 10^{-19}\ \text{C})} \\ \vec{\mathbf{E}} &= \left(5.58 \times 10^{-11}\ \frac{\text{N}}{\text{C}},\ \text{downward}\right)\end{aligned}$$

23.31 The field line diagram of this charge distribution can be found in Figure 23.16 of the text:

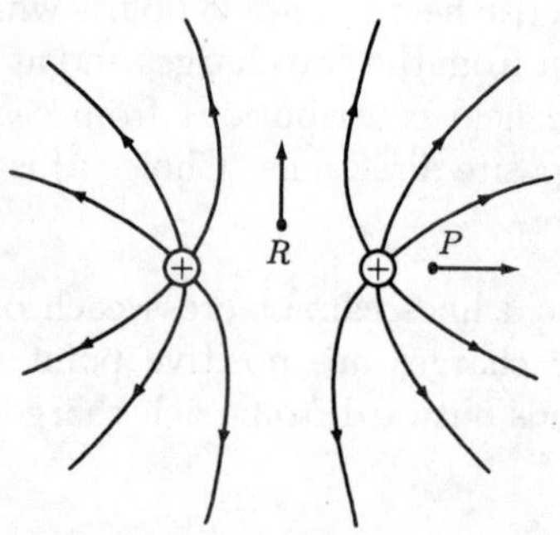

We drew the direction of $\vec{\mathbf{E}}$ at points R and P based upon the field line diagram, since the direction of the field line at any given point shows the direction of $\vec{\mathbf{E}}$ at that point.
At R, $\vec{\mathbf{E}}$ is upward
At P, $\vec{\mathbf{E}}$ is to the right.

23.35 The things to remember are:

a) The charges are of equal magnitude so an equal number of lines will be entering/leaving any given charge.

b) We are told that the charges are placed on a line as opposed to a line segment, so the line is either infinite or very long. These two facts taken together tells us that there is no one special charge; the field line diagram must be symmetric about each and every charge. In fact, since any given charge has only charges of opposite sign at either end, the field line diagram must be mirror symmetric about each and every charge.

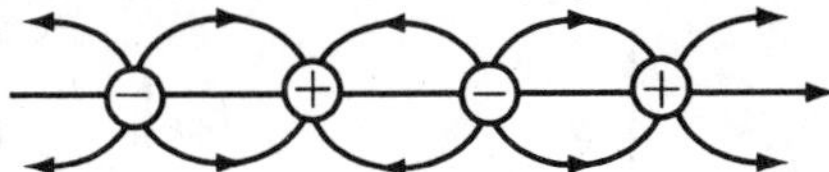

23.39 The electric field vectors have been drawn at the center of the pentagon, O. While it's not immediately obvious that they sum to zero, we can use geometric arguments to show $\vec{\mathbf{E}}(O) = 0$, without resorting to breaking everything down into components.

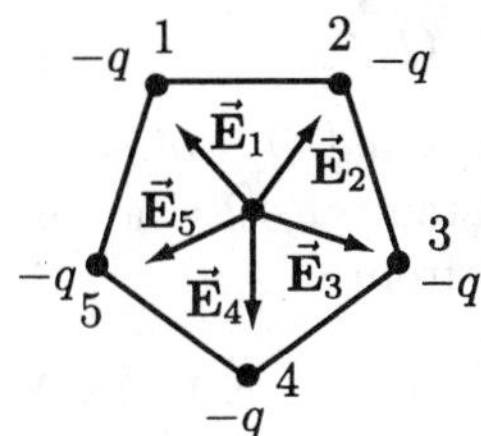

First of all, since O is equidistant from each corner of the pentagon, the magnitudes of all the fields at O are equal. Since each interior angle of the pentagon is equal (a property of all regular polygons) and each vector points away from a corner, the angle between any two vectors which are next to each other = interior angle of pentagon = $\frac{2\pi}{5}$.

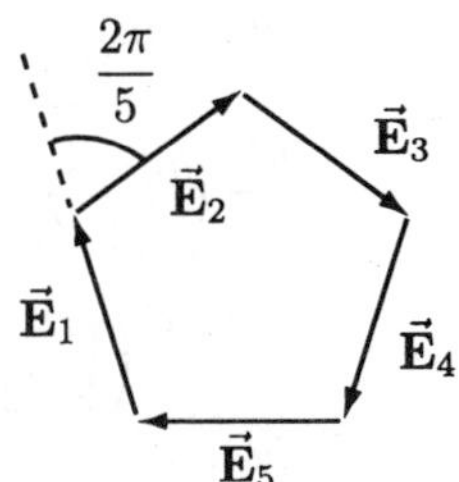

Now if we imagine adding all these vectors by sliding the tail of one to the head of the previous one, the outside angle is also $\frac{2\pi}{5}$—that is, each new vector added turns $\frac{2\pi}{5}$ CW from the previously added vector.

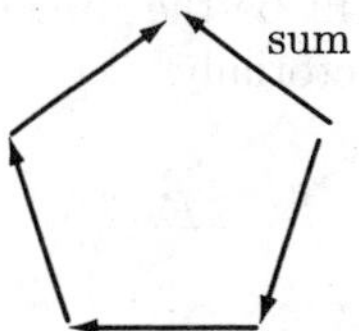

Since there are 5 such vectors to be added, the last vector added completes a pentagon of vectors, as shown in the diagram. Thus, the vector sum is zero and $\vec{\mathbf{E}}(O) = 0$. If one charge is removed, we remove the corresponding electric field vector from the vector sum. Then the sum, as shown, points away from the missing charge. The test charge accelerates directly away from the empty corner.

23.43 a) The magnitude of the electric field at P due to Q is:

$$\begin{aligned} E_1(P) &= \frac{kQ}{r^2} \\ &= \frac{\left(\frac{9.0\times10^9\ \mathrm{N\cdot m^2}}{\mathrm{C^2}}\right)(3.0\ \mu\mathrm{C})}{8.0\ \mathrm{m}^2} \\ &= 3375\ \frac{\mathrm{N}}{\mathrm{C}} \end{aligned}$$

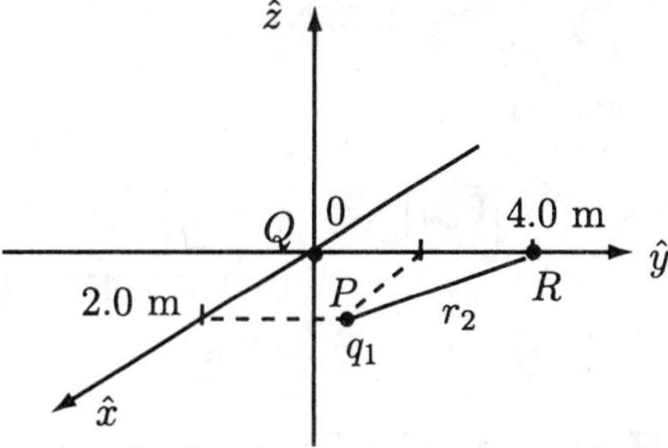

So the force on q due to the field produced by Q is:

$$F_1 = qE_1(P) = (1.0\ \mu\mathrm{C})(3375\ \mathrm{N/C}) = 3.4\ \mathrm{mN}$$

b) The distance between R and P is:

$$r_2 = \sqrt{(2.0\ \mathrm{m})^2 + (2.0\ \mathrm{m})^2} = \sqrt{8}\ \mathrm{m}$$

P

$r_1 = \sqrt{8.0\ \mathrm{m}}$

2.0 m

Q

2.0 m

Since the magnitude of the 2nd charge is equal to the magnitude of the 1st charge and the distance r_1 is equal to r_2, the magnitude of $E(P)$ due to $-Q$ will be equal to the magnitude of $E(P)$ due to $+Q$. The thing that will be different will be the direction. We must add the fields vectorially.

$$E_2(P) = 3375\ \frac{\text{N}}{\text{C}}$$

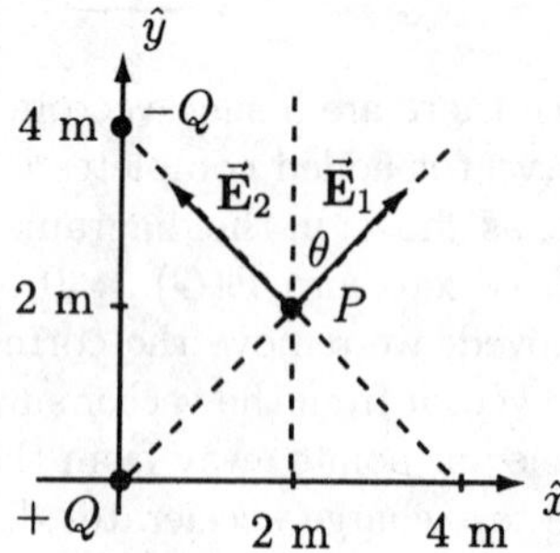

From the figure we can see that the horizontal components of $\vec{\mathbf{E}}_1$, $\vec{\mathbf{E}}_2$ cancel while the vertical components add, so:

$$\vec{\mathbf{E}}_{\text{net}}(P) = \vec{\mathbf{E}}_1 + \vec{\mathbf{E}}_2 = \frac{2kQ}{r^2}\cos\theta\hat{\mathbf{j}}$$

where $\theta = 45°$ and $r = r_1 = r_2 = \sqrt{8}$ m.

$$\begin{aligned}\vec{\mathbf{E}}_{\text{net}} &= \frac{2(9.0\times10^9\ \text{N}\cdot\text{m}^2/\text{C}^2)(3.0\ \mu\text{C})}{8.0\ \text{m}^2}\left(\frac{1}{\sqrt{2}}\right)\hat{\mathbf{j}} \\ &= 4773\ \text{N/C}\hat{\mathbf{j}}\end{aligned}$$

and so,

$$\begin{aligned}\left|\vec{\mathbf{F}}_{\text{net}}\right| &= q\left|\vec{\mathbf{E}}_{\text{net}}\right| \\ &= (1.0\ \mu\text{C})\,(4773\ \text{N/C}) \\ &= 4.8\ \text{mN}\end{aligned}$$

23.47 a) Since point P is at the center of the square, it's equidistant from each of the 4 charges, so

$$\left|\vec{\mathbf{E}}_1(P)\right| = \left|\vec{\mathbf{E}}_2(P)\right| = \left|\vec{\mathbf{E}}_3(P)\right| = \left|\vec{\mathbf{E}}_4(P)\right|$$

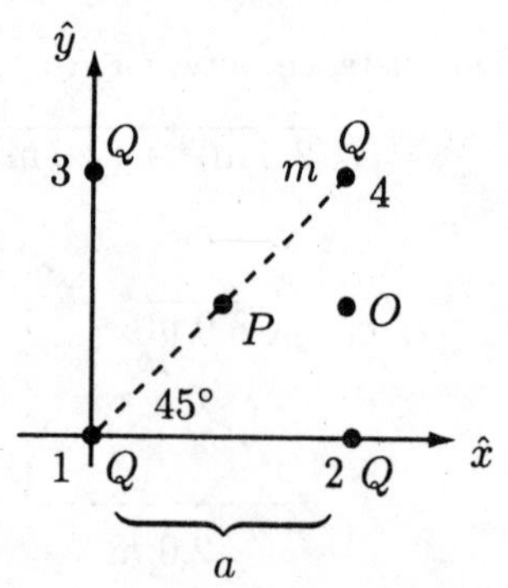

However, Q_1 and Q_4 lie on a straight line which passes through P, so $\vec{\mathbf{E}}_1(P)$ points in the exact opposite direction that $\vec{\mathbf{E}}_4(P)$ does:

$$\vec{\mathbf{E}}_1(P) = -\vec{\mathbf{E}}_4(P).$$

Therefore the vector sum

$$\vec{\mathbf{E}}_1(P) + \vec{\mathbf{E}}_4(P) = 0$$

the electric fields at P due to Q_1 and Q_4 sum to zero. By the same argument,

$$\vec{\mathbf{E}}_2(P) + \vec{\mathbf{E}}_3(P) = 0$$

so

$$\vec{\mathbf{E}}_{\text{net}}(P) = \sum_{n=1}^{4} E_n(P) = 0$$

b) We'll calculate E_{net} at Q_4, which we'll call point m. We cannot include $\vec{\mathbf{E}}_4(m)$ in this calculation since Q_4 cannot feel its own electric field—so Q_4 cannot contribute to $\vec{\mathbf{E}}_{\text{net}}(m)$.

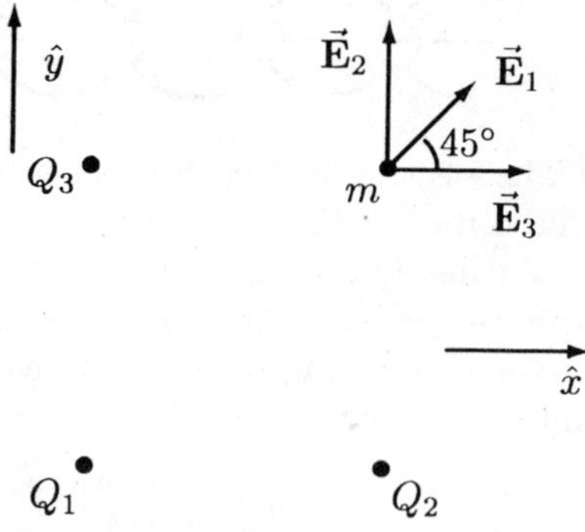

Since Q_3 and Q_2 are both a distance a and Q_1 is a distance $a\sqrt{2}$ (the diagonal of a square of side a) from m,

$$\begin{aligned}\vec{\mathbf{E}}_{\text{net}}(m) &= \vec{\mathbf{E}}_1(m) + \vec{\mathbf{E}}_2(m) + \vec{\mathbf{E}}_3(m) \\ &= \frac{kQ}{(a\sqrt{2})^2}\left[\hat{\mathbf{i}}\sin 45° + \hat{\mathbf{j}}\cos 45°\right] \\ &\quad + \frac{kQ}{a^2}\hat{\mathbf{j}} + \frac{kQ}{a^2}\hat{\mathbf{i}} \\ &= \hat{\mathbf{i}}\frac{kQ}{a^2}\left\{\frac{1}{2\sqrt{2}} + 1\right\} + \hat{\mathbf{j}}\frac{kQ}{a^2}\left\{\frac{1}{2\sqrt{2}} + 1\right\}\end{aligned}$$

so the field strength at m is:

$$\begin{aligned}E_{\text{net}}(m) &= \left|\vec{\mathbf{E}}_{\text{net}}(m)\right| = \sqrt{E_x^2 + E_y^2} \\ &= \frac{kQ}{a^2}\sqrt{\left\{\frac{1}{2\sqrt{2}} + 1\right\}^2 + \left\{\frac{1}{2\sqrt{2}} + 1\right\}^2} \\ &= \frac{kQ\sqrt{2}}{a^2}\left(\frac{1}{2\sqrt{2}} + 1\right) \\ &= \frac{kQ}{2a^2}\left(1 + 2\sqrt{2}\right)\end{aligned}$$

and it points outward along the diagonal of the square.

c) Point O is equidistant from Q_2 and Q_4, so

$$\left|\vec{\mathbf{E}}_2(O)\right| = \left|\vec{\mathbf{E}}_4(O)\right|$$

but $\vec{\mathbf{E}}_2(O)$ and $\vec{\mathbf{E}}_4(O)$ point in opposite directions, so they sum to zero:

$$\vec{\mathbf{E}}_2(O) + \vec{\mathbf{E}}_4(O) = 0$$

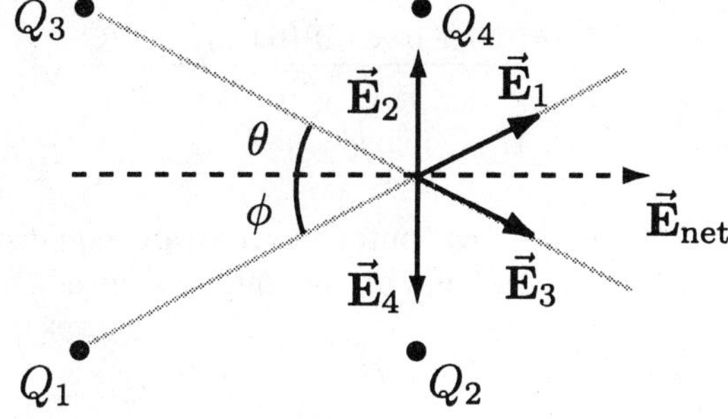

From the figure, since point O is a midpoint of a side of the square, $\theta = \phi$ so the vertical components of $\vec{\mathbf{E}}_1$ and $\vec{\mathbf{E}}_3$ cancel while their horizontal components add.

$$\theta = \phi = \tan^{-1}\left(\frac{\frac{1}{2}a}{a}\right) = 26.6°$$

and,

$$h = \sqrt{a^2 + \left(\frac{1}{2}a\right)^2} = a\sqrt{\frac{5}{4}} = \frac{1}{2}a\sqrt{5}$$

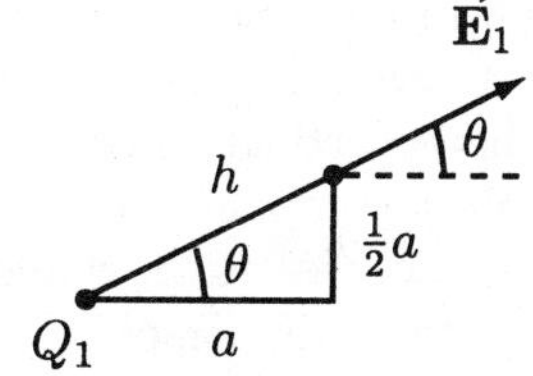

$$\begin{aligned}\left|\vec{\mathbf{E}}_{net}(O)\right| &= E_1(O)\cos\theta + E_3(O)\cos\theta \\ &= \frac{kQ}{\left(\frac{1}{2}a\sqrt{5}\right)^2}\cos(26.6°) \\ &\quad + \frac{kQ}{\left(\frac{1}{2}a\sqrt{5}\right)^2}\cos(\theta) \\ &= \frac{8kQ}{5a^2}\cos(\theta)\end{aligned}$$

Now

$$\sec^2\theta = \frac{1}{\cos^2\theta} = 1 + \tan^2\theta$$

so

$$\cos\theta = \frac{1}{\sqrt{1+\frac{1}{4}}} = \frac{2}{\sqrt{5}}$$

Thus

$$\begin{aligned}E_{net}(O) &= \left(\frac{8kQ}{5k^2}\right)\frac{2}{\sqrt{5}} = \frac{16kQ}{5\sqrt{5}a^2} \\ &= \frac{16}{25}\sqrt{5}\frac{kQ}{a^2}\end{aligned}$$

and $\vec{\mathbf{E}}_{net}$ is $\perp$ to the side.

23.51 Let's calculate the field due to the charges at each corner in pairs across the diagonals.

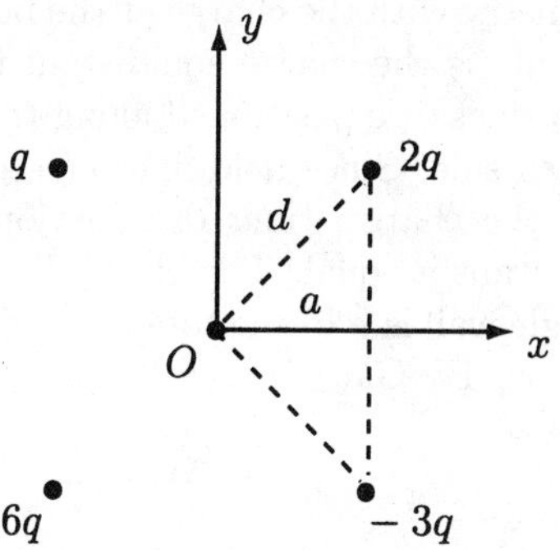

Note:

$$d^2 = a^2 + a^2 = 2a^2$$

q:

$$\vec{\mathbf{E}} = \left(\frac{kq}{d^2},\ \text{toward} - 3q\right) = \frac{kq}{d^2}\left(\frac{\hat{\mathbf{i}} - \hat{\mathbf{j}}}{\sqrt{2}}\right)$$

$-3q$:

$$\vec{\mathbf{E}} = \left(\frac{3kq}{d^2},\ \text{toward} - 3q\right)$$

The sum of these is

$$\vec{\mathbf{E}}_{sum} = \left(\frac{4kq}{d^2},\ \text{toward} - 3q\right)$$

$2q$:

$$\vec{\mathbf{E}} = \left(\frac{2kq}{d^2},\ \text{toward } 6q\right)$$

$6q$:

$$\vec{\mathbf{E}} = \left(\frac{6kq}{d^2},\ \text{away from } 6q\right)$$

The sum of these is

$$\vec{\mathbf{E}}_{sum} = \left(\frac{4kQ}{d^2},\ \text{toward } 2q\right)$$

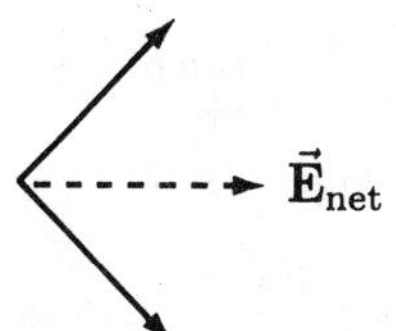

Thus the net electric field at O is the sum of two vectors of equal magnitude, each at 45° to the x-axis

$$\left|\vec{\mathbf{E}}_{\text{net}}\right| = 2\left|\vec{\mathbf{E}}_{\text{sum}}\right|\cos 45° = \sqrt{2}\cdot\frac{4kq}{d^2}$$

and

$$\vec{\mathbf{E}}_{\text{net}} = \frac{2\sqrt{2}kq}{a^2}\hat{\mathbf{i}}$$

23.55 Field lines go out from the charge and emerge from the box's walls. There is a great deal of symmetry with the charge at the box's center—not only is the charge equidistant from each of the 6 sides, it's positioned along the center line of each side. Since field lines emerge uniformly from the charge, their distribution on each of the 6 walls is equal. Therefore, the flux through a single wall is $\frac{1}{6}$ of the total flux emerging from the box. By Gauss's Law,

$$\Phi_{e,\text{net}} = \frac{Q_{\text{in}}}{\epsilon_0} = \frac{Q}{\epsilon_0}$$

so

$$\Phi_{\text{wall}} = \frac{1}{6}\Phi_{e,\text{net}} = \frac{Q}{6\epsilon_0}$$

If the charge is off center, field line distributions are no longer equal over each of the 6 walls. In this scenario, we can state the net flux through the box's surface, (by Gauss's Law it's simply $\frac{Q_{\text{in}}}{\epsilon_0}$) but we cannot easily find the flux through any given wall.

23.59

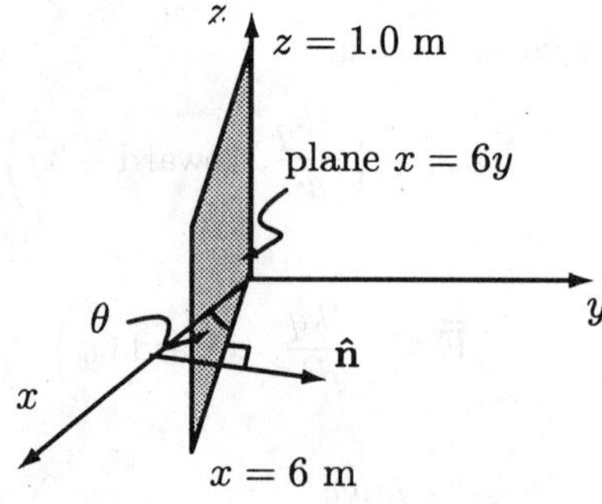

$$\vec{\mathbf{E}} = (6.0\text{ N/C})\hat{\mathbf{i}} + (7.0\text{ N/C})\hat{\mathbf{j}}$$

To compute the flux we need the normal to the plane. It is $\perp$ to the line $x = 6y$ or $y = \frac{x}{6}$, which makes an angle θ with the x-axis where

$$\tan\theta = \frac{1}{6}$$

$\hat{\mathbf{n}}$ makes an angle $90° - \theta$ with the x-axis, so

$$\frac{n_x}{n_y} = \tan\theta = \frac{1}{6}$$

thus

$$\hat{\mathbf{n}} = \frac{\hat{\mathbf{i}} + 6\hat{\mathbf{j}}}{\sqrt{37}}$$

Since $\vec{\mathbf{E}}$ is constant, we may calculate the flux as

$$\begin{aligned}\Phi_E &= \int \vec{\mathbf{E}}\cdot\hat{\mathbf{n}}\,dA = \vec{\mathbf{E}}\cdot\hat{\mathbf{n}}A \\ &= \left((6.0\text{ N/C})\hat{\mathbf{i}} + (7.0\text{ N/C})\hat{\mathbf{j}}\right)\cdot\left(\frac{\hat{\mathbf{i}}+6\hat{\mathbf{j}}}{\sqrt{37}}\right)(6.0\text{ m}\\ &= \frac{(6.0 + 6\times 7.0)(6.0)\text{m}^2\cdot\text{N/C}}{\sqrt{37}} \\ &= 47\text{ N}\cdot\text{m}^2/\text{C}\end{aligned}$$

23.63 Since the two outer charges are equidistant from point P and are of equal magnitude,

$$\left|\vec{\mathbf{E}}_1(P)\right| = \left|\vec{\mathbf{E}}_3(P)\right|$$

but $\vec{\mathbf{E}}_1(P)$ and $\vec{\mathbf{E}}_3(P)$ point in opposite directions and therefore sum to zero.

$\vec{\mathbf{E}}_3$ P $\vec{\mathbf{E}}_1$

⊕ ←⊕→ ⊕
1 2 3

Since $\vec{\mathbf{E}}_{\text{net}}(P) = 0$ there is no force acting on charge 2, that is, charge 2 is in equilibrium. If the 3 charges were negative, both $\vec{\mathbf{E}}_1(P)$ and $\vec{\mathbf{E}}_3(P)$ remain constant in magnitude but change sign. They will still sum to zero and Q_2 will still be in equilibrium. If Q_2 is displaced either up or down, $\vec{\mathbf{E}}_1(P+\Delta y)$ and $\vec{\mathbf{E}}(P+\Delta y)$ will both have vertical components which add to a non-zero y-component of $\vec{\mathbf{E}}_{\text{net}}$. In this case Q_2 will feel a net electric field and will therefore accelerate in the direction of $\vec{\mathbf{E}}_{\text{net}}$ away from the two other charges.

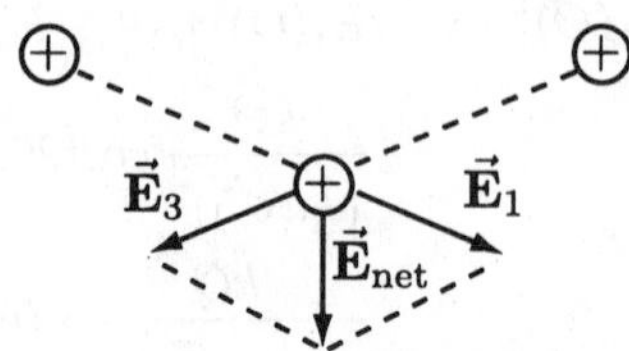

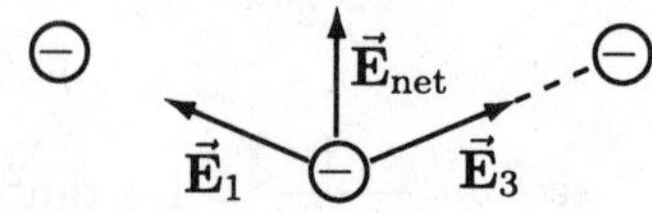

If the charges are negative, Q_2 will accelerate in the direction opposite of $\vec{\mathbf{E}}_{\text{net}}$ which is away from the system. Therefore, whether the changes are + or −, *the equilibrium is an unstable one.*

Now suppose Q_2 is opposite in sign to the other 2 charges. In the first figure, $\vec{\mathbf{E}}_{\text{net}}(P)$ is still zero whether Q_1, Q_2 are both positive or negative. The system is still in equilibrium. In figures 2 and 3 we saw that Q_2 accelerated away from the system. By changing the sign of Q_2, it would accelerate in the opposite direction in each case—back towards the system.

However, a displacement along the line of the 3 charges increases the magnitude of the attractive force toward the nearer charge while decreasing the other force. Thus there is a net force away from equilibrium. This situation is also unstable. This entire discussion was dependent only on the sign of Q_2; not its magnitude. The magnitude would change acceleration magnitude, but not its direction; *the system will behave the same no matter how great or small Q_2 gets.*

23.67 The electric field at P is the sum of the electric fields due to the two charges.

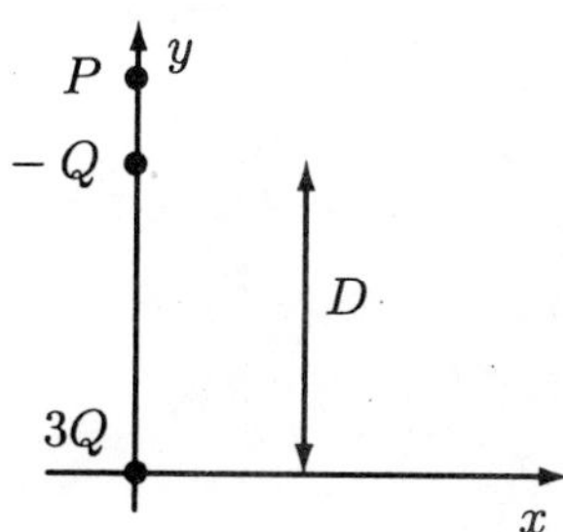

Both fields have only y components. Thus

$$\begin{aligned}
\vec{\mathbf{E}}_{3Q} &= \frac{k(3Q)}{y^2}\hat{\mathbf{j}} \\
\vec{\mathbf{E}}_{-Q} &= -\frac{kQ}{(y-D)^2}\hat{\mathbf{j}} \\
\vec{\mathbf{E}}_{\text{net}} &= kQ\left[\frac{3}{y^2} - \frac{1}{(y-D)^2}\right]\hat{\mathbf{j}} \\
&= kQ\frac{3\left(y^2 - 2yD + D^2\right) - y^2}{y^2(y-D)^2}\hat{\mathbf{j}} \\
&= kQ\frac{\left(2y^2 - 6yD + 3D^2\right)}{y^2(y-D)^2}\hat{\mathbf{j}}
\end{aligned}$$

The field has its maximum value where $\frac{dE}{dy} = 0$ or

$$\begin{aligned}
-\frac{6}{y^3} + \frac{2}{(y-D)^3} &= 0 \\
3(y-D)^3 &= y^3 \\
(y-D) &= \left(\frac{1}{3}\right)^{\frac{1}{3}} y
\end{aligned}$$

$$y\left(1 - \left(\frac{1}{3}\right)^{1/3}\right) = D$$

$$y = \frac{D}{1 - \left(\frac{1}{3}\right)^{1/3}} = 3.3D$$

At this point the field lines approach each other before separating again.

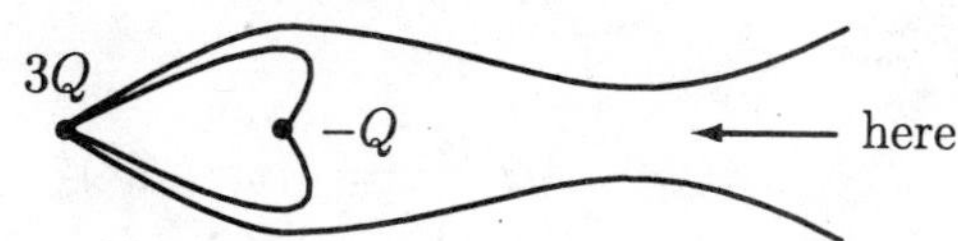

23.69 We may begin with Eqn. (23.9)

$$\vec{\mathbf{E}} = \frac{2kQy}{(a^2+y^2)^{3/2}}\hat{\mathbf{j}}$$

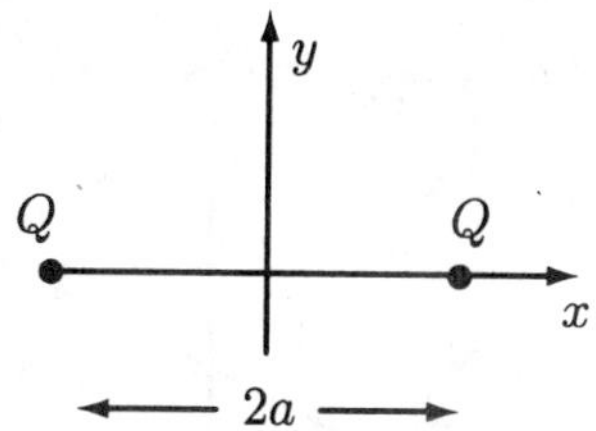

Thus $\left|\vec{\mathbf{E}}\right|$ is a max where $\frac{dE}{dy} = 0$

$$\frac{dE}{dy} = \frac{2kQ}{(a^2+y^2)^{\frac{3}{2}}} - \frac{3}{2}\frac{(2y)(2kQy)}{(a^2+y^2)^{5/2}} \overset{\text{set}}{\equiv} 0$$

The maximum occurs at

$$\begin{aligned}
1 &= \frac{3y^2}{a^2+y^2} \\
a^2 + y^2 &= 3y^2 \\
2y^2 &= a^2 \\
y &= \pm\frac{a}{\sqrt{2}}
\end{aligned}$$

23.71 The $\vec{\mathbf{E}} = 0$ point is the point at which the field lines begin to bend away from each other.

−Q +Q −Q

From the figure above, we see that $\vec{\mathbf{E}}$ is zero at points on the y-axis, so let's calculate the field there. Using Eqn. 23.9 we find the field due to the two negative changes

$$E_{y-} = -\frac{2kQy}{(y^2+L^2)^{\frac{3}{2}}}$$

The field due to the + charge is $E_{y+} = \frac{kQ}{y^2}$. Thus

$$E_{\text{net},y} = kQ\left[\frac{1}{y^2} - \frac{2y}{(y^2+L^2)^{\frac{3}{2}}}\right]$$

This is zero where:

$$2y^3 = \left(y^2+L^2\right)^{\frac{3}{2}}$$

or

$$\begin{aligned} y2^{1/3} &= (y^2+L^2)^{\frac{1}{2}} \\ y^2 \cdot 2^{2/3} &= y^2+L^2 \\ y^2\left[2^{2/3}-1\right] &= L^2 \end{aligned}$$

$$y = \frac{\pm L}{\sqrt{2^{2/3}-1}} = \pm 1.3\ L$$

This is the point marked P in the diagram.

23.75 Let P have coordinates (x, y).

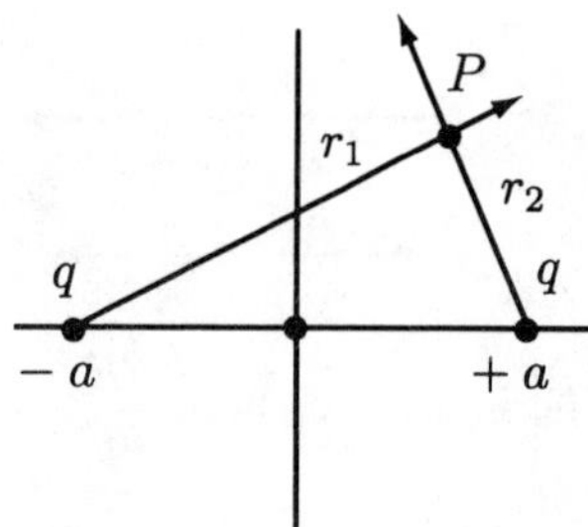

Field due to charge at $-a$:

$$\begin{aligned} \left|\vec{\mathbf{E}}\right| &= \frac{kq}{r_1^2} = \frac{kq}{(x+a)^2+y^2} \\ E_x &= \frac{kq(x+a)}{((x+a)^2+y^2)^{3/2}} \\ E_y &= \frac{kqy}{[(x+a)^2+y^2]^{3/2}} \end{aligned}$$

Similarly, the charge at $+a$ creates $\vec{\mathbf{E}}$ at P with components

$$\begin{aligned} E_x &= \frac{kq(x-a)}{[(x-a)^2+y^2]^{\frac{3}{2}}} \\ E_y &= \frac{kqy}{[(x-a)^2+y^2]^{3/2}} \end{aligned}$$

The total electric field at P has components:

$$\begin{aligned} E_x &= kq\left[\frac{x+a}{[(x+a)^2+y^2]^{3/2}} + \frac{x-a}{[(x-a)^2+y^2]^{3/2}}\right] \\ &= kq\left[\frac{(x+a)r_-^3 + (x-a)r_+^3}{r_+^3 r_-^3}\right] \end{aligned}$$

$$\begin{aligned} E_y &= kqy\left[\frac{1}{[(x+a)^2+y^2]^{3/2}} + \frac{1}{[(x-a)^2+y^2]^{3/2}}\right] \\ &= kqy\left[\frac{r_-^3+r_+^3}{(r_+r_-)^3}\right] \end{aligned}$$

Thus

$$\begin{aligned} \frac{dy}{dx} &= \frac{E_y}{E_x} \\ &= y\frac{\left[r_-^3+r_+^3\right]}{x\left(r_-^3+r_+^3\right)+a\left(r_-^3-r_+^3\right)} \\ &= \frac{y}{x+a\left(\frac{r_-^3-r_+^3}{r_-^3+r_+^3}\right)} \end{aligned}$$

where

$$r_- = \sqrt{(x-a)^2+y^2}$$

is the distance from the charge at $+a$, and r_+ is defined similarly.

Chapter 24

Static Electric Fields: Applications

24.1 From the result of Exercise 24.1, the electric field outside the ball is the same for a point charge located at the *center* of the ball.

$$\begin{aligned}\vec{\mathbf{E}} &= \frac{kQ}{r^2}\hat{\mathbf{r}} \\ &= \frac{(9.0 \times 10^9\ \mathrm{N \cdot m^2/C^2})\,(16\ \mathrm{nC})\,\hat{\mathbf{r}}}{(.21\ \mathrm{m})^2} \\ &= 3.3 \times 10^3\ \mathrm{N/C}\hat{\mathbf{r}}\end{aligned}$$

[Note: $r = 16\ \mathrm{cm} + 5\ \mathrm{cm} = 21\ \mathrm{cm}$]

24.3 From the result of Example 24.2, for an infinitely long filament,

$$\begin{aligned}\left|\vec{\mathbf{E}}\,(r = .50\,\mathrm{m})\right| &= \frac{\lambda}{2\pi r \epsilon_0} = \frac{2\lambda k}{r} \\ &= \frac{2(175\ \mu\mathrm{C/m})\left(9.0\times10^9\ \mathrm{N{\cdot}m^2/C^2}\right)}{.50\ \mathrm{m}} \\ &= 6.3 \times 10^6\ \mathrm{N/C}\end{aligned}$$

direction: radially outward from axis along which filament lies.

24.5 From the result for an infinite plane of uniform charge density given in Example 24.5

$$\begin{aligned}E\,(z) &= \frac{\sigma}{2\epsilon_0} \\ &= \frac{4.6 \times 10^{-3}\ \mathrm{C/m^2}}{2\,(8.85 \times 10^{-12}\ \mathrm{C^2/N \cdot m^2})} \\ &= 2.6 \times 10^8\ \mathrm{N/C}\end{aligned}$$

and $\vec{\mathbf{E}}$ is normal to the plate. Note that the result is independent of height above the plate.

24.7 From the result of Example 24.5,

$$\begin{aligned}E_{\mathrm{plane}} &= \frac{\sigma}{2\epsilon_0} \\ &= \frac{(175\ \mu\mathrm{C/m^2})}{2\,(8.85 \times 10^{-12}\ \mathrm{C^2/N \cdot m^2})} \\ &= 9.89 \times 10^6\ \mathrm{N/C}\end{aligned}$$

The plane's charge is negative and the proton's charge is positive—the proton will be accelerated toward the plane at an acceleration of

$$\begin{aligned}a &= \frac{qE}{m} \\ &= \frac{(1.602 \times 10^{-19}\ \mathrm{C})\,(9.89 \times 10^6\ \mathrm{N/C})}{1.673 \times 10^{-27}\ \mathrm{kg}} \\ &= 9.47 \times 10^{14}\ \mathrm{m/s^2}\end{aligned}$$

It will reach the plane in a time

$$\begin{aligned}\Delta t &= \sqrt{\frac{2\mathrm{s}}{a}} = \sqrt{\frac{2\,(.100\ \mathrm{m})}{(9.47 \times 10^{14}\ \mathrm{m/s^2})}} \\ &= 1.45 \times 10^{-8}\ \mathrm{s}\end{aligned}$$

24.13 The result of Example 24.1 is that $\vec{\mathbf{E}}$ due to a uniformly charged spherical shell is the same as for a point charge of equal net charge as the shell, located at the shell's center. Since $\vec{\mathbf{E}}$ points inward, the charge is negative.

$$\begin{aligned}E_r\,(r = R_E) &= \frac{kQ_{\mathrm{tot}}}{R_E^2} \\ \Rightarrow Q_{\mathrm{tot}} &= \frac{R_E^2 E_r\,(r = R_E)}{k}\end{aligned}$$

$$= \frac{\left(6.375 \times 10^6\ \text{m}\right)^2 (-100\ \text{N/C})}{9.0 \times 10^9\ \text{N} \cdot \text{m}^2/\text{C}^2}$$
$$= -.5\ \text{MC}$$

Assuming that the charge is uniformly distributed,

$$\sigma = \frac{Q_{\text{tot}}}{A_{\text{Earth}}} = \frac{-.5\ MC}{4\pi R_E^2} = -.9\ \text{nC/m}^2$$

24.17 Let O be the center of an equilateral triangle and consider one side of the triangle, say, AB. Each piece of AB contributes to the field at O due to AB. Consider some piece of AB a distance x from point P which is at the center of AB.

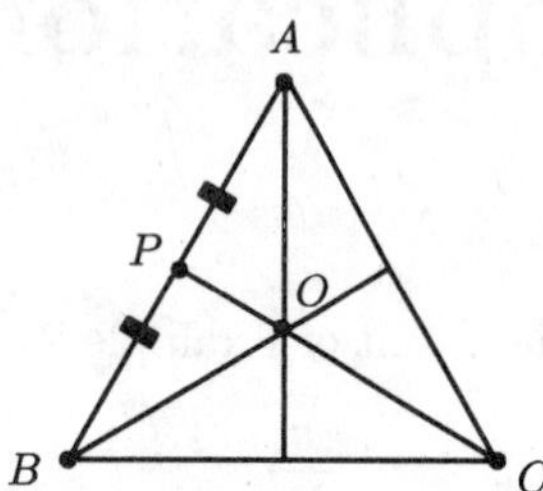

The contribution of the piece to $\vec{\mathbf{E}}$ at O is some vector $d\vec{\mathbf{E}}_1$ which points along the straight line between the piece itself and O. For every such piece there exists another piece also a distance x from P which also contributes to $\vec{\mathbf{E}}$.

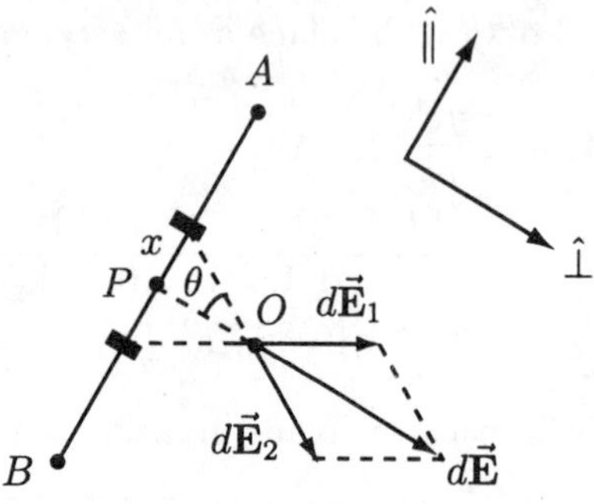

Because both pieces are equidistant from both P and O, it can be seen that the $\hat{\parallel}$ components of $d\vec{\mathbf{E}}_1$ and $d\vec{\mathbf{E}}_2$ cancel and the $\hat{\perp}$ components add. The result is the vector sum

$$d\vec{\mathbf{E}} = 2d\vec{\mathbf{E}}_1 \cos\theta = 2d\vec{\mathbf{E}}_2 \cos\theta$$

which lies along the $\perp$ bisector of AB. We can pick *any* piece of AB and there will be a corresponding piece equidistant from the line's middle whose contribution to $\vec{\mathbf{E}}$ will cancel the 1st piece's $\hat{\parallel}$ component and add to the first piece's $\hat{\perp}$ component. The result is that $\vec{\mathbf{E}}$ due to each of the three rods at O will be equal in magnitude to $\vec{\mathbf{E}}$ due to the other two rods because all three rods are 1) equal in length, charge density and distance to O and 2) directed along the $\perp$ bisector of the rod, away from the rod. Thus, $\vec{\mathbf{E}}$ due to the three rods at point O would be the sum of the three vectors:

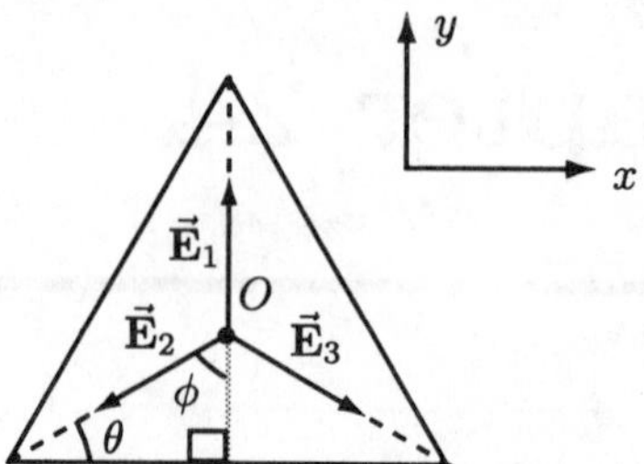

In an equilateral triangle, the $\perp$ bisector also bisects the opposite angle. Therefore $\theta = \frac{1}{2}(60°) = 30°$. This means that $\phi = 60°$ and the angle between $\vec{\mathbf{E}}_2$ and $\vec{\mathbf{E}}_3$ is 120°. Using the drawn coordinate system, the vector sum is:

$$\vec{\mathbf{E}}_{\text{tot}} = E_1\hat{\mathbf{j}} - E_2\cos 60°\hat{\mathbf{j}} - E_3\cos 60°\hat{\mathbf{j}} \\ +E_3\sin 60°\hat{\mathbf{i}} - E_2\sin 60°\hat{\mathbf{i}}$$

Since $\left|\vec{\mathbf{E}}_1\right| = \left|\vec{\mathbf{E}}_2\right| = \left|\vec{\mathbf{E}}_3\right|$, this sum equals zero.

24.21 Using Eq. (24.5),

$$\vec{\mathbf{E}}_{\text{hoop}}(z) = \frac{kQz}{(z^2 + a^2)^{3/2}}\hat{\mathbf{k}}$$

on the axis of each hoop, with $a = R$.

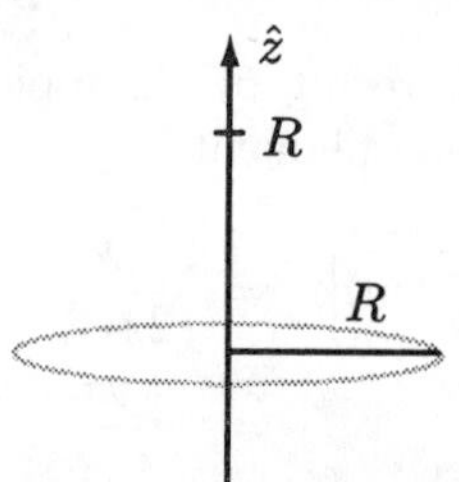

$$E(z = .65\ \text{m}) = \frac{\left(9.0\times10^9\ \text{N}\cdot\text{m}^2/\text{C}^2\right)(6.3\ \mu\text{C})(.65\ \text{m})}{\left[(.65\ \text{m})^2+(.65\ \text{m})^2\right]^{\frac{3}{2}}}$$
$$= 4.7 \times 10^4\ \text{N/C}$$

Using the result of Example 24.3, $\vec{\mathbf{E}}$ at points P on the bisector of a straight current filament,

$$E_{\text{bisector}} = \frac{kQ}{y\sqrt{y^2 + \frac{1}{4}\ell^2}}$$

so

$$E_{\text{bisector}}\left(y = \sqrt{2}R\right) = \frac{kQ}{\sqrt{2}R\sqrt{2R^2 + \frac{1}{4}(2\pi R)^2}}$$
$$= 2.7 \times 10^4\ \text{N/C}$$

$\perp$ to and away from the filament. The second result is smaller because much of the charge is at a distance $> \sqrt{2}R$ from P.

24.25 The charged rod can be modeled as a collection of different line elements of length dx and charge $dq = \lambda\, dx$.

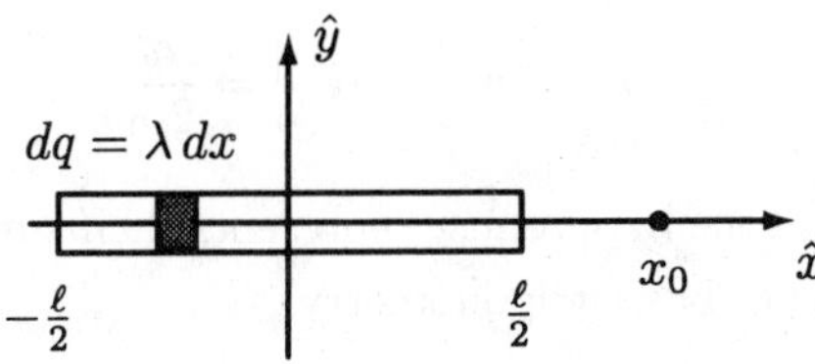

Each element can be located by the coordinate x. The distance between any typical element dq and the point x_0 is $x_0 - x$. Therefore, the electric field at x_0 due to a typical element dq is:

$$d\vec{\mathbf{E}} = \frac{k\,dq}{r^2}\hat{\mathbf{i}} = \frac{k\lambda\,dx}{(x_0 - x)^2}\hat{\mathbf{i}}$$

To integrate the entire rod, we let x vary from $-\frac{1}{2}\ell$ to $+\frac{1}{2}\ell$.

$$\begin{aligned}
\vec{\mathbf{E}}(x_0) &= \int_{-\ell/2}^{\ell/2} \frac{k\lambda\hat{\mathbf{i}}}{(x_0 - x)^2} dx \\
&= k\lambda\hat{\mathbf{i}} \int_{-\ell/2}^{\ell/2} \frac{dx}{(x_0 - x)^2} \\
u &= x_0 - x \\
du &= -dx \\
\vec{\mathbf{E}}(x_0) &= k\lambda\hat{\mathbf{i}} \int_{x_0-\ell/2}^{x_0+\ell/2} u^{-2}\, du \\
&= -\frac{k\lambda\hat{\mathbf{i}}}{u}\Bigg|_{x_0-\frac{1}{2}\ell}^{x_0+\frac{1}{2}\ell} \\
&= \frac{k\lambda\hat{\mathbf{i}}}{x_0 - \frac{1}{2}\ell} - \frac{k\lambda\hat{\mathbf{i}}}{x_0 + \frac{1}{2}\ell} \\
&= \frac{k\lambda\ell}{x_0^2 - \frac{1}{4}\ell^2}\hat{\mathbf{i}}
\end{aligned}$$

24.29 Step 1: The semicircle can be modeled as a collection of differential arc lengths, ds. The problem statement chose the coordinates to be used for us—each element will be identified by the θ coordinate which is the angle to the positive x-axis.

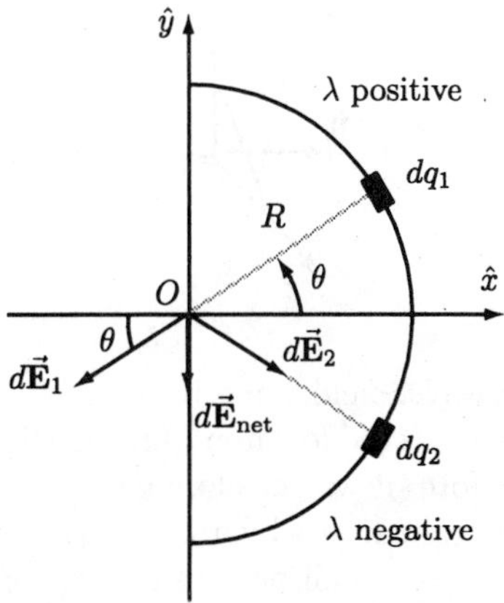

Step 2: Each differential arc length element has a length $ds = R\,d\theta$ and a charge of

$$dq = \lambda(\theta)\; ds = \lambda_0 \sin\theta R\, d\theta$$

Step 3: Because of the mirror symmetry about the x-axis, the horizontal components of $d\vec{\mathbf{E}}$ due to any two elements located at $+\theta$ and $-\theta$ will cancel and the vertical components will add as shown in the diagram. (Note that dq is negative for all elements for which $\theta < 0$!) therefore the net $\vec{\mathbf{E}}(0)$ will be in the $-\hat{y}$ direction and we will want to sum only the vertical component of the contributions to $\vec{\mathbf{E}}(0)$.

$$\begin{aligned}
d\vec{\mathbf{E}} &= -\frac{k\,dq}{r^2}\sin\theta\hat{\mathbf{j}} = -\frac{k(\lambda_0 \sin\theta R\,d\theta)}{R^2}\sin\theta\hat{\mathbf{j}} \\
&= -\frac{k\lambda_0 \sin^2\theta\, d\theta}{R}\hat{\mathbf{j}}
\end{aligned}$$

Step 4: θ will vary from $-\frac{\pi}{2}$ to $+\frac{\pi}{2}$.

Step 5:

$$\begin{aligned}
\vec{\mathbf{E}}(0) &= \int_{\text{all }\theta} d\vec{\mathbf{E}}(\theta) \\
&= \hat{\mathbf{j}} \int_{-\frac{\pi}{2}}^{\frac{\pi}{2}} -\left(\frac{k\lambda_0 \sin^2\theta}{R}\right) d\theta \\
&= -\hat{\mathbf{j}}\frac{k\lambda_0}{R} \int_{-\frac{\pi}{2}}^{\frac{\pi}{2}} \sin^2\theta\, d\theta
\end{aligned}$$

Using $\sin^2\theta = \frac{1}{2}(1 - \cos 2\theta)$

$$\begin{aligned}
\vec{\mathbf{E}}(0) &= -\frac{k\lambda_0}{2R}\hat{\mathbf{j}} \int_{-\frac{\pi}{2}}^{\frac{\pi}{2}} [1 - \cos(2\theta)]\, d\theta \\
&= -\frac{k\lambda_0}{2R}\hat{\mathbf{j}} \left[\theta - \frac{\sin(2\theta)}{2}\right]_{-\frac{\pi}{2}}^{\frac{\pi}{2}} \\
\vec{\mathbf{E}} &= -\frac{\pi k\lambda_0}{2R}\hat{\mathbf{j}}
\end{aligned}$$

24.33 Since there is a net positive charge on the surface of the tetrahedron, field lines extend from the surface outward.

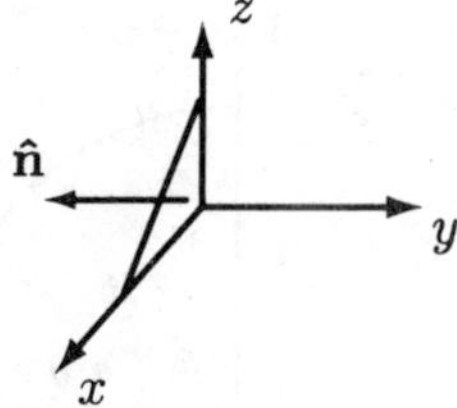

As always, field lines begin ⊥ to the charged surface; even for non-uniform field patterns. Therefore, if we get close enough to the face of the tetrahedron which lies in the xz plane, the field lines will all be $\|$ to the vector $\hat{\mathbf{n}}$ drawn in the figure. Since $\hat{\mathbf{n}} = -\hat{\mathbf{j}}$, choice **b** is the correct answer.

24.37 We are looking at the plate's field at a distance which is much smaller than the radius of the plate—an entire order of magnitude smaller. The figure shows, in scale, what the situation looks like:

If P doesn't get too close to the edges of the charged plate, the plate will look like an infinite plane charge. So, to a good approximation,

$$\vec{\mathbf{E}}(P) = \frac{\sigma}{2\epsilon_0}\hat{\mathbf{k}} = \frac{(4.3\times10^{-9}\ \mathrm{C/m^2})\,\hat{\mathbf{k}}}{2\,(8.85\times10^{-12}\ \mathrm{C^2/N\cdot m^2})} = (240\ \mathrm{N/C})\,\hat{\mathbf{k}}$$

24.41 From the text, the result for the electric field due to an infinite plane is

$$\vec{\mathbf{E}}_{\text{plane}} = \frac{\sigma}{2\epsilon_0}\hat{\mathbf{n}}$$

where $\hat{\mathbf{n}}$ is a unit vector normal to the plane. Consider the two intersecting planes:

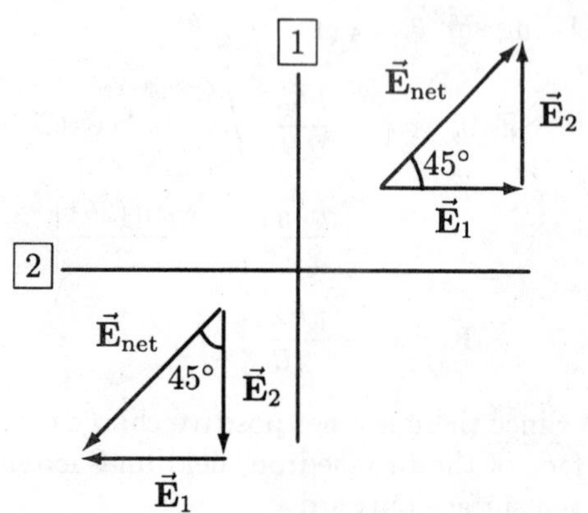

Since $\vec{\mathbf{E}}_1$ is independent of distance to plane $\boxed{1}$ and $\vec{\mathbf{E}}_2$ is independent of distance to plane $\boxed{2}$, their sum, $\vec{\mathbf{E}}_{net}$ will be independent of the distance to either plane. Furthermore since $\vec{\mathbf{E}}_1$ is ⊥ to $\boxed{1}$, $\vec{\mathbf{E}}_2$ is ⊥ to $\boxed{2}$ and $\left|\vec{\mathbf{E}}_1\right| = \left|\vec{\mathbf{E}}_2\right|$, $\vec{\mathbf{E}}_{\text{net}}$ will always point 45° from each plane at every point in space (see figure.) The vector diagrams are isosceles right triangles. The two legs represent the individual fields,

$$\left|\vec{\mathbf{E}}_1\right| = \left|\vec{\mathbf{E}}_2\right| = \frac{\sigma}{2\epsilon_0}$$

and the hypotenuse represents their sum, $\left|\vec{\mathbf{E}}_{\text{net}}\right|$. From basic trigonometry,

$$\left|\vec{\mathbf{E}}_{\text{net}}\right| = \frac{E_1}{\sin 45^\circ} = \sqrt{2}E_1 = \frac{\sigma\sqrt{2}}{2\epsilon_0}$$

24.45 Step 1: The charge distribution is cylindrically symmetric so it makes sense to model the system as a collection of differential ring charges. Cylindrical coordinates is a natural system of coordinates to use for this geometry.

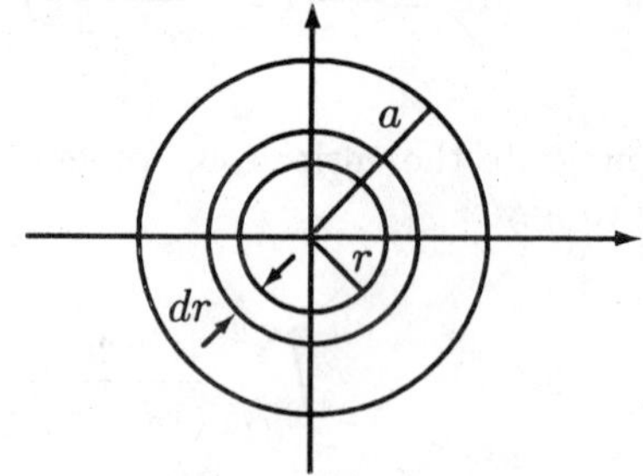

Step 2: Each element can be uniquely identified by the r coordinate and has an area of

$$dA = 2\pi r\,dr$$

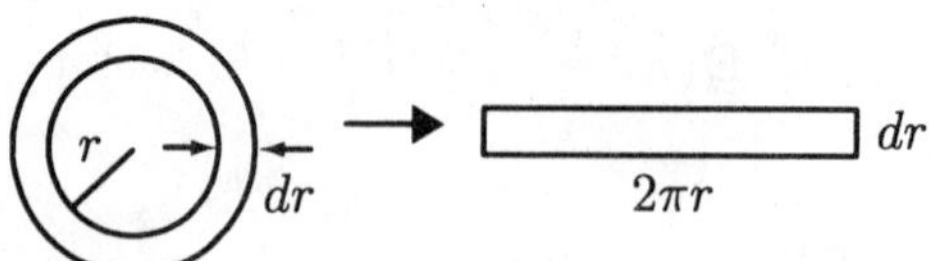

The charge of such a typical element would be given by:

$$dq = \sigma(r)\,dA = \frac{2\pi\sigma_0 r^3\,dr}{a^2}$$

Step 3: From Eq. (24.5), the electric field produced by the ring charge at some coordinate r is:

$$d\vec{\mathbf{E}} = \frac{k\,(dq)\,z}{(z^2+r^2)^{\frac{3}{2}}}\hat{\mathbf{k}} = \frac{2\pi k\sigma_0 r^3\,z\,dr}{a^2\,(z^2+r^2)^{\frac{3}{2}}}\hat{\mathbf{k}}$$

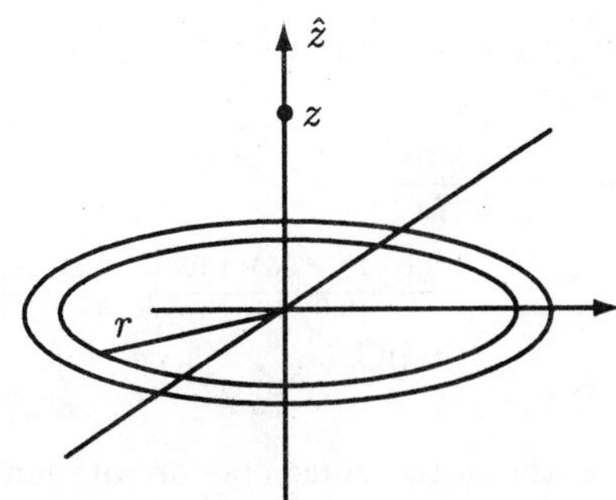

Step 4: To integrate the entire disk, r must vary from zero to a.

Step 5:

$$\begin{aligned}
\vec{\mathbf{E}}(z) &= \frac{2\pi k\sigma_0 z\hat{\mathbf{k}}}{a^2}\int_0^a \frac{r^3}{[z^2+r^2]^{\frac{3}{2}}}dr \\
u &= z^2+r^2 \\
du &= 2r\,dr \\
r^2 &= u-z^2 \\
&= \frac{\pi k\sigma_0 z\hat{\mathbf{k}}}{a^2}\int_{z^2}^{z^2+a^2} u^{-\frac{3}{2}}\left[u-z^2\right]du \\
&= \frac{\pi k\sigma_0 z\hat{\mathbf{k}}}{a^2}\int_{z^2}^{z^2+a^2} \left(u^{-\frac{1}{2}}-z^2u^{-\frac{3}{2}}\right)du \\
&= \frac{\pi k\sigma_0 z\hat{\mathbf{k}}}{a^2}\left[2u^{\frac{1}{2}}+2z^2u^{-\frac{1}{2}}\right]_{z^2}^{z^2+a^2}
\end{aligned}$$

$$\begin{aligned}
&= \frac{2\pi k\sigma_0 z\hat{\mathbf{k}}}{a^2}\left[\sqrt{z^2+a^2}+\frac{z^2}{\sqrt{z^2+a^2}}-z-z\right] \\
&= \frac{2\pi k\sigma_0 z\hat{\mathbf{k}}}{a^2}\left[\frac{z^2+a^2+z^2}{\sqrt{z^2+a^2}}-2z\right] \\
&= \frac{2\pi k\sigma_0 z\hat{\mathbf{k}}}{a^2}\left[\frac{2z^2+a^2}{\sqrt{z^2+a^2}}-2z\right]
\end{aligned}$$

Check: For $a \ll z$, the term in square brackets becomes $2z-2z=0$, so $\vec{\mathbf{E}} \to 0$, as expected.

24.49 The result for $\vec{\mathbf{E}}$ inside a uniformly charged sphere is:

$$\vec{\mathbf{E}} = \frac{Qr}{4\pi R^3\epsilon_0}\hat{\mathbf{r}} = \frac{\rho\vec{\mathbf{r}}}{3\epsilon_0}$$

where $Q=\frac{4}{3}\pi R^3\rho$. We solved a problem similar to this when calculating center of mass. The trick is to treat the larger sphere as if it didn't have a hole and superimpose the solution for an oppositely charged sphere of radius $\frac{R}{2}$ centered at O'.

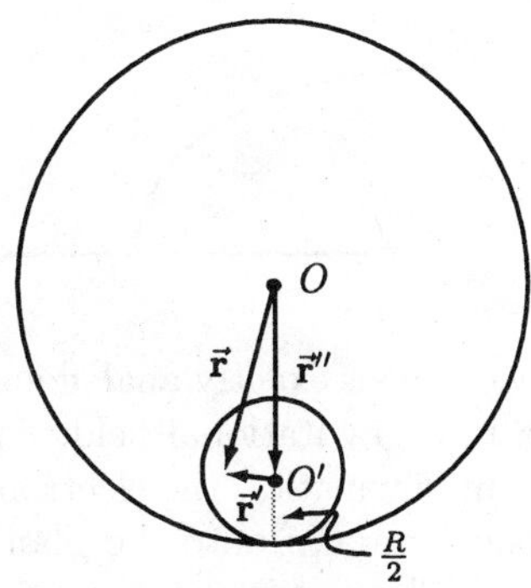

$\vec{\mathbf{E}}$ due to the larger sphere at some point within the region of the smaller sphere is:

$$\vec{\mathbf{E}}_1 = \frac{\rho\vec{\mathbf{r}}}{3\epsilon_0}$$

$\vec{\mathbf{E}}$ due to the sphere of negative charge centered at O' is:

$$\vec{\mathbf{E}}_2 = -\frac{\rho\vec{\mathbf{r}}'}{3\epsilon_0}$$

Superimposing the solution for the two spheres,

$$\vec{\mathbf{E}}_{\text{net}} = \vec{\mathbf{E}}_1+\vec{\mathbf{E}}_2 = \frac{\rho\vec{\mathbf{r}}}{3\epsilon_0}-\frac{\rho\vec{\mathbf{r}}'}{3\epsilon_0}$$

But from the diagram $\vec{\mathbf{r}}-\vec{\mathbf{r}}'=\vec{\mathbf{r}}''$, a vector from the large sphere to the center of the hole. Thus

$$\vec{\mathbf{E}} = \left(\frac{\rho}{3\epsilon_0}\right)\vec{\mathbf{r}}''$$

and is uniform in the hole.

24.53 Field lines extend upward in the $\hat{y}$-direction (above the plane).

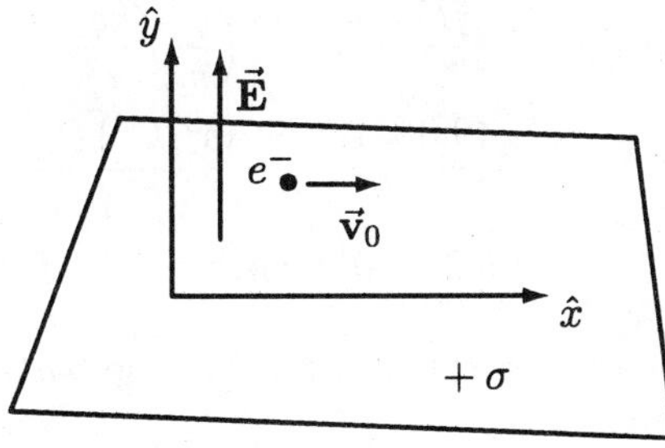

Since the electron is negatively charged it feels a force in the direction opposite $\vec{\mathbf{E}}$, in the $-\hat{y}$-direction. Specifically,

$$\vec{\mathbf{F}} = -eE\hat{\mathbf{j}} = -\frac{e\sigma}{2\epsilon_0}\hat{\mathbf{j}}$$

and $\vec{\mathbf{F}}$ is constant in both direction and magnitude.

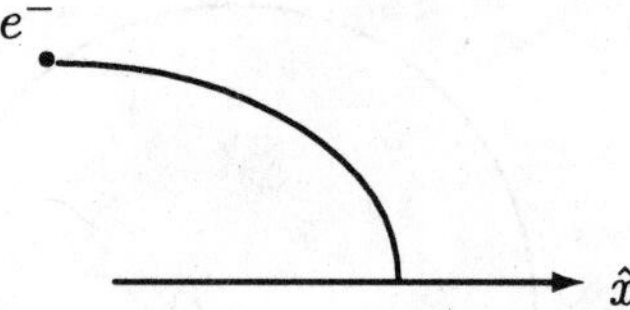

What we have is exactly analagous to projectile motion in a gravitational field. From what we learned in kinematics, the electron will follow a parabolic arc down onto the plane, in the $-\hat{y}$-direction. Choice e is the correct choice.

a) is incorrect. While the speed in the x-direction will remain constant, the electron will eventually collide with the sheet and "stick."

b) is incorrect. There is no force on the electron in the x-direction due to the sheet.

c) is incorrect. There is no force on the electron in the x-direction due to the sheet.

d) is incorrect. The force is *attractive* (toward the plane).

24.57 By N3L, the forces exerted by each particle on the other are equal so the magnitude of the identical particles' accelerations are also equal.

$r = .10$ m

q_1 q_2

Since both particles are positively charged, they will accelerate away from each other on a line which passes through them both. The magnitude of their initial acceleration, by Eq. (24.10) of the text is:

$$
\begin{aligned}
a &= \frac{q_1 E_{\text{due to 2}}}{m} = \left(\frac{q_1}{m}\right)\frac{kq_2}{r^2} \\
&= \frac{(9.0\times 10^9\ \text{N}\cdot\text{m}^2/\text{C}^2)\,(17\ \text{nC})^2}{(5.0\times 10^{-3}\ \text{kg})\,(.10\ \text{m})^2} \\
&= 5.2\times 10^{-2}\ \text{m/s}^2
\end{aligned}
$$

The acceleration decreases as the particles separate.

24.61 Applying N2L for centripetal motion,

$$F = \frac{mv^2}{r}$$

Here, the force F is the Coulomb force between 2 charged particles.

$$\frac{k\,|q|\,Q}{r^2} = \frac{mv^2}{r}$$

Solving for the unknown charge, q

$$
\begin{aligned}
|q| &= \frac{rm^2}{kQ} \\
&= \frac{(2.0\times10^{-2}\ \text{m})(1.0\times10^{-3}\ \text{kg})(5.0\ \text{m/s})^2}{(9.0\times10^9\ \text{Nm}^2/\text{C}^2)(1.0\ \mu\text{C})} \\
&= 56\ \text{nC}
\end{aligned}
$$

Since the force must be an attractive force, the two charges have opposite signs. Thus $q = -56$ nC.

24.65 From Example 24.2 of the text, the electric field due to an infinite filament of charge is

$$E_{\text{line}} = \frac{\lambda}{(2\pi\epsilon_0 r)}$$

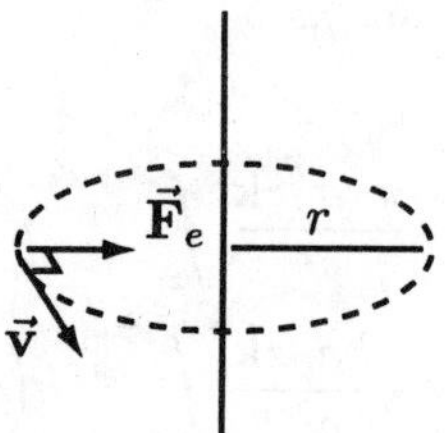

Applying N2L for centripetal motion,

$$
\begin{aligned}
F_e &= \frac{mv^2}{r} \\
qE_{\text{line}} &= \frac{mv^2}{r}
\end{aligned}
$$

(The force is inward, as required, since the particle and the filament have opposite signs of charge.) Solving for the charge's speed,

$$
\begin{aligned}
v &= \left(\frac{qrE_{\text{line}}}{m}\right)^{\frac{1}{2}} = \left(\frac{q\lambda}{2\pi m\epsilon_0}\right)^{\frac{1}{2}} \\
&= \left[\frac{(15\ \text{nC})(1.0\ \mu\text{C/m})}{2\pi(1.5\times10^{-3}\ \text{kg})(8.85\times10^{-12}\ \text{C}^2/\text{N}\cdot\text{m}^2)}\right]^{1/2} \\
&= 0.42\ \text{m/s}
\end{aligned}
$$

24.71 The charge distribution is positive above the x-axis and negative below the x-axis, being most positive on the $+y$-axis and most negative on the $-y$-axis. The net charge of the filament is zero and it should vaguely remind you of a dipole because, in a way, it is! We make use of the symmetry of the system to draw field lines (Figure 1.)

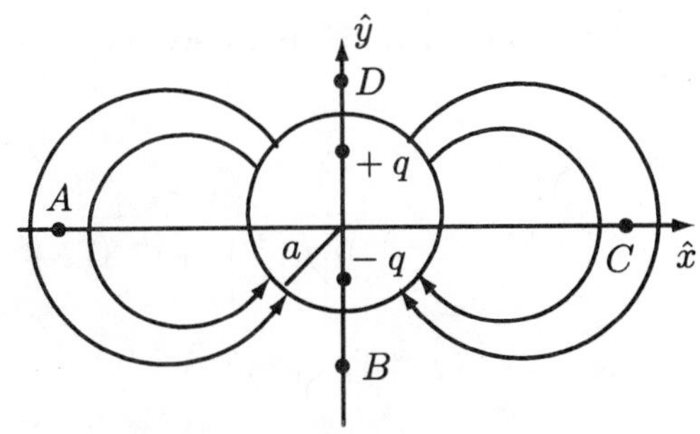

Then we can see that at points A and C the field points in the $-y$-direction.

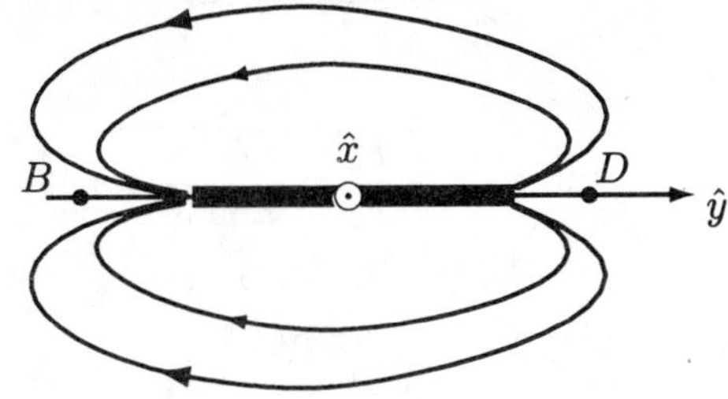

Figure 2 shows a change of viewpoint. Again, the field is dipolar. From Figure 2 we can see the field will be in the y-direction at both points B and D.

24.73 Example 24.11 calculates the electric field on the axis of a dipole

$$E = \frac{2kp}{x^3}$$

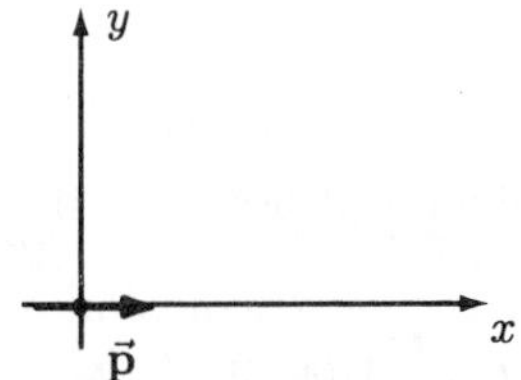

The dipole moment of the needle is

$$\begin{aligned} p &= q\ell = (1.0\,\text{nC})\,(2.0\,\text{mm}) \\ &= 2.0 \times 10^{-12}\,\text{C}\cdot\text{m} \end{aligned}$$

so:

$$\begin{aligned} E &= \frac{2\,(9.0\times 10^{9}\,\text{N}\cdot\text{m}^2/\text{C}^2)\,(2.0\times 10^{-12}\,\text{C}\cdot\text{m})}{(.33\,\text{m})^3} \\ &= 1.0\,\text{N/C} \end{aligned}$$

24.77 From Example 24.1 of the text, uniformly charged shells produce an electric field outside the shell which is identical to the electric field due to a point charge with the same total charge, located at the center of the shell.

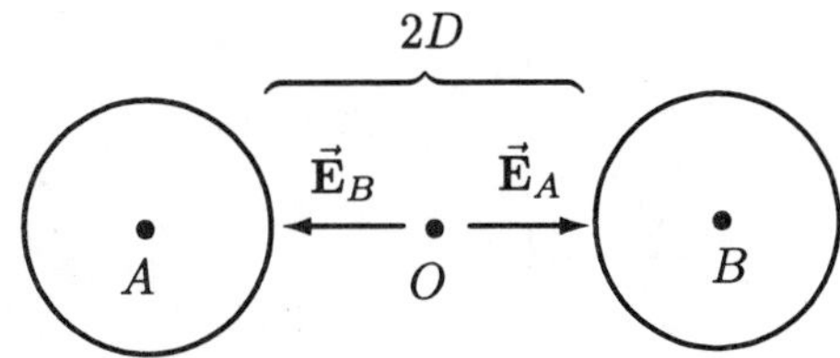

Therefore, the field at O is equal to the field due to two point charges Q located at A and B. Since O is equidistant from A and B and all three points are colinear, $\vec{\mathbf{E}}_B$ and $\vec{\mathbf{E}}_A$ are equal in magnitude and opposite in direction at O. Therefore,

$$\vec{\mathbf{E}}\,(0) = \vec{\mathbf{E}}_A + \vec{\mathbf{E}}_B = 0$$

Also, from Example 24.1 from the text we learned that the electric field inside a uniformly charged shell is zero so $\vec{\mathbf{E}}$ at the center of each shell is due solely to the other shell: ($\vec{\mathbf{E}} = \frac{kQ}{9D^2}$ directly away from the other shell).

24.81 The Coulomb force provides the acceleration necessary for centripetal motion.

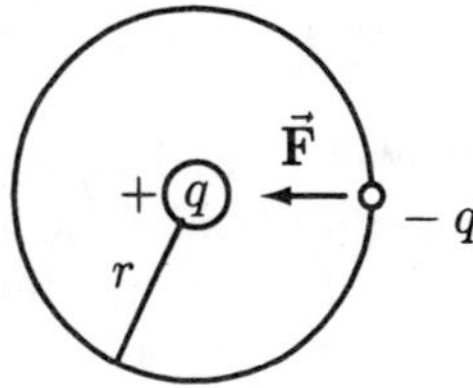

Applying N2L to the electron,

$$\begin{aligned} F &= \frac{mv^2}{r} \\ eE &= \frac{mv^2}{r} \\ \frac{ke^2}{r^2} &= \frac{mv^2}{r} \end{aligned}$$

Multiplying through by $\frac{r^3}{ke^2}$,

$$r = \frac{mv^2r^2}{ke^2} = \frac{L^2}{mke^2}$$

where $L = mvr$ is the electron's angular momentum

24.85 The vertical infinite plane produces a horizontal uniform electric field $\frac{\sigma}{2\epsilon_0}$ to the right.

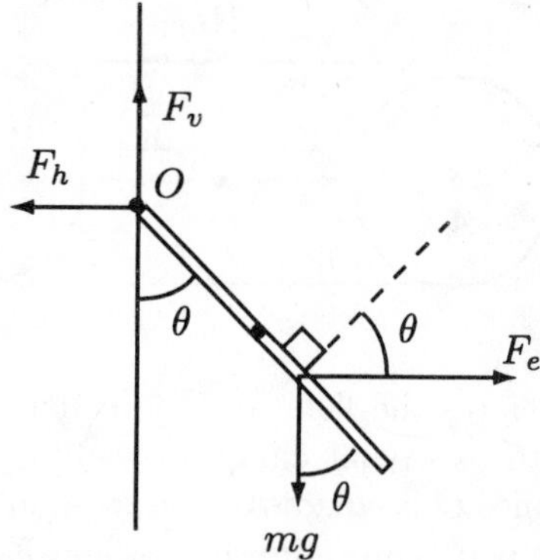

This field exerts an electric force on the hanging rod acting directly away from the plane (to the right) and at the rod's center of charge. Because the rod is uniformly charged, the center of charge coincides with the rod's center of mass at the rod's center. At this point we can get a qualitative feel for how large θ is. From the way we defined θ in the figure, θ can vary physically between 0 and π. However, if $\theta > \frac{\pi}{2}$ the torque about 0 due to Mg and F_e will be in the same direction and there would be no hope for the rod to be in equilibrium. So $\theta < \frac{\pi}{2}$. We break the hinge's force on the rod into Cartesian components:

$$F_{\text{hinge}} = \sqrt{F_{\text{h}}^2 + F_{\text{v}}^2}$$

The directions of F_{v} and F_{h} are determined by looking at the direction of the other two forces, Mg and F_e, and being mindful that we want force balance in each coordinate direction. Applying N2L,

$$\begin{aligned} \sum F_x &= F_e - F_{\text{h}} = 0 \\ \Rightarrow F_{\text{h}} &= F_e = \frac{\sigma}{2\epsilon_0} Q_{\text{rod}} \end{aligned}$$

where $\frac{\sigma}{2\epsilon_0} = E_{\text{plane}}$. Since the rod is uniformly charged,

$$\begin{aligned} Q_{\text{rod}} &= \lambda_{\text{rod}} L_{\text{rod}} \\ F_{\text{h}} &= F_e = \frac{\sigma}{2\epsilon_0} \lambda L \end{aligned}$$

$$\begin{aligned} &= \frac{(85 \times 10^{-6}\ \text{C/m}^2)(1.5 \times 10^{-6}\ \text{C/m})(1.2\ \text{m})}{2\,(8.85 \times 10^{-12}\ \text{C}^2/\text{N}\cdot\text{m}^2)} \\ &= 8.64\ \text{N} \end{aligned}$$

$$\begin{aligned} \sum F_y &= F_{\text{v}} - Mg = 0 \\ \Rightarrow F_{\text{v}} &= (0.75\ \text{kg})\,(9.8\ \text{m/s}^2) \\ &= 7.35\ \text{N} \end{aligned}$$

The net force that the hinge exerts on the rod is:

$$\begin{aligned} F_{\text{hinge}} &= \sqrt{F_{\text{h}}^2 + F_{\text{v}}^2} \\ &= \sqrt{(8.64\ \text{N})^2 + (7.35\ \text{N})^2} \\ &= 11\ \text{N} \end{aligned}$$

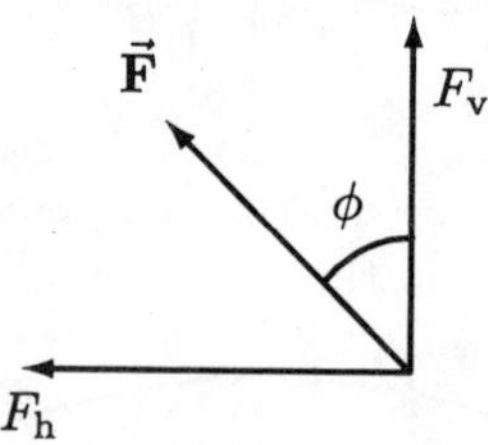

$$\begin{aligned} \tan\phi &= \frac{F_{\text{h}}}{F_{\text{v}}} = \frac{8.64\ \text{N}}{7.35\ \text{N}} = 1.18 \\ \Rightarrow \phi &= 50° \end{aligned}$$

So

$$F = (11\ \text{N, at } 50° \text{ to the vertical})$$

Now let's apply torque balance about point O.

$$\begin{aligned} \sum \vec{\tau}_0 &= \left(\frac{1}{2}L\right) F_e \cos\theta \hat{\odot} \\ &\quad + \left(\frac{1}{2}L\right) Mg \sin\theta \hat{\otimes} \\ &= 0 \end{aligned}$$

Dividing through by $\frac{1}{2}L\sin\theta$ and then solving for θ,

$$\theta = \tan^{-1}\left(\frac{F_e}{Mg}\right) = 50°$$

$\therefore$ the rod hangs at 50° with the vertical.

24.89 In writing the electric field at a point P due to a point charge at point P', we write:

$$\vec{\mathbf{E}} = \frac{kq}{R^3}\vec{\mathbf{R}}$$

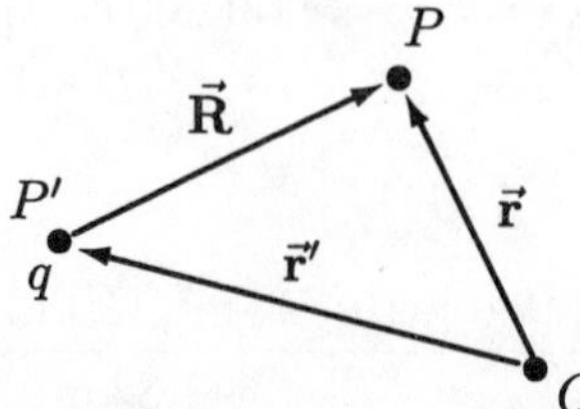

where $\vec{\mathbf{R}}$ is the vector from P' to P. Another way to express this is: if $\vec{\mathbf{r}}$ is the position vector of P and $\vec{\mathbf{r}}'$ is the position vector of P', then $\vec{\mathbf{R}} = \vec{\mathbf{r}} - \vec{\mathbf{r}}'$:

$$\vec{\mathbf{E}}_{\text{at } P} = \frac{kq}{R^3}\vec{\mathbf{R}} = \frac{kq\,(\vec{\mathbf{r}} - \vec{\mathbf{r}}')}{|\vec{\mathbf{r}} - \vec{\mathbf{r}}'|^3} \qquad (1)$$

where

$$\vec{\mathbf{r}} = x\hat{\mathbf{i}} + y\hat{\mathbf{j}} + z\hat{\mathbf{k}} = \text{position of } P$$

and

$$\vec{\mathbf{r}}' = x'\hat{\mathbf{i}} + y'\hat{\mathbf{j}} + z'\hat{\mathbf{k}} = \text{position of } P'$$

so:

$$\begin{aligned}\vec{\mathbf{r}} - \vec{\mathbf{r}}' &= (x - x')\,\hat{\mathbf{i}} + (y - y')\,\hat{\mathbf{j}} + (z - z')\,\hat{\mathbf{k}} \\ &= \vec{\mathbf{R}}\end{aligned}$$

If we dot (1) first with $\hat{\mathbf{i}}$ and then with $\hat{\mathbf{j}}$ we get the components of $\vec{\mathbf{E}}$ at P.

$$\begin{aligned}E_x(P) &= \frac{kq}{|\vec{\mathbf{r}} - \vec{\mathbf{r}}'|^3}(x - x') \\ E_y(P) &= \frac{kq}{|\vec{\mathbf{r}} - \vec{\mathbf{r}}'|^3}(y - y')\end{aligned}$$

For the problem at hand, the dipoles are both located in the xy-plane so there is no z-component of the field and therefore F_z will be zero also. A charge q' located at P will experience a force $\vec{\mathbf{F}}$ with components:

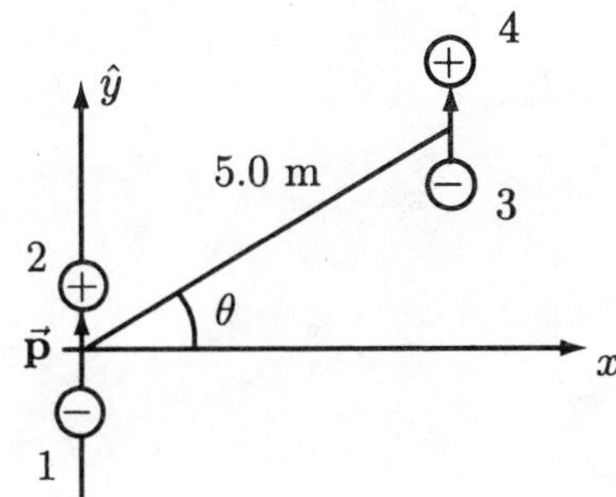

$$\begin{aligned}F_x(P) &= q'E_x(P) = \frac{kqq'(x - x')}{|\vec{\mathbf{r}} - \vec{\mathbf{r}}'|^3} \\ F_y(P) &= q'E_y(P) = \frac{kqq'(y - y')}{|\vec{\mathbf{r}} - \vec{\mathbf{r}}'|^3}\end{aligned}$$

For this problem we will have two charges acting on two other charges, as illustrated below. We will have eight force components for each θ

$$\begin{matrix} F_{x,\,1\text{ on }3} & F_{y,\,1\text{ on }3} \\ F_{x,\,1\text{ on }4} & F_{y,\,1\text{ on }4} \\ F_{x,\,2\text{ on }3} & F_{y,\,2\text{ on }3} \\ F_{x,\,2\text{ on }4} & F_{y,\,2\text{ on }4} \end{matrix}$$

Let's do one component for $\theta = 30°$ and use it as a model for how the computer will do the other components. For $\theta = 30°$ we'll do $F_{x,\,2\text{ on }3}$.

$$\begin{aligned}\vec{\mathbf{r}}_1' &= -0.05\text{ m}\hat{\mathbf{j}} \\ \vec{\mathbf{r}}_2' &= +0.05\text{ m}\hat{\mathbf{j}} \\ \vec{\mathbf{r}}_3 &= (5.0\text{ m}\cos 30°)\,\hat{\mathbf{i}} + (5.0\text{ m}\sin 30° - .05\text{ m})\,\hat{\mathbf{j}} \\ \vec{\mathbf{r}}_4 &= [5.0\text{ m}\cos 30°]\,\hat{\mathbf{i}} + [5.0\text{ m}\sin 30° + .05\text{ m}]\,\hat{\mathbf{j}}\end{aligned}$$

$$\begin{aligned}\vec{\mathbf{R}}_{2\text{ on }3} &= \vec{\mathbf{r}}_3 - \vec{\mathbf{r}}_2' \\ &= 5.0\text{ m}\cos 30°\hat{\mathbf{i}} \\ &\quad + [5.0\text{ m}\sin 30° + .10\text{ m}]\,\hat{\mathbf{j}}\end{aligned}$$

so,

$$F_{x,\,2\text{ on }3} = \frac{kqq'}{\left|\vec{\mathbf{R}}\right|^3}(x_3 - x_2)$$

$$= \frac{(9.0\times10^9\text{ N}\cdot\text{m}^2/\text{C}^2)(10^{-5}\text{ C})(-10^{-5}\text{ C})}{\left[(5.0\text{ m}\cos 30°)^2 + (5.0\text{ m}\sin 30° - .10\text{ m})^2\right]^{\frac{3}{2}}}(5.0\text{ m}\cos 30°)$$

$$= -3.21 \times 10^{-2}\text{ N}$$

In this problem you want to keep θ as a variable in your expressions.

c) The result of Example 24.15 was

$$F_y = \frac{-6kp^2}{y^4} = -8.6 \times 10^{-5}\text{ N}$$

(rotated 90°) in good agreement with the spreadsheet result.

24.93 Step 1: The hemispherical shell can be modelled as a collection of differential ring charges. The geometry is spherical and we choose the coordinates accordingly.

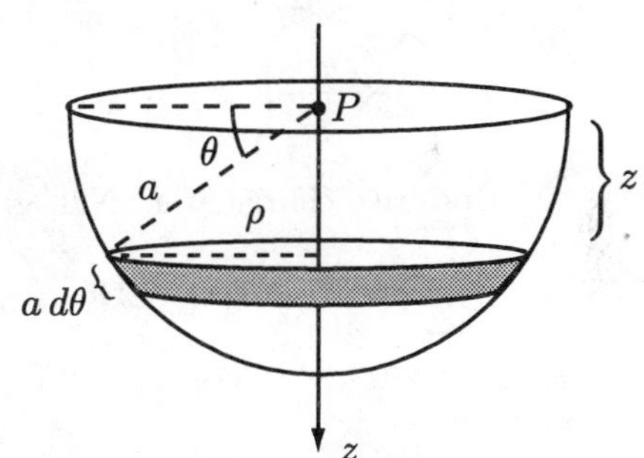

Step 2: We could label the ring charges by a z coordinate, but its a bit tricky getting a correct expression for dq—if you're interested in the details you can refer back to the center of mass chapter where it's done. Instead, we'll uniquely identify each ring by the θ coordinate, measured downward from the origin as shown in the figure. The charge of one such ring is

$$dq = \sigma\,da = \sigma 2\pi\rho\,(a\,d\theta) = 2\pi\sigma a\rho\,d\theta$$

Step 3: Using Eq. (24.5), the electric field at P due to one of these rings is:

$$\begin{aligned}d\vec{\mathbf{E}} &= -\frac{k\,dq\,(z)}{\left((z)^2 + \rho^2\right)^{\frac{3}{2}}}\hat{\mathbf{k}} \\ &= -\frac{2\pi k\sigma\rho a\,(z)}{\left[(z)^2 + \rho^2\right]^{\frac{3}{2}}}d\theta\hat{\mathbf{k}}\end{aligned}$$

Now we need to write all the variables in terms of the variable of integration, θ. From the figure and some trig we have:

$$\rho = a\cos\theta$$

and

$$z = a\sin\theta$$

Plugging in:

$$\begin{aligned} d\vec{\mathbf{E}} &= -\frac{2\pi k\sigma_0 a^3\cos\theta\sin\theta}{a^3}\hat{\mathbf{k}}\,d\theta \\ &= -2\pi k\sigma_0\cos\theta\sin\theta\hat{\mathbf{k}}\,d\theta \end{aligned}$$

Step 4: θ varies between 0 and $\frac{\pi}{2}$.

Step 5:

$$\begin{aligned} \vec{\mathbf{E}}(P) &= -2\pi k\sigma_0\hat{\mathbf{k}}\int_0^{\frac{\pi}{2}}\cos\theta\sin\theta\,d\theta \\ &= -\pi k\sigma_0\hat{\mathbf{k}}\int_0^{\frac{\pi}{2}}\sin(2\theta)\,d\theta \\ &= +\frac{\pi k\sigma_0\hat{\mathbf{k}}}{2}\cos(2\theta)\Big|_0^{\frac{\pi}{2}} \\ &= -\frac{\pi k\sigma_0}{2}\hat{\mathbf{k}} \end{aligned}$$

If we put a positive charge at P, N2L gives:

$$\begin{aligned} \sum F_z &= -F_{\text{elec}} + W = ma_z \\ -qE(p) + mg &= ma_z \\ -\frac{\pi k\sigma_0}{2}q + mg &= ma_z \end{aligned}$$

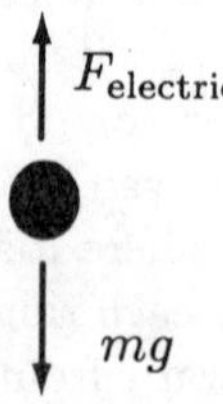

If q is in equilibrium, a_z is zero and we can solve for m.

$$m = \frac{\pi k\sigma_0 q}{2g}$$

If the charge is displaced upward (downward), the electric field becomes smaller (larger) in which case the particle's weight force is larger than (smaller than) F_{elec}. In either case, the particle accelerates back towards P after a vertical displacement.

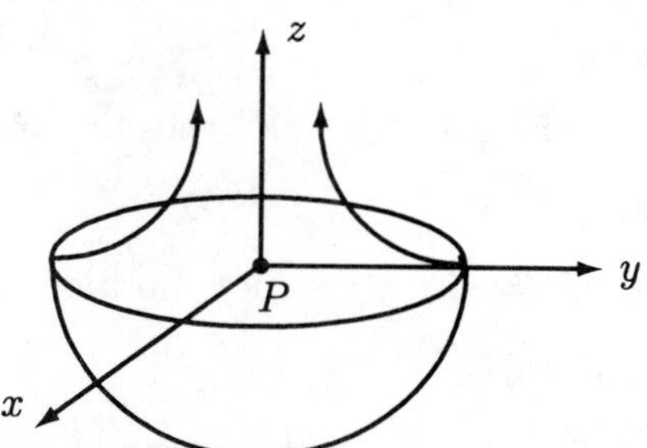

Looking at a field line diagram for the hemishell, field lines go inward and upward (it should be similar to the ring charge.) Therefore, $\vec{\mathbf{E}}$ in the xy-plane points to P. If q were displaced horizontally from P, a horizontal electric force would accelerate it back towards P. The equilibrium is stable for all displacements.

Chapter 25

Electric Potential Energy

25.1 The potential energy of the system is

$$\begin{aligned} U &= \frac{kQ^2}{d} \\ &= \frac{(9.0 \times 10^9\ \mathrm{N \cdot m^2/C^2})\left(1.6 \times 10^{-19}\ \mathrm{C}\right)^2}{(5.0 \times 10^{-9}\ \mathrm{m})} \\ &= 4.6 \times 10^{-20}\ \mathrm{J} \end{aligned}$$

25.3 The work required is given by

$$W = Q\Delta V = (6\ \mathrm{C})\,(100\ \mathrm{V}) = 600\ \mathrm{J}$$

25.5 a) Only the electric field is zero at the center of the triangle, C.

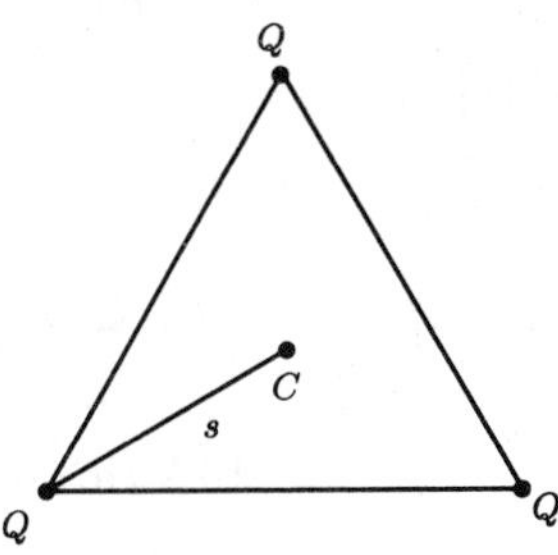

The electric potential is $\frac{3kQ}{s}$ at point C; because all the charges have the same sign the potential can only be zero at ∞. The electric fields produced by the 3 charges at C are three vectors of equal magnitude at 120° angles. Their sum is zero.

25.7

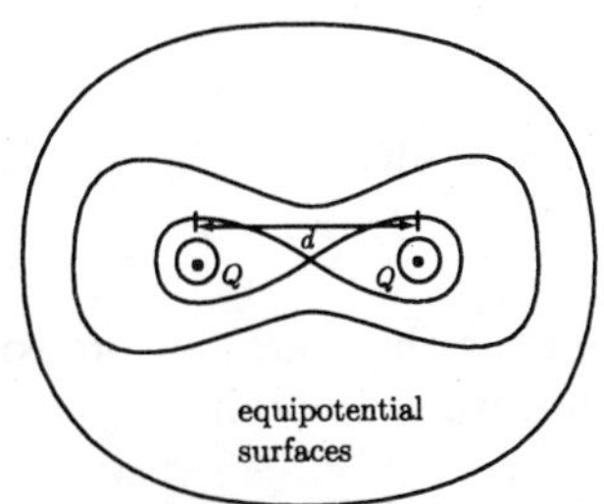

25.9 Since every charge element is the same distance a from P,

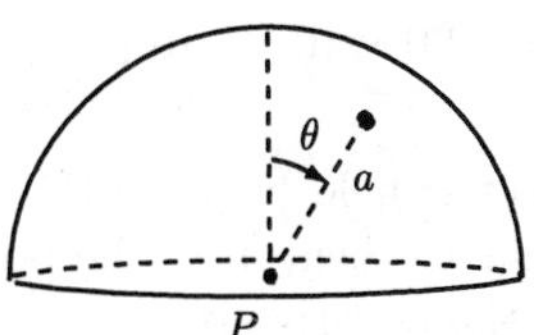

$$dV_P = k\frac{dQ}{a} \quad \text{and} \quad V(P) = \frac{kQ}{a}$$

25.11 Zero. The electric field is zero within a conductor. The charge distribution on the surface serves to enforce $\vec{\mathbf{E}} = 0$ inside.

25.13 From Example 25.15,

$$\begin{aligned} \frac{Q_1}{R_1} &= \frac{Q_2}{R_2} \\ \Rightarrow Q_2 &= \frac{R_2}{R_1}Q_1 \\ &= \left(\frac{20\ \mathrm{cm}}{10\ \mathrm{cm}}\right) Q_1 = 2Q_1 \end{aligned}$$

Therefore b) is the correct answer.

25.15 The work done equals the change in the charge's potential energy. Since both initial and final distances of q from Q are the same, the potential energy is the same in the initial and final states, so the net work done is zero.

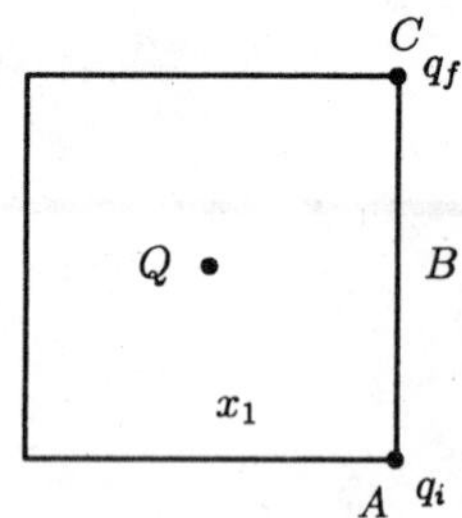

Work is done moving from A to B, as q gets close to Q, but negative work is done moving from B to C. The net result is zero.

25.19 Use conservation of energy to find the minimum distance d_{min}.

	Before		After	
	KE	PE	KE	PE
Energy:	$\frac{1}{2}mv^2$	0	0	$\frac{kqQ}{d_{\text{min}}}$

Then, setting the energy before equal to the energy after, we can obtain d_{min}

$$\begin{aligned} d_{\text{min}} &= \frac{2kqQ}{mv^2} \\ &= \frac{2(9.0\times10^{9}\ \text{N}\cdot\text{m}^2/\text{C}^2)(1.0\times10^{-6}\ \text{C})(1.0\times10^{-9}\ \text{C})}{(1\times10^{-5}\ \text{kg})(3.0\times10^{5}\ \text{m}/\text{s})^2} \\ &= 2.0\times10^{-11}\ \text{m} \end{aligned}$$

25.23 The electric field between the plates runs perpendicular to them from the positively charged plate to the negative plate.

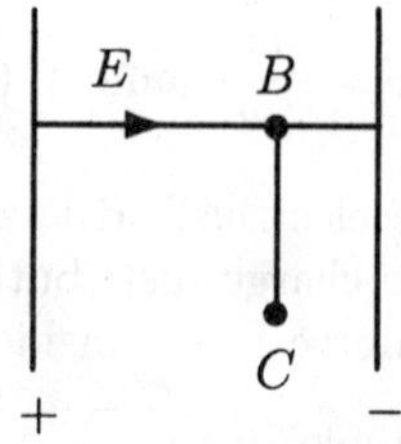

Since the line BC is parallel to the plates, it is perpendicular to $\vec{\mathbf{E}}$, so

$$V_B - V_C = -\int_C^B \vec{\mathbf{E}}\cdot d\vec{\ell} = 0$$

The potential at C is the *same* as at B.

25.25 The potential difference is given by

$$\Delta V = -\int_1^2 \vec{\mathbf{E}}\cdot d\vec{\ell}$$

independent of path between points 1 and 2.

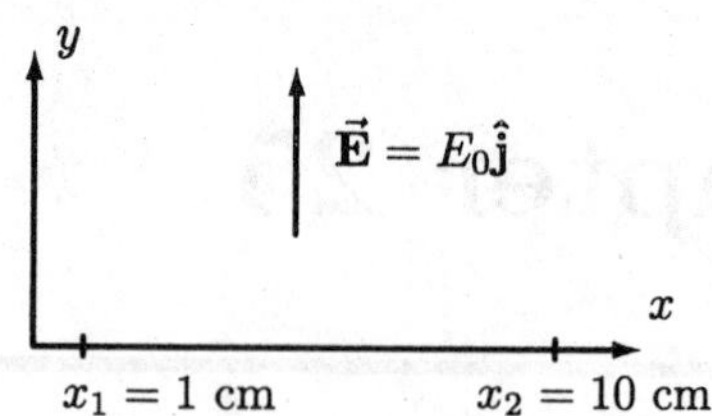

If we choose a path along the x-axis, then $\vec{\mathbf{E}}\cdot d\vec{\ell}=$ zero everywhere on the path, and hence $\Delta V = 0$. i.e, since the field is uniform and perpendicular to the x-axis, the potential is the same at $x_1 = 1$ cm as it is at $x_2 = 2$ cm.

25.29 The potential at point P is given by

$$\Delta V = -\int_{(0,0)}^{P} \vec{\mathbf{E}}\cdot d\vec{\ell}$$

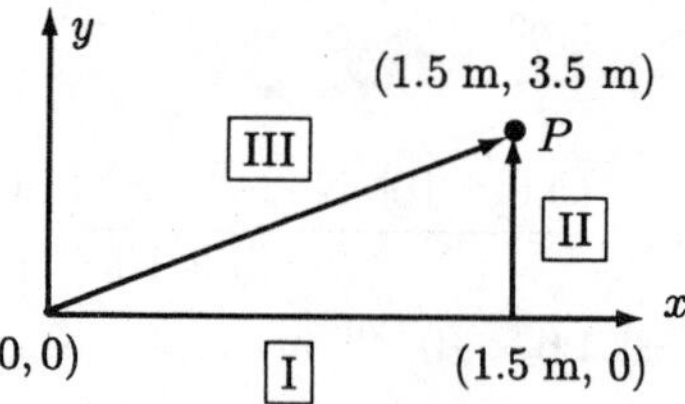

Since $\vec{\mathbf{E}}$ is conservative, $\int_{(0,0)}^{P}\vec{\mathbf{E}}\cdot d\vec{\ell}$ is independent of the path taken from the origin to point P. That means that path III is equivalent to path I and path II. Therefore, we may write

$$d\vec{\ell} = dx\,\hat{\mathbf{i}}$$

on path I and

$$d\vec{\ell} = dy\,\hat{\mathbf{j}}$$

on path II so

$$\begin{aligned} \Delta V &= V(P) - V(0) \\ &= -\int_{x=0}^{x=1.5\ \text{m}} E_x\,dx - \int_{y=0}^{y=3.5\ \text{m}} E_y\,dy \\ &= -\int_0^{1.5\ \text{m}} (16\ \text{V/m})\,dx \\ &\quad -\int_0^{3.5\ \text{m}} (8.5\ \text{V/m})\,dy \end{aligned}$$

Since $V(0) = 0$, we have that

$$\begin{aligned} V(P) &= -(16\ \text{V/m})\,x|_0^{(1.5\ \text{m})} - (8.5\ \text{V/m})\,y|_0^{(3.5\ \text{m})} \\ &= -(16\ \text{V/m})(1.5\ \text{m}) - (8.5\ \text{V/m})(3.5\ \text{m}) \\ &= -54\ V \end{aligned}$$

25.33 From Gauss's law, the electric field between the shells depends only on Q_1, the charge on the inner shell.

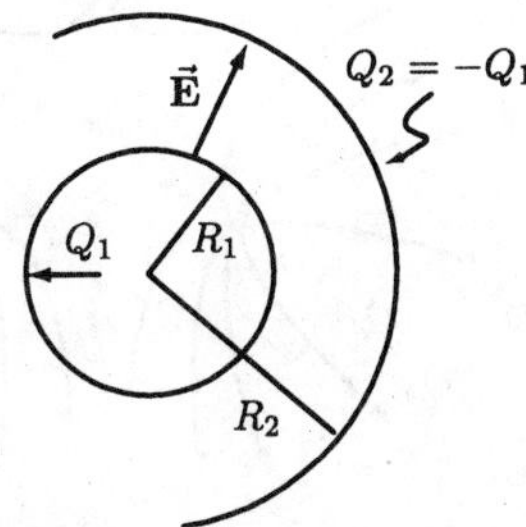

So, the potential between the shells varies in the same way as the potential due to a single charged shell. The potential difference is

$$\begin{aligned}\Delta V &= V_2 - V_1 = -\int_{R_1}^{R_2} \frac{kQ_1}{r^2}\,dr \\ &= \frac{kQ_1}{R_2} - \frac{kQ_1}{R_1} \\ &= kQ_1\left(\frac{1}{R_2} - \frac{1}{R_1}\right) \\ &= (9.00 \times 10^9\ \mathrm{N\cdot m^2/C^2})\,(3.27 \times 10^{-6}\ \mathrm{C}) \\ &\quad \times \left(\frac{1}{(0.300\ \mathrm{m})} - \frac{1}{(0.200\ \mathrm{m})}\right) \\ &= -4.91 \times 10^4\ \mathrm{V}\end{aligned}$$

25.37 There are 10 distinct pairs. To see this, consider the 5 charges:

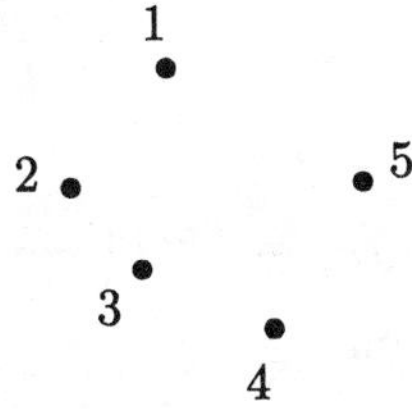

The distinct pairs are

$$\begin{aligned}&1-2, 1-3, 1-4, 1-5\\&2-3, 2-4, 2-5\\&3-4, 3-5\\&4-5\end{aligned}$$

25.41 The potential energy of the system of charges is given by the work required to assemble the changes (c.f., Equation 25.14)

$$U_{\text{system}} = \frac{1}{2}\sum_i Q_i V_i$$

Now, V_i is the potential evaluated at the location of the ith change. Each negative charge is at equal distance from a $+Q$ and $-Q$ charge, so the potential at its location is zero. Thus

$$U_{\text{system}} = \frac{1}{2}\left\{-2\frac{kQ^2}{a} + 0 + 0\right\}$$

$$U = -\frac{kQ^2}{a}$$

25.45 To place the third charge, q, at the third corner, we bring q in from infinity. The potential energy of the system is equal to the work required to assemble the three charges.

$$U = W_1 + W_2 + W_3$$

where,

$$\begin{aligned}W_1 &= 0\\ W_2 &= \frac{kQ^2}{\ell}\\ W_3 &= \frac{kqQ}{\ell} + \frac{kqQ}{\ell} = \frac{2kqQ}{\ell}\end{aligned}$$

so

$$U = \frac{kQ}{\ell}(Q + 2q)$$

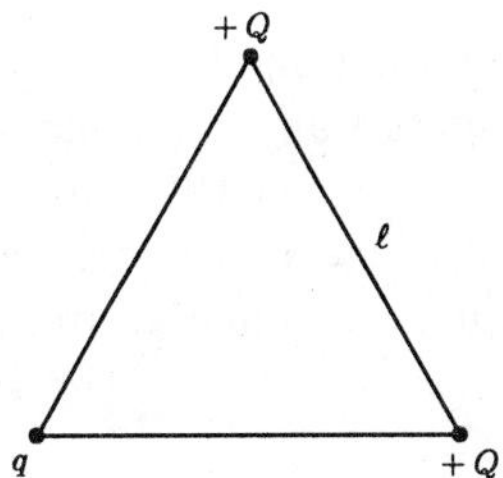

When q is subsequently released, it gains kinetic energy at the expense of the potential energy of the system. However, since the two charges $+Q$ remain fixed, the potential energy of the system is $\frac{kQ^2}{\ell}$ after q has moved a very far distance from the original configuration. Applying conservation of energy yields

	Before		After	
Energy:	KE	PE	KE	PE
	0	$\frac{kQ}{\ell}(Q+2q)$	$\frac{1}{2}mv^2$	$\frac{kQ^2}{\ell}$

Setting the energy of the system before the release of q equal to the system's total energy after q's release, we have

$$\frac{kQ}{\ell}(Q + 2q) = \frac{1}{2}mv^2 + \frac{kQ^2}{\ell}$$

The kinetic energy of the third charge is therefore,

$$\mathrm{K} = \frac{1}{2}mv^2 = \frac{2kqQ}{\ell}$$

25.47 Potential due to a dipole (in (r, θ) coordinates):

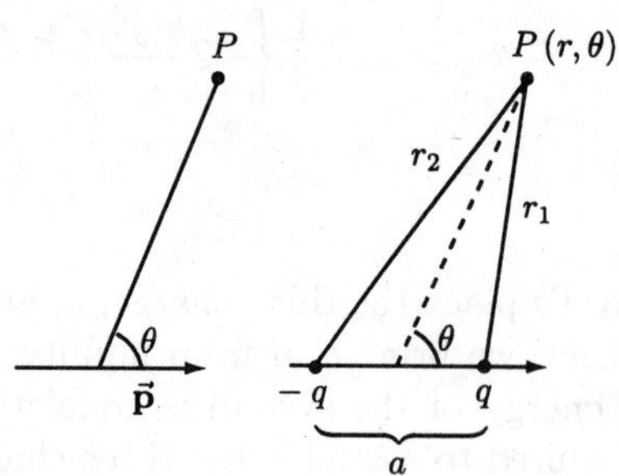

The potential at P is given by

$$V(P) = kq\left(\frac{1}{r_1} - \frac{1}{r_2}\right) = kq\frac{(r_2 - r_1)}{r_1 r_2}$$

where

$$\begin{aligned} r_1 &= \sqrt{r^2 + \left(\frac{a}{2}\right)^2 - 2\frac{ra}{2}\cos\theta} \\ r_2 &= \sqrt{r^2 + \left(\frac{a}{2}\right)^2 + ra\cos\theta} \end{aligned}$$

For an ideal dipole, $p = qa$ and $a \to 0$; thus we may ignore any differences between r, r_1, and r_2 except when they are subtracted. To find $r_2 - r_1$, look at the triangle of distances, and drop a perpendicular from q to the $-q \to P$ line:

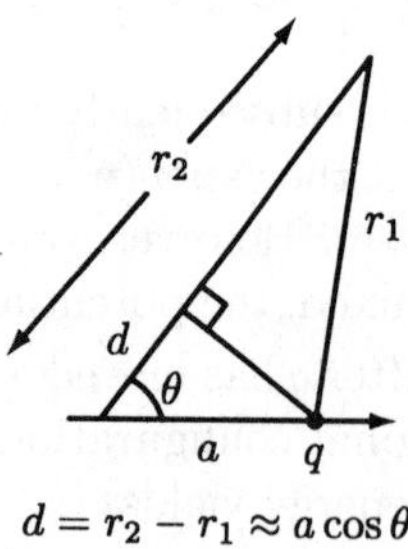

$d = r_2 - r_1 \approx a\cos\theta$

Thus

$$\begin{aligned} V(P) &= kq\frac{(r_2 - r_1)}{r_1 r_2} \\ &= kq\frac{a\cos\theta}{r^2} = kp\frac{\cos\theta}{r^2} \end{aligned}$$

where $p = qa \equiv$ dipole moment.

25.51 Near each charge, the equipotentials form circles around that charge. At large distances, the equipotentials form circles around the entire system.

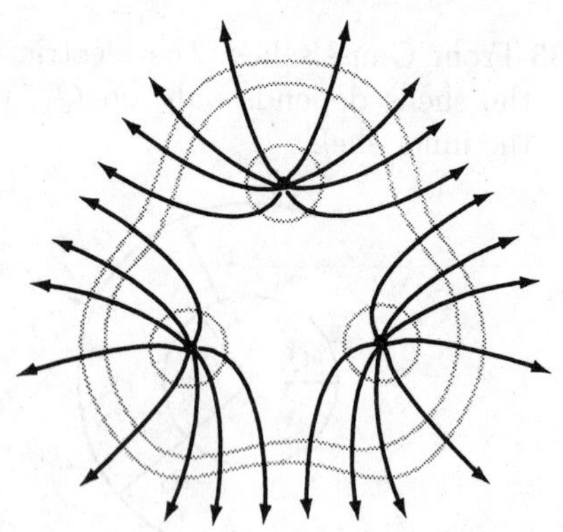

25.55 The net charge of the system is negative ($-1.5\ \mu$C) so there is a $V = 0$ equipotential surface surrounding the + charge. $V = 0$ for some $y < 0$ and some y between 0 and D.

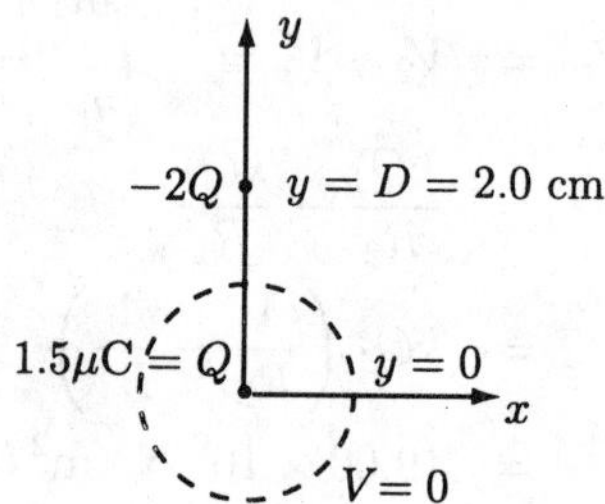

a) $0 < y < D$

$$\begin{aligned} V(y) &= kQ\left(\frac{1}{y} - \frac{2}{(D - y)}\right) = 0 \\ \Rightarrow 2y &= D - y \\ y &= \frac{1}{3}D = \frac{2}{3}\text{ cm} = 0.67\text{ cm} \end{aligned}$$

$y < 0$

$$\begin{aligned} V(y) &= kQ\left(\frac{1}{-y} - \frac{2}{(D - y)}\right) = 0 \\ \Rightarrow 2y &= y - D \\ y &= -D = -2.0\text{ cm} \end{aligned}$$

b) The field will be zero somewhere below the + charge, at $y < 0$

$$\begin{aligned} E_y(y) &= kQ\left(-\frac{1}{y^2} + \frac{2}{(D - y)^2}\right) = 0 \\ \Rightarrow 2y^2 &= (D - y)^2 \\ \pm\sqrt{2}y &= D - y \Rightarrow y = \frac{D}{1 \pm \sqrt{2}} \end{aligned}$$

Only the negative sign gives $y < 0$, so

$$\begin{aligned} y &= -\frac{D}{(\sqrt{2} - 1)} = -D\left(\sqrt{2} + 1\right) \\ &= -(2.4)\,D = -4.8\text{ cm} \end{aligned}$$

At this point the potential is

$$\begin{aligned} V(y) &= kQ\left(\frac{1}{|y|} - \frac{2}{|(D-y)|}\right) \\ &= kQ\left(\frac{1}{\frac{+D}{\sqrt{2}-1}} - \frac{2}{D+\frac{D}{\sqrt{2}-1}}\right) \\ &= \frac{kQ}{D}\left(\sqrt{2}-1-\frac{2}{\sqrt{2}+2}\right) \\ &= \frac{(9.0\times10^9\ \text{N}\cdot\text{m}^2/\text{C}^2)}{(2.0\times10^{-2}\ \text{m})}\left[\sqrt{2}-1-\frac{2}{\sqrt{2}+2}\right] \\ &\quad \times (1.5\times10^{-6}\ \text{C}) \\ &= -1.2\times10^5\ \text{V} \end{aligned}$$

25.59 The potential of a sphere of radius R is given by

$$V = \frac{kQ}{R}$$

the charge, Q, necessary to reach 10^4 V is therefore

$$\begin{aligned} Q &= \frac{RV}{k} = \frac{(15\times10^{-2}\ \text{m})(10^4\ \text{V})}{(9\times10^9\ \text{N}\cdot\text{m}^2/\text{C}^2)} \\ &= 2\times10^{-7}\ \text{C} = 0.2\ \mu\text{C} \end{aligned}$$

25.63 a) For $y \ll \text{L}$, treat L as infinite. The potential (c.f. Example 25.13) at y is then ($\text{L}\to\pm\infty$) given by

$$V(y) = 2k\lambda\ln\left(\frac{y_0}{y}\right)$$

where y_0 is a suitably chosen ($y_0 \neq 0$, $y_0 \neq \infty$) reference point where $V(y_0) = 0$.

b) For $y \gg \text{L}$, approximate by using Coulomb's law

$$V(y) = \frac{k(\lambda\text{L})}{y}$$

c) By exact integration, divide the rod into elements of length dx carrying charge $dq = \lambda\,dx$.

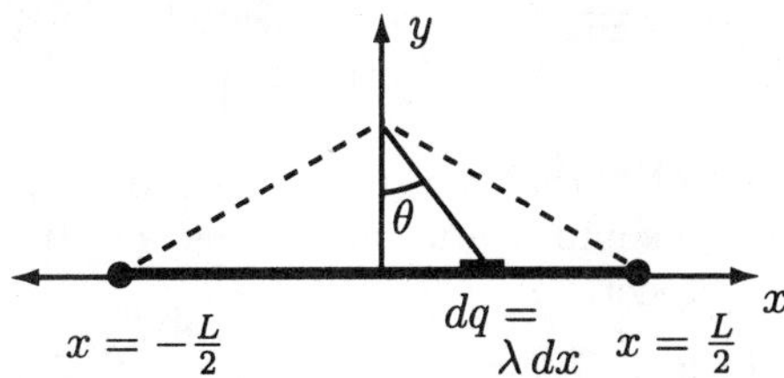

The potential due to one such segment is

$$\begin{aligned} dV(y) &= \frac{kdq}{\sqrt{x^2+y^2}} = k\lambda\frac{dx}{\sqrt{x^2+y^2}} \\ V(y) &= k\lambda\int_{-\frac{\text{L}}{2}}^{\frac{\text{L}}{2}}\frac{dx}{\sqrt{x^2+y^2}} \end{aligned}$$

Let

$$x = y\tan\theta,\ dx = y\sec^2\theta\,d\theta$$

$$\begin{aligned} V(y) &= k\lambda\int_{\tan^{-1}-\frac{\text{L}}{2y}}^{\tan^{-1}\frac{\text{L}}{2y}}\frac{\sec^2\theta\,d\theta}{\sec\theta} \\ &= k\lambda\ln\left[\sec\theta+\tan\theta\right]_{-\tan^{-1}\left(\frac{\text{L}}{2y}\right)}^{\tan^{-1}\frac{\text{L}}{2y}} \end{aligned}$$

Therefore,

$$V(y) = k\lambda\ln\left(\frac{\sqrt{\left(\frac{\text{L}^2}{4}\right)+y^2}+\frac{\text{L}}{2}}{\sqrt{\left(\frac{\text{L}^2}{4}\right)+y^2}-\frac{\text{L}}{2}}\right)$$

To check this answer, investigate the cases $y \ll \text{L}$ and $y \gg \text{L}$.

Case a): $y \ll \text{L}$ or $\frac{y}{\text{L}} \ll 1$. Rewrite $V(y)$ in terms of $\frac{y}{\text{L}}$:

$$V(y) = k\lambda\ln\left(\frac{\sqrt{1+\frac{4y^2}{\text{L}^2}}+1}{\sqrt{1+\frac{4y^2}{\text{L}^2}}-1}\right)$$

Now, if $\frac{2y}{\text{L}} \ll 1$,

$$\sqrt{1+\frac{4y^2}{\text{L}^2}} \approx 1+\frac{2y^2}{\text{L}^2}$$

and

$$\begin{aligned} V(y) &\approx k\lambda\ln\left(\frac{2\left(1+\frac{y^2}{\text{L}^2}\right)}{2\left(\frac{y^2}{\text{L}^2}\right)}\right) \\ &\approx 2k\lambda\ln\left(\frac{\text{L}}{y}\right) \end{aligned}$$

By comparison with result a) we see that the choice $y_0 = L$ matches the exact calculation with $V = 0$ at ∞. For any other choice of y_0, the two answers differ by a constant $2k\lambda\ln\left(\frac{\text{L}}{y_0}\right)$.

Case b): $y \gg \text{L}$ or $1 \gg \frac{\text{L}}{y}$. Rewrite $V(y)$ in terms of $\frac{\text{L}}{y}$:

$$\begin{aligned} V(y) &= k\lambda\ln\left(\frac{y\sqrt{1+\frac{\text{L}^2}{4y^2}}+\frac{\text{L}}{2}}{y\sqrt{1+\frac{\text{L}^2}{4y^2}}-\frac{\text{L}}{2}}\right) \\ &= k\lambda\ln\left(\frac{y+\frac{\text{L}}{2}}{y-\frac{\text{L}}{2}}\right) \\ &= k\lambda\ln\left(\frac{1+\frac{\text{L}}{2y}}{1-\frac{\text{L}}{2y}}\right) \\ &= k\lambda\ln\left(1+\frac{\text{L}}{2y}\right) - k\lambda\ln\left(1-\frac{\text{L}}{2y}\right) \end{aligned}$$

Since $\ln(1+x) = x - \frac{1}{2}x^2 + \ldots$, for $|x| < 1$, then for $\frac{L}{2y} \ll 1$, we have that

$$V(y) \approx k\lambda\left(\frac{L}{2y} - \left(-\frac{L}{2y}\right)\right) = k\frac{\lambda L}{y}$$

which is identical to our result in b).

25.67 The potential due to a thin hemispherical shell at its center is

$$V = \int \frac{k\,dq}{r} = \frac{k}{r}\int dq = \frac{kQ}{r}$$

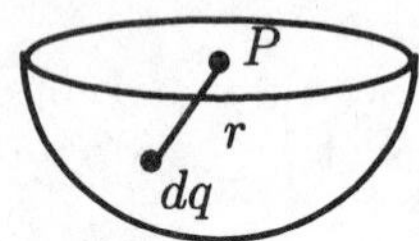

since every charge element is the same distance from P. Now we imagine our object to be composed of many such differential shells, each carrying charge

$$dq = 2\pi r^2 dr\,\rho$$

and contributing potential

$$dV = \frac{k\left(2\pi r^2\,dr\right)}{r}\rho = 2\pi k\rho r\,dr \text{ at } P$$

Thus

$$V = \int_a^b 2\pi k\rho r\,dr = 2\pi k\rho\frac{r^2}{2}\bigg|_a^b = \pi k\rho\left(b^2 - a^2\right)$$

25.71 a) With $V \to 0$ at ∞, the potential of each object may be approximated as that of an isolated sphere. Thus

$$V_s = \frac{Q_s}{r_s} > 0$$

and

$$V_b = \frac{Q_b}{r_b} < 0$$

In addition, $|Q_b| > |Q_s|$ and $r_b < r_s$, so $|V_b| > |V_s|$. The correct diagram is a).

b) The ball is placed inside the shell, so we may no longer consider the two objects to be isolated. Let's look at the field line diagram.

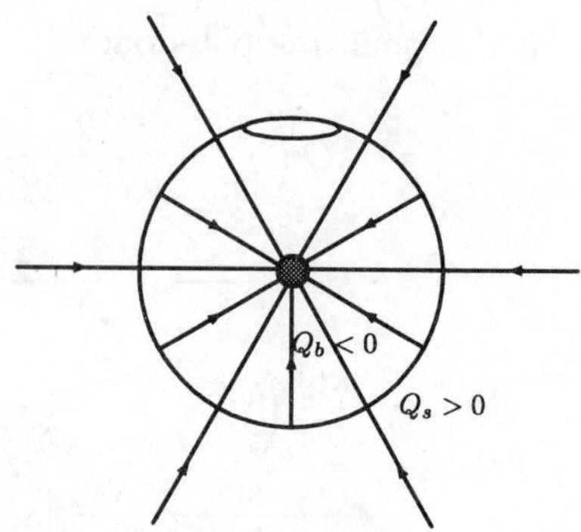

Since the *net* charge of the system is negative ($|Q_b| > Q_s$) field lines run inward to the system from infinity. All of the field lines leaving the positive charge on the shell run inward to the negative charge on the ball. Since potential decreases along field lines, the potential is negative on *both* the shell *and* the ball if it is zero at infinity. Since the net charge inside the shell's outer surface is less in magnitude than the charge on the ball,

$$|V_s| < |V_b|$$

The correct answer is d).

25.75 The object's field will be greatest near its narrow tip and will induce a positive charge density on the grounded conducting plane:

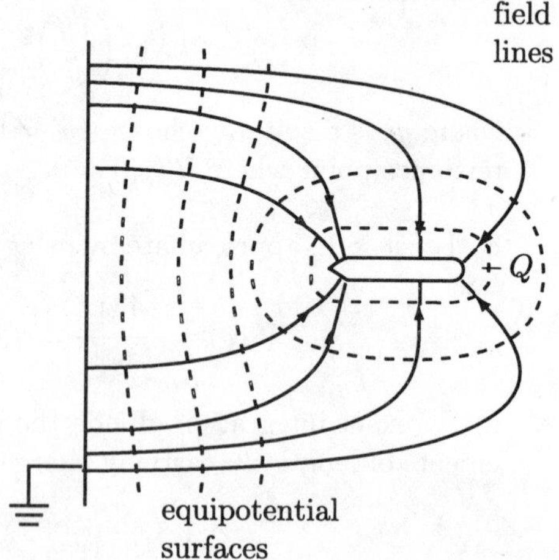

Notice that the grounded conducting plane forms an equipotential surface, as does the conducting object.

25.79 The electric field outside the copper cube near the center of the face is given by Eqn. (25.17)

$$\vec{\mathbf{E}} = 4\pi k\sigma\hat{\mathbf{n}} = 4\pi\left(9.00\times10^9\ \mathrm{N\cdot m^2/C^2}\right)\times\left(0.78\times10^{-6}\ \mathrm{C/m^2}\right) = 8.8\times10^4\ \mathrm{V/m} = 88\ \mathrm{kV/m}$$

Near the edges and corners of the copper cube, the radius of curvature becomes very small and therefore the field strength becomes larger there.

25.83 The potential energy is given by

$$U = QV$$

so, the required charge is

$$\begin{aligned} Q &= \frac{U}{V} = \frac{(15)\,(130 \times 10^6 \text{ J})}{(12 \text{ V})} \\ &= 163 \times 10^6 \text{ C} \end{aligned}$$

One battery will deliver

$$\begin{aligned} Q_B &= (5 \text{ C/s})\,(60 \text{ s/min})\,(60 \text{ min/h})\,(10 \text{ h}) \\ &= 0.18 \times 10^6 \text{ C} \end{aligned}$$

The number of batteries, N_B, required is

$$N_B = \frac{Q}{Q_B} = \frac{163 \times 10^6 \text{ C}}{0.18 \times 10^6 \text{ C}} = 900 \text{ batteries}$$

25.87 The net field between the plastic sheets is

$$E = (2\pi k\sigma) - (-2\pi k\sigma) = 4\pi k\sigma$$

The potential difference between the plates is therefore

$$\begin{aligned} \Delta V &= Ed = 4\pi k\sigma d \\ &= 4\pi\left(9.0 \times 10^9 \,\frac{\text{N}\cdot\text{m}^2}{\text{C}^2}\right)\left(1.3 \times 10^{-6}\,\frac{\text{C}}{\text{m}^2}\right) \\ &\quad \times \left(1.3 \times 10^{-3} \text{ m}\right) \\ &= 190 \text{ V} \end{aligned}$$

Notice that outside the system,

$$\vec{\mathbf{E}} = \vec{\mathbf{E}}_1 + \vec{\mathbf{E}}_2 = 0$$

b) The free charges inside the metal slab quickly distribute themselves to produce an opposing field, $\vec{\mathbf{E}}'$, so that the net electric field, $\vec{\mathbf{E}}_{\text{net}}$, inside the conductor is zero.

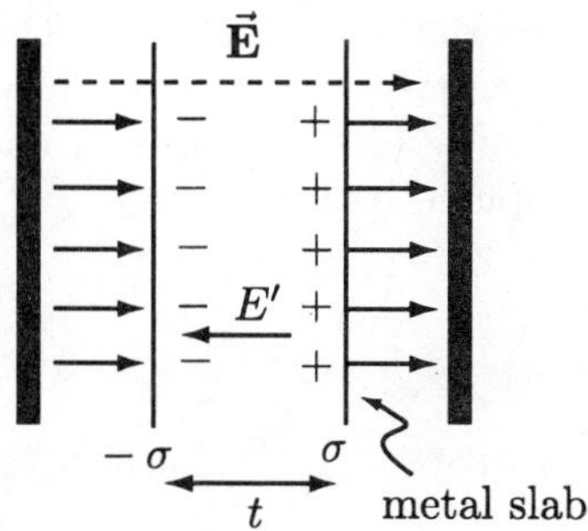

Inside the metal slab,

$$\vec{\mathbf{E}}_{\text{net}} = \vec{\mathbf{E}} + \vec{\mathbf{E}}' = 0 \Rightarrow \vec{\mathbf{E}}' = -\vec{\mathbf{E}}$$

Thus the surface density at the edges of the slab is σ and $-\sigma$, as shown. These charged sheets produce zero field to the left and right of the slab. Therefore, the field between the left-hand sheet and the slab is

$$E_{\text{left}} = E = 4\pi k\sigma$$

The field between the right-hand sheet and the slab is

$$E_{\text{right}} = E = 4\pi k\sigma$$

The potential difference between the sheets is now given by

$$\begin{aligned} \Delta V &= 4\pi k\sigma\,(d - t) \\ &= 4\pi\left(9.0 \times 10^9 \,\frac{\text{N}\cdot\text{m}^2}{\text{C}^2}\right)\left(1.3 \times 10^{-6}\,\frac{\text{C}}{\text{m}^2}\right) \\ &\quad \times (1.3 \text{ mm} - 1.0 \text{ mm}) \\ &= 44 \text{ V} \end{aligned}$$

25.91 Assume the dome is spherical. Then the potential on the dome is related to the charge on it by $V = \frac{kQ}{R}$. Let $\frac{dQ}{dt}$ denote the rate of charge transfer to the dome. Then the time, t, required to charge the dome to a given potential difference, ΔV, is given by

$$\begin{aligned} t &= \frac{Q}{\frac{dQ}{dt}} = \frac{VR}{k\left(\frac{dQ}{dt}\right)} \\ &= \frac{(3.0 \times 10^6 \text{ V})\,(2.0 \text{ m})}{(9.0 \times 10^9 \text{ N}\cdot\text{m}^2/\text{C}^2)\,(17 \times 10^{-6} \text{ C/s})} \\ &= 39 \text{ s} \end{aligned}$$

The power, P, required to operate the machine at potential difference ΔV is

$$\begin{aligned} P &= \Delta V\left(\frac{dQ}{dt}\right) = \left(3 \times 10^6 \text{ V}\right)\left(17 \times 10^{-6} \text{ C/s}\right) \\ &= 51 \text{ W} \end{aligned}$$

25.95 The Matlab program, LBPOT, (see appendix) was written to calculate the values of the potential, $V(P)$, at points P_1 and P_2. The values computed by LBPOT are

$$\begin{aligned} V(P_1) &= 1.45 \times 10^5 \text{ V} \\ V(P_2) &= -2.0 \times 10^5 \text{ V} \end{aligned}$$

25.99 a) First consider a spherical shell of charge.

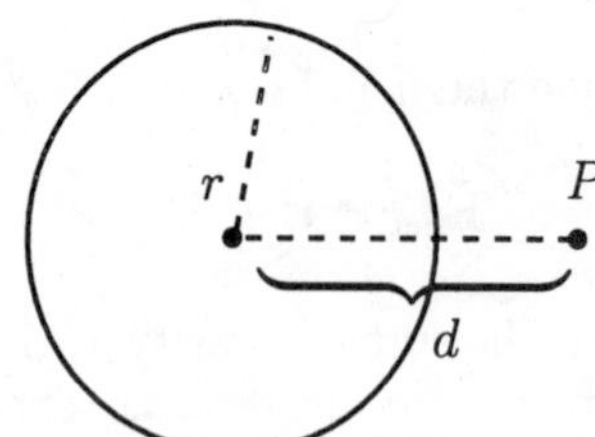

We know that the electric field outside the shell is

$$\vec{\mathbf{E}} = \frac{kQ}{r^2}\hat{\mathbf{r}}$$

and inside it is zero. Thus the potential inside the shell is constant $= \frac{kQ}{R}$ and outside it is $\frac{kQ}{r}$. Now consider the sphere to be composed of shells of radius r and thickness dr. Each carries charge

$$dQ = 4\pi r^2 dr\, \rho,$$

and produces potential

$$dV = \frac{k\left(4\pi r^2 \rho\, dr\right)}{r} = 4\pi k\rho r\, dr$$

at all points inside and

$$dV = \frac{k\left(4\pi r^2 \rho\, dr\right)}{d}$$

at point P distance $d > r$ from its center. Now we are ready to calculate the potential due to the whole sphere:

Point Outside:

$$\begin{aligned} V &= \int_0^R \frac{k4\pi r^2 \rho\, dr}{d} = \frac{4\pi k\rho}{d}\left(\frac{r^3}{3}\right)\Bigg|_0^R \\ &= \frac{4}{3}\frac{\pi k\rho R^3}{d} = \frac{kQ}{d} \end{aligned}$$

as expected.

Point Inside:

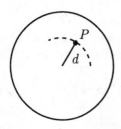

All shells with $r > d$ produce potential $4\pi k\rho r\, dr$ at P

$$\begin{aligned} V(P) &= \int_0^d 4\pi\rho k\frac{r^2}{d}\, dr + \int_d^R 4\pi k\rho r\, dr \\ &= 4\pi\rho k\left[\frac{1}{d}\frac{r^3}{3}\Bigg|_0^d + \frac{r^2}{2}\Bigg|_d^R\right] \\ &= 4\pi\rho k\left[\frac{d^2}{3} + \frac{R^2 - d^2}{2}\right] \\ &= 4\pi\rho k\left[\frac{R^2}{2} - \frac{d^2}{6}\right] \end{aligned}$$

or since the total charge on the sphere, Q, is given by:

$$Q = \frac{4\pi}{3}\rho R^3$$

we may write:

$$V(P) = \frac{kQ}{2}\frac{\left(3R^2 - d^2\right)}{R^3}$$

As a check, note that at $d = R$,

$$V(P(d = R)) = \frac{kQ}{2}\frac{\left(3R^2 - R^2\right)}{R^3} = \frac{kQ}{R}$$

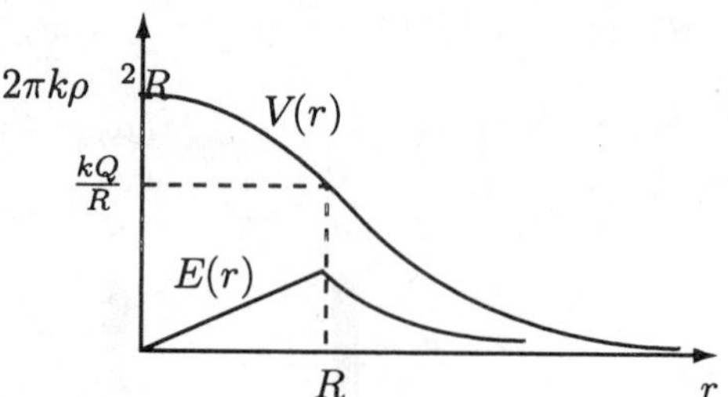

plots of the potential, $V(r)$, and the electric field $E(r)$ for part a).

b) The electric potential energy stored in the sphere may be obtained by generalizing equation (25.14) for a continuous distribution of charge:

$$U_{\text{system}} = \frac{1}{2}\sum_i Q_i V_i \qquad \text{(discrete)}$$

$$\Rightarrow \frac{1}{2}\int V\, dq \qquad \text{(continuous)}$$

With $dQ = 4\pi\rho r^2\, dr$ on a shell of radius r,

$$\begin{aligned} U_{\text{sphere}} &= \frac{1}{2}\int_{\text{sphere}} V(r)\, dQ \\ &= \frac{1}{2}k(4\pi\rho)^2\int_0^R\left(\frac{R^2}{2} - \frac{r^2}{6}\right)r^2\, dr \\ &= \frac{1}{4}k(4\pi\rho)^2\left[R^2\frac{r^3}{3}\Bigg|_0^R - \frac{1}{3}\left(\frac{r^5}{5}\right)\Bigg|_0^R\right] \end{aligned}$$

$$= \frac{4k(\pi\rho)^2}{3}\left(R^5 - \frac{1}{5}R^5\right)$$
$$= \frac{3}{5}\frac{kQ^2}{R}$$

(where $Q = 4\pi\rho\frac{R^3}{3}$.) We obtain the same result by calculating the work, W, required to assemble the sphere by adding successive thin spherical shells of charge:

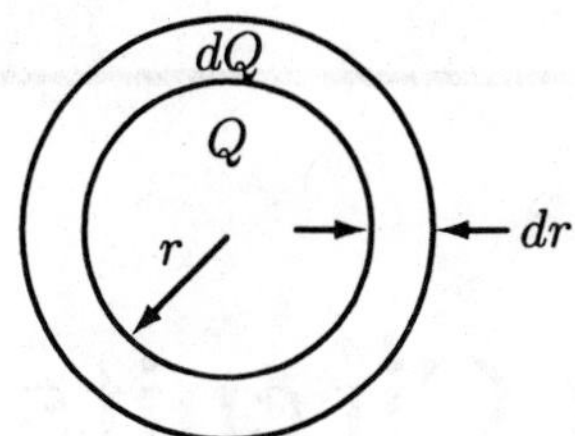

$$\begin{aligned} dW &= k\frac{Q\,dQ}{r} \\ &= \frac{k\left(\frac{4}{3}\pi\rho r^3\right)\left(4\pi r^2\rho\,dr\right)}{r} \\ &= \frac{k(4\pi\rho)^2}{3}r^4\,dr \\ W &= \frac{k(4\pi\rho)^2}{3}\int_0^R r^4\,dr \\ &= \frac{k(4\pi\rho)^2}{3}\frac{R^5}{5} = \frac{3}{5}\frac{kQ^2}{R} \end{aligned}$$

Chapter 26

Introduction to Electric Circuits

26.1 Because the current is constant, we can use the definition

$$I = \frac{\Delta Q}{\Delta t}$$

to give an exact answer:

$$\Delta Q = I\Delta t = (5.0 \text{ C/ s})(4 \text{ min}) = 1200 \text{ C}$$

26.3 Since the light bulb is the only load in the circuit, the batteries' full emf must drop across the light bulb. That is,

$$\Delta V_{\text{lightbulb}} = \mathcal{E} = 5.0 \text{ V}$$

Then, from the definition of power,

$$\begin{aligned} P_{\text{lightbulb}} &= \Delta V_{\text{lightbulb}} I \\ \Rightarrow I &= \frac{15 \text{ W}}{5 \text{ V}} = 3 \text{ A} \end{aligned}$$

Although the light bulb is a non-ohmic device, by the definition of resistance, $R = \frac{\Delta V}{I}$ must still hold for any given R.

$$R_{\text{lightbulb}} = \frac{\Delta V}{I} = \frac{5 \text{ V}}{3 \text{ A}} = 1.7 \ \Omega$$

26.5 From Eqn. 26.11, (note $\rho_1 = \rho_2$)

$$\begin{aligned} \frac{R_1}{R_2} &= \frac{\frac{\rho_1 \ell_1}{A_1}}{\frac{\rho_2 \ell_2}{A_2}} = \frac{\ell_1}{A_1}\frac{A_2}{\ell_2} \\ &= \frac{(1.0 \text{ m})}{\pi(1.0 \text{ mm})^2} \cdot \frac{\pi(.5 \text{ mm})^2}{(.75 \text{ m})} \\ &= \frac{1}{3} \end{aligned}$$

26.7 For a parallel combination, the equivalent resistance is

$$R_{\text{eq}} = \left(\frac{1}{36 \text{ k}\Omega} + \frac{1}{5.0 \text{ k}\Omega}\right)^{-1} = 4.4 \text{ k}\Omega$$

26.9 Using Kirchoff's loop rule:

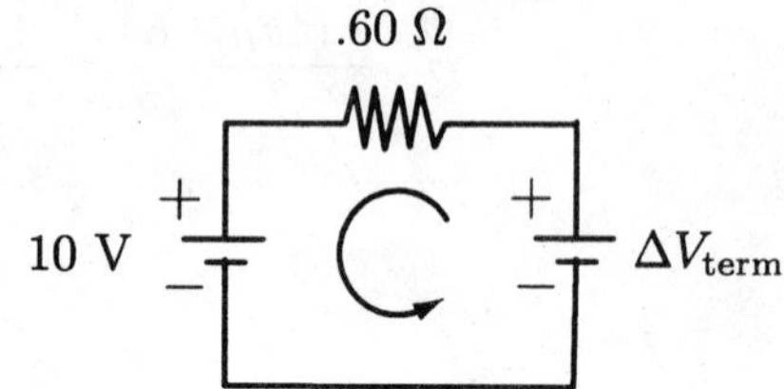

$$\begin{aligned} \sum \Delta V &= \Delta V_{\text{term}} - (.60 \ \Omega)(2.0 \text{ A}) - 10 \text{ V} \\ &= 0 \\ \Rightarrow \Delta V_{\text{term}} &= 10 \text{ V} + 1.2 \text{ V} = 11.2 \text{ V} \end{aligned}$$

26.11 From Eqn. 26.15, the condition of equilibrium is:

$$R_x = \frac{R_4}{R_3} R_1 = \frac{3.76 \text{ k}\Omega}{25.0 \text{ k}\Omega} 15.0 \ \Omega = 2.26 \ \Omega$$

26.13 Imagine the ring with an imaginary line passing through it as shown.

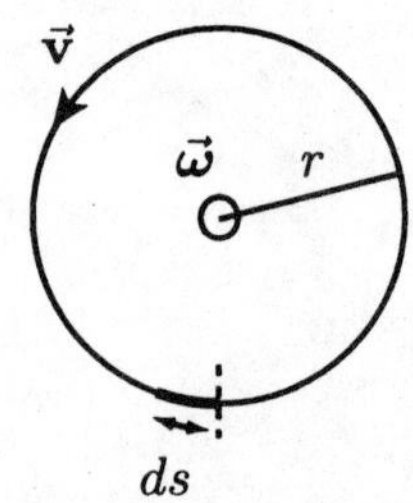

Let's calculate the amount of charge passing through the dashed line in a time dt. Suppose an arc length ds passes through the dashed line in a time dt. From the definition of charge density, a charge $dq = \lambda\, ds$ passes through the line in a time dt. The ring rotates at a speed $v = \omega r$, so:

$$dq = \lambda\, ds = \lambda v\, dt = \lambda \omega r\, dt$$

So the current produced by the rotating ring is:

$$I = \frac{dq}{dt} = \lambda \omega r$$

26.17 We can model the non-ideal battery as an ideal battery with a resistor connected in series. The value of the resistor's resistance is the battery's internal resistance. In the diagram shown, V_{AB} is the terminal voltage.

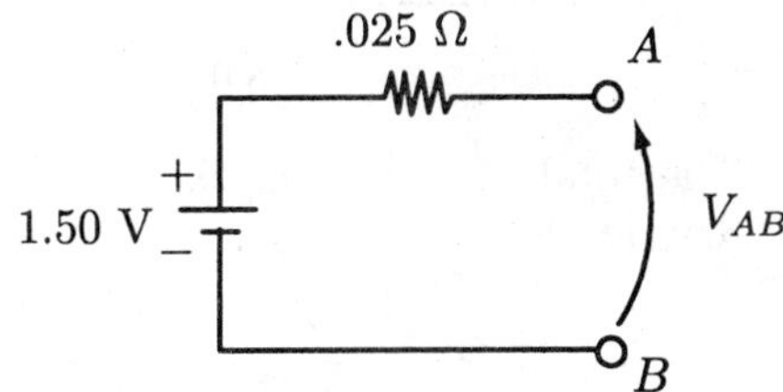

When delivering a current of 2.33 A, a potential of

$$\begin{aligned}\Delta V &= IR = (2.33 \text{ A})(.025) \\ &= 0.0582 \text{ V}\end{aligned}$$

is dropped over the .025 resistor. Thus,

$$V_{AB} = 1.50 \text{ V} - .0582 \text{ V} = 1.44 \text{ V}$$

26.21 From $P = I\Delta V$ and Ohm's Law, we get

$$\begin{aligned}P &= \frac{(\Delta V)^2}{R} \\ \Rightarrow \Delta V &= \sqrt{PR} = \sqrt{(25 \text{ W})(.15\ \Omega)} \\ &= 1.9 \text{ V}\end{aligned}$$

26.25 From Eqn. 26.11 and Table 26.1, the minimum diameter required is:

$$0.50 \times 10^3\ \Omega \stackrel{\text{set}}{\equiv} \frac{\rho_{\text{Al}}\ell}{\frac{1}{4}\pi d_{\text{min}}^2}$$

$$\begin{aligned}\Rightarrow d_{\text{min}} &= \left[\frac{4(2.8 \times 10^{-8}\ \Omega \cdot \text{m})(150 \times 10^3 \text{ m})}{\pi(0.50 \times 10^3\ \Omega)}\right]^{1/2} \\ d_{\text{min}} &= 3.3 \text{ mm}\end{aligned}$$

26.29 From Eq. 26.11, the resistance per unit length of a wire is:

$$\frac{R}{\ell} = \frac{\frac{\rho\ell}{A}}{\ell} = \frac{\rho}{A}$$

Setting R/ℓ for copper and tungsten equal (ρ_{cu} and ρ_{w} and can be found in Table 26.1: $1.7 \times 10^{-8}\ \Omega\cdot$ m and $5.6 \times 10^{-8}\ \Omega\cdot$ m respectively.)

$$\begin{aligned}\frac{\rho_{\text{cu}}}{A_{\text{cu}}} &= \frac{\rho_{\text{w}}}{A_{\text{w}}} \\ \Rightarrow A_{\text{w}} &= \frac{\rho_{\text{w}}}{\rho_{\text{cu}}} A_{\text{cu}} \\ \Rightarrow d_{\text{w}} &= d_{\text{cu}}\sqrt{\frac{\rho_{\text{w}}}{\rho_{\text{cu}}}} \\ &= (.80 \times 10^{-3} \text{ m})\sqrt{\frac{5.6}{1.7}} \\ &= 1.5 \text{ mm}\end{aligned}$$

From the text,

$$E = \frac{I\rho}{A}$$

Note that $\frac{\rho}{A}$ is equal for both wires from the 1st part of the problem. Thus, *for both wires*, (using the copper wire numbers)

$$\begin{aligned}E &= \frac{(5.0 \times 10^{-3} \text{ A})(1.7 \times 10^{-8}\ \Omega \cdot \text{m})}{\pi(.40 \times 10^{-3} \text{ m})^2} \\ &= 1.7 \times 10^{-4} \text{ V/m}\end{aligned}$$

Note that we could've used the tungsten wire values and we would've gotten the same value of E.

26.33 When the switch is closed, the total circuit resistance goes from $R + R_{\text{bulb}}$ to $R_{\parallel} + R_{\text{bulb}}$. Since resistors placed in parallel decrease the circuit's resistance (ie: $R_{\parallel} = \frac{1}{2}R$),

$$R_{\parallel} + R_{\text{bulb}} = \frac{1}{2}R + R_{\text{bulb}} < R + R_{\text{bulb}}$$

That is, the circuit's total resistance decreases when the switch is closed. By Ohm's Law, if the resistance decreases, current has to increase (remember, the battery's emf remains constant). The light bulb's brightness depends on how much current passes through the bulb—the larger the current, the brighter the bulb. Since the circuit's current increases with the switch closed, the bulb burns brighter with the switch closed.

26.37 We may replace the two resistors in parallel with the equivalent resistance

$$R_{\parallel} = \left[\frac{1}{R_1} + \frac{1}{R_2}\right]^{-1}$$

This combination is in series with the third resistor. Thus:

$$\begin{aligned} R_{\text{eq}} &= 15.0\text{ k}\Omega + R_{\parallel} \\ &= 15.0\text{ k}\Omega + \left[\frac{1}{10.0\text{ k}\Omega} + \frac{1}{20.0\text{ k}\Omega}\right]^{-1} \\ &= 15.0\text{ k}\Omega + \frac{20.0}{3}\text{ k}\Omega = 21.7\text{ k}\Omega \end{aligned}$$

26.41 Let r be the battery's internal resistance.

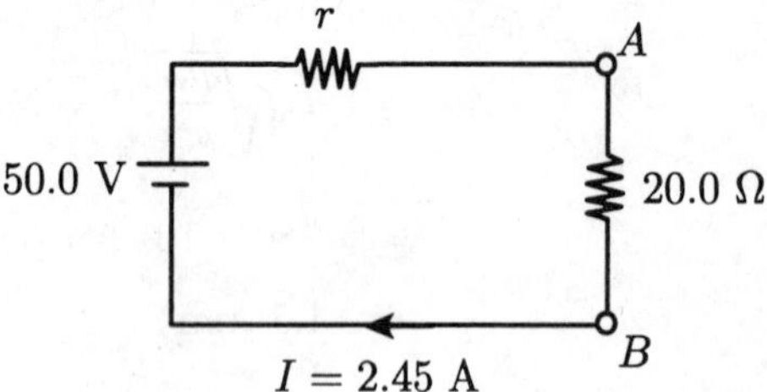

Points A and B represent the battery terminals. The circuit's total resistance is $R_{\text{eq}} = 20.0\ \Omega + r$. From the definition of resistance

$$I = \frac{50.0\text{ V}}{20.0\ \Omega + r} = 2.45\text{ A} \qquad (1)$$

Solving (1) for r gives:

$$r = \frac{50.0\text{ V}}{2.45\text{ A}} - 20.0\ \Omega = 0.4\ \Omega$$

If we used a voltmeter to measure the potential between points A and B (measuring the terminal voltage), we would measure

$$V_{AB} = IR = (2.45\text{ A})(20.0\ \Omega) = 49.0\text{ V}$$

26.45 Refer to the circuit diagram.

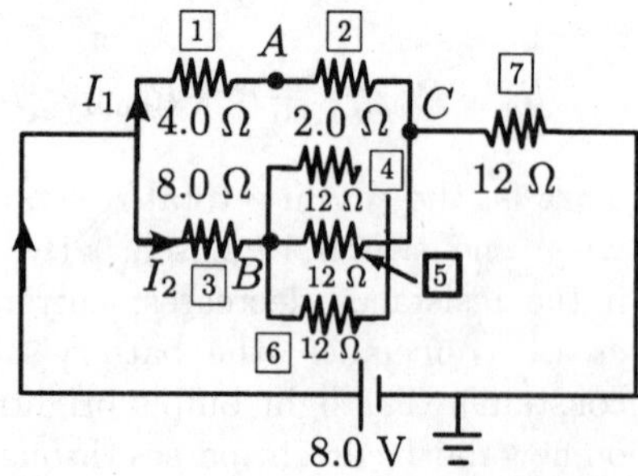

The series combination of resistors 1 and 2 is 6.0 Ω, while the parallel combination of the three 12 Ω resistors is

$$R = \left(\frac{1}{12\ \Omega} \times 3\right)^{-1} = 4.0\ \Omega.$$

So, an equivalent circuit is given by the second diagram.

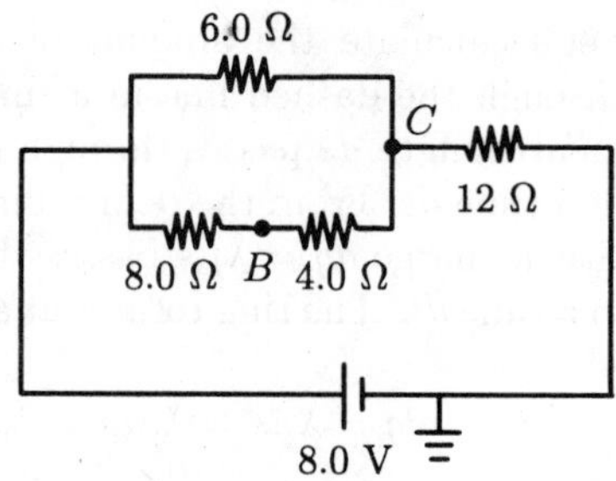

Now the 8 Ω and 4 Ω resistances give a 12.0 Ω combination (next diagram.)

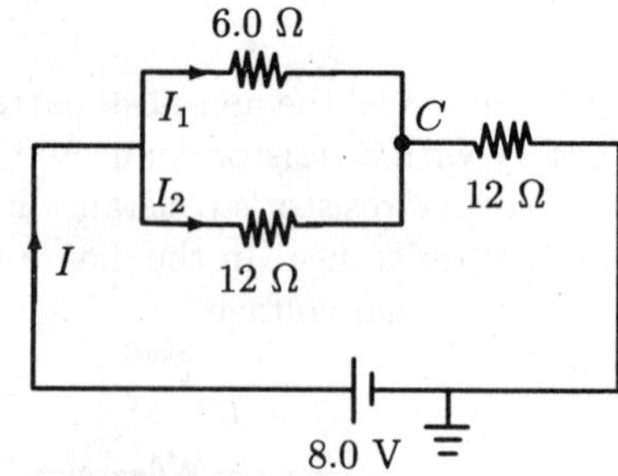

The total resistance in the circuit of the last diagram is

$$R = 12\ \Omega + \left[\frac{1}{6.0\ \Omega} + \frac{1}{12\ \Omega}\right]^{-1} = 16\ \Omega$$

So, the total current is

$$I = \frac{8.0\text{ V}}{16\ \Omega} = 0.50\text{ A}$$

The currents in the parallel arms of the last diagram are inversely proportional to the resistances in the arms. So,

$$I_1 = \frac{12\ \Omega(0.50\text{ A})}{12\ \Omega + 6.0\ \Omega} = 0.33\text{ A}$$

and

$$I_2 = \frac{6.0\ \Omega(0.50\text{ A})}{12\ \Omega + 6.0\ \Omega} = 0.17\text{ A}$$

I_2 splits into three equal currents through the three $12 - \Omega$ resistors each of these currents is

$$\frac{0.17\text{ A}}{3} = 0.057\text{ A}$$

So now the potential at A is

$$\begin{aligned} V_A &= 8.0\text{ V} - (4.0\ \Omega)(0.33\text{ A}) \\ &= 6.7\text{ V} \end{aligned}$$

At B,

$$\begin{aligned} V_B &= 8.0\text{ V} - (8.0\Omega)(0.17\text{ A}) \\ &= 6.6\text{ V} \end{aligned}$$

At C,

$$\begin{aligned} V_C &= +(12\ \Omega)(0.50\text{ A}) \\ &= 6.0\text{ V} \end{aligned}$$

26.49 Suppose that the current I through the resistor increases or remains the same. (This is equivalent to supposing that potential difference increases or remains the same.) Let's see if this guess gets us into trouble.

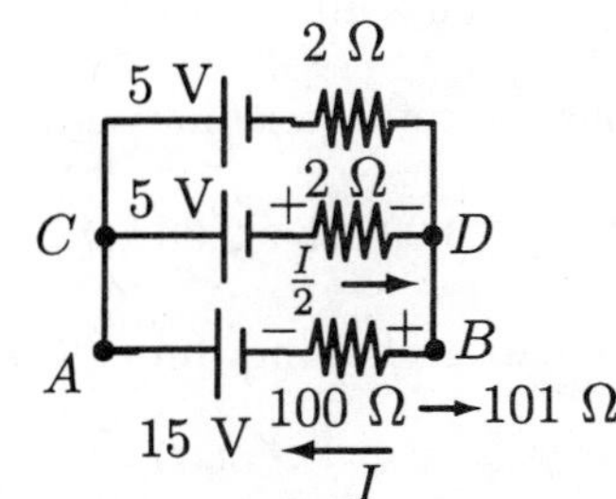

Since the potential difference across the emf is fixed, a greater *drop* across the resistor means a reduced potential difference $V_A - V_B$. But, if I increases, so must the currents through the other two resistors, which are each equal to $\frac{I}{2}$. Thus, the potential drop across either of the 2 Ω resistors increases, so that $V_C - V_D$ increases. But

$$V_C - V_D = V_A - V_B,$$

so $V_A - V_B$ must both increase and decrease—a contradiction. So, the current I must decrease, as must the potential change across the resistor.

Note: I decreases by a smaller fraction than R increases so $V_A - V_B$ still decreases, as does $V_C - V_D$.

26.53 Step 1: The circuit cannot be simplified by series || methods.

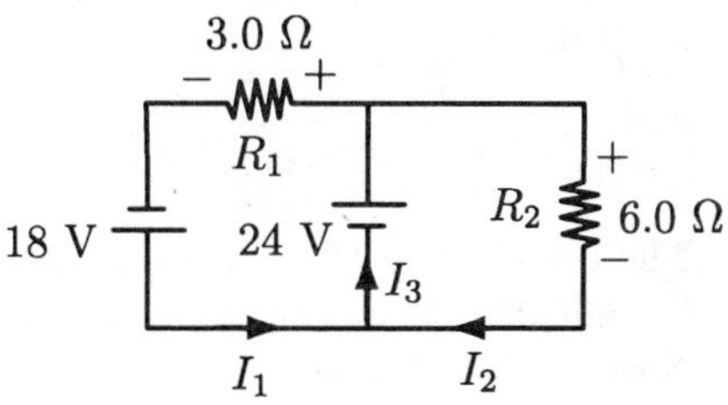

Steps 2 3 4:

$$I_3 = I_1 + I_2 \tag{2}$$

Step 5:
left:

$$\sum \Delta V = 18\text{ V} + 24\text{ V} - I_1R_1 = 0 \tag{3}$$

right:

$$\sum \Delta V = 24\text{ V} - I_2R_2 = 0 \tag{4}$$

Step 6: From (3):

$$I_1 = \frac{18\text{ V} + 24\text{ V}}{R_1} = 14\text{ A}$$

From (4):

$$I_2 = \frac{24\text{ V}}{R_2} = 4\text{ A}$$

From (2):

$$I_3 = I_1 + I_2 = 18\text{ A}$$

26.59 An educated guess would be that there is no current in the middle two branches because the top branch short circuits the middle two branches which have resistors. The bottom branch has no emf source to oppose the flow of current. We'll apply Kirchhoff's Rules to get a quantitative answer.

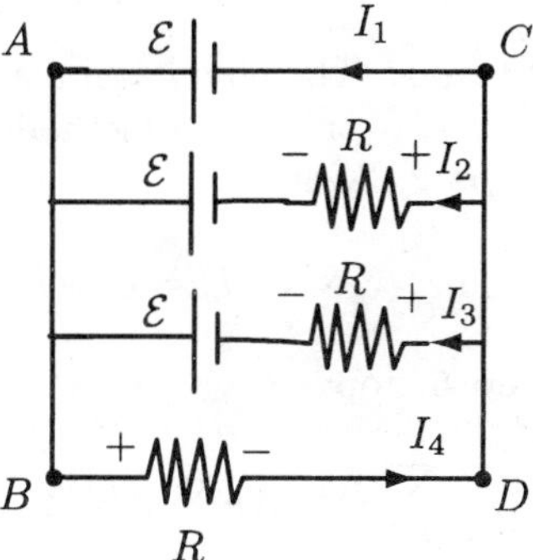

Step 1: The circuit cannot be simplified.
Steps 2, 3: Since $V_{AB} = V_{CD}$ and the middle two branches are symmetric, it would be a good guess that $I_2 = I_3$, but our personal style of solving these types of problems is to determine this from the loop equations in Step 5.
Step 4:

$$I_1 + I_2 + I_3 = I_4 \tag{5}$$

Step 5: top loop:

$$\begin{aligned}\sum \Delta V &= \mathcal{E} - \mathcal{E} + I_2R = 0\\ \Rightarrow I_2R &= 0\\ \Rightarrow I_2 &= 0\end{aligned}$$

middle loop:

$$\begin{aligned}\sum \Delta V &= \mathcal{E} - \mathcal{E} + I_3R - I_2R = 0\\ \Rightarrow I_3R &= I_2R\\ \Rightarrow I_3 &= 0\end{aligned}$$

bottom loop:

$$\sum \Delta V = \mathcal{E} - I_4R - I_3R = 0$$

Step 6: From the first two loop equations,

$$I_2 = I_3 = 0,$$

as expected. From (5),

$$\mathcal{E} = I_4 R \Rightarrow I_4 = \frac{\mathcal{E}}{R}$$

And by (5),

$$I_1 = I_4 = \frac{\mathcal{E}}{R}$$

Summarizing, we have:

Top arm	$I = \frac{\mathcal{E}}{R}$	to the left
2nd arm	$I = 0$	
3rd arm	$I = 0$	
Bottom arm	$I = \frac{\mathcal{E}}{R}$	to the right

26.63 Step 1: The circuit cannot be simplified by series and/or parallel methods.
Steps 2, 3, 4:

$$I_3 = I_1 + I_2 \qquad (6)$$

Step 5: top:

$$\sum \Delta V = 9\text{ V} + I_2 R_2 - I_1 R_1 = 0 \qquad (7)$$

bottom:

$$\sum \Delta V = 6\text{ V} - I_3 R_3 - I_2 R_2 = 0 \qquad (8)$$

Step 6: Use (6) to eliminate I_3 in (8). Solve for I_2.

$$I_2 = \frac{6\text{ V} - I_1 R_3}{R_2 + R_3} \qquad (9)$$

Now use (9) to eliminate I_2 in (7) and solve for I_1.

$$I_1 = \frac{(15\text{ V})\, R_2 + (9\text{ V})\, R_3}{R_1 R_2 + R_1 R_3 + R_2 R_3} = 2.5\text{ A}$$

We can plug the value for I_1 into Eqn. (9) and solve for I_2.

$$I_2 = \frac{6\text{ V} - (2.5\text{ A})\,(4\ \Omega)}{2\ \Omega + 4\ \Omega} = -\frac{2}{3}\text{ A}$$

Our sign convention for I_2 resulted in a negative answer—current is directed from a toward b. The easiest way to get I_3 is by Eqn. (6).

$$I_3 = I_1 + I_2 = 2.5\text{ A} - \frac{2}{3}\text{ A} = 1.8\text{ A}$$

Lastly,

$$V_{ba} = V_b - V_a = -I_2 R_2 = 1.3\text{ V}$$

26.67 *After* the meter has been connected, it forms a parallel combination with a 3 kΩ resistor, with equivalent resistance

$$R_{\parallel} = \left(\frac{1}{5 \times 10^5} + \frac{1}{3 \times 10^3}\right)^{-1} \Omega = 2.98\text{ k}\Omega$$

Thus the total resistance in the circuit is

$$3.00\text{ k}\Omega + 2.98\text{ k}\Omega = 5.98\text{ k}\Omega$$

The current flowing through the battery is:

$$I = \frac{\mathcal{E}}{R_{\text{eq}}} = \frac{5.00\text{ V}}{5.98\text{ k}\Omega} = 0.836\text{ mA}$$

From the result of problem 26.38, the current flowing through the meter is:

$$\begin{aligned} I_m &= \frac{3.00\text{ k}\Omega}{3.00\text{ k}\Omega + 5.00 \times 10^5\ \Omega}(0.836\text{ mA}) \\ &= 4.99 \times 10^{-6}\text{ A} \end{aligned}$$

Therefore, the potential dropped over R_m (which is the potential that the meter actually reads) is:

$$\begin{aligned} V_m &= I_m R_m \\ &= (4.99 \times 10^{-6}\text{ A})(5.00 \times 10^5\ \Omega) \\ &= 2.49\text{ V} \end{aligned}$$

Before the meter has been connected, half the battery's emf is dropped over the 1st 3 kΩ resistor and the other half is dropped over the 2nd 3 kΩ resistor

$$V_{3\text{ k}\Omega} = 2.50\text{ V}$$

The reason why $V_m \neq V_{3\text{ k}\Omega}$ is because the meter affects the circuit that it's hooked up to. Now matter how good the voltmeter is, it will always affect the circuit that it's hooked up to, and in turn, the reading will not be the "true value." Usually, this effect is small, but can sometimes be substantial. When the effect gets really bad, it's called *meter loading*.

26.71 If the switches are both open, resistors R_2 and R_3 are removed from the circuit.

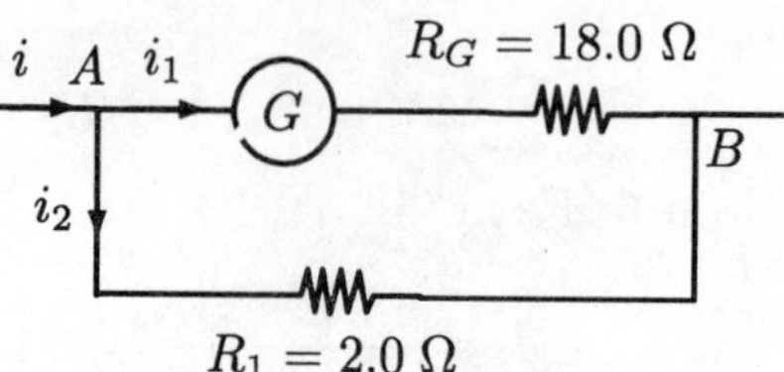

A full-scale deflection occurs when $i_1 = 50.0\ \mu$A. The maximum current that can be measured is i, where $i = i_1 + i_2$ and

$$i_1 R_G = V_{AB} = i_2 R_1$$

Thus

$$\begin{aligned} i &= i_1\left[1+\frac{R_G}{R_1}\right] \\ &= (50.0\ \mu\text{A})\left(1+\frac{18.0}{2.0}\right) \\ &= (50.0\ \mu\text{A})\left(\frac{20.0}{2.0}\right) \\ &= 500\ \mu\text{A} \\ i_{\max} &= 0.50\ \text{mA} \end{aligned}$$

If S_1 is now closed, an additional resistance is connected in parallel with G, thus *reducing* the total resistance in the circuit and also reducing the fraction of the current that passes through G. *The ammeter becomes less sensitive*, but also perturbs the system less.

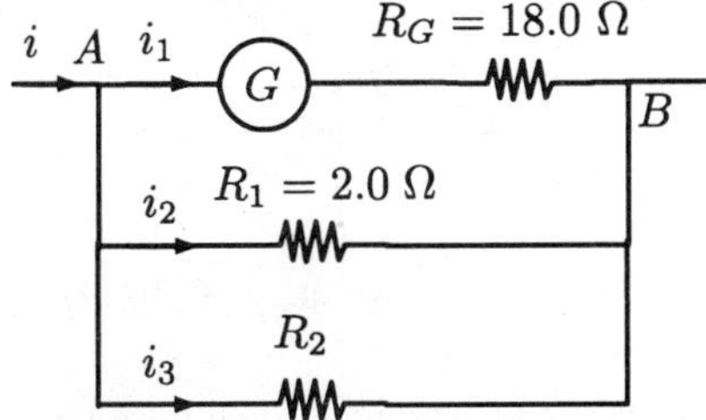

To obtain a factor of 10 in sensitivity, we need $i_{\max} = 5.00$ mA, while i, remains 50.0 μA. So

$$i_1 + i_2 + i_3 = 5.00\ \text{mA}$$

$$V_{AB} = i_1 R_G = i_2 R_1 = i_3 R_2$$

so

$$\begin{aligned} i_2 &= \frac{R_G}{R_1} i_1 \\ i_3 &= \frac{R_G}{R_2} i_1 \end{aligned}$$

Thus

$$i_1\left[1+\frac{R_G}{R_1}+\frac{R_G}{R_2}\right] = i_{\max}$$

$$(50.0\ \mu\text{A})\left(1+\frac{18.0}{2.0}+\frac{18.0\ \Omega}{R_2}\right) = 5.00\ \text{mA}$$

$$\begin{aligned} \frac{18.0\ \Omega}{R_2} &= \left(\frac{5.00\ \text{mA}}{50.0\ \mu\text{A}}\right) - \frac{20.0}{2.0} \\ &= 100 - 10.0 = 90.0 \end{aligned}$$

Thus

$$R_2 = \frac{18.0\ \Omega}{90} = 0.20\ \Omega$$

Finally, when the 2nd switch is closed:

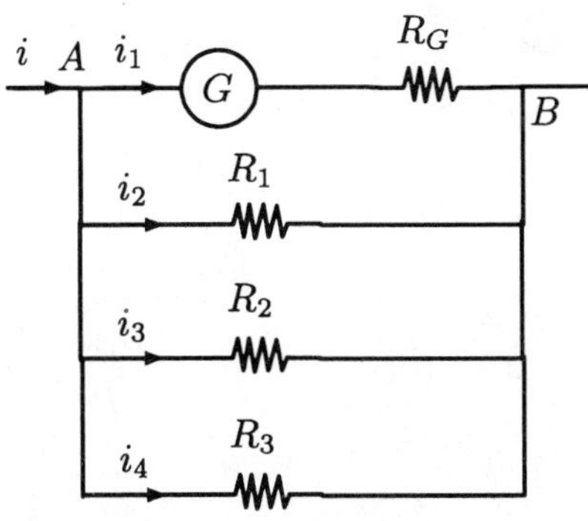

Now we want $i_{\max} = 50.0$ mA, and, proceeding as before,

$$i_4 = \frac{R_G}{R_3} i_1$$

and

$$\begin{aligned} i_{\max} &= i_1 + i_2 + i_3 + i_4 \\ &= i_1\left[1+\frac{R_G}{R_1}+\frac{R_G}{R_2}+\frac{R_G}{R_3}\right] \\ \frac{i_{\max}}{i_1} &= \frac{50.0\ \text{mA}}{50.0\ \mu\text{A}} \\ &= 1+\frac{18.0}{2.0}+\frac{18.0}{0.20}+\frac{18.0\ \Omega}{R_3} \\ \frac{18.0\ \Omega}{R_3} &= 1.00\times10^3 - [1+9.0+90.0] \\ &= 1.00\times10^3 - 0.100\times10^3 \\ R_3 &= \frac{18.0\ \Omega}{0.90\times10^3} = 0.020\ \Omega \end{aligned}$$

26.75 From Ohm's Law,

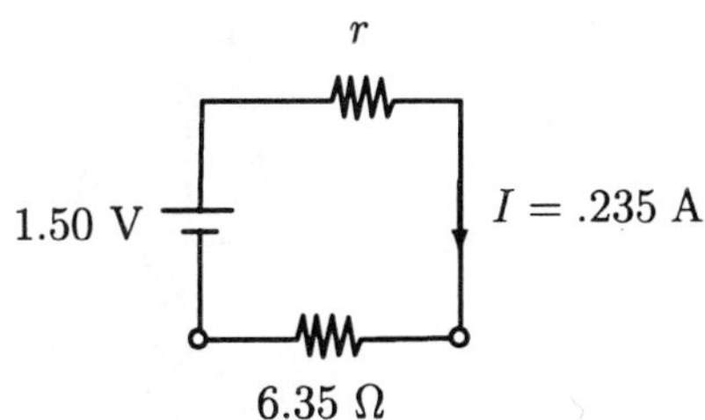

$$\begin{aligned} R_{\text{total}} &= \frac{V}{I} \\ r + 6.35\ \Omega &= \frac{1.50\ \text{V}}{0.235\ \text{A}} \Rightarrow r = 0.03\ \Omega \end{aligned}$$

26.79 Since the network is semi-infinite, it makes no difference if we start at A and B or at C and D.

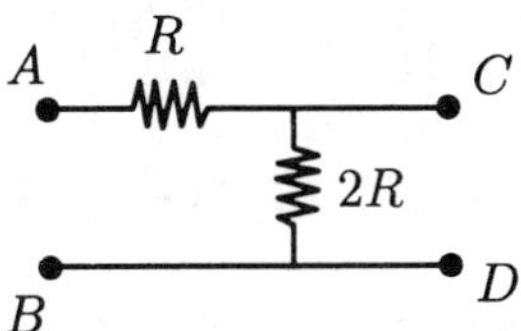

R_{AB} is the series combination of R and [the parallel combination of $2R$ and R_{CD}]

$$R_{AB} = R + \left[\frac{1}{2R} + \frac{1}{R_{CD}}\right]^{-1}$$

$$R_{\text{net}} = R + \left[\frac{R_{\text{net}} + 2R}{2R\,R_{\text{net}}}\right]^{-1}$$

$$= R + \frac{2R\,R_{\text{net}}}{R_{\text{net}} + 2R}$$

$$(R_{\text{net}} - R)(R_{\text{net}} + 2R) = 2R\,R_{\text{net}}$$

$$R_{\text{net}}^2 - R\,R_{\text{net}} - 2R^2 = 0$$

$$(R_{\text{net}} - 2R)(R_{\text{net}} + R) = 0$$

Since R_{net} cannot be negative:

$$R_{\text{net}} = 2R$$

26.83 The circuit is geometrically symmetric about the plane through points A, C, B and E.

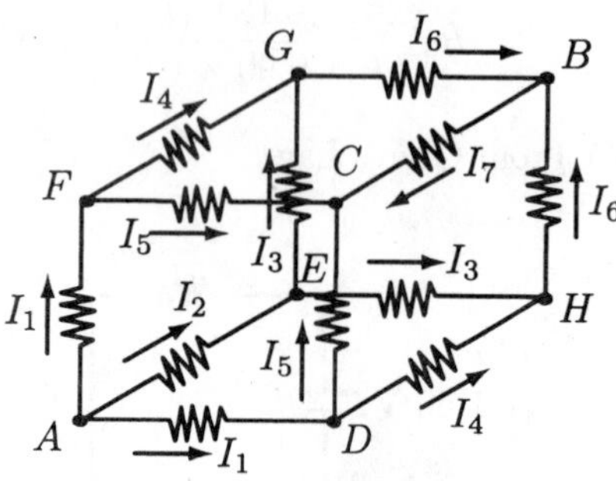

That fact reduces the problem to the 7 current variables shown in the figure, since currents symmetrically located with respect to plane $ACBE$ must be equal. Now, we know that reversing the battery would just reverse the currents' directions, but not change their magnitudes. This same reversal can be done by rotating the system about an axis perpendicular to the face $ADCF$ so that points A and C are interchanged. (That is, the resistor array would look exactly the same but with currents flowing toward the lower left corner of the face rather than away.) By this reasoning, $I_5 = I_1$ and $I_4 = 0$. Then the cubical set of resistors is equivalent to the circuit:

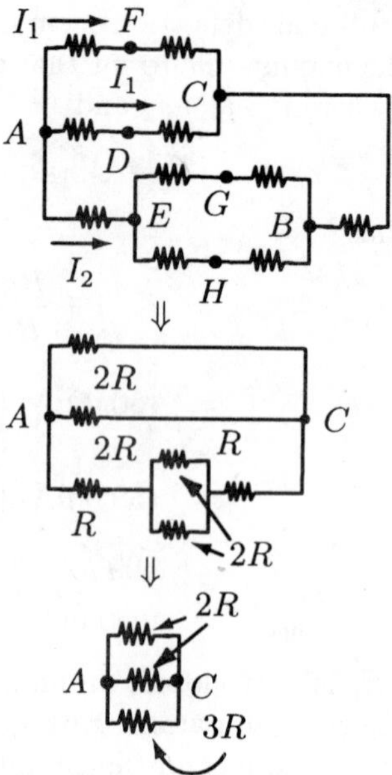

So

$$\frac{1}{R_{\text{eff}}} = \frac{2}{2R} + \frac{1}{3R} = \frac{8}{6R}$$

so $R_{\text{eff}} = \frac{3}{4}R$. So the total current through the battery is

$$I_{\text{total}} = \frac{\mathcal{E}}{R_{\text{eff}}} = \frac{1.0\ \text{V}}{\frac{3}{4}(10\ \Omega)} = 0.13\ \text{A}$$

26.87 a) At junction A, each loop current symbolizes an amount of current that enters junction A and continues through arm AB.

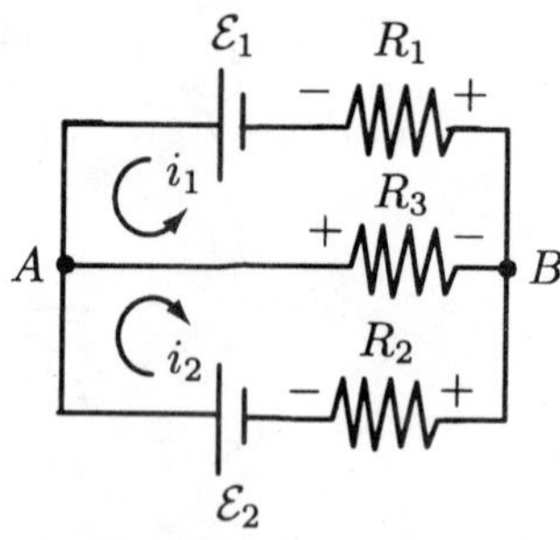

Thus, the current through the arm containing R_1 and $\mathcal{E}_1$ is i_1; the current through R_2 and $\mathcal{E}_2$ is i_2 and the current through R_3 is $i_1 + i_2$. Thus, currents entering A automatically add up to those leaving A and Kirchhoff's loop rule is satisfied.

b) Write one loop equation for each loop current:

$$\mathcal{E}_1 - R_3(i_1 + i_2) - i_1 R_1 = 0;$$

$$\mathcal{E}_2 - R_3(i_1 + i_2) - i_2 R_2 = 0$$

So

$$i_2 = \frac{\mathcal{E}_1 - i_1 (R_1 + R_3)}{R_3}$$

and

$$\mathcal{E}_2 - [\mathcal{E}_1 - i_1 (R_1 + R_3)] \frac{(R_3 + R_2)}{R_3} - i_1 R_3 = 0$$

Then,

$$-\mathcal{E}_2 + \mathcal{E}_1 \left(\frac{R_3+R_2}{R_3}\right) = \frac{i_1}{R_3} \left[(R_1 + R_3)(R_3 + R_2) - R_3^2\right]$$

$$\Rightarrow i_1 = \frac{(R_3 + R_2)\mathcal{E}_1 - R_3\mathcal{E}_2}{R_1R_2 + R_1R_3 + R_2R_3}$$

And,

$$i_2 = \frac{\mathcal{E}_1}{R_3} - \frac{R_1 + R_3}{R_3}\frac{(R_3 + R_2)\mathcal{E}_1 - R_3\mathcal{E}_2}{R_1R_2 + R_1R_3 + R_2R_3}$$

$$= \frac{\mathcal{E}_1[R_1R_2+R_1R_3+R_2R_3-(R_1+R_3)(R_3+R_2)]+\mathcal{E}_2R_3(R_r+R_3)}{R_3(R_1R_2+R_1R_3+R_2R_3)}$$

$$= \frac{[\mathcal{E}_2 (R_1 + R_3) - R_3\mathcal{E}_1]}{(R_1R_2 + R_1R_3 + R_2R_3)}$$

26.91 a) When M is a voltmeter of extremely high resistance, no current flows from A to B through the meter, and

$$\Delta V_{\text{open}} = I_N R_N$$

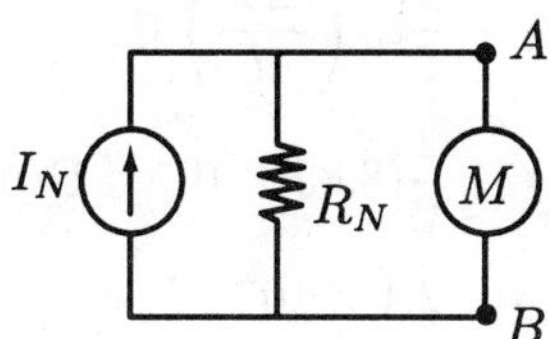

Similarly, when M is an ammeter of extremely low resistance, all the current flows through M. Thus

$$I_{\text{short}} = I_N$$

and

$$R_N = \frac{\Delta V_{\text{open}}}{I_{\text{short}}}$$

b) When M is a voltmeter of high resistance, no current flows through M: $i = 0$.

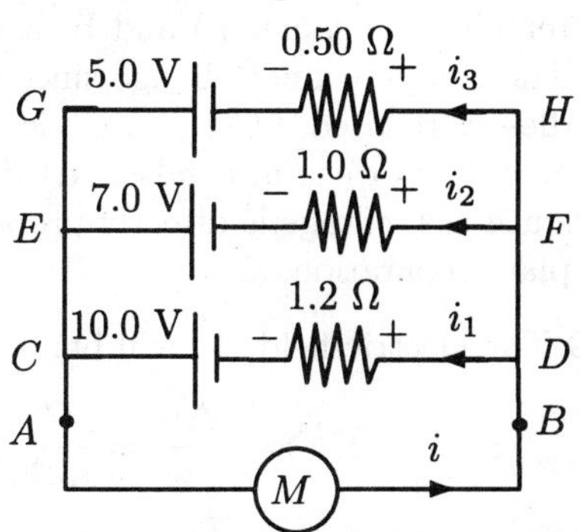

Thus applying Kirchhoff's rules:
Junction rule

$$i_1 + i_2 + i_3 = 0$$

Loop rule $CDFE$

$$-10.0\text{ V} + i_1 (1.2\ \Omega) - i_2 (1.0\ \Omega) + 7.0\text{ V} = 0$$

$$1.2i_1 - i_2 = 3.0\text{ A}$$

$$\Rightarrow i_2 = 1.2i_1 - 3.0\text{ A}$$

Loop rule $CDHG$

$$-10.0\text{ V} + (1.2\ \Omega)\, i_1 - (0.50\ \Omega)\, i_3 + 5.0\text{ V} = 0$$

$$1.2i_1 - 0.50i_3 = 5.0\text{ A}$$

$$\Rightarrow i_3 = 2.4i_1 - 10\text{ A}$$

$$i_1 + (1.2i_1 - 3\text{ A}) + (2.4i_1 - 10\text{ A}) = 0$$

$$4.6i_1 = 13\text{ A} \Rightarrow i_1 = 2.83\text{ A}$$

Then $i_2 = 0.3913$ A and $i_3 = -3.217$ A. Thus

$$\Delta V_{\text{open}} = 10.0\text{ V} - (1.2\ \Omega)(2.83\text{ A}) = 6.61\text{ V}$$

Next we let M be a very low resistance ammeter, and apply K's rules again.

$$i = i_1 + i_2 + i_3$$

We apply the loop rule as follows. Loop $CABD$

$$0 = 10.0\text{ V} - (1.2\ \Omega)\, i_1 \Rightarrow i_1 = 8.33\text{ A}$$

$EABF$

$$0 = 7.0\text{ V} - (1.0\ \Omega)\, i_2 \Rightarrow i_2 = 7.00\text{ A}$$

$GABH$

$$0 = 5.0\text{ V} - (0.50\ \Omega)\, i_3 \Rightarrow i_3 = 10.0\text{ A}$$

Thus

$$i = I_N = (8.33 + 7.00 + 10.0)\text{A} = 25.3\text{ A}$$

Then

$$R_N = \frac{\Delta V_{\text{open}}}{I_{\text{short}}} = \frac{6.61\text{ V}}{25.3\text{ A}} = 0.26\ \Omega$$

Chapter 27

Capacitance and Electrostatic Energy

27.1 The charge on the capacitor is

$$\begin{aligned} Q &= C\Delta V = (10^{-11}\text{ F})(12\text{ V}) \\ &= 1.2\times 10^{-10}\text{ C} \\ &= 120\text{ pC} \end{aligned}$$

27.3 The separation, d, is

$$\begin{aligned} d &= \frac{\epsilon_0 A}{C} \\ &= \frac{(8.85\times10^{-12}\text{ C}^2/\text{N}\cdot\text{m}^2)\pi(4.25\times10^{-2}\text{ m})^2}{(16\times10^{-12}\text{ F})} \\ &= 3.14\times 10^{-3}\text{ m} \end{aligned}$$

or $d = 3.1$ mm

27.5 The stored energy is given by

$$\begin{aligned} U &= \frac{1}{2}C(\Delta V)^2 \\ &= \frac{1}{2}(3.9\times 10^{-9}\text{ F})(6.0\text{ V})^2 \\ &= 7.0\times 10^{-8}\text{ J} \end{aligned}$$

27.7 For parallel connection:

$$C_{\text{eq}} = C_1 + C_2 = 1.8\text{ pF} + 4.7\text{ pF} = 6.5\text{ pF}$$

27.9 The capacitance is:

$$C = \kappa\epsilon_0\frac{A}{d}$$

$\kappa = 2.5$ for Polystyrene (Table 27.1), so

$$\begin{aligned} C &= \frac{(2.5)(8.85\times 10^{-12}\text{ F/m})(11\times 10^{-4}\text{ m}^2)}{(0.74\times 10^{-3}\text{ m})} \\ &= 3.3\times 10^{-11}\text{ F} \\ &= 33\text{ pF} \end{aligned}$$

27.11 The energy density is given by

$$u = \frac{1}{2}\epsilon_0 E^2$$

with $E = \dfrac{\Delta V}{d}$

$$\begin{aligned} u &= \frac{1}{2}\epsilon_0\left(\frac{\Delta V}{d}\right)^2 \\ &= \frac{1}{2}(8.85\times 10^{-12}\text{ F/m})\frac{(1.5\text{ V})^2}{(2.2\times 10^{-3}\text{ m})^2} \\ &= 2.1\times 10^{-6}\text{ J}\cdot\text{m}^{-3} \end{aligned}$$

27.15 Only (**e**) occurs. The capacitance is $C = \frac{\epsilon_0 A}{d}$ and *decreases* as d increases and the potential difference is given by

$$\Delta V = Ed = \left(\frac{\sigma}{\epsilon_0}\right)d$$

Therefore the stored charge

$$Q = C\Delta V = \frac{\epsilon_0 A}{d}\left(\frac{\sigma}{\epsilon_0}\right)d = \sigma A$$

remains the same. **a**) and **b**) are both untrue. The charge cannot change since insulated handles were used. Clearly, C decreases if d increases so **c**) is also false. **d**) doesn't happen since the voltage is directly proportional to the plate separation, d.

27.19 The electric field, E, will be:

$$E = \frac{\Delta V}{d} = \frac{C}{\epsilon_0 A}\Delta V$$

$$
\begin{aligned}
&= \frac{(28 \times 10^{-12}\ \text{F})(12\ \text{V})}{(8.85 \times 10^{-12}\ \text{C}^2/\text{N}\cdot\text{m}^2)(420 \times 10^{-4}\ \text{m}^2)} \\
&= 904\ \text{V/m} \\
&= 0.90\ \text{kV/m}
\end{aligned}
$$

27.23 Let $C_1 = 3C_2$ and $Q_1 = Q_2 = Q$. Then:

$$
\begin{aligned}
C_1 &= \frac{Q_1}{V_1} = \frac{Q}{V_1} \\
C_2 &= \frac{Q_2}{V_2} = \frac{Q}{V_2}
\end{aligned}
$$

and so $C_1 = 3C_2$ implies that $\frac{Q}{V_1} = \frac{3Q}{V_2}$ or that $\frac{V_2}{V_1} = 3$. The ratio of the stored energies is determined by:

$$
\begin{aligned}
U_1 &= \frac{1}{2}\frac{Q^2}{C_1} \\
U_2 &= \frac{1}{2}\frac{Q^2}{C_2}
\end{aligned}
$$

therefore

$$\frac{U_2}{U_1} = \frac{C_1}{C_2} = 3$$

27.25 To form a 6 μF module from six 5 μF capacitors, connect *five* of them in series together and one of the capacitors in parallel with the five in series:

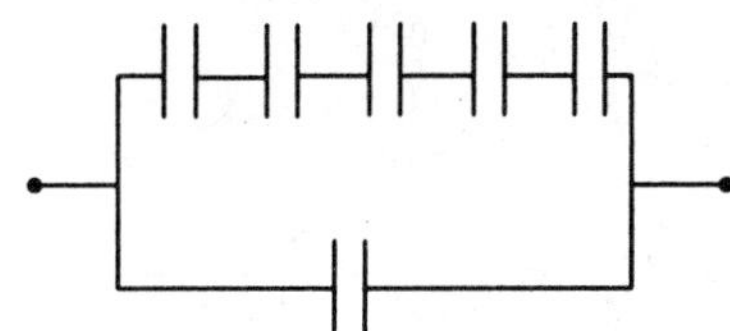

$$
\begin{aligned}
C_{\text{eq}} &= 5\ \mu\text{F} + \frac{1}{\frac{1}{5\ \mu\text{F}} + \frac{1}{5\ \mu\text{F}} + \frac{1}{5\ \mu\text{F}} + \frac{1}{5\ \mu\text{F}} + \frac{1}{5\ \mu\text{F}}} \\
&= 5\ \mu\text{F} + \frac{1}{5\left(\frac{1}{5\ \mu\text{F}}\right)} \\
&= 5\ \mu\text{F} + 1\ \mu\text{F} = 6\ \mu\text{F}
\end{aligned}
$$

27.27 From Figure 27.24, we can see that the capacitors are connected in parallel:

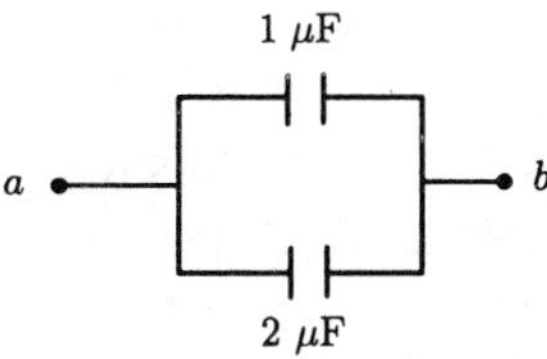

$$C_{\text{eq}} = C_1 + C_2 = 1\ \mu\text{F} + 2\ \mu\text{F} = 3\ \mu\text{F}$$

27.29 When capacitors are connected *in series*, the total potential difference across the bank is divided between them, thus reducing the potential difference across each individual capacitor.

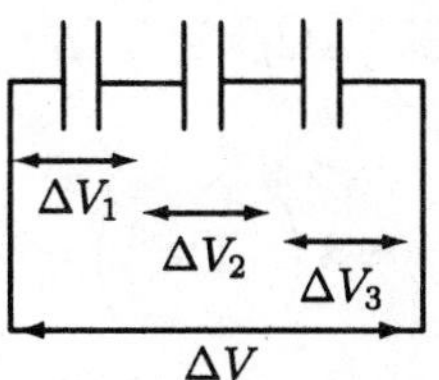

$$\Delta V = \Delta V_1 + \Delta V_2 + \Delta V_3$$

The equivalent capacitance is given by:

$$
\begin{aligned}
\frac{1}{C_{\text{eq}}} &= \frac{1}{C_1} + \frac{1}{C_2} + \frac{1}{C_3} \\
&= \frac{1}{2\ \mu\text{F}} + \frac{1}{4\ \mu\text{F}} + \frac{1}{8\ \mu\text{F}} \\
&= \frac{4+2+1}{8\ \mu\text{F}} \\
&= \frac{7}{8\ \mu\text{F}} \\
C_{\text{eq}} &= \frac{8}{7}\ \mu\text{F}
\end{aligned}
$$

27.33 When connected in parallel, the same potential difference $\mathcal{E}$ is across each capacitor, and each stores change $Q_i = C\mathcal{E}$.

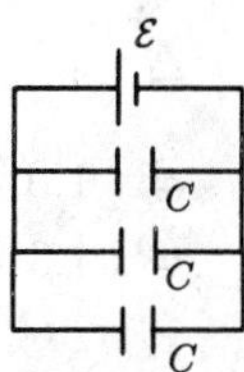

The total stored energy is

$$U_i = 3\left(\frac{1}{2}C\mathcal{E}^2\right) = \frac{3}{2}C\mathcal{E}^2$$

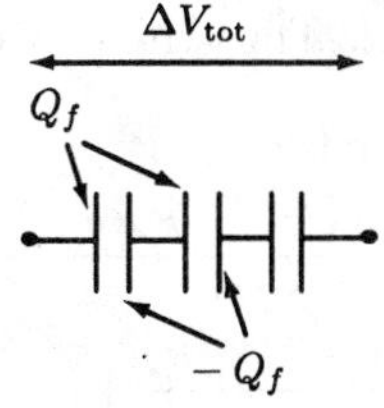

When connected in series, the same charge Q resides on each capacitor. But the same charge $Q_i = C\mathcal{E}$ was on each capacitor in the first place, so *no change* occurs. The total potential difference is now the sum of the potential differences across each capacitor.

$$\Delta V_{tot} = 3\Delta V_i = 3\mathcal{E}$$
$$Q_f = Q_i = C\mathcal{E}$$

and

$$U_f = 3\left(\frac{1}{2}C\mathcal{E}^2\right) = \frac{3}{2}C\mathcal{E}^2 = U_i$$

27.37 The electric field in the dielectric is reduced by the polarization.

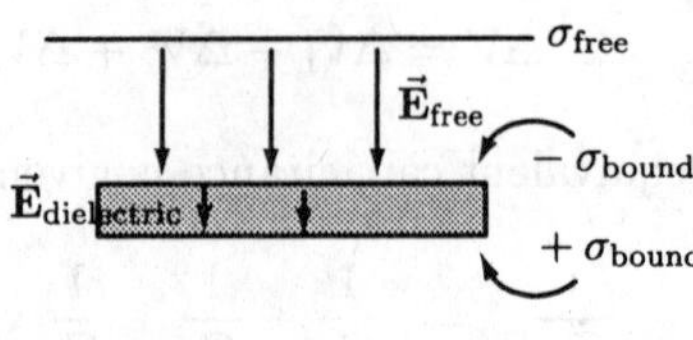

However, unlike the case of a conductor, the electric field inside the dielectric is *not zero.* Thus not every field line from the positive capacitor plate terminates on the dielectric surface, and $|\sigma_{bound}| < |\sigma_{free}|$. The answer is **b**).

27.41 For a concentric spherical shell:

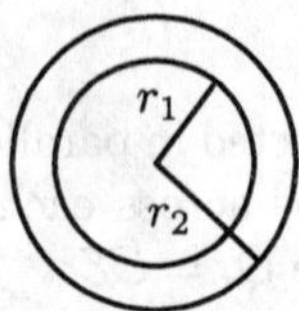

$$C = 4\pi\epsilon_0\kappa\frac{r_1 r_2}{r_2 - r_1}$$
$$r_1 = 1.0 \text{ cm}$$
$$r_2 = 1.3 \text{ cm}$$

$$C = \left(\frac{1}{9.0\times10^9 \text{ m/F}}\right)(4.7)\frac{(1.0\times10^{-2}\text{ m})(1.3\times10^{-2}\text{ m})}{0.3\times10^{-2}\text{ m}}$$
$$= 2.26\times10^{-11}\text{ F}$$
$$= 20 \text{ pF}$$

27.45 The capacitor may be modeled as two capacitors in parallel. Each has plate area $\frac{A}{2}$ and separation d. Thus the capacitance is:

$$C = C_1 + C_2$$
$$= \frac{\kappa_1\epsilon_0\left(\frac{A}{2}\right)}{d} + \frac{\kappa_2\epsilon_0\left(\frac{A}{2}\right)}{d}$$
$$= \frac{\epsilon_0 A}{d}\left(\frac{\kappa_1+\kappa_2}{2}\right)$$
$$= \frac{(8.85\times10^{-12}\text{ C}^2/\text{N·m}^2)(25\times10^{-4}\text{ m}^2)}{(2\times10^{-3}\text{ m})}\left(\frac{1.5+2.5}{2}\right)$$
$$= 2.2\times10^{-11}\text{ F}$$

27.47 a) The capacitances are:

$$C_1 = \kappa_1\frac{\epsilon_0 A}{d_1}$$
$$= \frac{(3.1)(8.85\times10^{-12}\text{ C}^2/\text{N·m}^2)(16\times10^{-4}\text{ m}^2)}{(2\times10^{-3}\text{ m})}$$
$$= 2.19\times10^{-11}\text{ F} = 22\text{ pF}$$
$$C_2 = \frac{\epsilon_0 A}{d_2}$$
$$= \frac{(8.85\times10^{-12}\text{ C}^2/\text{N}\cdot\text{m}^2)(16\times10^{-4}\text{ m}^2)}{(3\times10^{-3}\text{ m})}$$
$$= 4.72\times10^{-12}\text{ F} = 4.7\text{ pF}$$
$$C_{eq} = C_1 + C_2$$

so the total stored charge, Q, is:

$$Q = (C_1+C_2)V$$
$$= (2.662\times10^{-11}\text{ F})(12\text{ V})$$
$$= 31.94\times10^{-11}\text{ C}$$
$$= 320\text{ pC}$$

The stored energy is:

$$U = \frac{Q^2}{2C_{eq}} = 1.92\times10^{-9}\text{ J} = 1.9\text{ nJ}$$

b) After the battery has been removed, and the Plexiglas slab removed from C_1 and inserted into C_2, the new capacitances will be:

$$C_1' = \frac{\epsilon_0 A}{d_1}$$
$$= 7.06\times10^{-12}\text{ F} = 7.1\text{ pF}$$
$$C_2' = \frac{\kappa\epsilon_0 A}{\kappa d_2 - d_1(\kappa-1)}$$
$$= 8.61\times10^{-12}\text{ F} = 8.6\text{ pF}$$

Because the battery has been disconnected, the total charge, Q, stored in the circuit remains unchanged. Therefore, the voltage across each capacitor is:

$$V = \frac{Q}{C_1'+C_2'} = 20.38\text{ V}$$

The charges on the capacitors are:

$$Q_1 = C_1'V = (7.1\text{ pF})(20.4\text{ V})$$
$$= 0.14\text{ nC}$$
$$Q_2 = C_2'V = (8.6\text{ pF})(20.4\text{ V})$$
$$= 0.18\text{ nC}$$

The stored energy is:

$$U' = \frac{Q^2}{2(C_1'+C_2')} = 3.3\text{ nJ}$$

The work done by the system as the slab is moved is

$$\begin{aligned} W &= U - U' \\ &= 1.92 \times 10^{-9}\ \text{J} - 3.26 \times 10^{-9}\ \text{J} \\ &= -1.34 \times 10^{-9}\ \text{J} \end{aligned}$$

Thus work 1.3 nJ is done on the system in moving the Plexiglas slab.

27.51 The energy density is:

$$\begin{aligned} u_E &= \frac{1}{2}\epsilon_0 E^2 = \frac{\epsilon_0}{2}\left(\frac{Q}{4\pi\epsilon_0 r^2}\right)^2 \\ &= \frac{Q^2}{32\pi^2\epsilon_0 r^4} \\ &= \frac{(6 \times 10^{-6}\ \text{C})^2}{32\pi^2\,(8.85 \times 10^{-12}\ \text{F/m})\,(0.1\ \text{m})^4} \\ &= 100\ \text{J/m}^3 \end{aligned}$$

27.57 At P with $|x_p| < \ell$, the electric field due to $-Q$ is

$$\vec{\mathbf{E}}_- = -\frac{kQ}{(x+\ell)^2}\hat{\mathbf{i}}$$

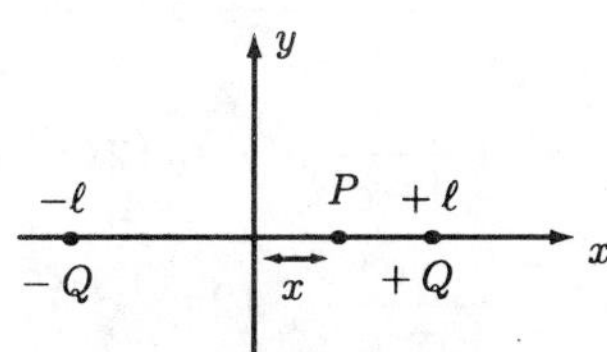

and due to $+Q$ is

$$\vec{\mathbf{E}}_+ = -\frac{kQ}{(\ell-x)^2}\hat{\mathbf{i}}$$

Thus

$$\begin{aligned} \vec{\mathbf{E}} &= \vec{\mathbf{E}}_+ + \vec{\mathbf{E}}_- \\ &= -kQ\left[\frac{1}{(x+\ell)^2} + \frac{1}{(\ell-x)^2}\right]\hat{\mathbf{i}} \\ &= -\frac{kQ}{(\ell^2-x^2)^2}\left[2\left(\ell^2+x^2\right)\right]\hat{\mathbf{i}} \end{aligned}$$

Thus

$$\begin{aligned} U_E &= \frac{1}{2}\epsilon_0 E^2 = \frac{1}{2}\epsilon_0(k^2Q^2)\frac{4(\ell^2+x^2)^2}{(\ell^2-x^2)^4} \\ &= \frac{1}{2\pi}kQ^2\frac{(\ell^2+x^2)^2}{(\ell^2-x^2)^4} \end{aligned}$$

At P with $|x_p| > \ell$

$$\vec{\mathbf{E}}_+ = \frac{kQ}{(x-\ell)^2}\hat{\mathbf{i}}$$

and

$$\vec{\mathbf{E}}_- = -\frac{kQ}{(x+\ell)^2}\hat{\mathbf{i}}$$

$$\vec{\mathbf{E}} = \vec{\mathbf{E}}_+ + \vec{\mathbf{E}}_- = kQ\hat{\mathbf{i}}\left[\frac{1}{(x-\ell)^2} - \frac{1}{(x+\ell)^2}\right]$$

and

$$\begin{aligned} u_E &= \frac{1}{2}\epsilon_0 E^2 = \frac{1}{2}\epsilon_0(kQ)^2\frac{(4x\ell)^2}{(x^2-\ell^2)^4} \\ &= \frac{2k}{\pi}Q^2\frac{(x\ell)^2}{(x^2-\ell^2)^4} \end{aligned}$$

at $x = 2\ell$

$$\begin{aligned} u_E &= \frac{2kQ^2}{\pi}\frac{(2\ell^2)^2}{(4\ell^2-\ell^2)^4} \\ &= \frac{2k}{\pi}Q^2\frac{4}{81\ell^4} \\ &= \frac{8}{81\pi}\frac{kQ^2}{\ell^4} \end{aligned}$$

27.61

Forsterite	Paraffin
$\kappa = 6.2$	$\kappa = 2.2$
Dielectric strength	Dielectric strength
$9.4\ \frac{\text{MV}}{\text{m}}$	$9.8\ \frac{\text{MV}}{\text{m}}$

The charge on a capacitor is

$$Q = C\Delta V$$

Since the capacitors have the same dimensions

$$\frac{C_F}{C_P} = \frac{\kappa_F}{\kappa_P} = \frac{6.2}{2.2}$$

and

$$\Delta V_{\text{max}} = E_{\text{max}} d$$

So

$$\frac{\Delta V_{\text{max},\text{F}}}{\Delta V_{\text{max},\text{P}}} = \frac{E_{\text{max},\text{F}}}{E_{\text{max},\text{P}}} = \frac{9.4\ \frac{\text{MV}}{\text{m}}}{9.8\ \frac{\text{MV}}{\text{m}}}$$

Thus

$$\begin{aligned} \frac{Q_{\text{max},\text{F}}}{Q_{\text{max},\text{P}}} &= \frac{C_F\Delta V_{\text{max},\text{F}}}{C_P\Delta V_{\text{max},\text{P}}} \\ &= \frac{6.2}{2.2}\cdot\frac{9.4}{9.8} = 2.7 \end{aligned}$$

The Forsterite filled capacitor can hold 2.7 times as much charge as the paraffin.

27.65 $U = \frac{1}{2}C\Delta V^2$ So

$$\begin{aligned} U &= \frac{1}{2}C\left(E_{\text{max}}d\right)^2 \\ &= \frac{1}{2}\epsilon_0\kappa\frac{A}{d}E_{\text{max}}^2 d^2 \\ &= \frac{1}{2}\epsilon_0\kappa E_{\text{max}}^2 V \end{aligned}$$

If $E_{\text{max}} = 10\ \frac{\text{MV}}{\text{m}}$ and $\kappa_{\text{max}} \sim$ few hundred (for ceramics) then

$$\begin{aligned} V_{\text{min}} &= \frac{2U}{\epsilon_0\kappa_{\text{max}}E_{\text{max}}^2}\frac{2(1\text{ MJ})}{9\times10^{-12}\text{ C}^2/\text{N}\cdot\text{m}^2(\text{few }100)\left(10\ \frac{\text{MV}}{\text{m}}\right)^2} \\ &\sim \frac{10^{11}\times10^6\text{ J}\cdot\text{N}\cdot\text{m}^2/\text{C}^2}{100\times10^{14}\left(\frac{\text{V}}{\text{m}}\right)^2} \\ &= \frac{10^{17}}{10^{16}}\text{ m}^3 \\ V_{\text{min}} &\simeq 10\text{ m}^3 \end{aligned}$$

27.69 The electric field on the x-axis is the superposition of the fields due to the two charges.

y

$-1\ \mu\text{C}$ $+2\ \mu\text{C}$ x

$$\begin{aligned} x<0 \quad & E_x = \left(-\frac{2}{(x-1\text{ cm})^2}+\frac{1}{x^2}\right)k\,(1\ \mu\text{C}) \\ 0<x<1\text{ cm} \quad & E_x = \left(-\frac{2}{(x-1\text{ cm})^2}-\frac{1}{x^2}\right)k\,(1\ \mu\text{C}) \\ x>1\text{ cm} \quad & E_x = \left(\frac{2}{(x-1\text{ cm})^2}-\frac{1}{x^2}\right)k\,(1\ \mu\text{C}) \end{aligned}$$

The energy density is

$$u\,(x) = \frac{1}{2}\epsilon_0 E_x^2$$

The "center of charge" for this system is at

$$x_{\text{CQ}} = \frac{0\,(-1)+(1\text{ cm})\,(2)}{2-1} = 2\text{ cm}$$

Thus we should put the 1 μC charge at $x = 2$ cm. The electric field due to this charge is

$$E_x = \frac{k\,(1\ \mu\text{C})}{(x-2\text{ cm})^2}$$

The results are shown in the figures in the appendix. The single charge result is good to better than 20% for $|x| > 8.25$ cm.

27.73 Uniform sphere of charge, radius R

We may use Gauss' law to find the electric field everywhere inside and outside the sphere. We expect the field lines to be radial, so we choose a spherical Gaussian surface of radius r. Then

$$\begin{aligned} 4\pi r^2 E_r &= \Phi_E = \frac{Q}{\epsilon_0} = \frac{4}{3}\frac{\pi r^3\rho}{\epsilon_0},\ r<R \\ \Rightarrow E_r &= \frac{\rho r}{3\epsilon_0},\ r<R \\ E_r &= \frac{-e}{(4\pi r^2\epsilon_0)},\ r>R \end{aligned}$$

The charge density is

$$\rho = \frac{-e}{\frac{4}{3}\pi R^3},$$

so

$$E_r = -\frac{er}{4\pi\epsilon_0 R^3},\ r<R$$

Now we calculate the total electrostatic energy by summing up the energy in spherical shells:

$$\begin{aligned} dU &= u\,dV = \frac{\epsilon_0 E^2}{2}4\pi r^2 dr \\ &= \left(\frac{-er}{4\pi\epsilon_0 R^3}\right)^2\epsilon_0\left(2\pi r^2\,dr\right),\ r<R \\ &= \frac{e^2}{16\pi^2\epsilon_0 R^6}\left(2\pi r^4\,dr\right),\ r<R \end{aligned}$$

and

$$\begin{aligned} u\,dV &= \frac{e^2}{(4\pi r^2)^2}\frac{1}{\epsilon_0}2\pi r^2 dr,\ r>R \\ &= \frac{e^2}{8\pi r^4}\frac{r^2\,dr}{\epsilon_0},\ r>R \end{aligned}$$

Thus

$$\begin{aligned} U &= \int dU = \int u\,dV \\ &= \int_0^R \frac{e^2}{16\pi R^6}\frac{2r^4}{\epsilon_0}dr + \int_R^\infty \frac{e^2}{8\pi\epsilon_0 r^2}dr \\ &= \frac{e^2}{8\pi\epsilon_0 R^6}\left[\frac{r^5}{5}\bigg|_0^R\right] + \frac{e^2}{8\pi\epsilon_0}\left[-\frac{1}{r}\bigg|_R^\infty\right] \\ &= \frac{e^2}{8\pi\epsilon_0}\left[\frac{R^5}{5R^6}+\frac{1}{R}\right] = \frac{6}{5}\frac{e^2}{8\pi\epsilon_0 R} \end{aligned}$$

If $U = mc^2$, then

$$\begin{aligned} R &= \frac{6}{5}\frac{e^2}{8\pi\epsilon_0 mc^2} = \frac{3}{5}\frac{ke^2}{mc^2} \\ &= 1.7\times10^{-15}\text{ m} \end{aligned}$$

Thus the factor $\frac{1}{2}$ we calculated for the spherical shell becomes $\frac{3}{5}$ for a sphere.

Chapter 28

Static Magnetic Fields

28.1 At point C, $\vec{\mathbf{v}}$ is parallel to $\vec{\mathbf{B}}$ so the magnetic force is zero.

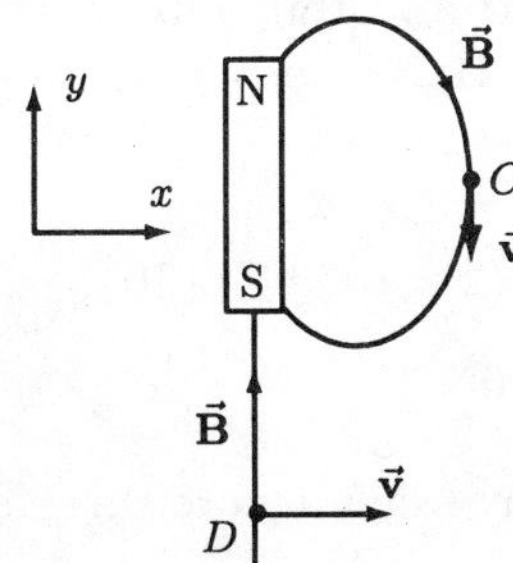

At point D, the magnetic field is in $+y$ direction. Since,

$$\vec{\mathbf{F}}_{\text{mag}} = q\vec{\mathbf{v}} \times \vec{\mathbf{B}} = -ev\hat{\mathbf{i}} \times B\hat{\mathbf{j}} = -evB\hat{\mathbf{k}}$$

the force will be in the negative z direction, i.e. into the page.

28.3 The magnetic force, $\vec{\mathbf{F}}_{\text{mag}} = q\vec{\mathbf{v}} \times \vec{\mathbf{B}}$, will be greatest if the charge moves perpendicular to $\vec{\mathbf{B}}$, i.e., in the y-z-plane.

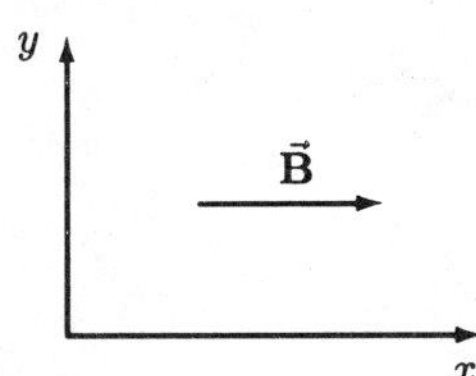

The maximum force is given by:

$$\begin{aligned} F_{\text{max}} &= qvB \\ &= (0.65 \times 10^{-6}\ \text{C})(375\ \text{m/s})(0.48\ \text{T}) \\ &= 1.2 \times 10^{-4}\ \text{N} \end{aligned}$$

28.5 From Equation (28.5),

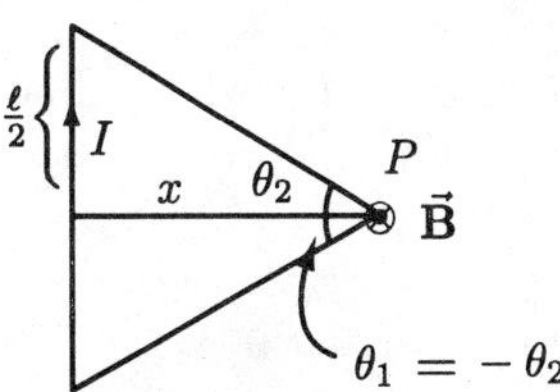

the magnetic field has magnitude given by:

$$\left|\vec{\mathbf{B}}\right| = \frac{\mu_0 I}{4\pi x}\left(\sin\theta_2 - \sin\theta_1\right)$$

Here, $|\theta_2| = |\theta_1| = |\theta|$; and,

$$\sin\theta = \frac{\left(\frac{\ell}{2}\right)}{\sqrt{x^2 + \frac{\ell^2}{4}}}$$

therefore:

$$\begin{aligned} \left|\vec{\mathbf{B}}\right| &= \frac{\mu_0 I}{4\pi x}\left[\frac{\left(\frac{\ell}{2}\right)}{\left(x^2 + \frac{\ell^2}{4}\right)^{\frac{1}{2}}} - \frac{\left(-\frac{\ell}{2}\right)}{\left(x^2 + \frac{\ell^2}{4}\right)^{\frac{1}{2}}}\right] \\ &= \frac{\mu_0 I}{4\pi x}\frac{\ell}{\left(x^2 + \frac{\ell^2}{4}\right)^{\frac{1}{2}}} \\ &= \frac{(4\pi\times10^{-7}\ \text{N/A}^2)(10^{-3}\ \text{A})(17\times10^{-2}\ \text{m})}{4\pi(5\times10^{-2}\ \text{m})\left((5\times10^{-2}\ \text{m})^2 + \frac{(17\times10^{-2}\ \text{m})^2}{4}\right)^{\frac{1}{2}}} \\ &= (3.4\times10^{-9}\ \text{T}) \end{aligned}$$

28.7 The magnetic moment, m, is given by:

$$m = IA$$

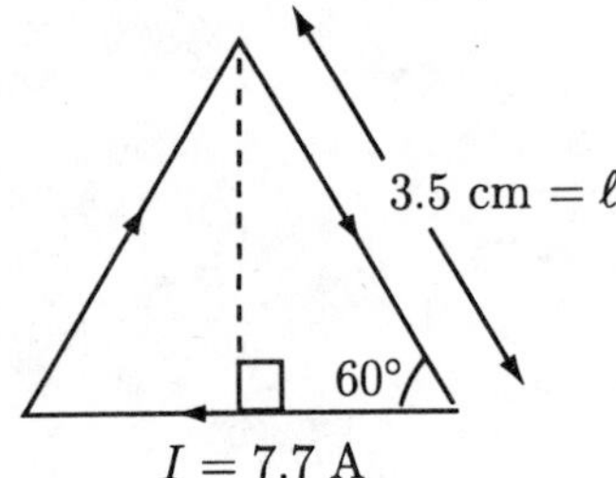

where A is the area of the loop.

$$A = \frac{1}{2}\ell\left(\frac{\ell\sqrt{3}}{2}\right)$$

$$m = (7.7\text{ A})((3.5\times10^{-2}\text{ m})^2\frac{\sqrt{3}}{4}$$
$$= 4.1\times10^{-3}\text{ A}\cdot\text{m}^2$$

28.9 The circulation about each wire is given by Ampère's Law:

$$\mathcal{C} = \oint \vec{\mathbf{B}}\cdot d\vec{\ell} = \mu_0 I$$

The direction of I is determined by the right hand rule. $\mathcal{C}_1$ encloses all 3 wires. The net current is

$$2I - I - I = 0$$
$$\mathcal{C}_1 = 0$$

$\mathcal{C}_2$ encloses the two outer wires but not the central one. The net current is $2I - I = I$. Thus

$$\mathcal{C}_2 = \mu_0 I$$

$\mathcal{C}_3$ encloses only one wire with current $2I$:

$$\mathcal{C}_3 = 2\mu_0 I$$

28.11 Applying Ampère's Law to a circle surrounding the central wire,

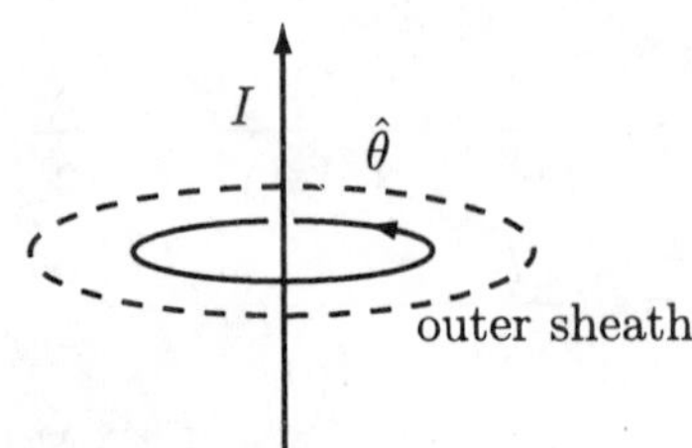

as in Example 28.7,

$$\vec{\mathbf{B}} = \frac{\mu_0 I}{2\pi r}\hat{\theta}$$
$$= \frac{(4\pi\times10^{-7}\text{ N/A}^2)(5.5\times10^{-3}\text{ A})}{2\pi(0.75\times10^{-2}\text{ m})}\hat{\theta}$$
$$= (1.5\times10^{-7}\text{ T})\hat{\theta}$$

Outside the coaxial cable, the magnetic field is zero, since the net current through a circle surrounding the entire cable is also zero.

28.13 Since the north pole of the magnet points in the y-direction, $\vec{\mathbf{B}}$ is in the y-direction also.

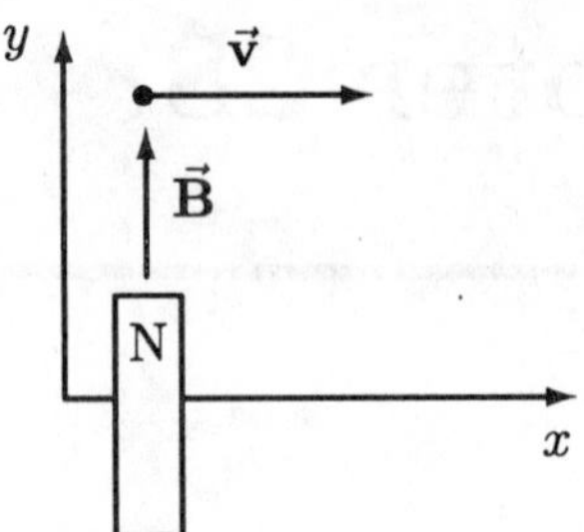

Now

$$\vec{\mathbf{F}}_{\text{mag}} = q\vec{\mathbf{v}}\times\vec{\mathbf{B}} = qv\hat{\mathbf{i}}\times B\hat{\mathbf{j}}$$

and since $\hat{\mathbf{i}}\times\hat{\mathbf{j}} = +\hat{\mathbf{k}}$, but the particle accelerates in the $-z$-direction, the charge on the particle must be *negative.*

28.17 The shuttle's orbit is approximately perpendicular to $\vec{\mathbf{B}}$, so the force is approximately

$$F = QvB_E$$
$$= (2\times10^{-6}\text{ C})\left(10^4\ \frac{\text{m}}{\text{s}}\right)(3\times10^{-5}\text{ T})$$
$$= 6\times10^{-7}\text{ N}$$

NASA may safely ignore this effect.

28.21 Since $\vec{\mathbf{F}} = q\vec{\mathbf{v}}\times\vec{\mathbf{B}}$, then

$$\left|\vec{\mathbf{F}}\right| = qvB\sin\theta$$

$$= 26(1.62\times10^{-19}\text{ C})\left(1800\ \frac{\text{m}}{\text{s}}\right)(0.057\text{ T})\sin 30^\circ$$
$$= 2.2\times10^{-16}\text{ N}$$

and the direction of $\vec{\mathbf{F}}$ is out of the page.

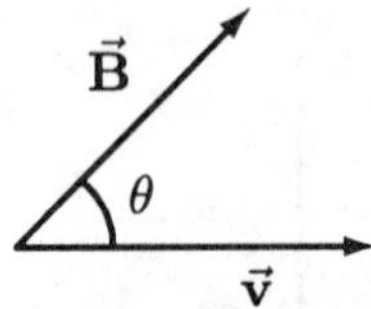

28.25 a) Since the magnetic moment is proportional to the area, A, enclosed by a loop,

$$m = IA$$

and since a circle encloses the maximum area for a given length, you ought to use a circular shape of loop.

b) Wrapping the wire in a coil of N turns might seem to be a clever way of increasing the moment since $m = NIA$. But for a given length of wire, wrapping the wire in a coil reduces the original loop's radius and hence its area. The wire should be wrapped in a single turn.

$$m = NIA = N(I\pi r^2)$$

But

$$\ell = 2\pi r N$$

is the total length of wire so

$$r = \frac{\ell}{2\pi N}$$

Thus

$$m = NI\pi\left(\frac{\ell}{2\pi N}\right)^2 \propto \frac{1}{N}$$

and is maximum for $N = 1$.

28.29 The field strength on the axis of a loop of wire is:

$$\begin{aligned} B &= \frac{\mu_0 I a^2}{2\left(z^2+a^2\right)^{\frac{3}{2}}} \qquad \text{(Eqn. 28.7)} \\ &= \frac{(4\pi \times 10^{-7}\ \text{N/A}^2)(66\ \text{A})\,(0.45\ \text{m})^2}{2\left((.15\ \text{m})^2 + (.45\ \text{m})^2\right)^{\frac{3}{2}}} \\ &= 7.9 \times 10^{-5}\ \text{T} \end{aligned}$$

28.33 By equation (28.6), and superposition, the net field *midway* between the wires is:

$$\vec{\mathbf{B}} = \vec{\mathbf{B}}_1 + \vec{\mathbf{B}}_2 = \frac{\mu_0 I}{2\pi r_\perp}\hat{\mathbf{k}} + \frac{\mu_0 I}{2\pi r_\perp}(-\hat{\mathbf{k}}) = 0$$

(The z-direction is defined in the figure.)

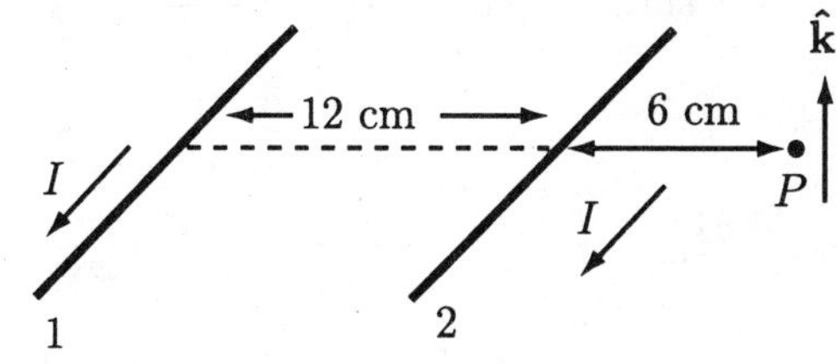

b) 6.0 cm to the right of wire 2,

$$\begin{aligned} \vec{\mathbf{B}} &= \frac{\mu_0 I\hat{\mathbf{k}}}{2\pi(6.0\times10^{-2}\ \text{m} + 12\times10^{-2}\ \text{m})} \\ &\quad + \frac{\mu_0 I(\hat{\mathbf{k}})}{2\pi(6.0\times10^{-2}\ \text{m})} \\ &= \frac{\mu_0 I}{2\pi(6.0\times10^{-2}\ \text{m})}\frac{4}{3}(+\hat{\mathbf{k}}) \\ &= \frac{(4\pi\times10^{-7}\ \text{N/A}^2)(25\ \text{A})}{2\pi(6\times10^{-2}\ \text{m})}\frac{4}{3}(+\hat{\mathbf{k}}) \\ &= (1.1\times10^{-4}\ \text{T})\hat{\mathbf{k}} \end{aligned}$$

28.37 Given a square loop of side $2a$ which carries a current I counter-clockwise around the loop, we want to find the magnetic field at a fixed point on the z-axis.

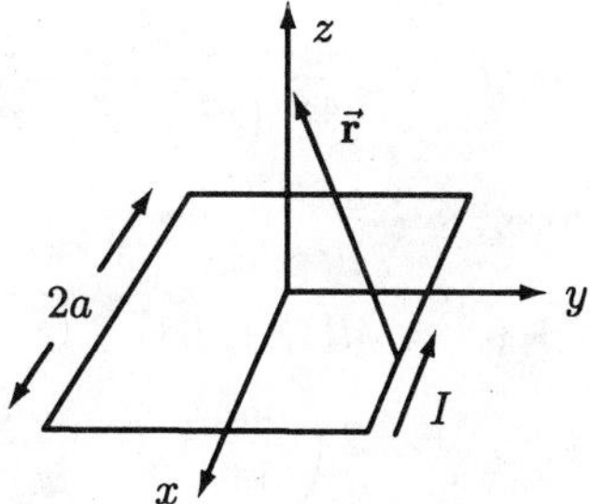

To do so, we must consider the contribution from each segment of the loop:

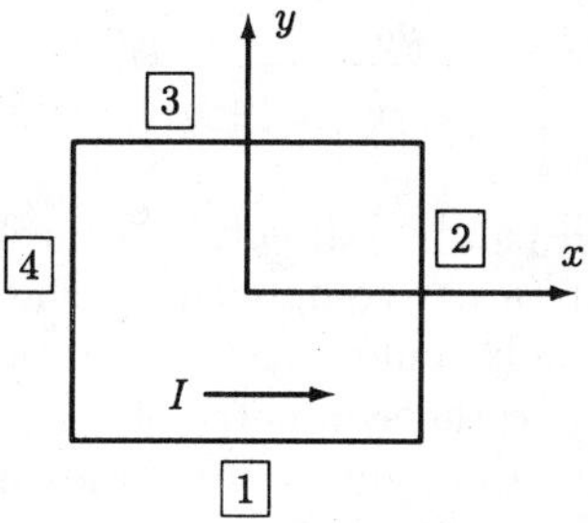

Label each segment of the loop; then apply the Biot-Savart law to each segment.

1:

$$-a \le x \le a,\ y = -a$$

2:

$$x = a, -a \le y \le a$$

3:

$$-a \le x \le a,\ y = a$$

4:

$$x = a, -a \le y \le a$$

Consider, for example, segment 3 ($-a \le x \le a,\ y = a$). For this leg of the loop,

$$\begin{aligned} I\,d\vec{\ell} &= -I\,dx\hat{\mathbf{i}} \\ \vec{\mathbf{r}} &= z\hat{\mathbf{k}} - (x\hat{\mathbf{i}} + a\hat{\mathbf{j}}) \end{aligned}$$

Then, from the Biot-Savart law, the contribution to the magnetic field at z due to an element of current flowing along segment 3 is:

$$\begin{aligned} d\vec{\mathbf{B}}_3 &= \frac{\mu_0 I}{4\pi}\frac{d\vec{\ell}\times\vec{r}}{r^3} \\ &= \frac{\mu_0 I}{4\pi}\frac{a\,dx\hat{\mathbf{k}} + z\,dx\hat{\mathbf{j}}}{(x^2+a^2+z^2)^{\frac{3}{2}}} \end{aligned}$$

The contribution to the magnetic field at z from a current element on the opposite side of the loop $(-a \le x \le a, y = -a)$

$$\begin{aligned} d\vec{\ell} &= +dx\hat{\mathbf{i}}, \vec{\mathbf{r}} = z\hat{\mathbf{k}} - (x\hat{\mathbf{i}} - a\hat{\mathbf{j}}) \\ d\vec{\mathbf{B}}_1 &= \frac{\mu_0 I}{4\pi} \frac{adx\hat{\mathbf{k}} - zdx\hat{\mathbf{j}}}{(x^2+a^2+z^2)^{\frac{3}{2}}} \end{aligned}$$

Therefore,

$$\begin{aligned} d\vec{\mathbf{B}}_{13} &= d\vec{\mathbf{B}}_1 + d\vec{\mathbf{B}}_3 \\ &= 2\left(\frac{\mu_0 I}{4\pi}\right) \frac{adx\hat{\mathbf{k}}}{(x^2+a^2+z^2)^{\frac{3}{2}}} \end{aligned}$$

is the contribution to the magnetic field at z due to current elements along segments 1 and 3. The source terms

$$\frac{\mu_0 I}{4\pi} \frac{\pm z\, dx}{(x^2+a^2+z^2)^{\frac{3}{2}}}$$

in the $+\hat{\mathbf{j}}$ and $-\hat{\mathbf{j}}$-directions cancel exactly, leaving only a net component in the $+\hat{\mathbf{k}}$-direction. This result could have been anticipated by appealing to the symmetry of the problem. If we consider the figure below, we can resolve $d\vec{\mathbf{B}}_1$ and $d\vec{\mathbf{B}}_3$ each into two components: one, $d\vec{\mathbf{B}}_z$, along the z-axis and another, $dB_\perp$, at right angles to the z-axis.

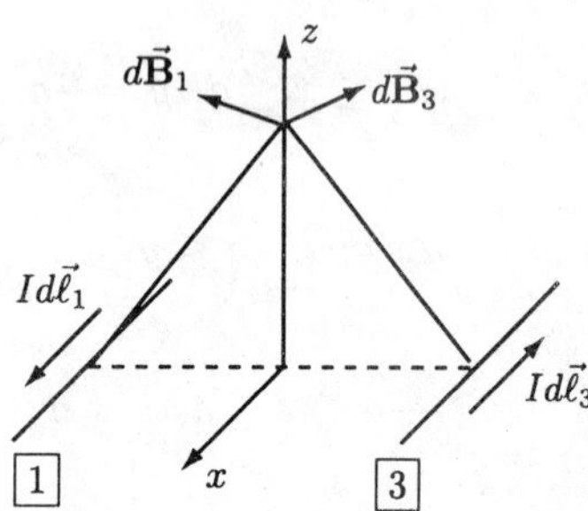

Since

$$I\, d\ell_1 = -I\, d\ell_3, (dB_1)_\perp = -(dB_3)_\perp$$

and thus the perpendicular components cancel, leaving only the sum of the parallel components:

$$d\vec{\mathbf{B}}_{13} = \frac{2\left(\frac{\mu_0 I}{4\pi}\right) a\, dx\hat{\mathbf{k}}}{(x^2+a^2+z^2)^{\frac{3}{2}}}$$

Integrating $d\vec{\mathbf{B}}_{13}$ yields the magnetic field at z due to segments 1 and 3:

$$\vec{\mathbf{B}}_{13} = \frac{\mu_0 I}{2\pi} \int_{-a}^{a} \frac{a\, dx}{(x^2+a^2+z^2)^{\frac{3}{2}}} \hat{\mathbf{k}}$$

Let

$$\begin{aligned} x &= \sqrt{a^2+z^2} \tan\theta \\ dx &= \sqrt{a^2+z^2} \sec^2\theta\, d\theta \end{aligned}$$

Then

$$\begin{aligned} x^2+a^2+z^2 &= (a^2+z^2)(1+\tan^2\theta) \\ &= (a^2+z^2)\sec^2\theta \end{aligned}$$

Thus

$$B_z = \frac{\mu_0 I a}{2\pi} \frac{\sqrt{a^2+z^2}}{(a^2+z^2)^{\frac{3}{2}}} \int_{-\alpha}^{\alpha} \frac{\sec^2\theta\, d\theta}{\sec^3\theta}$$

where

$$\tan\alpha = \frac{a}{\sqrt{a^2+z^2}}$$

Thus

$$\begin{aligned} B_z &= \frac{\mu_0 I a}{2\pi(a^2+z^2)} \int_{-\alpha}^{\alpha} \cos\theta\, d\theta \\ &= \frac{\mu_0 I a}{2\pi(a^2+z^2)} 2\sin\alpha \end{aligned}$$

Now

$$\begin{aligned} \sec^2\alpha &= 1+\tan^2\alpha = 1 + \frac{a^2}{a^2+z^2} \\ &= \frac{2a^2+z^2}{a^2+z^2} \end{aligned}$$

so

$$\cos^2\alpha = \frac{a^2+z^2}{2a^2+z^2}$$

and thus

$$\sin\alpha = \sqrt{1-\cos^2\alpha} = \frac{a}{\sqrt{2a^2+z^2}}$$

So

$$\vec{\mathbf{B}}_{13} = \left(\frac{\mu_0 I}{\pi}\right) \frac{a^2}{(a^2+z^2)(2a^2+z^2)^{\frac{1}{2}}} \hat{\mathbf{k}}$$

By similar arguments, the magnetic field at z due to segments 2 and 4 is:

$$\begin{aligned} \vec{\mathbf{B}}_{24} &= 2\left(\frac{\mu_0 I}{4\pi}\right) \int_{-a}^{a} \frac{a\, dy}{(y^2+a^2+z^2)^{\frac{3}{2}}} \hat{\mathbf{k}} \\ &= \left(\frac{\mu_0 I}{\pi}\right) \frac{a^2}{(a^2+z^2)(2a^2+z^2)^{\frac{1}{2}}} \hat{\mathbf{k}} \end{aligned}$$

Therefore, the net magnetic field at z is:

$$\begin{aligned} \vec{\mathbf{B}}(z) &= \vec{\mathbf{B}}_{13} + \vec{\mathbf{B}}_{24} \\ &= \frac{2\mu_0 I}{\pi} \frac{a^2}{(a^2+z^2)(2a^2+z^2)^{\frac{1}{2}}} \hat{\mathbf{k}} \end{aligned}$$

For

$$2a = 3.5 \times 10^{-2}\ \text{m}, I = 4.77\ \text{A}$$

and $z = 7 \times 10^{-2}$ m, we have:

$$B(z = 7\ \text{cm}) = 3.02 \times 10^{-6}\ \text{T}$$

The magnetic moment of this loop is

$$\begin{aligned}\vec{\mathbf{m}} &= IA\hat{\mathbf{k}} = (4.77\ \text{A})\left(3.5 \times 10^{-2}\ \text{m}\right)^2\hat{\mathbf{k}} \\ &= \left(5.84 \times 10^{-3}\ \text{A}\cdot\text{m}^2\right)\hat{\mathbf{k}}\end{aligned}$$

And using the dipole approximation we have:

$$\begin{aligned}B_z &= \frac{2\,k_m \cdot m}{z^3} \\ &= \frac{(2 \times 10^{-7}\frac{\text{T}\cdot\text{m}}{\text{A}})\,(5.84 \times 10^{-3}\ \text{A}\cdot\text{m}^2)}{(7 \times 10^{-2}\ \text{m})^3} \\ &= 3.41 \times 10^{-6}\ \text{T}\end{aligned}$$

which gives a 13% error from the actual result.

28.41 Short solenoid:

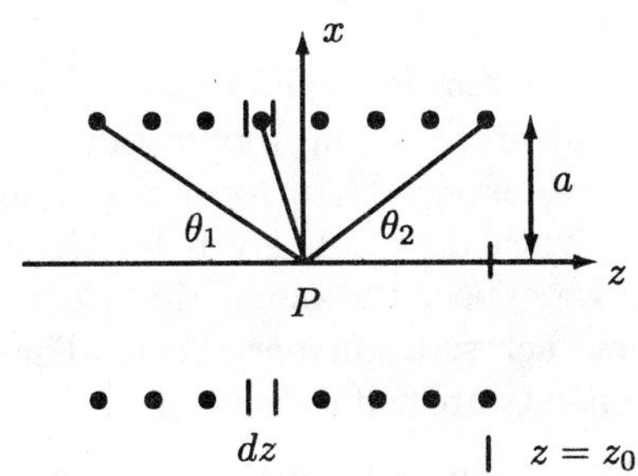

Treat each segment as a current loop of radius a and current

$$I\left(\frac{N}{L}\right)\,dz = In\,dz$$

where L is the length of the solenoid. Let $z = 0$ at P; then the magnetic field strength at P due to a single segment of the solenoid of width dz is given by Eqn. (28.7):

$$\begin{aligned}d\vec{\mathbf{B}}(P) &= \frac{\mu_0 nI\,dz}{2}\frac{a^2}{\left((-z)^2 + a^2\right)^{\frac{3}{2}}}\hat{\mathbf{k}} \\ &= \frac{\mu_0 nI\,dz}{2}\frac{a^2}{(z^2 + a^2)^{\frac{3}{2}}}\hat{\mathbf{k}}\end{aligned}$$

Denote one end of the solenoid by $z = z_0 (z_0 > 0)$, so the limits of integration will be

$$z_0 - L \le z \le z_0$$

The resultant field at P is therefore:

$$\begin{aligned}\vec{\mathbf{B}}(P) &= \frac{\mu_0 nI}{2}a^2\int_{z_0-L}^{z_0}\frac{dz}{(z^2 + a^2)^{\frac{3}{2}}}\hat{\mathbf{k}} \\ &= \frac{\mu_0 nI}{2}\left.\frac{z}{\sqrt{z^2 + a^2}}\right|_{z_0-L}^{z_0}\hat{\mathbf{k}} \\ &= \frac{\mu_0 nI}{2}\left\{\frac{z_0}{\sqrt{z_0^2 + a^2}} - \frac{(z_0 - L)}{\sqrt{(z_0 - L)^2 + a^2}}\right\}\hat{\mathbf{k}}\end{aligned}$$

From the figure,

$$\begin{aligned}\cos\theta_2 &= \frac{z_0}{\sqrt{z_0^2 + a^2}} \\ \cos\theta_1 &= \frac{|(z_0 - L)|}{\sqrt{(z_0 - L)^2 + a^2}} \\ &= -\frac{(z_0 - L)}{\sqrt{(z_0 - L)^2 + a^2}}\end{aligned}$$

therefore

$$\begin{aligned}\vec{\mathbf{B}}(P) &= \frac{\mu_0 nI}{2}(\cos\theta_1 + \cos\theta_2)\hat{\mathbf{k}} \\ &= 2\pi k_m nI(\cos\theta_1 + \cos\theta_2)\hat{\mathbf{k}}\end{aligned}$$

where $k_m = \frac{\mu_0}{4\pi}$. Consider the field at the center of the 10 cm long solenoid

$$\begin{aligned}B &= 2\pi k_m nI(2\cos\theta)\hat{\mathbf{k}} \\ &= \mu_0 nI\cos\theta \\ &= B_{\text{infinite}}\cos\theta\end{aligned}$$

where

$$\cos\theta = \frac{5\,\text{cm}}{\sqrt{(5\,\text{cm})^2 + (2\,\text{cm})^2}} = 0.93$$

A 7% error is introduced by assuming an infinite solenoid.

28.45 According to Gauss' law for the magnetic field the magnetic flux through any closed surface is zero. Thus the flux through this box is *zero*.

28.49 Within a wire of uniform current density, $\hat{\mathbf{j}}$,

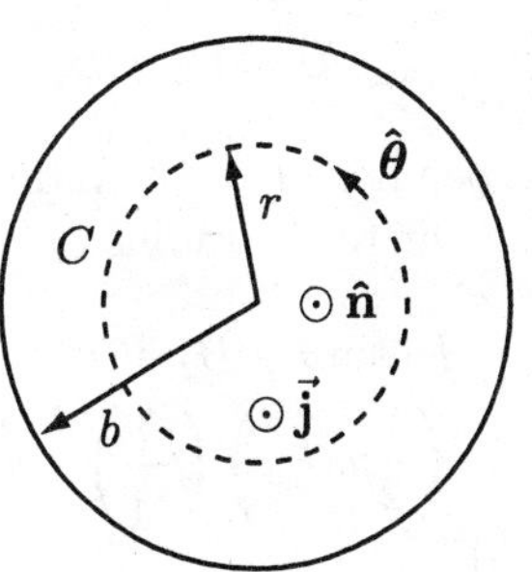

Ampères law gives:

$$\mathcal{C} = \oint_C \vec{\mathbf{B}} \cdot d\vec{\ell} = \mu_0 \int_S \hat{\mathbf{j}} \cdot \hat{\mathbf{n}}\, dA$$

We expect $\vec{\mathbf{B}}$ to follow circles around the center of the wire, so

$$\begin{aligned} 2\pi r B_\theta &= \mu_0 j \pi r^2 \\ B_\theta &= \frac{1}{2}\mu_0 j r \end{aligned}$$

at a distance $r \leq b$ from the center of the wire. Superposition allows us to think of the magnetic field within the hole as arising from two current densities:

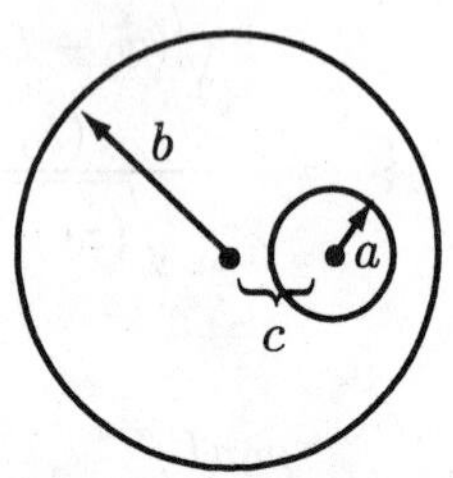

1) A uniform current density, $\hat{\mathbf{j}}$, carried by the cylinder of radius b.

2) A uniform current density, $-\hat{\mathbf{j}}$, carried by a cylinder of radius a.

The magnetic field due to each of 1) and 2) forms circles around the center of each cylinder, and has magnitude $(\frac{1}{2}\mu_0 j) \times$(distance from center of cylinder). At point P within the hole, the magnetic field is the sum of two contributions, as shown:

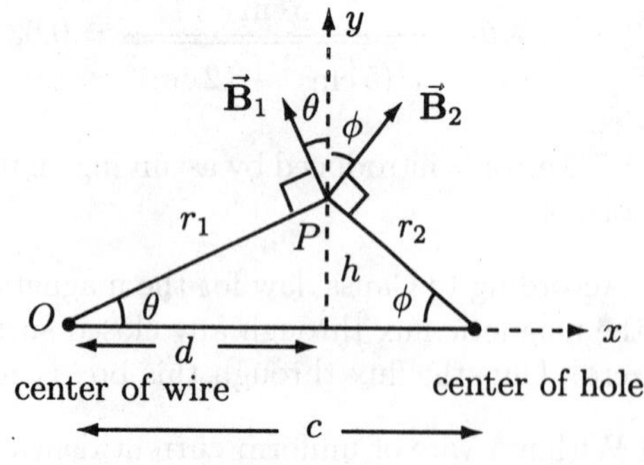

$$\vec{\mathbf{B}} = \vec{\mathbf{B}}_1 + \vec{\mathbf{B}}_2$$

$\vec{\mathbf{B}}_1$ is perpendicular to the radius r_1 and $\vec{\mathbf{B}}_2$ is perpendicular to the radius r_2.

$$\begin{aligned} B_x &= -B_1 \sin\theta + B_2 \sin\phi \\ &= -\left(\frac{1}{2}\mu_0 j r_1\right)\left(\frac{h}{r_1}\right) + \left(\frac{1}{2}\mu_0 j r_2\right)\left(\frac{h}{r_2}\right) \\ &= 0 \end{aligned}$$

and

$$\begin{aligned} B_y &= B_1 \cos\theta + B_2 \cos\phi \\ &= \left(\frac{1}{2}\mu_0 j r_1\right)\left(\frac{d}{r_1}\right) + \left(\frac{1}{2}\mu_0 j r_2\right)\left(\frac{c-d}{r_2}\right) \\ &= \frac{1}{2}\mu_0 j c = \text{constant.} \end{aligned}$$

Thus within the hole $\vec{\mathbf{B}}$ is perpendicular to the line from the center of the cylinder to the center of the hole and has magnitude $\frac{1}{2}\mu_0 j c$.

28.53

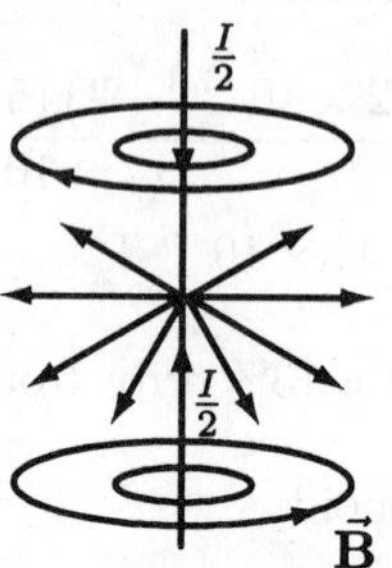

a) The system has cylindrical symmetry about the two wires. A long way from the conducting sheet we expect $\vec{\mathbf{B}}$ to form circles around the wires, with direction given by Oersted's rule. Also we expect the magnetic field direction to reverse across a current sheet. These two requirements are self consistent.

b) At a radius r from the wire we can draw a curve that follows an arc of a circle above and below the sheet, and is $\perp$ to the sheet on the other two sides.

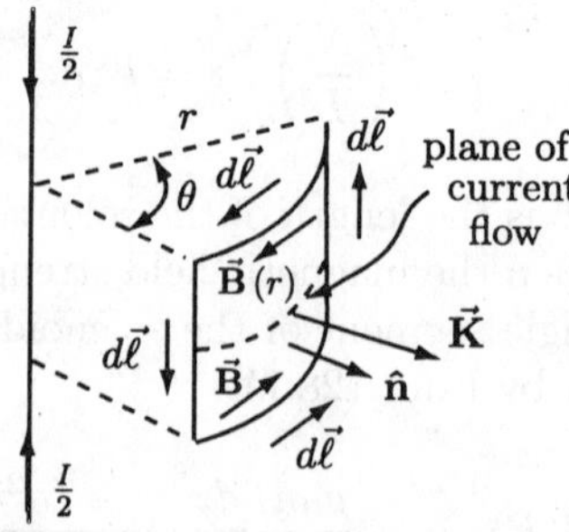

Then $\vec{\mathbf{B}} \parallel d\vec{\ell}$ on the two arcs and $\vec{\mathbf{B}} \perp d\vec{\ell}$ on the two straight sides. The surface current $\vec{\mathbf{K}}$ exists in a plane that cuts through the middle of the curve along the line shown. $\vec{\mathbf{K}}$ is $\parallel$ to $\hat{\mathbf{n}}$ everywhere on the line. Applying Ampère's law:

$$2B(r)r\theta = \int \vec{\mathbf{K}} \cdot \hat{\mathbf{n}}\, d\ell(\mu_0) = (r\theta)K\mu_0$$

when K is the surface current density in the sheet. The total current must be the same at every r:

$$K(2\pi r) = I \Rightarrow K = \frac{I}{2\pi r}$$

Thus

$$B = \mu_0 \frac{K}{2} = (\mu_0)\frac{I}{4\pi r} = \frac{\mu_0 I}{4\pi r}$$

c) Apply Ampère's law to a circle around the upper wire.

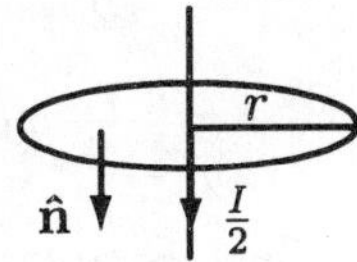

$$2\pi r B(r) = \mu_0 \left(\frac{I}{2}\right) \Rightarrow B(r) = \frac{\mu_0 I}{4\pi r},$$

as in part (b).

d) The system has the same cylindrical symmetry that we noted in part a),

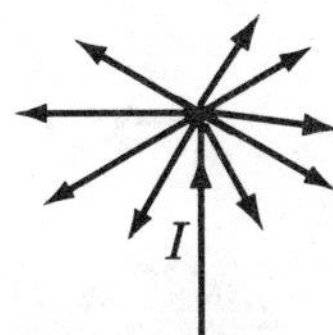

but we have lost the (anti-)symmetry above and below the plane. Since there is no long wire above the plane, Ampère's law applied to circles around the symmetry axis shows $\vec{\mathbf{B}} = 0$ on the side without the wire. On the side with the wire,

$$B(r) = \frac{\mu_0 I}{2\pi r}$$

twice as large as before. We can show that these results are consistent with Ampères law applied to the curved rectangle we used in part b). Now $\vec{\mathbf{B}} \cdot d\vec{\ell} \neq 0$ only on the lower side of the rectangle.

$$\begin{aligned} B(r)(r\theta) &= r\theta k(\mu_0) = \frac{\mu_0 I}{2\pi r}(r\theta) \\ \Rightarrow B(r) &= \frac{\mu_0 I}{2\pi r} \end{aligned}$$

as above.

28.57 Yes. Since the magnetic poles do not correspond to the geographic poles, there are some locations between the two where a compass, pointing toward *magnetic* north, is pointing *away* from geographic north, i.e. it points south.

28.61 At the center of the loop,

$$\vec{\mathbf{B}} = \frac{\mu_0 I}{2a}\hat{\mathbf{k}}$$

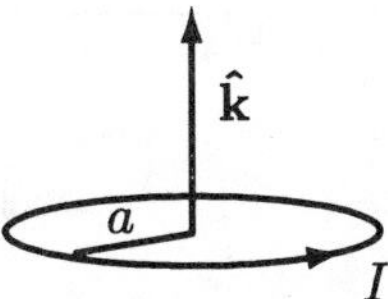

The velocity of the charged particle is: $\vec{\mathbf{v}} = v\hat{\mathbf{k}}$, which is parallel to $\vec{\mathbf{B}}$. The magnetic force acting on the charged particle is zero, since the cross product of two parallel vectors is *zero*.

$$\vec{\mathbf{F}} = q\vec{\mathbf{v}} \times \vec{\mathbf{B}} = \frac{qv\mu_0 I}{2a}(\hat{\mathbf{k}} \times \hat{\mathbf{k}}) = 0$$

Therefore no force acts on the particle as it passes through the center of the current loop.

28.67 a) For the plane current sheets where the two current densities are parallel, we label their direction $\hat{\mathbf{j}}$ as shown in the figure.

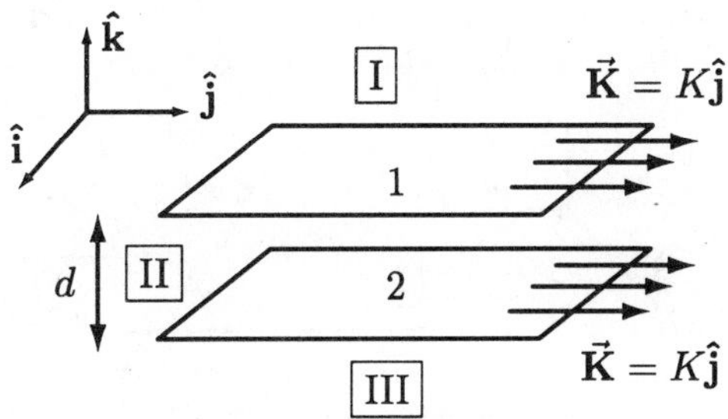

We use the principal of superposition, and the result of Example 28.9. In Region I (above both sheets) the two contributions to $\vec{\mathbf{B}}$ are in the same direction:

$$\vec{\mathbf{B}} = \vec{\mathbf{B}}_1 + \vec{\mathbf{B}}_2 = 2\left(\frac{\mu_0}{2}\right)K\hat{\mathbf{i}} = \mu_0 K\hat{\mathbf{i}}$$

In Region II (between the sheets) the two contributions are equal and opposite:

$$\vec{\mathbf{B}} = -\frac{\mu_0 K}{2}\hat{\mathbf{i}} + \frac{\mu_0 K}{2}\hat{\mathbf{i}} = 0$$

In Region III (below the sheets) again the two contributions are in the same direction, but opposite the direction in Region I.

$$\vec{\mathbf{B}} = -\frac{\mu_0 K}{2}\hat{\mathbf{i}} - \frac{\mu_0 K}{2}\hat{\mathbf{i}} = -\mu_0 K\hat{\mathbf{i}}$$

b) Anti-parallel:

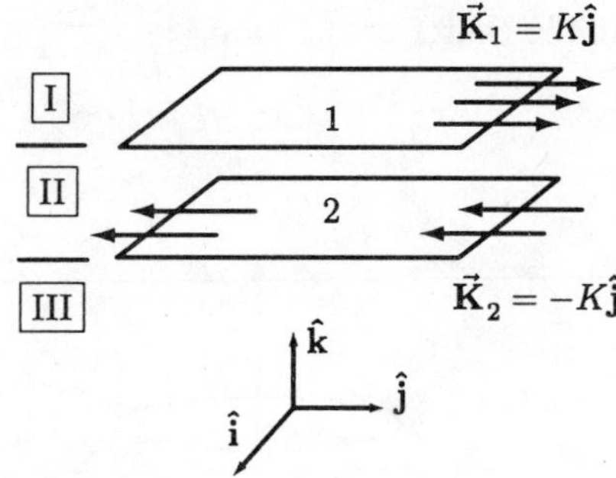

In Region I the two contributions are now opposite:

$$\vec{\mathbf{B}} = \vec{\mathbf{B}}_1 + \vec{\mathbf{B}}_2 = \frac{\mu_0}{2}K\hat{\mathbf{i}} + \frac{\mu_0}{2}K(-\hat{\mathbf{i}}) = 0$$

In Region II the two contributions are in the same direction:

$$\begin{aligned}\vec{\mathbf{B}} &= \frac{\mu_0}{2}K(-\hat{\mathbf{i}}) + \frac{\mu_0}{2}K(-\hat{\mathbf{i}}) \\ &= -\mu_0 K\hat{\mathbf{i}}\end{aligned}$$

In Region III the two contributions are again opposite:

$$\vec{\mathbf{B}} = \frac{\mu_0}{2}K(-\hat{\mathbf{i}}) + \frac{\mu_0}{2}K(\hat{\mathbf{i}}) = 0$$

c) In each region the two magnetic field contributions are at right angles.

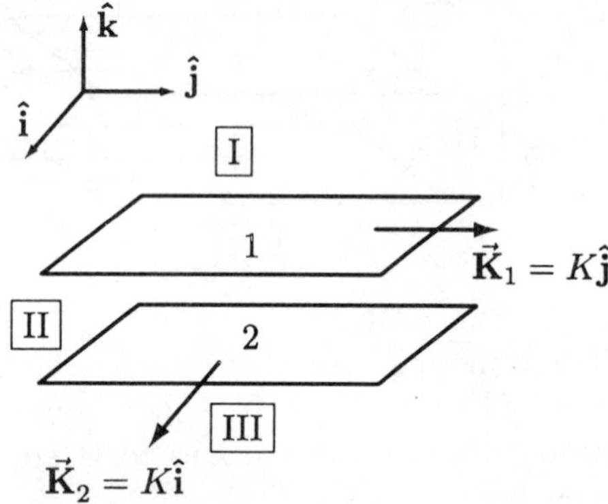

Above both sheets (Region I)

$$\begin{aligned}\vec{\mathbf{B}}_1 &= \frac{\mu_0}{2}K\hat{\mathbf{i}} \\ \vec{\mathbf{B}}_2 &= \frac{\mu_0}{2}K(-\hat{\mathbf{j}})\end{aligned}$$

So

$$\begin{aligned}\vec{\mathbf{B}} &= \frac{\mu_0}{2}K(\hat{\mathbf{i}} - \hat{\mathbf{j}}) \\ \left|\vec{\mathbf{B}}\right| &= \frac{\sqrt{2}}{2}\mu_0 K\end{aligned}$$

Between the sheets (region II)

$$\begin{aligned}\vec{\mathbf{B}}_1 &= \frac{\mu_0}{2}K(-\hat{\mathbf{i}}) \\ \vec{\mathbf{B}}_2 &= \frac{\mu_0}{2}K(-\hat{\mathbf{j}})\end{aligned}$$

So

$$\begin{aligned}\vec{\mathbf{B}} &= -\frac{\mu_0}{2}K(\hat{\mathbf{i}} + \hat{\mathbf{j}}) \\ \left|\vec{\mathbf{B}}\right| &= \mu_0\frac{\sqrt{2}}{2}K\end{aligned}$$

Below the sheets, (region III)

$$\begin{aligned}\vec{\mathbf{B}}_1 &= \frac{\mu_0}{2}K(-\hat{\mathbf{i}}) \\ \vec{\mathbf{B}}_2 &= \frac{\mu_0}{2}K\hat{\mathbf{j}} \\ \vec{\mathbf{B}} &= \frac{\mu_0}{2}K(-\hat{\mathbf{i}} + \hat{\mathbf{j}}) \\ \left|\vec{\mathbf{B}}\right| &= \frac{\sqrt{2}}{2}\mu_0 K\end{aligned}$$

28.69 The Earth's field is approximately a dipole field so we may use the expression for the magnetic field due to an ideal dipole (c.f. Exercise 24.6)

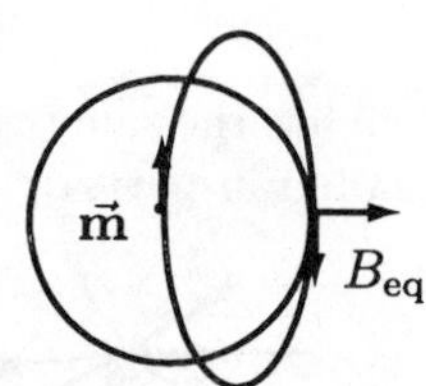

$$\begin{aligned}B_{eq} &= \frac{k_m m}{R^3} \\ \Rightarrow m &= \frac{R^3 B_{eq}}{k_m} \\ &= \frac{(3\times 10^{-5}\ \mathrm{T})(6400\times 10^3\ \mathrm{m})^3}{10^{-7}\ \mathrm{N/A^2}} \\ &= 7.9\times 10^{22}\ \mathrm{A\cdot m^2}\end{aligned}$$

For a current loop:

$$m = IA$$

with

$$A = \pi\left(\frac{R}{2}\right)^2 = \frac{\pi R^2}{4}$$

Then

$$B_{eq} \sim k_m I\frac{\left(\frac{\pi R^2}{4}\right)}{R^3} = \frac{k_m I\pi}{4R}$$

Thus

$$\begin{aligned}I \sim \frac{4RB_{eq}}{k_m\pi} &= \frac{4(6400\times 10^3\ \mathrm{m})(3\times 10^{-5}\ \mathrm{T})}{(10^{-7}\ \mathrm{N/A^2})\pi} \\ &= 2\times 10^9\ \mathrm{A}\end{aligned}$$

28.73 $d = R = 12.00$ cm

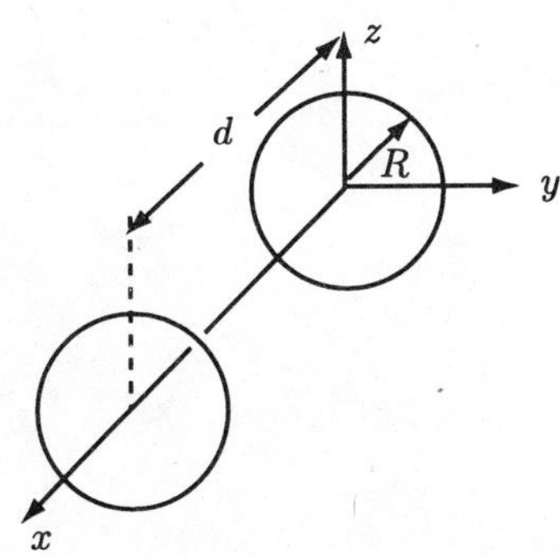

$$
\begin{aligned}
B(x) &= \frac{\mu_0 N I R^2}{2}\left\{\frac{1}{(R^2+x^2)^{3/2}} + \frac{1}{\left[(d-x)^2+R^2\right]^{3/2}}\right\} \\
&= \frac{\mu_0 N I R^2}{2}\left\{\frac{1}{(R^2+x^2)^{3/2}} + \frac{1}{\left((d-x)^2+R^2\right)^{3/2}}\right\}
\end{aligned}
$$

since $d = R$. Now,

$$
\begin{aligned}
\frac{\mu_0 I R^2}{2} &= \frac{(4\pi\times 10^{-7}\ \mathrm{N/A^2})(7.75\ \mathrm{A})\left(12\times 10^{-2}\ \mathrm{m}\right)^2}{2} \\
&= (7.01\times 10^{-8}\ \mathrm{N\cdot m^2/A})
\end{aligned}
$$

Sample Calculations:

1) $x = 0$ cm

$$
B(x=0) = (7.01\times 10^{-8})\left\{\frac{1}{R^3} + \frac{1}{2^{3/2}R^3}\right\}\frac{\mathrm{N\cdot m^2}}{\mathrm{A}}
$$

$$
\begin{aligned}
&= (7.01\times 10^{-8})\left\{578.7\ \mathrm{m^{-3}} + 204.6\ \mathrm{m^{-3}}\right\}\frac{\mathrm{N\cdot m^2}}{\mathrm{A}} \\
&= (7.01\times 10^{-8})(783.3)\ \frac{\mathrm{N}}{\mathrm{m\cdot A}} \\
&= 5.49\times 10^{-5}\ \mathrm{T}
\end{aligned}
$$

since $1\ \dfrac{\mathrm{N}}{\mathrm{m\cdot A}} = 1$ T.

2) $x = 6.0$ cm

$$B(x = 6.0\times 10^{-2}\ \mathrm{m})$$

$$
\begin{aligned}
&= (7.01\times 10^{-8})\left\{\frac{1}{\left[R^2+x^2\right]^{\frac{3}{2}}} + \frac{1}{\left[(d-x)^2+R^2\right]^{\frac{3}{2}}}\right\} \\
&= (7.01\times 10^{-8}\ \tfrac{\mathrm{N\cdot m^2}}{\mathrm{A}})\left\{\frac{1}{\left[(12\times 10^{-2}\ \mathrm{m})^2+(6\times 10^{-2}\ \mathrm{m})^2\right]^{\frac{3}{2}}}\right. \\
&\quad \left. + \frac{1}{\left[(6\times 10^{-2}\ \mathrm{m})^2+(12\times 10^{-2}\ \mathrm{m})^2\right]^{\frac{3}{2}}}\right\} \\
&= 5.81\times 10^{-5}\ \mathrm{T}
\end{aligned}
$$

As a check, recall that mid-way between two Helmholtz coils (see Example 28.5):

$$
\begin{aligned}
B &= \frac{8\mu_0 I}{(5)^{3/2}R} \\
&= \frac{8(4\pi\times 10^{-7}\ \mathrm{N/A^2})(7.75\ \mathrm{A})}{(5)^{3/2}(12\times 10^{-2}\ \mathrm{m})} \\
&= 5.81\times 10^{-5}\ \mathrm{T}
\end{aligned}
$$

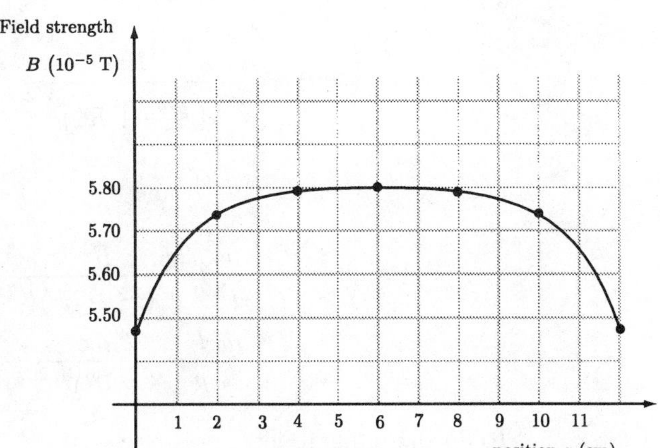

x(cm)	$B(x)$(T)
0.0	5.49×10^{-5}
2.0	5.73×10^{-5}
4.0	5.80×10^{-5}
6.0	5.81×10^{-5}
8.0	5.80×10^{-5}
10.0	5.73×10^{-5}
12.0	5.49×10^{-5}

28.77

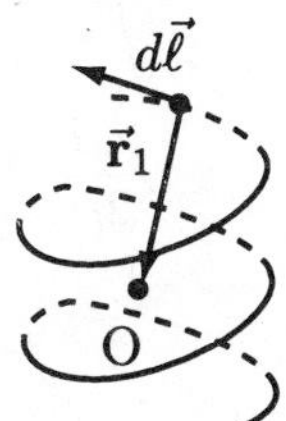

For observation point at the origin, O, the Biot-Savart law gives

$$d\vec{\mathbf{B}}\,(\text{origin}) = \frac{\mu_0 I}{4\pi}\frac{d\vec{\ell}\times\vec{\mathbf{r}}_1}{r_1^3}$$

where

$$
\begin{aligned}
\vec{\mathbf{r}}_1 &= -\vec{\mathbf{r}}\,(\theta) = -\frac{p\theta}{2\pi}\hat{\mathbf{k}} - R\left[\cos\theta\hat{\mathbf{i}} + \sin\theta\hat{\mathbf{j}}\right] \\
d\vec{\ell} &= \frac{p\,d\theta}{2\pi}\hat{\mathbf{k}} + R\left[-\sin\theta\,d\theta\,\hat{\mathbf{i}} + \cos\theta\,d\theta\,\hat{\mathbf{j}}\right]
\end{aligned}
$$

Therefore, the z component of $d\vec{\ell}\times\vec{\mathbf{r}}_1$ is

$$
\begin{aligned}
\left(d\vec{\ell}\times\vec{\mathbf{r}}_1\right)_z &= R^2\left(\sin^2\theta + \cos^2\theta\right)d\theta \\
&= R^2\,d\theta
\end{aligned}
$$

and, so

$$dB_z\,(\text{origin}) = \frac{\mu_0 I}{4\pi} \frac{R^2\,d\theta}{\left(\frac{p^2\theta^2}{4\pi^2} + R^2\right)^{3/2}}$$

Let $x = \frac{p\theta}{2\pi}$, then $dx = \frac{p\,d\theta}{2\pi}$, and

$$\begin{aligned} dB_z\,(\text{origin}) &= \frac{\mu_0}{4\pi} I \frac{2\pi}{p} \frac{R^2\,dx}{(x^2+R^2)^{3/2}} \\ &= \frac{\mu_0}{2} \frac{I}{p} \frac{R^2\,dx}{(x^2+R^2)^{3/2}} \end{aligned}$$

To do the integral, let

$$\begin{aligned} x &= R\tan\phi \\ dx &= R\sec^2\phi\,d\phi \end{aligned}$$

$$\begin{aligned} \int_{-\infty}^{\infty} \frac{R^2\,dx}{(x^2+R^2)^{3/2}} &= \int_{-\pi/2}^{+\pi/2} \frac{R^2\left(R\sec^2\phi\,d\phi\right)}{R^3\left(\sec^3\phi\right)} \\ &= \int_{-\pi/2}^{+\pi/2} \cos\phi\,d\phi \\ &= \sin\phi \Big|_{-\pi/2}^{+\pi/2} = 2 \end{aligned}$$

Thus we have

$$\begin{aligned} B_z\,(\text{origin}) &= \int dB_z \\ &= \int_{-\infty}^{+\infty} \frac{\mu_0 I}{2p} \frac{R^2\,dx}{(x^2+R^2)^{3/2}} \\ &= \frac{\mu_0 I}{2p}(2) = \frac{\mu_0 I}{p} \end{aligned}$$

Chapter 29

Static Magnetic Fields: Applications

29.1 Assume uniform magnetic field $B = 0.3$ nT. Electron of energy $K = 10^{-19}$ J has speed

$$v = \sqrt{\frac{2K}{m}}$$

where for an electron,

$$m = 9.11 \times 10^{-31} \text{ kg}$$

Thus

$$\begin{aligned} v &= \sqrt{\frac{2K}{m}} = \sqrt{\frac{2\left(10^{-19} \text{ J}\right)}{9.11 \times 10^{-31} \text{ kg}}} \\ &= 4.7 \times 10^{5} \text{ m/s} \end{aligned}$$

The Larmor radius, where $q = 1.6 \times 10^{-19}$ C, is given by

$$\begin{aligned} r &= \frac{mv}{qB} = \frac{\left(9.11 \times 10^{-31} \text{ kg}\right)\left(4.7 \times 10^{5} \text{ m/s}\right)}{\left(1.6 \times 10^{-19} \text{ C}\right)\left(0.3 \times 10^{-9} \text{ T}\right)} \\ &= 9 \times 10^{3} \text{ m} \end{aligned}$$

Cyclotron frequency is

$$\begin{aligned} \omega_c &= \frac{qB}{m} = \frac{\left(1.6 \times 10^{-19} \text{ C}\right)\left(0.3 \times 10^{-9} \text{ T}\right)}{9.11 \times 10^{-31} \text{ kg}} \\ &= 53 \text{ rad/s} \end{aligned}$$

29.3 Larmor radius given by

$$r = \frac{mv}{qB}$$

Given:

$$\begin{aligned} r &= \frac{\text{diameter}}{2} = \frac{4 \text{ in.}}{2} \\ &= 2 \text{ in.} \times \frac{2.54 \times 10^{-2} \text{ m}}{1 \text{ in.}} = 0.051 \text{ m} \end{aligned}$$

Use known kinetic energy to find speed v

$$K = \frac{1}{2}mv^2$$

Deuteron:

$$m = 2\left(1.67 \times 10^{-27} \text{ kg}\right) = 3.34 \times 10^{-27} \text{ kg}$$

Singly ionize:

$$\begin{aligned} q &= 1.6 \times 10^{-19} \text{ C} \\ K &= 80 \text{ keV} \\ &= \left(80 \times 10^{3} \text{ eV}\right) \times \left(\frac{1.6 \times 10^{-19} \text{ J}}{1 \text{ eV}}\right) \\ &= 1.28 \times 10^{-14} \text{ J} \\ &= \frac{1}{2}mv^2 \\ \Rightarrow v &= \sqrt{\frac{2K}{m}} = \sqrt{\frac{2\left(1.28 \times 10^{-14} \text{ J}\right)}{3.34 \times 10^{-27} \text{ kg}}} \\ &= 2.8 \times 10^{6} \text{ m/s} \\ r &= \frac{mv}{qB} \\ \Rightarrow B &= \frac{mv}{rq} \\ &= \frac{\left(3.34 \times 10^{-27} \text{ kg}\right)\left(2.8 \times 10^{6} \text{ m/s}\right)}{\left(0.0508 \text{ m}\right)\left(1.6 \times 10^{-19} \text{ C}\right)} \\ &= 1.1 \text{ T} \end{aligned}$$

29.5 Since the magnetic field is uniform, we may use Eqn. (29.6). The vector $\vec{\ell}$ joining the ends of the wire segment lies along the x-axis and has length

$$\ell = L$$

Thus

$$\begin{aligned}\vec{\mathbf{F}} &= I\vec{\ell}\times\vec{\mathbf{B}} = IL\hat{\mathbf{i}}\times B_0\left(\hat{\mathbf{j}}+\hat{\mathbf{k}}\right)\\ &= ILB_0\left(\hat{\mathbf{k}}-\hat{\mathbf{j}}\right)\\ &= (0.23\text{ A})(0.52\text{ m})(0.75\text{ T})\left(\hat{\mathbf{k}}-\hat{\mathbf{j}}\right)\\ &= (0.090\text{ N})\left(\hat{\mathbf{k}}-\hat{\mathbf{j}}\right)\end{aligned}$$

29.7

$$\begin{aligned}\frac{dF}{d\ell} &= \frac{\mu_0 I_1 I_2}{2\pi d}\\ I_1 &= I_2 = 1.7\text{ A},\ d = 0.38\text{ m}\\ \frac{dF}{d\ell} &= \frac{(4\pi\times10^{-7}\text{ T}\cdot\text{m/A})(1.7\text{ A})(1.7\text{ A})}{2\pi(0.38\text{ m})}\\ &= 1.5\times10^{-6}\text{ N/m}\end{aligned}$$

At B, magnetic field due to A is out of the page. $d\vec{\ell}_B$ is up the page. Therefore, $d\vec{\mathbf{F}}_B$ is to the right. So

$$\frac{d\vec{\mathbf{F}}}{d\ell} = (1.5\ \mu\text{N/m, to the right})$$

29.9 Hall potential given by

$$|V_H| = v_d BL$$

where

$$\begin{aligned}v_d &= \text{drift velocity}\\ B &= 1.3\text{ T}\\ L &= \text{diameter of wire}\\ &= 1.0\text{ mm} = 1.0\times10^{-3}\text{ m}\end{aligned}$$

Need v_d.

$$v_d = \frac{I}{neA}$$

where

$$\begin{aligned}I &= \text{current} = 2.5\times10^{-3}\text{ A}\\ e &= 1.6\times10^{-19}\text{ C}\\ A &= \text{cross-section area of wire}\\ &= \pi r^2 = \pi\left(\frac{d}{2}\right)^2 = \frac{\pi}{4}d^2\\ &= \left(\frac{\pi}{4}\right)(1.0\times10^{-3}\text{ m})^2\\ &= 7.85\times10^{-7}\text{ m}^2\end{aligned}$$

$$\begin{aligned}n &= \text{number of conduction electrons per unit volume}\\ &= 8.5\times10^{28}\text{ m}^{-3}\ \text{(See Ex. 26.4)}\end{aligned}$$

so

$$\begin{aligned}v_d &= \frac{I}{neA}\\ &= \frac{2.5\times10^{-3}\text{ A}}{(8.5\times10^{28}\text{ m}^{-3})(1.6\times10^{-19}\text{ C})(7.85\times10^{-7}\text{ m}^2)}\\ &= 2.3\times10^{-7}\text{ m/s}\end{aligned}$$

so

$$\begin{aligned}|V_H| &= v_d BL\\ &= (2.3\times10^{-7}\text{ m/s})(1.3\text{ T})(1.0\times10^{-3}\text{ m})\\ &= 3.0\times10^{-10}\text{ V}\\ &= 0.30\text{ nV}\end{aligned}$$

29.11 Given: for platinum,

$$H = 1.00000\times10^2\text{ A/m}$$

Find B and M. From Table 29.2,

$$\begin{aligned}\chi_m &= 2.8\times10^{-4}\\ B &= \mu_0(1+\chi_m)H\\ M &= \chi_m H\\ &= (2.8\times10^{-4})(100.000\text{ A/m})\\ &= 0.028\text{ A/m}\end{aligned}$$

$$\begin{aligned}B &= \mu_0(1+\chi_m)H\\ &= (4\pi\times10^{-7}\text{ T}\cdot\text{m/A})(1+0.00028)(100.000\text{ A}\\ &= 1.25699\times10^{-4}\text{ T}\end{aligned}$$

29.13

$$\begin{aligned}\vec{\mathbf{v}} &= v\left(+\hat{\mathbf{j}}\right)\ \text{up the page}\\ \vec{\mathbf{B}} &= B\left(-\hat{\mathbf{k}}\right)\ \text{into page}\\ q &= -e\\ \vec{\mathbf{F}} &= q\vec{\mathbf{v}}\times\vec{\mathbf{B}}\\ &= (-e)\left(v\hat{\mathbf{j}}\right)\times\left(-B\hat{\mathbf{k}}\right)\\ &= +evB\hat{\mathbf{i}}\ \rightarrow \text{Force is to right}\end{aligned}$$

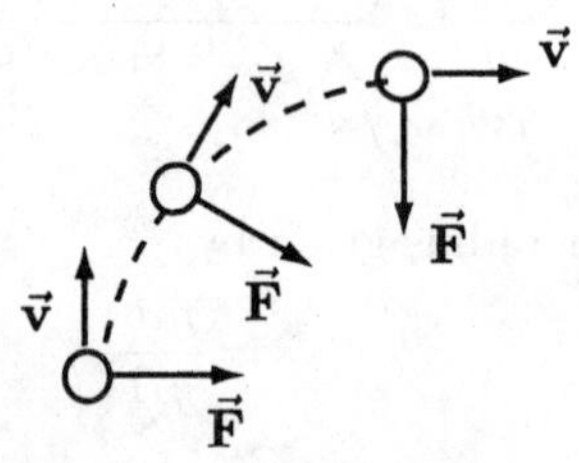

Subsequent motion will be on a clockwise circle.

29.17 At the equator $\vec{\mathbf{B}}$ is roughly parallel to the Earth's surface.

At the poles, $\vec{\mathbf{B}}$ is roughly normal to the Earth's

surface.
At the equator, an incoming cosmic ray (charged particle) is deflected sideways and so is less likely to strike the Earth.
At the poles, an incoming charged particle will be undeflected, and so be more likely to strike the Earth.

29.21 From ex. 29.10, a dipole field has the form

$$B \propto \frac{1}{r^3}$$

At equator, $B_1 = \frac{A}{r_e^3}$, where

$$\begin{aligned} A &= \text{constant} \\ r_e &= 6.37 \times 10^6 \text{ m} = \text{Earth's radius} \end{aligned}$$

At $r = 3.4 r_e$,

$$B = \frac{A}{(3.4 r_e)^3} = \frac{B_1}{(3.4)^3}$$

As given, $B_1 = 3.12 \times 10^{-5}$ T, so

$$B_2 = 7.94 \times 10^{-7} \text{ T}$$

Proton:

$$\begin{aligned} m &= 1.67 \times 10^{-27} \text{ kg} \\ q &= 1.6 \times 10^{-19} \text{ C} \\ v &= 3.0 \times 10^5 \text{ m/s} \end{aligned}$$

$$\begin{aligned} r &= \frac{mv}{qB} \\ &= \frac{(1.67 \times 10^{-27} \text{ kg})(3.0 \times 10^5 \text{ m/s})}{(1.6 \times 10^{-19} \text{ C})(7.94 \times 10^{-7} \text{ T})} \\ &= 3.9 \times 10^3 \text{ m} = 3.9 \text{ km} \end{aligned}$$

Cyclotron frequency,

$$\begin{aligned} f_c &= \frac{\omega_c}{2\pi} = \frac{qB}{2\pi m} \\ &= \frac{(1.6 \times 10^{-19} \text{ C})(7.94 \times 10^{-7} \text{ T})}{2\pi (1.67 \times 10^{-27} \text{ kg})} \\ &= 12 \text{ Hz} \end{aligned}$$

Electron:

$$\begin{aligned} |q| &= 1.6 \times 10^{-19} \text{ C} \\ m &= 9.11 \times 10^{-31} \text{ kg} \end{aligned}$$

As given,

$$v = 3.0 \times 10^5 \text{ m/s}$$

Larmor radius

$$\begin{aligned} r &= \frac{mv}{qB} \\ &= \frac{(9.11 \times 10^{-31} \text{ kg})(3.0 \times 10^5 \text{ m/s})}{(1.6 \times 10^{-19} \text{ C})(7.94 \times 10^{-7} \text{ T})} \\ &= 2.2 \text{ m} \end{aligned}$$

Cyclotron frequency:

$$\begin{aligned} f_c &= \frac{\omega_c}{2\pi} = \frac{qB}{2\pi m} \\ &= \frac{(1.6 \times 10^{-19} \text{ C})(7.94 \times 10^{-7} \text{ T})}{2\pi (9.11 \times 10^{-31} \text{ kg})} \\ &= 22 \text{ kHz} \end{aligned}$$

29.25 The distance between beam entrance and detector is the diameter of the particle's circular path.

$$\begin{aligned} r &= \frac{d}{2} = \frac{0.30 \text{ m}}{2} \\ &= 0.15 \text{ m} \\ B &= 1.3 \text{ T} \\ m &= 78 (1.66 \times 10^{-27} \text{ kg}) \\ &= 1.29 \times 10^{-25} \text{ kg} \\ q &= 1.6 \times 10^{-19} \text{ C} \end{aligned}$$

$$\begin{aligned} r &= \frac{mv}{qB} \\ \Rightarrow v &= \frac{qrB}{m} \\ &= \frac{(1.6 \times 10^{-19} \text{ C})(0.15 \text{ m})(1.3 \text{ T})}{1.29 \times 10^{-25} \text{ kg}} \\ &= 2.4 \times 10^5 \text{ m/s} \end{aligned}$$

$$\begin{aligned} K &= \frac{1}{2} mv^2 \\ &= \frac{1}{2} (1.29 \times 10^{-25} \text{ kg})(2.4 \times 10^5 \text{ m/s})^2 \\ &= 3.76 \times 10^{-15} \text{ J} \\ &= 23 \text{ keV} \end{aligned}$$

But $K = q\Delta V$, where $q = e$, so

$$\Delta V = 23 \text{ kV}$$

29.29 The electrons follow a circular path from the entrance hole to the exit hole, at speed v. The radius of the circle is the Larmor radius

$$r = \frac{mv}{eB}$$

and the center of the circle lies on the x-axis (see figure.)

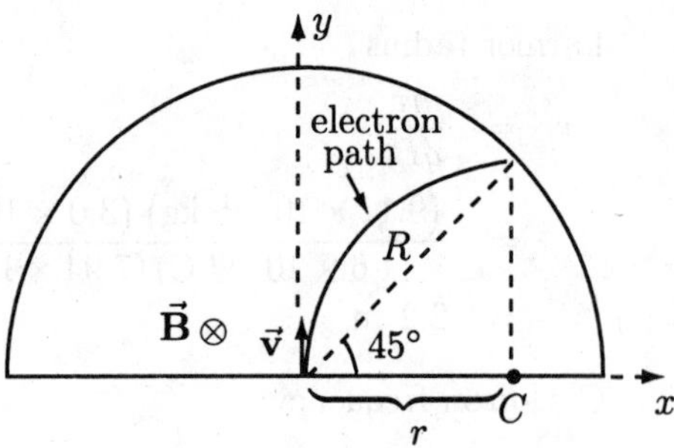

Thus the circle is described by the equation

$$(x-r)^2 + y^2 = r^2$$

Since the point $x = \frac{R}{\sqrt{2}}$, $y = \frac{R}{\sqrt{2}}$ lies on the circle, we have

$$\left(\frac{R}{\sqrt{2}} - r\right)^2 + \left(\frac{R}{\sqrt{2}}\right)^2 = r^2$$

$$\frac{R^2}{2} - \frac{2Rr}{\sqrt{2}} + r^2 + \frac{R^2}{2} = r^2$$

$$\Rightarrow r = \frac{R}{\sqrt{2}}$$

and so the particle exits the hole moving horizontally, then

$$\frac{mv}{eB} = \frac{R}{\sqrt{2}}$$

$$\Rightarrow v = \frac{eBR}{m\sqrt{2}}$$

$$= \frac{(1.6\times10^{-19}\text{ C})(5.4\times10^{-4}\text{ T})(0.15\text{ m})}{(9.11\times10^{-31}\text{ kg})\sqrt{2}}$$

$$= 1.0\times10^{7}\text{ m/s}$$

29.33 In the x-z-plane, $\vec{\mathbf{B}}$ due to I_1 is in the $\pm\hat{\mathbf{j}}$ direction.

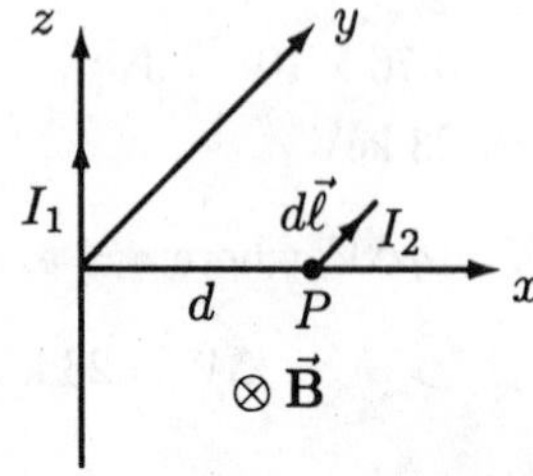

Since $\vec{\mathbf{B}}$ is parallel to the y-axis, it cannot exert any force along the y-axis. So, there is no y-component of force regardless of orientation.

29.37 The magnetic field produced by the long wire forms circles around the wire.

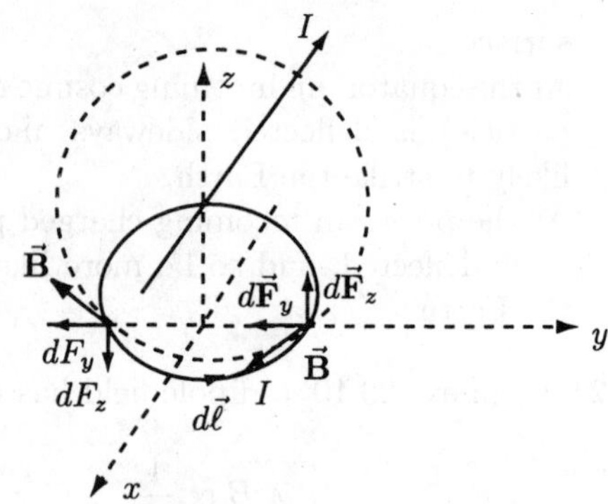

In the y-z-plane, at the position of the loop, $\vec{\mathbf{B}}$ has both y and z components. At two current elements on opposite sides of the loop, B_y is the same but B_z has opposite values. Similarly, $d\vec{\ell}$ has both x and y components, the y component being the same at both locations but the x-component reversing. The force on an element of loop,

$$d\vec{\mathbf{F}} = I\,d\vec{\ell}\times\vec{\mathbf{B}}$$

$$= I\left(d\ell_x\hat{\mathbf{i}} + d\ell_y\hat{\mathbf{j}}\right)\times\left(B_y\hat{\mathbf{j}} + B_z\hat{\mathbf{k}}\right)$$

$$= I\left(\underset{\text{c}}{d\ell_x}\underset{\text{s}}{B_y}\,\hat{\mathbf{k}} + \underset{\text{c}}{d\ell_x}\underset{\text{c}}{B_z}\left(-\hat{\mathbf{j}}\right) + \underset{\text{s}}{d\ell_y}\underset{\text{c}}{B_z}\,\hat{\mathbf{i}}\right)$$

[c = changes, s = same]
The only term which has the same sign on both sides of the loop is the middle term, and it contributes $d\vec{\mathbf{F}}$ in the $-y$-direction. [$d\ell_x$ is + when B_z is + and vice versa.] Thus the net force is in the $-y$-direction. The net torque is due to force components that change direction across the loop, it is in the + x-direction (see figure). Thus the correct answer is e).

29.41 $m = NIA$

$$= (6)(1.1\text{ A})(0.060\text{ m}\times0.12\text{ m})$$

$$= 4.75\times10^{-2}\text{ A}\cdot\text{m}^2$$

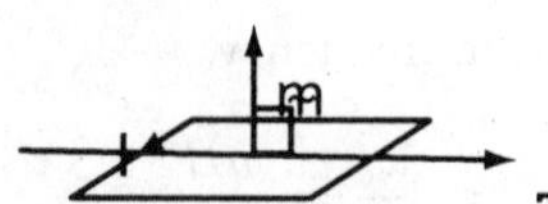

If the loop's plane is parallel to $\vec{\mathbf{B}}$, then its normal, and hence $\vec{\mathbf{m}}$, is perpendicular to $\vec{\mathbf{B}}$.

$$\vec{\tau} = \vec{\mathbf{m}}\times\vec{\mathbf{B}}$$

$B = 0.25$ T, $\theta = 90°$

$$\tau = mB\sin\theta$$

$$= \left(4.75\times10^{-2}\text{ A}\cdot\text{m}^2\right)(0.25\text{ T})$$

$$= 1.2\times10^{-2}\text{ N}\cdot\text{m}$$

in direction $\vec{\mathbf{m}}\times\vec{\mathbf{B}}$.

29.45 On side AC,

$$\vec{\ell}_{AC} = L\left[\cos(60^\circ)\,\hat{\mathbf{i}} + \sin(60^\circ)\,\hat{\mathbf{j}}\right]$$

On CB

$$\vec{\ell}_{CB} = L\left[\cos(60^\circ)\,\hat{\mathbf{i}} - \sin(60^\circ)\,\hat{\mathbf{j}}\right]$$

On BA:

$$\vec{\ell}_{BA} = -L\hat{\mathbf{i}}$$

and

$$\vec{\mathbf{B}} = B\hat{\mathbf{j}}$$

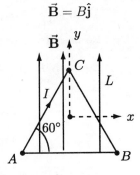

Thus the force on side AC is:

$$\begin{aligned} \vec{\mathbf{F}}_1 &= I\vec{\ell}_{AC} \times \vec{\mathbf{B}} \\ &= ILB\cos(60^\circ)\left(\hat{\mathbf{i}} \times \hat{\mathbf{j}}\right) \end{aligned}$$

since $\hat{\mathbf{j}} \times \hat{\mathbf{j}} = 0$

$$\vec{\mathbf{F}}_1 = \frac{1}{2}ILB\hat{\mathbf{k}}$$

On side CB:

$$\begin{aligned} \vec{\mathbf{F}}_2 &= I\vec{\ell}_{CB} \times \vec{\mathbf{B}} \\ &= ILB\cos(60^\circ)\left(\hat{\mathbf{i}} \times \hat{\mathbf{j}}\right) \\ &= \frac{1}{2}ILB\hat{\mathbf{k}} \end{aligned}$$

On side BA:

$$\begin{aligned} \vec{\mathbf{F}}_3 &= I\vec{\ell}_{BA} \times \vec{\mathbf{B}} \\ &= ILB\cos(60^\circ)\left(-\hat{\mathbf{i}} \times \hat{\mathbf{j}}\right) \\ &= -ILB\hat{\mathbf{k}} \end{aligned}$$

$$\begin{aligned} \vec{\mathbf{F}}_{\text{net}} &= \vec{\mathbf{F}}_1 + \vec{\mathbf{F}}_2 + \vec{\mathbf{F}}_3 \\ &= \frac{1}{2}ILB\hat{\mathbf{k}} + \frac{1}{2}ILB\hat{\mathbf{k}} - ILB\hat{\mathbf{k}} \end{aligned}$$

Thus $\vec{\mathbf{F}}_{\text{net}} = 0$.

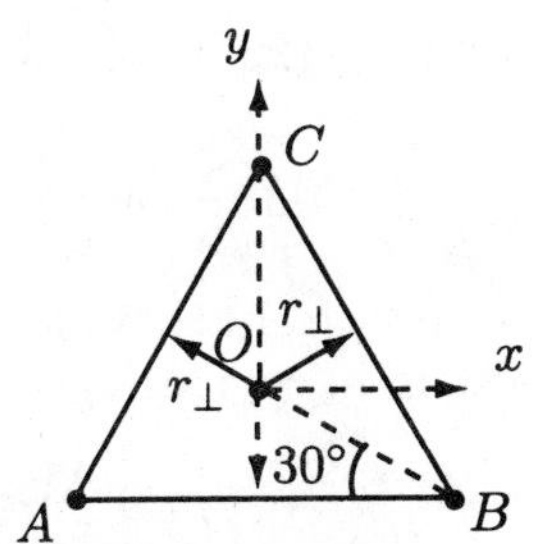

The forces acting on all three sides of the triangle are in the $\pm z$-direction, and act at equal perpendicular distances $r_\perp$ from O, the center of the triangle, where

$$r_\perp = \frac{L}{2}\tan 30^\circ = \frac{L}{2}\left(\frac{1}{\sqrt{3}}\right) = \frac{\sqrt{3}L}{6}$$

For side AC:

$$\begin{aligned} \vec{\tau}_1 &= r_\perp\left(-\cos 30^\circ\hat{\mathbf{i}} + \sin 30^\circ\hat{\mathbf{j}}\right) \times \vec{\mathbf{F}}_1 \\ &= \frac{\sqrt{3}L}{6}\left(-\hat{\mathbf{i}}\frac{\sqrt{3}}{2} + \frac{\hat{\mathbf{j}}}{2}\right) \times \left(\frac{1}{2}ILB\right)\hat{\mathbf{k}} \\ &= \frac{\sqrt{3}L^2}{12}IB\left[\frac{\sqrt{3}}{2}\hat{\mathbf{j}} + \frac{\hat{\mathbf{i}}}{2}\right] \end{aligned}$$

Similarly, for side BC:

$$\begin{aligned} \vec{\tau}_2 &= r_\perp\left(\cos 30^\circ\hat{\mathbf{i}} + \sin 30^\circ\hat{\mathbf{j}}\right) \times \vec{\mathbf{F}}_2 \\ &= \frac{\sqrt{3}L}{6}\left(\frac{\sqrt{3}}{2}\hat{\mathbf{i}} + \frac{\hat{\mathbf{j}}}{2}\right) \times \left(\frac{1}{2}ILB\right)\hat{\mathbf{k}} \\ &= \frac{\sqrt{3}L^2}{12}IB\left[-\frac{\sqrt{3}}{2}\hat{\mathbf{j}} + \frac{\hat{\mathbf{i}}}{2}\right] \end{aligned}$$

and for side BA:

$$\begin{aligned} \vec{\tau}_3 &= r_\perp\left(-\hat{\mathbf{j}}\right) \times \vec{\mathbf{F}}_3 \\ &= \frac{\sqrt{3}L}{6}\left(-\hat{\mathbf{j}}\right) \times (-ILB)\,\hat{\mathbf{k}} \\ &= \frac{\sqrt{3}L^2IB}{6}\hat{\mathbf{i}} \end{aligned}$$

Thus the total torque is

$$\begin{aligned} \vec{\tau} &= \vec{\tau}_1 + \vec{\tau}_2 + \vec{\tau}_3 \\ &= \frac{\sqrt{3}L^2IB}{6}\left[\frac{\sqrt{3}}{4}\hat{\mathbf{j}} + \frac{\hat{\mathbf{i}}}{4} - \frac{\sqrt{3}}{4}\hat{\mathbf{j}} + \frac{\hat{\mathbf{i}}}{4} + \hat{\mathbf{i}}\right] \\ &= \frac{\sqrt{3}L^2IB}{6}\left[\frac{1}{2} + 1\right]\hat{\mathbf{i}} = \frac{\sqrt{3}L^2IB}{4}\hat{\mathbf{i}} \end{aligned}$$

$$\begin{aligned} \vec{\mathbf{m}} &= IA\hat{\mathbf{n}} \\ &= I\left(\frac{1}{2}\right)(L)\left(\frac{\sqrt{3}}{2}L\right)\left(-\hat{\mathbf{k}}\right) \\ &= -\frac{\sqrt{3}}{4}IL^2\hat{\mathbf{k}} \end{aligned}$$

$$\begin{aligned} \vec{\tau} &= \vec{\mathbf{m}} \times \vec{\mathbf{B}} \\ &= \left(-\frac{\sqrt{3}}{4}IL^2\hat{\mathbf{k}}\right) \times \left(B\hat{\mathbf{j}}\right) \\ &= \frac{\sqrt{3}}{4}IL^2B\hat{\mathbf{i}} \end{aligned}$$

29.49 The magnetic force $-e\vec{\mathbf{v}}_D \times \vec{\mathbf{B}}$ drives negative charge to one side of the strip, producing a Hall field $\vec{\mathbf{E}}_{\mathrm{H}}$ along the side of the strip of length w.

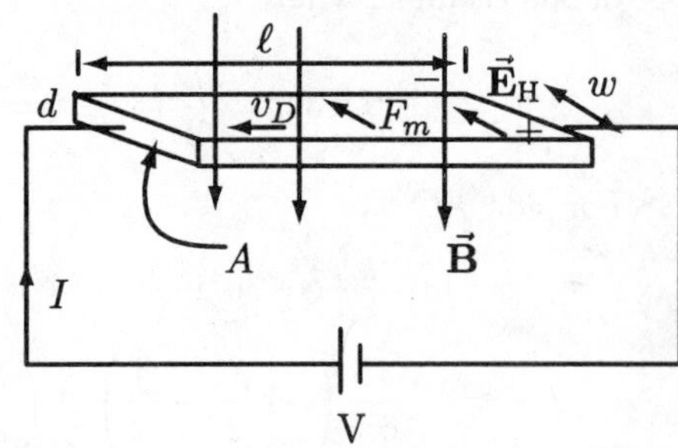

Thus

$$|\Delta V_{\mathrm{H}}| = E_{\mathrm{H}} w = v_D B w$$

Now

$$v_D = \frac{I}{neA} = \frac{I}{newd}$$

and

$$V = IR$$

with

$$\begin{aligned} R &= \rho\frac{\ell}{A} = \rho\frac{\ell}{dw} \\ I &= \frac{V}{\frac{\rho\ell}{dw}} = \frac{V\,dw}{\rho\ell} \\ |\Delta V_{\mathrm{H}}| &= \left(\frac{I}{newd}\right) Bw \\ &= \frac{B}{ned}\frac{V\,dw}{\rho\ell} \\ &= \frac{BVw}{ne\rho\ell} \end{aligned}$$

ΔV_{H} is proportional to V, w, and B. ΔV_{H} is inversely proportional to ρ and ℓ. ΔV_{H} has no dependence on d.

29.53 The magnetic force on a conduction electron is

$$\begin{aligned} \vec{\mathbf{F}}_m &= -e\vec{\mathbf{v}}_D \times \vec{\mathbf{B}} \\ &= -ev_D\hat{\mathbf{k}} \times \left(B_{\mathrm{horiz}}\hat{\mathbf{i}} - B_{\mathrm{vert}}\hat{\mathbf{j}}\right) \\ &= +ev_D\left(-B_{\mathrm{horiz}}\hat{\mathbf{j}} - B_{\mathrm{vert}}\hat{\mathbf{i}}\right) \end{aligned}$$

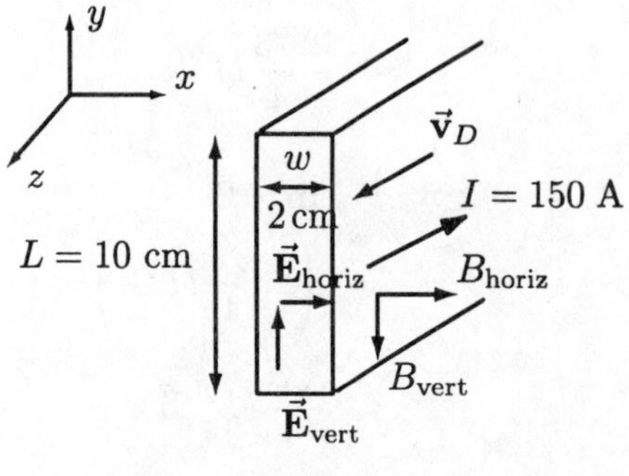

$$\begin{aligned} I &= 150\ \mathrm{A} \\ B_{\mathrm{horiz}} &= 3.0 \times 10^{-5}\ \mathrm{T} \\ w &= 2.0\ \mathrm{cm} = 0.020\ \mathrm{m} \\ B_{\mathrm{vert}} &= +2.7 \times 10^{-5}\ \mathrm{T} \\ L &= 10.0\ \mathrm{cm} = 0.100\ \mathrm{m} \\ n &= 8 \times 10^{28}\ \mathrm{m}^{-3} \\ e &= 1.6 \times 10^{-19}\ \mathrm{C} \end{aligned}$$

Since

$$-e\vec{\mathbf{E}}_{\mathrm{H}} = \vec{\mathbf{F}}_e = -\vec{\mathbf{F}}_m$$

then $\vec{\mathbf{E}}_{\mathrm{H}}$ also has two components

$$\vec{\mathbf{E}}_{\mathrm{H}} = v_D B_{\mathrm{vert}}\hat{\mathbf{i}} + v_D B_{\mathrm{horiz}}\hat{\mathbf{j}}$$

$$\begin{aligned} |\Delta V_{\mathrm{H}}|_{\mathrm{vert}} &= E_{\mathrm{vert}} L = v_D B_{\mathrm{horiz}} L \\ &= \frac{I}{neLw} B_{\mathrm{horiz}} L = \frac{IB_{\mathrm{horiz}}}{new} \\ &= \frac{(150\ \mathrm{A})\left(3.0\times10^{-5}\ \mathrm{T}\right)}{(8\times10^{28}\ \mathrm{m}^{-3})(1.6\times10^{-19}\ \mathrm{C})(0.02\ \mathrm{m})} \\ &= 17.6 \times 10^{-12}\ \mathrm{V} \end{aligned}$$

$$\begin{aligned} |\Delta V_{\mathrm{H}}|_{\mathrm{horiz}} &= E_{\mathrm{horiz}} w = \frac{I}{neLw} B_{\mathrm{vert}} w \\ &= \frac{I}{neL} B_{\mathrm{vert}} \\ &= \frac{(150\ \mathrm{A})\left(2.7\times10^{-5}\ \mathrm{T}\right)}{(8\times10^{28}\ \mathrm{m}^{-3})(1.6\times10^{-19}\ \mathrm{C})(0.10\ \mathrm{m})} \\ &= 3.2\ \mathrm{pV} \end{aligned}$$

Measured corner to corner, the potential difference would be

$$18\ \mathrm{pV} + 3\ \mathrm{pV} = 21\ \mathrm{pV}$$

29.55 a) For this part, let's ignore any effect on the main circuit due to allowing current flow in the shunts. Then the drift speed of electrons along the slab is

$$v_d = \frac{\left|\vec{\mathbf{j}}\right|}{ne} = \frac{I}{neLw}$$

The corresponding magnetic force on the electrons is balanced by the Hall electric field:

$$E_{\mathrm{H}} = v_d B = \frac{IB}{neLw}$$

The resulting Hall potential is

$$\Delta V_{\mathrm{H}} = E_{\mathrm{H}} \cdot L = \frac{IB}{new}$$

With respect to 1 kΩ, we may neglect resistance of the strip in computing I'. Then

$$I' = \frac{\Delta V_{\mathrm{H}}}{R}$$

and the power extracted is

$$
\begin{aligned}
P_{\mathrm{H}} &= N(I')^2 R \\
&= \frac{N}{R}\left(\frac{IB}{new}\right)^2
\end{aligned}
$$

$$
\begin{aligned}
&= \frac{100}{1.0\ \mathrm{k}\Omega}\left[\frac{(5.0\ \mathrm{A})(1.0\ \mathrm{T})}{\left(\frac{8.5\times10^{28}}{\mathrm{m}^3}\right)(1.6\times10^{-19}\ \mathrm{C})(1.0\times10^{-3}\ \mathrm{m})}\right]^2 \\
&= 1.4\times10^{-14}\ \mathrm{W}
\end{aligned}
$$

Hardly a practical power source!

b) Now, if a Hall current flows across the strip, it will give rise to a Hall effect in the direction of the main current that opposes the electric field due to the battery. The main current flow is reduced slightly. Let I be the current that actually flows in the main circuit and I' the current in each shunt wire. Let s be the length of the strip in the direction of the main current flow. Then the drift velocity component of electrons across the strip is

$$
v_{d\perp} = \frac{\left|\hat{\mathbf{j}}_{\mathrm{Hall}}\right|}{ne} = \frac{NI'}{news}
$$

The corresponding Hall potential is

$$
\mathcal{E}_{\mathrm{H},\perp} = v_{d\perp} B s = \frac{NI'B}{new}
$$

So, the potential difference driving current in the main circuit is $\mathcal{E} - \mathcal{E}_{\mathrm{H},\perp}$. The current in the main circuit is

$$
I = \frac{\mathcal{E} - \mathcal{E}_{\mathrm{H},\perp}}{r}
$$

where r is the resistance of the slab. The power output of the battery is

$$
\begin{aligned}
P_{\mathrm{batt}} &= \mathcal{E} I = \frac{\mathcal{E}}{r}(\mathcal{E} - \mathcal{E}_{\mathrm{H},\perp}) \\
&= \frac{(\mathcal{E} - \mathcal{E}_{\mathrm{H},\perp})^2}{r} + \frac{\mathcal{E}_{\mathrm{H},\perp}}{r}(\mathcal{E} - \mathcal{E}_{\mathrm{H},\perp}) \\
&= I^2 r + \mathcal{E}_{\mathrm{H},\perp} I \\
&= I^2 r + \frac{NI'B}{new} I
\end{aligned}
$$

Using the relation derived in part a) between I and I'.

$$
\begin{aligned}
P_{\mathrm{batt}} &= I^2 r + \frac{NI'B}{new}\frac{new}{B} R I' \\
&= I^2 r + N(I')^2 R
\end{aligned}
$$

= Power dissipated in slab + Power dissipated in shunts. Q.E.D.

29.59

$$
\begin{aligned}
\vec{\tau} &= \vec{\mathbf{m}} \times \vec{\mathbf{B}} \\
\tau &= mB\sin\theta
\end{aligned}
$$

From example 12.17 and equation 12.12, the precession frequency is

$$
\Omega = \frac{\tau}{L} = \frac{mB\sin\theta}{L} = \frac{B\sin\theta}{\left(\frac{L}{m}\right)}
$$

29.63 The wire's magnetic field circles the wire. The solenoid's field is parallel to the wire. In general, $\vec{\mathbf{B}} \parallel d\vec{\ell}$ so $\vec{\mathbf{F}}_{\mathrm{net}} = 0$. However, we are told the solenoid is "loosely wound".

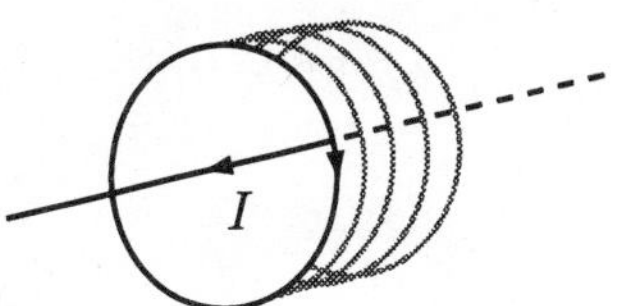

If we take into account the effect of the wire's field on the longitudinal component of the solenoid's current, the net effect is a set of forces that attempt to expand the solenoid, i.e., the forces are radially outward.

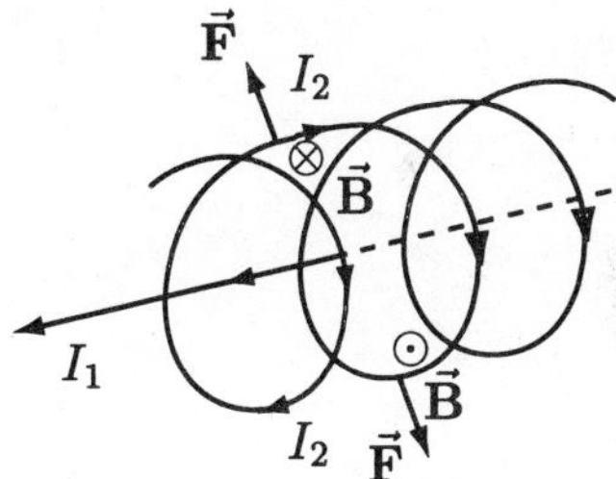

29.67 $F = I\ell B$

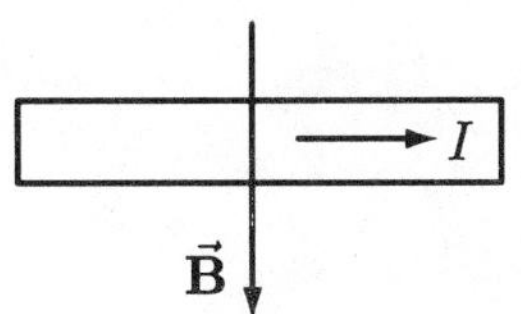

$$
\begin{aligned}
\rightarrow I &= \frac{F}{\ell B} \\
&= \frac{(1\times10^5\ \mathrm{N})}{(3\ \mathrm{m})(10^{-5}\ \mathrm{T})} \\
&= 0.33\times10^{10}\ \mathrm{A} \\
&= 3\times10^9\ \mathrm{A}
\end{aligned}
$$

$$
\begin{aligned}
P &= I^2 R = (3\times10^9\ \mathrm{A})^2(1\ \Omega) \\
&= 10^{19}\ \mathrm{W}
\end{aligned}
$$

This is ridiculously high, and unrealistic.

29.71 Let B_E = horizontal component of Earth's magnetic field. The magnetic field produced by the coil

$$B_{\rm coil} = \frac{\mu_0 NI}{2r}$$

where r = radius of coil. It is perpendicular to to the plane of the coil at its center.

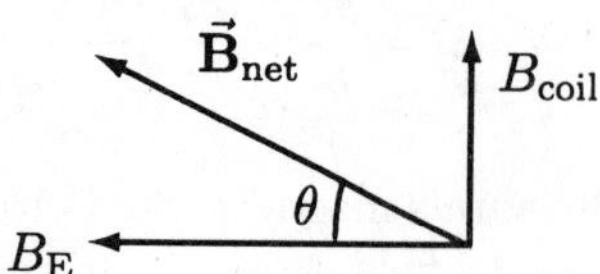

The needle aligns with the net field. Thus

$$\tan\theta = \frac{B_{\rm coil}}{B_E} = \frac{\mu_0 NI}{2rB_E}$$

Chapter 30

Dynamic Fields

30.1 The average emf developed in each loop of the coil is

$$\mathcal{E} = \left|\frac{d\Phi}{dt}\right| = \frac{(2.40 - 1.60)\text{ Wb}}{0.15\text{ s}} = 5.3\text{ V}$$

The average emf developed in the coil of 150 loops is

$$150(5.3\text{ V}) = 0.80\text{ kV}$$

30.3 The induced emf is

$$\begin{aligned}\mathcal{E} &= -\frac{d\Phi}{dt} \\ &= -\frac{d}{dt}\left[(1\text{ mWb})\left(At^4 + \cos\omega t\right)\right] \\ &= -(1\text{ mWb})\left[4At^3 - \omega\sin\omega t\right] \\ &= (4\text{ mV})\left\{(100)\sin\left[(400\text{ rad/s})\,t\right] \right. \\ &\qquad \left. - \left(10^{-4}/\text{s}^3\right)t^3\right\}\end{aligned}$$

30.5 The induced emf is

$$\begin{aligned}|\mathcal{E}| &= \left|\frac{d\Phi_B}{dt}\right| = \frac{d(BA)}{dt} = B\frac{dA}{dt} \\ &= B\ell\frac{dx}{dt} = B\ell v\end{aligned}$$

For the smaller loop,

$$\mathcal{E} = (0.5\text{ T})(0.1\text{ m})(10\text{ m/s}) = 0.5\text{ V}$$

For the larger loop,

$$\mathcal{E} = (0.5\text{ T})(0.2\text{ m})(10\text{ m/s}) = 1\text{ V}$$

That is, $\mathcal{E}_2 = 2\mathcal{E}_1$. The heat losses are

$$\frac{\mathcal{E}^2}{R} = \frac{(0.5\text{ V})^2}{10\ \Omega} = 0.025\text{ W}$$

in the smaller loop and

$$\frac{(0.5\text{ V})^2}{10\ \Omega} = 0.1\text{ W}$$

in the larger. That is, $P_2 = 4P_1$. The forces on the loops are $BI\ell$, or

$$(0.5\text{ T})\left(\frac{0.5\text{ V}}{10\ \Omega}\right)(0.2\text{ m}) = 2.5\times10^{-3}\text{ N}$$

in the smaller loop, and

$$(0.5\text{ T})\left(\frac{1\text{ V}}{10\ \Omega}\right)(0.2\text{ m}) = 1.0\times10^{-2}\text{ N}$$

in the larger. That is,

$$F_2 = 4F_1$$

It could easily be predicted that the larger force would be needed on the larger loop, since a greater emf would clearly be developed across the larger loop.

30.7 The maximum torque developed is

$$\begin{aligned}\tau &= NIAB \\ &= (50)(2.5\text{ A})(1.0\times10^{-2}\text{ m})^2(0.75\text{ T}) \\ &= 9.4\times10^{-3}\text{ N}\cdot\text{m}\end{aligned}$$

30.9 The induced electric field

$$|E| = \frac{\left|\frac{dB}{dt}\right| r}{2}$$

where B is the magnetic field inside the solenoid and r the distance from the axis. Since $B = \mu_0 nI$,

$$\frac{dB}{dt} = \mu_0 n\frac{dI}{dt}$$

Thus

$$
\begin{aligned}
|E| &= \frac{\mu_0 n \left|\frac{dI}{dt}\right| r}{2} \\
&= \frac{\left(4\pi\times10^{-7}\ \frac{\text{Wb}}{\text{A}\cdot\text{m}}\right)\left(25\times10^{2}\ \frac{\text{turns}}{\text{m}}\right)\left(15\ \frac{\text{A}}{\text{s}}\right)\left(1.0\times10^{-2}\ \text{m}\right)}{2} \\
&= 0.24\ \text{mV/m}
\end{aligned}
$$

30.11 When a conductor moves between the poles of a magnet, eddy currents are generated which give a retarding force. Thus a force is needed to keep the plate moving. [This form of force is derived in *Digging Deeper*: Force due to Eddy currents.]

$$F = \frac{vB^2V}{\rho} = \frac{vB^2\pi r^2 w}{\rho}$$

[Dimensions of field = 3 cm radius circle]

$$
\begin{aligned}
w &= 0.002\ \text{m} \\
B &= 0.3\ \text{T} \\
v &= 2\ \text{m/s} \\
\rho &= 1.7\times10^{-8}\ \Omega\cdot\text{m (copper @ 20° C)}
\end{aligned}
$$

So

$$
\begin{aligned}
F &= \frac{(2\ \text{m/s})(0.3\ \text{T})^2\pi\left(3\times10^{-2}\ \text{m}\right)^2\left(2\times10^{-3}\ \text{m}\right)}{1.7\times10^{-8}\ \Omega\cdot\text{m}} \\
&= 60\ \text{N}
\end{aligned}
$$

30.13 Consider an Amperian loop of radius r = 1.0 cm and use Maxwell's version of Ampère's Law.

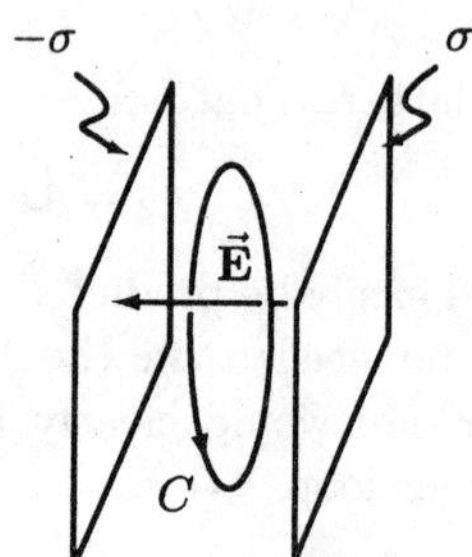

$\vec{\mathbf{B}}$ circles around the displacement current, so

$$
\begin{aligned}
\oint \vec{\mathbf{B}}\cdot d\vec{\ell} &= \mu_0\epsilon_0 \frac{d\Phi_E}{dt} \\
&= \mu_0\epsilon_0 \frac{d}{dt}\left(E\pi r^2\right)
\end{aligned}
$$

or

$$\oint \vec{\mathbf{B}}\cdot d\vec{\ell} = \mu_0\epsilon_0\pi r^2 \frac{dE}{dt} \text{ and } E = \frac{\sigma}{\epsilon_0}$$

(see § 27.1)

$$\oint \vec{\mathbf{B}}\cdot d\vec{\ell} = \frac{\mu_0\epsilon_0\pi r^2}{\epsilon_0}\frac{d\sigma}{dt}$$

so that

$$B\left(2\pi r\right) = \mu_0\pi r^2 \frac{d\sigma}{dt}$$

solving for B, we get

$$
\begin{aligned}
B &= \frac{\mu_0 r}{2}\frac{d\sigma}{dt} \\
&= \frac{\left(4\pi\times10^{-7}\ \frac{\text{Wb}}{\text{A}\cdot\text{m}}\right)\left(1\times10^{-2}\ \text{m}\right)}{2}\left(35\times10^{-3}\ \text{C/m}^2\cdot\text{s}\right) \\
&= 0.22\ \text{nT}
\end{aligned}
$$

the field circles as shown in the diagram.

30.17 At first, the field of the magnet through the center of the loop is directed along the positive x-axis.

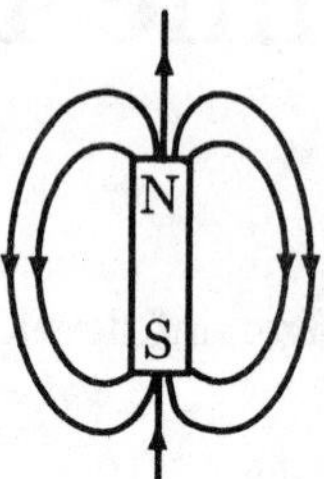

At the end, the field is directed along the $-x$-direction.

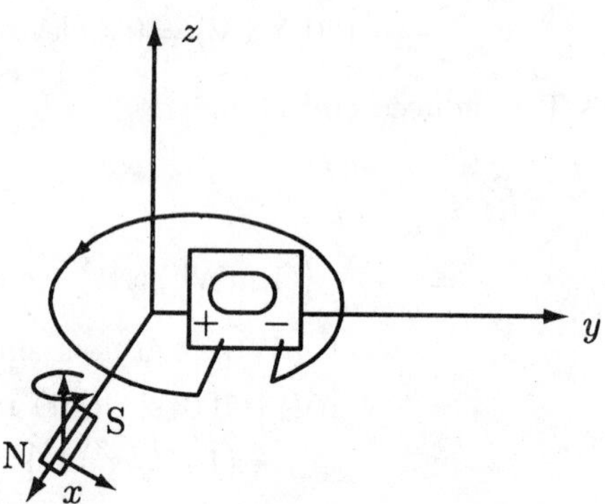

In between, the field is nearly parallel to the loop plane. Taking flux through the loop as + when $\vec{\mathbf{B}}$ points toward the loop, the flux loop changes from − to +, i.e., it increases.

As the flux changes, the induced emf must cause a current to flow that tries to keep the flux constant. The current that does this is in the counterclockwise direction, into the + terminal of the meter. Since the magnetic flux in the loop is increasing throughout the 180° rotation, current continues to flow in the meter's positive sense. At the end of the motion the magnet stops ⇒ current goes to 0 ⇒ **a)** and **e)** are incorrect because they describe emf of both signs.

b) is incorrect because the flux does change

in the loop and we presume the meter has adequate sensitivity to detect the emf.

d) gives an induced field of the wrong direction and is incorrect $\therefore$ **c)** is correct.

30.21 Since the area is constant and perpendicular to the magnetic field, we have

$$\begin{aligned} |\mathcal{E}| &= \left|\frac{d\Phi_B}{dt}\right| = \frac{d}{dt}|BA| = A\frac{d|B|}{dt} \\ &= \pi\left(2.50\times10^{-2}\ \text{m}\right)^2(0.305\ \text{T/s}) \\ &= 0.599\ \text{mV} \end{aligned}$$

30.25 The loop on right, with resistance $= 0.030\ \Omega$ and radius

$$a = 1.0\ \text{cm} = 10^{-2}\ \text{m},$$

is fixed. We move the other loop and the magnetic flux through the stationary loop changes, and induces an emf about the stationary loop. The moving loop carries current

$$I = 0.15\ \text{A}$$

and has radius

$$r = 10.0\ \text{cm} = 0.10\ \text{m}$$

For a current carrying loop, the magnetic field on the axis has magnitude

$$B = \frac{\mu_0 I r^2}{2\left(x^2+r^2\right)^{3/2}}$$

direction $\perp$ to plane of the loop. If $r^2 \ll x^2$ then

$$B \approx \frac{\mu_0 I r^2}{2x^3}$$

The flux through the small loop is

$$\begin{aligned} \Phi &= \pi a^2 B \\ &= \frac{\pi\mu_0 I r^2 a^2}{2x^3} \end{aligned}$$

and

$$\frac{d\Phi}{dt} = -\frac{3\pi\mu_0 I r^2 a^2}{2x^4}\frac{dx}{dt}$$

The current is:

$$i = \frac{\mathcal{E}}{R}$$

The magnitude of

$$|i| = \frac{3\pi\mu_0 I r^2 a^2}{2x^4 R}\left|\frac{dx}{dt}\right|$$

using

$$\begin{aligned} x &= 0.50\ \text{m} \\ a &= 0.01\ \text{m} \\ r &= 0.10\ \text{m} \\ I &= 0.15\ \text{A} \\ R &= 0.03\ \Omega \\ \frac{dx}{dt} &= 2.2\ \text{m/s} \\ \mu_0 &= 0.4\pi\times10^{-6}\ \Omega\cdot\text{s/m} \end{aligned}$$

thus

$$\begin{aligned} |i| &= \frac{3\pi\left(0.4\pi\times10^{-6}\frac{\Omega\cdot\text{s}}{\text{m}}\right)(0.10\ \text{m})^2(0.01\ \text{m})^2\left(2.2\frac{\text{m}}{\text{s}}\right)(0.15\ \text{A})}{2(0.50\ \text{m})^4(0.03\ \Omega)} \\ &= 1.0\times10^{-9}\ \text{A} \end{aligned}$$

The direction will be such as to try to reduce the flux through the loop.

By the right hand rule, it will have to cause a clockwise current flow in the same loop (as viewed on figure 30.39.)

30.27 The resistance of the loop is

$$\begin{aligned} R &= \frac{\rho\ell}{A} \\ &= \frac{(1.7\times10^{-8}\ \Omega\cdot\text{m})(9.0\times10^{-2}\ \text{m})}{\pi(0.5\times10^{-3}\ \text{m})^2} \\ &= 1.9\ \text{m}\Omega \end{aligned}$$

The area of the loop is

$$\begin{aligned} A &= (L\cos60^\circ)(L\sin60^\circ) \\ &= \frac{L^2\sqrt{3}}{4} \end{aligned}$$

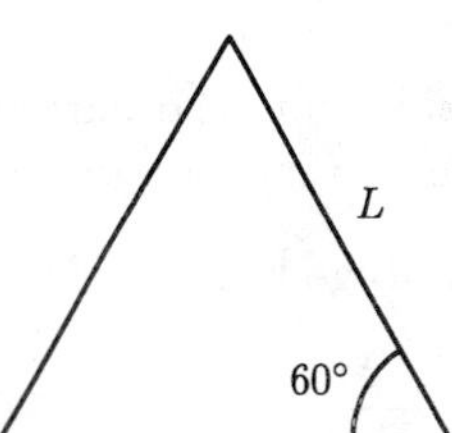

The positive direction of current is clockwise as viewed from positive z. The current in the loop is

$$\begin{aligned} I &= \frac{\left|\frac{d\Phi}{dt}\right|}{R} \\ &= \frac{A_{\text{loop}}\frac{dB}{dt}}{R} \\ &= \frac{(1.5\times10^{-2}\ \text{m})^2\left(\sqrt{3}\right)\frac{d}{dt}\left\{B_0\left[\left(\frac{t}{t_0}\right)^2-\left(\frac{t}{t_0}\right)^4\right]\right\}}{R} \\ &= \frac{(1.5\times10^{-2}\ \text{m})^2\left(\sqrt{3}\right)(0.33\ \text{T})(2)\left[\frac{t}{t_0^2}-2\frac{t^3}{t_0^4}\right]}{1.95\ \text{m}\Omega} \\ &= (0.13\ \text{A})\left\{\left[\frac{t}{(1.0\ \text{s})}\right]-2\left(\frac{t}{(1.0\ \text{s})}\right)^3\right\} \end{aligned}$$

[Note: we used the intermediate result 1.95 mΩ for R.]

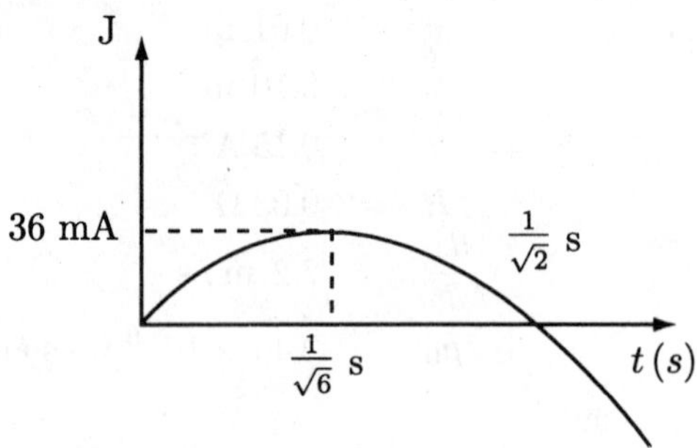

The maximum current occurs when $\frac{dI}{dt} = 0$, or $1 - \left(\frac{6}{s^2}\right) t^2 = 0$ or

$$t = \frac{1}{\sqrt{6}} \text{ s} = 0.41 \text{ s}$$

The maximum current value is

$$(0.132 \text{ A}) \left[0.41 - 2\,(0.41)^3\right] = 36 \text{ mA}$$

The power dissipated at this time is

$$I^2R = (36 \text{ mA})^2(1.95 \text{ m}\Omega) = 2.5\ \mu\text{W}$$

The power is zero when $\frac{d}{dt}\vec{\mathbf{B}} = 0$.

$$\begin{aligned} t - \left(\frac{2}{s^2}\right) t^3 &= 0 \\ 1 - \left(\frac{2}{s^2}\right) t^2 &= 0 \\ t &= \frac{1 \text{ s}}{\sqrt{2}} = 0.71 \text{ s} \end{aligned}$$

After this time the current reverses direction, and the power increases again.

30.31 The voltage is $\mathcal{E} = nBA\omega$, so

$$\begin{aligned} n &= \frac{\mathcal{E}}{BA\omega} \\ &= \frac{8.50\times10^3 \text{ V}}{(1.25 \text{ T})(15.0\times10^{-2} \text{ m})(35.0\times10^{-2} \text{ m})(2\pi)(60.0 \text{ Hz})} \\ &= 344 \text{ turns} \end{aligned}$$

30.35 The situation is shown below

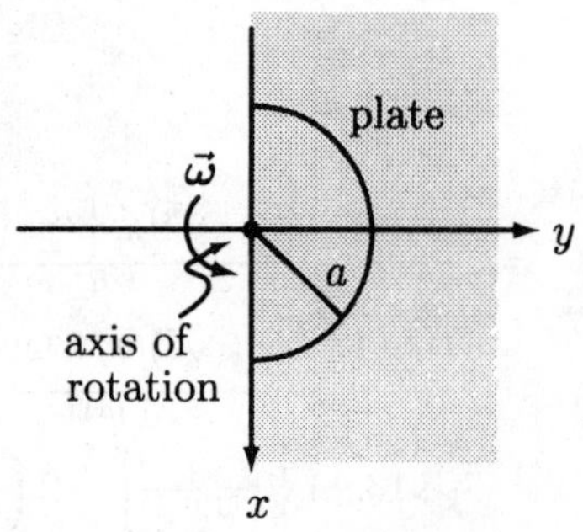

The area of the wire loop that is in the magnetic field is

$$\begin{aligned} A_1 &= \left(1 - \frac{\omega t}{\pi}\right)\frac{\pi a^2}{2} \quad 0 < t < \frac{\pi}{\omega} \\ A_2 &= \left(\frac{\omega t}{\pi} - 1\right)\frac{\pi a^2}{2} \quad \frac{\pi}{\omega} < t < \frac{2\pi}{\omega} \end{aligned}$$

With normal parallel to $\hat{\mathbf{k}}$,

$$\begin{aligned} \Phi_1 &= A_1 B \\ \Phi_2 &= A_2 B \end{aligned}$$

Then the positive sense for $\mathcal{E}$ is counterclockwise in the figure.

$$\begin{aligned} \mathcal{E}_1 &= -\frac{d\Phi_1}{dt} = \frac{\omega a^2}{2}B \quad 0 < t < \frac{\pi}{\omega} \\ \mathcal{E}_2 &= -\frac{d\Phi_2}{dt} = -\frac{\omega a^2}{2}B \quad \frac{\pi}{\omega} < t < \frac{2\pi}{\omega} \end{aligned}$$

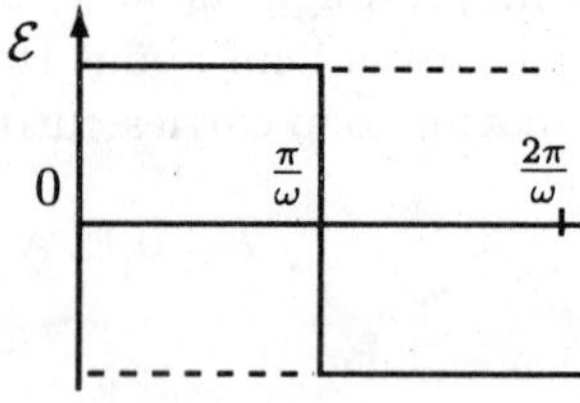

30.39 For all these cases, we know that the voltmeter reading ΔV is

$$\Delta V = vB_{\text{vert}}\ell$$

where v is the velocity of the conductor relative to the voltmeter.

Since the metal rod is on the boat, the engineer on the boat reads no voltage across it. Motion through the field induces equal potential difference in the rod and the meter leads. The engineer on the buoy, however, reads

$$(12 \text{ m/s})(10^{-3} \text{ T})(1 \text{ m}) = 0.1 \text{ mV}$$

since there is no induced emf in the meter wires to counteract the emf induced in the rod.

If the engineer on the boat tosses his leads in the water so that the line between them is perpendicular to v, he similarly reads 0.1 mV. Now, it is induced emf in the meter leads that causes the meter reading. The engineer on the buoy reads 0 V if he tosses his leads into the water, since his velocity relative to the water is zero, and there is no induced emf in either rod or meter leads.

30.41 With the velocity and magnetic field vectors as shown, magnetic force on the electrons is toward the stationary end of the rod.

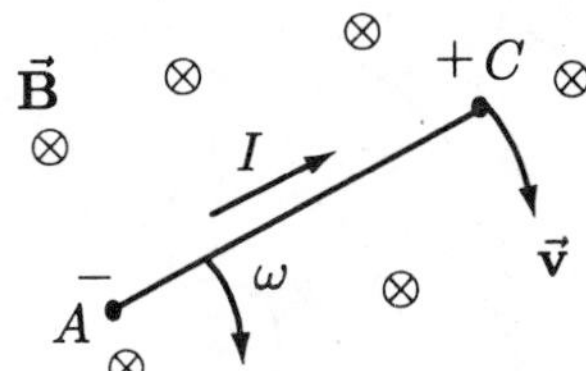

The potential difference $V_A - V_C$ is therefore negative. Since the magnetic forces are proportional to the speed of the rod, they are proportional to ω, and so is the potential difference. So **b)** is the correct answer.

30.45 The potential difference between the two ends of the conducting cable is (cf. Example 30.8)

$$\begin{aligned}\Delta V &= vB\ell \\ &= (7\times10^3\ \mathrm{m/s})(2\times10^{-5}\ \mathrm{T})(1\times10^3\ \mathrm{m}) \\ &= 140\ \mathrm{V}\end{aligned}$$

Note that with the cable pointing radially outward, its velocity is perpendicular to the magnetic field lines. Since $\vec{\mathbf{B}}$ points north and $\vec{\mathbf{v}}$ east, $\vec{\mathbf{v}}\times\vec{\mathbf{B}}$ points out from the Earth. $\vec{\mathbf{F}}_e$ opposes $\vec{\mathbf{F}}_m$ in equilibrium (see figure), thus $\vec{\mathbf{E}}$ points inward. That is, the satellite end is at higher potential.

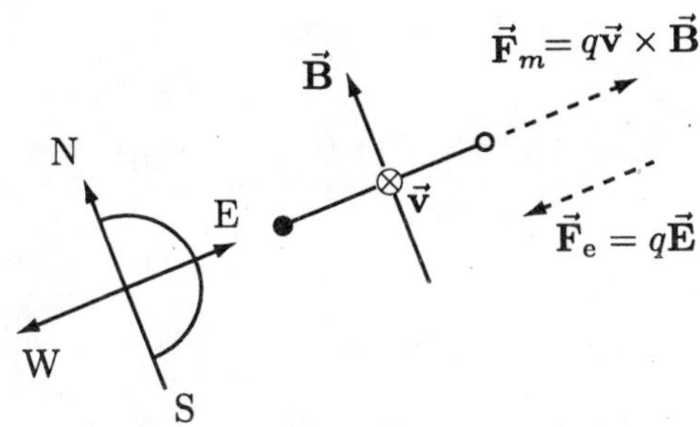

Even though B is rather small, v and ℓ are large enough that the potential difference is fairly big.

30.49 Since $\vec{\mathbf{E}}' = \vec{\mathbf{E}}+\vec{\mathbf{v}}\times\vec{\mathbf{B}}$, the electric field applied in the lab is

$$\begin{aligned}\vec{\mathbf{E}} &= \vec{\mathbf{E}}' - \left(\vec{\mathbf{v}}\times\vec{\mathbf{B}}\right) \\ &= (0.550\ \mathrm{kV/m})\hat{\mathbf{i}} - (3.00\times10^4\ \mathrm{m/s})(\hat{\mathbf{i}}+\hat{\mathbf{j}}) \\ &\qquad \times(5.50\times10^{-3}\ \mathrm{T})(\hat{\mathbf{j}}+\hat{\mathbf{k}}) \\ &= (0.550\ \mathrm{kV/m})\hat{\mathbf{i}} - (165\ \mathrm{V/m})(\hat{\mathbf{i}}-\hat{\mathbf{j}}+\hat{\mathbf{k}}) \\ &= (385\ \mathrm{V/m})\hat{\mathbf{i}} + (165\ \mathrm{V/m})\left(\hat{\mathbf{j}}-\hat{\mathbf{k}}\right)\end{aligned}$$

(*Remember*: $\hat{\mathbf{i}}\times\hat{\mathbf{j}}=\hat{\mathbf{k}}$, $\hat{\mathbf{i}}\times\hat{\mathbf{k}}=-\hat{\mathbf{j}}$, $\hat{\mathbf{j}}\times\hat{\mathbf{k}}=\hat{\mathbf{i}}$, and $\hat{\mathbf{j}}\times\hat{\mathbf{j}}=0$)

30.53 We may assume that the magnetic field between the poles of the electromagnet is uniform. Thus we have cylindrical symmetry, and may use Faraday's law to calculate the induced electric field.

To know if we are measuring $\vec{\mathbf{E}}$ inside or outside the poles, we must find the radius R of the poles:

$$A = \pi R^2$$

or

$$R = \sqrt{\frac{A}{\pi}} = \sqrt{\frac{0.30\ \mathrm{m}^2}{\pi}} = 0.31\ \mathrm{m}$$

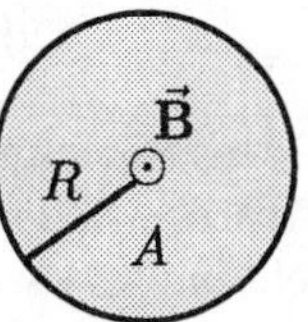

When $r = 0.20$ m, we are inside the poles, and

$$\begin{aligned}\left|\oint \vec{\mathbf{E}}\cdot d\vec{\ell}\right| &= 2\pi r\,|E| = \left|\frac{d\Phi_B}{dt}\right| \\ &= \pi r^2\left|\frac{dB}{dt}\right|\end{aligned}$$

or

$$\begin{aligned}|E| &= \frac{r\left|\frac{dB}{dt}\right|}{2} \\ &= \frac{(0.20\ \mathrm{m})\,(0.020\ \mathrm{T/s})}{2} \\ &= 2.0\ \mathrm{mV/m}\end{aligned}$$

When $r = 0.50$ m, we are outside the poles, and

$$\left|\oint \vec{\mathbf{E}}\cdot d\vec{\ell}\right| = 2\pi r\,|E| = A\left|\frac{dB}{dt}\right|$$

or

$$\begin{aligned}|E| &= \frac{A\left|\frac{dB}{dt}\right|}{2\pi r} = \frac{(0.30\ \mathrm{m}^2)\,(0.020\ \mathrm{T/s})}{(2\pi)\,(0.50\ \mathrm{m})} \\ &= 1.9\ \mathrm{mV/m}\end{aligned}$$

As a check, when $r = R$, we get the same field using both equations above.

30.57 For $y < 0$,

$$\vec{\mathbf{B}} = B_0\hat{\mathbf{i}}$$

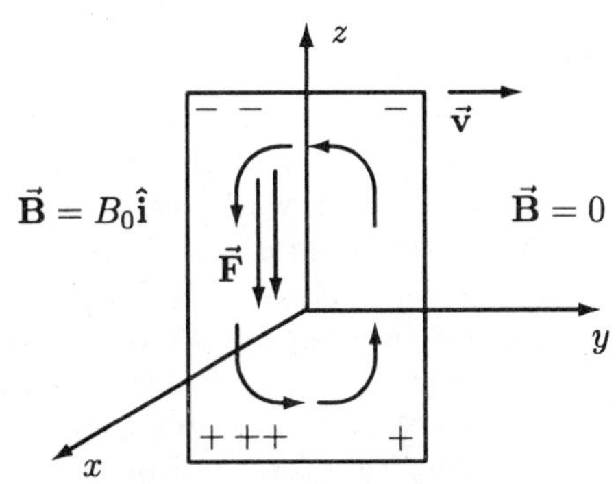

The force on charges is

$$\vec{\mathbf{F}} = q\vec{\mathbf{v}} \times \vec{\mathbf{B}}$$

For $y > 0$,

$$\vec{\mathbf{F}} = 0$$

since $\vec{\mathbf{B}} = 0$.

$$\begin{aligned} y < 0 \quad \vec{\mathbf{F}} &= qvB\hat{\mathbf{j}} \times \hat{\mathbf{i}} \\ &= -qvB\hat{\mathbf{k}} \end{aligned}$$

For $y < 0$ there is a current set up that creates a charge distribution on the plate. Thus there is an $\vec{\mathbf{E}}$ field in the $+\hat{\mathbf{k}}$ direction that drives current in the $\hat{\mathbf{k}}$ direction for $y > 0$ as shown.

As the plate is pulled from the field, then the current causes a force on the plate in the $-y$ direction that opposes its motion.

As the plate is pulled further out of the field, the force will decrease since less current in the plate is exposed to the field.

I is not parallel to $\vec{\mathbf{v}} \times \vec{\mathbf{B}}$ everywhere because charges on the plate create the necessary $\vec{\mathbf{E}}$ to drive current in closed loops through the plate.

30.61

$$\begin{aligned} t &= 1.0 \text{ cm} \\ r_0 &= 1.5 \text{ cm} \\ R &= 0.5 \text{ m} \\ \omega &= 2\pi \cdot 1500 \text{ min}^{-1} \\ &= 25 \times 2\pi \text{ s}^{-1} \\ \vec{\mathbf{B}} &= 0.25 \text{ T into page} \end{aligned}$$

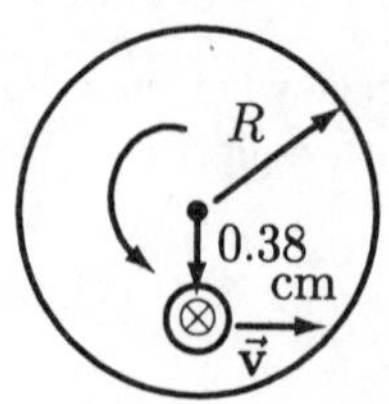

a) The motion of the conductor forces charge out of the field area into the surrounding metal.

Since $\vec{\mathbf{F}} = q\vec{\mathbf{v}} \times \vec{\mathbf{B}}$, the magnetic field forces a current flow towards the center of the disk. This current returns to the field area throughout the remainder of the disk. Since the disk does not have zero resistance, one expects currents to flow closer to the field to return instead of taking the long way around (i.e., current follows the path of least resistance.)

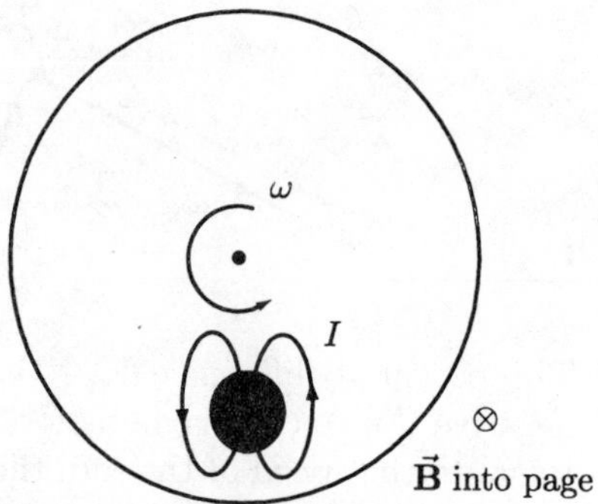

With this direction of current in the field, the wheel slows since $d\vec{\mathbf{F}} = I\, d\vec{\ell} \times \vec{\mathbf{B}}$ gives a force opposing the motion. The force on the conductor slab contained in the field is

$$F \approx \frac{VB_{\text{av}}^2 v}{\rho}$$

where V is the volume where $B \neq 0$. The torque is

$$\begin{aligned} Fr &= \tau \\ &\cong \frac{rVB^2v}{\rho} \end{aligned}$$

where r is the distance from center of the poles to the pivot. Now $v = \omega r$, so

$$\begin{aligned} V &= \pi r_0^2 t \\ &= \left(1.5 \times 10^{-2}\,\text{m}\right)^2 (0.01\,\text{m}) \\ &= 7.07 \times 10^{-6} \text{ m}^3 \end{aligned}$$

$$\begin{aligned} \tau &= \frac{(0.38\,\text{m})^2(1500\text{rpm})(2\pi)\left(\frac{1\,\text{min}}{60\ \text{s}}\right)(0.25\text{T})^2\left(7.07\times10^{-6}\ \text{m}\right.}{2.8\times10^{-8}\ \Omega\cdot\cdot\,\text{m}} \\ &= 400\ \text{N}\cdot\text{m} \end{aligned}$$

c) Newton's 2nd Law applied to the disk gives

$$\tau = I_i\alpha = I_i\frac{d\omega}{dt}$$

where $I_i \equiv$ rotational inertia of the disk, or

$$I_i\frac{d\omega}{dt} + \frac{r^2VB^2}{\rho}\omega = 0$$

which has solution

$$\omega = \omega_0 e^{-\frac{VB^2t}{\rho I_i}}$$

After each time segment $\frac{\rho I_i}{VB^2}$ (the "time constant"), the angular velocity of the disk is reduced by a factor of $\frac{1}{e}$. Thus the wheel is almost at rest after 5 time constants

$$5\tau_c = 5\left(\frac{\rho I_i}{B^2V}\right)$$

$$= \frac{5\rho}{B^2 \pi r_0^2 t} \frac{MR^2}{2}$$
$$= \frac{5\rho}{2B^2 \cdot \pi r_0^2 t} \rho_0 t \cdot \pi R^2 R^2$$
$$= \frac{5\rho \rho_0}{2B^2 r_0^2} R^4$$

where $\rho_0 =$ density of Al

$= 2.7 \times 10^3 \text{ kg} / \text{m}^3$ Table 13.1,

Putting in numbers

$$5\tau_c = \frac{5(2.8\times10^{-8}\ \Omega\text{·m})(2.7\times10^3\ \text{kg}/\text{m}^3)(0.50\ \text{m})^4}{2(0.25\ \text{T})^2(0.38\ \text{m})^2}$$
$$\sim 1\text{ ms}$$

30.65 The electric field in a region is

$$\vec{\mathbf{E}} = \left(69.5\text{ kV/m}\cdot\text{s}^2\right) t^2 \left(3\hat{\mathbf{i}} + 4\hat{\mathbf{j}}\right)$$

Use the E_y, component to calculate the displacement current through a circle in the x-z-plane (choose $\hat{\mathbf{n}} = \hat{\mathbf{j}}$). Since $\vec{\mathbf{E}}$ is uniform,

$$\Phi_E = \int \vec{\mathbf{E}} \cdot \hat{\mathbf{n}}\, da = E_y A$$
$$= \left(69.5 \times 10^3\text{ kV/m}\cdot\text{s}^2\right)\left(t^2\right)(4) A$$
$$\frac{d\Phi_E}{dt} = 8\left(69.5 \times 10^3\text{ kV/m}\cdot\text{s}^2\right) At$$
$$A = \pi r^2 = \pi (0.15\text{ m})^2 = (0.071)\text{ m}^2$$
$$I_d = \epsilon_0 \frac{d\Phi_E}{dt}$$
$$\epsilon_0 = 8.85 \times 10^{-12}\text{ F/m}$$
$$I_d = \left(3.48 \times 10^{-7}\text{ A/s}\right) t$$

At $t = 0.5$ sec, $I_d = 0.17\ \mu$A.
At $t = 7.7$ sec, $I_d = 2.7\ \mu$A.

30.69 Charge Q accumulates on the sphere

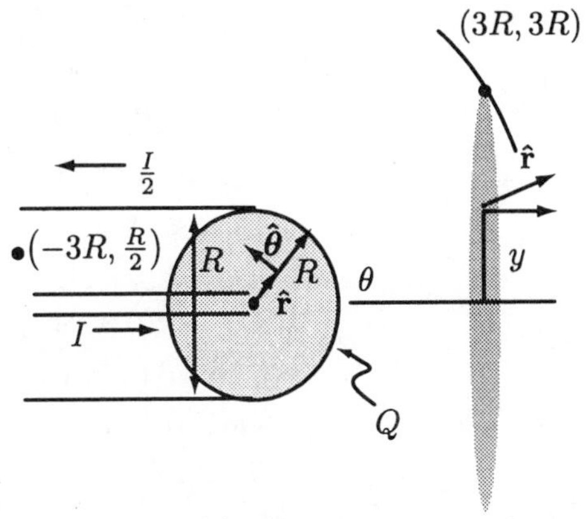

$$\Rightarrow \vec{\mathbf{E}} = \begin{cases} \frac{Q}{4\pi\epsilon_0} \frac{1}{(x^2+y^2)} \hat{\mathbf{r}} & \text{outside sphere} \\ 0 & \text{inside sphere} \end{cases}$$

at $(3R, 3R)$

$$\vec{\mathbf{E}} = \frac{Q}{4\pi\epsilon_0} \frac{1}{\left((3R)^2 + (3R)^2\right)} \hat{\mathbf{r}}$$
$$= \frac{Q}{4\pi\epsilon_0} \frac{1}{18R^2} \hat{\mathbf{r}}$$

We want the flux through a surface bounded by a circle at $x = 3R$. For ease of calculation, we pick a plane circle of radius $3R$. First we calculate the flux through a circular ring of thickness dy:

$$d\Phi = 2\pi y\, dy \vec{\mathbf{E}} \cdot \hat{\mathbf{i}}$$
$$= 2\pi y\, dy \frac{Q}{4\pi\epsilon_0} \frac{1}{[(x^2) + y^2]} \cdot \frac{x}{(x^2+y^2)^{1/2}}$$
$$= \frac{Qx}{2\epsilon_0} \frac{y\, dy}{(x^2+y^2)^{3/2}}$$

Then the flux through the whole circle is

$$\Phi = \int d\Phi = \frac{Q}{2\epsilon_0} x \int_0^{y_{\max}} \frac{y\, dy}{(x^2+y^2)^{3/2}}$$
$$= \frac{Qx}{2\epsilon_0} \left[- \frac{1}{(x^2+y^2)^{1/2}} \Bigg|_0^{y_{\max}} \right]$$
$$= \frac{Q}{2\epsilon_0} \left[1 - \frac{x}{\sqrt{x^2 + y_{\max}^2}} \right]$$

with $x = 3R$, $y_{\max} = 3R$

$$\Phi_e = \frac{Q}{2\epsilon_0} \left(1 - \frac{1}{\sqrt{2}}\right)$$

and

$$\frac{d\Phi_e}{dt} = \frac{1}{2\epsilon_0} \frac{dQ}{dt} \left(1 - \frac{1}{\sqrt{2}}\right)$$

$$I_d = \epsilon_0 \frac{d\Phi_e}{dt} = \frac{1}{2}\left(\frac{I}{2}\right)\left(1 - \frac{1}{\sqrt{2}}\right)$$
$$= \frac{I}{4}\left(1 - \frac{1}{\sqrt{2}}\right)$$

Using Ampère-Maxwell

$$\oint \vec{\mathbf{B}} \cdot d\vec{\ell} = \mu_0 (I + I_d) = \mu_0 I_d \text{ at } (3R, 3R)$$

$\vec{\mathbf{B}}$ forms circles around the cable axis

$$2\pi \cdot 3R\, B_\theta = \mu_0 I_d$$

$$\begin{aligned} B_\theta &= \frac{\mu_0}{6\pi R} \cdot \frac{I}{4}\left(1 - \frac{1}{\sqrt{2}}\right) \\ \vec{\mathbf{B}} &= \frac{\mu_0 I}{24\pi R}\left(1 - \frac{1}{\sqrt{2}}\right)\hat{\theta} \\ \hat{\theta} &\equiv \text{right-handed around } x\text{-axis} \end{aligned}$$

For $(-3R, 3R)$ we add in the current flow in a loop

$$\oint \vec{\mathbf{B}} \cdot d\vec{\ell} = \mu_0 (I + I_d)$$

At this location I_d is in the opposite direction i.e.,

$$I_d = -\frac{I}{4}\left(1 - \frac{1}{\sqrt{2}}\right)$$

but the conductor current is $\frac{I}{2}$ in the direction to the right

$$\begin{aligned} \oint \vec{\mathbf{B}} \cdot d\vec{\ell} &= \frac{\mu_0 I}{2} + \mu_0 I_d \\ 2\pi (3R) B &= \frac{\mu_0 I}{2} - \frac{\mu_0 I}{4}\left(1 - \frac{1}{\sqrt{2}}\right) \\ &= \frac{\mu_0 I}{4}\left(1 + \frac{1}{\sqrt{2}}\right) \\ \Rightarrow \vec{\mathbf{B}} &= \frac{\mu_0 I}{24\pi R}\left(1 + \frac{1}{\sqrt{2}}\right)\hat{\theta} \end{aligned}$$

For the third case, the result is similar except that the Amperian loop contains current I not $\frac{I}{2}$ and the displacement current is different

$$\begin{aligned} I_d &= \epsilon_0 \frac{d\Phi_e}{dt} \\ &= -\frac{1}{2}\left(\frac{I}{2}\right)\left(1 - \frac{3R}{(3R)^2 + (0.5R)^2}\right) \\ &= -\frac{I}{4}\left[1 - \frac{3}{\sqrt{9.25}}\right] \end{aligned}$$

We use the Ampere-Maxwell law, with I_d to the left

$$\begin{aligned} \oint \vec{\mathbf{B}} \cdot d\vec{\ell} &= \mu_0 (I + I_d) \\ &= \mu_0 I - \mu_0 \frac{I}{4}\left(1 - \frac{3}{\sqrt{9.25}}\right) \\ &= \frac{3}{4}\mu_0 I\left(1 + \frac{1}{\sqrt{9.25}}\right) \\ 2\pi\left(\frac{R}{2}\right) B &= \frac{3}{4}\mu_0 I\left(1 + \frac{1}{\sqrt{9.25}}\right) \end{aligned}$$

and noting the direction around the axis

$$\begin{aligned} \vec{\mathbf{B}} &= \frac{3\mu_0 I}{4\pi R}\left(1 + \frac{1}{\sqrt{9.25}}\right)\hat{\theta} \\ &= 0.32\,\mu_0 I\,\hat{\theta} \end{aligned}$$

30.73 The area penetrated by the field is A, the magnetic field is decreasing $\frac{dB}{dt} = 0.250$ T/s.

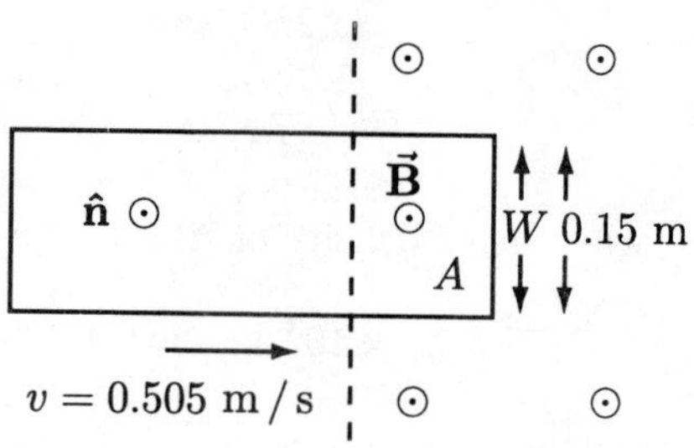

With $\hat{\mathbf{n}}$ outward from the plane of the diagram,

$$\Phi(t) = AB$$

Then

$$\frac{d\Phi}{dt} = B\frac{dA}{dt} + A\frac{dB}{dt} \quad (1)$$

A and B are functions of t, the width is

$$W \Rightarrow \frac{dA}{dt} = Wv$$

At $t = 0$, $A = 0$

$$\Rightarrow A(t) = Wvt$$

and

$$B(t) = B(0) + \left(\frac{dB}{dt}\right)t$$

putting into (1) gets us

$$\begin{aligned} \frac{d\Phi}{dt} &= \left[B(0) + \left(\frac{dB}{dt}\right)t\right]Wv + (Wvt)\left(\frac{dB}{dt}\right) \\ &= Wv\left[B(0) + 2\frac{dB}{dt}t\right] \end{aligned}$$

Now

$$\mathcal{E} = -\frac{d\Phi}{dt}$$

at $t = 1.00$ s

$$\begin{aligned} \mathcal{E} &= -[1.500\text{ T} - 2(0.250\text{ T/s})(1.00\text{ s})] \\ &\quad \cdot (0.15\text{ m})(0.505\text{ m/s}) \\ \mathcal{E}(1.0) &= -75.8\text{ mV (i.e., clockwise)} \end{aligned}$$

at $t = 5.00$ s

$$\begin{aligned} \mathcal{E} &= -[1.500\text{ T} - 2(0.250\text{ T/s})(5.00\text{ s})] \\ &\quad \cdot (0.15\text{ m})(0.505\text{ m/s}) \\ \mathcal{E}(5.00) &= +75.8\text{ mV (i.e., counterclockwise)} \end{aligned}$$

30.77 Refer to Figure 30.49

$Q = 1.3$ pC $\qquad A = 2.0 \times 10^{-2}$ m^2

$d = 1.0 \times 10^{-4}$ m

given

$$\begin{aligned} n &= 4 \times 10^4 \text{ turns/m} \\ \frac{dI}{dt} &= 1.0 \times 10^4 \text{ A/s} \end{aligned}$$

center of capacitor at $r = 0.10$ m from center of solenoid

a) For a parallel plate capacitor

$$\begin{aligned} \left|\vec{\mathbf{E}}_s\right| &= \frac{\sigma}{\epsilon_0} = \frac{Q}{A\epsilon_0} \\ &= \frac{1.3 \times 10^{-12}\text{ C}}{(2.0 \times 10^{-2}\text{ m}^2)(8.854 \times 10^{-12}\text{ F}\cdot\text{m}^{-1})} \\ \left|\vec{\mathbf{E}}_s\right| &= 7.3 \text{ V/m} \end{aligned}$$

b) $\Delta V \equiv$ potential difference

$$\begin{aligned} &= -\int_A^B \vec{\mathbf{E}}_c \cdot d\vec{\ell} \\ &= (7.34\text{ V/m})\left(1.0 \times 10^{-4}\text{ m}\right) \\ \Delta V &= 7.3 \times 10^{-4}\text{ V} \end{aligned}$$

c) The magnetic field inside a solenoid is

$$B = \mu_0 n I$$

out of page in Figure 30.49.

Since the magnetic field is changing, there must be an induced emf and hence field inside the solenoid.

The $\vec{\mathbf{E}}$-field is circular about the axis

$$\oint \vec{\mathbf{E}} \cdot d\vec{\ell} = -\frac{d\Phi}{dt}$$

$\odot\ \vec{\mathbf{B}} \qquad \hat{\boldsymbol{\theta}}$

around a circle of radius r, and area πr^2

$$\begin{aligned} E_\theta 2\pi r &= -\pi r^2 \cdot \mu_0 n \frac{dI}{dt} \\ E_\theta &= -\frac{\mu_0 n r}{2}\frac{dI}{dt} \end{aligned}$$

The center of the capacitor plates is at $r = 0.10$ m from the axis of the solenoid.

$$\begin{aligned} E_\theta &= (0.1\text{ m})\left(\frac{0.4}{2}\pi \times 10^{-6}\ \Omega\cdot\text{s/m}\right) \\ &\quad \cdot \left(4.0 \times 10^4\text{ m}^{-1}\right)\left(1.0 \times 10^4\text{ A/s}\right) \\ \left|\vec{\mathbf{E}}\right| &= 25\text{ V/m} \end{aligned}$$

d) The induced $\vec{\mathbf{E}}$ is clockwise, while the static $\vec{\mathbf{E}}$-field is counterclockwise as the plates are drawn $\Rightarrow$ the total $\vec{\mathbf{E}}$-field is the vector sum

$$\begin{aligned} E_{\text{result}} &= (25.1 - 7.34)\text{ V/m} \\ &= 18\text{ V/m (clockwise)} \end{aligned}$$

e) To balance the induced field, the static field must be 25 V/m counterclockwise $\Rightarrow$

$$25\text{ V/m} = \frac{Q}{A\epsilon_0}$$

so $Q \propto E$, thus we have

$$\begin{aligned} Q &= \frac{25.1\text{ V/m}}{7.34\text{ V/m}} \times 1.3 \times 10^{-12}\text{ C} \\ &= 4.4 \times 10^{-12}\text{ C} \\ \Rightarrow Q &= 4.4\text{ pC} \end{aligned}$$

Note that $\Delta V \equiv$ potential difference

$$= -\int_A^B \vec{\mathbf{E}}_c \cdot d\vec{\ell}$$

where $\vec{\mathbf{E}}_c \equiv$ coulomb static field. This is independent of the induced emf; there would still be a potential difference.

30.81 We need to apply both Faraday's law and Newton's 2nd law to the motion of the rod.

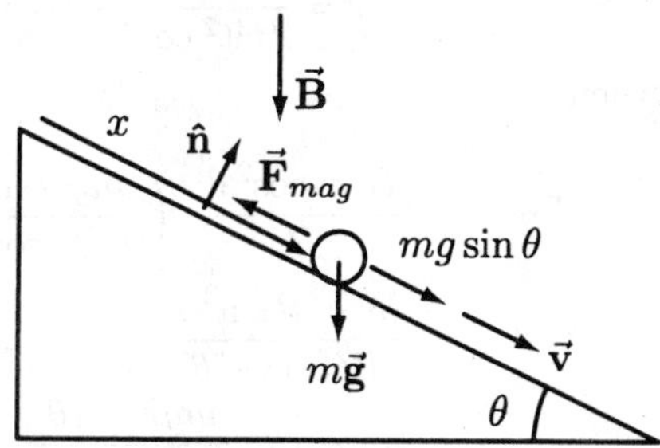

The emf around the loop is

$$\mathcal{E} = -\frac{d\Phi}{dt}$$

with

$$\Phi = A\vec{\mathbf{B}}\cdot\hat{\mathbf{n}} = -\ell x B\cos\theta$$

Then

$$\mathcal{E} = lvB\cos\theta$$

The current about the loop is thus $\frac{\mathcal{E}}{R}$

$$I = \frac{lvB\cos\theta}{R}$$

The magnetic force on the rod is

$$\begin{aligned}\vec{\mathbf{F}}_m &= I\vec{\ell}\times\vec{\mathbf{B}} = I\ell\hat{\mathbf{j}}\times B\left(-\cos\theta\hat{\mathbf{k}}+\sin\theta\hat{\mathbf{i}}\right)\\ &= -I\ell B\left(\cos\theta\hat{\mathbf{i}}+\sin\theta\hat{\mathbf{k}}\right)\end{aligned}$$

The x-component is

$$F_{m,x} = -\frac{\ell^2 vB^2\cos^2\theta}{R}$$

In equilibrium the gravitational force component down the plane balances the magnetic force component up the plane:

$$mg\sin\theta = \frac{B^2\ell^2 v}{R}\cos^2\theta$$

Thus the speed of the rod is

$$v = \frac{mgR\sin\theta}{B^2\ell^2\cos^2\theta}$$

The gravitational energy is lost by the rod at a rate

$$P_g = mg\sin\theta v$$

Heat is dissipated in the resistor at a rate

$$\begin{aligned}P_R &= I^2R = \frac{\ell^2v^2B^2}{R^2}\cos^2\theta R\\ &= \frac{\ell^2B^2\cos^2\theta}{R}v^2\end{aligned}$$

where

$$v = \frac{mgR\sin\theta}{B^2\ell^2\cos^2\theta}$$

then

$$\begin{aligned}P_R &= \frac{B^2\ell^2\cos^2\theta}{R}\cdot\left(\frac{mgR\sin\theta}{B^2\ell^2\cos^2\theta}\right)^2\\ &= \frac{m^2g^2R\sin^2\theta}{B^2\ell^2\cos^2\theta}\\ &= mg\sin\theta\cdot\frac{mgR\sin\theta}{B^2\ell^2\cos^2\theta}\\ &= mg\sin\theta v\\ &= P_g\end{aligned}$$

30.85 We have that

$$\text{emf} = -\frac{d\Phi}{dt}$$

To get the flux, we have to integrate the emf over time. Use the trapezoidal rule. The value of flux at $t = 2.0$ s is the value of the integral of $-$(emf) evaluated between $t = 0$ and $t = 2.0$ s.

time t (sec)	emf (V)	delta Φ
0.00	1.00	
0.10	1.20	−0.11
0.20	1.42	−0.13
0.30	1.64	−0.15
0.40	1.88	−0.18
0.50	2.12	−0.20
0.60	2.38	−0.23
0.70	2.65	−0.25
0.80	2.94	−0.28
0.90	3.24	−0.31
1.00	3.55	−0.34
1.10	3.89	−0.37
1.20	4.25	−0.41
1.30	4.63	−0.44
1.40	5.04	−0.48
1.50	5.47	−0.53
1.60	5.95	−0.57
1.70	6.46	−0.62
1.80	7.02	−0.67
1.90	7.63	−0.73
2.00	8.29	−0.80

Flux at $t = 2.0$ s is the sum of the third column: −7.80 Weber.

Chapter 31

Introduction to Time-dependent Circuits

31.1 This is an RC circuit.

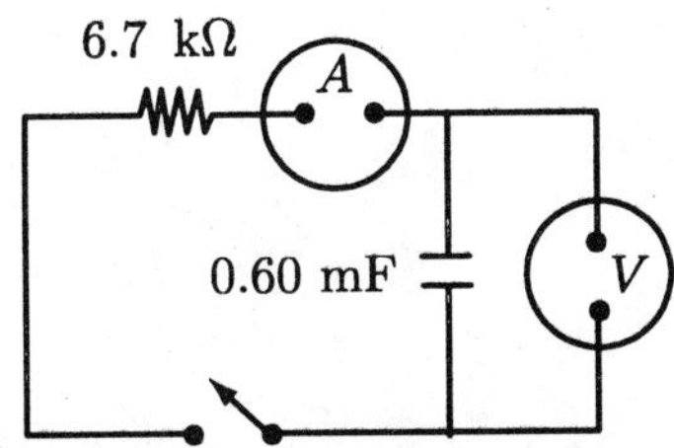

We know that at a time t after the switch is closed, the potential difference across the capacitor will be

$$\Delta V(t) = \frac{Q_0}{C} e^{-t/RC}$$

and the current will be

$$I(t) = \frac{Q_0}{RC} e^{-t/RC} = \frac{\Delta V}{R}$$

If $Q_0 = 0.27$ C, the readings at $t = 6.0$ s will be

$$\begin{aligned} V &= \frac{0.27 \text{ C}}{0.60 \times 10^{-3} \text{ F}} e^{-6.0 \text{ s}/\left(6.7\times10^3 \ \Omega\right)\left(0.60\times10^{-3} \text{ F}\right)} \\ &= 0.10 \text{ kV} \end{aligned}$$

and

$$I = \frac{0.10 \text{ kV}}{6.7 \text{ k}\Omega} = 0.015 \text{ A} = 15 \text{ mA}$$

The time constant of the system is

$$RC = \left(6.7 \times 10^3 \ \Omega\right)\left(0.60 \times 10^{-3} \text{ F}\right) = 4.0 \text{ s}$$

so the current at $t = 6.0$ s has decreased significantly from

$$I_0 = \frac{Q_0}{RC} = 67 \text{ mA}$$

and the voltage from $\Delta V_0 = \frac{Q_0}{RC} = 450$ V.

31.3 This is an RC circuit in which the capacitor is being charged.

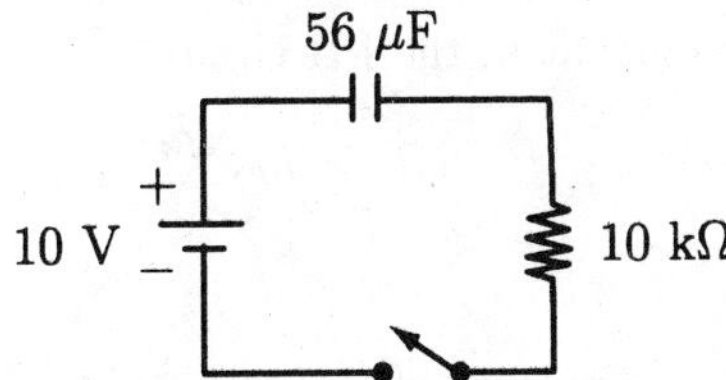

The charge on the capacitor at time t after the switch is closed is

$$Q(t) = C\mathcal{E}\left(1 - e^{-t/RC}\right)$$

For large t, we have

$$Q = C\mathcal{E} = (56 \ \mu\text{F})(10 \text{ V}) = 0.56 \text{ mC}$$

We also know that

$$I(t) = \frac{dQ}{dt} = \frac{\mathcal{E}}{R} e^{-t/RC}$$

so that for large t, $I = 0$. This makes sense since the charged capacitor opposes the emf of the battery. For the charge to reach 95% of its final value, a time of

$$t = 3\tau = 3RC = 3\,(10\text{ k}\Omega)\,(56\ \mu\text{F}) = 1.7\text{ s}$$

is needed. The initial current is

$$I_0 = \frac{\mathcal{E}}{R} = \frac{10\text{ V}}{10\text{ k}\Omega} = 1.0\text{ mA}$$

31.5 The long straight wire has magnetic field

$$B = \frac{\mu_0 I}{2\pi r}$$

at a distance r from the wire. Thus the magnetic energy density at a distance r from the wire is

$$\begin{aligned} U_B &= \frac{B^2}{2\mu_0} = \left(\frac{\mu_0 I}{2\pi r}\right)^2 \cdot \frac{1}{2\mu_0} \\ &= \left[\frac{4\pi\times10^{-7}\ \frac{\text{Wb}}{\text{A}\cdot\text{m}}\,(1.7\text{ A})}{2\pi(0.15\text{ m})}\right]^2 \cdot \frac{1}{2\left(4\pi\times10^{-7}\ \frac{\text{Wb}}{\text{A}\cdot\text{m}}\right)} \\ &= 2.0\ \mu\text{J/m}^3 \end{aligned}$$

31.7 Since there is no emf source, we know the current must be decreasing in time.

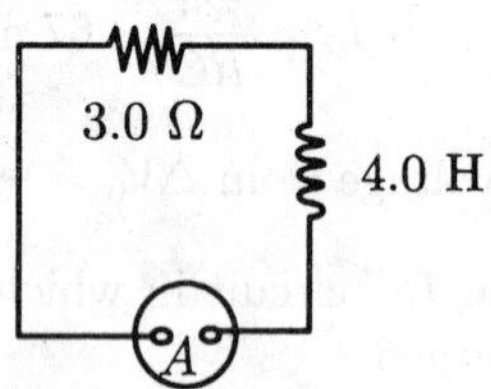

The current in the LR circuit is

$$I(t) = I_0 e^{-Rt/L}$$

So

$$\begin{aligned} I_0 &= I(t)e^{+Rt/L} = (0.050\text{ A})e^{\frac{(3.0\Omega)(4.0\text{ s})}{(4.0\text{ H})}} \\ &= 1.0\text{ A} \end{aligned}$$

We could have guessed this answer by noting

$$\tau = \frac{L}{R} = \frac{4}{3}\text{ s};$$

so $3\tau = 4$ s, and the current at this time is 5% of I_0.

31.9 This is an LC circuit. It oscillates at a frequency of

$$\begin{aligned} f &= \frac{\omega}{2\pi} = \frac{1}{2\pi\sqrt{LC}} \\ &= \frac{1}{2\pi\sqrt{(56\times10^{-6}\text{ F})\,(3.0\times10^{-3}\text{ H})}} \\ &= 390\text{ Hz} \end{aligned}$$

If the capacitor is uncharged at $t = 0$, all the energy of the circuit is stored in the inductor. This energy is

$$\begin{aligned} E &= \frac{1}{2}LI^2 = \frac{1}{2}\,(3.0\times10^{-3}\text{ H})\,(17\times10^{-3}\text{ A})^2 \\ &= 0.43\ \mu\text{J} \end{aligned}$$

Thus *total energy is conserved* as the system oscillates; it is stored in turn between the capacitor and the inductor, or shared by them. *The current oscillates at* 390 Hz.

31.13 We know that the time constant $\tau = RC$ in an RC circuit. We also know

$$Q = Q_0 e^{-t/\tau} = Q_0 e^{-t/(RC)}$$

Thus if $\dfrac{Q}{Q_0} = \dfrac{1}{2}$, we have

$$\begin{aligned} e^{-t/(RC)} &= \frac{1}{2} \\ t &= -RC\ln\left(\frac{1}{2}\right) = 0.69\ RC \end{aligned}$$

31.17 For the following circuit, Kirchhoff's loop rule gives:

$$-\frac{Q}{C} - IR + \mathcal{E} = 0$$

or

$$\frac{dQ}{dt} + \frac{Q}{RC} = \frac{\mathcal{E}}{R}$$

The solution is $Q\,(t) = C\mathcal{E} + Ae^{-t/(RC)}$ where A is a constant. Since $Q_0 = 2C\mathcal{E}$,

$$A = C\mathcal{E}$$

and

$$Q\,(t) = C\mathcal{E}\left(1 + e^{-t/(RC)}\right)$$

The current

$$I\,(t) = \frac{dQ}{dt} = -\frac{\mathcal{E}}{R}e^{-t/RC}$$

The minus sign means that current actually flows in the opposite sense from what we labeled positive, which makes sense since the initial potential difference across the capacitor is

$$\frac{Q_0}{C} = \frac{2C\mathcal{E}}{C} = 2\mathcal{E}$$

31.21 Finding the solution to this problem involves some trial and error. We can quickly eliminate the possibility of resistors in series and capacitors in parallel, since this would give a time constant of

$$[(56+22)\,\mu\text{F}]\cdot[(18+33)\,\text{k}\Omega] = 4.0\text{ s}$$

Also capacitors in series, resistors in parallel would result in

$$RC < (18\text{ k}\Omega)\,(22\,\mu\text{F}) = 0.40\ s$$

The solution is capacitors in parallel, resistors in parallel:

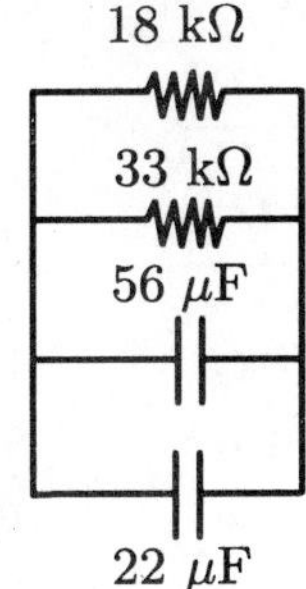

Then

$$\begin{aligned} R &= \frac{1}{\frac{1}{18}+\frac{1}{33}}\text{ k}\Omega = 11.6\text{ k}\Omega \\ C &= 56\,\mu\text{F} + 22\,\mu\text{F} = 78\,\mu\text{F} \end{aligned}$$

and

$$RC = (11.6\text{ k}\Omega)\,(78\,\mu\text{F}) = 0.91\text{ s}.$$

31.25 The magnetic energy in a solenoid of N turns, area A, current I, and length ℓ, is

$$U = \frac{1}{2}\mu_0\frac{N^2A}{\ell}I^2$$

So a solenoid with length 2ℓ will have half the magnetic energy of one with length ℓ.

31.29 The total electromagnetic energy density contained in a magnetic field B and electric field E is

$$\begin{aligned} u &= \frac{1}{2}\left(\epsilon_0E^2 + \frac{B^2}{\mu_0}\right) \\ &= \frac{1}{2}\left[(8.85\times10^{-12}\text{ C}^2/\text{N}\cdot\text{m}^2)\,(100\text{ V/m})^2 \right. \\ &\qquad \left. + \frac{(4\times10^{-5}\text{ T})^2}{4\pi\times10^{-7}\,\frac{\text{Wb}}{\text{A}\cdot\text{m}}}\right] \\ &= 0.6\text{ mJ/m}^3 \end{aligned}$$

The fraction due to the electric field is

$$\begin{aligned} \frac{\epsilon_0E^2}{B^2/\mu_0} &= \frac{\mu_0\epsilon_0E^2}{B^2} \\ &= (4\pi\times10^{-7}\text{ N/A}^2)(8.85\times10^{-12}\text{ C}^2/\text{N}\cdot\text{m}^2) \\ &\quad\cdot\left[\frac{100\text{ V/m}}{4\times10^{-5}\text{ T}}\right]^2 \\ &= 7\times10^{-5} \end{aligned}$$

31.31 We know that the magnetic field at point P is

$$B = \frac{\mu_0I}{4\pi R}\left(1+\frac{d}{\sqrt{d^2+R^2}}\right)$$

and the electric field is

$$E = \frac{1}{4\pi\epsilon_0}\frac{Q}{d^2+R^2}$$

$$Q = 3.0\text{ nC}$$

I d P R

So the total electromagnetic energy density at point P is

$$\begin{aligned} u &= \frac{1}{2}\left(\epsilon_0E^2+\frac{B^2}{\mu_0}\right) \\ &= \frac{1}{2}\left(\frac{Q^2}{16\pi^2\epsilon_0(d^2+R^2)^2}+\frac{\mu_0I^2}{16\pi^2R^2}\left(1+\frac{d}{\sqrt{d^2+R^2}}\right)^2\right) \end{aligned}$$

Substituting $Q = 3.0$ pC, $d = 1.5$ cm, $R = 2.0$ cm, and $I = 6.0$ mA, we get

$$u = 9.16\times10^{-9}\text{ J/m}^3$$

Note that

$$\frac{B^2}{\mu_0} = 1.8\times10^{-9}\text{ J/m}^3$$

while

$$\epsilon_0E^2 = 1.6\times10^{-8}\text{ J/m}^3$$

so that the electric field due to the point charge makes a larger contribution to the energy density than the magnetic field due to the current.

31.35 a) The inductance is the flux through the solenoids divided by the current (Eqn. 31.11)

$$L = \frac{\Phi_m}{I}$$

If the solenoids are wound in the *same sense*, both produce magnetic field in the same direction. If they are wound in opposite senses, they produce field in opposite directions. In the first case, the flux through the solenoids is greater for a given I than in the second case: Thus *the inductance is greater when the coils are wound in the same sense.*

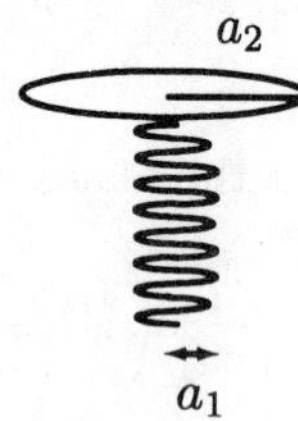

b) Inside the outer solenoid

$$B = \mu_0 n I$$

where

$$n = \frac{N}{\ell}$$

In the inner region (inside *both* solenoids) the magnetic field is the sum of the fields produced by the two solenoids

$$B = 2\mu_0 n I$$

The flux through each turn of the outer coil is

$$\begin{aligned} \Phi_{1,o} &= \pi a_1^2 2\mu_0 n I + \pi\left(a_2^2 - a_1^2\right)\mu_0 n I \\ &= \pi\mu_0 n I\left(a_2^2 + a_1^2\right) \end{aligned}$$

And the flux through the whole outer coil is

$$\Phi_o = N\pi\mu_0 n I\left(a_2^2 + a_1^2\right)$$

The flux through each turn of the inner coil is

$$\Phi_{1,i} = \pi a_1^2(2\mu_0 n I)$$

and the flux through the whole inner coil is

$$\Phi_i = 2N\pi a_1^2 \mu_0 n I$$

Thus the flux through the whole device is

$$\Phi_B = N\pi\mu_0 n I\left(a_2^2 + a_1^2 + 2a_1^2\right)$$

And thus

$$\begin{aligned} L &= \frac{\Phi_B}{I} = Nn\pi\mu_0\left(a_2^2 + 3a_1^2\right) \\ &= \frac{N^2\pi\mu_0}{\ell}\left(a_2^2 + 3a_1^2\right) \end{aligned}$$

31.43 In an LRC circuit, doubling the inductance doubles τ, since $\tau = \dfrac{2L}{R}$. A larger inductor can store more energy, which takes longer to dissipate. The change in ω_1 depends on the relations of L, R, and C. Changing L to $2L$ means ω_1 changes from

$$\left(\frac{1}{LC} - \frac{R^2}{4L^2}\right)^{\frac{1}{2}} \text{ to } \left(\frac{1}{2LC} - \frac{R^2}{16L^2}\right)^{\frac{1}{2}}$$

We wish to know if ω_1 will increase; this will be true if

$$\begin{aligned} \left(\frac{1}{LC} - \frac{R^2}{4L^2}\right)^{\frac{1}{2}} &< \left(\frac{1}{2LC} - \frac{R^2}{16L^2}\right)^{\frac{1}{2}} \\ \frac{1}{LC} - \frac{R^2}{4L^2} &< \frac{1}{2LC} - \frac{R^2}{16L^2} \\ \frac{4L^2 - R^2CL}{4L^3C} &< \frac{16L^2 - 2R^2CL}{32L^3C} \\ 32L^2 - 8R^2CL &< 16L^2 - 2R^2CL \\ 16L^2 &< 6R^2CL \\ L &< \frac{3}{8}R^2C \end{aligned}$$

otherwise, ω_1 decreases.

31.47 Here

$$\tau = \frac{L}{R} = \frac{1.0}{10}\text{ s} = 0.1\text{ s}$$

and we are assuming

$$I_0 = \frac{\mathcal{E}}{R} = 0.20\text{ A}$$

since the switch has been in position 1 for a long time.

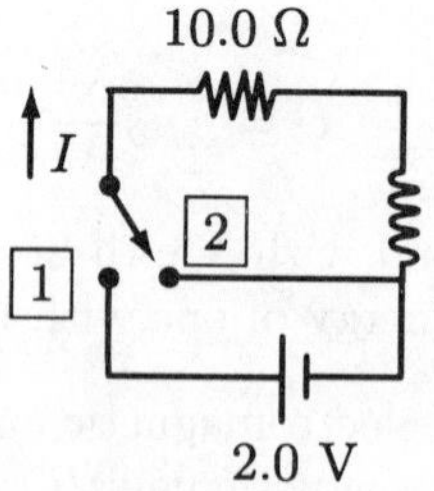

0.30 s(3τ) after the switch is put into position 2, current will drop by 95%; so

$$I = (0.05)(0.2\text{ A}) = 1.0 \times 10^{-2}\text{ A}$$

Current direction is as shown.

31.49 The current in the LR circuit satisfies

$$I(t) = \frac{\mathcal{E}}{R}\left(1 - e^{-Rt/L}\right)$$

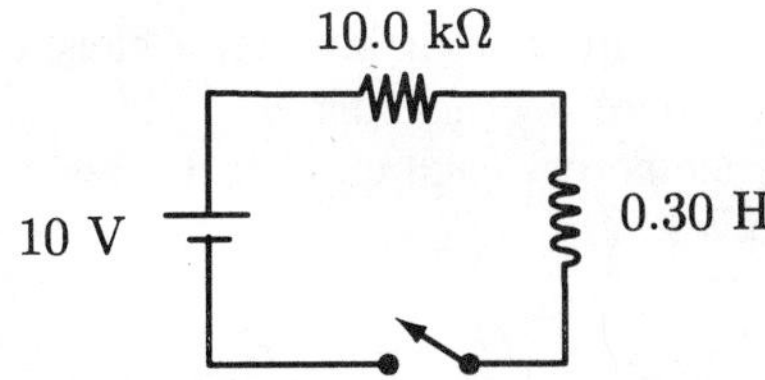

We know $I_0 = 0$ since

$$\tau = \frac{L}{R} = 3.0 \times 10^{-5}\text{ s}$$

and the switch was open for several seconds. So $I_0 = 0$,

$$I_f = \frac{\mathcal{E}}{R} = \frac{10.0\text{ V}}{1.00 \times 10^4\ \Omega} = 1.00\text{ mA}$$

and it requires

$$3\tau = 3\frac{L}{R} = 9.0 \times 10^{-5}\text{ s} = 0.090\text{ ms}$$

for the current to reach 95% of its final value.

31.53 The period

$$\tau = \frac{1}{f} = \frac{2\pi}{\omega} = 2\pi\sqrt{LC}$$

We can quickly eliminate the cases where the capacitors are in series and the inductors are in parallel, since

$$2\pi\sqrt{LC} < 2\pi\sqrt{(22\text{ mH})\,(2.2\ \mu\text{F})} = 1.4\text{ ms}$$

and where the capacitors are in parallel and the inductors in series, when

$$\tau = 2\pi\sqrt{(78\text{ mH})\,(7.8\ \mu\text{F})} = 4.9\text{ ms}$$

It is also easy to see that the result is the same if the components are all in series or if the capacitors and inductors are both in parallel, since

$$\left(\frac{1}{\frac{1}{5.6}+\frac{1}{2.2}}\right)(78) = (7.8)\left(\frac{1}{\frac{1}{22}+\frac{1}{56}}\right)$$

When we check one of these cases, we indeed get $\tau = 2.2$ ms.

31.57 We know that the inductance of a solenoid with radius r, length ℓ_s and N turns is

$$L = \frac{\mu_0 \pi r^2 N^2}{\ell_s}$$

If the length of the wire is ℓ_w, we know $N = \dfrac{\ell_w}{2\pi r}$. Also,

$$\ell_s = Nd$$

where d is the diameter of the wire. The resistance of the wire is

$$R = \frac{4\ell_w \rho}{\pi d^2}$$

So

$$\begin{aligned}\tau &= \frac{L}{R} \\ &= \mu_0 \pi r^2 \left(\frac{\ell_w^2}{4\pi^2 r^2}\right)\left(\frac{2\pi r}{\ell_w d}\right)\left(\frac{\pi d^2}{4\ell_w \rho}\right) \\ &= \frac{\mu_0 \pi r d}{8\rho}\end{aligned}$$

depends only on the radius of the coil and the diameter and resistivitity of the wire.

31.61 When S_1 and S_2 are closed, the charge on C_1 is

$$\mathcal{E}C_1 = (18\text{ V})(18\ \mu\text{F}) = 0.32\text{ mC}$$

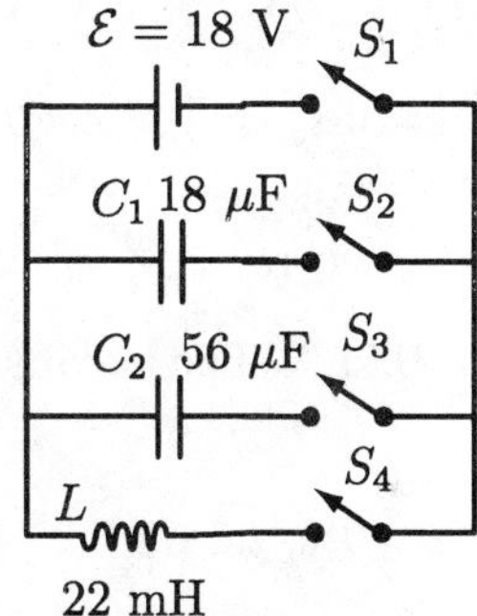

The charge on C_2 is of course still 0. Next we have

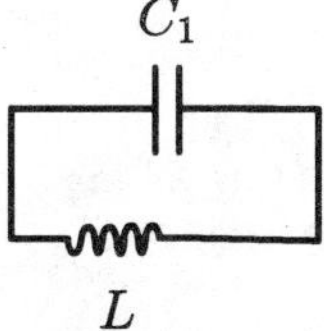

For this circuit,

$$Q(t) = Q_a \cos\omega t + Q_b \sin\omega t$$

Since $Q(0) = Q_0 = 0.32$ mC,

$$Q_a = Q_0$$

Since $I(0) = 0$ because of the inductor, $Q_b = 0$. Thus

$$Q(t) = Q_0 \cos\omega t$$

and

$$|I| = \omega Q_0 \sin\omega t$$

This reaches a maximum at $\omega t = \frac{\pi}{2}$ or

$$\begin{aligned} t &= \frac{\pi}{2\omega} = \frac{\pi}{2}\sqrt{LC_1} \\ &= \frac{\pi}{2}\sqrt{(22 \times 10^{-3})(18 \times 10^{-6})}\text{ s} \\ &= 1.0\text{ ms} \end{aligned}$$

This value is

$$\omega Q_0 = \frac{0.32 \times 10^{-3}}{\sqrt{(22 \times 10^{-3})(18 \times 10^{-6})}}\text{ A} = 0.51\text{ A}$$

Finally we have

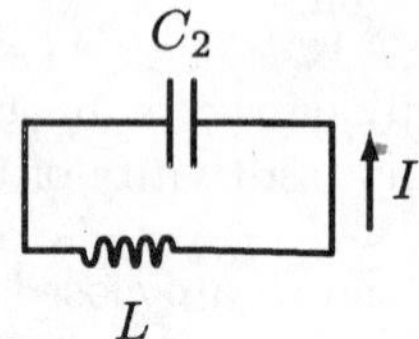

Again

$$Q(t) = Q_a \cos \omega t + Q_b \sin \omega t$$

Since $Q(0) = 0, Q_a = 0$. Since $I(0) = I_0, \omega Q_b = I_0$. So

$$Q(t) = \frac{I_0}{\omega} \sin \omega t$$

where $I_0 = 0.51$ A. The maximum charge on C_2 is

$$\begin{aligned} \frac{I_0}{\omega} &= I_0\sqrt{LC_2} \\ &= (0.51\text{ A})\sqrt{(22\text{ mH})(56\ \mu\text{F})} \\ &= 0.57\text{ mC} \end{aligned}$$

this occurs at $\omega t = \frac{\pi}{2}$ or

$$t = \frac{\pi}{2\omega} = \frac{\pi}{2}\sqrt{LC_2} = 1.7\text{ ms}$$

after the opening of S_2 and closing of S_3, or 2.7 ms after $t = 0$.

31.65 A long time after the switch is closed,

(i) there will be a steady current through the resistor

(ii) the capacitor is uncharged,

(iii) the potential drop across the inductor is zero and

(iv) $I = \frac{\mathcal{E}}{R}$.

$\mathcal{E}$ R L C

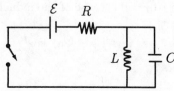

31.69 When the switch has been closed for a long time there are no currents. The potential difference across each capacitor equals $\mathcal{E}$, so $Q_1 = \mathcal{E}C_1$ and $Q_2 = \mathcal{E}C_2$.

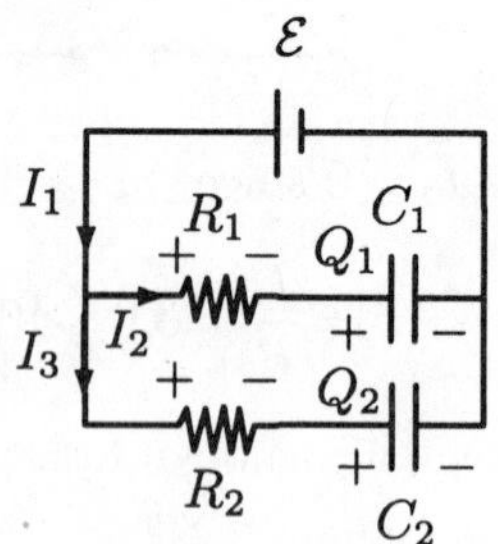

Kirchhoff's rules give:
Junction rule:

$$I_1 = I_2 + I_3$$

Loop rule: Top loop

$$-I_2 R_1 - \frac{Q_1}{C_1} + \mathcal{E} = 0$$

Outside loop:

$$\mathcal{E} - I_3 R_2 - \frac{Q_2}{C_2} = 0$$

$I_3 = \frac{dQ_2}{dt}$, so

$$\mathcal{E} = \frac{dQ_2}{dt} R_2 + \frac{Q_2}{C_2}$$

Also $I_2 = \frac{dQ_1}{dt}$; so

$$\frac{dQ_1}{dt} R_1 + \frac{Q_1}{C_1} = \mathcal{E}$$

$$\begin{aligned} Q_1(t) &= Ae^{-t/(R_1C_1)} + \mathcal{E}C_1 \\ Q_1(0) &= 0,\ A = -\mathcal{E}C_1 \\ Q_1(t) &= \mathcal{E}C_1\left(1 - e^{-t/(R_1C_1)}\right) \end{aligned}$$

Since Q_2 satisfies the same eqn. with $R_1 \to R_2$ and $C_1 \to C_2$, then:

$$Q_2(t) = \mathcal{E}C_2\left(1 - e^{-t/(R_2C_2)}\right)$$

The current supplied by the battery is I_1:

$$\begin{aligned} I_1 &= I_2 + I_3 \\ &= \frac{dQ_1}{dt} + \frac{dQ_2}{dt} \\ &= \frac{\mathcal{E}}{R_1} e^{-t/(R_1C_1)} + \frac{\mathcal{E}}{R_2} e^{-t/(R_2C_2)} \end{aligned}$$

and $\to 0$ as $t \to \infty$ as we predicted. When the switch is opened again, the circuit is

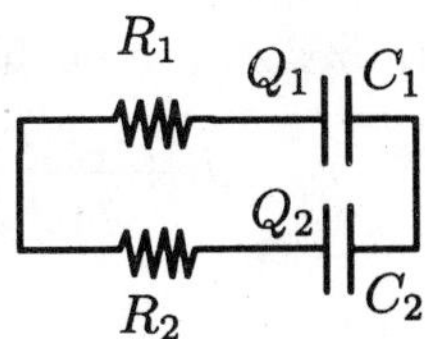

In the final state we expect the potential difference across each capacitor to be the same:

$$\frac{Q_{1f}}{C_1} = \frac{Q_{2f}}{C_2}$$

But after a long time the potential difference across each capacitor is $\mathcal{E}$, the same. Thus, when the switch is opened, *nothing happens*.

31.73 a) Applying Kirchhoff's rules:

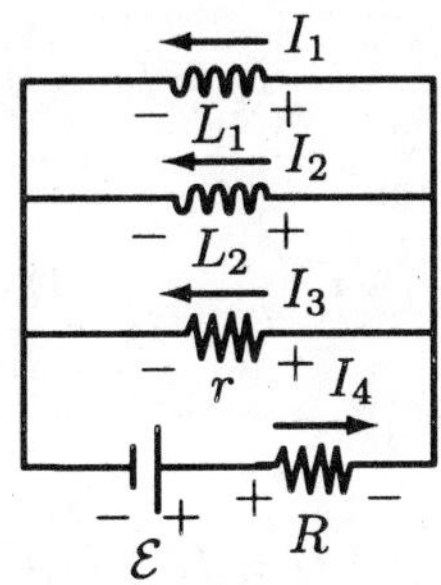

Junction rule:

$$I_4 = I_1 + I_2 + I_3$$

loop rule:

$$\mathcal{E} - I_4 R = I_3 r = L_2 \frac{dI_2}{dt} = L_1 \frac{dI_1}{dt}$$

Initial behavior Immediately after the switch is closed:

$$\begin{aligned} I_1 &= I_2 = 0 \\ I_3 &= I_4 = \frac{\mathcal{E}}{R+r} \end{aligned}$$

as $t \to \infty$, there is no resistance in the top two arms, so $I_3 \to 0$ and $I_4 = \frac{\mathcal{E}}{R}$. The current divides between I_1 and I_2. From the loop rule,

$$L_2 I_2 = L_1 I_1 + \text{const}$$

Since both are zero at $t = 0$, the const $= 0$ and

$$L_2 I_2 = L_1 I_1$$

at all times. Particularly, at $t \to \infty$,

$$\begin{aligned} I_1 + I_2 &= I_1 \left(1 + \frac{L_1}{L_2}\right) = \frac{\mathcal{E}}{R} \\ \Rightarrow I_1 &\to \frac{\mathcal{E}}{R} \frac{L_2}{(L_1 + L_2)} \end{aligned}$$

and

$$I_2 \to \frac{\mathcal{E}}{R} \frac{L_1}{(L_1 + L_2)}$$

From the junction equation:

$$\frac{dI_4}{dt} = \frac{dI_1}{dt} + \frac{dI_2}{dt} + \frac{dI_3}{dt}$$

From the loop rule

$$-R\frac{dI_4}{dt} = r\frac{dI_3}{dt} = L_2 \frac{d^2 I_2}{dt^2} = L_1 \frac{d^2 I_1}{dt^2}$$

Substituting these into the junction rule we may solve for I_3

$$-\frac{r}{R}\frac{dI_3}{dt} = \frac{r}{L_1} I_3 + \frac{r}{L_2} I_3 + \frac{dI_3}{dt}$$

or

$$\begin{aligned} \frac{dI_3}{dt}\left[1 + \frac{r}{R}\right] &= -I_3 \left[\frac{r}{L_1} + \frac{r}{L_2}\right] \\ \frac{dI_3}{dt} &= -I_3 r \left[\frac{L_1 + L_2}{L_1 L_2}\right] \frac{R}{(r+R)} \\ &= -I_3 \frac{R_\parallel}{L_\parallel} \end{aligned}$$

where

$$L_\parallel = \left(\frac{1}{L_1} + \frac{1}{L_2}\right)^{-1}$$

is the parallel combination of L_1 and L_2 and

$$R_\parallel = \left(\frac{1}{r} + \frac{1}{R}\right)^{-1}$$

is the parallel combination of r and R. Thus

$$I_3 = A + Be^{-t/\tau}$$

with $\tau = \frac{L_\parallel}{R_\parallel}$. Applying the initial conditions

$$I_3(0) = \frac{\mathcal{E}}{R+r} = A + B$$

The long term behavior is

$$I_3(t \to 0) = A = 0$$

thus

$$I_3 = \frac{\mathcal{E}}{R+r} e^{-t/\tau}$$

Then

$$\begin{aligned} I_4 &= -\frac{r}{R} I_3 + \frac{\mathcal{E}}{R} \\ &= -\frac{r}{R}\frac{\mathcal{E}}{(R+r)} e^{-t/\tau} + \frac{\mathcal{E}}{R} \end{aligned}$$

$$I_4 = \frac{\mathcal{E}}{R}\left[1 - \left(\frac{r}{R+r}\right)e^{-t/\tau}\right]$$

as $t \to \infty$, $I_4 \to \frac{\mathcal{E}}{R}$ as $t \to 0$

$$\begin{aligned} I_4(0) &= \frac{\mathcal{E}}{R} - \frac{r}{R}\frac{\mathcal{E}}{(R+r)} \\ &= \frac{\mathcal{E}}{R}\left[\frac{R}{R+r}\right] = \frac{\mathcal{E}}{R+r} \end{aligned}$$

as expected. Then

$$\frac{dI_2}{dt} = \frac{r}{L_2}I_3 = \frac{r}{L_2}\frac{\mathcal{E}}{(R+r)}e^{-t/\tau}$$

Thus

$$I_2 = -\frac{r}{L_2}\frac{\mathcal{E}}{(R+r)}\tau e^{-t/\tau} + \text{const}$$

Since $I_2 = 0$ at $t = 0$,

$$\begin{aligned} \text{const} &= \frac{r\mathcal{E}}{L_2(R+r)}\tau \\ &= \frac{r\mathcal{E}}{L_2(R+r)}\frac{L_1L_2}{(L_1+L_2)}\frac{(R+r)}{rR} \end{aligned}$$

Thus

$$I_2 = \frac{\mathcal{E}}{R}\left(\frac{L_1}{L_1+L_2}\right)\left(1 - e^{-t/\tau}\right)$$

and then

$$I_1 = \frac{\mathcal{E}}{R}\frac{L_2}{(L_1+L_2)}\left(1 - e^{-t/\tau}\right)$$

and these relations agree with the long term behavior we found above.

b) Short term behavior.

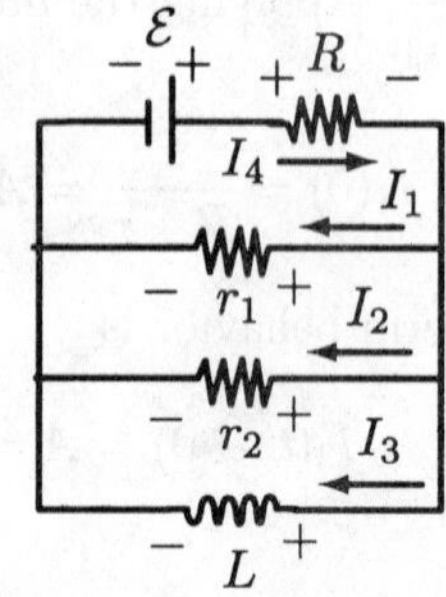

At $= 0$, there is no current through the inductor

$$I_3 = 0$$

r_1 and r_2 form a parallel combination with resistance

$$r_{\parallel} = \frac{r_1r_2}{r_1+r_2}$$

and the total resistance in the circuit is

$$R + r_{\parallel} = R + \frac{r_1r_2}{r_1+r_2} = R_s$$

The current

$$I_4 = \frac{\mathcal{E}}{R_s} = I_1 + I_2$$

and

$$\begin{aligned} I_1r_1 &= I_2r_2 \\ \Rightarrow I_1 &= \frac{\mathcal{E}}{R_s}\frac{r_2}{(r_1+r_2)} \\ I_2 &= \frac{\mathcal{E}}{R_s}\frac{r_1}{(r_1+r_2)} \end{aligned}$$

Long term behavior:

$$I_1 = I_2 \to 0$$

as $t \to \infty$, and

$$I_4 = I_3 = \frac{\mathcal{E}}{R}$$

Junction rule

$$I_4 = I_1 + I_2 + I_3$$

Loop rule

$$\mathcal{E} - I_4R = I_1r_1 = I_2r_2 = L\frac{dI_3}{dt}$$

Solve for I_3:

$$\left(\mathcal{E} - L\frac{dI_3}{dt}\right)\frac{1}{R} = \frac{L}{r_1}\frac{dI_3}{dt} + \frac{L}{r_2}\frac{dI_3}{dt} + I_3$$

$$\frac{dI_3}{dt}\left[\frac{L}{r_1} + \frac{L}{r_2} + \frac{L}{R}\right] + I_3 = \frac{\mathcal{E}}{R}$$

$$\left.\begin{aligned} I_{3,I} &= \tfrac{\mathcal{E}}{R} \\ I_{3,H} &= Ae^{-t/\tau} \\ \tau &= \tfrac{L}{r_1} + \tfrac{L}{r_2} + \tfrac{L}{R} \end{aligned}\right\} I_3 = \frac{\mathcal{E}}{R} + Ae^{-t/\tau}$$

which has the right form as $t \to \infty$. At $t = 0$,

$$\frac{\mathcal{E}}{R} + A = 0 \Rightarrow A = -\frac{\mathcal{E}}{R}$$

$$I_3 = \frac{\mathcal{E}}{R}(1 - e^{-t/\tau}), \tau = L\left(\frac{1}{r_1} + \frac{1}{r_2} + \frac{1}{R}\right)$$

Then

$$\begin{aligned} I_2 &= \frac{L}{r_2}\frac{dI_3}{dt} = \frac{L}{r_2}\cdot\frac{\mathcal{E}}{R}\frac{1}{\tau}e^{-t/\tau} \\ &= \frac{\mathcal{E}}{R}\frac{L}{r_2}\frac{r_1r_2R}{L(r_2R + r_1R + r_1r_2)}e^{-t/\tau} \\ I_2 &= \frac{\mathcal{E}r_1}{(r_2R + r_1R + r_1r_2)}e^{-t/\tau} \end{aligned}$$

Similarly

$$I_1 = \frac{\mathcal{E} r_2}{(r_2 R + r_1 R + r_1 r_2)} e^{-t/\tau}$$

and

$$\begin{aligned} I_4 &= I_1 + I_2 + I_3 \\ &= \frac{\mathcal{E}}{R}(1 - e^{-t/\tau}) + \frac{\mathcal{E}(r_1 + r_2)}{(r_1 + r_2)R + r_1 r_2} e^{-t/\tau} \\ &= \frac{\mathcal{E}}{R} - e^{-t/\tau}\mathcal{E}\frac{R(r_1 + r_2) + r_1 r_2 - R(r_1 + r_2)}{R\left[(r_1 + r_2)R + r_1 r_2\right]} \\ &= \frac{\mathcal{E}}{R}\left[1 - \frac{r_1 r_2}{R(r_1 + r_2) + r_1 r_2} e^{-t/\tau}\right] \end{aligned}$$

which has the expected behavior as $t \to 0$ and $t \to \infty$.

31.77 The inductance of the combination is $L = \frac{\Phi_B}{I}$.

First assume both coils are wound in the same sense. Then: A current I through coil one produces flux $\Phi_1 = L_1 I$ through coil 1 and $\Phi_{12} = MI$ through coil 2. Similarly a current I through coil 2 produces flux $L_2 I = \Phi_2$ through coil 2 and flux $\Phi_{21} = MI$ through coil 1. The total flux through the combination is:

$$\Phi_{\text{tot}} = L_1 I + L_2 I + 2MI$$

Then

$$L_{\text{eff}} = \frac{\Phi_{\text{tot}}}{I} = L_1 + L_2 + 2M$$

If the coils are wound in opposite senses, then the fields produced by the two coils are in opposite directions. Then

$$\Phi_{12} = -MI$$

and

$$L_{\text{eff}} = L_1 + L_2 - 2M$$

Finally, then

$$L_{\text{eff}} = L_1 + L_2 \pm 2M$$

When the inductors are in parallel,

$$I = I_1 + I_2$$

(Kirchhoff's junction rule.) The potential difference across each element is the same:

$$V = L_1\frac{dI_1}{dt} \pm M\frac{dI_2}{dt} = L_2\frac{dI_2}{dt} \pm M\frac{dI_1}{dt}$$

Thus

$$I_1(L_1 \mp M) = I_2(L_2 \mp M)$$

So

$$I = I_1\left[1 + \frac{L_1 \mp M}{L_2 \mp M}\right] = I_1\left(\frac{L_1 + L_2 \mp 2M}{L_2 \mp M}\right)$$

Thus

$$V = L_1\frac{dI}{dt}\left(\frac{L_2 \mp M}{L_1 + L_2 \mp 2M}\right) + M\frac{dI}{dt}\left(\frac{L_1 \mp M}{L_1 + L_2 \mp 2M}\right)$$

and

$$\begin{aligned} L_{\text{eff}} &= \frac{V}{\frac{dI}{dt}} \\ &= \frac{L_1(L_2 \mp M) \pm M(L_1 \mp M)}{L_1 + L_2 \mp 2M} \\ &= \frac{L_1 L_2 - M^2}{L_1 + L_2 \mp 2M} \end{aligned}$$

where the 2 signs account for the two possible senses of the windings. Notice that the $-$ sign corresponds to windings in the same sense and $+$ to opposite windings in this case. When $M \to 0$ this reduces to the result of problem 45.

31.81 Junction rule

$$I = I_1 + I_2 \tag{1}$$

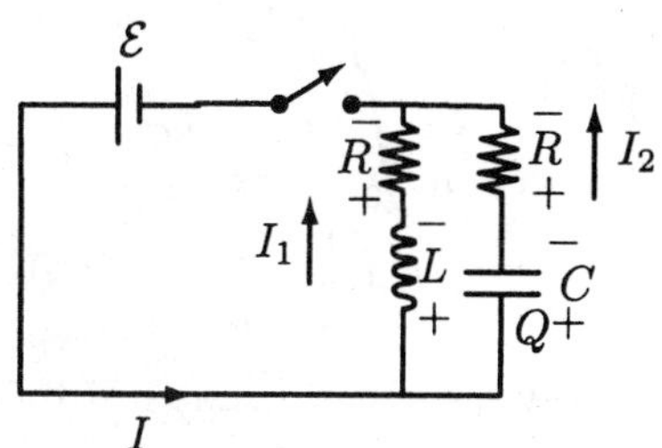

Loop rule, Left loop

$$\mathcal{E} = L\frac{dI_1}{dt} + I_1 R \tag{2}$$

Outside loop

$$\mathcal{E} = \frac{Q}{C} + I_2 R \tag{3}$$

At $t = 0$, $I_1 = 0, Q = 0$ and

$$I_2 = I = \frac{\mathcal{E}}{R}$$

As $t \to \infty$

$$\begin{aligned} I &= \text{constant} = \frac{\mathcal{E}}{R} = I_1 \\ I_2 &= 0 \\ Q &= \mathcal{E}C \end{aligned}$$

Since $I_2 = \frac{dQ}{dt}$, Eqn. (3) becomes:

$$\begin{aligned} \mathcal{E} &= \frac{Q}{C} + R\frac{dQ}{dt} \\ \frac{dQ}{dt} + \frac{Q}{RC} &= \frac{\mathcal{E}}{R} \end{aligned}$$

Thus

$$Q = \mathcal{E}C(1 - e^{-t/RC})$$

and

$$I_2 = \frac{\mathcal{E}}{R}e^{-t/RC}$$

From Eqn (2):

$$\begin{aligned} \frac{dI_1}{dt} + \frac{R}{L}I_1 &= \frac{\mathcal{E}}{L} \\ I_1 &= \frac{\mathcal{E}}{R}(1 - e^{-tR/L}) \end{aligned}$$

Thus

$$\begin{aligned} I &= I_1 + I_2 \\ &= \frac{\mathcal{E}}{R}\left(1 + e^{-t/RC} - e^{-tR/L}\right) \end{aligned}$$

31.85 The energy stored in the capacitor is

$$\begin{aligned} E_C(T) &= \frac{1}{C}\int_0^T Q\frac{dQ}{dt}\,dt \\ &= \frac{1}{C}\int_0^T \left[C\mathcal{E}\left(1 - e^{-t/(RC)}\right)\right] \\ &\quad \cdot\frac{\mathcal{E}}{R}e^{-t/(RC)}\,dt \\ &= \frac{\mathcal{E}^2}{R}\int_0^T \left(e^{-t/(RC)} - e^{-2t/(RC)}\right)dt \\ &= \mathcal{E}^2 C\left(\frac{1}{2} + \frac{1}{2}e^{-2T/(RC)} - e^{-T/(RC)}\right) \end{aligned}$$

The energy dissipated in the resistor is

$$\begin{aligned} E_R(T) &= \int_0^T I^2 R\,dt \\ &= \int_0^T \left(\frac{\mathcal{E}}{R}e^{-t/(RC)}\right)^2 R\,dt \\ &= \frac{\mathcal{E}^2}{R}\int_0^T e^{-2t/(RC)}\,dt \\ &= \frac{1}{2}\mathcal{E}^2 C\left(1 - e^{-2T/(RC)}\right) \end{aligned}$$

The energy drawn from the battery is

$$\begin{aligned} E_B(T) &= \mathcal{E}\int_0^T \frac{dQ}{dt}\,dt = \mathcal{E}\int_0^T \frac{\mathcal{E}}{R}e^{-t/(RC)}\,dt \\ &= \frac{\mathcal{E}^2}{R}\int_0^T e^{-t/(RC)}\,dt \\ &= \mathcal{E}^2 C\left(1 - e^{-T/(RC)}\right) \end{aligned}$$

Clearly at any time T,

$$E_C(T) + E_R(T) = E_B(T)$$

31.89 The sign conventions for currents are given in the figure, so we begin by writing Kirchhoff's rule equations. There are no junction equations.

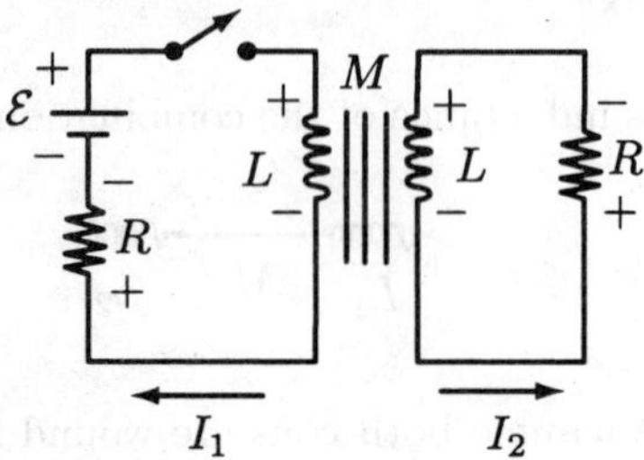

In loop 1:

$$\mathcal{E} - L\frac{dI_1}{dt} - \frac{MdI_2}{dt} - I_1 R = 0$$

(Here, we assume the coils are wrapped around their axes so that positive I_2 causes flux in the same direction as positive I_1.) In loop 2:

$$\frac{+LdI_2}{dt} + \frac{MdI_1}{dt} + I_2 R = 0$$

With $M = \frac{L}{2}$, the equations become:

$$\begin{aligned} \frac{\mathcal{E}}{R} &= \frac{L}{R}\frac{dI_1}{dt} + \frac{L}{2R}\frac{dI_2}{dt} + I_1 \\ 0 &= LR\frac{dI_2}{dt} + \frac{L}{2R}\frac{dI_1}{dt} + I_2 \end{aligned}$$

Now, let $Z_1 \equiv I_1 + I_2$ and $Z_2 \equiv I_1 - I_2$. Then, adding the two equations gives:

$$\frac{\mathcal{E}}{R} = 3\frac{L}{2R}\frac{dZ_1}{dt} + Z_1$$

Subtracting them gives:

$$\frac{\mathcal{E}}{R} = \frac{L}{2R}\frac{dZ_2}{dt} + Z_2$$

These are both the exponential equation. The initial state has

$$I_1(0) = I_2(0) = 0$$

so

$$Z_1(0) = Z_2(0) = 0$$

The final state has $I_2 = 0$ and $I_1 = \frac{\mathcal{E}}{R}$, so

$$Z_1(\infty) = Z_2(\infty) = \frac{\mathcal{E}}{R}$$

Thus, the solutions are

$$Z_1(t) = \frac{\mathcal{E}}{R}\left[1 - e^{-2Rt/3L}\right]$$

and

$$Z_2(t) = \frac{\mathcal{E}}{R}\left[1 - e^{-2Rt/L}\right]$$

Then,

$$\begin{aligned} I_1(t) &= \frac{Z_1 + Z_2}{2} \\ &= \frac{\mathcal{E}}{R}\left[1 - \frac{1}{2}e^{-Rt/3L} - \frac{1}{2}e^{-2Rt/L}\right] \end{aligned}$$

and

$$\begin{aligned} I_2(t) &= \frac{Z_1 - Z_2}{2} \\ &= \frac{\mathcal{E}}{2R}\left[e^{-2Rt/L} - e^{-2Rt/3L}\right] \end{aligned}$$

With the given numbers,

$$\frac{\mathcal{E}}{R} = \frac{12\ \text{V}}{1.6\ \text{k}\Omega} = 7.5\ \text{mA}$$

and

$$\frac{R}{L} = \frac{1.6\ \text{k}\Omega}{3.0\ \text{mH}} = 5.33 \times 10^5\ /\text{s}$$

Then $e^{-2Rt/3L} = 0.701$, and $e^{-2Rt/L} = 0.344$ at $t = 1.0\ \mu\text{s}$. So

$$\begin{aligned} I_1(1.0\ \mu\text{s}) &= (7.50\ \text{mA})\left[1 - \frac{1}{2}(0.701 + 0.344)\right] \\ &= 3.6\ \text{mA} \\ I_2(1.0\ \mu\text{s}) &= \left(\frac{7.50\ \text{mA}}{2}\right)[0.344 - 0.701] \\ &= -1.3\ \text{mA} \end{aligned}$$

Chapter 32

Introduction to AC Circuits

Notation: $\mathcal{V}$ = voltage phasor, $\mathcal{I}$ = current phasor.

32.1 The reactance of the capacitor, X_C, is related to the angular frequency ω by the equation:

$$|X_C| = \frac{1}{\omega C}$$

where $\omega = 2\pi f$ and $f = 60.0$ Hz, then:

$$\begin{aligned} |X_C| &= \frac{1}{(2\pi f)(C)} \\ &= \frac{1}{(6.28)(60.0\text{ Hz})(5.6\times 10^{-6}\text{ F})} \end{aligned}$$

thus

$$|X_C| = 470\ \Omega$$

Notice that the unit of reactance is the same as the unit of resistance.

32.3 $P = 60$ W, $V_{\text{rms}} = 120$ V. The power P of the bulb is related to the voltage by $P = VI$ where I is the current. The definition of resistance:

$$I = \frac{V}{R}$$

so that

$$P = (V)(\frac{V}{R}) = \frac{V^2}{R}$$

Thus averaging over time, the resistance is given by

$$R = \frac{\langle V^2\rangle}{\langle P\rangle} = \frac{(120V_{\text{rms}})^2}{60\text{ W}} = 240\ \Omega$$

32.5 $f = 0.5\times 10^3$ Hz, $L = 5\times 10^{-3}$ H, $R = 10.0\ \Omega$ and $V_0 = 10.0$ V.

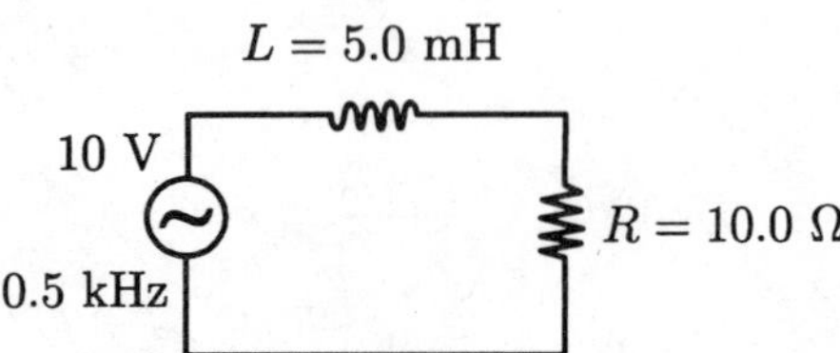

The angular frequency ω is related to the signal frequency $(0.5\times 10^3$ Hz) by

$$\omega = 2\pi f = 3.14\times 10^3\ \text{rad/s}$$

The inductive reactance X_L is

$$\begin{aligned} X_L &= \omega L \\ &= (3.14\times 10^3\text{ Hz})\,(5\times 10^{-3}\text{ H}) \\ &= 15.7\,\Omega \end{aligned}$$

The impedance Z of the circuit is

$$\begin{aligned} Z &= \sqrt{R^2 + X_L^2} \\ &= \sqrt{(10.0\ \Omega)^2 + (15.7\ \Omega)^2} \\ &\approx 18.6\,\Omega \end{aligned}$$

The current amplitude is found from Eqn. 32.18:

$$I_0 = \frac{V_0}{Z} = \frac{10.0\text{ V}}{18.6\ \Omega} = 0.54\text{ A}$$

The average power is found by using the rms current

$$I_{\text{rms}} = \frac{I_0}{\sqrt{2}}$$

and the resistance R in the relation

$$\begin{aligned} P_{\text{ave}} &= I_{\text{rms}}^2 \times R = (0.38\text{ A})^2\,(10.0\ \Omega) \\ &= 1.4\text{ W} \end{aligned}$$

32.7 The current as a function of time is related to the circuit parameters by the equation:

$$I(t) = \frac{V_0}{Z}\cos(\omega t - \phi)$$

where

$$I_0 = \frac{V_0}{Z}$$

is the current amplitude and ϕ is the phase angle. First, compute the parameters in the equation $I(t)$.

$$V_0 = \left(\sqrt{2}\right) V_{\rm rms} = (1.414)(120\ \text{V}) = 169.7\ \text{V}$$

To compute the impedance Z, we need the inductive reactance

$$\begin{aligned} X_L &= \omega L = (2\pi)(60\ \text{Hz})(2.2 \times 10^{-3}\ \text{H}) \\ &= 0.829\ \Omega \\ Z &= \sqrt{R^2 + X_L^2} \\ &= \sqrt{(5.60\ \Omega)^2 + (0.829\ \Omega)^2} \\ &= 5.66\ \Omega \end{aligned}$$

The phase angle ϕ is

$$\phi = \tan^{-1}\left(\frac{X_L}{R}\right)$$

i.e.,

$$\phi = \tan^{-1}\left(\frac{0.829}{5.60}\right) = 8.4^\circ = 0.147\ \text{rad}$$

and

$$\begin{aligned} I(t) &= \left(\frac{169.7}{5.66}\right)\cos(\omega t - 0.147\ \text{rad}) \\ &= (30\ \text{A})\cos(\omega t - 0.147\ \text{rad}) \end{aligned}$$

The amplitude has 2 significant figures.

32.9 See figure.

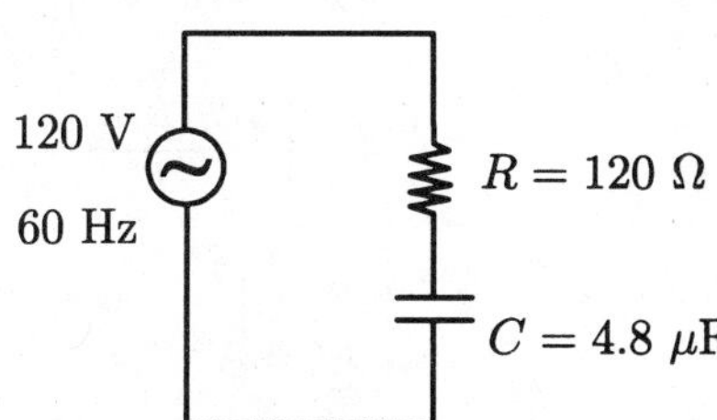

The capacitative reactance is

$$\begin{aligned} X_C &= \frac{1}{2\pi f C} = \frac{1}{2\pi(60)(4.8 \times 10^{-6}\ \text{F})} \\ &= 553\ \Omega \end{aligned}$$

The impedance is given by

$$\begin{aligned} Z &= \sqrt{R^2 + X_C^2} \\ &= \sqrt{(120\ \Omega)^2 + (553\ \Omega)^2} \\ &= 566\ \Omega \end{aligned}$$

The current I in the circuit is

$$I = \frac{V}{Z}$$

The voltage drop across the resistor is

$$V_R = IR$$

and across the capacitor is

$$V_C = IX_C$$

The capacitor voltage lags the current by 90°. From the phasor diagram,

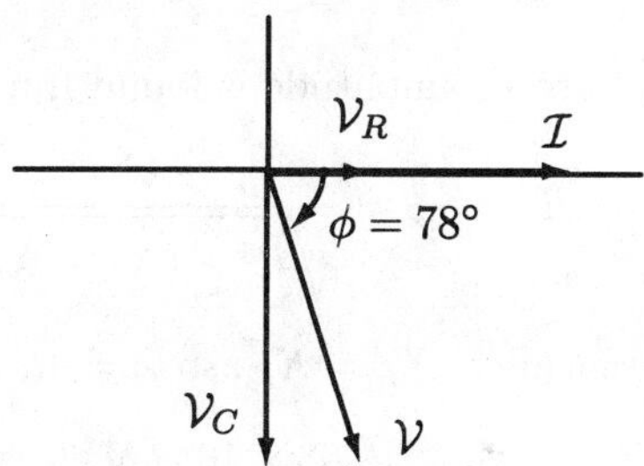

$$\begin{aligned} \tan\phi &= \frac{V_C}{V_R} = \frac{X_C}{R} = \frac{552.6\ \Omega}{120\ \Omega} \\ &= 4.605 \end{aligned}$$

So $\phi = 77.8^\circ$ and

$$\cos\phi = \frac{V_R}{V} = \frac{R}{Z}$$

The power is

$$\begin{aligned} P &= \frac{\langle V^2 \rangle}{Z}\cos\phi = \frac{(120\ \text{V})^2 R}{Z^2} \\ &= \frac{(120\ \text{V})^2 (120\ \Omega)}{(566\ \Omega)^2} \\ &= 5.4\ \text{W} \end{aligned}$$

32.11 First find expressions for L and R. We know that Q can be obtained from

$$Q = \frac{\omega_0 L}{R}$$

from which

$$L = \frac{QR}{\omega_0}$$

At resonance, $X_L = X_C$, i.e.,

$$\omega_0 L = \frac{1}{\omega_0 C}$$

from which

$$L = \frac{1}{\omega_0^2 C}$$

Eliminate L by equating the equations for L to obtain

$$\begin{aligned} R &= \frac{1}{\omega_0 QC} = \frac{1}{2\pi f_0 CQ} \\ &= \frac{1}{(6.28)(610 \times 10^3 \text{ Hz})(56 \times 10^{-9} \text{ F})(60)} \\ &= 0.078\ \Omega \\ L &= \frac{(60)(0.078\ \Omega)}{2\pi f_0} \\ &= \frac{4.7}{(6.28)(610 \times 10^3 \text{ Hz})} \\ &= 1.2 \times 10^{-6} \text{ H} \\ &= 1.2\ \mu\text{H} \end{aligned}$$

The current amplitude is found from

$$I_0 = \frac{V_0}{Z} = \frac{V_0}{\sqrt{R^2 + (X_L - X_C)^2}}$$

at resonance, $X_L = X_C$, so $Z = R$, and

$$I_0 = \frac{V_0}{R} = \frac{(0.68 \times 10^{-3} \text{ V})}{0.078\ \Omega} = 8.7 \text{ mA}$$

32.13 The reactance of an inductor is given by:

$$X_L = \omega L$$

therefore, the larger inductance $2L$ has the larger reactance by a factor of 2 i.e.,

$$X_{2L} = \omega(2L) = 2\omega L = 2X_L$$

For a given rate of change of current a larger inductor produces a large potential difference:

$$\Delta V = L\frac{dI}{dt}$$

Thus $|X| = \frac{V_0}{I_0}$ also increases as L increases.

32.17 The important consideration is the light travel time $\tau = \frac{L}{c}$ for the system. We need

$$f < \frac{1}{\tau} = \frac{c}{L}$$

a) For the United States, $L \cong 5000$ km. Thus

$$f < \frac{3 \times 10^8 \text{ m/s}}{5 \times 10^6 \text{ m}} \sim 60 \text{ Hz}$$

b) For Denver, $L \cong 30$ km

$$\begin{aligned} f &< \frac{3 \times 10^8 \text{ m/s}}{30 \times 10^3 \text{ m}} = 10^4 \text{ Hz} \\ &= 10 \text{ kHz} \end{aligned}$$

c) In a 2-m-diameter supercomputer

$$\begin{aligned} f &< \frac{3 \times 10^8 \text{ m/s}}{2 \text{ m}} = 1.5 \times 10^8 \text{ Hz} \\ &= 150 \text{ MHz} \end{aligned}$$

d) In a microprocessor chip:

$$\begin{aligned} f &< \frac{3 \times 10^8 \text{ m/s}}{10^{-2} \text{ m}} = 3 \times 10^{10} \text{ Hz} \\ &= 30 \text{ GHz} \end{aligned}$$

32.21 The average power delivered to the hair dryer is related to the voltage and the resistance by:

$$\langle P \rangle = \frac{\langle V^2 \rangle}{R} = \frac{V_{\text{rms}}^2}{R}$$

The resistance is therefore,

$$R = \frac{V_{\text{rms}}^2}{\langle P \rangle} = \frac{(120 \text{ V})^2}{1200 \text{ W}} = 12\ \Omega$$

32.25 To apply the definition (32.6) of V_{rms}, we first need $V(t)$. For the triangle wave we have

$$\begin{aligned} V(t) &= \frac{V_0 t}{\frac{T}{4}}, & 0 \le t < \frac{T}{4} \\ &= 2V_0 - \frac{V_0 t}{\frac{T}{4}}, & \frac{T}{4} < t < \frac{3T}{4} \\ &= \frac{V_0(t-T)}{\frac{T}{4}}, & \frac{3T}{4} < t < T \end{aligned}$$

Thus

$$\begin{aligned} \langle V^2 \rangle &= \frac{1}{T}\left[\int_0^{\frac{T}{4}} \left(\frac{4V_0 t}{T}\right)^2 dt \right. \\ &\quad + \int_{\frac{T}{4}}^{\frac{3T}{4}} \left(2V_0 - \frac{4V_0 t}{T}\right)^2 dt \\ &\quad \left. + \int_{\frac{3T}{4}}^{T} \left[\frac{4V_0(t-T)}{T}\right]^2 dt \right] \\ &= \frac{V_0^2}{T}\left[\frac{16}{T^2}\frac{t^3}{3}\Big|_0^{\frac{T}{4}} \right. \\ &\quad + \frac{1}{3}\left(2 - \frac{4t}{T}\right)^3 \left(-\frac{T}{4}\right)\Big|_{\frac{T}{4}}^{\frac{3T}{4}} \\ &\quad \left. + \frac{1}{3}\frac{T}{4}\left(\frac{4(t-T)}{T}\right)^3\Big|_{\frac{3T}{4}}^{T} \right] \end{aligned}$$

$$= \frac{V_0^2}{T}\left[\frac{T}{4\cdot 3} + \left(\frac{-T}{4\cdot 3}\right)(-1-1) + \frac{T}{3\cdot 4}\left[0-(-1)\right]\right]$$

$$= \frac{V_0^2}{3\cdot 4}(1+2+1) = \frac{V_0^2}{3}$$

Thus

$$V_{\text{rms}} = \frac{V_0}{\sqrt{3}}$$

Since $(-V)^2 = V^2$, V_{rms} for the rectified triangle wave would be the same $\frac{V_0}{\sqrt{3}}$.

32.29 The power increases as ω increases.

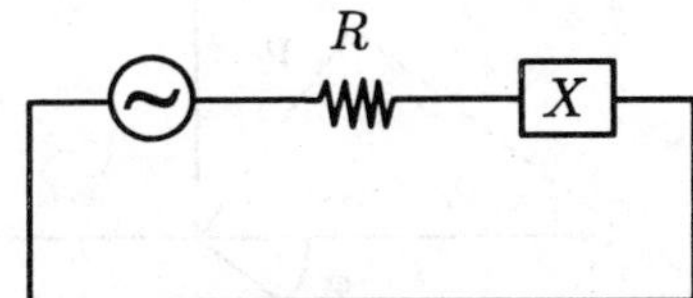

We may immediately conclude that X is not a resistor, since power is independent of frequency if the circuit contains only resistance (Eqn. 32.7) In a 2-component circuit, the average power is

$$\langle P\rangle = \frac{V_{\text{rms}}^2}{Z}\cos\phi = V_{\text{rms}}^2 \cdot \frac{R}{Z^2}$$

(Eqn. 32.20), where $Z^2 = R^2 + X^2$ is a function of frequency. If $\langle P\rangle$ increases as ω increases, then Z^2 decreases as ω increases. *Thus X is a capacitor*:

$$X = \frac{1}{\omega c}$$

decreases as ω increases.

32.33 The requirement is that the total reactance $X_L - X_C = 0$, or $X_L = X_C$. Then

$$\omega L = \frac{1}{\omega C}$$

and

$$\omega^2 LC = 1$$

solving for C gives

$$C = \frac{1}{\omega^2 L} = \frac{1}{(2\pi f)^2 L}$$

with $f = 60$ Hz and $L = 8.0$ H

$$C = \frac{1}{(2\pi \times 60\text{ Hz})^2(8.0\text{ H})} = 0.88\ \mu\text{F}$$

32.37 The inductive and capacitative reactances are given by $X_L = \omega L$ and so

$$V_L = I\omega L$$
$$X_C = \frac{1}{\omega C}$$

and so

$$V_C = \frac{I}{\omega}$$

The ratio of the potential difference across an inductor to that across the capacitor is then

$$\frac{V_L}{V_C} = \frac{I\omega L}{\frac{I}{\omega C}} = \omega^2 LC$$

Therefore, the potential differences are only equal at resonance, i.e., when

$$\omega^2 LC = 1$$

or

$$\omega = \frac{1}{\sqrt{LC}}$$
$$f = \frac{\omega}{2\pi} = \frac{1}{2\pi\sqrt{LC}}$$
$$= \frac{1}{2\pi}\frac{1}{\sqrt{(1.9\times 10^{-3}\text{ H})(53\times 10^{-6}\text{ F})}}$$
$$= 0.50\text{ kHz}$$

The woofer should be connected to the capacitor since the potential difference is greater across the capacitor at low frequencies; and the tweeter should be connected to the inductor since the potential difference across the inductor is greater at high frequencies (see plot.)

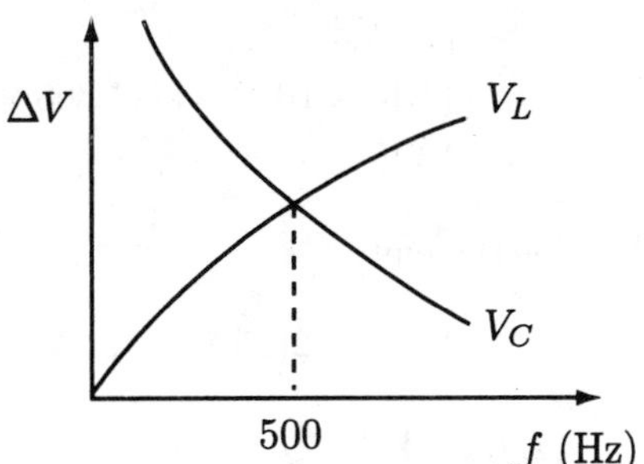

32.41 See figure.

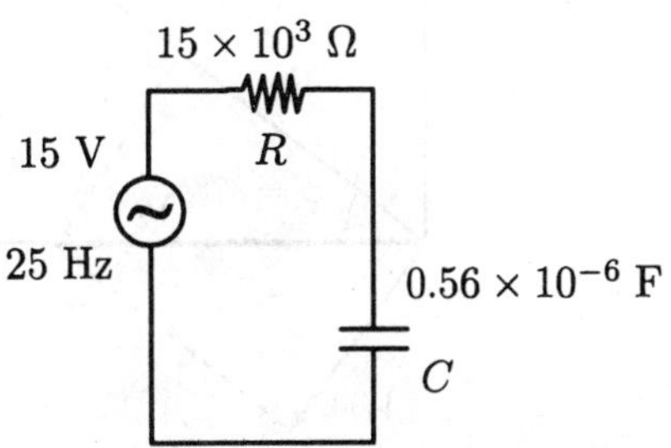

The capacitance reactance is

$$\begin{aligned} X_C &= \frac{1}{\omega C} \\ &= \frac{1}{(2\pi)(25\text{ Hz})(.56\times 10^{-6}\text{ F})} \\ &= 11.4\text{ k}\Omega \end{aligned}$$

The impedance Z is therefore

$$Z = \sqrt{R^2 + X_C^2} = 18.8\text{ k}\Omega$$

and so the current amplitude is

$$I_0 = \frac{V_0}{Z} = \frac{\sqrt{2}(15\text{ V})}{18.8\text{ k}\Omega} = 1.1\text{ mA}$$

The magnitude of the phase shift is given by

$$\begin{aligned} \tan|\phi| &= \frac{|X_c|}{R} \\ &= \left(\frac{11.4\text{ k}\Omega}{15\text{ k}\Omega}\right) \\ \phi &= -37.2^\circ = .21\pi\text{ rad} \\ I(t) &= I_0\cos(\omega t - \phi) \\ &= 1.1\times 10^{-3}\text{ A}\cos(50\pi t + 0.21\pi) \end{aligned}$$

The voltage phasors for the resistor and capacitor have magnitude

$$\begin{aligned} |\mathcal{V}_R| &= |\mathcal{I}|\,R \\ &= (1.13\times 10^{-3}\text{ A})\,(15\times 10^3\ \Omega) \\ &= 17\text{ V} \end{aligned}$$

and

$$\begin{aligned} |\mathcal{V}_c| &= |\mathcal{I}|\,|X_c| \\ &= (1.13\times 10^{-3}\text{ A})\,(11.368\times 10^3\ \Omega) \\ &= 13\text{ V} \end{aligned}$$

while for the emf

$$|\mathcal{V}_\mathcal{E}| = \sqrt{2}(15\text{ V}) = 21\text{ V}$$

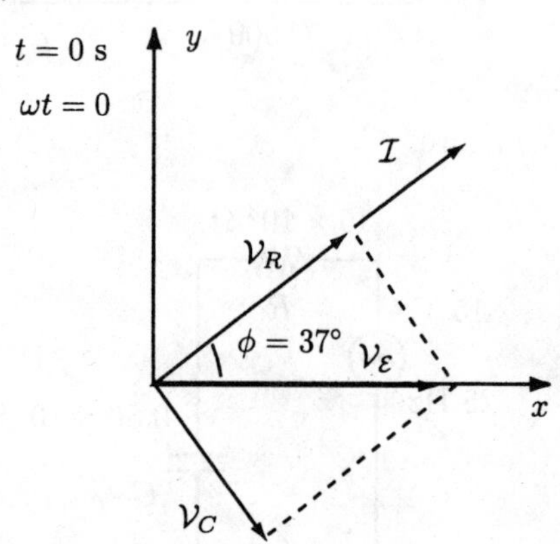

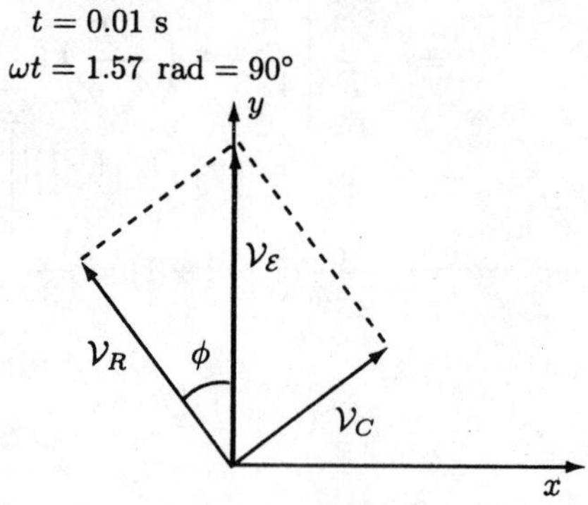

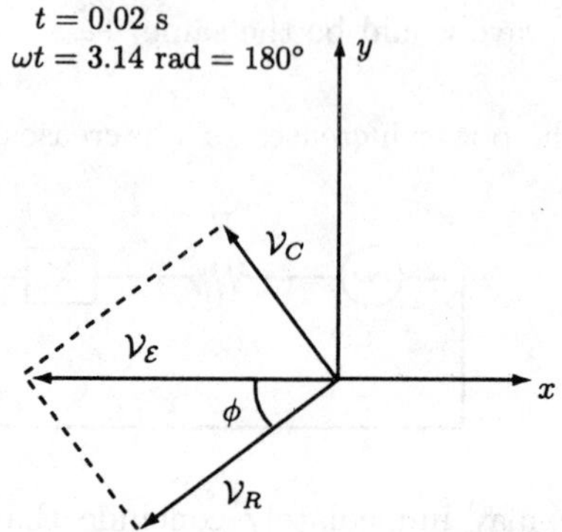

32.45 See figure.

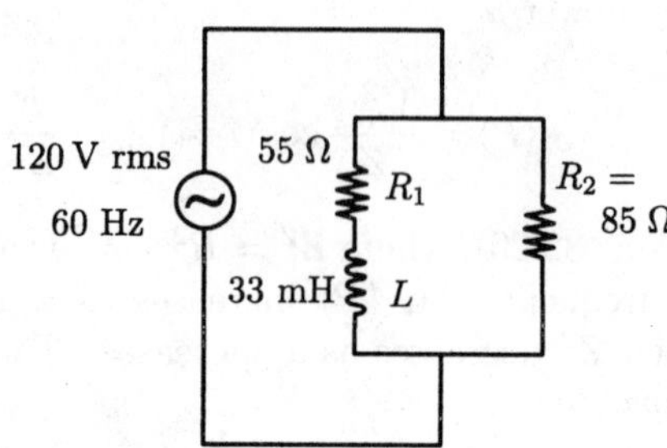

The circuit may be redrawn by replacing the series combination of R_1 and L with a single impedance Z_1:

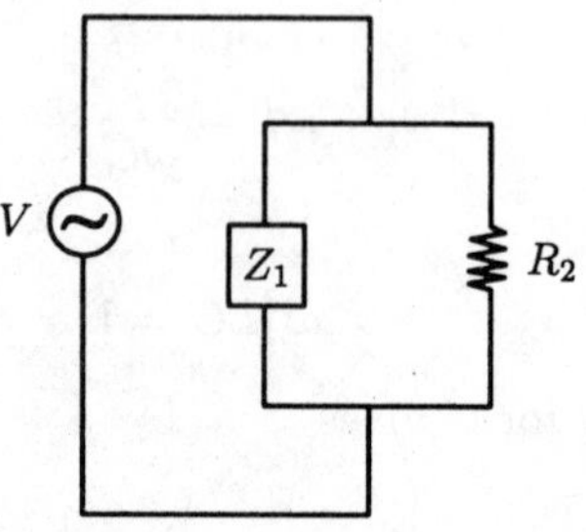

where

$$Z_1 = \sqrt{R_1^2 + X_L^2}$$

The phasor diagram for the series combination of R_1 and L is given by

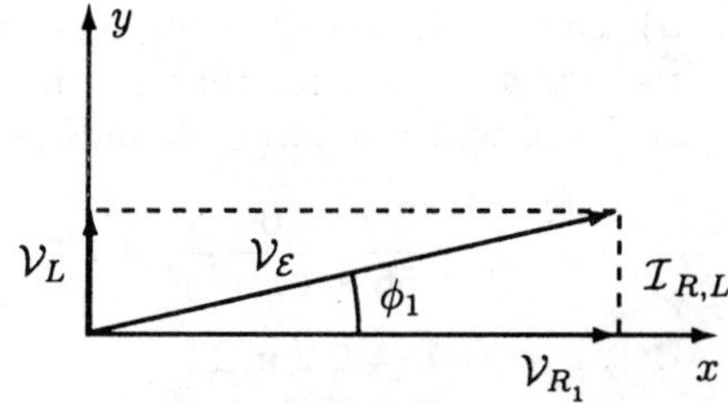

$$\begin{aligned}\phi_1 &= \tan^{-1}\left(\frac{\mathcal{V}_L}{\mathcal{V}_{R_1}}\right) = \tan^{-1}\left(\frac{X_L}{R_1}\right) \\ &= \tan^{-1}\left(\frac{\omega L}{R_1}\right) \\ &= \tan^{-1}\frac{2\pi\,(60\text{ Hz})\,(33\times10^{-3}\text{ H})}{(55\ \Omega)} \\ &= 12.7^\circ\end{aligned}$$

$$\begin{aligned}Z_1 &= \sqrt{R_1^2+\omega^2L^2} \\ &= \sqrt{(55\ \Omega)^2+(2\pi\,(60\text{ Hz})\,(33\times10^{-3}\text{ H}))^2} \\ &= 56.4\ \Omega\end{aligned}$$

By Kirchhoff's junction rule, the total current through the parallel combination must equal the phasor sum of the individual current phasors. The current through the impedance Z, has magnitude

$$|\mathcal{I}_{R,L}| = \frac{|\mathcal{V}_\mathcal{E}|}{Z_1}$$

and

$$|\mathcal{I}_{R_2}| = \frac{|\mathcal{V}_\mathcal{E}|}{R_2}$$

On a phasor diagram these two currents are represented as

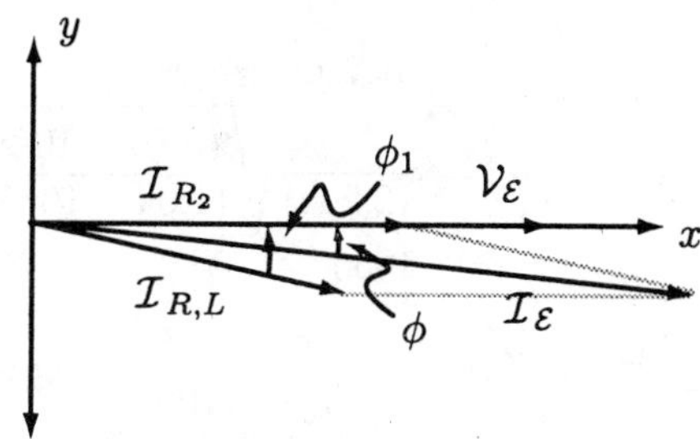

The total current is the vector sum of $\mathcal{I}_{R_2}$ and $\mathcal{I}_{R,L}$:

$$\begin{aligned}|\mathcal{I}| &= \sqrt{(|\mathcal{I}_{R,L}|\cos\phi_1+|\mathcal{I}_{R_2}|)^2+|\mathcal{I}_{R,L}|^2\sin^2\phi_1} \\ &= \sqrt{|\mathcal{I}_{R,L}|^2+|\mathcal{I}_{R_2}|^2+2\,|\mathcal{I}_{R,L}|\,|\mathcal{I}_{R_2}|\cos\phi_1} \\ &= \mathcal{V}_\mathcal{E}\sqrt{\frac{1}{Z_1^2}+\frac{1}{R_2^2}+\frac{2}{R_2Z_1}\cos\phi_1}\end{aligned}$$

The phase of the current with respect to the EMF, $\mathcal{V}_\mathcal{E}$, is given by

$$\tan|\phi| = \frac{|\mathcal{I}_{R,L}|\sin\phi_1}{|\mathcal{I}_{R_2}|+|\mathcal{I}_{R,L}|\cos\phi_1}$$

where ϕ will be negative since the current lags $\mathcal{V}_\mathcal{E}$. Thus

$$\begin{aligned}\tan|\phi| &= \frac{\left(\frac{1}{Z_1}\right)\sin\phi_1}{\frac{1}{R_2}+\frac{1}{Z_1}\cos\phi_1} \\ &= \frac{\sin\phi_1}{\frac{Z_1}{R_2}+\cos\phi_1}\end{aligned}$$

putting in numerical values

$$\begin{aligned}\cos\phi_1 &= \cos(12.7^\circ) = 0.975 \\ \sin\phi_1 &= \sin(12.7^\circ) = 0.221\end{aligned}$$

The current phasor's magnitude is

$$\begin{aligned}|\mathcal{I}| &= (170\text{ V}) \\ &\quad\times\sqrt{\frac{1}{(56.4\ \Omega)^2}+\frac{1}{(85\ \Omega)^2}+\frac{2(0.975)}{(56.4\ \Omega)(85\ \Omega)}} \\ &= 4.98\text{ A}\end{aligned}$$

$$\begin{aligned}\tan|\phi| &= \frac{(0.221)}{\frac{(56.4\ \Omega)}{(85\ \Omega)}+(0.975)} = 0.135 \\ \phi &= -7.7^\circ\end{aligned}$$

The power consumed by both the television and the vacuum cleaner is

$$\begin{aligned}\langle P(t)\rangle &= \frac{1}{2}\mathcal{V}\cdot\mathcal{I} \\ &= \frac{1}{2}\mathcal{V}_\mathcal{E}\,|\mathcal{I}|\cos(\phi) \\ &= \frac{1}{2}(170.0\text{ V})\,(4.98\text{ A})\cos(-7.7^\circ) \\ &= 420\text{ W}\end{aligned}$$

The power consumed by the TV set is

$$\begin{aligned}\langle P_{\text{TV}}\rangle &= \frac{1}{2}|\mathcal{I}_R|\,\mathcal{V}_\mathcal{E} \\ &= \frac{1}{2}\frac{\mathcal{V}_\mathcal{E}^2}{R_1} \\ &= \frac{1}{2}\frac{(170.0\text{ V})^2}{(85\ \Omega)} \\ &= 170\text{ W}\end{aligned}$$

The average power consumed by the vacuum cleaner is

$$\begin{aligned}\langle P_{\text{vac}}\rangle &= \frac{1}{2}|\mathcal{I}_{R,L}|\,\mathcal{V}_\mathcal{E}\cos\phi_1 \\ &= \frac{1}{2}\frac{\mathcal{V}_\mathcal{E}^2}{Z_1}\cos\phi_1 \\ &= \frac{1}{2}\frac{(170.0\text{ V})^2}{(56.4\ \Omega)}\cos(12.7^\circ) \\ &= 250\text{ W}\end{aligned}$$

Since

$$\langle P(t)\rangle = 420 \text{ W}$$

the respective fractions are given by

$$\begin{aligned} f_{\text{TV}} &= \frac{170 \text{ W}}{420 \text{ W}} = 0.40 = 40\% \\ f_{\text{vac}} &= \frac{250 \text{ W}}{420 \text{ W}} = 0.60 = 60\% \end{aligned}$$

Therefore, 40% of the power is consumed by the TV set and 60% of the power is consumed by the vacuum cleaner.

32.49 If $C \to 4C$ i.e., if the capacitance is quadrupled, then

$$\begin{aligned} \omega_0 &= \frac{1}{\sqrt{LC}} \to \frac{1}{\sqrt{L(4C)}} \\ &= \frac{1}{2}\omega_0 \\ \tau &= 2\frac{L}{R} \end{aligned}$$

remains unchanged

$$Q = \frac{1}{R}\sqrt{\frac{L}{C}} \to \frac{1}{R}\sqrt{\frac{L}{4C}} = \frac{1}{2}Q$$

Thus ω_0 and Q are each halved, τ is unchanged.

32.53 Since $\omega_0 = \frac{1}{\sqrt{LC}}$, to get the maximum resonant frequency we need the minimum values of L and C. $L = 0.75$ mH and $C = 56$ pF.

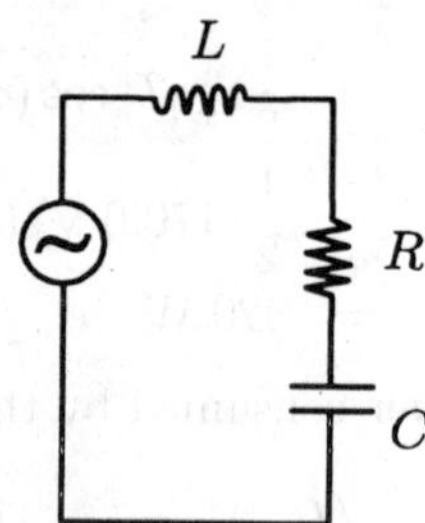

a)

$$\begin{aligned} f_0 &= \frac{1}{2\pi\sqrt{LC}} \\ &= \frac{1}{2\pi\sqrt{(.75 \times 10^{-3} \text{ H})(56 \times 10^{-12} \text{ F})}} \\ &= 777 \text{ kHz} = .78 \text{ MHz} \end{aligned}$$

The Q value is related to L and R by the expression

$$\begin{aligned} Q_0 &= \frac{\omega_0 L}{R} = \frac{2\pi f_0 L}{R} \\ &= \frac{(6.28)(777 \text{ kHz})(0.75 \times 10^{-3} \text{ H})}{0.95 \; \Omega} \\ &= 3850 \approx 3900 \end{aligned}$$

b) The maximum Q is obtained when the $\frac{L}{C}$ ratio is a maximum, that is, the largest Q is achieved with the larger L and the smaller C.

$$\frac{L}{C} = \frac{2.2 \times 10^{-3} \text{ H}}{56 \times 10^{-12} \text{ F}} = 3.9 \times 10^7$$

$L = 2.2$ mH

$C = 56$ μF

$R = 0.95 \; \Omega$

Then

$$\begin{aligned} Q &= \frac{1}{R}\sqrt{\frac{L}{C}} \\ &= \frac{1}{0.95 \; \Omega}\sqrt{3.9 \times 10^7} \; \Omega \\ &= 6600 \end{aligned}$$

The resonant frequency for this circuit is

$$f_0 = \frac{1}{2\pi\sqrt{LC}} = 0.45 \text{ MHz}$$

32.59 The capacitance required to tune the circuit is

$$\begin{aligned} C &= \frac{1}{\omega_0^2 L} \\ &= \frac{1}{4\pi^2 (660 \times 10^3 \text{ Hz})^2 (0.56 \times 10^{-3} \text{ H})} \\ &= 1.04 \times 10^{-10} \text{ F} \end{aligned}$$

The Q value is

$$\begin{aligned} Q &= \frac{1}{R}\sqrt{\frac{L}{C}} \\ &= \frac{1}{(2.2 \; \Omega)}\sqrt{\frac{(0.56 \times 10^{-3} \text{ H})}{(1.04 \times 10^{-10} \text{ F})}} \\ &= 1056 \end{aligned}$$

Thus

$$Q = 1100$$

Now the capacitor voltage is given by Equation (32.28):

$$V_C(\omega) = \frac{V_0 Q}{\sqrt{\left(\frac{\omega}{\omega_0}\right)^2 + Q^2\left[\left(\frac{\omega}{\omega_0}\right)^2 - 1\right]^2}}$$

Since,

$$\frac{\omega}{\omega_0} = \frac{f}{f_0} = \frac{600 \text{ Hz}}{660 \text{ Hz}} = 0.909$$

We have

$$\begin{aligned}\frac{V_C(\omega)}{V_0 Q} &= \frac{1}{\sqrt{(0.909)^2 + (1056)^2\left[(0.909)^2 - 1\right]^2}} \\ &= 5.5\times 10^{-3}\end{aligned}$$

The unwanted input signal is only 0.55% of the value of the selected signal.

32.63 The maximum energy stored in the inductor is given by the current amplitude

$$U_{L,\max} = \frac{1}{2}L\,I_{\max}^2 = \frac{1}{2}L\left(\frac{V_0}{Z}\right)^2.$$

The maximum energy stored in the capacitor is given by the voltage amplitude

$$U_{C,\max} = \frac{1}{2}C\,V_{\max}^2 = \frac{1}{2}C\left[X_C\frac{V_0}{Z}\right]^2$$

So, the ratio

$$\begin{aligned}\frac{U_{C,\max}}{U_{L,\max}} &= \frac{\frac{1}{2}CX_C^2\frac{V_0^2}{Z^2}}{\frac{1}{2}L\frac{V_0^2}{Z^2}} = \frac{C}{L}\left(\frac{1}{\omega C}\right)^2 \\ &= \frac{1}{\omega^2 LC} = \frac{\omega_0^2}{\omega^2}\end{aligned}$$

as required.

Now, the emf of the generator is

$$\mathcal{E}(t) = V_0\cos\omega t$$

and the current in the circuit is

$$I(t) = \frac{V_0}{Z}\cos(\omega t - \phi)$$

We desire the work done by the generator between the time of maximum capacitor charge

$$I(t) = 0 \Longleftrightarrow \omega t - \phi = \frac{\pi}{2}$$

and the time of max current

$$\omega t - \phi = \pi$$

We expect that

$$\begin{aligned}W_{\text{generator}} &= W_{\text{resistor}} + \Delta U_{\text{stored}} \\ &= W_{\text{resistor}} + U_{L,\max} - U_{C,\max}\end{aligned}$$

Now

$$U_{L,\max} - U_{C,\max} = \frac{V_0^2}{2Z^2}\left[L - \frac{1}{\omega^2 C}\right]$$

Now

$$\begin{aligned}W_{\text{generator}} &= \int_{\omega t=\phi+\frac{\pi}{2}}^{\omega t=\phi+\pi} \mathcal{E}(t)\,I(t)\,dt \\ &= \frac{V_0^2}{Z}\int_{\omega t=\phi+\frac{\pi}{2}}^{\phi+\pi}\cos\omega t\cos(\omega t-\phi)\,dt \\ &= \frac{V_0^2}{\omega Z}\int_{\phi+\frac{\pi}{2}}^{\phi+\pi}\cos u\cos(u-\phi)\,du\end{aligned}$$

$$\begin{aligned}&= \frac{V_0^2}{\omega Z}\int_{\phi+\frac{\pi}{2}}^{\phi+\pi}\left[\cos^2 u\cos\phi + \cos u\sin u\sin\phi\right]du \\ &= \frac{V_0^2}{\omega Z}\left[\cos\phi\left[\frac{u}{2}+\frac{\sin 2u}{4}\right] + \frac{1}{2}\sin\phi\sin^2 u\right]_{\phi+\frac{\pi}{2}}^{\phi+\pi} \\ &= \frac{V_0^2}{\omega Z}\left[\cos\phi\left[\frac{\pi}{4}+\frac{\sin(2\phi+2\pi)-\sin(2\phi+\pi)}{4}\right]\right. \\ &\qquad \left. + \frac{1}{2}\sin\phi\left[\sin^2(\phi+\pi) - \sin^2\left(\phi+\frac{\pi}{2}\right)\right]\right] \\ &= \frac{V_0^2}{\omega Z}\left[\frac{\pi}{4}\cos\phi + \frac{1}{2}\sin 2\phi\cos\phi\right. \\ &\qquad \left. + \frac{1}{2}\sin\phi\left(\sin^2\phi - \cos^2\phi\right)\right] \\ &= \frac{V_0^2}{\omega Z}\left[\frac{\pi}{4}\cos\phi + \frac{1}{2}\sin\phi\right]\end{aligned}$$

Now, note that the first term is $\frac{V_0^2}{2}\frac{\cos\phi}{Z}\frac{\pi}{2\omega}$, which is the average power dissipated by the resistor multiplied by $\frac{1}{4}$ period; that is, the work dissipated by the resistor. Since

$$\tan\phi = \frac{X_L - |X_C|}{R} \qquad \text{(Eqn. 32.25)}$$

we have

$$\sin\phi = \frac{X_L - |X_C|}{\sqrt{R^2 + (X_L - |X_C|)^2}} = \frac{\omega L - \frac{1}{\omega}C}{Z}$$

So, the second term in $W_{\text{generator}}$ is

$$\begin{aligned}\frac{V_0^2}{\omega Z}\cdot\frac{\omega}{2}\frac{\left(L - \frac{1}{\omega^2 C}\right)}{Z} &= \frac{V_0^2}{2Z^2}\left(L - \frac{1}{\omega^2 C}\right) \\ &= U_{L,\max} - U_{C,\max}\end{aligned}$$

as required.

32.67 The resistors form a balanced Wheatstone bridge. So, the *steady state behavior* results in *no potential differences across the capacitor.* There could be a transient effect due to some initial charge on the capacitor.

32.71 Since the two arms of the circuit have a common voltage phasor, equal to that of the AC generator, draw the voltage phasor at a time when it is along the x-axis.

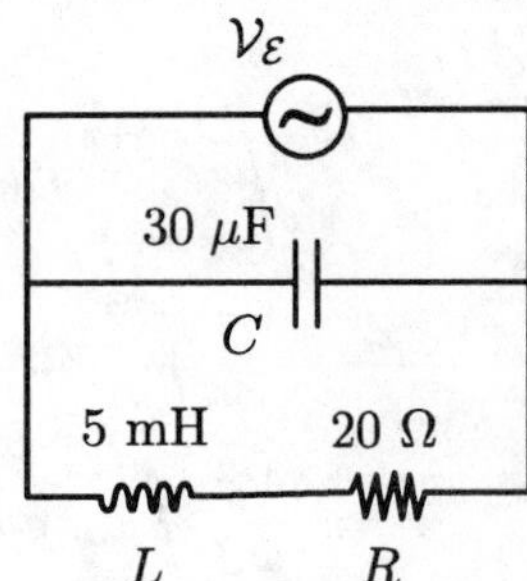

The capacitor current leads the voltage phasor, so its phasor lies along the $+y$-axis.

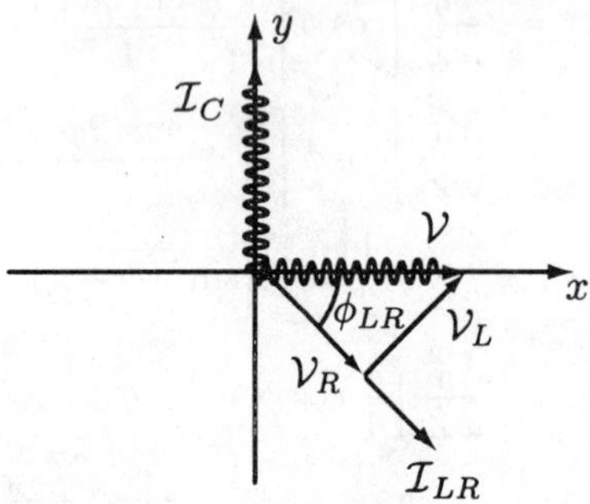

The current in the LR circuit arm will lag the voltage so that the sum of the resistor and inductor voltage phasors equals the generator voltage. The inductor voltage leads the current.

We have

$$\begin{aligned} |\mathcal{I}_C| &= \frac{V_0}{|X_C|} \\ |\mathcal{V}_R| &= R|\mathcal{I}_{LR}| \\ |\mathcal{V}_L| &= X_L\,|\mathcal{I}_{LR}| \\ \tan\phi_{L,R} &= \frac{X_L}{R} \end{aligned}$$

and

$$\begin{aligned} V_0^2 &= |\mathcal{V}_R|^2 + |\mathcal{V}_L|^2 \\ &= \left(R^2 + X_L^2\right)|\mathcal{I}_{LR}|^2 \end{aligned}$$

Now, the net current in the circuit is the phasor sum of $\mathcal{I}_C$ and $\mathcal{I}_{LR}$.

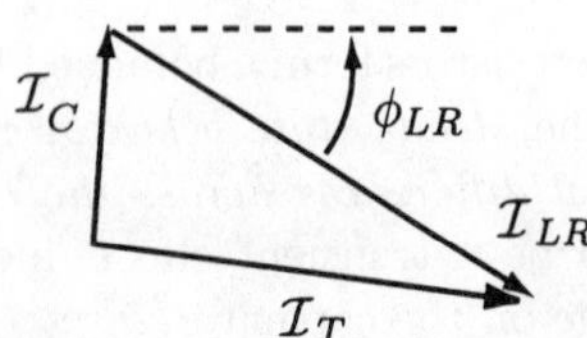

So,

$$|\mathcal{I}_T|^2 = |\mathcal{I}_C|^2 + |\mathcal{I}_{LR}|^2 - 2\,|\mathcal{I}_C|\,|\mathcal{I}_{LR}|\sin\phi_{LR}$$

So,

$$\begin{aligned} |\mathcal{I}_T|^2 &= \frac{V_0^2}{X_C^2} + \frac{V_0^2}{R^2+X_L^2} \\ &\quad - \frac{2V_0^2}{|X_C|\sqrt{R^2+X_L^2}} \cdot \frac{X_L}{\sqrt{R^2+X_L^2}} \\ &= V_0^2\left[\omega^2C^2 + \frac{1}{R^2+\omega^2L^2} - \frac{2\omega^2 LC}{R^2+X_L^2}\right] \\ &= V_0^2\left[\omega^2C^2 + \frac{1-2\omega^2LC}{R^2+\omega^2L^2}\right] \\ &= \frac{V_0^2}{R^2+\omega^2L^2}\left[\begin{array}{c} R^2\omega^2C^2 + \left(\omega^2LC\right)^2 \\ -2\omega^2LC + 1 \end{array}\right] \\ &= \frac{V_0^2}{R^2+\omega^2L^2}\left[R^2\omega^2C^2 + \left(\omega^2LC-1\right)^2\right] \end{aligned}$$

So the impedance is

$$Z = \frac{V_0}{|\mathcal{I}_T|} = \left\{\frac{R^2+\omega^2L^2}{R^2\omega^2C^2 + \left(\omega^2LC-1\right)^2}\right\}^{1/2}$$

Now,

$$\begin{aligned} \omega &= 2\pi f = 2\pi\,(330\text{ Hz}) \\ &= 2.07\times10^3\text{ rad}/\text{s} \end{aligned}$$

So,

$$\begin{aligned} \omega C &= \left(2.07\times10^3\text{ rad}/\text{s}\right)\left(3\times10^{-5}\text{ F}\right) \\ &= 6.2\times10^{-2}\text{ F/s} \\ &= 6.2\times10^{-2}\ \Omega^{-1} \end{aligned}$$

and

$$\begin{aligned} \omega L &= \left(2.07\times10^3\text{ rad}/\text{s}\right)\left(5\times10^{-3}\text{ H}\right) \\ &= 10.4\text{ H/s} \\ &= 10.4\ \Omega \end{aligned}$$

So,

$$\begin{aligned} &Z \\ &= \left\{\frac{400\ \Omega^2+(10.4\ \Omega)^2}{(400\Omega^2)(6.2\times10^{-2}\Omega^{-1})^2+[(10.4\Omega)(6.2\times10^{-2}\Omega^{-1})-1]^2}\right\}^{\frac{1}{2}} \\ &= 17.4\ \Omega \end{aligned}$$

(20 Ω to 1 sig. fig. accuracy of data.)

32.75 a) The resonance frequency is $\omega_0 = \frac{1}{\sqrt{LC}}$, with the capacitance depending on the fluid level. Modeling the capacitor as a parallel combination of the air filled and fluid filled parts, we have

$$\begin{aligned} C &= \frac{\epsilon_0 w\,(\ell - x)}{d} + \frac{\kappa\epsilon_0 w x}{d} \\ &= \frac{\epsilon_0 w}{d}\left[\ell + (\kappa-1)\,x\right] \end{aligned}$$

So

$$\begin{aligned} \omega_0^2 &= \frac{1}{LC} \\ \Rightarrow C &= \frac{1}{\omega_0^2 L} = \frac{\epsilon_0 w}{d}\left[\ell + (\kappa - 1)\,x\right] \end{aligned}$$

So

$$x = \left(\frac{1}{\kappa - 1}\right)\left[\frac{d}{\omega_0^2 \epsilon_0 L w} - \ell\right]$$

$$\begin{aligned} &= \frac{1.0\times10^{-3}\text{ m}}{(1.3)\omega_0^2(8.85\times10^{-12}\text{ F/m})(2.5\times10^{-6}\text{ H})(5.0\times10^{-2}\text{ m})} - \frac{0.200\text{ m}}{1.3} \\ &= \left(6.95\times10^{14}\text{ m}/\text{s}^2\right)\omega_0^{-2} - 0.15\text{ m} \end{aligned}$$

b) The current amplitude in the circuit is

$$\begin{aligned} I &= \frac{V}{Z} = \frac{V_0}{\sqrt{R^2 + \left(\omega L - \frac{1}{\omega C}\right)^2}} \\ &= \frac{V_0 \omega C}{\sqrt{\frac{\omega^2 R^2 C^2}{\omega_0^2 LC} + \left(\omega^2 LC - 1\right)^2}} \\ &= \frac{V_0 \omega C\left(\frac{1}{\omega_0\sqrt{LC}}\right)\frac{\sqrt{L/C}}{RQ}}{\sqrt{\frac{\omega^2}{\omega_0^2}\frac{1}{Q^2} + \left(\frac{\omega^2}{\omega_0^2} - 1\right)^2}} \end{aligned}$$

With $Q = \frac{1}{R}\sqrt{\frac{L}{C}}$. So

$$I = \frac{V_0}{R}\frac{\frac{\omega}{\omega_0}}{\sqrt{\frac{\omega^2}{\omega_0^2} + Q^2\left(\frac{\omega^2}{\omega_0^2} - 1\right)^2}}$$

We are asked for the value of Q which results in a 1% error in ω_0 when there is a 1% error in measuring I. So, let

$$\frac{I}{I_{\text{max}}} = \frac{RI}{V_0} = f = 0.99$$

and let $\frac{\omega}{\omega_0} = f$ as well, and solve for Q. This will be the minimum Q that meets the given condition. So

$$\begin{aligned} f &= \frac{f}{\sqrt{f^2 + Q^2\left(f^2 - 1\right)^2}} \\ &\Rightarrow\; f^2 + Q^2\left(f^2 - 1\right)^2 = 1 \\ &\Rightarrow\; Q^2 = \frac{1 - f^2}{\left(f^2 - 1\right)^2} \\ &= \frac{1}{1 - f^2}\ \text{(since } f^2 < 1\text{)} \end{aligned}$$

So

$$Q_{\text{min}} = \frac{1}{\sqrt{1 - f^2}} = 7.1$$

Now

$$\begin{aligned} Q_{\text{min}} &= \frac{1}{R_{\text{max}}}\sqrt{\frac{L}{C}} \\ \Rightarrow R_{\text{max}} &= \frac{1}{Q_{\text{min}}}\sqrt{\frac{L}{C_{\text{min}}}} \\ &= \frac{1}{7.1}\sqrt{\frac{2.5\times10^{-6}\text{ H}}{\left(8.85\times10^{-12}\,\frac{\text{F}}{\text{m}}\right)(0.20\text{ m})(0.050\text{ m})/(1.0\times10^{-3}\text{ m})}} \\ &= 24\ \Omega \end{aligned}$$

32.79 The parallel circuit arms have a common voltage phasor, which we draw along the x-axis of the first phasor diagram.

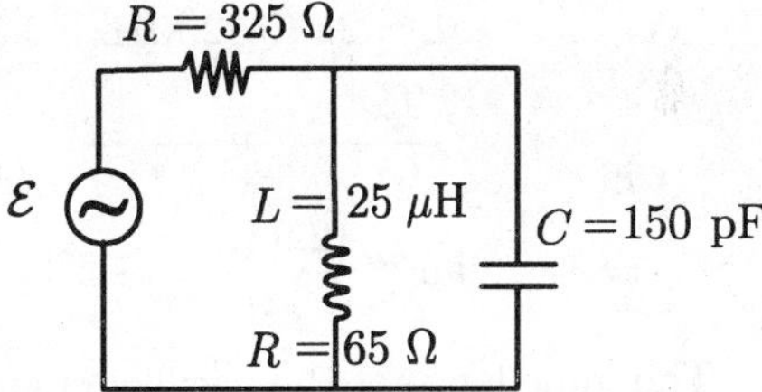

The capacitor current $\mathcal{I}_c$ leads $\mathcal{V}_{\|}$ by 90°, and the inductor current lags $\mathcal{V}_{\|}$ by an amount that ensures that the voltage across its inductance and internal resistance add to give $\mathcal{V}_{\|}$.

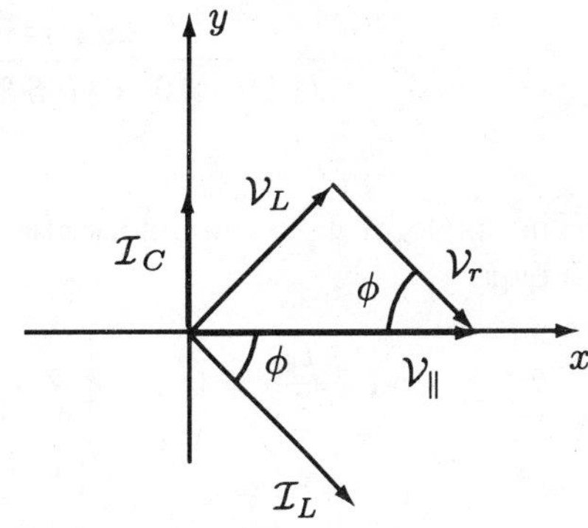

We have

$$\begin{aligned} |\mathcal{I}_C| &= \frac{|\mathcal{V}_{\|}|}{|X_C|} \\ \phi &= \tan^{-1}\frac{X_L}{r} \\ \mathcal{V}_{\|}^2 &= \mathcal{V}_L^2 + \mathcal{V}_r^2 \\ &= |\mathcal{I}_L|^2\left\{X_L^2 + r^2\right\} \\ \Rightarrow Z_L &= \frac{|\mathcal{V}_{\|}|}{|\mathcal{I}_L|} = \sqrt{X_L^2 + r^2} \end{aligned}$$

To add the current phasors, use their Cartesian components:

$$\mathcal{I}_{\parallel y} = |\mathcal{V}_{\parallel}|\left\{\frac{1}{X_C} - \frac{\sin\phi}{Z_L}\right\}$$
$$\mathcal{I}_{\parallel x} = |\mathcal{V}_{\parallel}|\frac{\cos\phi}{Z_L}$$

So

$$|\mathcal{I}_{\parallel}| = |\mathcal{V}_{\parallel}|\left\{\left(\frac{1}{X_C} - \frac{\sin\phi}{Z_L}\right)^2 + \frac{\cos^2\phi}{Z_L^2}\right\}^{1/2} = |\mathcal{V}_{\parallel}|\left\{\frac{1}{X_C^2} + \frac{1}{Z_L^2} - \frac{2X_L}{X_C Z_L^2}\right\}^{1/2}$$

Now for the numbers:

$$X_C = \frac{1}{\omega C} = [2\pi\,(0.82\text{ MHz})\,(150\text{ pF})]^{-1} = 1.29\times 10^3\ \Omega$$
$$X_L = \omega L = 2\pi\,(0.82\text{ MHz})\,(2.5\ \mu\text{H}) = 12.9\ \Omega$$
$$Z_L = \sqrt{(65\ \Omega) + (12.9\ \Omega)} = 66.3\ \Omega$$
$$\phi = \tan^{-1}\frac{12.9\ \Omega}{65\ \Omega} = 11.2^\circ$$

The impedance of the parallel circuit is

$$Z_{\parallel} = \frac{|\mathcal{V}_{\parallel}|}{|\mathcal{I}_{\parallel}|} = \left\{\frac{1}{X_C^2} + \frac{1}{Z_L^2} - \frac{2X_L}{X_C Z_L^2}\right\}^{-\frac{1}{2}}$$
$$= \left\{\left[1.29\times 10^3\ \Omega\right]^{-2} + [66.3\ \Omega]^{-2} - \frac{2\,(12.9\ \Omega)}{(1.29\times 10^3\ \Omega)\,(66.3\ \Omega)^2}\right\}^{-\frac{1}{2}}$$
$$= 66.9\ \Omega$$

The angle of $\mathcal{I}_{\parallel}$ counterclockwise from the x-axis is

$$\theta = \tan^{-1}\frac{\mathcal{I}_{\parallel y}}{\mathcal{I}_{\parallel x}} = \tan^{-1}\left\{\frac{\frac{1}{X_C} - \frac{\sin\phi}{Z_L}}{\frac{\cos\phi}{Z_L}}\right\}$$
$$= \tan^{-1}\left\{\frac{\left[\frac{Z_L}{X_C} - \sin\phi\right]}{\cos\phi}\right\}$$
$$= \tan^{-1}\left\{\frac{\left[\frac{66.3\ \Omega}{1.29\times 10^3\ \Omega} - \sin(11.2^\circ)\right]}{\cos(11.2^\circ)}\right\}$$
$$= -8.3^\circ$$

That is, the current lags the voltage.

Now, we do the phasor diagram for the series connection, which has the common current phasor we described in the first part of the calculation.

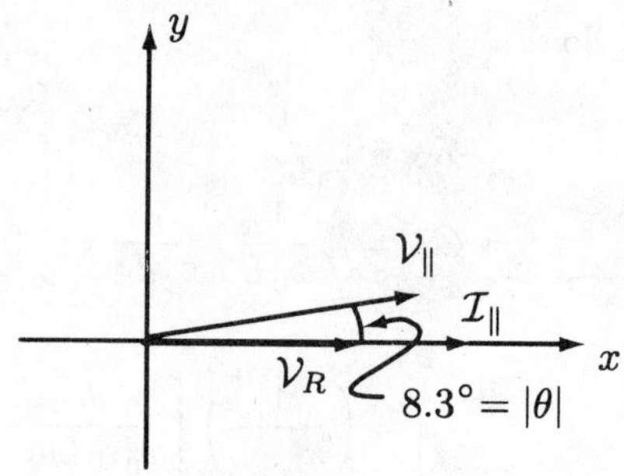

The antenna voltage phasor is the sum of phasors for the resistor and the parallel combination.

$$\mathcal{V}_{\text{ant}} = \mathcal{V}_{\parallel} + \mathcal{V}_R$$

So

$$|\mathcal{V}_{\text{ant}}|^2 = \left[|\mathcal{V}_R| + |\mathcal{V}_{\parallel}|\cos|\theta|\right]^2 + \left[|\mathcal{V}_{\parallel}|\sin|\theta|\right]^2$$
$$= |\mathcal{V}_R|^2 + |\mathcal{V}_{\parallel}|^2 + 2\,|\mathcal{V}_R|\,|\mathcal{V}_{\parallel}|\cos|\theta|$$
$$= |\mathcal{I}_{\parallel}|\left\{R^2 + Z_{\parallel}^2 + 2RZ_{\parallel}\cos\theta\right\}$$

The ratio of capacitor voltage to antenna voltage is

$$\frac{V_C}{V_{\text{ant}}} = \frac{|\mathcal{V}_{\parallel}|}{|\mathcal{V}_{\text{ant}}|} = \frac{|\mathcal{I}_{\parallel}|\,Z_{\parallel}}{|\mathcal{I}_{\parallel}|\left\{R^2 + Z_{\parallel}^2 + 2RZ_{\parallel}\cos\theta\right\}^{1/2}}$$
$$= \frac{66.9\ \Omega}{\left\{(325\ \Omega)^2 + (66.9\ \Omega)^2 + 2(325\ \Omega)(66.9\ \Omega)\cos(8.3^\circ)\right\}^{1/2}}$$
$$= 0.17$$

32.83 The figure shows a circuit diagram that models the jumping ring system. The AC generator is in series with the resistance, and inductance of the coil/iron core as well as with the mutual inductance of coil and ring.

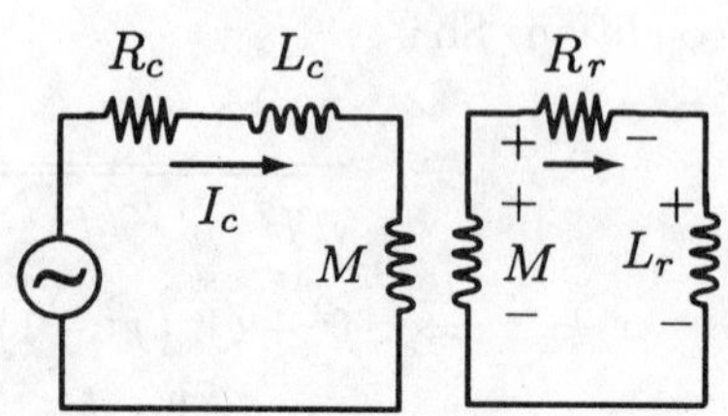

The ring is modeled as a resistance in series with an inductance and the mutual inductance.

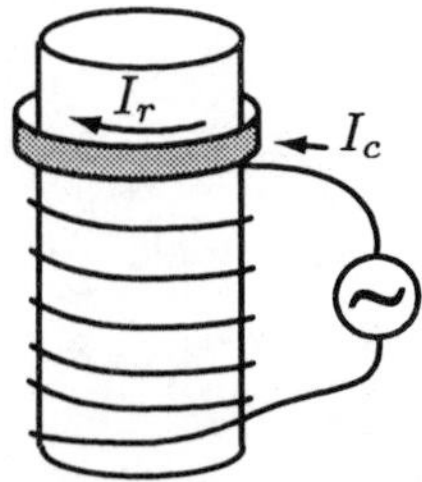

We don't care about phase shift and amplitude relations in the generator circuit; it is coil current as a function of time that influences the ring. Amplitude and phase relations between coil current and ring current are what count. So, let the coil current be

$$I_c = I_0 \cos \omega t$$

where the relation of "positive" current to the system is shown in the second figure. Increasing I_c would then result in increasing upward flux through the ring, resulting in emf in the sense shown for positive current I_r. So, in the circuit diagram the sign convention for positive potential change across the mutual inductance in the ring circuit is the same as the positive sense of ring current, as shown. With these sign conventions, the equation for current in the ring circuit is

$$+M\frac{dI_c}{dt} - RI_r - L\frac{dI_r}{dt} = 0$$

So,

$$\begin{aligned} L_r\frac{dI_r}{dt} + R_r I_r &= MI_0\left(-\omega \sin \omega t\right) \\ &= \omega M I_0 \cos\left(\omega t + \frac{\pi}{2}\right) \end{aligned}$$

The mutual inductance acts like an applied voltage in the ring circuit which leads the coil current by $\frac{\pi}{2}$. The phasor diagram is drawn at a time when the coil current phasor is along the x-axis.

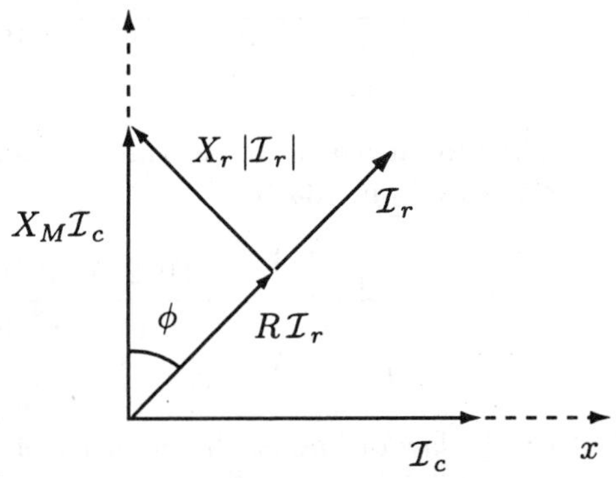

The ring current has the value such that the sum of voltage phasors across the ring resistance and ring inductance equals the "driving voltage" phasor $X_m\mathcal{I}_c$.

Thus

$$X_m^2\,|\mathcal{I}_c|^2 = \left(R_r^2 + X_r^2\right)|\mathcal{I}_r|^2$$

and

$$\phi = \tan^{-1}\frac{X_r}{R_r}$$

So, the current in the ring is given by

$$I_r = \frac{X_m I_0}{\sqrt{R_r^2 + X_r^2}}\cos\left(\omega t + \frac{\pi}{2} - \tan^{-1}\frac{X_r}{R_r}\right)$$

Now the magnetic field produced by the coil is proportional to the current in the coil and is generally upward for positive coil current.

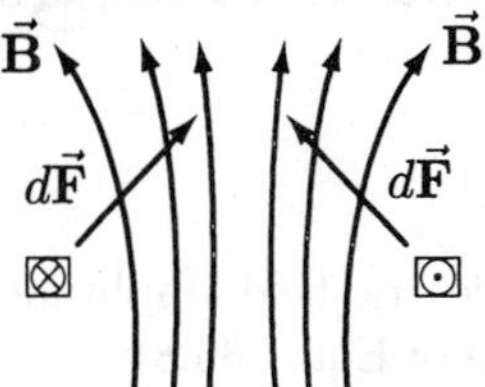

The figure shows field lines of $\vec{\mathbf{B}}$ and the direction of current in two diametrically opposite elements of the ring at a time when both I_c and I_r are positive.

Forces on the two elements $\left(I_r d\vec{\ell} \times \vec{\mathbf{B}}\right)$ have a net upward component and balancing horizontal components. So, the net upward force acting on the ring is proportional to the product I_cI_r. The average upward force component is proportional to the average of I_cI_r:

$$\begin{aligned} \langle F_{\text{up}}\rangle &= \langle I_cI_r\rangle \\ &\propto \left\langle \cos\left(\omega t + \tfrac{\pi}{2}\right)\cos\left(\omega t + \tfrac{\pi}{2} - \tan^{-1}\tfrac{X_r}{R_r}\right)\right\rangle \end{aligned}$$

$$\begin{aligned} &= \left\langle \cos\left(\omega t + \frac{\pi}{2}\right)\left[\cos\left(\omega t + \frac{\pi}{2}\right)\cos\left(\tan^{-1}\frac{X_r}{R_r}\right)\right.\right. \\ &\qquad \left.\left. + \sin\left(\omega t + \frac{\pi}{2}\right)\sin\left(\tan^{-1}\frac{X_r}{R_r}\right)\right]\right\rangle \\ &= \left[\left\langle \cos^2\left(\omega t + \frac{\pi}{2}\right)\right\rangle \cos\left(\tan^{-1}\frac{X_r}{R_r}\right) + 0\right] \\ &= \frac{1}{2}\cos\left(\tan^{-1}\frac{X_r}{R_r}\right) \\ &= \frac{1}{2}\frac{R_r}{R_r^2 + \omega^2 L_r^2} \end{aligned}$$

We find an average upward force, as required.

Chapter 33

Electromagnetic Waves

33.1 The electric field amplitude may be determined by use of Eqn. (33.8):

$$B_0 = \frac{E_0}{c}$$

or

$$\begin{aligned} E_0 &= cB_0 \\ &= (3.00 \times 10^8 \text{ m/s})(1.56 \times 10^{-8} \text{ T}) \\ &= 4.68 \text{ V/m} \end{aligned}$$

33.3 The magnetic field amplitude, B_0, is given by:

$$B_0 = \frac{E_0}{c}$$

Note, the unit vector $\hat{\mathbf{E}}$ is $\frac{(\hat{\mathbf{x}} + \hat{\mathbf{y}})}{\sqrt{2}}$, so

$$\begin{aligned} B_0 &= \frac{(6.0 \text{ V/m})\sqrt{2}}{(3.0 \times 10^8 \text{ m/s})} \\ &= 2.8 \times 10^{-8} \text{ T} \end{aligned}$$

Since the direction of propagation is in the $+z$-direction, $\vec{\mathbf{B}}$ must lie in the x-y-plane (since EM waves are transverse waves). Furthermore, $\vec{\mathbf{B}}$ must be perpendicular to $\vec{\mathbf{E}}$. Therefore

$$\vec{\mathbf{B}} = (2.8 \times 10^{-8} \text{ T}) \frac{(-\hat{x} + \hat{y})}{\sqrt{2}}$$

33.5 The magnitude of the Poynting vector is given by c.f. Example (33.2):

$$\begin{aligned} \left|\vec{\mathbf{S}}\right| &= \frac{\left|\vec{\mathbf{E}} \times \vec{\mathbf{B}}\right|}{\mu_0} \\ &= \frac{E_0 B_0}{\mu_0} \left|\cos^2(kx - \omega t)\right| \\ \left\langle\left|\vec{\mathbf{S}}\right|\right\rangle &= \frac{E_0^2}{2c\mu_0} \end{aligned}$$

Since

$$\left\langle\left|\vec{\mathbf{S}}\right|\right\rangle = 2 \times 10^{-8} \text{ W/m}^2$$

we have

$$\begin{aligned} E_0 &= \sqrt{2c\mu_0 \left\langle\left|\vec{\mathbf{S}}\right|\right\rangle} \\ &= \sqrt{2\left(3 \times 10^8 \tfrac{\text{m}}{\text{s}}\right)\left(4\pi \times 10^{-7} \tfrac{\text{H}}{\text{m}}\right)\left(2 \times 10^{-8} \tfrac{\text{W}}{\text{m}^2}\right)} \\ &= 4 \times 10^{-3} \text{ V/m} \end{aligned}$$

33.7 The time-averaged magnitude of $\vec{\mathbf{S}}$ is given by c.f. Example 33.2:

$$\begin{aligned} \left\langle\left|\vec{\mathbf{S}}\right|\right\rangle &= \frac{E_0^2}{2c\mu_0} \\ &= \frac{(3.0 \text{ V/m})^2}{2(3 \times 10^8 \text{ m/s})(4\pi \times 10^{-7} \text{ H/m})} \\ &= 0.012 \text{ W/m}^2 \end{aligned}$$

The rate at which momentum is transported is given by Eqn. (33.10):

$$\begin{aligned} \left|\frac{d\vec{\mathbf{p}}}{dt}\right| &= \frac{\left\langle\left|\vec{\mathbf{S}}\right|\right\rangle A}{c} = \frac{(0.012 \text{ W/m}^2)(1.5 \text{ m}^2)}{(3 \times 10^8 \text{ m/s})} \\ &= 6.0 \times 10^{-11} \text{ N} \end{aligned}$$

33.9 By the law of Malus, the intensity of the transmitted light is given by

$$I_t = I_0 \cos^2\theta$$

$$\begin{aligned} &= \left(7.75\ \text{W/m}^2\right)\cos^2(20°) \\ &= 6.84\ \text{W/m}^2 \end{aligned}$$

33.11 From Eqn. (33.12) we have that

$$\tan\theta_B = \frac{n_2}{n_1}$$

From Table (16.2) we have

$$\begin{aligned} n_1 &= 1.00\ \text{(air)} \\ n_2 &= 2.42\ \text{(diamond)} \end{aligned}$$

Therefore,

$$\theta_B = \tan^{-1}\left(\frac{2.42}{1.00}\right) = 67.5°$$

33.13 The cutoff wavelength (TE mode) is

$$\begin{aligned} \lambda_c &= 2a = 2\left(3.56\times10^{-2}\ \text{m}\right) \\ &= 7.12\times10^{-2}\ \text{m} \end{aligned}$$

then, the cutoff frequency is

$$\begin{aligned} f_c &= \frac{c}{\lambda_c} \\ &= \frac{\left(3.00\times10^{8}\ \text{m/s}\right)}{\left(7.12\times10^{-2}\ \text{m}\right)} \\ &= 4.21\times10^{9}\ \text{Hz} \end{aligned}$$

33.15 Since $\vec{\mathbf{E}}\times\vec{\mathbf{B}} \parallel \hat{\mathbf{k}}$, and since

$$-\hat{\mathbf{z}}\times\hat{\mathbf{y}} = \hat{\mathbf{x}}$$

$\vec{\mathbf{E}}$ must be in the $-\hat{z}$ direction.

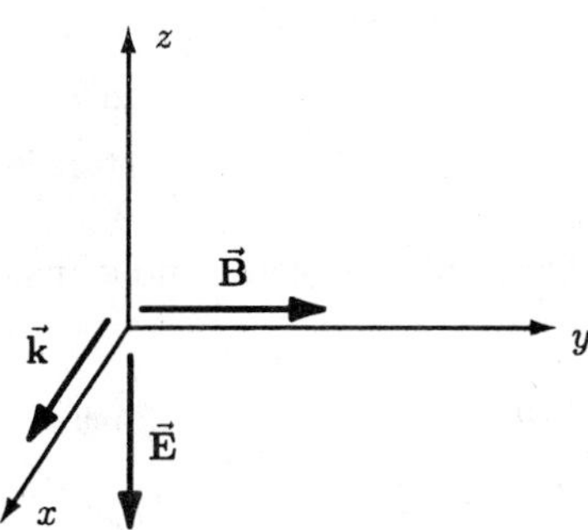

33.19 The wave vector magnitude is

$$\begin{aligned} k &= \frac{2\pi}{\lambda} = \frac{2\pi}{c}f \\ &= \frac{2\pi\left(16.5\times10^{6}\ \text{Hz}\right)}{\left(3.00\times10^{8}\ \text{m/s}\right)} \\ &= 0.346\ \text{rad/m} \end{aligned}$$

Thus

$$\vec{\mathbf{k}} = (0.346\ \text{rad/m})\,\hat{\mathbf{x}}$$

The magnetic field amplitude is

$$\begin{aligned} B_0 &= \frac{E_0}{c} \\ &= \frac{(0.792\ \text{V/m})}{\left(3.00\times10^{8}\ \text{m/s}\right)} \\ &= 2.64\times10^{-9}\ \text{T} \end{aligned}$$

33.23 Since $\vec{\mathbf{k}} = (\hat{\mathbf{x}}+2\hat{\mathbf{y}})\left(\frac{1}{m}\right)$, we have

$$k = \left|\vec{\mathbf{k}}\right| = \sqrt{(1)^2+(2)^2} = \frac{\sqrt{5}}{m}$$

Now, since $\omega = ck$ (in a vacuum) and

$$f = (2\pi)^{-1}\omega$$

the frequency f is

$$\begin{aligned} f &= \frac{ck}{2\pi} \\ &= \frac{\left(3\times10^{8}\ \text{m/s}\right)\left(\frac{\sqrt{5}}{m}\right)}{2\pi} \\ &= 107\times10^{6}\ \text{Hz} \end{aligned}$$

The electric field vector amplitude is given by

$$\begin{aligned} \vec{\mathbf{E}}_0 &= c\vec{\mathbf{B}}_0\times\hat{\mathbf{k}} \\ &= \left(3.00\times10^{8}\ \text{m/s}\right)\left(1.43\times10^{-10}\ \text{T}\right) \\ &\quad \cdot(2\hat{\mathbf{x}}-\hat{\mathbf{y}}+\hat{\mathbf{z}})\times\frac{(\hat{\mathbf{x}}+2\hat{\mathbf{y}})}{\sqrt{5}} \\ &= \left(1.92\times10^{-2}\ \text{V/m}\right)(-2\hat{\mathbf{x}}+\hat{\mathbf{y}}+5\hat{\mathbf{z}}) \end{aligned}$$

33.27 Since the required power is given by

$$\left(\frac{dU}{dt}\right) = (0.01)\left|\vec{\mathbf{S}}\right|A$$

the area A is

$$\begin{aligned} A &= \frac{\left(\frac{dU}{dt}\right)}{(0.01)\left|\vec{\mathbf{S}}\right|} \\ &= \frac{\left(1\times10^{9}\ \text{W}\right)}{(0.01)\left(1.4\times10^{3}\ \text{W/m}^2\right)} \\ &= 7\times10^{7}\ \text{m}^2 \end{aligned}$$

If we denote the area of the (perfectly efficient) collector on the surface of the earth by A_c, then the required intensity is

$$\begin{aligned} I &= \left(\frac{dU}{dt}\right)\frac{(1.00)}{A_c} = \frac{\left(1\times10^{9}\ \text{W}\right)}{\left(2.5\times10^{4}\ \text{m}^2\right)} \\ &= 40\times10^{3}\ \text{W/m}^2 \end{aligned}$$

Since

$$I = \frac{E_0^2}{2c\mu_0}$$

the electric field amplitude is

$$\begin{aligned} E_0 &= \sqrt{2c\mu_0 I} \\ &= \sqrt{2\left(3\times10^8\ \tfrac{\text{m}}{\text{s}}\right)\left(4\pi\times10^{-7}\ \tfrac{\text{H}}{\text{m}}\right)\left(40\times10^3\ \tfrac{\text{W}}{\text{m}^2}\right)} \\ &= 5.5\times10^3\ \text{V/m} \end{aligned}$$

33.31

$$\begin{aligned} F &= Ma \\ &= \left(3.0\times10^4\ \text{kg}\right)(0.01)\left(9.8\ \text{m}/\text{s}^2\right) \\ &= 3\times10^3\ \text{N} \end{aligned}$$

is the required force. Now, the force due to radiation pressure is

$$\begin{aligned} F_{\text{rad}} &= P_{\text{rad}}A = \frac{|\vec{\mathbf{S}}|}{c}A \\ &= 3\times10^3\ \text{N} \end{aligned}$$

but the power is given by

$$P = \left|\vec{\mathbf{S}}\right| A$$

so

$$\begin{aligned} P &= \left(3\times10^3\ \text{N}\right)c \\ &= \left(3\times10^3\ \text{N}\right)\left(3\times10^8\ \text{m}/\text{s}\right) \\ &= 9\times10^{11}\ \text{W} \end{aligned}$$

33.35 Consider a grain of radius r and density $\rho = 3\times10^3\ \text{kg}/\text{m}^3$, with the Sun's light incident on it.

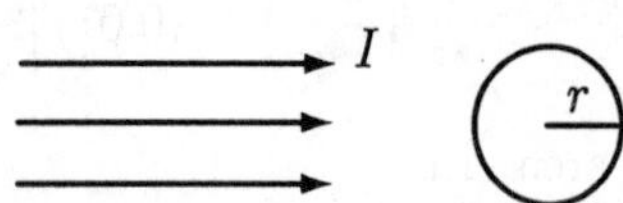

The radiation pressure force is

$$\begin{aligned} F_{\text{rad}} &= \frac{I}{c}\left(\pi r^2\right) \\ &= \frac{L}{4\pi R^2 c}\left(\pi r^2\right) \end{aligned}$$

where R is the distance from the sun to the grain, L is the Sun's luminosity and πr^2 is the effective cross-section of the grain. If $F_{\text{rad}} > F_{\text{grav}}$, the grain will be blown out of the solar system, i.e.,

$$\frac{L}{4\pi R^2 c}\left(\pi r^2\right) > \frac{GM_{\text{sun}}}{R^2}\rho\left(\frac{4\pi}{3}r^3\right)$$

therefore

$$r < \frac{3}{16\pi}\frac{L}{GM_{\text{sun}}c\rho}$$

$$\begin{aligned} r &= \frac{3}{16\pi} \\ &\quad\times\frac{\left(3.9\times10^{26}\ \text{W}\right)}{\left(6.7\times10^{-11}\ \frac{\text{m}^3}{\text{s}^2\cdot\text{kg}}\right)\left(2\times10^{30}\ \text{kg}\right)\left(3\times10^8\ \frac{\text{m}}{\text{s}}\right)\left(3\times10^3\ \ldots\right.} \\ &= 0.19\times10^{-6}\ \text{m} \end{aligned}$$

Therefore

$$R_c = 0.19\ \mu\text{m}$$

Note to instructors: This solution (and the problem) neglects the possible complicating effects caused by the fact that this value is less than the wavelength of light ($\lambda \sim 0.5\ \mu\text{m}$).

33.39 Yes, set the filter's transmission axis horizontal. Since light reflected from the lake surface will be partially polarized with $\vec{\mathbf{E}}$ parallel to the lake surface, turning the transmission axis of the camera's polarizing filter horizontal will reduce the intensity of unreflected light by half and reduce the intensity of reflected light by a smaller factor. Thus, reflected light is enhanced relative to unreflected light.

33.43 Light with incident angle 58° is 100% polarized, implies that $\theta_B = 58°$ for the air/glass interface.

air	glass
$n_1 = 1.0$	$n_2 = ?$

$$\begin{aligned} \tan\theta_B &= \frac{n_2}{n_1} \\ \Rightarrow n_2 &= n_1\tan\theta_B \\ &= (1.0)\tan 58° = 1.6 \end{aligned}$$

33.47 The light is reflected and transmitted as shown:

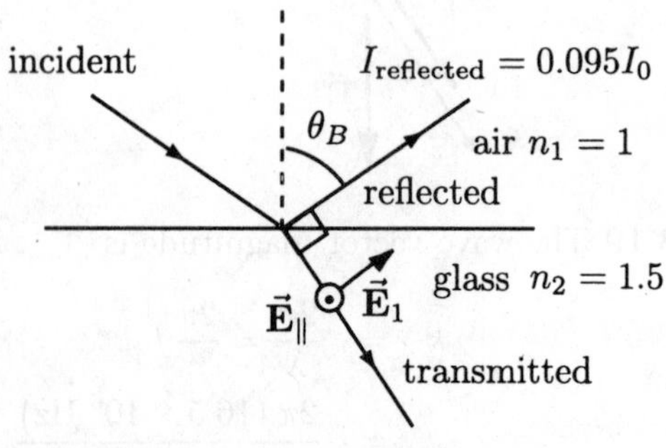

a) $\theta_B = \tan^{-1}\left(\dfrac{n_2}{n_1}\right) = 0.98\ \text{rad} = 56°$

b) The reflected ray is 100% polarized, parallel to the surface.

c) The transmitted light has total intensity

$$(1-0.095)\,I_0 = 0.905I$$

Light of intensity

$$I_{\|} = (0.50-0.095)\,I_0 = 0.405 I_0$$

is polarized parallel to the surface, while light of intensity $I_1 = 0.50 I_0$ is in the other polarization. Thus

$$\begin{aligned} p_{\%} &= 100\left|\frac{0.50I_0 - 0.405I_0}{0.50I_0 + 0.405I_0}\right| \\ &= 10.5\% \text{ polarized} \end{aligned}$$

33.55 The transverse magnetic mode will have field lines with similar patterns to those in Figure 33.29. Since both arise from superposing waves bouncing off the wave guide surfaces. The difference may be thought of as arising from different polarizations of the bouncing waves in the two cases. So, the electric and magnetic fields will switch roles, with $\vec{\mathbf{B}}$ parallel to the guide surfaces and $\vec{\mathbf{E}}$ following loop-like field lines. But, the different boundary conditions on $\vec{\mathbf{E}}$ and $\vec{\mathbf{B}}$ impose a shift on the pattern. $\vec{\mathbf{B}}$ parallel to the surfaces can have large amplitude at the surfaces, while $\vec{\mathbf{E}}$ has to be perpendicular to the guide surfaces. That means that the guide surfaces have to cut the loop-like patterns of $\vec{\mathbf{E}}$ through the middle, rather than containing complete loops, as in Figure 33.29. The resulting field pattern is shown in the figure below. Also shown are the pattern of charge and current densities in the guide surfaces.

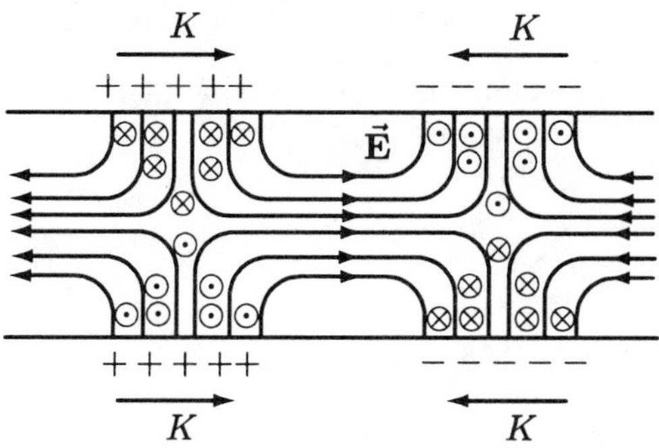

33.59 The cutoff wavelength is related to the separation by

$$\begin{aligned} \lambda_c &= 2a = 2\left(15\times10^{-2}\text{ m}\right) \\ &= 0.30\text{ m} \end{aligned}$$

Thus, the cutoff frequency is

$$\begin{aligned} f_c &= \frac{c}{\lambda_c} = \frac{3\times10^{8}\text{ m/s}}{0.30\text{ m}} \\ &= 1\times10^{9}\text{ Hz} = 1.0\text{ GHz} \end{aligned}$$

33.63 $\vec{\mathbf{E}} = E_0\cos\left(\frac{\pi y}{a}\right)\sin\omega t\hat{\mathbf{z}}$

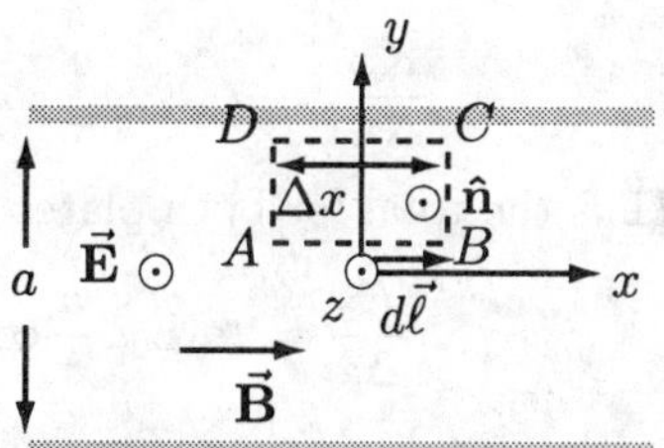

Note that Amperian path $ABCD$ is entirely within the cavity; it does not pass through the conducting surface.

a) (displacement current per unit area) j_d:

$$\begin{aligned} j_d &= \epsilon_0\frac{\partial\vec{\mathbf{E}}}{\partial t}\cdot\hat{\mathbf{n}} \\ j_d &= \epsilon_0E_0\omega\cos\left(\frac{\pi y}{a}\right)\cos\omega t \\ \int\vec{\mathbf{B}}\cdot d\vec{\ell} &= \mu_0I_d = \mu_0\iint j_d\,dx\,dy \end{aligned}$$

$$B_x(y)\,\Delta x - B_x\left(\tfrac{a}{2}\right)\Delta x$$
$$= \mu_0\epsilon_0E_0\omega\cos\omega t\,\Delta x\int_y^{a/2}\cos\left(\tfrac{\pi y}{a}\right)dy$$
$$B_x(y) - B_x\left(\tfrac{a}{2}\right)$$
$$= \mu_0\epsilon_0E_0\omega\,(\cos\omega t)\left(\frac{a}{\pi}\right)\sin\frac{\pi y}{a}\bigg|_y^{a/2}$$
$$B_x(y) - B_x\left(\tfrac{a}{2}\right)$$
$$= \mu_0\epsilon_0E_0\omega\,(\cos\omega t)\left(\frac{a}{\pi}\right)\left(\sin\frac{\pi}{2} - \sin\frac{\pi y}{a}\right)$$
$$\Rightarrow B_x(y) = -\mu_0\epsilon_0E_0\omega\frac{a}{\pi}\cos\omega t\sin\frac{\pi y}{a}$$

b) $\vec{\mathbf{B}}$ is zero within the conductor (see Section 33.42, *Waveguides*)

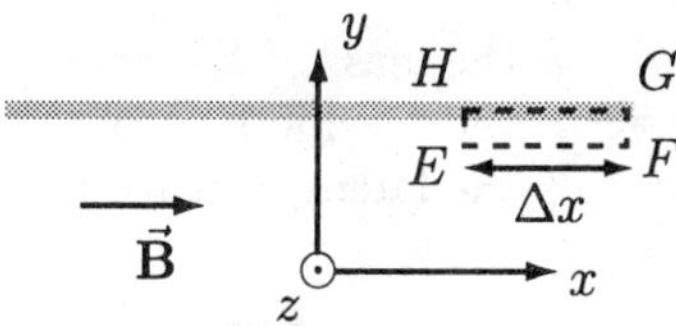

Note that the upper segment GH of the Amperian path EFGH, falls within the conducting plate. The lower segment, EF, is located at $y = \frac{a}{2}$, just below the conducting plate.

$$\int\vec{\mathbf{B}}\cdot d\vec{\ell} = \mu_0K_s\Delta x$$

where K_s = surface current

$$B_x\left(\frac{a}{2}\right)\Delta x = \mu_0 K_s \Delta x$$
$$\Rightarrow \vec{\mathbf{K}}_s = \frac{-\epsilon_0\mu_0 E_0\omega}{\mu_0}\frac{a}{\pi}\cos\omega t\,\Delta x\,\hat{\mathbf{z}}$$

For each plate, the surface current per unit length is

$$\frac{\vec{\mathbf{K}}_s}{\Delta x} = -\epsilon_0 E_0\omega\frac{a}{\pi}\cos\omega t\,\hat{\mathbf{z}}$$

Thus the total for both plates is

$$\frac{K_s}{\Delta x} = 2\epsilon_0 E_0\omega\frac{a}{\pi}\cos\omega t$$

in the $-z$-direction. The total displacement current in cavity (per unit length)

$$\frac{I_d}{\Delta x} = \int_{-a/2}^{a/2}\frac{\epsilon_0\partial\vec{\mathbf{E}}}{\partial t}\cdot\hat{\mathbf{n}}\,dy$$
$$= \epsilon_0 E_0\omega\cos\omega t\int_{-a/2}^{a/2}\cos\frac{\pi y}{a}\,dy$$

since $\hat{\mathbf{n}} = \hat{\mathbf{z}}$ and is parallel to $\vec{\mathbf{E}}$.

$$\frac{I_d}{\Delta x} = \epsilon_0 E_0\omega(\cos\omega t)\left(\frac{a}{\pi}\right)\sin\frac{\pi y}{a}\bigg|_{-a/2}^{a/2}$$
$$= \epsilon_0 E_0\omega\frac{a}{\pi}\cos\omega t\left(\sin\left(\frac{\pi}{2}\right)-\sin\left(-\frac{\pi}{2}\right)\right)$$
$$= 2\epsilon_0 E_0\omega\frac{a}{\pi}\cos\omega t$$

I_d is in the $+z$ direction. Thus

$$\left(\frac{K_s}{\Delta x}\right) = -\left(\frac{I_d}{\Delta x}\right)$$

33.67 The laser of power P must transfer enough energy (per second) to boil a 1 mm^3 $(= 10^{-9}\text{ m}^3)$ volume of water initially at 300 K. The energy required is found from

$$E = mc\Delta T + mL$$

where

$m =$ the mass of the water
$c =$ the specific heat of water, and
$L =$ the latent heat (of vaporization) of water

$$\Rightarrow E = m(c\Delta T + L)$$
$$= \rho V(c\Delta T + L)$$

where $\rho =$ the density of water. Thus

$$E = (10^3\text{ kg/m}^3)\frac{\pi}{4}(10^{-9}\text{ m}^3)$$
$$\times[4186\text{ J/kg}\cdot\text{K}(373\text{ K} - 300\text{ K}) + 3.26\times10^5\text{ J/kg}]$$
$$= 0.5\text{ J}$$

So the laser must have a power $\frac{1}{2}$W. The force exerted by the radiation is

$$F = \frac{dp}{dt} = A\frac{S}{c} = \frac{P}{c} = 2\times10^{-9}\text{ N}$$

33.73 Since electric field lines in the TE mode are parallel to the plates 3 cm apart, they are perpendicular to the ends of the finite guide whatever the value of b.

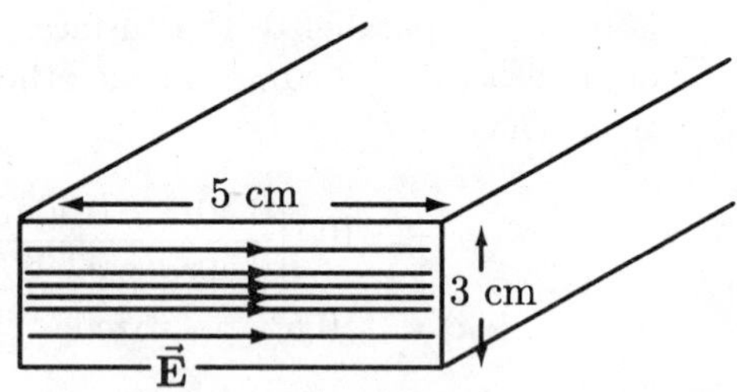

Thus $\vec{\mathbf{E}}$ automatically satisfies its boundary condition at the ends of the finite guide. Since $\vec{\mathbf{B}}$ field lines in this mode are in planes parallel to the new boundaries, $\vec{\mathbf{B}}$ also automatically satisfies its boundary condition. Thus the finite boundaries do not affect propagation of the mode.

Waves polarized parallel to 5 cm-sides are cut off at

$$\lambda_c = 2a = 6\text{ cm}$$

Waves polarized parallel to the 3 cm-sides are cut off at

$$\lambda_c = 2b = 10\text{ cm}$$

$$f = 4\times10^9\text{ Hz}$$
$$\Rightarrow \lambda = \frac{c}{f} = \frac{3\times10^8\text{ m/s}}{4\times10^9\text{ /s}}$$
$$= 7.5\times10^{-2}\text{ m} = 7.5\text{ cm}$$

$$2a < \lambda < 2b$$

$\Rightarrow$ Only light polarized parallel to the 3 cm-sides propagates.

Chapter 34

Relativity and Spacetime

34.1 The rocket ship has maximum length, ℓ, in its own rest frame. In a rest frame with respect to which it is moving, the asteroid's rest frame, the rocket ship will be shorter. In that frame, the rocket ship's length ℓ', will be given by

$$\begin{aligned}\ell' &= \frac{\ell}{\gamma} = \ell\sqrt{1-\beta^2} \\ &= 0.10\text{ km}\left(1-(0.70)^2\right)^{1/2} \\ &= 0.10\text{ km}\,(1-0.49)^{1/2} \\ &= 0.10\text{ km}\,(0.71) = 0.071\text{ km} \\ &= 71\text{ m}\end{aligned}$$

34.3 The time period between two events, the sending of signals, is shortest in the spaceship's rest frame where it is $\Delta t_1 = 12$ h. In the Earth's frame, where the spaceship is moving, it is given by $\Delta t = 12$ h 3 min $= 12.05$ h. Since

$$\begin{aligned}\Delta t_2 &= \gamma\Delta t_1 = \frac{\Delta t_1}{\sqrt{1-\beta^2}} \\ \sqrt{1-\beta^2} &= \frac{\Delta t_1}{\Delta t_2} \\ 1-\beta^2 &= \left(\frac{\Delta t_1}{\Delta t_2}\right)^2 \\ \beta &= \left[1-\left(\frac{\Delta t_1}{\Delta t_2}\right)^2\right]^{1/2} \\ &= \left[1-\left(\frac{12\text{ h}}{12.05\text{ h}}\right)^2\right]^{1/2} \\ &= [1-0.991]^{1/2} = 0.091\end{aligned}$$

Since $\beta = \frac{v}{c}$,

$$\begin{aligned}v &= 0.091\,c \\ &= 2.7\times 10^7\text{ m/s}\end{aligned}$$

We see that this effect occurs for either direction of motion, since $\beta = -0.091$ is also a valid solution.

34.5 The space-time interval between two events is given by

$$(\Delta s)^2 = c^2(\Delta t)^2 - (\Delta x)^2$$

For the two events, E_1 at $t = 0, x = 3.50$ m and E_2 at $t = 1.3\times 10^{-9}$ s, $x = -0.75$ m,

$$\begin{aligned}(\Delta s)^2 &= 9.00\times 10^{16}\text{ m}^2/\text{s}^2\left(1.3\times 10^{-9}\text{ s}-0\right)^2 \\ &\quad -(-0.75-3.50\text{ m})^2 \\ &= 9.00\times 10^{16}\text{ m}^2/\text{s}^2\left(1.69\times 10^{-18}\text{ s}^2\right) \\ &\quad -18.1\text{ m}^2 \\ &= 0.15\text{ m}^2 - 18.1\text{ m}^2 \\ &= -18\text{ m}^2\end{aligned}$$

Since $(\Delta s)^2$ is negative, $(\Delta x)^2$ is greater than $c^2(\Delta t)^2$ and the interval is *spacelike*.

34.7 The space station is the unprimed frame. In the unprimed frame the first rocket has velocity

$$\frac{dx}{dt} = v_1 = 0.85\,c$$

The second rocket has velocity v_2 which may be taken as $v_2 = \beta c = 0.73c$. Then we want to find

$\frac{dx'}{dt'}$, the speed of the first rocket in the frame of the second rocket.

$$\frac{dx'}{dt'} = \frac{\gamma(dx - \beta c dt)}{\gamma\left(dx - \frac{\beta}{c}dt\right)} = \frac{\frac{dx}{dt} - \beta c}{1 - \frac{\beta}{c}\frac{dx}{dt}}$$
$$= \frac{0.85c - 0.73c}{1 - (0.85)(0.73)} = 0.32c$$

34.9 Since $K = (\gamma - 1)mc^2$ and $mc^2 = 938$ MeV for a proton, we have

$$\gamma - 1 = \frac{K}{mc^2} = \frac{450 \text{ MeV}}{938 \text{ MeV}} = 0.48$$

and $\gamma = 1.48$. We can use either $E = K + E_0$ or $E = \gamma mc^2$ to find that $E = 1390$ MeV. Then,

$$pc = \sqrt{E^2 - m^2c^4}$$
$$= \sqrt{(1388 \text{ MeV})^2 - (938 \text{ MeV})^2}$$
$$= 1020 \text{ MeV}$$

So $p = 1020$ MeV/c. Its energy momentum invariant is

$$E^2 - p^2c^2 = m^2c^4 = (938 \text{ MeV})^2$$

34.13 Since the cruiser is moving relative to the freighter, the freighter sees it as having length $\frac{L}{\gamma}$. That is why the doors on the freighter match the ends of the cruiser perfectly in the freighter's coordinate system. If we call one end of the cruiser O and the other end X in the cruiser's system, then the ends are O' and X' in the freighter's system. We are told that

$$t'_{O'} = t'_{x'} = 0$$

Then

$$t_O = \gamma(0 - 0) = 0$$

and

$$t_x = \gamma\left(0 - \frac{\beta}{c}x\right)$$

Thus, in the cruiser's frame, the astronaut at the front must leap first, $\frac{\gamma\beta x}{c}$ seconds earlier.

34.17 A test of time dilation involves comparison of measurements at two different events. In one reference frame, both events occur at the same spatial place and can be measured by a single observer. In the other reference frame, the events occur at two different spatial locations, so two observers from that frame are required, one at each position. The minimum number of observers is three. If the experimenters wish to test the symmetry of time-dilation, they need measurements of two pairs of events such that one pair of events occurs at the same place in each frame. Four observers are needed for this test.

34.21 Since $\Delta t'_2 = \gamma \Delta t_1$,

$$x' = v\Delta t_2 = 0.95c\gamma\Delta t_1$$
$$= \frac{(0.95)(3.0 \times 10^8 \text{ m/s})(2.9 \times 10^{-10} \text{ s})}{\sqrt{1 - (0.95)^2}}$$
$$= 0.26 \text{ m}$$

34.25 You observe $L' = \frac{L}{\gamma} = 417$ m and $\tau' = \gamma\tau = 3.0\ \mu$s, since it is moving with respect to you. Since

$$\frac{L'}{\tau'} = v = 1.39 \times 10^8 \text{ m/s}$$

The length measured by the rocket crew is:

$$L = \gamma(417 \text{ m})$$
$$= \frac{417 \text{ m}}{\sqrt{1 - \left(\frac{1.39}{3.00}\right)^2}} = 471 \text{ m}$$

34.29 We want to find

$$\frac{\Delta\ell}{\ell_0} = \frac{\ell}{\ell_0} - \frac{\ell_0}{\ell_0} = \frac{\ell}{\ell_0} - 1$$

The molecule's speed is $v = \omega r$. So,

$$\ell = \frac{\ell_0}{\gamma} = \frac{\ell_0}{\sqrt{1 - \frac{\omega^2 r^2}{c^2}}}$$

But $\omega r \ll c$, so we may expand $\left(1 - \frac{\omega^2 r^2}{c^2}\right)^{-\frac{1}{2}}$:

$$\left(1 - \frac{\omega^2 r^2}{c^2}\right)^{-\frac{1}{2}} = 1 - \frac{1}{2}\frac{\omega^2 r^2}{c^2} + \cdots$$

Thus

$$\frac{\Delta\ell}{\ell_0} = 1 - \frac{1}{2}\frac{\omega^2 r^2}{c^2} - 1 = -\frac{1}{2}\frac{\omega^2 r^2}{c^2}$$
$$= -\frac{1}{2}\left(\frac{10^5 \text{ rad/s}\,(0.1 \text{ m})}{3 \times 10^8 \text{ m/s}}\right)^2$$
$$= -6 \times 10^{-10}$$

34.33 The diagram below shows events P and Q with their past and future light cones.

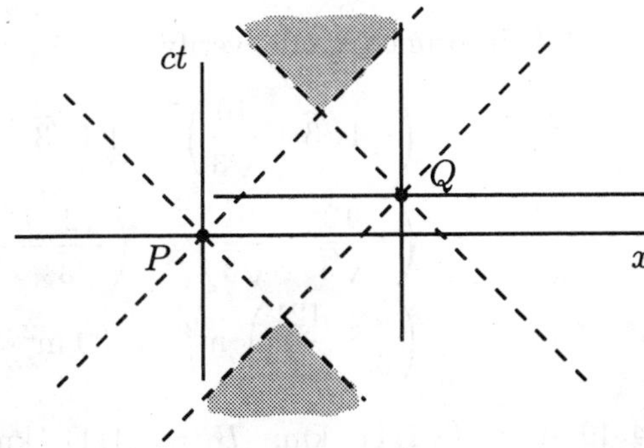

The grids show where their pasts and futures overlap. For the past of Q to overlap the future of P it would have to be shifted toward Q until the light cone right boundaries overlapped and then a bit more. Then the separation of the events would become timelike. So no, the past of one cannot intersect the future of the other as long as they have spacelike separation.

34.37 We find

$$\begin{aligned}(\Delta s)^2 &= (ct_2 - ct_1)^2 - (x_2 - x_1)^2 \\ &= (37.56 - 0.45)^2\,\text{m}^2 - (23.49 - 16.29)^2\,\text{m}^2 \\ &= 1325\text{m}^2 \\ \Delta s &= 36.4\text{ m}\end{aligned}$$

Here Δs represents c times the time between the events in a reference frame where they occur at the same place. Proper length does not have any meaning because it is impossible to find a frame in which both events occur at the same time.

34.41 The space-time diagram below shows two points A and B that are separated by a spacelike interval. The axis, x', that is drawn through them is the required axis.

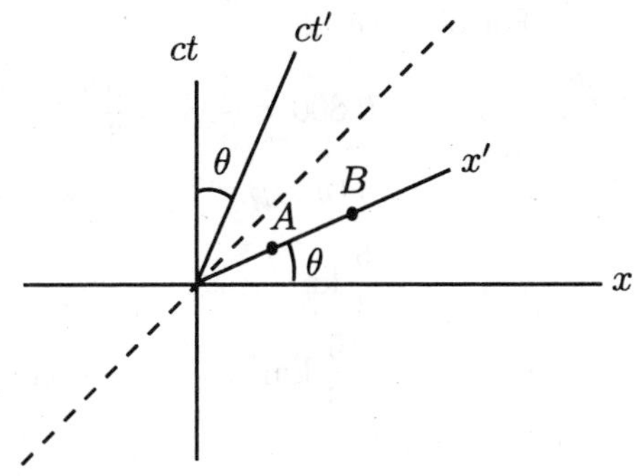

34.43 We need to calculate the relative velocity both ways with the Newtonian and the relativistic formulas, and find the difference.

$$\begin{aligned}V_{N,R} &= v_1 + v_2 \\ V_{E,R} &= \frac{v_1 + v_2}{1 + \frac{v_1 v_2}{c^2}}\end{aligned}$$

$$\begin{aligned}\frac{V_{N,R} - V_{E,R}}{V_{E,R}} &= \frac{V_{N,R}}{V_{E,R}} - 1 \\ &= \frac{(v_1 + v_2)}{(v_1 + v_2)}\left(1 + \frac{v_1 v_2}{c^2}\right) - 1 \\ &= \frac{v_1 v_2}{c^2}\end{aligned}$$

The magnitude of the fractional error is

$$\begin{aligned}\frac{v_1 v_2}{c^2} &= \left(\frac{1800 \times 10^3\text{ m}}{3600\text{ s}}\right)^2 \frac{1}{(3 \times 10^8\text{ m/s})^2} \\ &= 2.8 \times 10^{-12}\end{aligned}$$

The absolute error is

$$\begin{aligned}\Delta v &= (2.8 \times 10^{-12})(1800\text{ km/h}) \\ &= 5.0 \times 10^{-9}\text{ km/h}\end{aligned}$$

34.47 Given A is on the time axis of the 'moving' frame. The ct' and x' axes are symmetrically placed about the line $x = ct$.

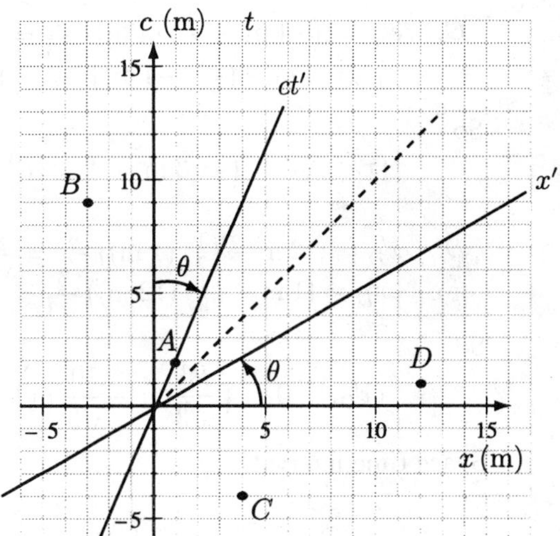

From the figure we have

$$\tan\theta = \beta = \frac{1}{2},\ \beta = \frac{1}{2}$$

We can read off the coordinates of A, B, C, D in the lab frame

$$\begin{aligned}A\!:&\quad ct_A = 2\text{ m} \\ &\quad x_A = 1\text{ m} \\ B\!:&\quad ct_B = 9\text{ m} \\ &\quad x_B = -3\text{ m} \\ C\!:&\quad ct_C = -4\text{ m} \\ &\quad x_C = 4\text{ m} \\ D\!:&\quad ct_D = 1\text{ m} \\ &\quad x_D = 12\text{ m}\end{aligned}$$

Use the Lorentz transformation to convert from lab frame to moving or prime frame

$$\left.\begin{array}{l} x' = (x - \beta ct)\,\gamma \\ ct' = (ct - \beta x)\,\gamma \end{array}\right\} \gamma = \frac{1}{\sqrt{1-\left(\frac{1}{2}\right)^2}} = \sqrt{\frac{1}{\frac{3}{4}}} = \frac{2}{\sqrt{3}}$$

$$\begin{aligned} A:\quad x'_A &= (x_A - \beta ct_A)\,\gamma \\ &= \left(1\text{ m} - \left(\tfrac{1}{2}\right)(2\text{ m})\right)\tfrac{2}{\sqrt{3}} = 0 \\ ct'_A &= (ct_A - \beta x_A)\,\gamma \\ &= \left(2\text{ m} - \left(\tfrac{1}{2}\right)(1\text{ m})\right)\tfrac{2}{\sqrt{3}} = \sqrt{3}\text{ m} \end{aligned}$$

So

$$(x'_A, ct'_A) = \left(0\text{ m}, \sqrt{3}\text{ m}\right)$$

$$\begin{aligned} B:\quad ct'_B &= \left(9\text{ m} - \left(\tfrac{1}{2}\right)(-3\text{ m})\right)\tfrac{2}{\sqrt{3}} = 7\sqrt{3}\text{ m} \\ x'_B &= \left(-3\text{ m} - \left(\tfrac{1}{2}\right)(9\text{ m})\right)\tfrac{2}{\sqrt{3}} = -5\sqrt{3}\text{ m} \end{aligned}$$

So

$$(x'_B, ct'_B) = \left(-5\sqrt{3}\text{ m},\ 7\sqrt{3}\text{ m}\right)$$

$$\begin{aligned} C:\quad ct'_C &= \left(-4\text{ m} - \left(\tfrac{1}{2}\right)(4\text{ m})\right)\tfrac{2}{\sqrt{3}} = -4\sqrt{3}\text{ m} \\ x'_B &= \left(4\text{ m} - \left(\tfrac{1}{2}\right)(-4\text{ m})\right)\tfrac{2}{\sqrt{3}} = 4\sqrt{3}\text{ m} \end{aligned}$$

So

$$(x'_C, ct'_C) = \left(4\sqrt{3}\text{ m},\ -4\sqrt{3}\text{ m}\right)$$

$$\begin{aligned} D:\quad x'_D &= \left(12\text{ m} - \left(\tfrac{1}{2}\right)(1\text{ m})\right)\tfrac{2}{\sqrt{3}} = \tfrac{23}{\sqrt{3}}\text{ m} \\ cx'_D &= \left(1\text{ m} - \left(\tfrac{1}{2}\right)(12\text{ m})\right)\tfrac{2}{\sqrt{3}} = -\tfrac{10}{\sqrt{3}}\text{ m} \end{aligned}$$

So

$$(x'_D, ct'_D) = \left(\frac{23}{\sqrt{3}}\text{ m}, -\frac{10}{\sqrt{3}}\text{ m}\right)$$

Space time intervals:

$$s^2 = c^2(\Delta t)^2 - (\Delta x)^2$$

for AB in lab frame ("timelike")

$$\begin{aligned} s^2_{AB} &= (9\text{ m} - 2\text{ m})^2 - (-3\text{ m} - 1\text{ m})^2 \\ &= 49\text{ m}^2 - 16\text{ m}^2 = 33\text{ m}^2 \end{aligned}$$

CD in lab frame ("spacelike")

$$\begin{aligned} s^2_{CD} &= (1+4)^2\text{ m}^2 - (12-4)^2\text{ m}^2 \\ &= 25\text{ m}^2 - 64\text{ m}^2 = -39\text{ m}^2 \end{aligned}$$

for AB in primed frame, verify

$$\begin{aligned} s^{2\prime}_{AB} &= \left(7\sqrt{3} - \sqrt{3}\right)^2\text{ m}^2 - \left(-5\sqrt{3}\right)^2 \\ &= \left(6^2\cdot 3 - 5^2\cdot 3\right)\text{ m}^2 = (36-25)(3)\text{ m}^2 \\ &= 33\text{ m}^2 = s^2_{AB} \end{aligned}$$

CD in primed frame, verify

$$\begin{aligned} s^{2\prime}_{CD} &= \left(-4\sqrt{3} + \frac{10}{\sqrt{3}}\right)^2 - \left(4\sqrt{3} - \frac{23}{\sqrt{3}}\right)^2 \\ &= \left(-\frac{12}{\sqrt{3}} + \frac{10}{\sqrt{3}}\right)^2 - \left(\frac{12}{\sqrt{3}} - \frac{23}{\sqrt{3}}\right)^2\text{ m}^2 \\ &= \left(\frac{4}{3} - \frac{121}{3}\right)\text{ m}^2 = -39\text{ m}^2 = s^2_{CD} \end{aligned}$$

34.49 $A = (-1,1)$ km; $B = (1,1)$ km; $C = (-1,-1)$ km; $D = (-1,-1)$ km; $E = (0,0)$ km

For observer 1,

$$\begin{aligned} \beta_x &= 0.800 = \frac{4}{5},\ \gamma = 1.667 = \frac{5}{3} \\ x_1 &= \gamma(x - \beta ct) = \gamma x = \frac{5}{3}x \\ y_1 &= y \\ t_1 &= \gamma\left(t - \frac{\beta x}{c}\right) = -\gamma\beta\frac{x}{c} \\ &= -\frac{5}{3}\cdot\frac{4}{5}\frac{x}{c} = -\frac{4}{3}\frac{x}{c} \end{aligned}$$

Therefore,

$$\begin{aligned} x_{1A} &= -\frac{5}{3}\text{ km},\ x_{1B} = \frac{5}{3}\text{ km}, \\ x_{1C} &= -\frac{5}{3}\text{ km},\ x_{1D} = \frac{5}{3}\text{ km} \\ x_{1E} &= 0\text{ km} \end{aligned}$$

$A_1 = \left(-\frac{5}{3}, 1\right)$ km; $B_1 = \left(\frac{5}{3}, 1\right)$ km; $C_1 = \left(-\frac{5}{3}, -1\right)$ km; $D_1 = \left(\frac{5}{3}, -1\right)$ km; $E_1 = (0,0)$ km

$$\begin{aligned} t_{1A} &= -\left(-\frac{4}{3}\frac{10^3}{3\times 10^8}\right) = 4.44\times 10^{-6}\text{ s} = t_{1C} \\ t_{1B} &= t_{1D} = -4.44\times 10^{-6}\text{ s} = t_{1C} \\ t_{1E} &= 0\text{ s} \end{aligned}$$

Sequence: (B,D), E, (A,C).

For observer 2,

$$\begin{aligned} \beta_y &= 0.800 = \frac{4}{5},\ \gamma = \frac{5}{3} \\ y_2 &= \gamma(y - \beta ct) = \gamma y,\ t_2 = -\beta\gamma\frac{y}{c} \\ y_{2A} &= \frac{5}{3}\text{ km},\ y_{2B} = \frac{5}{3}\text{ km} \\ y_{2C} &= -\frac{5}{3}\text{ km},\ y_{2D} = -\frac{5}{3}\text{ km} \\ y_{2E} &= 0 \end{aligned}$$

$A_2 = \left(-1, \frac{5}{3}\right)$ km; $B_2 = \left(1, \frac{5}{3}\right)$ km; $C_2 = \left(-1, -\frac{5}{3}\right)$ km; $D_2 = \left(1, -\frac{5}{3}\right)$ km

$$\begin{aligned} t_{2A} &= t_{2B} = -4.44\times 10^{-6}\text{ s} \\ &= t_{2C} = t_{2D} = 4.44\times 10^{-6}\text{ s} \\ t_{2E} &= 0\text{ s} \end{aligned}$$

Sequence: (A,B), E, (C,D).

34.51 Let the star Betelgeuse define the unprime frame. In the primed system moving with the ship, each of the components moves away at $\beta' = 0.60 = \frac{3}{5}$.

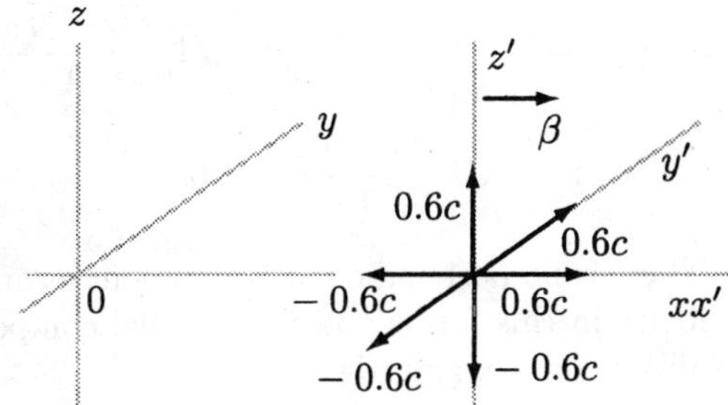

Given $\beta = \frac{4}{5}$. In vector form

$$\vec{\beta} = \frac{4}{5}\hat{\mathbf{x}}$$
$$\vec{\beta}' = \pm\frac{3}{5}\hat{\mathbf{x}},\ \pm\frac{3}{5}\hat{\mathbf{y}},\ \pm\frac{3}{5}\hat{\mathbf{z}}$$

The $\hat{y}$ and $\hat{z}$ motion is $\perp$ to the relative velocity of the two inertial frames; answers will be the same for both.

a) We use equations 34.9 (with β replaced with $-\beta$ to give the solution for unprimed velocity components in terms of the primed components)

$$u_x = \frac{u' + \beta c}{1 + \beta u'_x c}$$
$$u_y = \frac{u'_y}{\gamma\,(1 + \beta u'_x c)}$$
$$u_z = \frac{u'_z}{\gamma\,(1 + \beta u'_x c)}$$

thus

$$u_x = \frac{\left(\pm\frac{3}{5} + \frac{4}{5}\right)c}{\left(1 \pm \frac{4}{5}\cdot\frac{3}{5}\right)} = \frac{(\pm 3 + 4)\,c}{\left(5 \pm \frac{12}{5}\right)}$$
$$= \begin{cases} \frac{35}{37}c \\ \frac{5}{13}c \end{cases}$$

We calculate u_y; for each of these cases $u'_x \equiv 0$

$$\gamma = \frac{1}{\sqrt{1 - \left(\frac{4}{5}\right)^2}} = \frac{1}{\left(1 - \frac{16}{25}\right)^{1/2}} = \frac{5}{3}$$

So

$$u_y = \frac{3}{5}\left(\pm\frac{3}{5}c\right) = \pm\frac{9}{25}c$$

similarly

$$u_z = \pm\frac{9}{25}c$$

b) Next, consider the relative velocities of modules with respect to one another. Forget about Betelgeuse; modules move out at $\pm 0.6c$ on each of the axes in the ship frame.
Antiparallel: The ship becomes the unprimed frame, one of the modules becomes the primed frame. In this case we can use equations 34.9 directly. Here

$$\beta = 0.6$$
$$u_x = -0.6c$$

The relative speed is the speed relative to the primed frame:

$$u'_x = \frac{u_x - \beta c}{1 - \frac{\beta u_x}{c}} = \frac{-\frac{3}{5}c - \frac{3}{5}c}{1 - \frac{\left(\frac{3}{5}\right)\left(-\frac{3}{5}c\right)}{c}}$$
$$= \frac{-\frac{6}{5}}{1 + \frac{9}{25}}$$
$$|u'_x| = 0.88c$$

c) For the perpendicular case, transform the $\perp$ and $\|$ velocity components to the prime frame.

$$\beta = 0.6$$
$$\gamma = \frac{5}{4}$$

$$u'_x = \frac{0 - \beta c}{1 - (0.6)\,(0)} = -0.6c$$
$$u'_\perp = \frac{-0.6c}{1 - (0.6)\,(0)} \cdot \frac{1}{\gamma}$$
$$= -\frac{4}{5}\cdot\frac{3}{5}c = -\frac{12}{25}c$$

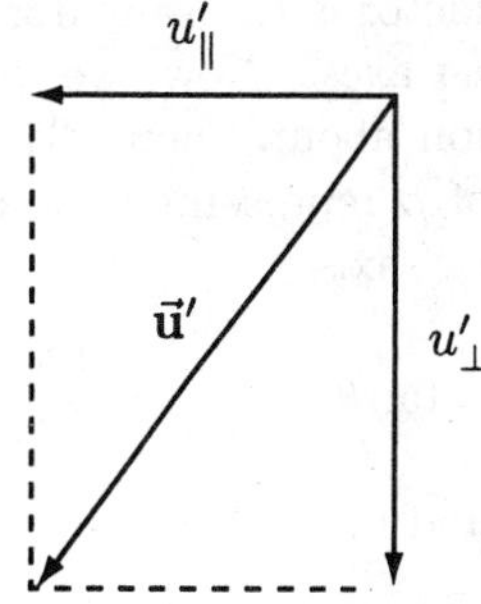

Relative speed $= u' = \sqrt{\left(u'_\|\right)^2 + \left(u'_\perp\right)^2}$ using Pythagoras

$$u' = \sqrt{\left(\frac{9}{25}\right) + \left(\frac{144}{625}\right)}c$$
$$= 0.77c$$

34.53 Consider two events O and A on the ct-axis separated by a time interval Δt, represented by the length OA in the diagram. Event B is simultaneous with event A. (The line AB is parallel to the x-axis.)

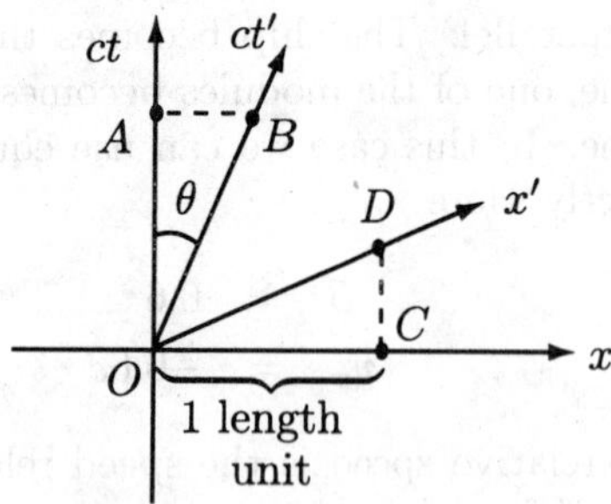

Since events O and B both occur at $x' = 0$, the time interval between O and B in the prime frame is $\frac{\Delta t}{\gamma}$. So, the longer length OB represents the shorter time interval. The scale on the ct axis is $\frac{AO}{\Delta t}$, while that on the ct' axis is

$$\frac{\overline{OB}}{\frac{\Delta t}{\gamma}} = \frac{\overline{AO}\sqrt{1+\tan^2\theta}}{\frac{\Delta t}{\gamma}} = \frac{\overline{AO}}{\Delta t}\frac{\sqrt{1+\beta^2}}{\sqrt{1-\beta^2}}$$

So, to represent a given space-time interval requires a geometrical length on the prime axis that is longer by a factor $\sqrt{(1+\beta^2)/(1-\beta^2)}$. The geometry for scales on the spatial axes is identical. Events O and C represent the length of an object at rest in the unprime frame. Events O and D represent the length of that object as measured in the prime frame, shorter by a factor γ but represented by the greater geometrical length

$$\overline{OD} = \overline{OC}\sqrt{1+\beta^2}$$

The scale factor is the same for spatial axes as for the time axes. Now, we are asked a specific question about Figure 34.11, so we need the value of β represented there. We measure β from the x'-axis

$$\beta = \tan\theta = \frac{\text{rise}}{\text{run}} = \frac{2.5 \text{ cm}}{4.0 \text{ cm}} = \frac{5}{8}$$

The predicted ratio of scales is thus

$$\sqrt{\frac{1+\beta^2}{1-\beta^2}} = \sqrt{\frac{1+\left(\frac{5}{8}\right)^2}{1-\left(\frac{5}{8}\right)^2}} = \sqrt{\frac{89}{39}} = 1.5$$

Measuring from Figure 34.11, 4.9 cm represents 4 m on the x'-axis, while 3.4 cm represent 4 m on the x-axis. The measured scale ratio is $\frac{4.9 \text{ cm}}{3.4 \text{ cm}} = 1.4$ With uncertainty of the measurements, the two results agree. So, the length on the diagram representing 1 μs on the ct' axis or 300 m on the x'-axis is 1.45 ± 0.05 cm.

34.55 We need to set $\beta = 1$ and have an arbitrary u_x and $u_\perp$. Then

$$u'_x = \frac{u_x - \beta c}{1 - \frac{\beta u_x}{c}} = \frac{u_x - c}{c - u_x} \cdot c = -c$$

$$u'_\perp = \frac{u_\perp}{\gamma\left(1 - \frac{\beta u_x}{c}\right)} = \sqrt{1-\beta^2}\frac{u_\perp}{1 - \frac{u_x}{c}}$$

$$= \sqrt{1-1}\frac{u_\perp}{1 - \frac{u_x}{c}} = 0$$

There is no perpendicular component to any velocity in this frame and the parallel component always has magnitude c.

34.59 $u'_\parallel = \frac{u_\parallel - \beta c}{1 - \frac{\beta u_\parallel}{c}}$, $t' = \gamma\left(t - \frac{\beta x}{c}\right)$

$$du'_\parallel = \frac{du_\parallel}{1 - \frac{\beta u_\parallel}{c}} - \frac{(u_\parallel - \beta c)}{\left(1 - \frac{\beta u_\parallel}{c}\right)^2}\left(-\frac{\beta du_\parallel}{c}\right)$$

$$dt' = \gamma\left(dt - \frac{\beta dx}{c}\right)$$

$$du'_\parallel = \frac{du_\parallel}{1 - \frac{\beta u_\parallel}{c}} - \frac{\beta\,(\beta c - u_\parallel)\,du_\parallel}{c\left(1 - \frac{\beta u_\parallel}{c}\right)^2}$$

$$= du_\parallel \frac{\left\{1 - \frac{\beta u_\parallel}{c} - \beta^2 + \frac{\beta u_\parallel}{c}\right\}}{\left(1 - \frac{\beta u_\parallel}{c}\right)^2}$$

$$= \frac{du_\parallel}{\left(1 - \frac{\beta u_\parallel}{c}\right)^2}(1-\beta^2)$$

$$= \frac{1}{\gamma^2}\frac{du_\parallel}{\left(1 - \frac{\beta u_\parallel}{c}\right)^2}$$

$$a'_\parallel = \frac{du'_\parallel}{dt'} = \frac{\frac{1}{\gamma^2}\frac{du_\parallel}{\left(1 - \frac{\beta u_\parallel}{c}\right)^2}}{\gamma\,(dt - \beta\,dx)}$$

$$= \frac{1}{\gamma^3}\frac{a_\parallel}{\left(1 - \frac{\beta u_\parallel}{c}\right)^3}$$

$$\left(\text{Note}: u_\parallel = \frac{dx}{dt}\right)$$

$$u'_\perp = \frac{u_\perp}{\gamma\left(1 - \frac{\beta u_\parallel}{c}\right)}$$

$$du'_\perp = \frac{du_\perp}{\gamma\left(1 - \frac{\beta u_\parallel}{c}\right)} - \frac{u_\perp}{\gamma}\frac{\left(-\beta\frac{du_\parallel}{c}\right)}{\left(1 - \frac{\beta u_\parallel}{c}\right)^2}$$

$$= \frac{1}{\gamma}\frac{1}{\left(1 - \frac{\beta u_\parallel}{c}\right)^2}$$

$$\times\left(du_\perp\left(1-\frac{\beta u_\parallel}{c}\right)+\beta\frac{u_\perp}{c}du_\parallel\right)$$

and

$$\begin{aligned} a'_\perp &= \frac{du'_\perp}{dt'} \\ &= \frac{1}{\gamma^2}\frac{1}{\left(1-\frac{\beta u_\parallel}{c}\right)^2}\frac{\left[du_\perp\left(1-\frac{\beta u_\parallel}{c}\right)+\frac{\beta u_\perp}{c}du_\parallel\right]}{dt-\frac{\beta\, dx}{c}} \\ &= \frac{1}{\gamma^2}\frac{1}{\left(1-\frac{\beta u_\parallel}{c}\right)^2}\frac{\left[a_\perp\left(1-\frac{\beta u_\parallel}{c}\right)+\frac{\beta u_\perp}{c}a_\parallel\right]}{1-\frac{\beta u_\parallel}{c}} \\ &= \frac{1}{\gamma^2}\frac{1}{\left(1-\frac{\beta u_\parallel}{c}\right)^3}\left[\left(1-\frac{\beta u_\parallel}{c}\right)a_\perp+\frac{\beta u_\perp}{c}a_\parallel\right] \end{aligned}$$

34.63 When a high-energy beam is directed onto a stationary target, the system formed by a single beam particle and its target has a sizable linear momentum that must be conserved in the lab frame. So, a large fraction of the beam particle's energy ends up as kinetic energy of the reaction products. For a highly relativistic beam in which $\gamma \gg 1$, the energy wasted as outgoing kinetic energy becomes prohibitive. Using beams with equal and opposite velocities results in zero net momentum of the two colliding beam/target particles; thus all of their energy is available for producing massive reaction products.

34.67 The binding energy, 13.6 eV, is very small compared with the total mass of the proton and electron. Therefore we can obtain a reasonable approximation to three significant figures by taking the sum of rest energies of the electron and proton. If we call the binding energy ΔE, the fraction we want is

$$\begin{aligned} &\frac{\Delta E}{(m_p+m_e)\,c^2} \\ &= \frac{(13.6\text{ eV})(1.6\times10^{-19}\text{ J/eV})}{(1.0079\text{ u}+0.00055\text{ u})(1.66\times10^{-27}\,\frac{\text{kg}}{\text{u}})(3.00\times10^8\text{ m/s})^2} \\ &= 1.44\times10^{-8} \end{aligned}$$

34.69 a) For an electron with

$$\begin{aligned} K &= 75.0\text{ eV} = 1.20\times10^{-17}\text{ J} \\ v^2 &= \frac{2K}{m} = \frac{2\,(1.2\times10^{-17}\text{ J})}{9.11\times10^{-31}\text{ kg}} \\ &= 2.64\times10^{13}\text{ m}^2/\text{s}^2 \end{aligned}$$

so

$$v = 5.14\times10^6\text{ m/s}$$

and a non-relativistic approach is justified. Its momentum is

$$\begin{aligned} p &= mv = (9.11\times10^{-31}\text{ kg})\,(5.14\times10^6\text{ m/s}) \\ &= 4.68\times10^{-24}\,\frac{\text{kg}\cdot\text{m}}{\text{s}} \end{aligned}$$

The energy-momentum invariant always has a value equal to

$$m_e^2c^4 = 6.72\times10^{-27}\text{ J}^2$$

b) The 6.9 GeV electron has the same energy-momentum invariant, so

$$p^2c^2 = E^2 - m^2c^4$$

Since

$$\begin{aligned} K &= E - E_0 = E - mc^2 \\ p^2c^2 &= K^2 + 2mc^2K + m^2c^4 - m^2c^4 \\ &= K^2 + 2mc^2K \end{aligned}$$

with

$$\begin{aligned} K &= (6.9\times10^9\text{ eV})\,(1.60\times10^{-19}\text{ J/eV}) \\ &= 1.10\times10^{-9}\text{ J} \\ p^2c^2 &= (1.10\times10^{-9}\text{ J}) \\ &\quad\cdot(1.10\times10^{-9}\text{ J}+2\,(8.20\times10^{-14}\text{ J})) \\ &= (1.10\times10^{-9}\text{ J})^2 \end{aligned}$$

We see that $pc = 1.10\times10^{-9}$ J and that we would have been justified in neglecting the energy associated with the rest mass of this relativistic electron. Then, $E \approx K$, and $E = pc$.

$$\begin{aligned} p &= \frac{E}{c} = \frac{1.10\times10^{-9}\text{ J}}{3.00\times10^8\text{ m/s}} \\ &= 3.7\times10^{-18}\text{ kg}\cdot\text{m/s} \end{aligned}$$

34.73 In rest frame of galley

10^8 GeV

proton pea

The proton collides with the pea inelastically; $\Rightarrow$ use conservation of momentum

$$\vec{p}_{\text{proton}} + \vec{p}_{\text{pea}} = \vec{p}\,'_{\text{proton}} + \vec{p}\,'_{\text{pea}}$$

The pea is initially at rest, the proton in the final state has the same velocity as the pea. Since $m_{\text{pea}} \ggg m_{\text{proton}}$ neglect the final momentum of the proton from the energy invariant

$$E^2 - p^2c^2 = m^2c^4$$

for the proton

$$\begin{aligned} p^2c^2 &= E^2 - m^2c^4 \\ p^2 &= \frac{E^2}{c^2} - \frac{m^2c^4}{c^2} \\ p &= \frac{1}{c}\left[E^2 - m^2c^4\right] \\ &= \frac{1}{c}\left[\left(10^8 \text{ GeV}\right)^2 - \left(938 \text{ MeV}\right)^2\right]^{1/2} \\ &\approx 10^8 \text{ GeV/c} \end{aligned}$$

$$\begin{aligned} p_{\text{prot}} &= \frac{10^{17} \text{ eV}}{(3.0 \times 10^8 \text{ m/s})} \times 1.602 \times 10^{-19} \text{ J/eV} \\ &= 5.3 \times 10^{-11} \text{ kg} \cdot \text{m/s} \end{aligned}$$

Thus

$$p_{\text{pea}} = 5.3 \times 10^{-11} \text{ kg} \cdot \text{m/s}$$

Because of its much larger mass, the resulting speed of the pea is non-relativistic.

$$\begin{aligned} v &= \frac{p}{m} = \frac{5.3 \times 10^{-11} \text{ kg} \cdot \text{m/s}}{5.0 \times 10^{-3} \text{ kg}} \\ v_{\text{pea}} &\cong 1.0 \times 10^{-8} \text{ m/s} \end{aligned}$$

One would not know the difference in the galley if a cosmic ray hit the pea or not.

34.77 a) Since

$$u' = \frac{u - \beta c}{1 - \frac{\beta u}{c}}$$

for relative motion along a line

$$\begin{aligned} \gamma_{u'} &= \frac{1}{\left[1 - \left(\frac{u'}{c}\right)^2\right]^{1/2}} \\ &= \frac{1}{\left[1 - \frac{1}{c^2}\left(\frac{u-\beta c}{1-\frac{\beta u}{c}}\right)^2\right]^{1/2}} \\ &= \frac{1}{\left[\left(1 - \frac{\beta u}{c}\right)^2 - \frac{(u-\beta c)^2}{c^2}\right]^{1/2}}\left(1 - \frac{\beta u}{c}\right) \\ &= \frac{1}{\left[1 - \beta^2 - \frac{u^2}{c^2} + \frac{\beta^2 u^2}{c^2}\right]^{1/2}}\left(1 - \frac{\beta u}{c}\right) \\ &= \frac{1}{\left[1 - \beta^2\right]^{1/2}\left[1 - \frac{u^2}{c^2}\right]^{1/2}}\left(1 - \frac{\beta u}{c}\right) \\ &= \gamma_u \gamma \left(1 - \frac{\beta u}{c}\right) \end{aligned}$$

b) Now

$$\begin{aligned} E' &= \gamma_{u'} mc^2 \\ &= \gamma\gamma_u\left(1 - \frac{\beta u}{c}\right) mc^2 \\ &= \gamma\left(\gamma_u mc^2 - \beta\gamma_u muc\right) \\ &= \gamma\left(E - \beta pc\right) \end{aligned}$$

as required, and

$$\begin{aligned} p'c &= \gamma'_u mu'c \\ &= \gamma\gamma_u\left(1 - \frac{\beta u}{c}\right) mc\frac{u - \beta c}{1 - \frac{\beta u}{c}} \\ &= \gamma\left[-\beta\gamma_u mc^2 + \gamma_u muc\right] \\ &= \gamma\left[-\beta E + pc\right] \end{aligned}$$

as required.

34.79 Given an *elastic* collision of two *identical* particles. Let the particles have mass m and speed $\pm v$ along the x-axis.

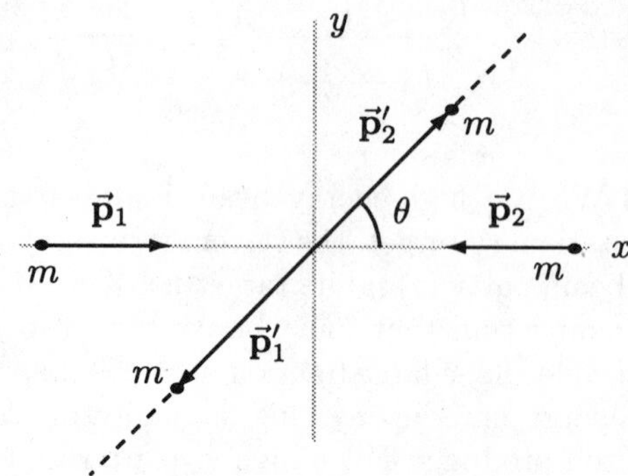

Prior to the collision we have $v_1 + v_2 = 0$ which says since the particles are identical that

$$\begin{aligned} \gamma_1 m\vec{v}_1 + \gamma_2 m\vec{v}_2 &= 0 \\ \gamma_1 &= \gamma_2 \\ \vec{p}_1 + \vec{p}_2 &= 0 \end{aligned}$$

Momentum is conserved in all collisions, so

$$\vec{p}'_1 + \vec{p}'_2 = 0$$

after the collision. Define a new x-axis x' rotated at an angle θ with respect to x; this axis defines the linear motion of the particles in the c-of-m frame and since

$$\vec{p}'_1 + \vec{p}'_2 = 0$$

then

$$\begin{aligned} \gamma'_1 m\vec{v}'_1 + \gamma'_2 m\vec{v}'_2 &= 0 \\ \gamma_1\vec{v}_1 + \gamma_2\vec{v}_2 &= 0 \qquad (1) \end{aligned}$$

Since

$$\gamma_i = \sqrt{\frac{1}{1 - \frac{v_i^2}{c^2}}}$$

then γ_1 and γ_2 are identical functions of v_1 and v_2 $\Rightarrow$ the solution of (1) that is possible is that

$$\vec{v}'_1 + \vec{v}'_2 = 0$$

$\therefore$ the particles will have equal speeds after the collision. Since kinetic energy is unchanged, the common speed must be the same as before the collision.

34.83 The average kinetic energy per particle is

$$\frac{3}{2}kT \equiv \text{KE}$$

for an ideal gas. But

$$\text{KE} = (\gamma - 1)\, mc^2$$

so

$$\frac{3}{2}kT = (\gamma - 1)\, mc^2 \approx \frac{1}{2}m\beta^2c^2$$

for $\beta \ll 1$

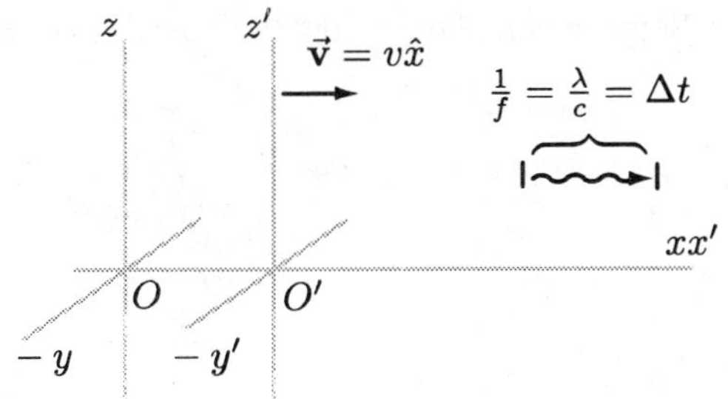

The line is broadened most by motions of atoms in the line of sight i.e., along our x-axis. In the rest frame of the particle, the time between two successive wave fronts is Δt. The signal travels at speed c

$$\Delta t = \frac{\lambda}{c}$$

In the primed frame the time interval between the two points in space-time depends on the Lorentz transformation.

$$t' = \gamma\left(t - \frac{v}{c^2}x\right)$$

so

$$\begin{aligned}
(t_2' - t_1') &= \Delta t' = \gamma\left(t_2 - t_1 - \frac{v}{c^2}(x_2 - x_1)\right) \\
&= \gamma\left(\Delta t + \frac{v}{c^2}(\lambda)\right) \\
\Delta t' &= \gamma\left(\Delta t + \frac{v}{c^2}\lambda\right) \\
&= \gamma\left(\frac{\lambda}{c} + \frac{v}{c^2}\lambda\right) \\
&= \gamma\frac{\lambda}{c}(1 + \beta)
\end{aligned}$$

the received frequency

$$f' = \frac{1}{\Delta t'}$$

and

$$\lambda f \equiv c \equiv \lambda' f'$$

thus

$$\begin{aligned}
f' &= f\frac{1}{\gamma}\left(\frac{1}{1+\beta}\right) \\
\frac{(f - f')}{f} &= \frac{1}{\gamma}\left(\frac{1}{1+\beta}\right) - 1 \\
\frac{\Delta f}{f} &= \frac{1}{\gamma}\left(\frac{1}{1+\beta}\right) - 1
\end{aligned}$$

For a non-relativistic gas

$$\begin{aligned}
\beta &\ll 1 \\
1 + \beta &\approx 1
\end{aligned}$$

and

$$\frac{1}{\gamma} \approx 1 - \frac{\beta^2}{2}$$

so

$$\frac{\Delta f}{f} = 1 - \frac{\beta^2}{2} - 1 \approx -\frac{\beta^2}{2}$$

Since

$$\frac{\beta^2}{2} = \frac{3}{2}\frac{kT}{mc^2} \text{ for } \beta \ll 1$$

then in this limit $T \to \Delta T$ and

$$\frac{1}{f}\frac{\Delta f}{\Delta T} \approx -\frac{3}{2}\frac{k}{mc^2}$$

for iron

$$\begin{aligned}
m &\approx 57 \times 1.6 \times 10^{-27}\text{ kg} \\
k &= 1.4 \times 10^{-23}\text{ J/K}
\end{aligned}$$

so

$$\begin{aligned}
\frac{1}{f}\frac{\Delta f}{\Delta T} &\approx -\left(\frac{3}{2}\right)\left(\frac{1.4 \times 10^{-23}\text{ J/K}}{57 \times 1.6 \times 10^{-27}\text{ kg}}\right) \\
&\quad \cdot \frac{1}{(3 \times 10^8\text{ m/s})^2} \\
&\cong -2.6 \times 10^{-15}\text{ K}^{-1}
\end{aligned}$$

which agrees well with Pound & Rebka.

34.87 Reference figures 34.7 and 34.6

a) The helicopter is at 0. The events simultaneous in the ground are not simultaneous as viewed at the helicopter. The distance signals take to travel $\overline{OB}$ is longer than taken to travel $\overline{OA}$.

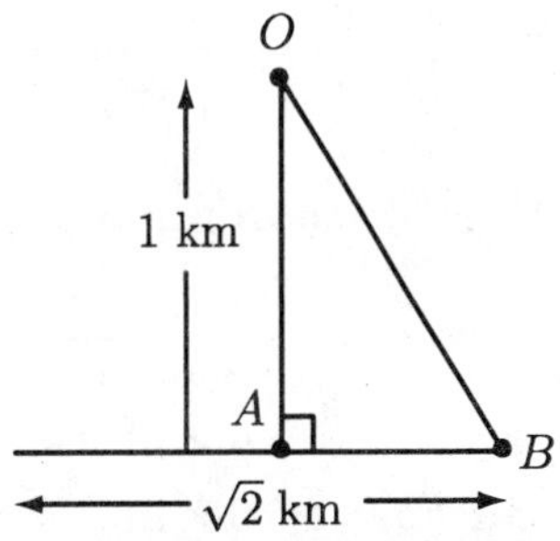

Using Pythagoras' theorem

$$(\overline{OA})^2 + (\overline{AB})^2 = (\overline{OB})^2$$

so

$$(\overline{OB}) = \left((\overline{AB})^2 + (\overline{OA})^2\right)^{1/2}$$

the time difference is

$$\begin{aligned}\Delta t &= \frac{1}{c}\left[(\overline{OB}) - (\overline{OA})\right] \\ &= \frac{1}{c}\left[\left((\overline{AB})^2 + (\overline{OA})^2\right)^{1/2} - (\overline{OA})\right]\end{aligned}$$

thus

$$\begin{aligned}\Delta t &= \frac{1}{c}\left\{\left[\frac{(OA)^2}{2} + (OA)^2\right]^{1/2} - (\overline{OA})\right\} \\ \Rightarrow \Delta t &= \left(\sqrt{\frac{3}{2}} - 1\right)\frac{(\overline{OA})}{c}\end{aligned}$$

or

$$\begin{aligned}\Delta t &= \left(\sqrt{\frac{3}{2}} - 1\right)\left(\frac{1\text{ km} \times 1000\text{ m/km}}{3 \times 10^8\text{ m/s}}\right) \\ &= 7.5 \times 10^{-7}\text{ sec}\end{aligned}$$

b) At time $t = 0$ as measured in the frame of the helicopter, the airplane flies past at

$$600\text{ km/h} = 166.7\text{ m/s}$$

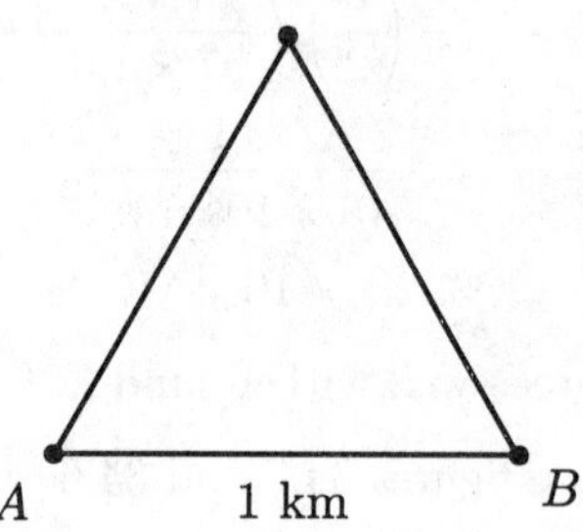

Simultaneous events in the helicopter frame are like $A + B$

$$\begin{aligned}\Delta t' &= \gamma\left(\Delta t - \frac{v}{c}\Delta x\right) \\ \gamma &= \sqrt{\frac{1}{1 - \left(\frac{166.7\text{ m/s}}{3\times 10^8\text{ m/s}}\right)^2}} \cong 1\end{aligned}$$

to a very good approximation, but

$$\begin{aligned}\frac{v}{c}\Delta x &= \frac{166.7\text{ m/s}}{3 \times 10^8\text{ m/s}} \times 10^3\text{ m} \\ &\doteq 5.6 \times 10^{-4}\text{ sec} \\ \Rightarrow \Delta t' &= 5.6 \times 10^{-4}\text{ sec}\end{aligned}$$

c) The error due to disagreement about simultaneity is a thousand times larger than that due to light-travel time.

34.91 If we let $\vec{\mathbf{v}} = \frac{d\vec{\mathbf{r}}}{dt}$ and define $\vec{\mathbf{u}} = \frac{d\vec{\mathbf{r}}}{d\tau}$, we can define a time-component for $\vec{\mathbf{u}}$ as

$$u_0 = \frac{d(ct)}{d\tau}$$

Further, since

$$\begin{aligned}c^2 d\tau^2 &= c^2 dt^2 - dx^2 \\ \left(\frac{dt}{d\tau}\right)^2 &= \frac{c^2}{c^2 - v^2} = \gamma^2\end{aligned}$$

so

$$u_0 = \gamma c \text{ and } \vec{\mathbf{u}} = \frac{dt}{d\tau}\frac{d\vec{\mathbf{r}}}{dt} = \gamma\vec{\mathbf{v}}$$

Now we can show that

$$\begin{aligned}\left(\frac{d(ct)}{d\tau}\right)^2 - \vec{\mathbf{u}}^2 &= c^2\gamma^2 - \gamma^2 v^2 \\ &= \gamma^2\left(c^2 - v^2\right) \\ &= \frac{c^2\left(1 - \frac{v^2}{c^2}\right)}{1 - \frac{v^2}{c^2}} = c^2\end{aligned}$$

Finally, since $E = \gamma mc^2 = \gamma E_0$,

$$\frac{E}{c} = \gamma mc = mc\frac{dt}{d\tau} = mu_0$$

Then with $\vec{\mathbf{p}} = \gamma m\vec{\mathbf{v}}$ or $\vec{\mathbf{p}} = m\vec{\mathbf{u}}$, the four-vector is $\boldsymbol{p} = (mu_0, m\vec{\mathbf{u}}) = m\boldsymbol{u}$.

Chapter 35

Light and Atoms

35.1 Photon of energy

$$\begin{aligned}
E &= 1.0 \text{ keV} = 1.0 \times 10^3 \text{ eV} \\
&= \frac{hc}{\lambda} \\
\Rightarrow \lambda &= \frac{hc}{E} \\
&= \frac{(6.626 \times 10^{-34} \text{ J}\cdot\text{s})(2.99 \times 10^8 \text{ m/s})}{(1.0 \times 10^3 \text{ eV})(1.6 \times 10^{-19} \text{ J/eV})} \\
&= 1.2 \times 10^{-9} \text{ m} = 1.2 \text{ nm} \\
E &= hf \\
\Rightarrow f &= \frac{E}{h} \\
&= \frac{(1.0 \times 10^3 \text{ eV})(1.6 \times 10^{-19} \text{ J/eV})}{6.626 \times 10^{-34} \text{ J}\cdot\text{s}} \\
&= 2.4 \times 10^{17} \text{ /s or } 2.4 \times 10^{17} \text{ Hz}
\end{aligned}$$

35.3 photo electric effect on silver (Ag)

$$\frac{1}{2}mv^2 = hf - \phi$$

ϕ = work function of Ag = 4.6 eV
energy per photon

$$\begin{aligned}
E &= hf = 15.0 \text{ eV} \\
\frac{1}{2}mv^2 &= hf - \phi = E - \phi \\
&= 15.0 \text{ eV} - 4.6 \text{ eV} \\
&= 10.4 \text{ eV}
\end{aligned}$$

= kinetic energy of emitted electrons.

35.5 Hydrogen atom undergoes transition

$$n' = 5 \rightarrow n = 2$$

the frequency of the emitted photon is given by

$$f_{nn'} = f_0\left(\frac{1}{n^2} - \frac{1}{n'^2}\right)$$

and the wavelength is related to the frequency by

$$\begin{aligned}
\lambda_{nn'} &= \frac{c}{f_{nn'}} = \frac{c}{f_0\left(\frac{1}{n^2} - \frac{1}{n'^2}\right)} \\
&= \frac{3.00 \times 10^8 \text{ m/s}}{(3.288 \times 10^{15} \text{ /s})\left(\frac{1}{2^2} - \frac{1}{5^2}\right)} \\
&= 4.34 \times 10^{-7} \text{ m} = 0.434 \,\mu\text{m}
\end{aligned}$$

35.7 We have a hydrogenic oxygen ($Z = 8$) atom. In other words, all of the electrons except one have been stripped from the atom. The remaining electron is in the ground state ($n = 1$). Stripping this last electron means taking it to a state $n' = \infty$. For hydrogenic atoms, an electron in the nth state has energy

$$\begin{aligned}
E_n &= \frac{-m\left(kZe^2\right)^2}{2\hbar^2}\frac{1}{n^2} \\
&= -Z^2\frac{E_0}{n^2}
\end{aligned}$$

the change in energy

$$\begin{aligned}
\Delta E &= E_{n'} - E_n = -Z^2\frac{E_0}{n'^2} - \left(-Z^2\frac{E_0}{n^2}\right) \\
&= Z^2E_0\left(\frac{1}{n^2} - \frac{1}{n'^2}\right) \\
&= 8^2(13.6 \text{ eV})\left(\frac{1}{1^2} - \frac{1}{\infty^2}\right)
\end{aligned}$$

$$= 8^2(13.6\text{ eV})(1-0)$$
$$= 870\text{ eV}$$

35.9 third shell:

$$n=3 \Rightarrow \ell = 0,1,2 \text{ and } -\ell \le m \le \ell$$

Angular momentum states:

$$\begin{array}{lll} n=3 & \ell=0 & m=0 \\ n=3 & \ell=1 & m=-1 \\ n=3 & \ell=1 & m=0 \\ n=3 & \ell=1 & m=1 \\ n=3 & \ell=2 & m=-2 \\ n=3 & \ell=2 & m=-1 \\ n=3 & \ell=2 & m=0 \\ n=3 & \ell=2 & m=1 \\ n=3 & \ell=2 & m=2 \end{array}$$

$\Rightarrow$ 9 states

Quantum states:
Two electrons can occupy *each* of these angular momentum states: one with $m_s = -\frac{1}{2}$ and one with $m_s = +\frac{1}{2}$. $\Rightarrow$ total number of electrons possible = 18

35.11 Momentum measured to 1 part in 10^3

$$\frac{\delta p}{p} = \frac{1}{10^3}$$
$$\delta p = \frac{p}{10^3} = \frac{\gamma m v}{10^3}$$

where

$$v = 1.5 \times 10^8\text{ m/s}$$

and

$$\gamma = \frac{2}{\sqrt{3}} = 1.15$$

From the uncertainty principle:

$$\delta x\, \delta p \ge \frac{\hbar}{2}$$
$$\Rightarrow \delta x \ge \frac{\hbar}{2\,\delta p}$$
$$\ge \frac{\hbar\, 10^3}{2\gamma m v}$$
$$\ge \frac{(1.055 \times 10^{-34}\text{ J}\cdot\text{s})(10^3)}{2\,(1.15)\,(9.11 \times 10^{-31}\text{ kg})(1.5 \times 10^8\text{ m/s})}$$
$$\ge 3.3 \times 10^{-10}\text{ m}$$
$$= 0.3\text{ nm}$$

35.13 The uncertainty in the time to transition is

$$\delta t = 6 \times 10^4\text{ s}$$

From the uncertainty principle:

$$\delta t\, \delta E \ge \frac{\hbar}{2\delta t}$$

So the uncertainty in the energy is:

$$\delta E \ge \frac{\hbar}{2\delta t}$$
$$\ge \frac{(1.05 \times 10^{-34}\text{ J}\cdot\text{s})}{2(6 \times 10^4\text{ s})}$$
$$\ge 8.8 \times 10^{-40}\text{ J}$$
$$= 5 \times 10^{-21}\text{ eV}$$

35.15 We need a photon with a minimum energy

$$E_{\text{min}} = 7.4\text{ eV}$$

This energy is related to the maximum wavelength

$$E_{\text{min}} = \frac{hc}{\lambda_{\text{max}}}$$
$$\Rightarrow \lambda_{\text{max}} = \frac{hc}{E_{\text{min}}}$$
$$= \frac{(6.626 \times 10^{-34}\text{ J}\cdot\text{s})(3.00 \times 10^8\text{ m/s})}{(7.4\text{ eV})(1.6 \times 10^{-19}\text{ J/eV})}$$
$$= 1.7 \times 10^{-7}\text{ m}$$
$$= 170\text{ nm}$$

35.19

$$\phi = \text{work function of sodium} = 2.75\text{ eV}$$
$$\lambda = 415\text{ nm} = 4.15 \times 10^{-7}\text{ m}$$

Then the electron's kinetic energy is:

$$\frac{1}{2}mv^2 = hf - \phi = \frac{hc}{\lambda} - \phi$$
$$= \frac{(6.626\times 10^{-34}\text{ J}\cdot\text{s})(3.00\times 10^8\text{ m/s})}{(4.15\times 10^{-7}\text{ m})(1.6\times 10^{-19}\text{ J/eV})} - 2.75\text{ eV}$$
$$= 0.24\text{ eV}$$

35.23 $\lambda = 4$ nm

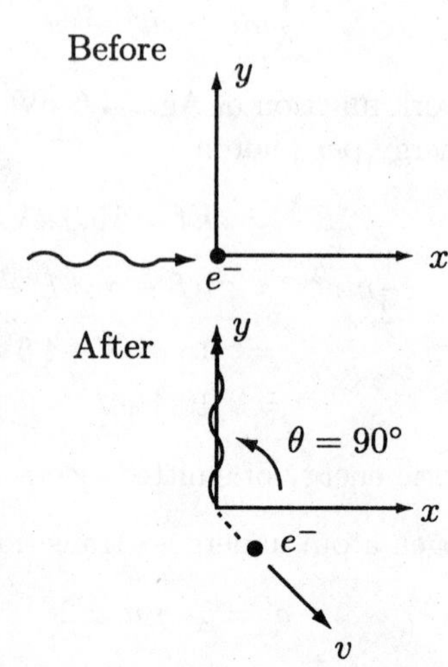

fractional change

$$\begin{aligned}\frac{\Delta\lambda}{\lambda_1} &= \lambda_c\frac{(1-\cos\theta)}{\lambda_1} \\ &= \frac{2.43\times10^{-3}\ \text{nm}(1-0)}{4.0\ \text{nm}} \\ &= 6.08\times10^{-4}\end{aligned}$$

35.27 Assuming that all X-rays directed toward the Earth actually reach the detector.

$$\frac{dn}{dt}=\frac{LA}{4\pi R^2E}$$

where

X-ray luminosity	$L=10^{26}$ W
area of detector	$A=900\ \text{cm}^2=.09\ \text{m}^2$
distance to source	$R=3\times10^{19}$ m
energy per photon	$E=6\times10^3$ eV

$$\begin{aligned}\frac{dn}{dt} &= \frac{LA}{4\pi R^2E} \\ &= \frac{10^{26}\ \text{W}(.09\ \text{m}^2)}{4\pi(3\times10^{19}\ \text{m})^2\cdot6\times10^3\ \text{eV}(1.6\times10^{-19}\ \text{J/eV})} \\ &= 0.8\ \text{photons per second}\end{aligned}$$

35.31 Change in photon energy = kinetic energy of electron

$$\begin{aligned}E_1-E_2 &= K \\ E_1 &= K+E_2 \\ \frac{hc}{\lambda_1} &= K+E_2 \\ \lambda_1 &= \frac{hc}{K+E_2}\end{aligned}$$

$$\begin{aligned}\lambda_1 &= \frac{(6.626\times10^{-34}\ \text{J}\cdot\text{s})(3\times10^8\ \text{m/s})}{(61\times10^3\ \text{eV}+150\times10^3\ \text{eV})(1.6\times10^{-19}\ \text{J/eV})} \\ &= 5.9\times10^{-12}\ \text{m} \\ &= 5.9\times10^{-3}\ \text{nm}\end{aligned}$$

35.35 A potential difference, V, can stop electrons when

$$\begin{aligned}eV &\geq \frac{1}{2}mv^2=hf-\phi \\ &\geq \frac{hc}{\lambda}-\phi \\ \Rightarrow\phi &\geq \frac{hc}{\lambda}-eV \\ \lambda &= 4.0\times10^{-7}\ \text{m} \\ V &= 0.80\ \text{V} \\ \phi &\geq \frac{(6.626\times10^{-34}\ \text{J}\cdot\text{s})(3\times10^8\ \text{m/s})}{4.0\times10^{-7}\ \text{m}} \\ &\quad -(1.6\times10^{-19}\ \text{C})(0.80\ \text{V}) \\ &\geq 3.7\times10^{-19}\ \text{J}=2.3\ \text{eV} \\ \lambda &= 3\times10^{-7}\ \text{m}\end{aligned}$$

$$\begin{aligned}eV &\geq \frac{hc}{\lambda}-\phi \\ V &\geq \frac{hc}{\lambda e}-\frac{\phi}{e}\end{aligned}$$

$$\begin{aligned}&= \frac{(6.626\times10^{-34}\ \text{J}\cdot\text{s})(3\times10^8\ \text{m/s})}{(3\times10^{-7}\ \text{m})(1.6\times10^{-19}\ \text{C})} \\ &\quad -\frac{3.7\times10^{-19}\ \text{J}}{1.6\times10^{-19}\ \text{C}}\end{aligned}$$

$$V\geq1.8\ V$$

35.39 Stopping potential V_s must be such that

$$\begin{aligned}eV_s &= \frac{1}{2}mv^2=hf-\phi \\ &= \frac{hc}{\lambda}-\phi \\ V_s &= \frac{hc}{\lambda e}-\frac{\phi}{e} \\ \phi &= 2.9\ \text{eV} \\ \lambda &= 320\ \text{nm}\end{aligned}$$

$$\begin{aligned}V_s &= \frac{(6.626\times10^{-34}\ \text{J}\cdot\text{s})(3\times10^8\ \text{m/s})}{(320\times10^{-9}\ \text{m})(1.6\times10^{-19}\ \text{C})} \\ &\quad -\frac{(2.9\,\text{eV})(1.6\times10^{-19}\ \text{J/eV})}{(1.6\times10^{-19}\ \text{C})} \\ &= 0.98\ \text{V} \\ \phi &= 2.9\ \text{eV} \\ \lambda &= 300\ \text{nm}\end{aligned}$$

$$\begin{aligned}V_s &= \frac{(6.626\times10^{-34}\ \text{J}\cdot\text{s})(3\times10^8\ \text{m/s})}{(300\times10^{-9}\ \text{m})(1.6\times10^{-19}\ \text{C})} \\ &\quad -\frac{(2.9\,\text{eV})(1.6\times10^{-19}\ \text{J/eV})}{(1.6\times10^{-19}\ \text{C})} \\ &= 1.2\ \text{V} \\ \Rightarrow\Delta V_s &= +0.2\ \text{V}\end{aligned}$$

35.43 Doubly ionized lithium is a hydrogenic atom with $Z=3$, so for any given pair of Bohr quantum numbers the photon frequency from lithium will be $Z^2=9$ times greater than the corresponding frequency for hydrogen. Put another way, for any hydrogen transition in the visible, say between levels n_1 and n_2, there will be a corresponding transition in lithium between levels $3n_1$ and $3n_2$. In lithium, however, transitions from different states, such as $3n_1-1$ to $3n_2$ will also be in the visible but will not have corresponding visible transitions in hydrogen. So, lithium will have the greater number of lines.

35.47 The photon frequencies are given by Eqn.(35.9):

$$f_{nn'} = f_0\left(\frac{1}{n^2} - \frac{1}{(n')^2}\right)$$

For $n' = 5 \to n = 3$

$$f_{nn'} = f_0\left(\frac{1}{9} - \frac{1}{25}\right) = 0.07 f_0$$

For $n' = 20 \to n = 4$,

$$f_{nn'} = f_0\left(\frac{1}{16} - \frac{1}{400}\right) = 0.06 f_0$$

So $n' = 5 \to n = 3$ has the greater frequency.

35.51 Short wavelength limit $n' = \infty$

$$\lambda_s = \frac{n^2}{R} = \frac{(N+1)^2}{R}, \quad n = N+1$$

α line:

$$\lambda_\alpha = \frac{1}{R\left(\frac{1}{n^2} - \frac{1}{n'^2}\right)}, \quad n = N, n' = N+1$$

$$\lambda_\alpha = \frac{1}{R}\left(\frac{1}{N^2} - \frac{1}{(N+1)^2}\right)^{-1}$$

Is $\lambda_s < \lambda_\alpha$ for any N?

$$\frac{(N+1)^2}{R} \overset{?}{<} \frac{1}{R}\left(\frac{1}{N^2} - \frac{1}{(N+1)^2}\right)^{-1}$$

$$(N+1)^2 \overset{?}{<} \frac{1}{\left(\frac{1}{N^2} - \frac{1}{(N+1)^2}\right)}$$

Let's see if this inequality holds for any Ns.
Arithmetic check:

N	inequality	True?
1	$4 < 1.333$	No
2	$9 < 7.199$	No
3	$16 < 20.57$	Yes
4	$25 < 44.4$	Yes

Algebraic check:

$$\frac{1}{(N+1)^2} \overset{?}{>} \frac{1}{N^2} - \frac{1}{(N+1)^2}$$

$$\frac{2}{(N+1)^2} > \frac{1}{N^2}$$

$$\frac{\sqrt{2}}{N+1} > \frac{1}{N}$$

$$\sqrt{2}N > N+1$$

$$N > \frac{1}{\sqrt{2}-1} = 2.4$$

For $n \geq 3$, $\lambda_s < \lambda_\alpha$.

35.55 The Bohr radius is $r_n = n^2 r_o$ ($r_o = .0529$ nm) $\therefore$ the $n = 2$ e^- is taken to be at a radius of

$$r_2 = 2^2 r_o = .2116 \text{ nm}$$

The speed of the e^- is given by

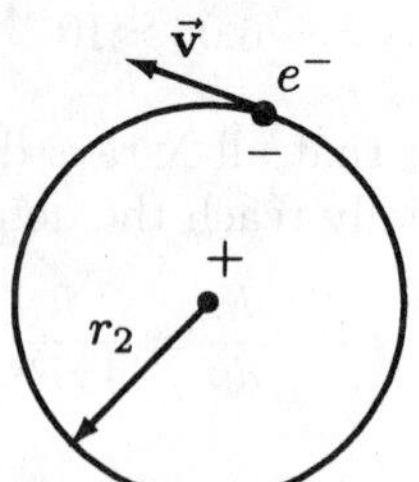

$$v_2 = \sqrt{\frac{ke^2}{mr_2}}$$

$$= \sqrt{\frac{9.0\times10^9 \frac{\text{N}\cdot\text{m}^2}{\text{C}^2}(1.6\times10^{-19}\text{ C})^2}{9.11\times10^{-31}\text{ kg}\cdot 0.2116\times10^{-9}\text{ m}}}$$

$$= 1.09\times10^6 \text{ m/s}$$

In 10^{-8} s the e^- will travel

$$d = (1.09\times10^6 \text{ m/s})(10^{-8}\text{ s}) = .011 \text{ m}$$

Each orbit is $2\pi r_2$ long so the e^- will make

$$\frac{.011 \text{ m}}{2\pi(.2116\times10^{-9}\text{ m})} = 8\times10^6 \text{ revolutions.}$$

35.59

$$r_n = \frac{n^2\hbar^2}{mke^2} = n^2 r_o$$

$r_o = .0529$ nm

$$r_{350} = (350)^2 \times .0529 \text{ nm} = 6480 \text{ nm}$$

$$PV = NkT$$

$$\frac{N}{V} = \rho = \frac{P}{kT}$$

$$\rho = \frac{1 \text{ atm}\cdot 1.013\times10^5 \text{ Pa/atm}}{1.38\times10^{-23}\text{ J/K}\cdot 300 \text{ K}}$$

$$= 2.4\times10^{25}\text{ molecules/m}^3$$

The intermolecular spacing is d

$$d^3 = \frac{1}{\rho}$$

$$d = 3.5 \text{ nm}$$

In the interstellar medium $\rho_i \approx 10^3\text{ m}^{-3}$ $\therefore$

$$d_i = \frac{1}{\sqrt[3]{\rho}} = 0.1 \text{ m} = 10^8 \text{ nm}$$

$$r_{350} \gg d, \; r_{350} \ll d_i$$

Thus such atoms are unlikely to occur naturally on Earth, but may occur in space.

35.65 $Z_{\text{Fe}} = 26$. The observed X-ray energy does not agree with the calculated energy because the Bohr theory does not take into account multiple electrons. An iron electron sees a $Z = 26$ nucleus and 25 other electrons.

$$\begin{aligned} E_{nn'} &= 13.6 \cdot Z_{\text{eff}}^2 \left(\frac{1}{n^2} - \frac{1}{n'^2} \right) \\ Z_{\text{eff}}^2 &= \frac{E_{nn'}}{13.6} \left(\frac{1}{n^2} - \frac{1}{n'^2} \right)^{-1} \\ &= \frac{6.4 \times 10^3 \text{ eV}}{13.6 \text{ eV}} \left(\frac{1}{1} - \frac{1}{2^2} \right)^{-1} \\ Z_{\text{eff}} &= 25 \end{aligned}$$

Z_{eff} is the effective charge internal to the transitioning electron.

35.67

$$\lambda = \frac{h}{p} = \frac{h}{mv}$$

electron:

$$\begin{aligned} \lambda &= \frac{6.626 \times 10^{-34} \text{ J} \cdot \text{s}}{(9.11 \times 10^{-31} \text{ kg})(100 \times 10^3 \text{ m/s})} \\ &= 7 \text{ nm} \end{aligned}$$

3 g pea:

$$\begin{aligned} \lambda &= \frac{6.626 \times 10^{-34} \text{ J} \cdot \text{s}}{(0.003 \text{ kg}) \cdot 10^5 \text{ m/s}} \\ &= 2 \times 10^{-36} \text{ m} \end{aligned}$$

proton:

$$\begin{aligned} \lambda &= \frac{6.626 \times 10^{-34} \text{ J} \cdot \text{s}}{(1.67 \times 10^{-27} \text{ kg}) \cdot (10^5 \text{ m/s})} \\ &= 4 \times 10^{-12} \text{ m} \end{aligned}$$

35.71 An alpha particle is two protons and two neutrons

$$\begin{aligned} \lambda &= \frac{h}{p} \\ p &= \sqrt{2mE} \\ \lambda &= \frac{h}{\sqrt{2mE}} \\ m_\alpha &= 2m_p + 2m_n \\ &= 2(1.675 \times 10^{-27} \text{ kg} + 1.673 \times 10^{-27} \text{ kg}) \\ &= 6.696 \times 10^{-27} \text{ kg} \\ E_\alpha &= 5.87 \times 10^6 \text{ eV} \cdot 1.6 \times 10^{-19} \text{ J/eV} \\ &= 9.392 \times 10^{-13} \text{ J} \\ \lambda &= \frac{6.626 \times 10^{-34} \text{ J} \cdot \text{s}}{(26.69 \times 10^{-27} \text{ kg} \cdot 9.392 \times 10^{-13} \text{ J})^{\frac{1}{2}}} \\ &= 5.9 \times 10^{-15} \text{ m} \end{aligned}$$

about $\frac{1}{3}$ the nuclear radius.

35.72 150 eV

35.73 There are four peaks between the center and the ring of iron atoms. So, we estimate that the $\frac{144 \text{ nm}}{2}$ radius of the ring is 4 wavelengths of the electrons. So, their energy is

$$\begin{aligned} K &\approx \frac{p^2}{2m} = \frac{\left(\frac{h}{\lambda}\right)^2}{2m} \\ &= \frac{(hc)^2}{2mc^2} \frac{1}{\lambda^2} \\ &= \frac{\left(1.24 \times 10^3 \text{ eV} \cdot \text{nm}\right)^2}{2(0.511 \times 10^6 \text{ eV}) \left(\frac{144}{8 \text{ nm}}\right)^2} \\ &= 5 \times 10^{-3} \text{ eV} \end{aligned}$$

35.75 $\lambda_B = \frac{h}{p}$ de Broglie wavelength

$$\begin{aligned} E^2 &= p^2c^2 + m^2c^4 \\ \Rightarrow p &= \frac{1}{c}\sqrt{E^2 - m^2c^4} \\ \lambda_B &= \frac{hc}{\sqrt{E^2 - m^2c^4}} \\ \lambda_c &= \frac{h}{mc} \text{ Compton wavelength} \end{aligned}$$

for $\lambda_B = \lambda_c$

$$\begin{aligned} \frac{h}{mc} &= \frac{hc}{\sqrt{E^2 - m^2c^4}} \\ E^2 - m^2c^4 &= m^2c^4 \\ E^2 &= 2m^2c^4 \\ E &= \sqrt{2}mc^2 = 0.723 \text{ MeV} \end{aligned}$$

35.79

$$\begin{aligned} \lambda &= \frac{hc}{E} \\ \delta\lambda &= hc(-1)(E^{-2})\delta E \\ &= -\frac{hc\delta E}{E^2} \end{aligned}$$

where

$$\begin{aligned} \delta E &\geq \frac{\hbar}{2\,\delta t} \\ \Rightarrow \delta\lambda &\geq \frac{2\pi\hbar^2 c}{E^2 2\,\delta t}\delta\lambda = \frac{\pi\hbar^2 c}{E^2\,\delta t} \\ &\geq \frac{\pi\left(1.055 \times 10^{-34} \text{ J} \cdot \text{s}\right)^2 (3 \times 10^8 \text{ m/s})}{[(2.48 \text{ eV})(1.6 \times 10^{-19} \text{ J/eV})]^2\, 10^{-8} \text{ s}} \\ &\geq 7 \times 10^{-15} \text{ m} = 7 \text{ fm} \end{aligned}$$

35.85

$$\lambda_d = \frac{h}{\gamma m v}$$

$$\frac{E_1}{c} = \frac{E_2}{c}\cos\theta + \gamma m v \cos\phi$$

$$0 = \frac{E_2}{c}\sin\theta - m v \sin\phi$$

when $\theta = 90°$

$$\frac{E_1}{c} = \gamma m v \cos\phi$$

$$\frac{E_2}{c} = \gamma m v \sin\phi$$

$$\Rightarrow \gamma^2 m^2 v^2 = \frac{E_1^2 + E_2^2}{c^2}$$

$$\gamma m v = \frac{1}{c}\left(E_1^2 + E_2^2\right)^{\frac{1}{2}}$$

$$\lambda_d = \frac{hc}{\left(E_1^2 + E_2^2\right)^{\frac{1}{2}}}$$

Now

$$E_1 = \frac{hc}{\lambda_1}$$

$$E_2 = \frac{hc}{\lambda_2}$$

$$\sqrt{E_1^2 + E_2^2} = \sqrt{h^2c^2\left[\frac{1}{\lambda_1^2} + \frac{1}{\lambda_2^2}\right]}$$

$$= hc\left[\frac{\sqrt{\lambda_2^2 + \lambda_1^2}}{\lambda_1\lambda_2}\right]$$

$$\Rightarrow \lambda_d = \frac{hc}{hc}\left(\frac{\lambda_1\lambda_2}{\sqrt{\lambda_1^2 + \lambda_2^2}}\right) = \frac{\lambda_1\lambda_2}{\sqrt{\lambda_1^2 + \lambda_2^2}}$$

and

$$\lambda_2 = \lambda_1 + \lambda_c(1 - \cos\theta)$$

where $\theta = 90°$

$$\lambda_2 = \lambda_1 + \lambda_c$$

case (a):

$$\lambda_1 \ll \lambda_c \Rightarrow \lambda_2 \approx \lambda_c \ \&\ \lambda_1^2 + \lambda_2^2 \approx \lambda_c^2$$

$$\lambda_d \approx \frac{\lambda_1\lambda_c}{\sqrt{\lambda_c^2}} \approx \lambda_1$$

case (b):

$$\lambda_1 \approx \lambda_c \Rightarrow \lambda_2 \approx 2\lambda_c \ \&\ \lambda_1^2 + \lambda_2^2 \approx 5\lambda_c^2$$

$$\Rightarrow \lambda_d \approx \frac{\lambda_1 2\lambda_c}{\sqrt{5\lambda_c^2}} \approx \lambda_1$$

case (c):

$$\lambda_1 \gg \lambda_c \Rightarrow \lambda_2 \approx \lambda_1$$

$$\lambda_1^2 + \lambda_2^2 \approx 2\lambda_1^2$$

$$\lambda_d \approx \frac{\lambda_1^2}{\sqrt{2\lambda_1^2}} \approx \lambda_1$$

35.89 1×10^5 J/mole = dissociation energy for a mole of silver bromide (AgBr). $N_A = 6 \times 10^{23}$ molecules of AgBr in one mole. The energy required to dissociate a single molecule is:

$$\text{Energy} = \left(1 \times 10^5 \ \frac{\text{J}}{\text{mole}}\right)\left(\frac{\text{mole}}{6 \times 10^{23} \ \text{molecules}}\right)$$

$$= 1.66 \times 10^{-19} \ \text{J/molecule}$$

$$\approx 1 \ \text{eV/molecule}$$

A photon with this energy has wavelength

$$\lambda = \frac{hc}{E}$$

$$= \frac{(6.626 \times 10^{-34} \ \text{J}\cdot\text{s})(3 \times 10^8 \ \text{m/s})}{(1 \ \text{eV})(1.6 \times 10^{-19} \ \text{J/eV})}$$

$$= 1.2 \times 10^{-6} \ \text{m} = 1.2 \ \mu\text{m}$$

which is therefore the maximum wavelength to which the plate is sensitive.

35.93 (See solutions to 35.91 and 35.92.) Positronium has a nuclear mass

$$m_N = m_e$$

$$d_n = \frac{n^2\hbar^2}{\mu(ke^2)}$$

where

$$\mu = \frac{m_e m_e}{m_e + m_e} = \frac{m_e}{2}$$

For $n = 1$

$$d_1 = \frac{2\hbar^2}{m_e k e^2}$$

$$= \frac{2\left(6.626\times10^{-34} \ \text{J}\cdot\text{s}\right)^2}{(9.11\times10^{-3} \ \text{kg})\left(8.99\times10^9 \ \frac{\text{N}\cdot\text{m}^2}{\text{C}^2}\right)(1.6\times10^{-19} \ \text{C})^2}$$

$$= 1.06 \times 10^{-10} \ \text{m}$$

$$= 0.106 \ \text{nm}$$

$$f_{nn'} = \frac{\mu\left(ke^2\right)^2}{4\pi\hbar^3}\left(\frac{1}{n^2} - \frac{1}{n'^2}\right)$$

$$\Rightarrow f_{12} = m_e\frac{\left(ke^2\right)^2}{8\pi\hbar^3}\left(\frac{1}{n^2} - \frac{1}{n'^2}\right)$$

$$= \frac{(9.11\times10^{-31} \ \text{kg})\left[\left(8.99\times10^9 \ \frac{\text{N}\cdot\text{m}^2}{\text{C}^2}\right)\left(1.6\times10^{-19} \ \text{C}\right)^2\right]^2\left[\frac{1}{1^2} - \frac{1}{2^2}\right]}{8\pi(1.055\times10^{-34} \ \text{J}\cdot\text{s})^3}$$

$$= 1.24 \times 10^{15} \ \text{Hz}$$

35.97 Ring of mass M, charge Q, radius r, spinning at frequency f.

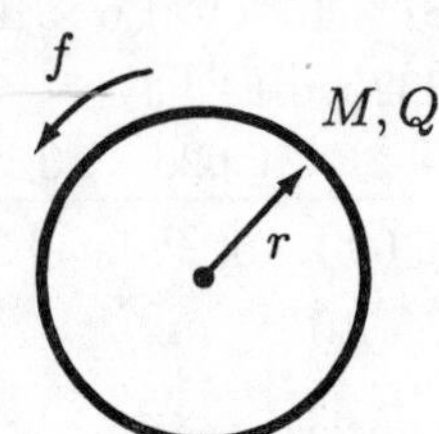

The magnetic moment is

$$\begin{aligned} m &= IA = (Qf)(\pi r^2) \\ \vec{\mathbf{L}} &= \vec{\mathbf{r}} \times \vec{\mathbf{p}} \\ L &= rMv = Mr^2(2\pi f) \\ \frac{m}{L} &= \frac{\pi r^2 Qf}{2\pi r^2 Mf} = \frac{Q}{2M} \\ m &= \frac{Q}{2M} L \end{aligned}$$

For spin

$$\begin{aligned} L_s &= \pm\frac{\hbar}{2} \\ m_s &= \frac{Q\hbar}{4M} = 4.63 \times 10^{-24}\ \mathrm{A \cdot m^2} \end{aligned}$$

For orbital angular momentum

$$\begin{aligned} L_{\text{orbit}} &= n\hbar \\ m_{\text{orbit}} &= n\frac{Q\hbar}{2M} \end{aligned}$$

for $n = 1$

$$m_{\text{orbit}} = 9.26 \times 10^{-24}\ \mathrm{A \cdot m^2}$$

The magnetic energy is

$$E_{\text{so}} = -\,\vec{\mathbf{m}}_{\text{spin}} \cdot \vec{\mathbf{B}}_{\text{orbit}}$$

where $\vec{\mathbf{B}}_{\text{orbit}}$ is due to $\vec{\mathbf{m}}_{\text{orbit}}$.

$$\begin{aligned} \vec{\mathbf{B}}_{\text{orbit}} &\approx -\frac{\mu_0}{4\pi}\frac{\vec{\mathbf{m}}_{\text{orbit}}}{r_1^3} \\ r_1 &= \frac{\hbar^2}{mke^2} \end{aligned}$$

$$\begin{aligned} E_{\text{so}} &= \pm\frac{\mu_0}{4\pi}\frac{1}{r_1^3} m_{\text{spin}} m_{\text{orbit}} \\ &\approx \pm\frac{4\pi \times 10^{-7}\ \mathrm{N/A^2}}{4\pi}\frac{1}{(5.3 \times 10^{-11}\ \mathrm{m})^3} \\ &\quad \cdot (4.63 \times 10^{-24}\ \mathrm{A \cdot m^2})(9.26 \times 10^{-24}\ \mathrm{A \cdot m^2}) \end{aligned}$$

$$E_{\text{so}} \approx 3 \times 10^{-23}\ \mathrm{J} = 1.8 \times 10^{-4}\ \mathrm{eV}$$

For the nucleus, $L_n = \frac{\hbar}{2}$, M_p = proton mass

$$\begin{aligned} m_h &= \pm\frac{Q\hbar}{4M_p} \\ &= \frac{(1.6 \times 10^{-19}\ \mathrm{C})(1.055 \times 10^{-34}\ \mathrm{J \cdot s})}{4(1.67 \times 10^{-27}\ \mathrm{kg})} \\ &= \pm 2.53 \times 10^{-27}\ \mathrm{A \cdot m^2} \end{aligned}$$

$$E_{\text{hyperfine}} = \frac{\mu_0}{4\pi r_1^3} m_h \cdot m_s \approx 5 \times 10^{-8}\ \mathrm{eV}$$

Chapter 36

Atomic Nuclei

36.1 Yttrium isotope containing 90 nucleons

$$\rightarrow n_p + n_n = A$$

or

$$Z + N = A$$

Yttrium's symbol is Y with $Z = 39 \Rightarrow$ symbol is ${}^{90}_{39}\text{Y}$ or just ${}^{90}\text{Y}$. For aluminum, the symbol is Al and $Z = 13$ for 27 nucleons, we get ${}^{27}_{13}\text{Al}$ or ${}^{27}\text{Al}$.

36.3 Deuterium nucleus has one proton + 1 neutron. From Table 36.1, the Deuterium mass is

$$\begin{aligned} m &= 2.0140 \text{ u} \\ &= (2.0140 \text{ u}) \left(931.49 \frac{\frac{\text{MeV}}{c^2}}{\text{u}} \right) \\ &= 1876.02 \frac{\text{MeV}}{c^2} \end{aligned}$$

and from Table 36.2

$$\begin{aligned} M_p &= 938.272 \frac{\text{MeV}}{c^2} \\ M_n &= 939.566 \frac{\text{MeV}}{c^2} \end{aligned}$$

then

$$\begin{aligned} \Delta E &= \left[\begin{array}{l} 1876.02 \frac{\text{MeV}}{c^2} \\ -(938.272 + 939.566) \frac{\text{MeV}}{c^2} \end{array} \right] \times c^2 \\ &= 1.8 \text{ MeV} \end{aligned}$$

36.5 Alpha decay of ${}^{183}\text{Au}$. We have in general for α-decay

$$^{A}Z \rightarrow^{(A-4)} (Z-2) + {}^{4}\text{He} + Q$$

From the periodic chart, for Au, $Z = 79$ then for

$$Z' = Z - 2 = 79 - 2 = 77$$

we get Ir (iridium). Putting this together,

$${}^{183}\text{Au} \rightarrow {}^{179}\text{Ir} + {}^{4}\text{He} + Q$$

36.7 For proton/antiproton annihilation, we have an analogous equation as for the electron/positron pair:

$$p^{+} + p^{-} \rightarrow 2\gamma$$

where 2γ are required to conserve the system momentum.

$$\begin{aligned} E_\gamma &= 2m_p c^2 = E_{\gamma_1} + E_{\gamma_2} \\ &= 2 \cdot \left(938.272 \frac{\text{MeV}}{c^2} \right) \times c^2 \\ &= 1876.544 \text{ MeV} \end{aligned}$$

This is the total energy of both γ rays. So each γ ray has energy 938 MeV.

36.9 For ${}^{23}\text{Ne}$, $T^{1/2} = 37.6$ s, $M = 15 \times 10^{-3}$ g $= 15 \times 10^{-6}$ kg. Now

$$R = \lambda N = \frac{\ln 2}{T_{1/2}} \times N$$

and N = mass of sample/mass of atom

$$N = \frac{15 \times 10^{-6} \text{ kg}}{(23) \times (1.66 \times 10^{-27} \text{ kg})} = 3.93 \times 10^{20}$$

Thus

$$\begin{aligned} R &= \left(\frac{\ln 2}{37.6 \text{ s}} \right) (3.93 \times 10^{20}) \\ &= 7.2 \times 10^{18} \text{ Bq.} \end{aligned}$$

Alternatively,

$$R = \frac{7.24 \times 10^{18}\ \text{Bq}}{3.7 \times 10^{10}\ \text{Bq/Ci}} = 2.0 \times 10^{8}\ \text{Ci}$$

36.11 ^{59}Co captures a neutron

$$^{59}\text{Co} + n \rightarrow {}^{60}\text{Co}^{*} \rightarrow {}^{60}\text{Co} + \gamma + Q$$

$$\begin{aligned} Q &= 931.49\ \frac{\text{MeV}}{\text{c}^2 \cdot \text{u}}\ [\text{reactants} - \text{products}]\ c^2 \\ &= 931.49\ \frac{\text{MeV}}{\text{c}^2 \cdot \text{u}} \\ &\quad \times [1.00866\ \text{u} + 58.93319\ \text{u} - 59.93388\ \text{u}]\ c^2 \\ &= 7.42\ \text{MeV} \end{aligned}$$

(masses from Table 36.1)

36.13 ^{235}U absorbs a neutron and produces ^{143}La + ^{87}Br + neutrons. The reaction is

$$^{235}\text{U} + n \rightarrow {}^{143}\text{La} + {}^{87}\text{Br} + xn + Q$$

where to conserve A (baryons)

$$\begin{aligned} 235 + 1 &\rightarrow 143 + 87 + x \\ x &= 235 + 1 - 143 - 87 \\ &= 6 \end{aligned}$$

$\Rightarrow$ 6 neutrons are produced, for a *net* gain of 5 neutrons. The Q for the reaction is

$$\begin{aligned} Q &= 931.49\ \frac{\text{MeV}}{c^2} \\ &\quad \times [235.04394 - 142.916 \\ &\quad - 86.9199 - 5(1.00866)] \\ &= 153.5\ \text{MeV} \end{aligned}$$

36.17 Fraction of lead isotopes:

^{204}Pb	1.4%
^{206}Pb	24.1%
^{207}Pb	22.1%
^{208}Pb	52.4%

Chemical atomic mass

$$\begin{aligned} &= (0.014)(204) + (0.241)(206) \\ &\quad + (0.221)(207) + (0.524)(208) \\ &= 207.241\ \text{u} \end{aligned}$$

The periodic table gives the chemical atomic weight as 207.2 u $\Rightarrow$ good agreement.

36.21 Tritium ^{3}H contains 1 p, 1 e, and 2 n.

$$\begin{aligned} \Delta E &\equiv \text{Binding energy} = c^2 \cdot \Delta M \\ &= 931.49\ \frac{\text{MeV}}{c^2} \\ &\quad \times [1.00727 + 2(1.00866) - 3.01605]\ c^2 \\ &\quad + 0.511\ \text{MeV} \\ &= 7.95\ \text{MeV} + 0.511\ \text{MeV} \end{aligned}$$

per nucleon, $\dfrac{\Delta E}{A} = 2.82$ MeV.

^{3}H contains 2 p, 2 e, 1 n. For ^{3}He the binding energy per nucleon is

$$\begin{aligned} \frac{\Delta E}{A} &= \frac{1}{3}\Big\{931.49\ \text{MeV} \times \\ &\quad [2 \times 1.00727 + 1.00866 - 3.01603] \\ &\quad + 2 \times 0.511\ \text{MeV}\Big\} \\ &= 2.57\ \text{MeV} \end{aligned}$$

^{3}H is more tightly bound than ^{3}He.

36.25 Equality of decay rates among the daughter species in a series requires that the geological sample in question has been undisturbed for a time much longer than the half-lives of the daughter species. If instead the sample has been disturbed more recently than the half-life of a particular species, that species will not have had time to build up its equilibrium number of atoms and so will be decaying at a lower rate than it is being produced by species earlier in the series.

The decay rate of the series' parent species is R_p and is essentially constant. The daughter species decays at a rate $N_d \lambda_d$, but starts at $N_d = 0$ when the sample is deposited in a sediment. The daughter species will increase in number at a rate

$$\frac{dN_d}{dt} = R_p - \lambda_d N_d$$

This equation is of the same form as Eqn. (31.7) and has a similar solution. The daughter species comes to equilibrium in a time $\sim \dfrac{1}{\lambda_d}$.

36.29 Suppose $n \rightarrow \gamma$. The only isotope that exists without containing neutrons in the nucleus is ^{1}H. Conversely, all elements containing neutrons would ultimately be unstable, as the normally stable nuclei would decay. We would have

$$\begin{aligned} A_z^A &\rightarrow A_z^{A-1} + \gamma + Q \\ A_z^{A-1} &\rightarrow A_z^{A-2} + \gamma + Q \\ A_z^{A-2} &\rightarrow A_z^{A-3} + \gamma + Q \\ &\vdots\ \text{etc} \end{aligned}$$

This continues until the net strong force is insufficient to overcome the Coulomb repulsion of the protons ⇒ each of these nuclei would eject protons or else completely split until a "stable" nucleus could be formed. But then the n-decay continues and all we would be left with is nuclei with 1 proton each (ie ^{1}H). Hence, the only possible *stable* nucleus would be ^{1}H.

36.33 For α decay,

$$^{A}Z \to \, ^{(A-4)}(Z-2) + \, ^4\text{He} + Q$$

Consulting the periodic chart:

$$\begin{array}{ll} ^{224}\text{Th}: & ^{224}\text{Th} \to \, ^{220}\text{Ra} + \, ^4\text{He} + Q \\ ^{211}\text{Bi}: & ^{211}\text{Bi} \to \, ^{207}\text{Tl} + \, ^4\text{He} + Q \\ ^{243}\text{Cm}: & ^{243}\text{Cm} \to \, ^{239}\text{Pu} + \, ^4\text{He} + Q \\ ^{221}\text{Fr}: & ^{221}\text{Fr} \to \, ^{217}\text{At} + \, ^4\text{He} + Q \\ ^{215}\text{At}: & ^{215}\text{At} \to \, ^{211}\text{Bi} + \, ^4\text{He} + Q \end{array}$$

36.37 ^{60}Co source.

$$\begin{aligned} T_{1/2} &= 5.2 \text{ y} \times 3.156 \times 10^7 \text{ s/y} \\ &= 1.641 \times 10^8 \text{ s} \end{aligned}$$

Initial activity $t = 0$

$$A(0) = 1100 \text{ Ci}$$

In Bq,

$$\begin{aligned} A(0) &= 1100 \text{ Ci} \times 3.7 \times 10^{10} \text{ Bq/Ci} \\ &= 4.1 \times 10^{13} \text{ Bq} \end{aligned}$$

Use

$$\begin{aligned} A(t) &= A(0)2^{-t/T_{1/2}}; \text{ at } t = 40 \text{ y} \\ &= \left(4.07 \times 10^{13}\right) 2^{-40/5.2} \text{ Bq} \\ A(40 \text{ y}) &= 2 \times 10^{11} \text{ Bq} \end{aligned}$$

Use

$$\begin{aligned} A_o &= \lambda N_o \\ &= \frac{\ln 2}{T_{1/2}} N_o \\ N_o &= \frac{T_{1/2}}{\ln 2} \times A_o \\ &= \left(\frac{1.64 \times 10^8 \text{ s}}{\ln 2}\right) 4.07 \times 10^{13} \text{ Bq} \\ N_o &= 9.6 \times 10^{21} \text{ atoms at } t = 0 \end{aligned}$$

36.41 Consider α decay of ^{152}Sm

$$^{152}\text{Sm} \to \, ^{148}\text{Nd} + \, ^4\text{He} + Q$$

Since ^{148}Nd is much heavier than ^{4}He, nearly all of the kinetic energy passes to the α-particle.

$$\begin{aligned} Q &= c^2 \left[\text{reactants} - \text{products}\right] \\ &= c^2 \Big[151.91976 \text{ u} - 4.00260 \text{ u} \\ &\quad - 147.9169 \text{ u}\Big] \times 931.49 \frac{\text{MeV}}{c^2 \cdot \text{u}} \\ &= 0.186 \text{ MeV} \end{aligned}$$

The α-particles are emitted with kinetic energy 0.186 MeV.

36.45 Start with ^{226}U, do 5 α-decays.

$$\begin{array}{l} ^{226}\text{U} \to \, ^{222}\text{Th} + \, ^4\text{He} + Q \\ ^{222}\text{Th} \to \, ^{218}\text{Ra} + \, ^4\text{He} + Q \\ ^{218}\text{Ra} \to \, ^{214}\text{Rn} + \, ^4\text{He} + Q \\ ^{214}\text{Rn} \to \, ^{210}\text{Po} + \, ^4\text{He} + Q \\ ^{210}\text{Po} \to \, ^{206}\text{Pb} + \, ^4\text{He} + Q \end{array}$$

The final isotope is ^{206}Pb.

36.49 The total activity of the sample is the sum of the activities of both isotopes:

$$\lambda_1 = \frac{\ln 2}{3.13 \text{ h}}, \quad \lambda_2 = \frac{\ln 2}{22 \text{ m}}$$

$$\begin{aligned} R &= R_1 + R_2 = \lambda_1 N_1 + \lambda_2 N_2 \\ R(t) &= \lambda_1 N_1(0) e^{-\lambda_1 t} + \lambda_2 N_2(0) e^{-\lambda_2 t} \\ &= R_1(0) e^{-\lambda_1 t} + R_2(0) e^{-\lambda_2 t} \\ &= R_1(0) 2^{-t/T_{1,1/2}} + R_2(0) 2^{-t/T_{2,1/2}} \end{aligned}$$

At $t = 0$

$$R(0) = 6.35 \times 10^7 \text{ Bq}$$

$t = 1$ h

$$R(1 \text{ h}) = 4.5 \times 10^7 \text{ Bq}$$

So

1) $R(0) = R_1(0) + R_2(0)$ and since

$$22 \text{ min} = \frac{22}{60} \text{ h} = 0.367 \text{ h}$$

$$R(1 \text{ h}) = R_1(0) 2^{-1/3.13} + R_2(0) 2^{-1/0.367}$$

2) $R(1) = 0.801 R_1(0) + 0.151 R_2(0)$
From 1)

$$R_2(0) = R(0) - R_1(0)$$

substitute into (2)

$$\begin{aligned} R(1) &= 0.801R_1(0) + 0.151\,(R(0) - R_1(0)) \\ &= 0.801R_1(0) - 0.151R_1(0) + 0.151R(0) \\ &= 0.65R_1(0) + 0.151R(0) \\ R_1(0) &= \frac{R(1) - 0.151R(0)}{0.65} \\ &= \frac{4.5\times10^7 - (0.151)(6.35\times10^7)}{0.65}\,\text{Bq} \\ R_1(0) &= 5.45\times10^7\ \text{Bq} \end{aligned}$$

thus

$$\begin{aligned} R_2(0) &= 6.35\times10^7 - 5.45\times10^7\ \text{Bq} \\ R_2(0) &= 0.90\times10^7\ \text{Bq} \end{aligned}$$

The initial activity of silver is 5.45×10^7 Bq and of palladium is 0.90×10^7 Bq. For palladium to contribute $\leq 0.1\%$ we must have $\frac{R_2(t)}{R_1(t)} \leq 0.001$ or

$$\begin{aligned} \frac{R_2(0)e^{-\lambda_2 t}}{R_1(0)e^{-\lambda_1 t}} &\leq 0.001 \\ \frac{R_2(0)}{R_1(0)}e^{-(\lambda_2-\lambda_1)t} &\leq 0.001 \\ 1000\frac{R_2(0)}{R_1(0)} &\leq e^{+(\lambda_2-\lambda_1)t} \\ \ln 1000 + \ln\frac{R_2}{R_1} &\leq +(\lambda_2-\lambda_1)t \\ 3\ln 10 + \ln\frac{R_2}{R_1} &\leq (\lambda_2-\lambda_1)t \end{aligned}$$

$$\begin{aligned} t &\geq \frac{1}{(\lambda_2-\lambda_1)}\left[3\ln 10 + \ln\frac{R_2}{R_1}\right] \\ t &\geq \frac{1}{\left(\frac{\ln 2}{22\text{ m}}\times\frac{60\text{ m}}{\text{h}} - \frac{\ln 2}{3.13\text{ h}}\right)} \cdot \left[3\ln 10 + \ln\frac{0.9}{5.45}\right] \\ t &\geq 3.1\ \text{hours.} \end{aligned}$$

$\Rightarrow$ The tech must wait at least 3.1 hours.

36.53 The amount of nuclide at time t is

$$\begin{aligned} N(t) &= N_o e^{-\lambda t} \\ R(t) &= R_o e^{-\lambda t} \\ R(t) &= R_o e^{-\lambda t} \equiv R \\ N_o &= \frac{R}{\lambda}e^{\lambda t_a} \end{aligned}$$

Find the minimum with respect to λ. Take $\frac{d}{d\lambda}$; set to 0

$$\begin{aligned} \frac{1}{R}\frac{dN_o}{d\lambda} &= -\frac{1}{\lambda^2}e^{\lambda t_a} + \frac{t_a}{\lambda}e^{\lambda t_a} \\ &= 0 \\ \Rightarrow \lambda &= \frac{1}{t_a} \end{aligned}$$

but

$$\lambda = \frac{\ln 2}{T_{1/2}}$$

thus

$$\frac{\ln 2}{T_{1/2}}t_a = 1$$

or

$$\begin{aligned} T_{1/2} &= t_a \ln 2 \\ R &= \lambda N_o e^{-\lambda t_a} \\ &= \frac{\ln 2}{T_{1/2}}N_o e^{-t\ln 2/T_{1/2}} \end{aligned}$$

but at minimum $\frac{t_a \ln 2}{T_{1/2}} = 1$

$$\begin{aligned} R &= \frac{\ln 2}{T_{1/2}}N_o e^{-1} \\ \Rightarrow N_o &= \frac{eRT_{1/2}}{\ln 2} \end{aligned}$$

36.59 Target atom absorbs n, final product is ^{65}Cu

$$\begin{aligned} \text{n} + {}^{A}Z &\rightarrow {}^{A+1}Z^* \\ {}^{A+1}Z^* &\rightarrow {}^{A+1}Z + \gamma \\ {}^{A+1}Z &\rightarrow {}^{A+1}(Z-1) + \text{e}^+ + \nu_\text{e} + Q \end{aligned}$$

from the above $A+1 = 65$ so $A = 64$ and $Z - 1 = 29$ or $Z = 30 \Rightarrow$ the target nucleus is ^{64}Zn.

36.61 $^2\text{H} + {}^2\text{H} \rightarrow {}^3\text{He} + \text{n} + Q$.

$$\begin{aligned} Q &= c^2 \times 931.49\ \text{MeV}/c^2\cdot\text{u} \\ &\quad [\text{reactants} - \text{products}] \\ &= 931.49\ \text{MeV/u} \\ &\quad [2\,(2.0141\ \text{u}) - 3.01603\ \text{u} - 1.000866\ \text{u}] \\ &= 3.27\ \text{MeV} \end{aligned}$$

36.65 Consider

$$\bar{\nu}_\text{e} + \text{p} \rightarrow \text{n} + \text{e}^+$$

The reaction goes forward if $Q > 0$ with no KE in the n or e$^+$

$$Q = c^2\,[\text{reactants} - \text{products}]$$

set

$$\begin{aligned} 0 &\leq E_{\bar{\nu}} + m_\text{p}c^2 - m_\text{n}c^2 - m_\text{e}c^2 \\ E_{\bar{\nu}} &\geq m_\text{n}c^2 + m_\text{e}c^2 - m_\text{p}c^2 \\ &\geq 939.566\ \text{MeV} + 0.511\ \text{MeV} \\ &\quad -938.272\ \text{MeV} \\ E_{\bar{\nu}} &\geq 1.805\ \text{MeV} \end{aligned}$$

36.69 The measured argon determines the amount of ^{40}K which has decayed. Let the number of ^{40}K nuclei in the rock originally be N_o. Then

$$N(t) = N_o e^{-\lambda t}$$

and

$$\frac{N(^{40}\text{Ar})}{N(^{40}\text{K})} = \frac{N_o\left(1-e^{-\lambda t}\right)}{N_o e^{-\lambda t}} = e^{-\lambda t} - 1$$

The measured argon, accounting for atmospheric contamination is

$$\begin{aligned} m(\text{Ar}) &= 2.39\times10^{-9}\text{ g} - 7.25\times10^{-10}\text{ g} \\ &= 1.67\times10^{-9}\text{ g} \end{aligned}$$

for each 0.1 g of K. The measured mass of ^{40}K is 1.18×10^{-5} of the total, or 1.18×10^{-6} g for each 0.1 g of K. Then

$$\frac{m(\text{Ar})}{m(^{40}\text{K})} = \frac{N(^{40}\text{Ar})(40\text{ u})}{N(^{40}\text{K})(40\text{ u})}$$

Setting the two ratios equal

$$e^{-\lambda t} - 1 = \frac{1.67\times10^{-9}\text{ g}}{1.18\times10^{-6}\text{ g}}$$

so

$$e^{\lambda t} = 1 + 0.0014$$

since

$$\begin{aligned} \lambda t &= \ln(1.0014) \\ &= 0.0014 \\ t &= \frac{0.0014}{\ln 2} T_{1/2} \\ &= (0.002)\left(1.28\times10^9\text{ y}\right) \\ &= 2.6\times10^6\text{ y} \end{aligned}$$

Lucy was 2.6 million years old.

36.73 Mean kinetic energy of a particle is

$$K = \frac{3}{2}kT$$

With 2 particles, set this to the electrostatic potential energy

$$2\times\frac{3}{2}kT = \frac{q_1 q_2}{4\pi\epsilon_0 r}$$

Given that $r = 2.0$ fm $= 2.0\times10^{-15}$ m, then

$$T = \frac{1}{3k}\frac{e^2}{4\pi\epsilon_0 r}$$

with

$$\begin{aligned} k &= 1.38\times10^{-23}\text{ J/K} \\ e &= 1.60\times10^{-19}\text{ C} \\ \frac{1}{4\pi\epsilon_0} &= 8.99\times10^9\text{ N}\cdot\text{m}^2/\text{C}^2 \end{aligned}$$

thus

$$\begin{aligned} T &= \frac{1\times8.99\times10^9\text{ N}\cdot\text{m}^2/\text{C}^2}{3\left(1.38\times10^{-23}\text{ J/K}\right)}\frac{\left(1.60\times10^{-19}\text{ C}\right)^2}{2\times10^{-15}\text{ m}} \\ &= 2.8\times10^9\text{ K} \end{aligned}$$

36.77 Given $m_{\text{sun}} \cong 2\times10^{30}$ kg. $r_* \equiv$ radius of neutron star $\simeq 10$ km.

$$\text{Density of neutron } * \equiv \rho_* = \frac{m_*}{V}$$

Assume a spherical volume;

$$V = \frac{4}{3}\pi r_*^3$$

thus

$$\begin{aligned} \rho_* &\simeq \frac{2\times10^{30}\text{ kg}}{\frac{4}{3}\pi\left(10\text{ km}\right)^3\times\left(10^3\text{ m/km}\right)^3} \\ &\cong 4.8\times10^{17}\text{ kg/m}^3 \\ &\simeq 5\times10^{17}\text{ kg/m}^3 \end{aligned}$$

Typical nucleus has $r \sim 1.2\text{ fm}\left(\text{A}^{1/3}\right)$. To minimize error, pick a relatively large nucleus, e.g. ^{235}U

$$r \sim 1.2\text{ fm}\,(235)^{1/3}$$

then

$$V = \frac{4}{3}\pi\left(1.2\text{ fm}\right)^3(235)$$

and to first order

$$m_{\text{nuc}} \simeq 235\times m_{\text{neutron}}$$

thus

$$\begin{aligned} \rho_{\text{nuc}} &\simeq \frac{235\times m_{\text{neutron}}}{\frac{4}{3}\pi\left(1.2\text{ fm}\right)^3(235)} \\ &= \frac{1.675\times10^{-27}\text{ kg}}{\frac{4}{3}\pi\left(1.2\times10^{-15}\right)^3} \\ &= 2.3\times10^{17}\text{ kg/m}^3 \end{aligned}$$

$\Rightarrow$ there is good agreement in considering a neutron star to be one big nucleus.

36.81 The binding energy of 2.226 MeV is much less than the rest energy ($\sim$ 400 MeV) of either neutron or proton, so we may conclude that both particles move non-relativistically. We analyze the collision in the usual way.

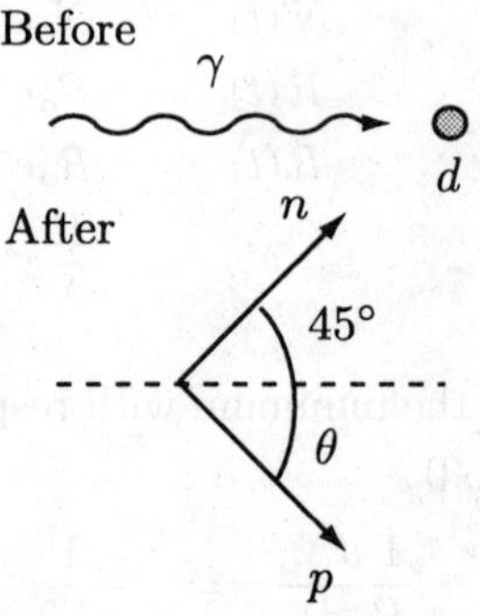

Momentum	Before		After
x-component	$\frac{E}{c}$	=	$p_n \cos 45° + p_p \cos\theta$
y-component	0	=	$p_n \sin 45° - p_p \sin\theta$
Energy	$E + m_d c^2$	=	$m_n c^2 + \frac{p_n^2}{2m_n} + m_p c^2 + \frac{p_p^2}{2m_p}$

Rearranging the energy equation:

$$E + (m_d - m_n - m_p)\, c^2 = \frac{p_n^2}{2m_n} + \frac{p_p^2}{2m_p}$$

The second term on the left-hand side of this equation is minus the binding energy

$$B = 2.226 \text{ MeV}$$

so

$$E - B = \frac{p_n^2}{2m_n} + \frac{p_p^2}{2m_p} = 2.774 \text{ MeV}$$

From the momentum equations we can eliminate θ:

$$\begin{aligned} p_p \sin\theta &= p_n \sin 45° \\ &= \frac{p_n}{\sqrt{2}} \\ p_p \cos\theta &= \frac{E}{c} - p_n \sin 45° \\ &= \frac{E}{c} - \frac{p_n}{\sqrt{2}} \end{aligned}$$

Square and add:

$$\begin{aligned} p_p^2 &= \frac{p_n^2}{2} + \frac{E^2}{c^2} + \frac{p_n^2}{2} - \frac{2E}{c}\frac{p_n}{\sqrt{2}} \\ &= p_n^2 + \frac{E^2}{c^2} - \sqrt{2}\frac{E}{c} p_n \end{aligned}$$

Substitute this result into the energy equation:

$$\begin{aligned} E - B &= \frac{p_n^2}{2m_n} \\ &\quad + \frac{1}{2m_p}\left(p_n^2 + \frac{E^2}{c^2} - \sqrt{2}\frac{E}{c} p_n\right) \\ &= \frac{p_n^2}{2}\left(\frac{1}{m_n} + \frac{1}{m_p}\right) \\ &\quad + \frac{E^2}{2m_p c^2} - \frac{\sqrt{2}}{2}\frac{E}{m_p c} p_n \end{aligned}$$

Let

$$\mu = \frac{m_n m_p}{m_n + m_p}$$

$$\begin{aligned} \frac{p_n^2}{2\mu} - \frac{\sqrt{2}}{2}\frac{E}{m_p c} p_n + \frac{E^2}{2m_p c^2}\left(\frac{\mu}{m_p}\right) + (B - E) &= 0 \\ p_n^2 - \sqrt{2}\left(\frac{\mu}{m_p}\right)\frac{E}{c} p_n + \frac{E^2}{c^2}\left(\frac{\mu}{m_p}\right) + (B - E)\, 2\mu &= 0 \end{aligned}$$

$$\begin{aligned} p_n &= \frac{\sqrt{2}}{2}\left(\frac{\mu}{m_p}\right)\frac{E}{c} \\ &\quad \pm \frac{1}{2}\sqrt{2\left(\frac{\mu}{m_p}\right)^2 \left(\frac{E}{c}\right)^2 - 4\left(\frac{E^2}{c^2}\frac{\mu}{m_p} + 2\mu\,(B - E)\right)} \\ &= \frac{\sqrt{2}}{2}\frac{\mu}{m_p}\frac{E}{c} \\ &\quad \pm \sqrt{\left(\frac{E}{c}\right)^2 \left(\frac{\mu}{m_p}\right)\left(\frac{\mu}{2m_p} - 1\right) + 2\mu\,(E - B)} \end{aligned}$$

Now

$$\begin{aligned} \frac{\mu}{m_p} &= \frac{m_n}{m_n + m_p} = \frac{1}{1 + \frac{m_p}{m_n}} \\ &= \frac{1}{1 + \frac{1.67262}{1.67493}} = 0.500345 \end{aligned}$$

so $2\mu = 938 \text{ MeV}/c^2$ and the second term in the square root is much the larger, and the square root itself is much larger than the term outside the root and only the + sign makes sense. So

$$p_n = \frac{1}{c}\left\{\frac{\sqrt{2}}{2}(0.500345)(5.000 \text{ MeV}) + \sqrt{\begin{array}{l}(5.000 \text{ MeV})^2 (0.5000345)\left(\frac{0.500345}{2} - 1\right) \\ +2\,(0.5000345)\,(938.272\text{MeV})\,(5.000 - 2.226) \text{ MeV}\end{array}}\right\}$$

$$\begin{aligned} p_n c &= 1.7690 \text{ MeV} \\ &\quad + \left[-9.3793\,(\text{MeV})^2 + 2.6046 \times 10^3\,(\text{MeV})^2\right]^{1/2} \\ &= 52.712 \text{ MeV} \end{aligned}$$

Then, the kinetic energy of the neutron is

$$\begin{aligned} K_n &= \frac{p_n^2}{2m_n} = \frac{p_n^2 c^2}{2m_n c^2} \\ &= \frac{(52.712 \text{ MeV})^2}{2\,(939.566 \text{ MeV})} \\ &= 1.479 \text{ MeV} \end{aligned}$$

Similarly, the proton kinetic energy is

$$K_p = \frac{p_p^2 c^2}{2m_p c^2} = \frac{p_n^2 c^2 + E^2 - \sqrt{2} E p_n c}{2m_p c^2}$$

$$= \frac{(52.712 \text{ MeV})^2 + (5.000 \text{ MeV})^2 - \sqrt{2}(5.000 \text{ MeV})(52.712 \text{ MeV})}{2(938.272 \text{ MeV})}$$

$$= 1.295 \text{ MeV}$$

36.85 Natural ^{238}U ore emits 1.26×10^4 α/s in the reaction

$$^{238}\text{U} \rightarrow {}^{234}\text{Th} + {}^{4}\text{He} + Q$$

From Fig. 36.29c, $T_{1/2}$ for ^{238}U is $T_{1/2} = 4.51 \times 10^9$ y.

$$R(t) = \lambda N(t)$$

with

$$\lambda = \frac{\ln 2}{T_{1/2}}$$

$$
\begin{aligned}
N(t) &= \frac{R}{\lambda} = \frac{R}{\ln 2} \cdot T_{1/2} \\
&= \frac{1.26 \times 10^4 \text{ s}^{-1}}{\ln 2} \\
& \quad \cdot \left(3.156 \times 10^7 \text{ s/y}\right)\left(4.51 \times 10^9 \text{ y}\right) \\
N &= \frac{1.79 \times 10^{21}}{\ln 2} = 2.59 \times 10^{21}
\end{aligned}
$$

$\Rightarrow$ the sample contains 2.59×10^{21} nuclei of ^{238}U.

^{226}Ra has a $T_{1/2} = 1600$ y. In the series, the intermediate species are in equilibrium $\Rightarrow$ ^{226}Ra and ^{238}U are decaying at the same rate $\left(1.26 \times 10^4 \text{ decays/s}\right)$. Thus the number of Ra nuclei is proportional to the half life

$$
\begin{aligned}
\Rightarrow N(\text{Ra}) &= \frac{N(\text{U})}{T(\text{U})} \cdot T(\text{Ra}) \\
&= 2.59 \times 10^{21} \times \frac{1600 \text{ y}}{4.51 \times 10^9 \text{ y}} \\
&= 9.24 \times 10^{14} \text{ Ra nuclei}
\end{aligned}
$$

The mass of radium is this value $\times$ 226.025 $\times$ 1.66×10^{-27} kg or $\sim 3.4 \times 10^{-10}$ kg.

Following the series, there are two ways to decay to ^{206}Pb. One way in β^- decay from ^{206}Tl is

$$
^{206}\text{Tl} \rightarrow \, ^{206}\text{Pb} + \beta^- + \bar{\nu}_e + Q
$$

with a half life of 4.2 min, $5 \times 10^{-5}\%$ of the decays to the final product ^{206}Pb are via this route $\Rightarrow$ $5 \times 10^{-5}\%$ of the original ^{238}U decays go this way, or 1 decay every $\frac{1}{0.0063 \text{ s}^{-1}}$, i.e., 1 every 159 seconds.

Chapter 37

Particle Physics

37.1 The time is given by Heisenberg's uncertainty principle, taking the mass-energy of the e^+/e^- pair as ΔE. So,

$$\begin{aligned} \Delta t_{\text{pair}} &\sim \frac{\hbar}{2\Delta E} = \frac{\hbar}{4m_e c^2} \\ &= \frac{1.0 \times 10^{-34}\ \text{J}\cdot\text{s}}{4(0.511 \times 10^6\ \text{eV})(1.60 \times 10^{-19}\ \text{J/eV})} \\ &= 3 \times 10^{-22}\ \text{s}. \end{aligned}$$

37.3 The Q value is the difference between the incoming mass-energy and outgoing mass-energy.

$$\begin{aligned} Q &= 2m_p c^2 - 4M_\pi c^2 \\ &= 2(9.38.27\ \text{MeV}) - 4(139.57\ \text{MeV}) \\ &= 1.3183\ \text{GeV}. \end{aligned}$$

37.5 Energy and momentum can be conserved when 2 particles $\rightarrow$ two particles. The particles on either side of the reaction have net half-integer spin, so angular momentum can be conserved. The reaction does not change electric charge, and leaves the quark number at 4, and the baryon number at 1. In the quark picture, the reaction is

$$d\bar{s} + uud \rightarrow u\bar{s} + udd,$$

so the quark flavor $u^2 d^2 \bar{s}$ is unchanged. The net isospin on either side of the reaction is

$$\frac{1}{2} + \left(-\frac{1}{2}\right) = 0$$

All the quantities conserved by the strong interaction are conserved in this reaction; so, it can go via the strong force.

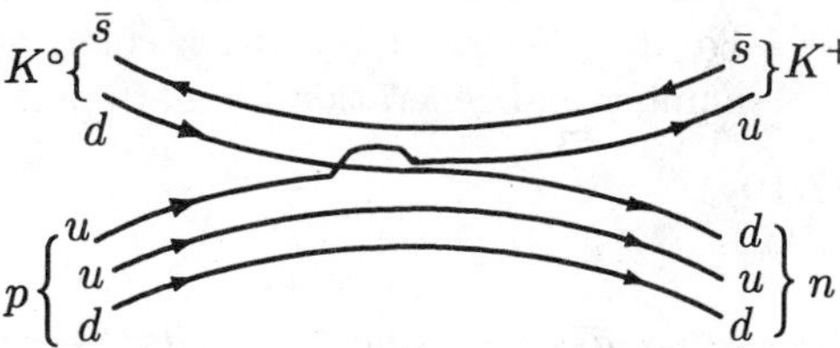

37.7 The particles can have values of I_3, differing by integer amounts, and ranging from a maximum of I to a minimum of $-I$:

$$I_3 = -\frac{3}{2},\ -\frac{1}{2},\ +\frac{1}{2},\ +\frac{3}{2}$$

$$\Delta^- \quad \Delta^0 \quad \Delta^+ \quad \Delta^{++}$$

In general, the number of possible isospin states is $N = 2I + 1$. Here, $N = 4$.

37.11 In the time δt that a massive virtual photon is allowed by Heisenberg's uncertainty principle to exist, it must be able to travel a distance over which Maxwell's equations are known to describe the behavior of things. Thus, any such known distance implies a limit on photon mass given by

$$\frac{\hbar}{2} \geq \delta E\, \delta t \overset{\text{set}}{=} M_{\text{photon}} c^2 \frac{\delta x}{c}$$

So

$$M_{\text{photon}} c^2 \leq \frac{\hbar c}{2\delta x} = \frac{hc}{4\pi \delta x}$$

From motors, $\delta x \sim 1$ m, and

$$M_{\text{photon}} c^2 \leq \frac{1.24 \times 10^3\ \text{eV}\cdot\text{nm}}{4\pi(1\ \text{m})} \approx 10^{-7}\ \text{eV}$$

From Earth's magnetic field, $\delta x \sim 10^{+7}$ m, and

$$M_{\text{photon}}c^2 \leq 10^{-14} \text{ eV}$$

Either of these numbers is indistinguishable from zero by any particle physics experiment we can presently imagine.

37.15 The quantum numbers are shown in the table.

	Before: $\overline{\Sigma^+} = \bar{u}\bar{u}\bar{s}$	After: $\bar{p} = \bar{u}\bar{u}\bar{d}$;	$\pi^\circ = S(\bar{u}u, \bar{d}d)$
Yes	Charge: -1	-1,	0
Yes	baryon #: -1	-1,	0
No	isospin I_3 : -1	$-\frac{1}{2}$,	0
No	Quark flavor: $\bar{s} \to \bar{d}$		
Yes	Energy: 1189.4 MeV	938.3 MeV	$+135.0$ MeV
		$Q > 0$	

Linear momentum is OK; $\bar{\text{p}}$ and π° have opposite momenta. Angular momentum: Incoming and out-going particles both have half-integer spin, so angular momentum can be conserved. We don't know about quark color, and lepton number and generation are not applicable.

37.19

$$\begin{array}{ccccccc} \Xi^- & + & p & \to & \Lambda & + & \Lambda \\ dss & + & uud & \to & uds & + & uds \end{array}$$

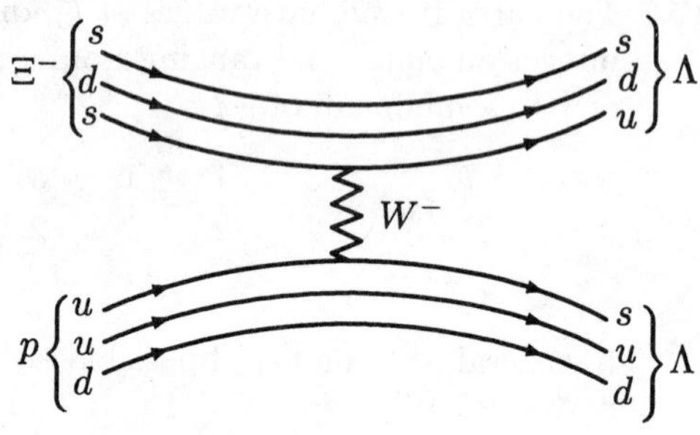

$$\begin{aligned} Q &= m_{\Xi^-} + m_p - 2m_\Lambda \\ &= [1321 + 938 - 2(1116)] \text{ MeV} \\ &= 27 \text{ MeV} \end{aligned}$$

37.23 Three quarks enter the reaction while four quarks and one anti quark leave, for a total of three.

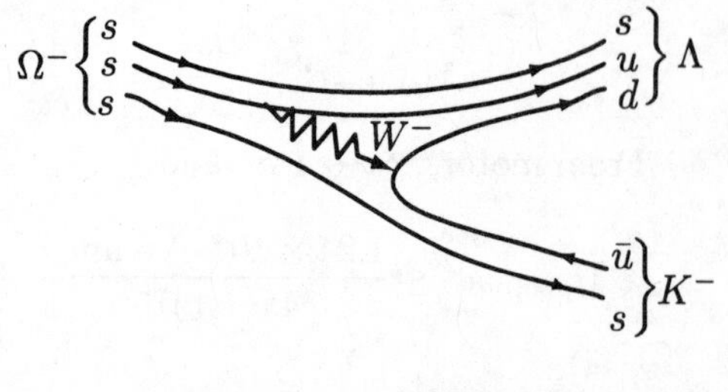

$$\begin{aligned} Q &= M_{\Omega^-} - M_\Lambda + M_{K^-} \\ &= (1672 - 1116 - 494) \text{ MeV} \\ &= 62 \text{ MeV} \end{aligned}$$

37.27 a)

$$\begin{array}{ccccc} \Omega^- & \to & \Xi^\circ & + & \pi^- \\ sss & \to & ssu & + & \bar{u}d \end{array}$$

quark flavor changes $\Rightarrow$ weak interaction

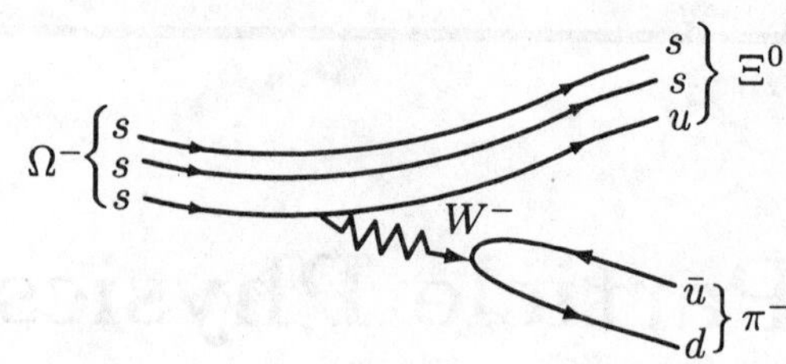

$$\begin{aligned} Q &= (1672 - 1315 - 139.6) \text{ MeV} \\ &= 217 \text{ MeV} \end{aligned}$$

b)

$$\begin{array}{ccccc} \Lambda_c^+ & \to & p & + & \bar{\text{K}}^\circ \\ udc & \to & uud & + & \bar{s}d \end{array}$$

quark flavor changes $\Rightarrow$ weak interaction

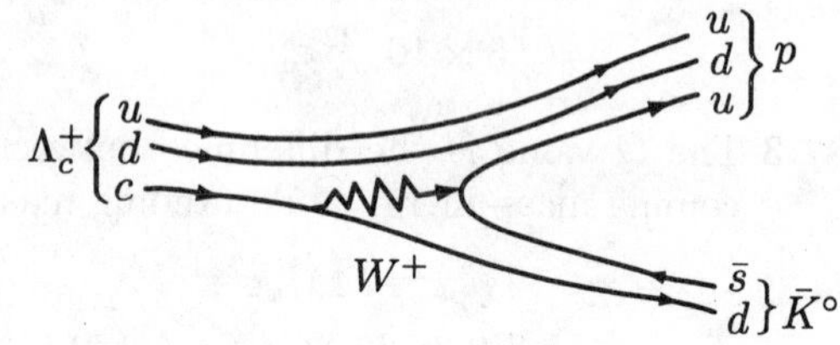

$$Q = (2285 - 938 - 498) \text{ MeV} = 849 \text{ MeV}$$

For Further Reading

Chapter 0

Cronin V. *The view from planet Earth: Man looks at the cosmos.* New York: Wm. Morrow and Co. 1983.

Davies, P. *God and the new physics.* New York: Simon and Schuster. 1983.

De Santillana, G. *The crime of Galileo.* New York: Time Inc. 1962.

Drake, S. *Galileo at work.* Chicago: University of Chicago Press. 1978.

———. The role of music in Galileo's experiments. *Scientific American.* Jun. 1975: 98

Ferris, T. *Coming of age in the Milky Way.* New York: Morrow. 1988.

Hanbury-Brown, R. *The wisdom of science.* Cambridge, Eng.: Cambridge University Press. 1986.

Hoyle, F. *From Stonehenge to modern cosmology.* San Francisco: W. H. Freeman. 1972.

Koestler, A. *The sleepwalkers.* New York: Macmillan. 1959.

Kuhn, T. *The Copernican revolution.* New York: Vintage. 1959.

McCloskey, M. Intuitive physics. *Scientific American.* Apr. 1983: 122.

Rocklin, G. I., ed. *Technology and social change.* San Francisco: Freeman.

Sky and Telescope magazine.

Chapter 1

Astin, A. V. Standards of measurement. Scientific American. Jun. 1968: 50.

Cohen, J. Psychological time. *Scientific American.* Nov. 1964: 116.

Crombie, A. C. Descartes. *Scientific American.* Oct. 1959: 160.

Gardner, M. Can time go backward? *Scientific American.* Jan. 1967: 98.

Layzer, D. The arrow of time. *Scientific American.* Dec. 1975: 56.

Ritchie-Calder, Lord. Conversion to the metric system. *Scientific American.* Jul. 1970: 17.

Taylor, B. N. New measurement standards for 1990. *Physics Today.* Aug. 1989: 23.

Taylor, B., D. Langenberg, and W. Parker. The fundamental physical constants. *Scientific American.* Oct. 1970: 62.

White, R. M. Measurement Technology: Immeasurably important. *Physics Today.* Apr. 1993: 55.

Chapter 2

Drake, S. Galileo's discovery of the law of fall. *Scientific American.* May 1973: 84.

Poggio, T., and C. Koch, Synapses that compute motion. *Scientific American.* May 1987: 46.

Rogers, T. A. The physiological effects of acceleration *Scientific American* Feb. 1962: 60.

Wallach, H. The perception of motion. *Scientific American.* Jul. 1959: 56.

Chapter 3

Drake, S., and J. MacLachlan. Galileo's discovery of the parabolic trajectory. *Scientific American.* Mar. 1975: 102.

Korfmann, M. The sling as a weapon. *Scientific American.* Oct. 1973: 34.

McEwen, E., R. Miller, and C. Bergman. Early bow design and construction *Scientific American.* Jun. 1991: 76.

Wallach, H. Perceiving a stable environment. *Scientific American.* May 1985: 118.

Interlude 2

Adams, J. L. *Conceptual blockbusting.* San Francisco: Freeman. 1974.

Gardner, M. *Aha! Insight.* New York: Scientific American/Freeman. 1978.

Polya, G. *How to solve it*, 2nd ed. Princeton, N.J.: Princeton University Press. 1971.

Price, J. J. Learning mathematics through writing: Some guidelines. *College Mathematics Journal*, **20**(5) Nov. 1989.

Wickelgren, W. A. *How to solve problems.* San Francisco: Freeman. 1974.

Chapter 4

Cohen, I. B. Isaac Newton. *Scientific American.* Dec. 1955: 73–83.

Da. C. Andrade, E. N. Robert Hooke. *Scientific American.* Dec. 1954: 94.

Krim, J. Friction at the atomic scale. *Scientific American* Oct. 1996: 74.

Palmer, F. Friction. *Scientific American.* Feb. 1951: 54–59.

Rabinowicz, E. Stick and Slip. *Scientific American.* May 1956: 109–119.

Sciama, D. Inertia. *Scientific American.* Feb. 1957: 99–109.

Vam, P. There's the rub: nanotribology reveals the atomic nature of friction. *Scientific American.* Jun. 1991: 30.

Essay 1

Becker, C. *The Declaration of Independence*: a study in the history of political ideas. New York: Vintage books. 1970.

Cohen, I. B. Newton's discovery of gravity. *Scientific American.* Mar. 1981: 166.

Drake, S. Newton's apple and Galileo's dialogue. *Scientific American.* Aug. 1980: 150.

Kuhn, T. *The Copernican revolution.* New York: Vintage Books. 1959.

Locke, J. *Of civil government, the second of two treatises of government, 1690.* Chicago: Henry Regnery. 1962.

Westfall, R. S. *Never at rest.* Cambridge, Eng.: Cambridge University Press. 1980.

Chapter 5

Artsutanov, Y. N. *Komsomolskaya Pravda.* Jul. 31, 1960.

Clarke, A. C. *The fountains of paradise.* New York: Ballentine/Del Rey Books. 1978.

Edwards, L. K. High speed tube transportation. *Scientific American.* Aug. 1965: 30.

Goldman, T., R. Hughes, and M. M. Nieto. Gravity and antimatter. *Scientific American.* Mar. 1988: 48.

Heiskanen, W. A. The Earth's gravity. *Scientific American.* Sep. 1955: 164.

Isaacs, J. D., H. Bradner, and G. E. Backus. Satellite elongation into a true sky hook. *Science.* Feb 11, 1966.

King, I. Globular clusters. *Scientific American.* Jun. 1985: 78.

Rubin, V. Dark matter in spiral galaxies. *Scientific American.* Jun. 1983: 96.

Sheffield, C. *The web between the worlds.* New York: Ace Science Fiction. 1979.

Will, C. *Was Einstein right?* New York: Basic Books. 1986.

Chapter 6

Alvarez, W., and F. Asaro. What caused the mass extinction? An extraterrestrial impact. *Scientific American.* Oct. 1990: 78.

Balsiger, H., H. Fechtig, and J. Geiss. A close look at Halley's comet. *Scientific American.* Sep. 1988: 96.

Beatty, J. K., and D. H. Levy. Awaiting the crash. *Sky and Telescope.* Jan. 1994: 40.

Charters, A. High speed impact. *Scientific American.* Oct. 1960: 128.

Grieve, R. Impact cratering on the Earth. *Scientific American.* Apr. 1990: 66.

Ley, W. Rockets. *Scientific American.* May 1949: 30.

Matthews, R. A rock watch for earthbound asteroids. News and Comment. *Science* **255**, Mar. 6, 1992: 1204.

Newgard, J., and M. Leroy. Nuclear rockets. *Scientific American.* May 1959: 46.

Niven, L., and J. Pournelle. *Lucifer's hammer.* Chicago: Playboy Press. 1977.

Struve, O. The great meteor of 1947. *Scientific American.* Jun. 1950: 42.

Wetherill, G. W. Apollo objects. *Scientific American.* Mar. 1979: 54.

Chapter 7

Bierlein, J. C. The journal bearing. *Scientific American.* Jul. 1975: 50.

Dibner, B. Moving the obelisk. *Scientific American.* Jun. 1951: 58.

Fitch, J. M., J. Templer, and P. Corcoran. The dimensions of stairs. *Scientific American.* Oct. 1974: 82.

Ginzberg, E. The mechanization of work. *Scientific American.* Sep. 1982: 66. (The entire September 1982 issue of *Scientific American* is devoted to mechanization.)

Moydenman, E. A. Bearings. *Scientific American.* Mar. 1966: 60.

Starr, C. Energy and power. *Scientific American.* Sep. 1971: 36.

Chapter 8

Scientific American has produced two special issues on the topic of energy: Sep. 1971 and Sep. 1990.

Classer, R., and L. Girifalco. Materials for energy utilization. *Scientific American.* Oct. 1986: 102.

Gibbons, J., P. Blair and H. Guin. Strategies for energy use. *Scientific American.* Sep. 1989: 136.

Goslino, J. M., and M. E. DeMont. Jet-propelled swimming in squids. *Scientific American.* Jan. 1985: 96.

Hubbard, H. M. The real cost of energy. *Scientific American.* Apr. 1991: 36.

Johnson, T. V., and L. A. Soderblom. Io. *Scientific American.* Dec. 1983: 56.

Ross, R., and P. Bornstein. Elastic fibers in the body. *Scientific American.* Jun. 1971: 44.

Rothschild, M., Y. Schlein, K. Parker, C. Neville, and S. Sternberg. The flying leap of the flea. *Scientific American.* Nov. 1973: 92.

Soedel, W., and V. Foley. Ancient catapults. *Scientific American.* Mar. 1979: 150.

Essay 2

Einstein, A., and L. Infeld. *The Evolution of Physics.* New York: Simon and Schuster 1967: ch. 3.

Chapter 9

Frohlich, C. The physics of somersaulting and twisting. *Scientific American.* Mar. 1980: 154.

Nichols, R. G. Satellites on a string. *Sky and Telescope.* Apr. 1987: 383–385.

Niven, L., and S. Barnes. *The descent of the Anansi.* New York: Tor Books. 1991.

Richardson, R. The discovery of Icarus. *Scientific American.* Apr. 1965: 106.

Urey, H. C. The origin of the Earth. *Scientific American.* Oct. 1952: 53.

Wetherill, G. W. Apollo objects. *Scientific American.* Mar. 1979: 54.

Whipple, F., and J. A. Hynek. Observations of satellite I. *Scientific American.* Dec. 1957: 37.

Chapter 10

Barger, V. D., and D. B. Cline. High energy scattering. *Scientific American.* Dec. 1967: 76.

Barnes, J., L. Hernquist, and F. Schweizer. Colliding galaxies. *Scientific American.* Aug. 1991: 40.

Benford, G. *Stars in shroud.* New York: Tor Books. 1984.

Binzel, R., M. A. Barucci, and M. Fulchignoni. The origin of the asteroids. *Scientific American.* Oct. 1991: 40.

Damask, A. C. Forensic physics of vehicle accidents. *Physics Today,* Mar. 1987: 36–44.

McHarris, W. C., and J. O. Rasmussen. High energy collisions between atomic nuclei. *Scientific American.* Jan. 1984: 58.

Walker, J. The amateur scientist. *Scientific American.* Sep. 1984: 215.

Zewail, A. H. The birth of molecules. *Scientific American.* Dec. 1990: 76.

Chapter 11

Condit, C. W. The wind bracing of buildings. *Scientific American.* Feb. 1974: 92

Shapiro, L. K., and H. I. Shapiro. Construction cranes. *Scientific American.* Mar. 1988: 72.

Steinman, D. B. Bridges. *Scientific American.* Nov. 1954: 64.

Chapter 12

Beams, J. W. Ultra-high speed rotation. *Scientific American.* Apr. 1961: 134.

Brody, H. The moment of inertia of a tennis racket. *The physics of sports.* New York: American Institute of Physics. 1992.

Frohlich, C. The physics of somersaulting and twisting. *Scientific American.* Mar. 1980: 154.

Howard, R. The rotation of the Sun. *Scientific American.* Apr. 1975: 106.

Kirkpatrick, P. Batting the ball. *The American Journal of Physics* **31** 1963: 606. Reprinted in *The physics of sports.* New York: American Institute of Physics. 1992.

Ostriker, J. The nature of pulsars. *Scientific American.* Jan. 1971: 48.

Pacini, F., and M. Rees. Rotation in high-energy astrophysics. *Scientific American.* Feb. 1973: 98.

Smylie, D. E., and L. Mansinha. The rotation of the Earth. *Scientific American.* Dec. 1971: 80.

Walker, J. *Roundabout—The physics of rotation in the everyday world.* New York: Freeman. 1979.

Whipple, F. The spin of comets. *Scientific American.* Mar. 1980: 124.

Essay 4: The bicycle

Brooks, A., A. Abbott, and D. Wilson. Human-powered watercraft. *Scientific American.* Dec. 1986: 120.

Drela, M, and J. Langford. Human-powered flight. *Scientific American.* Nov. 1985: 144.

Gross, A., C. Kyle., and D. Malewicki. The aerodynamics of human-powered land vehicles. *Scientific American.* Dec. 1983: 142.

Whitt, F. R., and D. G. Wilson. Bicycling science. Cambridge, Mass: MIT Press. 1982.

Wilson, S. S. Bicycle technology. *Scientific American.* Mar. 1973: 81. Reprinted in *Scientific technology and social change.* San Francisco: Freeman.

Chapter 13

Barker, J. A., and D. Henderson. The fluid phases of matter. *Scientific American.* Nov. 1981: 130.
Hess, F. The aerodynamics of boomerangs. Scientific American. Nov. 1968: 124.
Hodge, A . Siphons in Roman aqueducts. *Scientific American.* Jun. 1985: 114.
Hurt, H. *Aerodynamics for naval aviators.* U. S. Navy. 1960. Revised 1965.
Larrabee, E. The screw propeller. *Scientific American.* Jul. 1980: 134.
Lynch, D. K. Tidal bores. *Scientific American.* Oct. 1982: 146.
Niklas, K. Aerodynamics of wind pollination. *Scientific American.* Jul. 1987: 90.
Peterson, C. High performance parachutes. *Scientific American.* May 1990: 114.
Smith, N. Roman hydraulic technology. *Scientific American.* May 1978: 154.
Ward, P., L. Greenwald, and O. Greenwald. The buoyancy of the chambered nautilus. *Scientific American.* Oct. 1980: 190.
Wootton, R. The mechanical design of insect wings. *Scientific American.* Nov. 1990: 114.

Chapter 14

Bertsch, G. F. Vibrations of the atomic nucleus. *Scientific American.* May 1983: 62.
Itano, W. M., and N. E. Ramsey. Accurate measurement of time. *Scientific American.* Jul. 1993: 56.
Lippold, O. Physiological tremor. *Scientific American.* Mar. 1971: 65.
Lyons, H. Atomic clocks. *Scientific American.* Feb. 1957: 71.
Press, F. Resonant vibrations of the Earth. *Scientific American.* Nov. 1965: 28.
Sotter, J. G., and G. A. Flandro. Resonant combustion in rockets. *Scientific American.* Dec. 1968: 94.

Chapter 15

Anderson, D., and A. Dziewonski. Seismic tomography. *Scientific American.* Oct. 1984: 60.
Bascom, W. Ocean waves. *Scientific American.* Aug. 1959. Reprinted in *The physics of everyday phenomena.* San Francisco: W. H. Freeman. 1979.
Bernstein, J. Tsunamis. *Scientific American.* Aug. 1954: 60.
Kottick, E., K. Marshall, and T. Hendrickson. The acoustics of the harpsichord. *Scientific American.* Feb. 1991: 110.
Pierce, J. R. *The science of musical sound.* New York: Scientific American Books. 1983.
Rebbi, C. Solitons. *Scientific American.* Feb. 1979: 92
Rossing, T. D. The physics of kettledrums. *Scientific American.* Nov. 1982: 172.
Schelling, J. C. The physics of the bowed string. *Scientific American.* Jan. 1974: 87.
Weinreich, G. The coupled motions of piano strings. *Scientific American.* Jan. 1979: 118.

Chapter 16

Beranek, L. L. Noise. *Scientific American.* Dec. 1966: 66.
Bron, P., T. Sciascia, L. Linden, and J. Lettvin. The colors of things. *Scientific American.* Sep. 1986: 84.
Bryant, H. C., and N. Jarmine. The glory. *Scientific American.* Jul. 1974: 60.
Desurvire, E. Lightwave communications: The fifth generation. *Scientific American.* Jan. 1992: 114.
Devey, G., and P. Wells. Ultrasound in medical diagnosis. *Scientific American.* May 1978: 98.
Fletcher, N., and S. Thwaites. The physics of organ pipes. *Scientific American.* Jan. 1983: 94.

Fraser, A. B., and W. H. Mach. Mirages. *Scientific American.* Jan. 1976: 102.
Hudspeth. A. J. The hair cells of the inner ear. *Scientific American.* Jan. 1983: 54.
Mandoli, D., and W. Briggs. Fiber optics in plants. *Scientific American.* Aug. 1984: 90.
Nassau, K. The causes of color. *Scientific American.* Oct. 1980: 124.
Nussenzveig, H. M. The theory of the rainbow. *Scientific American.* Apr. 1977: 116.
Quate, C. The acoustic microscope. *Scientific American.* Oct. 1979: 62.
Sandage, A. The red shift. *Scientific American.* Sep. 1956: 170. Reprinted in *Cosmology* +1, San Francisco: Freeman.
Silverman, J., J. Mooney, and F. Shepherd. Infrared video cameras. *Scientific American.* Mar. 1992: 78.
Yariv, A. Guided wave optics. *Scientific American.* Jan. 1979: 64.
Lightwave communications. Special issue of *Physics Today.* May 1976.
A fish's view... in The Amateur Scientist. *Scientific American.* Mar. 1984.

Chapter 17

Abraham, E., C. Seaton and S. D. Smith. The optical computer. *Scientific American.* Feb. 1983: 85.
Alper, J. Antinoise creates the sounds of silence. *Science.* Apr. 26, 1991: 508.
Babcock, H. Diffraction gratings at the Mount Wilson Observatory. *Physics Today.* Jul. 1986: 34.
Brookner, E. Phased-array radars *Scientific American.* Feb. 1985: 94.
Bowmeister P., and G. Pincus. Optical interference coatings. *Scientific American.* Dec. 1970: 58.
Darragh, P., A. Gaskin, J. Sanders and D. Opals. *Scientific American.* Apr. 1976: 84.
Howells, M., J. Kirz, and D. Sayre. X-ray microscopes. *Scientific American.* Feb. 1991: 88.
Knudsen, V. O. Architectural acoustics. *Scientific American.* Nov. 1963: 78.
Levenson, M. D. Wavefront engineering for photolithography. *Physics Today.* Jul. 1993: 28.
Robinson, G., D. Perry, and R. Peterson. Optical interferometry of surfaces. *Scientific American.* Jul. 1991: 66.

Chapter 18

Chaisson, E. Early results from the Hubble space telescope. *Scientific American.* Jun. 1992: 44.
Hecht, E., and A. Zajac. Optics. Addison Wesley. Reading, Mass.: 1974.
Land, M. Animal eyes with mirror optics. *Scientific American.* Dec. 1978: 126.
Price, W. The photographic lens. *Scientific American.* Aug. 1976: 76.
Skinner, G. X-ray imaging with coded masks. *Scientific American.* Aug. 1988: 84.
Spiller, E., and R. Feder. The optics of long-wavelength X-rays. *Scientific American.* Nov. 1978: 70.
Tape, W. The topology of mirages. *Scientific American.* Jun. 1985: 120.
Thomas, D. E. Mirror images. *Scientific American.* Dec. 1980: 206.
Veldkamp, W., and T. McHugh. Binary optics. *Scientific American.* May 1992: 92.
Winston, R. Nonimaging optics. *Scientific American.* Mar. 1991: 76.
——— Bug eyed. *Scientific American.* Nov. 1990: 134.

Chapter 19

The September 1954 issue of *Scientific American* is entirely devoted to temperature and heat.
Jones, P., and T. Wigley. Global warming trends. *Scientific American.* Aug. 1990: 84.
Lavenda, B. Brownian motion. *Scientific American.* Feb. 1985: 70.
Milne, L., and M. Milne. Temperature and life. *Scientific American.* Feb. 1949: 46.
Proctor, W. G. Negative absolute temperatures. *Scientific American.* Aug. 1978: 90.

Chapter 20

Berry, S. When the melting and freezing points are not the same. *Scientific American*. Aug. 1990: 68.
Duncan, M., and D. Rouvray. Microclusters. *Scientific American*. Dec. 1989: 110.
Jayaroman, A. The diamond anvil high pressure cell. *Scientific American*. Apr. 1984: 54.
Kleppner, D. About Benjamin Thompson. *Physics Today*. Sep. 1992: 9.
Quinn, T. J. Temperature, 2nd ed. London: Academic Press. 1991.
Storey, K., and J. Storey. Frozen and alive. *Scientific American*. Dec. 1990: 92.
Turnbull, D. The undercooling of liquids. *Scientific American*. Jan. 1965: 38.
Walker, J., and C. Vause. Reappearing phases. *Scientific American*. May 1987: 98.
Thermometers and cooking. *Sunset* magazine. November 1986: 186.

Chapter 21

Eastman, G. Y. The heat pipe. *Scientific American*. May 1968: 38.
Gates, D. Heat transfer in plants. *Scientific American*. Dec. 1965: 76.
Leikind, B. J., and W. J. McCarthy. An investigation of firewalking. *The Skeptical Inquirer*. 1985: Vol. 10, 23.
McKenzie, D. The Earth's mantle. *Scientific American*. Sep. 1983: 66.
Oort, A. H. The energy cycle of the Earth. *Scientific American*. Sep. 1970: 54.
Pollack, H., and D. Chapman. The flow of heat from the Earth's interior. *Scientific American*. Aug. 1977: 60.
Rosenfeld, A., and D. Hafemeister. Energy efficient buildings. *Scientific American*. Apr. 1988: 78.
Sproull, R. The conduction of heat in solids. *Scientific American*. Dec. 1962: 92.
Velarde, M. and C. Normand. Convection. *Scientific American*. Jul. 1980: 93.
Ziman, J. The thermal properties of materials. *Scientific American*. Sep. 1967: 180.

Chapter 22

For a light-hearted rendition of thermodynamic theory, listen to *Thermodynamics*, by Flanders and Swann, Angel Records, 1961.

Andresen, B., P. Salamon, and R. Berry. Thermodynamics in finite time. *Physics Today*. Sep. 1984: 62
Angrist, S. W. Perpetual motion machines. *Scientific American*. Jan. 1968: 114.
Atkins, P. The second law. Scientific American Books. New York: Freeman 1984.
Bennett, C. H. Demons, engines and the second law. *Scientific American*. Nov. 1987: 108.
Bryant, L. Rudolf Diesel and his rational engine. *Scientific American*. Aug. 1969: 108.
Dyson, F. *Disturbing the Universe*. New York: Harper and Row. 1979.
Engel, L. The Philips air engine. *Scientific American*. Jul. 1948: 52.
Ferguson, E. The origins of the steam engine. *Scientific American*. Jan. 1964: 98.
Frautschi, S. Entropy in an expanding universe. *Science*. Vol. 217. Aug. 1982: 593.
Hossli, W. Steam turbines. *Scientific American*. Apr. 1969: 100.
Kohler, J. The sterling refrigeration cycle. *Scientific American*. Apr. 1965: 119.
Prigogine, I., G. Nicolis, and A. Babloyantz. Thermodynamics of evolution. *Physics Today*. Nov. 1972: 23 and Dec. 1972: 38.
Pynchon, T. Entropy, in *Slow Learner*. Boston: Little Brown. 1984.
Sandfort, J. The heat pump. *Scientific American*. May 1951: 54.
Schadewald, R. The perpetual quest. *Science 80*. Nov.
Summers, C. The conversion of energy. *Scientific American*. Sep. 1971: 148.
Taub, A. High compression. *Scientific American*. Feb. 1950: 16.
Waldrop, M. M. Inflation and the arrow of time. *Science*. Vol. 219. Mar. 1983: 1416.
Wheatley, J. and A. Cox. Natural engines. *Physics Today*. Aug. 1985: 50.

Wilson, D. Alternative automobile engines. *Scientific American.* Jul. 1978: 39.

Wilson, S. S. Sadi Carnot. *Scientific American.* Aug. 1981: 134.

Zoline, P. The heat death of the universe. In *The heat death of the universe and other stories.* Kingston, N.Y.: McPherson. 1988.

———. Toward the absolute zero. *Physics Today.* Dec. 1979.

Chapter 23

The Classical Period. In *Physics from AAPT Journals.* AAPT. 1985.

Finkbeiner, A. A universe in our image. *Sky and Telescope.* Aug. 1984: 107.

Gordon, M. J. The control of sex. *Scientific American.* Nov. 1958: 87. (Yes, this really is about electric fields!)

Kevles, D. Robert A. Millikan. *Scientific American.* Jan. 1979: 142.

Chapter 24

Ehrenreich, H. The force between molecules. *Scientific American.* Sep. 1967: 194.

Hailmeier, G. H. Liquid crystal display-devices. *Scientific American.* Apr. 1970: 100.

Steward, I. Gauss. *Scientific American.* Jul. 1977: 22.

Chapter 25

Fickett, A., C. Gellings, and A. Lovins. Efficient use of electricity. *Scientific American.* Sep. 1990: 64.

O'Neill, G. K. The spark chamber. *Scientific American.* Aug. 1962: 36.

Regan, D. Electrical responses evoked from the human brain. *Scientific American.* Dec. 1979: 134.

Rose, P., and A. Wittkower. Tandem van de Graaff accelerators. *Scientific American.* Aug. 1970: 24.

Shiers, G. Ferdinand Braun and the cathode ray tube. *Scientific American.* May 1974: 92.

Williams, E. R. The electrification of thunderstorms. *Scientific American.* Nov. 1988: 88.

Chapter 26

Cava, R. Superconductors beyond 1-2-3. *Scientific American.* Aug. 1990: 42.

De Santillana, G. Alessandro Volta. *Scientific American.* Jan. 1965: 82.

Ehrenreich, H. The electrical properties of materials. *Scientific American.* Sep. 1967: 194.

Fickett, A. P. Fuel-cell power plants. *Scientific American.* Dec. 1978: 70.

Heiblum, M., and L. Eastman. Ballistic electrons in semiconductors. *Scientific American.* Feb. 1987: 102.

Kaner, R., and A. MacDiarmid. Plastics that conduct electricity. *Scientific American.* Feb. 1988: 106.

Likharev, K., and T. Claeson. Single electronics. *Scientific American.* Jun. 1992: 80.

Matthias, B. T. Superconductivity. *Scientific American.* Nov. 1957: 92.

Shepherd, G. M., Microcircuits in the nervous system. *Scientific American.* Feb. 1978: 92.

Williams, L. P. Andre-Marie Ampere. *Scientific American.* Jun. 1989: 90.

Chapter 27

Harari, H. The structure of quarks and leptons. *Scientific American.* Apr. 1983: 56.

Kahlhamer, F. Energy storage systems. *Scientific American.* Dec. 1979: 56.

Trotter, D. Jr. Capacitors. *Scientific American.* Jul. 1988: 86.

Chapter 28

Blandford, R., and M. Begelman. Cosmic jets. *Scientific American.* May 1982: 116.
Bloxham, J., and D. Gubbins. The evolution of the Earth's magnetic field. *Scientific American.* Dec. 1989: 68.
Carrigan, R., Jr., Superheavy magnetic monopoles. *Scientific American.* Apr. 1982: 106.
Conn, R. The engineering of magnetic fusion reactors. *Scientific American.* Oct. 1983: 60.
———, V. Cheyanov, N. Inoue, and D. Sweetman. The international thermonuclear experimental reactor. *Scientific American.* Apr. 1992: 102.
Hones, E. W., Jr. The Earth's magnetotail. *Scientific American.* Mar. 1986: 40.
Parker, E. N. Magnetic fields in the cosmos. *Scientific American.* Aug. 1983: 44.
Rosensweig, R. Magnetic fluids. *Scientific American.* Oct. 1982: 136.
Verschuur, G. *Hidden attraction: the history and mystery of magnetism.* Oxford: Oxford University Press. 1993.

Chapter 29

Akasofu, S. I. The dynamic aurora. *Scientific American.* May 1989: 90.
Becker, J. J. Permanent magnets. *Scientific American.* Dec. 1970: 92.
Boebinger, G., A. Passner, and J. Bevk. Building world-record magnets. *Scientific American.* Jun. 1995: 59.
Chaudhari, P. Electronic and magnetic materials. *Scientific American.* Oct. 1986: 136.
Kolm, H., and R. Thornton Electromagnetic flight. *Scientific American.* Oct. 1973: 17.
Kryder, M. H. Data storage technologies for advanced computing. *Scientific American.* Oct. 1987: 16.
Laughlin, J. S. History of medical physics. *Physics Today.* Jul. 1983: 26.
Pykett, I. NMR imaging in medicine. *Scientific American.* May 1982: 78.
Shukman, R. NMR Spectroscopy of living cells. *Scientific American.* Jan. 1983: 86.

Chapter 30

Devons, S. The search for electromagnetic induction. *The Physics Teacher.* Dec. 1978: 625.
Elsasser, W. M. The Earth as a dynamo. *Scientific American.* May 1958: 44.
Meurig, J. T. Michael Faraday and the Royal Institution. Bristol, England: Adam Hilger. 1991.
Sharlin, H. I., From Faraday to the dynamo. *Scientific American.* May 1961: 107.
Shiers, G. The induction coil. *Scientific American.* May 1971: 80.
Wilson, M. Joseph Henry. *Scientific American.* Jul. 1954: 72.

Chapter 31

Rieder, W. Circuit breakers. *Scientific American.* Jan. 1971: 76.

Chapter 32

Barthold, L. O., and H. G. Pfeiffer. High voltage power transmission. *Scientific American.* May 1964: 38.
Cheney, M. Tesla—man out of time. Englewood Cliffs, N.J.: Prentice Hall. 1981.
Cohen, A. *Hi-fi loudspeakers and enclosures*, 2nd Ed. New York: Haydon. 1968. See especially the chapter on networks in multi-speaker systems.
Coltman, J. W. The transformer. *Scientific American.* Jan. 1988: 86.
Lampton, M. A three-way corner loudspeaker system. *Speaker builder.* Apr. 1982: 7.
Meindl, J. Microelectronic circuit elements. *Scientific American.* Sep. 1977: 70.
Skilling, H. Electrical engineering circuits. New York: Wiley. 1957.
Snowden, D. P. Superconductors for power transmission. *Scientific American.* Apr. 1972: 84.

Chapter 33

Ashkin, A. The pressure of laser light. *Scientific American.* Feb. 1972: 62.

Courvoisier, T. J-L. and E. I. Robson. The quasar 3C273. *Scientific American.* Jun. 1991: 50.

Dickinson, D. Cosmic masers. *Scientific American.* Jun. 1978: 90.

Elachi, C. Radar images of the Earth from space. *Scientific American.* Dec. 1982: 54.

Foster, K. and A. Guy. The microwave problem. *Scientific American.* Sep. 1986: 32.

Ginzton, E. The klystron. *Scientific American.* Mar. 1954: 84.

Henry, G. E. Radiation pressure. *Scientific American.* Jun. 1957: 99.

Mathis, J., B. Savage, and J. Cassinelli. A superluminous object in the large cloud of Magellan. *Scientific American.* Aug. 1984: 52.

Tsipis, K. Laser weapons. *Scientific American.* Dec. 1981: 51.

Wehner, R. Polarized light navigation by insects. *Scientific American.* Jul. 1976: 106.

Part VIII

McCormmach, R. *Night thoughts of a classical physicist.* Cambridge, Mass.: Harvard University Press. 1982.

Pais, A. *'Subtle is the Lord...', The science and the life of Albert Einstein.* Oxford, England: Oxford University Press. 1982.

Chapter 34

Benford, G. *Timescape.* New York: Simon and Schuster. 1980.

Burke, J. R., and F. Strode. Classroom exercises with the Terrell effect. *American Journal of Physics.* Oct. 1991: 912–915.

Einstein, A. Autobiographical notes. In P. A. Schilpp, *Albert Einstein: Philosopher-scientist.* Vol. 1, 3rd ed. La Salle, Ill.: Open Court 1969.

Epstein, L. *Relativity visualized.* San Francisco: Insight Press. 1983.

Feinberg, G. Particles that go faster than light. *Scientific American.* Feb. 1970: 68.

Gamow, G. *Mr. Tomkins in paperback.* Cambridge, England: Cambridge University Press. 1982.

McCrea, Sir W. Arthur Stanley Eddington. *Scientific American.* Jun. 1991: 92.

Taylor, E. and J. Wheeler. Spacetime physics, 2nd ed. New York: Freeman. 1992.

Whittaker, Sir E., and G. F. Fitzgerald. *Scientific American.* Nov. 1953: 93.

Chapter 35

Albert, D. Bohm's alternative to quantum mechanics. *Scientific American.* May 1994: 58.

Cassidy, D. Heisenberg, uncertainty and the quantum revolution. *Scientific American.* May 1992: 106.

Cohen, M. L., V. Heine, and J. C. Phillips. The quantum mechanics of materials. *Scientific American.* Jun. 1982: 82.

Da C. Andrade, E. The birth of the nuclear atom. *Scientific American.* Nov. 1956: 93

d'Espagnat, B. The quantum theory and reality. *Scientific American.* Nov. 1979: 158.

Dirac, P. A. M. The evolution of the physicists' picture of nature. *Scientific American.* May 1963: 45.

Feynmann, R. *The character of physical law.* London: British Broadcasting Corporation. 1965.

Gamow, G. The principle of uncertainty. *Scientific American.* Jan. 1958: 51.

———. The exclusion principle. *Scientific American.* Jul. 1959: 74.

Gutzwiller, M. Quantum chaos. *Scientific American.* Jan. 1992: 78.

Haensch, T. W., A. L. Schawlow, and G. W. Series. The spectrum of atomic hydrogen. *Scientific American.* 1979: 94.

Halliwell, J. Quantum cosmology and the creation of the universe. *Scientific American.* Dec. 1991: 76.

Horgan, J. Quantum philosophy. *Scientific American.* Jul. 1992: 94.

Hughes, R. I. G. Quantum logic. *Scientific American.* Oct. 1981: 202.

Kleppner, D., M. Littman, and M. Zimmerman. Highly excited atoms. *Scientific American.* May 1981: 130.

Nauenberg, M., C. Stroud, and J. Yeazell. The classical limit of an atom. *Scientific American.* Jun. 1994: 44.

Pool, R. Quantum pot watching. *Science* **246**. Nov. 17. 1989: 888.

Schrödinger, E. What is matter? *Scientific American.* Sep. 1953: 52.

Shimony, A. The reality of the quantum world. *Scientific American.* Jan. 1988: 46.

Weisskopf, N. How light interacts with matter. *Scientific American.* Sep. 1968: 60.

Chapter 36

Aftergood, S., D. Hafemeister, O. Prilutsky, J. Primack, and S. Rodionov. Nuclear power in space. *Scientific American.* Jun. 1991: 42.

Agnew, H. M. Gas-cooled nuclear power reactors. *Scientific American.* Jun. 1981: 64.

Armbruster, P., G. Munzenberg. Creating superheavy elements. *Scientific American.* May 1986: 66.

Baranger, M., and R. Sorensen. The size and shape of atomic nuclei. *Scientific American.* Aug. 1969: 58.

Bethe, H. A. The necessity of fission power. *Scientific American.* Jan. 1976: 21.

Clancy, T. *The sum of all fears.* New York: G. P. Putnam. 1991.

Cohen, B. The disposal of radioactive wastes from fission reactors. *Scientific American.* Jun. 1977: 21.

Cowan, G. A. A natural fission reactor. *Scientific American.* Jul. 1976: 36.

Curie, E. *Madame Curie.* V. Sheehan: translator. Garden City, N.Y.: Doubleday. 1937.

da C. Andrade, E. N. Carbon 14 and the prehistory of Europe. *Scientific American.* Oct. 1971: 63.

Fermi, L. *Atoms in the family.* Chicago: University of Chicago Press. 1954.

Golay, M. and N. Todreas. Advanced light-water reactors. *Scientific American.* Apr. 1990: 82.

Greiner, W., and A. Sandulescu. New radioactivities. *Scientific American.* Mar. 1990: 58.

Gutbrod, H., and H. Stocker. The nuclear equation of state. *Scientific American.* Nov. 1991: 58.

Häfele, W. Energy from nuclear power. *Scientific American.* Sep. 1990: 136.

Hedges, R., and J. Gowlett. Radiocarbon dating by accelerator mass spectroscopy. *Scientific American.* Jan. 1986: 100.

Learned, J. G. and D. Eichler. A deep sea neutrino telescope. *Scientific American.* Feb. 1981: 138.

McHarris, W. C. and J. O. Rasmussen. High energy collisions between atomic nuclei. *Scientific American.* Jan. 1984: 58.

Nero, Jr. A., Controlling indoor air pollution. *Scientific American.* May 1988: 42.

Rhodes, R. *The making of the atomic bomb.* New York: Simon and Schuster. 1986.

Rutherford, E., J. Chadwick, and C. D. Ellis. *Radiations from radioactive substances.* New York: Macmillan. 1930.

Chapter 37

Bloom, E., and G. Feldman. Quarkonium. *Scientific American.* May 1982: 66.

Close, F. *The cosmic onion: Quarks and the nature of the universe.* New York: American Institute of Physics. 1986.

Dodd, J. E. *The ideas of particle physics.* Cambridge, Eng.: Cambridge University Press. 1984.

Feldman, G., and J. Steinberger. The number of families of matter. *Scientific American.* Feb. 1991: 70.

Green, M. Superstrings. *Scientific American.* Sep. 1986: 48.

Jacob, M., and P. Landshoff. The inner structure of the proton. *Scientific American.* Mar. 1980: 66.

Lederman, L. The value of fundamental science. *Scientific American.* Nov. 1984: 40.

Myers, S., and E. Picasso. The LEP collider. *Scientific American.* Jul. 1990: 54.
Pickering, A. *Constructing quarks.* Chicago: University of Chicago Press. 1984.
Quigg, C. Elementary particles and forces. *Scientific American.* Apr. 1985: 84.
Rebbi, C. The lattice theory of quark confinement. *Scientific American.* Feb. 1983: 54.
Soergel, D., and N. Turok. Textures and cosmic structure. *Scientific American.* Mar. 1992: 52.